Security of
Self-Organizing Networks

MANET, WSN, WMN, VANET

Security of
Self-Organizing Networks

MANET, WSN, WMN, VANET

Edited by Al-Sakib Khan Pathan

CRC Press
Taylor & Francis Group
Boca Raton London New York

CRC Press is an imprint of the
Taylor & Francis Group, an **informa** business

AN AUERBACH BOOK

CRC Press
Taylor & Francis Group
6000 Broken Sound Parkway NW, Suite 300
Boca Raton, FL 33487-2742

First issued in paperback 2019

© 2011 by Taylor and Francis Group, LLC
CRC Press is an imprint of Taylor & Francis Group, an Informa business

No claim to original U.S. Government works

ISBN-13: 978-1-4398-1919-7 (hbk)
ISBN-13: 978-0-367-38352-7 (pbk)

Library of Congress Cataloging-in-Publication Data

Security of self-organizing networks : MANET, WSN, WMN, VANET / editor, Al-Sakib Khan Pathan.
  p. cm.
  Includes bibliographical references and index.
  ISBN 978-1-4398-1919-7 (hardcover : alk. paper)
  1. Ad hoc networks (Computer networks)--Security measures. 2. Self organizaing systems--Security measures. I. Pathan, Al-Sakib Khan.

TK5105.77.S43 2011
005.8--dc22   2010028807

Visit the Taylor & Francis Web site at
http://www.taylorandfrancis.com

and the CRC Press Web site at
http://www.crcpress.com

Contents

PART I GENERAL TOPICS: SECURITY OF WIRELESS AND SELF-ORGANIZING NETWORKS

PART II MOBILE *AD HOC* NETWORK AND VEHICULAR *AD HOC* NETWORK SECURITY

PART IV WIRELESS MESH NETWORK SECURITY

Preface

Various types of wireless networks and their applications can be seen in many places nowadays. With the increase in the number of Internet users, with the aid of wireless communications and demand of flexible anytime, anywhere networking, self-organizing wireless networks have already gained huge popularity among users. For wider support of wireless connectivity and developing easy-to-use technologies, substantial efforts are underway to reduce human intervention in the configuration, formation, and maintenance processes of these networks. This book is an attempt to address the security issues of four types of self-organizing networks (SONs): the Wireless Mobile *Ad hoc* Network (MANET), Wireless Sensor Network (WSN), Wireless Mesh Network (WMN), and Vehicular *Ad hoc* Network (VANET). Although various issues of these networks have been addressed extensively in different literature and numerous researchers around the globe are working on different aspects, here we have mainly focused on various facets of security in these networks.

As in an SON, hundreds and thousands of wireless devices can participate in a limited area with wireless communications but the information exchange among the devices needs appropriate privacy-, authenticity-, availability-, and nonrepudiation-ensuring mechanisms. Without proper security policy, any type of SON is exposed to a wide variety of security vulnerabilities and threats. Other than data security in such types of networks, there are also multiple factors that should be considered for ensuring overall security. For example, physical security of wireless devices is one of the important issues especially for WSNs and VANETs. Severe limitation of available device-resources (e.g., energy, radio, storage, processing, etc.) is a critical point for some networks. Again, security measures should be employed based on the special characteristics of a particular SON. All these issues related to security, problems, challenges, hopes, and solutions are discussed in various chapters of this book.

About the Contents of the Book

There are a total of 23 chapters in this book, which is divided into four parts. Different chapters in different parts address various security issues of SONs from different angles. The first part deals with general security topics. This part sets the base for the following chapters. Many terms are introduced in the chapters under Part I so that later they become easily accessible to readers. Part II deals with different security issues in MANETs and VANETs. For some of the chapters, the authors have proposed specific solutions to specific security problems. As VANET is a special class of MANET, we have included both of these networks under the same part. Part III discusses critical issues of

WSN security, and Part IV briefly touches on the security issues of WMNs. Most of these chapters are written in a tutorial manner. However, for some of the chapters, mathematical equations and detailed analysis are used for advanced readers. Hence, this book talks about relatively easier issues as well as analyzes some security-related issues in depth. Although in some cases, the contents of a chapter may overlap with some of the contents of another chapter, the repeated information might be helpful in clarifying specific points dealt with in that particular chapter. This is also useful in the sense that different people think of the same point from different perspectives. We hope that the book will really be helpful in giving a thorough and wide picture of the security aspects of MANETs, WSNs, WMNs, and VANETs.

What Not to Expect from the Book

This book is a not a basic tutorial on the security issues of SONs. Hence, it does not have detailed introductory information about security. Although some chapters contain some elementary information, those should not be considered adequate for a beginner. Users/readers need to have at least some basic knowledge about security in networking. Again, this book should not be taken as a detailed research report. Some chapters simply present a specific problem and its solution that might be helpful for graduate students, some talk about elementary information that might be useful for general readers, some discuss in-depth security issues that might be helpful for advanced readers, and some chapters talk about the latest updates in a particular research area that might assist researchers in determining their future research objectives.

Target Audience

The book is very useful for Master's- or PhD-level students working on security issues in these networks, for researchers, for faculty members at the university level, and for some industry professionals. Questions and their sample answers have been provided with each chapter so that the readers' understanding from a chapter can be tested. Supplementary materials have been provided so that each chapter can be taught in a classroom environment with presentation slides. The book's chapters have been placed in such a sequence that it can be helpful to readers if they are read sequentially or to other readers who skip some of the chapters as they like. Overall, this book can guide readers to the latest trends, issues, and aspects of security of MANETs, WSNs, WMNs, and VANETs.

Dr. Al-Sakib Khan Pathan

Acknowledgments

I am very much thankful to the Almighty for giving me strength, keeping me fit, and giving me this time to accomplish this work. My sincere thanks to the authors of different chapters of this book without whose invaluable contributions, this project could not be completed. All the authors have been cooperative on different occasions during the submission, review, and editing process of the book. I am very thankful to Richard O'Hanley who helped me at each step of the publication process. I express my earnest gratitude to him for giving me this opportunity to edit such a book. Also thanks to him for his contribution to this book of choosing a timely, concise, and catchy title. Finally, I would like to thank my parents; Abdus Salam Khan Pathan, Delowara Khanom, my loving wife Labiba Mahmud, my younger brother Dr. Mukaddim Pathan, and my elder sister Tahmina A. Khanam for their continuous support and encouragement while preparing the book.

Dr. Al-Sakib Khan Pathan

Editor

Al-Sakib Khan Pathan is an assistant professor at the Computer Science and Engineering Department of Bangladesh Rural Advancement Committee (BRAC) University, Bangladesh. He worked as a researcher at the Networking Lab, Department of Computer Engineering in Kyung Hee University, South Korea, where he received his PhD in 2009. He received his BSc degree in computer science and information technology from the Islamic University of Technology (IUT), Bangladesh, in 2003. He has served as a chair, organizing committee member, and technical program committee member in several international conferences/workshops such as HPCS 2010, ICA3PP 2010, WiMob'09 and 08, HPCC'09, IDCS'09 and 08. He is currently serving as an area editor of *IJCNIS*, associate editor of IASTED/ACTA Press *IJCA*, guest editor of several international journals, including Elsevier's *Mathematical and Computer Modelling*, and editor of a book. He also serves as a referee of a few renowned journals such as the *IEEE Transactions on Dependable and Secure Computing* (*IEEE TDSC*); the *IEEE Transactions on Vehicular Technology* (*IEEE TVT*); the *IEEE Communications Letters*; Elsevier's *Computer Communications, Computer Standards and Interfaces*; the *Computers & Electrical Engineering* journal; the *Journal of High Speed Networks* (*JHSN, IOS* Press); the *EURASIP Journal on Wireless Communications and Networking* (*EURASIP JWCN*); and the *International Journal of Communication Systems* (*IJCS*, Wiley). He is a member of IEEE and several other international organizations. His research interest includes wireless sensor networks, network security, and e-services technologies.

Contributors

Mozaffar Afaque
Department of Computer Science
  and Engineering
Indian Institute of Technology Kharagpur
Kharagpur, India

Johnson I. Agbinya
Centre for Real-Time Information Networks
University of Technology
Sydney, Australia

Syed Ishtiaque Ahmed
Department of Computer Science
  and Engineering
Bangladesh University of Engineering
  and Technology
Dhaka, Bangladesh

Tarem Ahmed
Department of Electrical and
  Electronic Engineering
BRAC University
Dhaka, Bangladesh

Nancy Alrajei
Department of Computer Science
  and Engineering
Oakland University
Rochester, Michigan

Hani Alzaid
Information Security Institute
Queensland University of Technology
Queensland, Australia

Sharifah Hafizah Syed Ariffin
Faculty of Electrical Engineering
Universiti Teknologi Malaysia
Skudai, Malaysia

Michel Barbeau
School of Computer Science
Carleton University
Ontario, Canada

Dan Chalmers
School of Informatics
University of Sussex
Brighton, United Kingdom

Gihwan Cho
Division of Electronics and
  Information Engineering
Chonbuk National University
Chonju, South Korea

John A. Clark
Department of Computer Science
University of York
York, United Kingdom

Zubair Muhammad Fadlullah
Graduate School of Information Sciences
Tohoku University
Sendai, Japan

Norsheila Fisal
Faculty of Electrical Engineering
Universiti Teknologi Malaysia
Skudai, Malaysia

Joaquin Garcia-Alfaro
School of Computer Science
Carleton University
Ontario, Canada

and

Computer Science and Multimedia
    Studies
Open University of Catalonia
Catalonia, Spain

M.S. Gaur
Department of Computer Engineering
Malaviya National Institute
    of Technology
Jaipur, India

Swapna Ghanekar
Department of Computer Science
    and Engineering
Oakland University
Rochester, Michigan

S.K. Ghosh
School of Information Technology
Indian Institute of Technology
    Kharagpur
Kharagpur, India

Stephen Glass
School of Information and Communication
    Technology
Griffith University

and

National Information and
    Communications Technology
    Australia (NICTA)
Queensland Research Laboratory
Queensland, Australia

Vikrant Gokhale
School of Information Technology
Indian Institute of Technology
    Kharagpur
Kharagpur, India

Sumit Goswami
DESIDOC, Defence Research
    & Development Organization
Delhi, India

Jyoti Grover
Department of Computer Engineering
Malaviya National Institute of Technology
Jaipur, India

Arobinda Gupta
Department of Computer Science
    and Engineering
Indian Institute of Technology Kharagpur
Kharagpur, India

Md. Abdul Hamid
Department of Information and
    Communications Engineering
Hankuk University of Foreign Studies
Kyonggi-do, South Korea

James Harbin
Department of Electronics
University of York
York, United Kingdom

Bing He
Department of Computer Science
University of Cincinnati
Cincinnati, Ohio

Md. Shariful Islam
Department of Computer Engineering
Kyung Hee University
Gyeonggi-do, South Korea

Evangelos Kranakis
School of Computer Science
Carleton University
Ontario, Canada

V. Laxmi
Department of Computer Engineering
Malaviya National Institute of Technology
Jaipur, India

Sungyoung Lee
Department of Computer Engineering
Kyung Hee University (Global Campus)
Gyeonggi-do, South Korea

Effie Makri
Department of Mathematics
University of the Aegean
Karlovassi, Greece

Yasir Arfat Malkani
School of Informatics
University of Sussex
Brighton, United Kingdom

Fatma Mili
Department of Computer Science
    and Engineering
Oakland University
Rochester, Michigan

Sudip Misra
School of Information Technology
Indian Institute of Technology Kharagpur
Kharagpur, India

Paul Mitchell
Department of Electronics
University of York
York, United Kingdom

Vallipuram Muthukkumarasamy
School of Information and
    Communications Technology
Griffith University
and
National Information and Communications
    Technology Australia (NICTA)
Queensland Research Laboratory
Queensland, Australia

Al-Sakib Khan Pathan
Department of Computer Science
    and Engineering
BRAC University
Dhaka, Bangladesh

and
Department of Computer Engineering
Kyung Hee University
Gyeonggi-do, South Korea

David Pearce
Department of Electronics
University of York
York, United Kingdom

Zeeshan Pervez
Department of Computer Engineering
Kyung Hee University (Global Campus)
Gyeonggi-do, South Korea

Marius Portmann
School of Information Technology
    and Electrical Engineering
University of Queensland

and

National Information and Communications
    Technology Australia (NICTA)
Queensland Research Laboratory
Queensland, Australia

Syed Muhammad Khaliq-ur-Rahman Raazi
Department of Computer Engineering
Kyung Hee University (Global Campus)
Gyeonggi-do, South Korea

Rumana Rahman
Department of Electrical and Electronic
    Engineering
BRAC University
Dhaka, Bangladesh

Rozeha A. Rashid
Faculty of Electrical Engineering
Universiti Teknologi Malaysia
Skudai, Malaysia

Ranga Reddy
U.S. Army—Communications Electronics
    Research, Development, and Engineering
    Center (CERDEC)
Fort Monmouth, New Jersey

Une Thoing Rosi
Department of Computer Science
and Engineering
United International University
Dhaka, Bangladesh

Kashif Saleem
Faculty of Electrical Engineering
Universiti Teknologi Malaysia
Skudai, Malaysia

Marcus Schöller
NEC Europe Ltd
Heidelberg, Germany

Jaydip Sen
Innovation Lab
Tata Consultancy Services
Kolkata, India

Sevil Şen
Department of Computer
Science
University of York
York, United Kingdom

Yannis C. Stamatiou
Department of Mathematics
University of Ioannina
Ioannina, Greece

and

Research Academic
Computer Technology
Institute
University of Patras
Patras, Greece

Tarik Taleb
NEC Europe Ltd
Heidelberg, Germany

Juan E. Tapiador
Department of Computer Science
University of York
York, United Kingdom

Thanh Dai Tran
Centre for Real-Time Information Networks
University of Technology
Sydney, Australia

Ian Wakeman
School of Informatics
University of Sussex
Brighton, United Kingdom

Gicheol Wang
Department of Computing and
Networking Resources
Supercomputing Center, KISTI
Daejeon, South Korea

Bin Xie
InfoBeyond Technology LLC
Louisville, Kentucky

Vikas Singh Yadav
Department of Computer
Science and Engineering
Indian Institute of Technology
Kharagpur
Kharagpur, India

Sharifah Kamilah Syed Yusof
Faculty of Electrical Engineering
Universiti Teknologi Malaysia
Skudai, Malaysia

David Zhao
U.S. Army—Communications Electronics
Research, Development, and Engineering
Center (CERDEC)
Fort Monmouth, New Jersey

GENERAL TOPICS—SECURITY OF WIRELESS AND SELF-ORGANIZING NETWORKS

Chapter 1

Secure Device Association
Trends and Issues

Yasir Arfat Malkani, Dan Chalmers, and Ian Wakeman

Contents

1.1 Introduction

More and more computing devices are coming into existence every day, which may vary in size, capabilities, mode of interaction, and so on. As a result we are moving toward a world in which computing is omnipresent. Many modern devices (e.g., smart printers, PDAs, smart phones, and cameras) support multiple communication channels and almost all of them use wireless technology in some form, such as Bluetooth, Infrared, Wibree, Zigbee, 802.11, IrDA, or ultrasound. Having wireless technology in these devices does not guarantee that all of these devices can also take advantage of Internet technology. However, those wireless-enabled devices that cannot connect to Internet can still take advantage of other colocated devices in the vicinity by forming short-term or long-term associations on *ad hoc* basis: for example, pairing a Bluetooth-enabled headset with a mobile phone or an MP3 player (short term) and pairing of a PDA with home devices in order to control them wirelessly (long term). Some other examples of pairing from everyday life include pairing of a Bluetooth keyboard with a Desktop computer, pairing of a laptop with an access point or a printer through the use of a WiFi or Bluetooth, and pairing of two mobile phones to exchange the music files or other data. Since wireless communication is susceptible to eavesdropping, thus one can easily launch man-in-the-middle (MiTM), denial-of-service (DoS), or bidding-down attacks to break the secure pairing process. Therefore, the main goal of secure pairing research is to provide assurance of the identity of the devices participating in the pairing process and to secure them from being victims of eavesdropping attacks, such as MiTM attack. Achieving this goal is a challenging problem from both the security and the usability or user interaction points of view.

Security challenges emerge due to the *ad hoc* and dynamic nature of mobile *ad hoc* networks (MANET), in which devices do not know each other *a priori*, but still need to develop spontaneous interactions between themselves. This precludes the idea of preshared secret keys. Further, traditional key exchange or key agreement approaches, such as Diffie–Hellman [1], are not applicable without modification in wireless environments due to their vulnerability to an MiTM attack.

From a usability point of view, since most of the device owners are nontechnical, they want minimal and easy interactions with their devices during the pairing process. They do not want to remember a list of PIN numbers or secret passwords to establish the secure communication channel between a pair of devices for several scenarios or situations. Since many users do not have a deep technical understanding of the risks of pairing and there is a substantial cognitive overhead in remembering the different kinds of steps of secure pairing for several categories of devices and situations, many users may either deactivate security of the devices or select an inappropriate pairing method, which may cause poor security. Therefore it is also a challenge to develop more general, standardized, and user-friendly interaction methods that might increase the usability of pairing schemes.

Some other challenges are due to the devices' heterogeneity in terms of their communication channels, user interfaces, power requirements, and sensing technology that make it hard to give a single or standard solution for secure pairing of devices.

As a result of these challenges, a wide community of researchers has proposed many protocols to deal with this issue. These protocols vary in the assumptions of required capabilities in the devices, required human intervention, and in the way they utilize out-of-band or location-limited side channels including physical, audio, visual, short-range wireless channels like Near-Field Communications (NFC), and also combination of these wherever possible. As a consequence, currently there exist many options for an ordinary user to establish a secure channel between the devices from entering pins and passwords to verifying hashes of public keys and pressing buttons simultaneously on the two devices. This notion contradicts with the usability goal of secure device association methods.

In this chapter, we discuss and analyze the different existing solutions of secure device association (pairing) and then discuss future directions by considering trade-offs among various existing approaches for device pairing.

1.2 Background

1.2.1 Attack Types in Device Association Model

Device association (also known as security initialization, first-connect or device pairing in the literature) can be referred as the process of establishing a secure channel between two unassociated human-operated devices over a short-range wireless channel, such as Bluetooth, Infrared, or 802.11. There are several kinds of possible security threats or attacks in device association scenarios. In this section, we describe them in brief.

1.2.1.1 Eavesdropping

The most significant risk in device association models is that the underlying communication channel is wireless (e.g., Bluetooth, 802.11, etc.), which is open to everyone including bona fide users as well as intruders or adversaries, and thus pairing partners cannot be physically secured the same way as two peers in a point-to-point wired network. In an eavesdropping attack, an adversary secretly listens to the conversation between pairing partners. The adversary's main goal is to obtain confidential information, including public/private keys, location information, contact details, data of commercial value, or even devices' capabilities. To reduce the risk of eavesdropping, the general solutions include encryption, and physically securing the medium (line of sight transmission, frequency hopping, etc.).

1.2.1.2 MiTM Attack

Simple eavesdropping is a passive attack, in which an adversary's goal is to steal some confidential information. However, active attacks are more dangerous, in which the main goal of an adversary is to fool the legitimate device(s) to associate with the adversary's device. An (MiTM) attack is the most widespread and well-known active attack against device pairing protocols. It is a kind of active eavesdropping, in which an adversary can fully intercept the messages moving in both directions, modify, or corrupt the message, store messages for later replay, or insert new messages. To successfully launch this attack, an adversary should be able to establish two independent connections with the victims. In the event of a successful attack, the victims believe that they are communicating with each other and the messages received by them are from a legitimate source; however, it is not the case. In fact, all conversation is passed through the adversary, who is able to illegitimately analyze and modify the real data, launch DoS attack, and even impersonate one partner to gain control

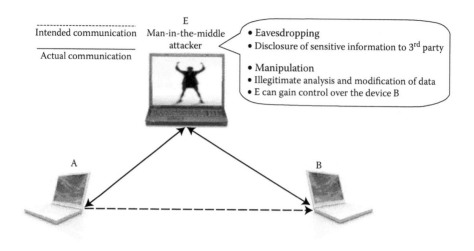

Figure 1.1 **MiTM attack scenario.**

over the victim's device(s) or gain access to data or resources. Figure 1.1 depicts the scenario of an MiTM attack.

1.2.1.3 DoS Attack

The general goal of an adversary launching a DoS attack is to prevent communication between wirelessly connected nodes. However, in the case of device pairing, a DoS attack prevents two legitimate pairing partners from establishing a secure channel. It is a general concept that this is the easiest attack that can be launched by an adversary in wireless environments. Since there has been less emphasis on the prevention of DoS attack in pairing scenarios, many of the pairing schemes are susceptible to DoS attack. For example, in pairing schemes that use audio as an out-of-band channel, an attacker can launch a DoS attack by creating noisy environment for the user/devices. The noisy environment may prevent the user from properly participating in the pairing process. In the case of visual out-of-band channels, this attack can be launched by manipulating the lights (dark, bright, flashing, etc.), so that bar codes, screens, and so on used to contain secure pairing information cannot be read. However, these kinds of DoS attacks can be recognized by the user, who can then try to eliminate them by changing the environment or by forcing the adversary not to do so in case of source detection.

1.2.1.4 Bidding-Down Attack

Bidding-down attack is possible in scenarios where a list of choices to establish a secure channel is available, and the selection of the best pairing protocol is negotiated based on some criteria, such as device capabilities or user preferences. In this kind of attack, the goal of an adversary is to fool (i.e., bid down) the intended pair able devices to use weaker security than is possible. For instance, when pairing two display and camera-equipped devices, an adversary could modify the capabilities of one of the devices into display-less and/or camera-less device (i.e., bidding down) to force a radio-based pairing protocol to be used, which is easier to intercept without being detected.

1.2.1.5 Compromised Devices

Compromised devices are a risk in any wireless system and are difficult to prevent at the protocol level. In the case of secure device pairing, it is possible that an adversary may install malicious code on the device(s). Then an adversary can access confidential information (e.g., shared secret) stored on the device or use it to get authorized access to other available services. Further, a compromised device could suggest pairing with only the adversary's device or could run a weak pairing protocol. It is the user's responsibility to eliminate the chance of this attack by some mechanism, such as deploying security software to detect the malicious code or to restrict the physical access of the device to only those people whom he/she trusts.

1.2.2 Device Association in Ad Hoc *Environments*

The problem of secure device association continues to be a very active area of research in *ad hoc* environments. The issue received significant attention from many researchers after Stajano et al. [2–4] highlighted the challenges inherent in secure device association. As a result, currently we have more than two dozen device association methods including their variations. Since overlapping material on device pairing has started to appear in the literature (e.g., [5–8]), we believe that the knowledge and understanding of existing methods is very important in order to propose new pairing methods that should be really *novel*. Considering this fact, in this section we present the survey of several approaches to device association along with a detailed comparative analysis (Section 1.2.3). We first present several schemes proposed by academia and then we also discuss efforts taken by the industry and standardization bodies.

1.2.2.1 Resurrecting Duckling Security Model

In their seminal *Resurrecting Duckling* paper [2] Stajano and Anderson presented a policy-based mother–duckling security model that played an important role in raising the issue of secure device association among a wide community of researchers. Their work [2–4] has been considered as the first effort toward secure transient association between devices for MANET and ubiquitous computing environments. The proposed mother–duckling model maps the relationships between devices. "Mother" is a master device that imprints a "duckling" that is a slave device. The slave device remains in one of the two states: imprinted or imprintable. The slave device is in the imprintable state at the beginning or bootstrapping time. However, it switches from imprintable to the imprinted (paired) state once it has got the shared secret from its master device. The slave remains in this state until its death (i.e., while it keeps the shared secret provided by its master device). In fact, the shared secret binds the slave device to its master device. As a consequence, the slave device remains faithful to the master device and obeys no one else. Since the shared secret is transferred from the master to the slave over a physical connection (such as using a cable) in plain-text form, the proposed approach does not require complex cryptographic methods, such as Diffie–Hellman [1]. The authors also highlighted the concept of device-control mechanisms, that is, how to gain control of personal devices, how to transfer or release control when needed, and how to regain control of the same devices.

1.2.2.2 Talking to Strangers

Balfanz et al. [9] extended Stajano and Anderson's work and proposed a two-phase authentication method for pairing of colocated devices using infrared as a location-limited side channel (also known as out-of-band channel). In their proposed solution, preauthentication information is exchanged over the infrared channel and then the user switches to the common wireless channel. Preauthentication data contains cryptographic material as well as the complete address of the device. The proposed method exploited public key cryptography in which devices exchange their public keys over an insecure wireless channel followed by exchanging the hashes of respective public keys over the location-limited side channel (i.e., infrared). Further, they are the first to introduce the concept of demonstrative identification (i.e., identification in the form of a representation of an object, for example, the printer in this room, the display in front of me, etc.) for authentication purposes in pairing process. Slightly different variations, of Balfanz et al. [9] approach are proposed in [10,11], which use laser and ultrasound as location-limited side channels to transfer the preauthentication data.

1.2.2.3 Device Association Using Visual Out-of-Band Channels

On the basis of the pairing protocol of Balfanz et al. [9], some other schemes are proposed through the use of audio and visual out-of-band channels. One such system is *Seeing-is-Believing* (SiB) [12]. SiB takes advantage of the common presence of cameras in modern handheld devices and utilizes two-dimensional bar codes for exchanging preauthentication data (i.e., public keys) between the devices. In the proposed approach, device A encodes cryptographic material into a two-dimensional bar code and displays it on the screen; then device B reads it through a camera to set up an authenticated channel. In the simplest case, SiB requires the first device (A) to have a display to show the 2D bar codes and the second device (B) a camera. Then the user is required to focus and place the camera of device B at the first device's (device A) screen properly to take a photograph of the displayed bar code. SiB supports several use cases based on the device capabilities. For example, when the first device has a camera and the other device has only a display, then only the first device (camera-equipped) can authenticate the other device—that is, the display-only device (1-way authentication). In the second use case, when both devices are camera and display equipped, then both the devices can authenticate each other by two protocol runs, one in each direction (2-way authentication). In another use case, when only one device has a camera and the other device has neither a camera nor a display, the user can then print a two-dimensional bar code on a sticker, containing the cryptographic material, and attach the sticker to the other (camera-less and display-less device) device. In this case, the user takes a photograph of the sticker and performs the SiB protocol as usual.

Another pairing method that uses a visual out-of-band channel is proposed by Saxena et al. [13]. To reduce the camera requirement in one of the pairing devices in SiB, they extended the work of McCune et al. [12] and proposed an improvement to it through the use of a simple light source, such as an LED, and short authenticated integrity checksums. In fact, they showed that mutual authentication can be achieved with a one-way visual channel, while SiB requires two visual channels, one in each direction (for full functionality). In the proposed scheme [13], device A needs to be equipped with a camera and device B with a single LED. Device A takes a video clip of a blinking pattern on device B's LED. Then the video clip is parsed to extract an authentication string.

1.2.2.4 Device Association Using Audio Out-of-Band Channels

Loud and Clear (L&C) [14] and *Human-Assisted Pure Audio Device Pairing* (HAPADEP) [15] use audio as an out-of-band channel to establish a secure channel between the devices. The main idea of the L&C [14] scheme is to encode the hash of the first device's public key into a MadLib sentence (i.e., grammatically correct but nonsensical sentence) and transmit it over a device-to-human channel using a speaker or a display. The second device also encodes the hash of the received public key from the first device into the MadLib sentence and transmits it over a device-to-human channel using a speaker or a display. The user is then responsible for comparing the two sentences and to accepting or rejecting the pairing. There are four variants of this approach: speaker to speaker, speaker to display, display to speaker, and display to display. In the first variant, the user is required to compare and verify the two sentences vocalized by the pairing candidate devices. In the second variant, the user is required to compare the vocalized MadLib sentence with the sentence displayed on the other device. In the third variant, the user is required to compare the displayed MadLib sentence on one device with the vocalized MadLib sentence from the other device. In the fourth variant, the user is required to compare the MadLib sentences displayed on both the devices. In all the variants, the user is responsible for accepting or rejecting the pairing based on the results of comparison.

Soriente et al. proposed HAPADEP [15], which is a follow-on from L&C [14]. Soriente et al. consider the problem of pairing two devices that have no common standard wireless communication channel, such as Bluetooth or WiFi, at the time of pairing. The proposed scheme uses only audio to exchange both public keys and hashes of public keys. The proposed system consists of two phases: key transfer and key verification. In the key-transfer phase, the first device (device A) encodes cryptographic material along with protocol messages into a fast audio codec and plays the resulting audio sequence. The other device (device B) records and decodes this audio sequence in order to obtain the key. This process is also repeated in reverse direction so that device A could get the key from device B. In the second phase, each device computes a hash of the received public key and encodes it into a pleasant audio sequence, such as a melody. Then the user is required to listen and compare the audio sequences played by both devices and accept or reject the pairing based on the results of comparison. This scheme is only applicable to those scenarios where both devices have a microphone and a speaker.

1.2.2.5 Device Association Using Accelerometers

Unlike the approaches described above, the idea of shaking devices together to pair them has become more common. *Smart-its-Friends* [5] is the first effort that proposed pairing of two devices using a common movement pattern and used accelerometers as an out-of-band channel. In this approach, two devices are held and shaken together simultaneously. Then common readings from the embedded accelerometers in the devices are utilized to establish the communication channel between the two devices. However, security has not been the major concern of Smart-its-Friends. The follow-on schemes to Smart-its-Friends are *Are You With Me* [8] and *Shake Well Before Use* [6]. In *Are You With Me* [8], the main goal was to show that accelerometer's data can be used to reliably determine that a set of devices are being carried by the same person. The authors showed that one can reliably determine whether the two devices are being carried by the same person or not using only eight seconds of walking data. However, one of the major limitations of the proposed system is that they require the user(s) to walk [8].

Mayrhofer and Gellersen [6] extended Holmquist et al.'s [5] approach and proposed two protocols to securely pair the devices. Both the proposed protocols exploit cryptographic primitives with accelerometer data analysis for secure device-to-device authentication. The first protocol uses public key cryptography and is more secure as compared to the second protocol, which is more efficient and computes a secret key directly from the accelerometer's data. In the second scheme, the user is required to hold and shake the devices together for approximately 20 s to generate a 128-bit shared secret [6]. Kirovski et al. proposed Martini Synch [16], another accelerometer-based approach to securely pair the devices that use the idea of joint fuzzy hashing [7].

1.2.2.6 Device Association Using Radio Signals

Another approach that requires shaking or moving patterns is *Shake Them Up* [17]. The authors suggest a movement-based technique for pairing two resource-constrained devices that involves shaking and twirling them in very close proximity to each other. In the proposed scheme, intended pairing partners are shaken together to exchange the radio packets and agree on a key one bit at a time relying on the attacker's inability to determine the source of each radio packet (i.e., sending device). Unlike accelerometer-based schemes, this approach exploits the source indistinguishability property of radio signals and does not require embedded accelerometers. Castelluccia and Mutaf [17] described the source indistinguishability as two parties Alice and Bob run the previously described key exchange protocol, but the eavesdropper should not be able to distinguish the packets sent by Alice from the packets sent by Bob. This source indistinguishability property requires that communication should be temporally and spatially indistinguishable. To achieve the temporal indistinguishability, the authors use a CSMA-based system. To achieve the spatial indistinguishability, the authors suggested that devices should be shaken and twirled in very close proximity to each other.

Varshavsky et al. [18] proposed *Amigo,* a proximity-based technique for secure pairing of colocated devices. They extended the Diffie–Hellman key exchange protocol with the addition of a key verification stage. The proposed approach utilizes commonality of radio signals from locally available wireless access points to establish the secure channel between the devices. Any attacker who is not physically very close would see a different pattern of access point signal strengths. Radio-based approaches to secure device association either require no or minimal hardware and user involvement during the pairing process. However, these schemes are not applicable in the scenarios where devices support only Bluetooth technology.

1.2.2.7 Device Association Using Biometric Data

Biometrics represents a common technique for identifying human beings. Owing to the success of biometric-based user authentication systems, researchers realized that many benefits could be achieved by combining biometrics with cryptography. As a consequence, Buhan et al. proposed two systems [19,20] that utilize biometric data to establish a secure channel between the devices. Both the proposed systems are based on the Balfanz et al. model [9], and biometrics is used as an out-of-band channel. In *Feeling-is-Believing* (FiB) [19]; Buhan et al. investigated the grip pattern and proposed to generate a shared secret key from biometric data using quantization and cryptanalysis. In *SAfE* [20], keys are extracted from images during the preauthentication phase, which are used for authentication in subsequent phase.

1.2.2.8 Button-Enabled Device Association (BEDA)

Soriente et al. [21] proposed BEDA. The main idea of the proposed approach is to transfer the short secret key from one device to the other using 'button-presses' and then use that key to authenticate the public keys of the devices. A short secret key (21-bits) is agreed between the two devices via one of its four variants. These variants are called button-to-button (B-to-B), display-to-button (D-to-B), short vibration-to-button (SV-to-B), and long vibration-to-button (LV-toB). In fact, the only difference between these variants is the way the first device (device A) transfers the bits of the generated short secret to the other device (device B). Bits of a short secret are encoded by the devices using the time interval between two events, such as a button-press-event. For example, the first and basic variant (i.e., B-to-B) involves the user simultaneously pressing buttons on both of the devices within certain random time intervals and each of these intervals are used to derive three bits of the short secret key. In the D-to-B variant, it requires the first device to have a display that emits visual signals by showing a blinking square on its screen. The user reacts to blinking square events by pressing the button on the other device. In the SV-to-B variant, it requires one of the devices to have vibration capability. It is similar to the D-to-B scheme; however, it transmits signals through short vibration events instead of blinking square. Finally, LV-to-B variant is also similar to the SV-to-B and D-to-B variants; however, in this scheme, instead of short vibration or blinking square events, signals are emitted through either the start or the end of a long vibration.

1.2.2.9 Bluetooth Pairing

Bluetooth [22] is a short-range wireless technology that allows modern devices—such as mobile phones, PDAs, cameras, and other handheld devices—to communicate with each other over a distance of up to 100 m. It works on a 2.4 GHz ISM band and is considered to be one of the simplest ways to wirelessly exchange information between two devices in close proximity. In order to establish a secure communication link between intended pairing devices, the user needs to go through the Bluetooth pairing setup procedure. In Bluetooth pairing, devices need to exchange a short passkey or PIN code to prove that the owners of both devices are agreed to pair the devices with each other. Below are the general steps involved in the Bluetooth pairing process (Figure 1.2):

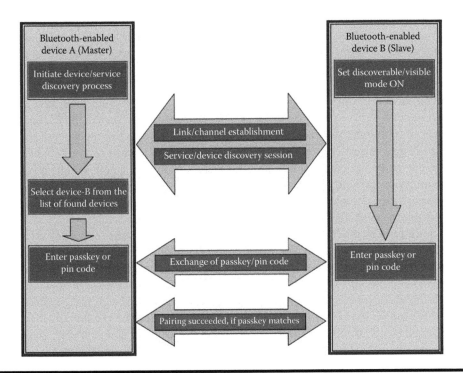

Figure 1.2   Bluetooth pairing process.

1. The pairing process starts when the first device (device A), such as Bluetooth-enabled mobile phone or PDA, searches for other Bluetooth-enabled devices in the vicinity. A list of Bluetooth devices found would be shown on the screen of device A. Note that only those devices can be found that are already in Bluetooth discoverable mode and their visibility option is turned ON.

2. Device A selects device B (such as other mobile phone or PDA) from the available list of devices. Then, device A asks the user to enter a PIN code or passkey. It could be any special code of your choice; however, it must be remembered, because it needs to be entered on the other device (device B). Note that in some of the resources/interface-constrained device scenarios, it is not possible to enter the passkey or PIN code. In that case, there is a fixed code, such as 0000, which the user is required to enter onto the other device.

3. Once the user has entered the passkey on device A, it sends it to device B.

4. If device B is not a resource-constrained device, it asks the user to enter the same PIN code or passkey; otherwise it simply uses its own standard/fixed passkey (e.g., 0000).

5. Finally, device B sends back the user-entered passkey to device A. If device B's passkey is the same as entered by device A, then automatically a trusted association takes place between the devices.

1.2.2.10 Device Association Using Near-Field Communication Technology

Near-Field Communication (NFC) is a short-range, high-frequency, low-bandwidth wireless connectivity standard defined by the NFC Forum [23]. Since NFC uses magnetic field induction to enable communication between devices, it allows users to securely pair the NFC-enabled devices by simply touching them together or holding them in very close proximity of up to 10 cm. NFC-enabled devices are capable of establishing a peer-to-peer network to exchange content and access services. It operates on a 13.56 MHz frequency with a data transfer rate of up to 424 kbit/s, with a bandwidth of 14 kHz. However, NFC in combination with other wireless technologies, such as Bluetooth or WiFi, can be used for exchanging a huge amount of data or can support longer communication.

In NFC, there are two kinds of devices—active devices that generate their own field, and passive devices that retrieve power from the field generated by active devices. NFC supports two basic modes of communication: active mode and passive mode. In active mode, both the devices generate their own magnetic field and require a power supply in each of them. While in passive mode, one of the devices (an active device) generates its magnetic field and the other devices (i.e., passive device, such as a contactless smart card) retrieve the power from the active device. There are many scenarios where NFC can be used. One such common scenario is the pairing of an NFC-enabled camera and a computer. In that scenario, the user could transfer all the photographs in the camera into his/her computer just touching them together or putting them in very close proximity. The touch mechanism makes it clear for the user which two devices are selected for intended association and takes away the burden of selecting the right devices (i.e., discovery and device identification) from a long list of available devices. Other possible applications/uses of NFC include smart posters, replacement of contactless credit cards with NFC-enabled mobile phones, and support services (through the use of voice clips) for the visually impaired people. WiFi protected setup (WPS) also incorporates one of the methods that use NFC as an out-of-band channel. Recently, there has been much greater availability of this technology in commercial devices including Nokia 6131, Motorola L7, SAGEM my700X Contactless, LG600V, and Samsung D500E.

1.2.2.11 Wireless Universal Serial Bus (WUSB) Association, WPS, and Windows Connect now-Net

The Wireless USB (WUSB) group was formed in 2004 to define the WUSB specifications that took about one year to complete. WUSB is a short-range (up to 10 m), high-bandwidth (110 Mbit/s) wireless radio communication technology, which is developed to simplify the process of establishing associations between a pair of wireless-enabled devices. The main goal of this technology is to replace wired USB. In WUSB, device A (i.e., the host device) and device B exchange connection host ID, connection device ID, and connection key during the association process. This information is utilized later on to set up secure communication between device A and the device B. WUSB supports two types of association models: cable association model and numeric association model. Device A or host device supports both the models; while the other device having only USB ports supports the cable association model, and the device with only a display supports the numeric association model. The cable association model utilizes a USB cable to perform the first-time association between a host and a device. Once the association has been completed, the cable is no longer needed and future communications with the device can be entirely wireless. In the numeric association model, the first-time association is performed over the ultra-wideband (UWB) radio. The WiFi Alliance officially launched WPSs in early 2007. The goal was to provide a standard and simple way for easy and secure establishment and configuration of wireless home networks. Another effort for standardization of secure device association is Microsoft's Windows Connect now-NET technology. It provides a way to set up secure wireless network, and works for both in-band wireless devices and out-of-band Ethernet devices.

1.2.3 Comparative Analysis of Device Association Methods

As described above, the issue of secure device association received significant attention from many researchers, after Stajano and Anderson in their seminal paper [2] highlighted the challenges inherent in secure device association. Since the secret key is transferred in plain-text form in their proposed approach, it is susceptible to dictionary attacks. It also requires the same physical interface in both the devices to transfer the secret, which makes such an approach inapplicable in scenarios where the devices do not have a common physical interface. Further, it is also difficult to carry the cables all the time. However, Resurrecting Duckling and Talking to Strangers both require minimal user interaction that is an advantage from usability point of view. The common drawback of Talking to Strangers [9] and some other similar approaches [10,11,24] (in terms of use of secondary-location-limited-side channel) is twofold: first, they need some kind of physical interface (e.g., IrDA, laser, ultrasound, etc.) for the preauthentication phase and are vulnerable to a passive eavesdropping attack in the location-limited side channels, for example, two remotes and one projector. Further, some of the location-limited side channels, such as infrared and laser, are highly vulnerable to denial of service (DoS) attacks. Those schemes that use audio and/or visual out-of-band channels [12–14] for secure device association also suffer from a few problems. For example, SiB [12] requires that one of the peers must be equipped with camera; while in L&C [14] a speaker and/or display is required. Camera-equipped devices are usually prohibited in high security areas; while the latter is not suitable for hearing-impaired users. Further, bar code scanning requires sufficient proximity and light in SiB; while L&C and HAPADEP [15] place some burden on the user for comparison of MadLib sentences and audible sequences, respectively. An adversary can easily subvert bar code stickers on devices in SiB to launch the successful attack, while ambient noise makes authentication either weak or difficult in L&C as well as in HAPADEP. For example in SiB, a user wants to pair his/her handheld device with a display-less printer to print a confidential document. Since the printer is display-less, a bar code sticker is attached to it. It is possible that an adversary subverts the bar code or swaps it with another printer available in the next building. In that scenario, once the pairing is established, and the user sends the document to the printer, it is printed by the adversary's printer in the next building. However, this scheme is more secure in the scenarios where both the devices are camera equipped and also have displays. Since [13] is a variation of SiB, so this scheme has some of the same limitations as SiB, such as requiring close proximity and a camera in at least one of the devices. Further, in the case of L&C and HAPADEP more research and development is required in the areas of speech engines, audio codec technology, as well as in L&C Dictionary. Moreover, L&C and HAPADEP also suffer from the fact that users cannot be forced to carefully listen to the audio played by the devices. It means a user who does not understand the importance of security might not pay proper attention to the sound played by the devices, and thus can easily ignore the verification stage, and may confirm a false match. Secure pairing of devices by shaking them together is an interesting approach. However, these schemes require embedded accelerometers in both the devices. Further, shaking devices together is always not possible, since there is a large variety of devices, such as printers, projectors, and laptops that cannot be held and shaken together simultaneously.

In contrast to the above approaches, AMIGO [18] and Shake Them Up [17] exploit radio signals to establish the secure channel. Since AMIGO uses the similarity of radio signals from the nearby access points, it is not applicable in scenarios, where the radio data is not available to process or where the wireless network is easy to eavesdrop on while remaining physically hidden to the bona fide users. Further, it is hard to identify the intended device in AMIGO when many other devices surround it, because in the proposed scheme, the calculated physical proximity is of coarse granular nature. Moreover, it is also a fact that in many developing countries 802.11-based wireless technology is less popular compared to Bluetooth technology that is common due to the widespread use of mobile phones. *Shake Them Up* is susceptible to attack by an eavesdropper who exploits the differences in the baseband frequencies of the two radio sources. Biometric-based solutions to device pairing are considered to be good from the usability point of view in which biometrics is used as an out-of-band channel. The reason is that biometric-based channels put little cognitive load on the users. However, the logic and calculations to accurately recognize the biometric patterns are a heavy burden on its applications. Since no two biometric measurements, even coming from the same user and using the same measurement setup are identical; the issues regarding the accuracy of recognition techniques still need more research and improvement. Another drawback of this approach is that it requires biometric readers in both the devices.

Bluetooth pairing requires the human operator to put the communicating partners into discovery mode. After discovery and selection of a device, the channel is secured by entering the same PIN or password into both devices that give rise to a number of usability and security issues [25,26]. For example, a short password or PIN number makes it vulnerable to dictionary or exhaustive search attacks. In [25] it was shown that an adversary can easily derive a 4-digit PIN from an eavesdropped communication during pairing process in less than 0.06 s on a common computer by mounting brute force attack. Further, in Bluetooth pairing an adversary can eavesdrop to break the security from a long distance using powerful antennas. As a consequence, the Bluetooth Special Interest Group (SIG) reacted to these concerns by creating Secure Simple Pairing. The protocol supports four association modes: passkey entry, numeric comparison, just works, and an out-of-band model. As far as NFC is concerned, it is extremely short-range technology as compared to other short-range technologies, such as Infrared and Bluetooth. Therefore in many scenarios, NFC is used in combination with Bluetooth, where NFC is used for pairing (Authenticating) a Bluetooth session used for the transfer of data. NFC setup time is much shorter than Bluetooth. NFC requires less than 0.2 ms to set up the connection; while Bluetooth requires approximately 6 s. Soriente et al. [27], described different possible types of attacks on NFC. For example, NFC offers no protection against eavesdropping and is also vulnerable to data corruption and data modifications. However, it is practically impossible to launch MiTM attack in NFC, especially when Active–Passive communication mode is used [27]. WUSB project is perceived to have failed at the end of 2008 after the withdrawal of Intel. Two major reasons that play a role in its failure are the need of a power supply cable for the WUSB devices and the consumption of a large amount of energy.

Some other efforts toward providing secure device association include Lokey [28], manual authentication [29], a generic framework [30], and NFC-based schemes [31]. LoKey uses SMS messages to authenticate key exchange over the Internet, which incurs substantial monetary cost and delay. Gehrmann et al. [29] proposed several manual schemes that enable handheld devices to authenticate their public keys by some kind of user interaction. In the proposed schemes, the user manually exchanges short message authentication codes between the devices. These short message authentication codes are strings of very short length, between 16 and 20 bits. For example, in one of the proposed method the user is required to compare the short strings displayed on the screens of intended pair able devices. While, in another case in which one of the device is display-less, the user is required to type the short string displayed on the first device onto the other device (i.e., display-less device).

In summary, each of the proposed schemes we surveyed has strengths and weaknesses—often in hardware requirements, strength against various attacks or usability in particular scenarios. We proposed in [30] a generic framework for secure device association. In the proposed system devices first register their capabilities with the directory service. Then, whenever two devices need to create an association, the client (device A) queries the directory service to discover and acquire the required information to initiate a secure pairing with the target device (device B). On the basis of the information from directory service, both the client (device A) and resource (device B) mutually execute a common pairing protocol. The protocol that is chosen can be selected to achieve mutually agreeable levels of security and usability within the constraints of the devices available and the scenario the users find themselves in.

1.3 Future Directions for Research

From the comparative analysis presented in the previous section, we can conclude that no one has yet devised the perfect pairing protocol. Pairing protocols vary in the strength of their security, the level of required user intervention, their susceptibility to environmental conditions and in the required physical capabilities of the devices, and the required proximity between the devices. It is therefore appropriate to investigate ways of integrating different pairing protocols within a general architecture for providing secure and usable pairing mechanisms for a large set of *ad hoc* scenarios.

Further, most of the prior work on secure device association considered the demonstrative approach (i.e., requires user involvement and/or manual efforts to identify the intended partner) for the identification and discovery of the intended pair able colocated device. For example, in SiB [12] and the Resurrecting Duckling Security Model [2], the discovery of the intended pair able device is performed manually; while in Talking to Strangers [9] communicating partners exchange their connectivity information over the secondary channel (i.e., infrared). However, in many situations automatic device discovery is required. If we continue to multiply the number of manuals or out-of-band discovery mechanisms, users will become confused about the selection of the device discovery method during pairing process. For instance, a user wanting to create an association of a mobile phone having microphone, speaker, camera, display, and infrared with another mobile phone having microphone, speaker, display, no camera, and no infrared might be confused about the varied types of manual or out-of-band possibilities of device discovery [13]. We therefore agree with the view proposed by Saxena et al. [13] that it should not be the user's responsibility to figure out how and which method to use for device discovery each time; instead an automatic device discovery should take place. One of the efforts toward this is taken by Malkani et al. [30,32]; however, this issue requires more attention from researchers.

Moreover, as described earlier, there is a large and growing literature on secure device association. However, some of the proposed techniques or protocols have not been implemented; while others are implemented and evaluated in a stand-alone manner without being compared with other related works. Examples of these include the Resurrecting Duckling Security Model [2], Talking to Strangers [9], AMIGO [18], Shake Well Before Use [6], some of the Saxena et al.'s proposed methods [33], and four variants of the BEDA [21] approach. It might be because of unavailability of such tools that provide a common platform to test the usability or security of these methods. This creates the need to design new tools such as simulators, benchmarks, and usability testing frameworks (e.g., [35,36]) that can be used to evaluate the existing as well as new pairing schemes. Finally, in Table 1.1, we have summarized the features of some of the device association methods described in this chapter.

1.4 Conclusions

Wireless networks are common-place nowadays and almost all modern computing devices support wireless communication in some form. These networks differ from more traditional computing systems due to the *ad hoc* and spontaneous nature of interactions among devices. These systems are prone to security risks, such as eavesdropping, and require different techniques as compared to traditional security mechanisms. Recently, secure device association has got significant attention from many researchers and a significant set of techniques and protocols have been proposed. More recently, numerous standardization and industrial bodies, (such as Microsoft, WiFi Alliance, Bluetooth SIG, and the USB Forum) have also recognized the significance of this problem and are working on specifying more general, usable, and secure procedures for device pairing. However, as we have shown in our survey of the state of the art, currently available schemes for secure device pairing vary in their security against different attacks, in the needed hardware capabilities, and in the necessary level of user attention. Some of these techniques consider devices equipped with infrared, laser, or ultrasound transceivers, while others require embedded accelerometers, cameras and/or LEDs, displays, and microphones and/or speakers. Some techniques exploit the knowledge of radio environment to securely pair the devices; others require the user's careful attention and significant manual intervention in the pairing process. However, less attention is paid toward the issue of a more generic or standard pairing mechanism or infrastructure that covers a large set of *ad hoc* scenarios. Finally, we attempted to highlight the gaps left by prior work and presented some future research directions using a survey and a comparative analysis of the existing methods for secure device association. Finally, we envision that in a world of heterogeneous devices and requirements, we need mechanisms to allow automated selection of the best device association protocols without requiring the user to have an in-depth knowledge of the minutiae of the underlying technologies. Further, these mechanisms should facilitate unobtrusive device identification, matching of pairing

Table 1.1   Features Summary of Device Association Methods

Pairing Method	Minimum Hardware or Equipment Required in Each of the Device		Human/User Effort Required	Out-of-Band/Location-Limited Secondary Channel
	Device A	Device B		
Resurrecting Duckling Security Model	A cable and the same physical inter-face (e.g., USB port) on both devices		Set up cable connection between the devices	Cable
Talking to Strangers	Infrared (IrDA) port on both devices		Set up infrared (IrDA) connection between the devices	Infrared (IrDA)
Smart-its-Friends	2D accelerometers on both devices		Move/shake devices together simultaneously until the response signal is received	Accelerometer/Motion/Tactile
Are You with Me?	2D accelerometers on both devices		Walk around to shake the devices (sensors) for a certain time period	Accelerometer/Motion
Shake Well Before Use	2D accelerometers on both devices		Move/shake devices together simultaneously until the response signal is received	Accelerometer/Motion/Tactile
SiB	Display	Photo camera	Properly place camera of device B at the displayed bar code on device A with sufficient proximity and take the photograph	Visual
L&C (Display–Speaker)	Display	Speaker	Compare the MadLib sentence displayed on the screen of device A with the vocalized MadLib sentence from device B	Combination of audio and visual

continued

Table 1.1 (continued) Features Summary of Device Association Methods

Pairing Method	Minimum Hardware or Equipment Required in Each of the Device		Human/User Effort Required	Out-of-Band/Location-Limited Secondary Channel
	Device A	*Device B*		
L&C (Speaker–Speaker)	Speaker	Speaker	Compare the two vocalized MadLib sentences from both of the devices	Audio
HAPADEP	Speaker	Microphone	Compare two audible sequences/melodies	Audio
Shake Them Up	802.11 network card/interface	802.11 network card/interface	Shake/twirl/move devices around until pairing is done or the response signal is received	Combination of 802.11 and motion
AMIGO	802.11 network card/interface	802.11 network card/interface	Shake/wave hand near the device until pairing is done or the response signal is received	Combination of 802.11 and tactile
BEDA (Button to button)	A single button on both devices		Press button on both devices simultaneously with random time intervals until the response signal is received	Tactile
BEDA (Display to button)	Display	A single button	Press and release the button on device B whenever the display of device A flashes	Tactile
BEDA (Short vibrations to button)	Vibration capability	A single button	Press and release the button on device B whenever device A vibrates	Tactile
BEDA (Long vibrations to button)	Vibration capability	A single button	Press and hold the button on device B while device A vibrates	Tactile

Source: Data from Malkani, Y. A. and L. D. Dhomeja, in *Proceedings of 5th IEEE International Conference on Emerging Technologies* (ICET-09). © IEEE 2009.

techniques to requirements, and chains of communication to bridge between devices of different capability and improved security by combining techniques where possible.

Acknowledgments

Some pieces of information are taken from the author's earlier published work [30,34].

Terminologies

Authentication
Device pairing
Eavesdropping attacks
Mobile/*Ad hoc* systems
Security initialization
Spontaneous interaction
Out-of-band channels

Questions and Sample Answers

1. Describe the term "device pairing."
   Device pairing is the task of establishing or bootstrapping a secure communication link between two devices in close proximity. To achieve this, the protocol must consider the absence of any prior common device context and trusted third party. Secure device association, security initialization, and secure first-connect are some of the alternative terms used to describe the process of device pairing.

2. What is meant by "out-of-band channel"?
   An out-of-band channel is a secondary communication channel. Such a channel usually has additional security guarantees (e.g., confidentiality or message integrity) that help to create a secure association between a pair of devices. In many cases, the additional security comes through the absence of vulnerability to attacks on the network and/or a requirement that engagement with the channel is physically visible to the users, and it might be as simple as direct person-to-person verbal exchange. Out-of-band channels are also known as location-limited side channels or constrained channels. One of the major uses of out-of-band channels is to transfer messages for authentication during the pairing process. Out-of-band channels can be categorized into two broad categories: input out-of-band channels and output out-of-band channels. The first category is usually used to enter some data into the device(s) during the pairing process, such as entering PIN code or passkey using a keypad. The latter category is used for verification purposes through the use of some output capability of the device, such as a display.

3. List any four common sources that represent input out-of-band channels.
   Keypad/Button, microphone, camera, and accelerometers.

4. List any four common sources that represent output out-of-band channels.
   Display, speaker, LED, and vibrators.

5. Why is the problem of secure device pairing challenging?

There are several reasons that make secure device pairing a challenging real-world problem. A list of some of the major reasons is given below:

1. *Wireless technology:* Devices involved in device pairing scenarios use wireless technology in some form and thus are susceptible to eavesdropping. As a consequence, it also opens doors for other security threats, such as MiTM attack.
2. *Ad hoc* and spontaneous interaction among the devices.
3. No preshared secret between the intended pairing partners.
4. Unavailability of centralized trusted third party.
5. Nonexistence of any offline or online security infrastructure, such as PKI.
6. Lack of common device context/capabilities between the devices.
7. Devices' heterogeneity in terms of communication channel, power requirements, and available sensor technology.

6. What are the major requirements for device pairing solutions?

When proposing a solution to a certain problem, one must need to consider its essential requirements. In the same way, there are also some major requirements that need significant attention when proposing/developing a solution for secure device pairing. A list of these requirements is given below:

1. *Usability:* This requirement states that the process of secure pairing should be easy to use and comprehensible by an ordinary (nontechnical) user.
2. *Security:* An attack against the pairing process should not be possible without an extensive preplanning and the use of very sophisticated equipment.
3. *No extra hardware:* The solution should avoid the addition of any extra hardware in the devices to properly carry out the pairing process.
4. *No additional interface:* This requirement states that the solution should use the same communication channel for both security initialization and further communication between the devices.
5. *Support for device heterogeneity:* The solution should support pairing in varied scenarios (use cases), with various device capabilities.

7. Write a detailed scenario of your choice that demonstrates the importance of secure device pairing in everyday life.

The scenario presented below clearly demonstrates the need and importance of secure device pairing in every day life.

Let us first introduce Angela who is working in a well-reputed organization. She organizes a meeting with representatives of some customers to give them a confidential briefing about a new product that her company is launching in the near future. The meeting is organized in a hotel equipped with modern smart devices, but which is unfamiliar to Angela. On the meeting day, Angela is getting late, so she leaves her office in hurry and forgets to print some important documents required during the meeting. When she reaches the hotel, she wants to pair her laptop with a nearby printer to print the documents, without having to gain special permissions on the hotel network or pass files to a receptionist. That she has been allowed into the room with the printer is sufficient credentials. Next she goes to the meeting room, where she wants to pair her laptop with the projector securely, since the presentation carries some sensitive data. In addition to preventing eavesdroppers on a connection expected to last for several hours, Angela's laptop selects a mechanism that allows her to demonstrate to the room that the data are coming from her laptop. After her meeting and before leaving, she needs to discuss a confidential issue with her boss. At this time, she wants to pair her Bluetooth-enabled headset with her mobile phone. Finally, when she finishes everything and needs to leave the hotel, she wants to provide the hotel with a signature stored on her work contactless smart ID card to use in authenticating their invoice.

8. Briefly describe the out-of-band association model of Bluetooth Secure Simple Pairing.
   The out-of-band association model of Bluetooth Secure Simple Pairing is designed to be used with several possible out-of-band channels, such as NFC technology. It addresses the two major requirements of device pairing, security and usability (or simplicity). It can be used in the scenarios where a demonstrative approach to device pairing is desired. For example, this association model allows the user to demonstratively discover (identify) the intended devices to establish the association between them. Since, in this model, cryptographic material is exchanged over the out-of-band channel between intended devices, thus the security of this model also relies on the type of out-of-band channels used during the pairing process. For example, when NFC is used as an out-of-band channel, it is hard to mount the MiTM attack due to the characteristics of the NFC channel. Further, the user's experience from usability point of view may also vary in several device pairing scenarios depending on the chosen out-of-band channel. For instance, in an NFC-based out-of-band channel, the user only needs to touch the two devices together to initiate the security relation between a pair of devices.

9. Name some of the short-range wireless data standards.
   1. Bluetooth 2.0/2.1/3.0
   2. NFC
   3. WiFi (IEEE 802.11)
   4. WUSB
   5. UWB
   6. Infrared (however, less common in modern devices)
   7. Wibree
   8. Zigbee—IEEE 802.15.4

10. What are the major features of NFC that make it an important technology for very short-range communication?
    Some of the major features of NFC that make it an important technology for short-range wireless communication are listed below:
    1. *Availability:* Recently, NFC has got significant attention from the industry, and become a rapidly growing technology. We advocate that NFC has found its place in the market, and currently a large number of NFC-enabled devices are available in the market.
    2. *Usability:* NFC provides user-friendly methods to establish the link between two NFC-enabled devices, such as simply touching a pair of devices or holding them in very close proximity.
    3. *Security against MiTM attack:* Owing to the characteristics of NFC, it is extremely difficult for an adversary to successfully mount the MiTM attack.
    4. *Support for variety of applications:* NFC has a number of applications, such as smart posters, easy payment methods for goods and ticketing, maintaining automatic attendance records for employees in an organization, and so on.
    5. *Compatibility:* NFC is compatible with other similar existing infrastructures, such as contactless infrastructure of ticketing and transportation.

Author's Biography

Yasir Arfat Malkani is a lecturer in the Institute of Mathematics and Computer Science (IMCS), University of Sindh, Jamshoro, Pakistan. Currently, he is a DPhil student and associate tutor at the University of Sussex, Brighton, United Kingdom. He was awarded the Vice Chancellor's silver medal for securing first position in MSc computer science at the University of Sindh in 2003. He was appointed as a research associate in the University of Sindh in 2004, and then as a lecturer in July 2005. He was awarded a PhD scholarship from the University of Sindh in 2006 to pursue his DPhil studies at the University of Sussex. His main area of research is pervasive computing. He studies the issue of establishing whether two devices are colocated and enabling secure communication based on evidence of colocation, without any other prior knowledge of each other. He has defined a framework and core protocol for such a system and implemented a basic prototype. Ongoing work is being undertaken in analyzing the security and usability of existing protocols, developing and testing features in his protocols, and building a more general implementation for evaluation.

Dan Chalmers is a senior lecturer in the Software Systems group of the School of Informatics at the University of Sussex, Brighton, United Kingdom. Before working at the University of Sussex he worked for Imperial College London and Ericsson Ltd. He has a BEng(Hons) in software engineering from University of Manchester Institute of Science and Technology (UMIST)—now a part of the University of Manchester, Manchester, United Kingdom, an MSc in advanced computing and a PhD, both from the Department of Computing, Imperial College London. His research focuses on the way knowledge of context (including resource limits, location, and other physical and social aspects of context) can be used to modify behavior and affect data display and configuration of systems.

Ian Wakeman is a senior lecturer in the Software Systems group of the School of Informatics at the University of Sussex, Brighton, United Kingdom. He has a BA in electrical and information sciences from Cambridge University, Cambridge, United Kingdom, an MS from Stanford University, California, and a PhD from University College London (UCL), London, United Kingdom. His research could be described as user-centered networking, investigating protocols and techniques to make computer networks work for people. This has spawned over 50 refereed papers in fields as diverse as congestion control for packetized video and programming languages for active networks and has more recently focused on trust-based approaches for network and system configuration in pervasive computing.

References

1. Diffie, W. and M. E. Hellman, New directions in cryptography. *IEEE Transactions on Information Theory*, 1976; IT-22(6): 644–654.
2. Stajano, F. and R. Anderson, The Resurrecting Duckling: Security issues for ad-hoc wireless networks, in *Security Protocols*. 2000. Springer: Berlin/Heidelberg, pp. 172–182.
3. Stajano, F., The Resurrecting Duckling–what next?, in Revised Papers from the *8th International Workshop on Security Protocols*. 2001, Springer: Berlin/Heidelberg, pp. 204–214.
4. Stajano, F. and R. Anderson, The Resurrecting Duckling: Security issues for ubiquitous computing. *Computer*, 2002; 35(4): 22–26.
5. Holmquist, L. E., et al., Smart-Its Friends: A technique for users to easily establish connections between smart artefacts, in *Proceedings of the 3rd International Conference on Ubiquitous Computing*. 2001, Springer: Berlin/Heidelberg, pp. 116–122.

6. Mayrhofer, R. and H. Gellersen, Shake Well Before Use: Authentication based on accelerometer data, in *5th International Conference on Pervasive Computing (Pervasive 2007)*. 2007. Toronto, Ontario, Canada.

7. Kirovski, D., M. Sinclair, and D. Wilson, The Martini Synch: Joint fuzzy hashing via error correction, in *Security and Privacy in Ad-hoc and Sensor Networks*. 2007. Springer: Berlin/Heidelberg, pp. 16–30.

8. Lester, J., B. Hannaford, and G. Borriello, Are You with Me?–using accelerometers to determine if two devices are carried by the same person, in *Pervasive Computing*. 2004. Springer: Berlin/Heidelberg pp. 33–50.

9. Balfanz, D., et al., Talking to strangers: Authentication in adhoc wireless networks, in *Symposium on Network and Distributed Systems Security (NDSS '02)*. 2002. San Diego, CA.

10. Mayrhofer, R., M. Hazas, and H. Gellersen, An authentication protocol using ultrasonic ranging, *Technical Report*. 2006, Lancaster University.

11. Mayrhofer, R. and M. Welch, A human-verifiable authentication protocol using visible laser light, in *the 2nd International Conference on Availability, Reliability and Security*, ARES 2007. Vienna, Austria.

12. McCune, J. M., A. Perrig, and M. K. Reiter, Seeing-is-believing: Using camera phones for human-verifiable authentication, in *IEEE Symposium on Security and Privacy*, 2005. Oakland, California, pp. 110–124.

13. Saxena, N., et al., Secure device pairing based on a visual channel, in *SP'06: Proceedings of the 2006 IEEE Symposium on Security and Privacy (S&P'06)*, 2006. IEEE Computer Society, Washington, DC, pp. 306–313.

14. Goodrich, M. T., et al. Loud and clear: Human-verifiable authentication based on audio, in *26th IEEE International Conference on Distributed Computing Systems, ICDCS 2006*. Lisbon, Portugal.

15. Soriente, C., G. Tsudik, and E. Uzun, HAPADEP: human assisted pure audio device pairing. *Cryptology ePrint Archive*, Report 2007/093, 2007.

16. Kirovski, D., M. Sinclair, and D. Wilson, The Martini Synch. *Technical Report* MSR-TR-2007-123, Microsoft Research, September 2007.

17. Castelluccia, C. and P. Mutaf, Shake them up!: A movement-based pairing protocol for CPU-constrained devices, in *Proceedings of the 3rd International Conference on Mobile Systems, Applications, and Services*. 2005, Seattle, Washington: ACM.

18. Varshavsky, A., et al., Amigo: Proximity-based authentication of mobile devices, in *Ubiquitous Computing, UbiComp 2007*, 2007. Innsbruck, Austria, pp. 253–270.

19. Buhan, I. R., J. M. Doumen, P. H. Hartel, and R. N. J. Veldhuis, Feeling is believing: A location limited channel based on grip pattern biometrics and cryptanalysis. Technical Report TR-CTIT-06-29, Centre for Telematics and Information Technology, University of Twente, Enschede. 2006.

20. Buhan, I., et al., Secure ad-hoc pairing with biometrics: SAfE, in *Proceedings of First International Workshop on Security for Spontaneous Interaction (IWSSI '07)*, 2007. Innsbruck, Austria.

21. Soriente, C., G. Tsudik, and E. Uzun, BEDA: Button-enabled device association, in *International Workshop on Security and Spontaneous Interaction (IWSSI '07)*. 2007 Innsbruck, Austria.

22. The Official Bluetooth® Technology Info Site: http://www.bluetooth.com

23. The Near Field Communication (NFC) Forum: http://www.nfc-forum.org/home

24. Spahic, A., et al., Pre-authentication using infrared, in *Privacy, Security and Trust within the Context of Pervasive Computing*, 2005. Springer: US, pp. 105–112.

25. Shaked, Y. and A. Wool, Cracking the Bluetooth PIN, in *MobiSys '05*, in *Proceedings of the 3rd International Conference on Mobile Systems, Applications, and Services*, 2005. Seattle, Washington: ACM.

26. Jakobsson, M. and S. Wetzel, Security weaknesses in Bluetooth, *Lecture Notes in Computer Science*, 2001. Springer: Berlin/Heidelberg, pp. 176–191.

27. Haselsteiner, E. and K. Breitfuß, Security in near field communication (NFC), in *Proceedings of Workshop on RFID Security*, July 2006. Graz, Austria, pp. 3–13.

28. Nicholson, A., et al., LoKey: Leveraging the SMS network in decentralized, end-to-end trust establishment, in *Pervasive Computing*. 2006. Springer: Berlin/Heidelberg, pp. 202–219.

29. Gehrmann, C. and C. J. Mitchell, Manual authentication for wireless devices. *RSA Cryptobytes*, 2004; 7(1): 29–37.

30. Malkani, Y. A., et al., Towards a general system for secure device pairing by demonstration of physical proximity, in *MWNS-09 Co-located with IFIP Networking 2009 Conference*, 2009. Shaker Verlag: Aachen, Germany, pp. 13–24.

31. Rekimoto, J., et al., Proximal interactions: A direct manipulation technique for wireless networking, in *INTERACT 2003*. 2003. Zurich, Switzerland.

32. Malkani, Y. A., D. Chalmers, and I. Wakeman, Towards a general system for secure device pairing by demonstration of physical proximity (Poster), in *UBICOMP Grand Challenge: Workshop on Ubiquitous Computing at a Crossroads: Art, Science, Politics and Design*. 6th and 7th January, 2009, Huxley Building, Imperial College, London.

33. Saxena, N. and J. Voris, Pairing devices with good quality output interfaces, in *International Workshop on Wireless Security and Privacy (WISP) (co-located with ICDCS)*, 2009. Montreal, Quebec, Canada.

34. Malkani, Y. A. and L. D. Dhomeja, Secure device association for ad hoc and ubiquitous computing environments, in *Proceedings of 5th IEEE International Conference on Emerging Technologies* (ICET-09), 2009. Islamabad, Pakistan.

35. Kostiainen, K., et al., Framework for comparative usability testing of distributed applications. *Technical Report NRC-TR-2007-005*, Nokia Research Center, Helsinki, Finland, 2007.

36. Mayrhofer, R., Towards an open source toolkit for ubiquitous device authentication, in *Proceedings of the Fifth IEEE international Conference on Pervasive Computing and Communications Workshops*, 2007. IEEE Computer Society, Washington, DC, pp. 247–254.

Chapter 2

Securing Route and Path Integrity in Multihop Wireless Networks

Stephen Glass, Marius Portmann,
and Vallipuram Muthukkumarasamy

Contents

2.1 Introduction

This chapter examines the problem of securing route integrity in multihop wireless networks such as mobile *ad hoc* networks (MANETs) and wireless mesh networks (WMNs). These networks are often based on the same technologies as conventional infrastructure-based networks. Standards such as WiFi (IEEE 802.11) and WiMax (IEEE 802.16) provide support for both infrastructure and multihop operation.

In an infrastructure wireless network, an ordinary network node is normally known as a Mobile Station (MS). The MS does not communicate directly with other MSs. Instead it communicates directly only with its Base Station (BS) and the BS forwards traffic appropriately. As a result, the BS and MS must be within radio communications range if the MS should have service. In contrast, nodes in a multihop wireless network communicate directly with their neighbors and route and forward traffic on their behalf. As long as the route exists between the source and destination, service can be maintained. A major advantage of multihop networks is that the region over which they provide coverage called the service area is larger than that seen for conventional infrastructure networks.

Multihop networks are also more flexible and resilient than their infrastructure equivalents because they allow multiple routes between communicating network nodes. These routes are discovered and maintained autonomously by the network nodes themselves and so multihop networks are said to be *self-organizing*. If a route to the destination becomes unworkable then an alternative route can be discovered and used. This means the network can often repair itself in the presence of changing topology, temporary interference, or deliberate radio-frequency jamming and gives rise to the description of multihop networks as being *self-healing*. Multihop wireless networks are attractive for applications where flexibility, resilience, and a large network service area are required.

2.1.1 IEEE 802.11 and IEEE 802.11s

In this discussion, particular attention is focused on multihop wireless networks conforming to the IEEE 802.11 standard [1]. This allows us to consider concrete examples of both security problems and their solutions. IEEE 802.11 wireless local area networks use inexpensive hardware and enable reliable and relatively high-bandwidth communications in radio spectrum that permits unlicensed access. From its inception, IEEE 802.11 has defined support for multihop networking in a mode of operation known as *ad hoc* or the Independent Basic Service Set (IBSS).

An IEEE 802.11 IBSS defines both the medium access control (MAC) and physical (PHY) layers but leaves unspecified the question of routing and many of the details needed to create a workable multihop network. Commercial WMNs have used IEEE 802.11 with proprietary extensions to deal

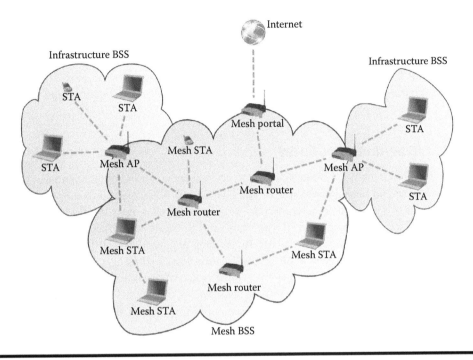

Figure 2.1   IEEE 802.11s network architecture.

with such issues as network formation, connection of the wireless network to the wired infrastructure, performance optimization, and routing. This has led to a lack of interoperability. IEEE 802.11s standards amendment aims to address these problems and specifies mandatory solutions to address the problems of interoperability in multihop networks [2]. IEEE 802.11s defines the Mesh Basic Service Set (MBSS), which is intended for small- to medium-scale WMNs with a maximum of 32 and 64 nodes.* Figure 2.1 shows network architecture in which network nodes are divided into different types depending on their functions:

- *Mesh Station (Mesh STA):* The most basic elements of an MBSS, these nodes participate in routing and forwarding traffic on behalf of their neighbors.
- *Mesh Router:* A mesh STA is dedicated to routing traffic. These nodes may have several wireless network interface controllers (WNICs) and additional power reserves.
- *Mesh AP:* A mesh STA which also provides access point (the term used for an 802.11 BS) functionality for infrastructure-mode 802.11 clients. The mesh AP can route frames between the infrastructure BSS and the MBSS.
- *Mesh Portal (MP):* A mesh STA which provides a bridge to a different kind of network. For example, a gateway to a WiMax or Universal Mobile Telecommunications System (UMTS) network which provides Internet access. MPs are important because they typically represent the destination for a significant amount of network traffic.

---

* The IEEE Working Group defines 32 as the intended maximum size of an MBSS but higher figures are often quoted in the literature.

In this description list, we have differentiated between mesh STAs and mesh routers. This is not a distinction made in the draft standard but it does help to clarify the difference between the client devices and network infrastructure. In a MANET, all nodes maybe mobile and there is no network infrastructure or back-haul network. In a WMN, mesh routers may provide a back-haul network to optimize the traffic flow. A more detailed overview of the 802.11s architecture and concepts is given by Hiertz et al. [3].

2.1.2 Implementation Support

In many early WNICs it was common to find that IBSS operation was not implemented or that only a subset of the required behavior was supported. The latter WNICs are said to support an "*ad hoc* demo" or "pseudo-IBSS" mode of operation. Most of these WNICs use a "FullMAC" architecture in which the WNIC contains a slave I/O processor and firmware to implement the whole IEEE 802.11 protocol stack. Upgrading to full IBSS mode often requires the WNIC to be provided with new firmware. In contrast, modern WNICs are often implemented using a "SoftMAC" design in which the WNIC is responsible only for timing critical aspects of the protocol and the host operating system implements the majority of the protocol behavior.

The implementation support for MBSS operations is significantly better than was the earlier support for IBSS. Even before the IEEE 802.11s standards amendment has been finalized implementations have been developed and support is already at an advanced stage. The GNU/Linux kernel, for example, has a software implementation of IEEE 802.11s that is based on the draft standard and which has tracked many changes as the standard develops. This means that SoftMAC WNICs can be used in an 802.11s mesh because the necessary support is already available in the operating system.

Another high-profile implementation of the IEEE 802.11s is the one laptop per child project.* This project has delivered over a million OLPC XO-1 laptops whose network protocol is based on an early draft of the IEEE 802.11s specification. Substantial revision of the draft standard has rendered the frame formats incompatible but newer XO-1s are intended to conform to IEEE 802.11s while interworking with existing XO-1s. Although the XO-1 is based on the GNU/Linux operating system it makes use of a FullMAC design for its network controller. This choice runs contrary to the current trend toward SoftMAC approaches but allows the WNIC to run even while the main CPU is asleep.

In modern WNICs, support for IBSS operation can be taken for granted and MBSS can be expected to be widely available at the time the standard is formally ratified. This was a substantial improvement during the early days of multihop networking. IEEE 802.11s mandates solutions to most of the problems that have led to proprietary extensions and hampered interoperability in existing multihop networks. There will, of course, be problems which the amendment does not address but the existence of standard implementations should encourage further work.

2.2 Background

In this section, we briefly introduce the background concepts of proactive and reactive routing and path-selection protocols in multihop wireless networks. We also describe how proactive and

---

* http://www.laptop.org/

reactive approaches have been combined to optimize routing and path-selection performance for the commonly occurring situations.

2.2.1 Routing Protocols

The purpose of routing protocols is to find the best route(s) between a source node and destination node in the network. It is important that this is done efficiently so that resources are not wasted carrying unnecessary information and reliably so that if a path exists it will be discovered. It is common to assume that the "best" route maybe one which has the fewest hops because such routes can be expected to have lower latency, provide fewer opportunities for problems and reach the destination more quickly. The best route, however, may not always be the shortest. Some links maybe highly congested or subjected to interflow interference from neighboring high-volume links. Many routing protocols allow for the use of *routing metrics* to assess the quality of a route and choose the best one. Hop count is possibly the most widely used metric but there are many other routing metrics that take into account the conditions of the wireless links such as expected transmission count (ETX), expected transmission time (ETT), and weighted cumulative ETT (WCETT).

2.2.1.1 Proactive Routing Protocols

Proactive or link state protocols ensure that every node in the network has up-to-date routes to all other nodes. Such protocols are attractive because they find optimal routes with no delays. The Open Link State Routing (OLSR) protocol is a proactive protocol that is widely used in wireless mesh network research. The protocol itself finds the optimal route based on the smallest hop count between network nodes and is fully described in Internet RFC 3626 [4]. The mechanism by which OLSR distributes link state updates throughout the network is particularly interesting.

In OLSR all network nodes periodically broadcast HELLO messages which identify the network node and its immediate neighbors. Receiving network nodes use this information to discover their immediate one-hop neighbors, whether a bidirectional link exists to that neighbor and identifies those network nodes which are in their two-hop neighborhood. Using the information from the HELLO messages every network node computes a set of multipoint relays (MPRs)—a subset of their neighbors through which they can reach all of their two-hop neighbors.

Route updates are broadcast across the network in the form of a topology control (TC) message. The MPRs are important because they are used to broadcast and rebroadcast TC messages minimizing the amount of traffic broadcast in routing updates. This substantially reduces the number of TC broadcasts compared to conventional flooding mechanism. However, this mechanism ensures that all network nodes will receive the update. An MPR can also choose just to send partial link-state announcements by notifying only links to those network nodes that have selected the MPR. This partial information is sufficient for the receivers to compute optimal routes to all network nodes.

Until draft 1.07 of IEEE 802.11s a variant of OLSR was an optional part of the draft standard. This was known as radio-aware OLSR (RA-OLSR) and modified the frequency of TC updates so that local nodes receive them more frequently than distant nodes. This makes sense because OLSR in general does not guarantee the integrity of routing information at any node and relies on frequent updates to minimize any node having unsynchronized routing information. RA-OLSR was, however, only an option and has now been removed from the standard.

2.2.1.2 Reactive Routing Protocols

Reactive or on-demand protocols work well in dynamic networks in which there are many mobile nodes and a rapidly changing network topology. The *Ad Hoc* On-demand Distance Vector (AODV) routing protocols are examples of a reactive routing protocol that is widely used in multihop wireless networks and is fully described in RFC 3561 [5]. Reactive routing protocols have a lower overhead than proactive protocols because they discover and maintain routes only when they are actually needed. This reduces traffic overheads for the network which need not manage routes that are never used. The disadvantage is that reactive protocols incur a delay when first establishing the connection during which the route is discovered.

The process for finding a route from a source node to a destination in AODV is illustrated in Figure 2.2. The process consists of two parts and begins when the source node broadcasts a Route Request (RREQ) packet to all of its immediate neighbors. Receivers will inspect the sequence number in the RREQ to determine if they have already seen the request before. If not they will update their routing table with an entry to the source and rebroadcast the request. Duplicate requests are simply ignored; this is known as *duplicate suppression* and prevents the presence of routing loops and the *counting to infinity* problem as well as reducing the number of rebroadcast messages. If the request has not previously been received then it is rebroadcasted. In Figure 2.2a, the arrows show how node *s* broadcasts the RREQ looking for a path to node *d* and the path taken as the broadcast advances through the network.

The second part of the process begins when the RREQ is received at node *d*. The destination replies to the source by sending a unicast Route Reply (RREP) along the path which the broadcast traversed to the destination. The route taken by the unicast reply is shown by the arrows in Figure 2.2b. As an optimization it is possible for a node receiving an RREP to reply on behalf of the final destination if it already has a routing table entry. In this case, the intermediate node sends the RREP to the source and a gratuitous unicast RREP to the actual destination. The latter ensures that the destination has routing table entries back to the intermediate node.

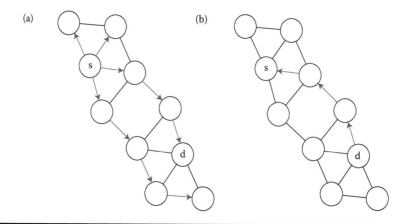

Figure 2.2   Routing of RREQ and RREP messages through network. (a) Broadcast Route Request (RREQ). (b) Unicast Route Reply (RREP).

2.2.2 Path-Selection Protocols

It is usually the routing protocol in the network layer which is responsible for ensuring that a frame reaches its final destination across multiplehops and multiple potential paths. Path selection is the act of finding a path at the MAC layer between a source and destination. It is directly analogous to routing but uses MAC addresses instead of IP addresses to route traffic. It is widely recognized that there are major advantages to having multihop wireless networks perform path selection in the MAC layer. This is because the MAC layer has timely and accurate information on network topology, link-quality, data rates, and queue lengths.

The Hybrid Wireless Mesh Protocol (HWMP) path-selection protocol is a mandatory new path-selection protocol introduced by IEEE 802.11s. It is a hybrid path-selection protocol in that it combines both proactive and reactive approaches to path selection. If a "root node" exists then HWMP uses proactive routing to find and maintain a route to it. Root nodes usually represent MPs that serve as gateways to non-802.11 networks. Proactively maintaining a path to a root node is, therefore, an optimization for one of the most likely traffic destinations. For all other nodes reactive path discovery is used exclusively. Reactive path discovery uses protocol primitives and rules from the AODV routing protocol. A detailed introduction to the 802.11s HWMP path selection protocol is provided by Bahr [6,7].

Although IEEE 802.11s mandates support for HWMP it also allows for alternative path selection algorithms to be implemented and also permits a "null" path-selection profile. The latter does not provide any path-selection algorithm at the MAC layer and so network nodes can communicate only with their direct neighbors. This allows for existing routing algorithms to be used at the network layer and so HWMP can be replaced with a traditional routing algorithm if desired.

2.3 Securing Routing and Path Selection

When multihop wireless networks are used in sensitive applications there must be robust security protocols that ensure secure operation. The goals of the security protocols should be to ensure the confidentiality, integrity, authenticity of network traffic, and preserve the availability of communications. Attacks intended to compromise routing integrity are a significant threat and can lead to a network wide loss of availability.

2.3.1 Threat Model

A threat model is the key to identifying and managing the key security threats faced by a secure system. The purpose of the threat model is to identify the sources of risk and select appropriate countermeasures. In the case of a multihop wireless network it is appropriate to divide the threat model into two and identify threats posed by outsiders and then threats posed by compromised nodes.

2.3.1.1 Threats Posed by Outsider Attacks

The goal of the security protocol is to divide the world into two: the trusted insiders and the untrusted outsiders. In the presence of an effective security protocol the outsider attacks are reduced to a handful of possibilities. The kind of attacks one could expect are to see signal jamming denial of service, traffic analysis, man-in-the-middle, and wormhole attacks. The self-healing property of multihop networks means that signal jamming has an effect in the immediate locality of the jammer but that any damage will be routed around to the extent possible. Traffic analysis and man-in-the-middle attacks are of limited utility to an adversary but the wormhole attack represents a singular opportunity to impact network availability and is described in more detail below.

2.3.1.2 Threats Posed by Compromised Nodes

Infrastructure-based networks often use purpose-built facilities to protect network equipment such as BSs. These facilities can provide physical protection and alarms to prevent and detect attempts to damage, access, or remove equipment. Many BSs are also protected from power problems using an uninterruptable power supply (UPS). This contrasts with multihop networks such as MANETs, WMNs, and WSNs where the lack of physical security for network nodes is a serious problem. Network nodes maybe distributed throughout areas where they are at risk from hostile adversaries. Ben-Salem and Hubaux describe four key threats arising from the lack of physical security [8]:

- Removal of network nodes
- Inspection of nodes to, for example, recover key material, routing tables or traffic transiting the node
- Modification of the internal state of a node
- Cloning and deployment of compromised nodes

These threats are particularly serious in that they expose the network to a possible hostile attack from a compromised node. Compromised nodes are a problem because they possess the cryptographic keys used by the security protocol to secure traffic and allow the hostile adversary to establish a foothold within the network. This opens the possibility of Byzantine attacks at a later stage when nodes under the control of the adversary collude to subvert the integrity of the network. An effective solution to the problems posed by Byzantine attacks remains an open research topic and generally no suitable mechanism exists to deal with this problem. Instead there is a need to rely on tamper-resistant and tamper-evident mechanisms to, for example, securely delete cryptographic keys if the equipment is interfered with. The security protocol maybe able to help by ensuring rapid revocation of credentials which are believed to be compromised.

2.3.2 Attacks against Routing and Path Selection

The other part of the threat model is the identification of the attacks to which the network maybe subjected. A malicious adversary can disrupt the route discovery process to affect the routing metrics, introduce gratuitous detours, attempt to create routing loops, and overflow routing tables. The following sections describe some of the more serious threats that are posed to the route integrity of a multihop wireless network.

2.3.2.1 Rushing Attacks

Rushing attacks are attacks against reactive routing protocols. The rushing attack subverts the route discovery process and increases the likelihood that the hostile node is included in a given route. This will allow the adversary to perform traffic analysis, conduct further attacks, and prevent routes from being established via other nodes. The rushing attack of the compromised node quickly forwards the route request messages to ensure that the RREQ (or PREQ for path-selection messages) messages from itself arrive earlier than do those from other nodes. The duplicate suppression mechanism will ensure that the duplicate requests arriving later from other nodes will be ignored [9]. The defense against this attack is in two parts: a secure neighbor discovery protocol and a modification to the route discovery logic of the routing protocol. At present these are not integrated into popular wireless routing protocols. This attack remains a potential threat especially when higher layer end-to-end higher layer security protocols are not in use.

2.3.2.2 Gray Holes and Black Holes

A black hole is a station which advertises its willingness to take part in a route but which forwards no traffic. As this is an attack against message forwarding and not route discovery this attack applies equally well to proactive and reactive routing or path-selection protocols. Black holes do not notify the sender of their failure to forward data and so the network's normal self-healing property is compromised. The gray hole is a more slippery variant of the black hole attack in which the compromised node conditionally decides on which traffic it will forward and which traffic it will not. In consequence it can be very difficult to discover.

2.3.2.3 Wormholes

The wormhole attack poses an extremely severe outsider threat to the routing integrity of the network [10]. A wormhole is a specialized man-in-the-middle attack in which the adversary connects the two otherwise distant regions of the network. At first sight, a wormhole appears beneficial because it optimizes traffic flow across the mesh but it does so using a link that is under the complete control of the hostile adversary. The presence of the wormhole subverts the network topology and thus undermines the network routing algorithms. Routes through the wormhole benefit from lower hop counts and other link-quality metrics than legitimate routes and increase the probability that traffic will be routed via the wormhole. The threat is that a wormhole permits an adversary to conduct active traffic analysis and large-scale denial-of-service attacks.

An example of the effect of a wormhole attack is shown in Figure 2.3. The first, Figure 2.3a, shows the network without a wormhole. The edges between nodes are a function of the number of shortest-path links carried by the node. Therefore, the links $\tilde{d}f$ and $\tilde{e}g$ carry more of the network's shortest paths than any of the other links in the network. The second Figure 2.3b, shows the effect on the topology when a wormhole is present. The wormhole is located within radio range of just two nodes c and i and all traffic from one is relayed to the other. This causes the legitimate stations link to establish the $\tilde{c}i$ link that traverses the wormhole and which is shown in light gray. In this example, the $\tilde{c}i$ link carries more of the shortest-path routes than any other link in the network. Thus it is evident that the presence of even a simple wormhole has the potential to substantially distort the network topology.

The wormhole attack can be devastatingly effective because the adversary enjoys complete control of the communication link and can inspect, inject, delay, delete, modify, and reorder traffic. Yet the adversary does not have to compromise the security protocol or any legitimate node for the attack to be effective. Even if the adversary does not understand any of the traffic that they are routing they can use the wormhole attack to substantially damage the network's routing integrity.

2.3.3 Defenses

There are several core mechanisms which have been proposed in the literature to ensure the integrity of routing in multihop wireless networks. These include the application of cryptographic techniques to route discovery and maintenance, the use of reputation-based schemes to choose routes from trusted parties and the use of location and distance information. These are described below.

2.3.3.1 Authenticated Routing

To address these risks to route integrity, a number of secure routing protocols have been proposed that use cryptography-based approaches to prevent many attacks. A survey of secure routing protocols is given by Hu and Perrig [11]. The use of cryptography allows for the authenticity and integrity of routing messages to be established and the nonrepudiation property allows for misbehaving nodes to be unambiguously identified.

A central division in the cryptographic techniques is whether the approach uses asymmetric or symmetric cryptography to sign messages. The latter is several orders of magnitude more efficient than the former and undesirable for high volumes of traffic. An example can be drawn from the Authenticated Routing for *Ad Hoc* Networks (ARAN) protocol in which the intermediate network nodes at every hop must compute and append a digital signature to the RREP/RREQ message [12]. The size of the RREQ/RREP in this protocol grows in direct proportion to the hop count and validating the message requires the verifier to obtain certificates for all the network nodes along the route. Some schemes avoid this overhead and sign messages only at their origin but in these cases they are unable to detect changes made at intermediate nodes.

Public-key cryptography is computationally expensive and so approaches which are less computationally intensive have been proposed. One of the most common is the use of hash chains which was introduced in the Secure *Ad Hoc* On-Demand Distance Vector (SEAD) protocol [13]. Hash chains are efficient and provide similar guarantees of authenticity and integrity to digital signatures but at a lower computational cost.

In practice, the IEEE 802.11s standard does not make use of cryptographic means to protect the integrity of routing control information. The close relationship between HWMP and AODV means that there is scope to follow the approach adopted by secure AODV (SAODV) [14] but for now there is no specific security mechanism for route control messages. Instead IEEE 802.11s defines a MAC layer security protocol for all network traffic and so routing is protected by virtue of the security protocol.

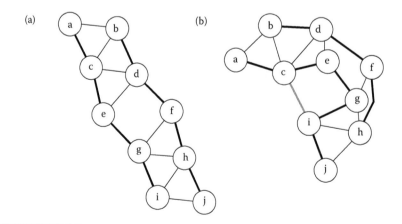

Figure 2.3 Effect of wormhole on network topology. (a) Original topology. (b) Topology with wormhole.

2.3.3.2 Pathrater/Watchdog

A novel approach to detecting misbehaving nodes (including gray holes, black holes, and wormholes) are reputation-based approaches such as the watchdog/pathrater protocol [15]. The first part of this scheme is watchdog which monitors neighbors to see that each of them is behaving honestly and not malfunctioning. This is achieved by using the WNICs to monitor the radio channel and observe the onward transmission of messages. To intercept transmissions from the next hop, the WNIC is placed into *promiscuous* mode in which all frames received at the antenna are passed to the kernel. In early WNICs this was a distinct mode of operation and prohibited normal WNIC operation. Modern WNICs are able to receive promiscuously and maintain normal operations at the same time using distinct logical devices on the same physical WNIC hardware. The watchdog may register a failure to forward which can be for a variety of reasons, for example: a malfunction in the forwarding node, network congestion, or collision at the watchdog receiver. Thus, the watchdog is prone to both false positives and false negatives.

The pathrater is responsible for evaluating the trustworthiness of neighbors and avoids paths via nodes that appear to behave improperly. The pathrater is tolerant of a certain level of failures from the watchdog to account for the problems identified above, which means that it will fail to identify a node that misbehaves only occasionally. Unfortunately, such reputation-based approaches are complicated in multichannel, multihop networks such as IEEE 802.11s supports. When a node employs multiple radios it may forward traffic using a radio channel that cannot be heard by the previous hop. A similar problem occurs when the security protocol uses different session keys for each link. In this case, the watchdog can confirm that it saw an onward transmission but cannot be sure about the content.

2.3.3.3 Packet Leashes

Hu et al. [10] have proposed the use of packet leashes as a defense against wormhole attacks. Packet leashes are intended to restrict the travel of a packet within a tightly defined geographical area. Despite the prefix "packet" this is a MAC layer mechanism as it applies on a per-hop basis. Packet leashes are of two varieties:

- *Geographical:* In which the sender appends their location to the outgoing packet. The receiver can make use of this location to compute a distance vector from the sender to its own location.
- *Temporal:* In which the sender appends a time stamp indicating the time of packet transmission (alternatively, the maximum time at which the packet will be valid). The receiver computes the difference between the packet time stamp and the actual time-of-arrival of the packet. Given this time difference and an estimate for the speed of travel of the signal the receiver can compute the distance to the sender.

It is important that the packet leash cannot simply be changed by the adversary conducting a wormhole attack. The leash must be protected by the security protocol to ensure authenticity. For both geographical and temporal leashes if the distance-vector between the sender and receiver is too large the receiver assumes that this is because a hostile adversary is conducting a wormhole attack. In that case, the node can take remedial measures such as invalidating the routing table entries that traverse the wormhole, blacklisting the node reached via the wormhole and searching for routes via nodes that are not blacklisted.

The necessary condition for the successful use of packet leashes is, therefore, either trustworthy geographical data or precisely synchronized clocks to restrict the travel of a packet within a defined geographical area. Hu et al. suggest that LORAN-C, WWVB, or receivers or atomic clocks can be used to satisfy the requirements for time synchronization and the Global Positioning System (GPS), and can be used to provide both time synchronization and location information.

Of these devices, GPS receivers are now probably the most inexpensive and commonly available to mobile computers and telephones. There are serious practical problems with using radio systems such as LORAN-C, WWVB, or GPS because both the low-frequency and GPS receivers do not work well when indoors and GPS is often problematic when the view of the sky is significantly obscured. A more significant objection to using radio signals as the basis for packet leashes is the inherent insecurity of the approach. Signals arriving at the receiver from the GPS satellites are very weak and even and a relatively low-power transmitter local to the receiver can be used to jam or spoof signals. This is exacerbated by the fact that LORAN-C, WWVB, and civilian-use GPS receivers all make use of unauthenticated signals. An adversary is, therefore, able to completely control the time and/or location reported by the receiver.

Despite the concerns at suggested sources of time synchronization and location information the packet leash proposal is valuable. The use of distance vectors as a means of detecting wormholes is sensible because the adversary cannot normally appear to be any closer than they are in reality. Neighbors in a wireless network are likely to be quite close and limiting the maximum distance between them limits the opportunity for an adversary to conduct a successful wormhole attack.

2.4 Future Directions for Research

The state-of-the-art for securing route integrity in the presence of routing wormholes covers a number of distinct approaches. There are centralized schemes for wormhole detection that adopt graph theoretic [16,17] and software visualization approaches [18]. These centralized approaches are well-suited to wireless sensor networks where each node is resource-limited and detection of an attack is unlikely to be handled by the nodes themselves. Decentralized defenses, in contrast, are ideally suited to MANETs and WMNs where the resources at each node are sufficient to allow for intrusion, detection and prevention, and autonomous action by network nodes to respond to problems will result in more timely action than can be achieved by relaying information to a central monitoring station. The following sections outline some of the decentralized defenses against wormhole attacks.

2.4.1 Secure MAC Protocols

IEEE 802.11 suffered from several serious flaws in its original security protocols which took several years to address. Without an effective security protocol it is impossible to partition the world into trusted insiders and untrustworthy outsiders and there is no defense against hostile adversaries. The IEEE 802.11 CCMP and Temporal Key Integrity Protocol (TKIP) security protocols have proven to be much more robust in practice and have been subject to significant theoretical scrutiny and model checking to verify their robustness [19]. Secure MACs for multihop networks seek to leverage the experience of security protocol design in single-hop networks. The new environment is not the same as the single-hop case because communications with trusted third parties must be appropriately secured and all stages are subject to eavesdropping and malicious interference.

One interesting secure MAC for multihop networks is MobiSEC [20]. MobiSEC extends the infrastructure-mode IEEE 802.11 security protocols to the WMN environment and uses only minimal changes to do so. Clients and mesh routers both authenticate to their peers using the normal IEEE 802.1X authentication and four-way handshake. Mesh routers then perform an additional authentication with a trusted key server in order to join the back-haul network. The additional overhead is justified as a one-off cost which is made only for higher capacity devices. MobiSEC claims fast hand-off and achieves this by using the same link encryption key across multiple links. This not only exposes traffic unnecessarily but makes it difficult to revoke credentials without rekeying all of the other link keys.

IEEE 802.11s takes a similar approach in extending the existing security protocols to the multihop environment. The existing preshared key authentication mechanism has been obsolete because of its inherent insecurity and replaced with a new scheme that uses a trusted third party for preshared key authentication. In most networks it is expected that the same IEEE 802.1X authentication mechanism will be used and, to ensure mutual authentication, it is run twice with each party assuming the role of both supplicant and authenticator. Using a well-known security protocol in this way maybe inefficient but it has the benefit of being well-understood. Kuhlman et al. have also subjected the IEEE 802.11s security protocol to the same kind of model checking that the base protocol and found no evidence of any significant security flaws [21].

2.4.2 Distance-Bounding Protocols

One defense against wormhole and man-in-the-middle attacks that appears very promising is the use of distance-bounding protocols. First introduced by Brands and Chaum the distance-bounding protocol seeks to fix an upper bound on the distance between legitimate parties by using precise timing of a cryptographic bit-commitment protocol in which bits are rapidly exchanged [22]. One distance-bounding protocol is SECTOR's mutual authentication with distance-bounding (MAD) protocol which can be used as a defense against wormhole and more general impersonation attacks [23]. MAD is an extension of the Brands/Chaum protocol in which the protocol is "doubled up"

$u \quad v$

— Initialization phase —

Generate random numbers $r \in \{0,1\}^\ell, r' \in \{0,1\}^{\ell'}$ Generate random numbers $s \in \{0,1\}^\ell, s' \in \{0,1\}^{\ell'}$

Compute commitment $c_u = H(r|r')$ $\xrightarrow{c_u}$ Compute commitment $c_v = H(s|s')$

$\xleftarrow{c_v}$

— **Distance-bounding phase** —

The bits of r are $r_1, r_2, \ldots, r_\ell$ the bits of s are $s_1, s_2, \ldots, s_\ell$

$\alpha_1 = r_1 \xrightarrow{\alpha_1}$

$\xleftarrow{\beta_1} \beta_1 = s_1 \oplus \alpha_1$

...

$\alpha_i = r_i \oplus \beta_{i-1} \xrightarrow{\alpha_i}$ measure delay between β_{i-1} and α_i

Measure delay between α_i and $\beta_i \xleftarrow{\beta_i} \beta_i = s_i \ominus \alpha_i$

...

$\alpha_\ell = r_\ell \oplus \beta_{\ell-1} \xrightarrow{\alpha_\ell}$ measure delay between $\beta_{\ell-1}$ and α_ℓ

Measure delay between α_ℓ and $\beta_\ell \xleftarrow{\beta_\ell} \beta_\ell = s_\ell \oplus \alpha_\ell$

— **Authentication phase** —

$s_i = \alpha_i \oplus \beta_i \ (i = 1, \ldots, \ell)$ $r_1 = \alpha_1$ and $r_i = \alpha_i \oplus \beta_{i-1} \ (i = 2, \ldots, \ell)$

$\mu_u = mac_{k_{uv}}(u|v|r_1|s_1|\ldots|r_\ell|s_\ell)$ $\mu_v = mac_{k_{uv}}(v|u|s_1|r_1|\ldots|s_\ell|r_\ell)$

$\xrightarrow{r'|\mu_u}$

$\xleftarrow{s'|\mu_v}$

Verify c_v and μ_v verify c_u and μ_u

Figure 2.4 Mutual authenticated with distance-bounding (MAD) protocol.

and each participant undertakes both the role of verifier and prover as shown in Figure 2.4. The key problem with distance-bounding protocols is that they require special hardware support to accomplish the high-speed, low-latency bit exchanges. Hancke and Kuhn suggest that submicrosecond times are necessary and suggest the use of ultra-wideband (UWB) transmission to achieve this in the case of RFID devices such as electronic passports [24]. In commercial IEEE 802.11 equipment this is not achievable because there is no provision for a single-bit exchange and the latency for even the smallest message round trip can take hundreds of microseconds. This is several orders of magnitude too large to be useful for distance-bounding protocols.

2.4.3 Secure Neighbor Discovery

The problem of the wormhole attack can be thought of as being a problem of authenticity—is the received frame really being received from the station that claims to send it. More generally this is the problem of secure neighbor discovery and any MAC protocol that can do this will be immune from wormhole, man-in-the-middle, and message replay attacks.

Korkmaz suggests using time-of-flight and signal-power models as part of a neighbor-verification protocol (NVP) [25]. NVP uses timing and power information to authenticate the exchanges. Unlike the secure neighbor verification protocol suggested to defend against rushing attacks the NVP protocol is not cryptographically secure. It is, nevertheless, a potential obstacle to hostile adversaries. The idea that physical aspects of the transmission form a unique "fingerprint" for a transceiver has been investigated by several authors.

An alternative approach is advocated by Eriksson in the form of the TrueLink protocol [26]. TrueLink is not intended to be a true distance-bounding protocol but can be used to establish the authenticity of neighboring nodes. The protocol has two phases. First, RTS/CTS packets are used to exchange nonces between nodes. The timing requirements of this exchange are such that a wormhole cannot relay the RTS/CTS packets. These nonces are then used to answer periodic authentication challenges that are not time critical but prove that the RTS/CTS nonces to be original. An advantage of this approach is that it requires only minor change required to the MAC protocol and can be used with standard hardware.

A MAC protocol modification very similar in spirit to that of TrueLink has been proposed by Glass et al. [27]. This protocol exploits the fact that wireless networks experience much higher bit error rates than do wired networks and in consequence use *positive acknowledgment*. Unicast data frames are explicitly acknowledged and the acknowledgment must arrive within strict time constraints. Even an adversary that conducts a wormhole attack at the PHY layer must contend for access to the radio spectrum at the exit of the wormhole. If the medium is busy the frame must be buffered and replayed at the first opportunity. The problem for the adversary is that the delays can be substantial and the transmitter will incur more retransmissions on links via the wormhole than to genuine neighbors.

An adversary can, however, avoid detection by link quality means if they send their own acknowledgment for any frame successfully received by the wormhole. For attacks where the round-trip latency introduced by the attacker exceeds the acknowledgment time-out it becomes necessary for the adversary to do so. Consider the example of Figure 2.5. In this case, the adversary waits until the first 10 octets of the frame have been seen before begging to relay the frame contents. In this case, the frame does not begin to be relayed back until after the ACKTimeout has expired.

A successful attack must, therefore, ensure that the ACK is received within the acknowledgment time-out period. In practical attacks this is achieved by having M itself to send an acknowledgment whenever it successfully receives a frame for A or B. All that needs to be done is to modify the WNIC's

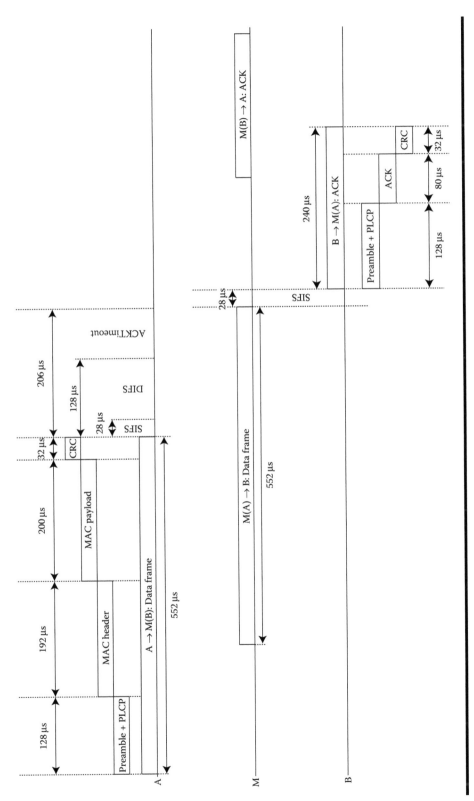

Figure 2.5   IEEE 802.11 DATA/ACK exchange timing diagram.

interface address. Reconfiguring the wireless interface in this way allows the adversary to meet the timing constraints but the adversary sends an acknowledgment *before* it has received an ACK from the final destination. This property can be used to expose the presence of a frame-relaying attack.

The technique used is to modify the MAC protocol so that a frame is occasionally not acknowledged on its first transmission. The sender and receiver share a secret key and agree on which frames to suppress the first acknowledgment by using a keyed hash function computed for the frame's body. The decision to suppress a frame is made when certain bits of the hash value are zeroes. The benefit to this scheme is that it has a very small impact on performance but has no false positives—the only way to trigger the mechanism is if someone *other* than the intended host generates an acknowledgment.

2.5 Conclusions

Multihop wireless networks face many security problems faced by infrastructure networks but the requirement to participate in routing and forwarding traffic on behalf of neighbors brings new threats. Of the threats discussed here the wormhole attack remains the most serious because it can be conducted by an outsider who does not need to undermine the security protocol but can damage the integrity of routing information and cause large-scale denial of service.

Terminologies

Clear-to-Send (CTS)
Counter CBC/MAC Protocol (CCMP)
Expected Transmission Count (ETX)
Expected Transmission Time (ETT)
Global Positioning System (GPS)
Hybrid Wireless Mesh Protocol (HWMP)
Independent Basic Service Set (IBSS)
Medium Access Control (MAC) layer
Mesh Basic Service Set (MBSS)
Mobile *Ad Hoc* network (MANET)
Multipoint Relays (MPRs)
Mutual Authentication with Distance-bounding (MAD)
Open Link State Routing (OLSR)
Request-to-Send (RTS)
Secure AODV (SAODV)
Secure *Ad Hoc* On-Demand Distance Vector (SEAD)
Temporal Key Integrity Protocol (TKIP)
Weighted Cumulative ETT (WCETT)
Wireless Mesh Networks (WMNs)

Questions and Sample Answers

1. Why multihop networks are said to be self-organizing?
   Multihop networks are also more flexible and resilient than their infrastructure equivalents because they allow multiple routes between communicating network nodes. These routes are

discovered and maintained autonomously by the network nodes themselves and so multihop networks are said to be *self-organizing*.

2. State the names of different types of nodes depending on their functions.

   Different types on network nodes depending on their functions:

   ■ *Mesh STA:* The most basic elements of an MBSS, these nodes participate in routing and forwarding traffic on behalf of their neighbors.

   ■ *Mesh Router:* A mesh STA which is dedicated to routing traffic. These nodes may have several WNICs and additional power reserves.

   ■ *Mesh AP:* A mesh STA which also provides access point (the term used for an 802.11 BS) functionality for infrastructure-mode 802.11 clients. The mesh AP can route frames between the infrastructure BSS and the MBSS.

   ■ *Mesh Portal:* A mesh STA which provides a bridge to a different kind of network. For example, a gateway to a WiMax or UMTS network which provides Internet access. MPs are important because they typically represent the destination for a significant amount of network traffic.

3. What is the purpose of a routing protocol?

   The purpose of routing protocols is to find the best route(s) between a source node and destination node in the network. It is important that this is done efficiently so that resources are not wasted carrying unnecessary information and reliably so that if a path exists it will be discovered.

4. Why proactive routing protocols are attractive?

   Proactive or link state protocols ensure that every node in the network has up-to-date routes to all other nodes. Such protocols are attractive because they find optimal routes with no delays.

5. What is OLSR?

   OLSR is a proactive routing protocol originally developed for Mobile *ad-hoc* Networks.

6. When do routing protocols work well?

   Reactive or on-demand protocols work well in dynamic networks in which there are many mobile nodes and a rapidly changing network topology.

7. What is HWMP?

   The Hybrid Wireless Mesh Protocol (HWMP) path-selection protocol is a mandatory new path-selection protocol introduced by IEEE 802.11s.

8. What is a threat model? How is it related to multihop wireless networks?

   A threat model is the key to identifying and managing the key security threats faced by a secure system. The purpose of the threat model is to identify the sources of risk and select appropriate countermeasures. In the case of a multihop wireless networks, it is appropriate to divide the threat model in two and identify threats posed by outsiders and then threats posed by compromised nodes.

9. What is the goal of a security protocol?

   The goal of a security protocol is to divide the world into two: the trusted insiders and the untrusted outsiders.

10. Mention some names of attacks against routing and path selection.

    Rushing attacks, gray holes and black holes, wormholes.

Author's Biography

Stephen Glass is a research engineer at National ICT Australia (NICTA) and a mature PhD candidate at Griffith University where he is investigating security of wireless mesh networks. His research interests include wireless networks, network security, secure software implementation, software-defined radio, and operating systems. Glass has an MSc and PostGraduate Diploma from the Open University in the United Kingdom. Contact him at stephen.glass@nicta.com.au.

Marius Portmann is a lecturer at the University of Queensland, Brisbane, and a researcher at the National ICT Australia (NICTA), where he is working on wireless mesh networks for public safety applications. His research interests include wireless networks, peer-to-peer systems, and network security. Portmann has a PhD in electrical engineering from the Swiss Federal Institute of Technology (ETH), Zürich. He is a member of the IEEE. Contact him at marius.portmann@nicta.com.au.

Vallipuram Muthukkumarasamy is a senior lecturer in the School of Information and Communication Technology, Griffith University. His research interests include Security in Wireless Networks, Intrusion Detection and Prevention Systems, Sensor Network Security, Information Assurance in e-Government Models, and Adaptive Equalization. Muthukkumarasamy has a PhD in Communications from the University of Cambridge. He is a member of the Institution of Electrical and Electronic Engineers. Contact him at v.muthu@griffith.edu.au.

References

1. LAN/MAN Standards Committee of the IEEE Computer Society. IEEE Standard for Information technology—Telecommunications and information exchange between systems—Local and metropolitan area networks—Specific requirements. Part 11: Wireless LAN MAC and Physical Layer (PHY) Specifications, 802.11-2007 edition, June 2007.
2. LAN/MAN Standards Committee of the IEEE Computer Society. IEEE P802.11s™/D3.03 Draft standard for information technology—Telecommunications and information exchange between systems—Local and metropolitan area networks—Specific requirements—Part 11: wireless LAN MAC and physical layer (PHY) specifications amendment 10: Wireless Mesh Networking, D3.03 edition, July 2009.
3. G. R. Hiertz, S. Max, R. Zhao, D. Denteneer, and L. Berlemann. Principles of IEEE 802.11s, in *Proceedings of 16th International Conference on Computer Communications and Networks ICCCN 2007*, pp. 1002–1007. Institution of Electrical and Electronics Engineers, August 2007.
4. T. H. Clausen and P. Jacquet. *RFC 3626: Optimized Link State Routing Protocol (OLSR)*, Internet RFC, October 2003. http://tools.ietf.org/html/rfc3626.
5. C. E. Perkins, E. M. Belding-Royer, and S. Das, *RFC 3561: Ad hoc On-demand Distance Vector (AODV) Routing*, Internet RFC, June 2003. Available from http://tools.ietf.org/html/rfc3561.
6. M. Bahr, J. Wang, and X. Jia, Routing in wireless mesh networks, in Y. Zhang, J. Luo, and H. Hu, eds., *Wireless Mesh Networking*, pp. 113–146, ACM Press, Auerbach, 2007.
7. M. Bahr, Update on the hybrid wireless mesh protocol of IEEE 802.11s, in *IEEE International Conference on Mobile Ad hoc and Sensor Systems (MASS 2007)*, pp. 1–6, Pisa, Italy, October 2007.
8. N. B. Salem and J.-P. Hubaux, Securing wireless mesh networks. *Wireless Communications*, 13(2), 50–55, 2006.
9. Y.-C. Hu, A. Perrig, and D. B. Johnson, Rushing attacks and defense in wireless ad hoc network routing protocols, in *WiSe '03: Proceedings of the 2003 ACM Workshop on Wireless Security*, pp. 30–40, ACM Press, New York, NY, USA, 2003.
10. Y.-C. Hu, A. Perrig, and D. B. Johnson, Wormhole attacks in wireless networks. *IEEE Journal on Selected Areas in Communications*, 24(2), 370–380, 2006.

11. Y.-C. Hu and A. Perrig, A survey of secure wireless ad hoc routing. *IEEE Security & Privacy Magazine*, 2(3), 28–39, 2004.
12. K. Sanzgiri, B. Dahill, B. N. Levine, C. Shields, and E. M. Belding-Royer. A secure routing protocol for ad hoc networks, in *ICNP '02: Proceedings of the 10th IEEE International Conference on Network Protocols*, pp. 78–89, IEEE Computer Society, Washington, DC, USA, 2002.
13. Y.-C. Hu, D. B. Johnson, and A. Perrig, SEAD: Secure efficient distance vector routing in mobile wireless ad hoc networks, in *Fourth IEEE Workshop on Mobile Computing Systems and Applications (WMCSA '02)*, pp. 3–13, June 2002.
14. M. G. Zapata, Secure ad hoc on-demand distance vector routing. *SIGMOBILE Mobile Computing and Communications Review*, 6(3), 106–107, 2002.
15. S. Marti, T. J. Giuli, K. Lai, and M. Baker, Mitigating routing misbehavior in mobile ad hoc networks, in *MobiCom '00: Proceedings of the 6th Annual International Conference on Mobile Computing and Networking*, pp. 255–265, ACM Press, New York, NY, USA, 2000.
16. L. Lazos, R. Poovendran, C. Meadows, P. Syverson, and L. W. Chang, Preventing wormhole attacks on wireless ad hoc networks: A graph theoretic approach. *IEEE Wireless Communications and Networking Conference (WCNC 2005)*, 2, 1193–1199, 2005.
17. R. Poovendran and L. Lazos, A graph theoretic framework for preventing the wormhole attack in wireless ad hoc networks. *Wireless Networks*, 13(1), 27–59, 2007.
18. W. Wang and B. Bhargava, Visualization of wormholes in sensor networks, in *WiSe '04: Proceedings of the 3rd ACM Workshop on Wireless Security*, pp. 51–60, ACM Press, New York, NY, USA, 2004.
19. C. He, M. Sundararajan, A. Datta, A. Derek, and J. C. Mitchell, A modular correctness proof of IEEE 802.11i and TLS, in *CCS '05: Proceedings of the 12th ACM Conference on Computer and Communications Security*, pp. 2–15, ACM Press, New York, NY, USA, 2005.
20. F. Martignon, S. Paris, and A. Capone, MobiSEC: A novel security architecture for wireless mesh networks, in *Q2SWinet '08: Proceedings of the 4th ACM Symposium on QoS and Security for Wireless and Mobile Networks*, pp. 35–42, ACM Press, New York, NY, USA, 2008.
21. D. Kuhlman, R. Moriarty, T. Braskich, S. Emeott, and M. V. Tripunitara, A correctness proof of a mesh security architecture, in *IEEE 21st Computer Security Foundations Symposium (CSF '08)*, pp. 315–330, IEEE Computer Society, June 23–25, 2008.
22. S. Brands and D. Chaum, Distance-bounding protocols, in *EUROCRYPT '93: Workshop on the Theory and Application of Cryptographic Techniques on Advances in Cryptology*, pp. 344–359, Secaucus, NJ, USA, 1993. Springer-Verlag, New York.
23. S. Čapkun, L. Buttyán, and J.-P. Hubaux, SECTOR: Secure tracking of node encounters in multi-hop wireless networks, in *SASN '03: Proceedings of the 1st ACM Workshop on Security of Ad Hoc and Sensor Networks*, pp. 21–32, ACM Press, New York, NY, USA, 2003.
24. G. P. Hancke and M. G. Kuhn, An RFID distance bounding protocol, in *First International Conference on Security and Privacy for Emerging Areas in Communications Networks*, 2005. SecureComm 2005., pp. 67–73, September 2005.
25. T. Korkmaz, Verifying physical presence of neighbors against replay-based attacks in wireless ad hoc networks, in *ITCC '05: Proceedings of the International Conference on Information Technology: Coding and Computing (ITCC'05)–Volume II*, pp. 704–709, IEEE Computer Society, Washington, DC, 2005.
26. J. Eriksson, S. V. Krishnamurthy, and M. Faloutsos, TrueLink: A practical countermeasure to the wormhole attack in wireless networks, in *14th Annual IEEE Conference on Network Protocols (ICNP 2006)*, pp. 75–84. IEEE Computer Society, 2006.
27. S. Glass, V. Muthukkumarasamy, and M. Portmann, Detecting man-in-the-middle and wormhole attacks in wireless mesh networks, in *IEEE 23rd International Conference on Advanced Information Networking and Applications (AINA-09)*, May 2009, Bradford, UK.

Chapter 3

Handling Security Threats to the RFID System of EPC Networks

Joaquin Garcia-Alfaro, Michel Barbeau, and Evangelos Kranakis

Contents

3.1 Introduction

Passive radio frequency identification (RFID) is a wireless communication technology that allows the automatic identification of objects, animals, and persons through radio waves. Passive RFID tags are electronic labels without self-power supply. They are energized by the electromagnetic field of radio frequency (RF) front-end devices (hereinafter referred as RFID readers). The radio spectrum used in RFID systems varies from low-frequency (LF) and high-frequency (HF) bands (typically 125 kHz and 13.56 MHz) to ultra-high-frequency (UHF) bands (typically 868 MHz in Europe, 915 MHz in North America, and 950 MHz in Japan). Distances from which the RFID tags can be interrogated vary with the frequency band. It may vary from a few centimeters, while using LF and HF, to a few meters, while using UHF.

Although no single technology is ideal for all applications [1], most of the modern RFID systems seem to be moving toward increasing the integration of long-distance passive tags into self-organizing wireless applications. This is the case with the modern electronic product code (EPC) Gen2 tags. They are becoming truly pervasive in wireless network applications, such as Mobile Wireless *Ad Hoc* Networks (MANETs), Wireless Sensor Networks (WSNs), and Vehicular *Ad Hoc* Networks (VANETs) [2]. Tags are potentially the targets of attack against their security and this raises major concerns. The objective of this chapter is to analyze some of these concerns and survey solutions that handle them.

3.1.1 Background

The EPC technology originates from the MIT's Auto-ID Center (now called the Auto-ID Labs). It had been further developed by different working groups at EPCglobal Inc. [3]. It is a layered service-oriented architecture to link objects, information, and organizations via Internet technologies. At the lowest layer, an identification system based on passive RFID tags and readers provides the means to access and identify objects in motion. This system possesses two primary interfaces: the Class 1 Generation 2 UHF Air Interface Protocol Standard (Gen2 for short) and Low Level Reader Protocol (LLRP). The former defines the physical and logical requirements for RFID readers (or interrogators) and passive tags (or labels). The latter specifies the air interface and interactions between its instances.

The next layer consists of a middleware composed of several services (such as filtering, fusion, aggregation, and correlation of events) that perform real-time processing of tag event data and collect the identifier of objects interrogated by RFID readers at different time points and locations. Data gathered by sensors, such as temperature and humidity, can also be aggregated at the middleware layer within tag events. The middleware forwards the complete set of events to a local repository where they are persistently stored (e.g., into a relational or XML database). The Reader Protocol (RP) and Reader Management (RM) interfaces define the interactions between a device capable of reading/writing RFID tags and the middleware. The middleware relies on a second interface called Application Level Event (ALE) for interaction with other applications (e.g., repository managers). At the top of the architecture, the EPC Information Services (EPCISs) offer the means to access the data stored in EPC network repositories. These EPCISs are implemented using standard Web technologies such as the Simple Object Access Protocol (SOAP) and Web Services Description Language (WSDL). Two additional services are defined for accessing the EPCIS of a given EPC network by external applications: a lookup service binding object identifiers and EPCISs, called the Object Name Service (ONS); and a EPC discovery service (EPCDS) to perform searches with high-level semantics (i.e., similar to Web engines for Web page browsing).

Security attacks can target the different services of the EPC network architecture. They may succeed if weaknesses within the underlying technologies are not handled properly. The exchange of information between EPC tags and readers, for example, is carried out via wireless channels that do not posses basic security attributes such as authenticity, integrity, and availability. This situation allows attackers to misuse the RFID service of an EPC network and perform unauthorized activities such as eavesdropping, rogue scanning, cloning, location tracking, and tampering of data. The attacker motivation for performing these activities is potentially high. The attacker can obtain financial gains (e.g., offering services for corporate espionage purposes). Mechanisms at the RFID level of the EPC architecture must be applied to mitigate these security risks.

The implementation of new security features in EPC tags faces several challenges, the main one being cost. The total cost of an EPC tag was estimated in [4] to be less than 10 cents per unit. The goal is to maintain a low cost. Other challenges include compatibility regulations, power consumption, and performance requirements [5]. In this chapter, we analyze threats to the security of the exchange of information between RFID readers and tags. Some of them need to be handled by appropriate countermeasures. Our threat analysis is based on a methodology proposed by the European Telecommunications Standards Institute (ETSI). It proposes the ranking of threats depending on their likelihood of occurrence, their possible impact on targeted systems, and the risk they represent [6] for corporate systems. The results of our analysis are intended for leading further research and developments of security of EPC-based technologies. We also study countermeasures for threats ranked at the critical or major level. We discuss the benefits and drawbacks associated with the surveyed solutions.

Section 3.2 outlines the methodology used for our analysis of threats. Section 3.3 presents the identified threats and their risk assessments. Section 3.4 surveys traditional security defenses for RFID solutions. Section 3.5 discusses some directions and trends for further research.

3.2 Threat Analysis Methodology

We define a threat as the objective of an attacker to violate security properties of a target system, such as authenticity, integrity, and availability. We define the attacker as an agent that is exploiting a vulnerability of the targeted system to carry out the threat. The exploitation of the vulnerability is defined as the attack. The security officer of the target system must put in place countermeasures to reduce the risk of the undesirable activities associated with all the threats. Given the difficulty of implementing countermeasures for every possible threat against a system, it is crucial for security officers to identify threats with potentially high impact and insure the presence of countermeasures. This is indeed the objective of the threat analysis.

The methodology we use is based on a framework proposed by the ETSI [6]. ETSI identifies three levels of threats: critical, major, and minor. Each level depends on estimated values for the likelihood of occurrence of the threat and its potential impact on a given system. The likelihood of a threat (*cf.* Figure 3.1a) is determined by the motivation for an attacker to carry out an attack

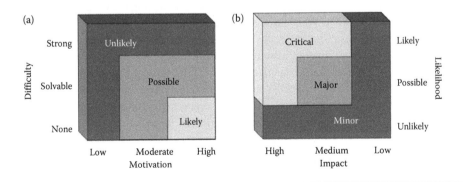

Figure 3.1 Likelihood and risk functions: (a) likelihood of a threat, (b) risk evaluation function.

associated to the threat versus the technical difficulties that must be resolved by the attacker to effectively implement the attack. The three levels of likelihood are: (1) *likely*, if the targeted system is almost assured of being victimized, given a high attacker motivation (e.g., financial gains as a result of selling private information or disrupting network services) and lack of technical difficulties (e.g., a precedent for the attack already exists); (2) *possible*, if the motivation for the attacker is moderate (e.g., limited financial gains) and technical difficulties are potentially solvable (e.g., the required theoretical and practical knowledge for implementing the attack is available); and (3) *unlikely*, in case there is little motivation for perpetrating the attack (e.g., few or no financial gains resulting from the attack) or if significant technical difficulties and obstacles must be overcome (e.g., theoretical or practical elements for perpetrating the attack are still missing).

The impact of a threat evaluates the potential consequences on the system when the threat is successfully carried out. The following three categories are identified: (1) *low*, if the consequences of the attack can be quickly repaired without suffering from financial losses; (2) *medium*, if the consequences are limited in time but might result in few financial losses; and (3) *high*, if the attack results in substantial financial loss and/or law violations. The risk of a threat is ranked in [6] as *minor*, if it is unlikely to happen and it has low or medium potential impact, or if it is possible but with low potential impact. A threat is ranked as *major* if it is likely but has low potential impact, if it is possible and has medium potential impact, or if it is unlikely but has high potential impact. A threat is ranked as *critical* if it is likely and has high or medium potential impact, or if it is possible and has high potential impact. Through our experience with the ETSI methodology, we have observed that several threats are overclassified as major, when they would better be ranked as minor. We have slightly adapted the risk function in order to focus on truly critical or major threats. Figure 3.1b presents the adapted risk function. A threat is ranked as *major* when its likelihood is *possible* and its potential impact is *medium*. A threat is ranked as *minor* when it is *unlikely* to happen or when its potential impact is *low*. Minor risk threats typically require no countermeasures. Major and critical threats need to be handled with appropriate countermeasures. Moreover, critical threats should be addressed with the highest priority.

3.3 Evaluation of Threats

The communication channel between the components of the RFID system of an EPC network, that is, tags and readers, is a potentially insecure wireless channel. It is fair to assume that most of the threats on EPC configurations are going to target this level. We analyze threats targeting basic security features such as authenticity, integrity, and availability during the exchange of data between a RFID tag and a RFID reader. We assume that attackers may only act from outside when trying to exploit the wireless channel between tags and readers, for example, the lack of authentication between these elements. We therefore assume that attackers do have physical access neither to the components of the system nor to the organization itself. The reason we ignore direct physical access is because we assume the presence of other security mechanisms in the organization (e.g., physical access control and surveillance of workers). Attackers, however, may have access to information about the system and its components or services. We summarize in Table 3.1 the results of our evaluation.

3.3.1 Authenticity Threats

The EPC Gen2 standard is designed to balance cost and functionality [4]. However, security features on board Gen2 tags are minimal. They protect message integrity via 16-bit Cyclic Redundancy

Table 3.1 Evaluation of Threats

Threats	Motivation	Difficulty	Likelihood	Impact	Risk
Eavesdropping, rogue scanning	High	Solvable	Possible	High	Critical
Cloning of tags, location tracking	Moderate	Solvable	Possible	Medium	Major
Tampering of data	Moderate	Solvable	Possible	High	Critical
Destruction of data, denial of service	Moderate	Solvable	Possible	Medium	Major
Malware	Moderate	Strong	Unlikely	Medium	Minor

Codes (CRC) and generate 16-bit pseudorandom strings. Their memory, very limited, is separated into four independent blocks: reserved memory, EPC data, Tag Identification (TID), and user memory. The absence of strong authentication on the tags opens the door to malicious readers that can impersonate legal readers and perform eavesdropping attacks. Figure 3.2 shows a simplified description of the steps of the Gen2 protocol for product inventory. In Step 1, the reader queries the tag and selects one of the following options: select, inventory, or access [3]. Figure 3.2 represents the execution of an inventory query. It assumes that a select operation has been completed in order to single out a specific tag from the population of tags. When the tag receives the inventory query, it returns a 16-bit random string denoted as RN16 in Step 2. This random string is temporarily stored in the tag memory. The reader replies to the tag in Step 3 with a copy of the random string, as an acknowledgment. If the echoed string matches the copy of RN16 stored in the tag memory, the tag enters the acknowledged state and returns the EPC.

Let us observe that any compatible Gen2 reader can access the EPC. The traffic between tags and readers flows through nonauthenticated wireless channels. Illegitimate collection of traffic might be slightly protected by reducing the transmission power or by sheltering the area. It is, although, theoretically possible to conduct eavesdropping attacks. We define forward eavesdropping as the passive collection of queries and commands sent from readers to tags; and backward eavesdropping as the passive collection of responses sent from tags to readers. Although the range for backward eavesdropping could be only of a few meters [3], and probably irrelevant for a real eavesdropping attack, the distance at which an attacker can eavesdrop the signal of an EPC reader can be much longer. In

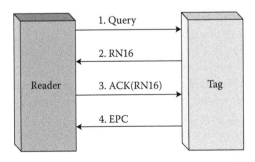

Figure 3.2 Inventory protocol of a Gen2 tag.

ideal conditions, for example, readers configured to transmit at maximum output power, the signal could be received from tens of kilometers away. Analysis attacks inferring sensitive information from forward eavesdropping, for example, analysis of the pseudorandom sequences generated by the tags (denoted as RN16 in Figure 3.2), are hence possible. Replay attacks enabled by this inferred data are also possible. The absence of a strong authentication process also enables scanning attacks. Although, the distance at which an attacker can perform scanning is considerably shorter than the distance for forward eavesdropping. The use of special hardware (e.g., highly sensitive receivers and high gain antennas) could enable rogue scanning attacks.

We can conclude that outsiders equipped with Gen2 compatible readers and special hardware can theoretically eavesdrop the communication between readers and tags; or scan objects in motion if they successfully manage to place their readers at appropriate distances. According to [3], the information stored on an EPC tag is limited to an identification number. No additional data beyond the number itself is conveyed in the EPC. Additional information associated with the code must be retrieved from an EPCIS. However, an attacker accessing these data may determine types and quantities of items in a supply chain and sell the information to competitors or thieves. An attacker can obtain information from the EPC, that is, the manufacturer and product number. This information may be used for corporate espionage purposes by competitors, or for other attacks against other services of the EPC infrastructure. Clearly, the motivation of an attacker to carry out this threat must be rated as high, since attackers can sell their services to competitors, thieves, or any other individual looking for the objects tagged in the organization. The difficulties for performing both eavesdropping and rogue scanning, as shown by the example depicted by Figure 3.2, are solvable. This level of motivation and degree of difficulty lead to a likelihood that is possible. Regarding the potential impact of these threats (e.g., disclosure of information considered by the organization as confidential or trade secrets), it is high, since it may have serious consequences for an organization if an attacker offers the malicious service to competitors or to thieves. These threats are assessed as critical and need to be handled by appropriate countermeasures.

Using the codes eavesdropped or scanned by unauthorized readers, an attacker may successfully clone the tags by conducting, for example, skimming attacks. Indeed, an attacker can simply dump data and responses from a given tag, and program it into a different device. The objective of the attacker for performing the cloning of tags is the possibility for counterfeiting. The attacker may create fake EPC tags that contain data and responses of real tags and sell these counterfeit tags for profit. The forgery of legal tags can be performed without physical access to the organization. We rank the motivation of attackers to carry out the attacks associated with this threat as moderate since they can obtain some financial gain by offering this service to third parties. Current EPC specifications do not include any mechanism for Gen2 compatible readers to verify if they communicate with genuine or fraudulent tags. We thus rate the difficulties associated to this threat as solvable. This level of motivation and degree of difficulty lead to a likelihood that is possible. Regarding the potential impact of this threat, it is medium and thus the threat is assessed as major.

The lack of a strong authentication process in Gen2 tags also has consequences to the privacy of tagged object bearers. Indeed, interrogations of Gen2 tags give attackers unique opportunities for the collection of personal information (and without the consent of the bearer). This can have serious consequences, such as location tracking or surveillance of the object bearers. An attacker can distinguish any given tag by just taking into account the EPC number. Following a reasoning similar to the one used for the cloning threat leads to ranking the risk of the location tracking threat as major. This threat, as well as the cloning threat, must be handled by appropriate countermeasures.

3.3.2 Integrity and Availability Threats

Gen2 tags are required to be writable [3]. They must also implement an access control routine, based on the use of 32-bit passwords, to protect the tags from unauthorized activation of the writing process. Other operations, also protected by 32-bit passwords, can be used in order to permanently lock or disable this operation. Although the writable feature of Gen2 tags is very interesting, it is also one of the least exploited features in current EPC scenarios (due probably to the lack of a strong authentication process, as reported in the previous section). Writable tags are hence locked in most of today's EPC applications. This option will, however, be extremely important in future EPC applications, especially on those self-organizing-based scenarios, where the addition of complementary information into the memory of the tags will require the unlocking of the writing process (e.g., to store routing parameters, locations, or time stamps). It is therefore important to analyze the risk of a tampering attack to the data stored by Gen2 tags, if they can be accessed in write mode from a wireless channel that does not guarantee strong authentication. Figure 3.3 presents a simplified description of protocol steps for requesting and accessing the writing process that modifies the memory of a Gen2 tag. We assume that a select operation has been completed, in order to single out a specific tag from the population of tags. It is also assumed that an inventory query has been completed and that the reader has a valid RN16 identifier (*cf.* Figure 3.2, Steps 2 and 3) to communicate and request further operations from the tag. Using this random sequence (*cf.* Figure 3.2, Step 5), the reader requests a new descriptor (denoted as *Handle* in the following steps). This descriptor is a new random sequence of 16 bits that is used by the reader and tag. Indeed, any command requested by the reader must include this random sequence as a parameter in the command. All the acknowledgments sent by the tag to the reader must also include this random sequence. Once the reader obtains the *Handle* descriptor in Step 6, it acknowledges by sending it back to the tag as a parameter of its query (*cf.* Step 7). To request the execution of the writing process, the reader needs first to be granted access by supplying the 32-bit password that protects the

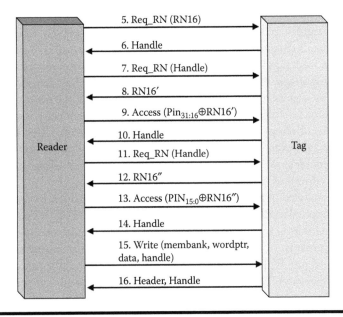

5. Req_RN (RN16)

6. Handle

7. Req_RN (Handle)

8. RN16′

9. Access ($Pin_{31:16} \oplus RN16'$)

10. Handle

11. Req_RN (Handle)

12. RN16″

13. Access ($PIN_{15:0} \oplus RN16''$)

14. Handle

15. Write (membank, wordptr, data, handle)

16. Header, Handle

Reader

Tag

Figure 3.3   Writing protocol of a Gen2 tag.

writing routine. This password is actually composed of two 16-bit sequences, denoted in Figure 3.3 as $PIN_{31:16}$ and $PIN_{15:0}$. To protect the communication of the password, the reader obtains in Steps 8 and 12, two random sequences of 16 bits, denoted in as *RN16'* and *RN16''*. These two random sequences *RN16'* and *RN16''* are used by the reader to blind the communication of the password toward the tag. In Step 9, the reader blinds the first 16 bits of the password by applying an XOR operation (denoted by the symbol ⊕ in Figure 3.3) with the sequence *RN16'*. It sends the result to the tag, which acknowledges the reception in Step 10. Similarly, the reader blinds the remaining 16 bits of the password by applying an XOR operation with the sequence *RN16''*, and sends the result to the tag in Step 13. The tag acknowledges the reception in Step 14 by sending a new *Handle* to the reader. By using the latter, the reader requests the writing operation in Step 15, which is executed and acknowledged by the tag in Step 16.

An attacker can find the 32-bit password that protects the writing routine. It suffices to intercept sequences RN16' and RN16'', in Steps 8 and 12, and to apply the XOR operation to the contents of Steps 9 and 13. Other techniques to retrieve similar passwords have also been reported in the literature. For example, in [7] the authors present a mechanism to retrieve passwords by simply analyzing the radio signals sent from readers to tags. Although the proof-of-concept implementation of this technique is only available for Gen1 tags [3], the authors state that Gen2 tags are equally vulnerable. The technical difficulties for setting up attacks to retrieve the password are therefore ranked as solvable. The likelihood of the tampering threat is classified as possible. Regarding the impact, it is ranked, because of some extreme scenarios, as high. For example, in the context of a pharmaceutical supply chain, corrupting data in the memory of EPC tags can be very dangerous: the supply of medicines with wrong information, or delivered to the wrong patients, can lead to situations where a sick person could take the wrong drugs. In these circumstances, the combination of likelihood and impact of the tampering threat lead to critical risk. The threat needs therefore to be addressed by appropriate countermeasures.

Let us note that these attacks enabled by retrieving the passwords, that protect both writing and self-destruction routines of Gen2 tags [3], can be used as models to analyze the risk of threats like destruction of data or denial of service [8]. Tag information can also be destroyed by devices that send strong electromagnetic pulses. Devices, such as the RFID-zapper [9], have been presented in the literature. We can also include here denial of service attacks consisting of jamming channels or flooding channels between tags and readers by sending a large number of requests and responses. For performing a jamming attack, the attacker uses powerful transmitters to generate noise in the range of frequencies used by readers and tags. In any case, the technical difficulties are ranked as solvable. The motivation of attackers to carry out these threats is rated as moderate, since they can obtain some financial gain by offering their malicious services. The likelihood of these two threats is hence classified as possible. However, since the impact of these threats represents to the victim temporal disruption of its operations rather than great financial losses, we rate the impact as medium, and so the risk of these two threats as major.

The final threat analyzed in this section, related to attacks to the integrity and availability of the back-end servers connected to RFID readers, was initially reported in [10]. Rieback et al. uncover the possibility of using malware to attack back-end databases. Their approach classifies such malware into three categories: (1) exploits, (2) worms, and (3) viruses. The exploits are attacks carried out within the information stored into RFID tags. They target the security of middleware services connecting readers to back-end databases. Worms and viruses are attacks that spread themselves over new RFID tags by using network connections (in the case of worms) or connectionless self-replication strategies (in the case of viruses). The malware reported in [10] exploits the trust relationship between back-end databases and the information sent by readers—obtained in turn from malicious tags. Rieback et al. consider that even if there is a very tiny window for storing information into an RFID tag, traditional attacks against information systems (e.g., buffer overflows and SQL injection attacks) might be condensed into a small string of bits harmful enough to break the security of a system. The authors present a proof-of-concept that uses tags carrying an SQL injection attack that compromises the security of the back-end layer of an *ad hoc* RFID setup. The work presented in [10] is interesting and relevant. However, we think that the likelihood for those threats must be rated as unlikely, since no real-world vulnerabilities on the filtering and collection of middleware services specified by EPCglobal can be exploited at the moment. We conclude that even if the impact is potentially serious, due to its unlikely degree of likelihood, the threat is assessed as minor.

3.4 Survey of RFID Security Defences

Research on security countermeasures for RFID technologies can be divided into two categories: (1) hardware-based security primitives for RFID tags, and (2) software protocols using the hardware-based primitives. We review in this section a nonexhaustive list of contributions in both categories.

3.4.1 Hardware-Based Primitives

According to research presented in [4], the cost of passive EPC tags should not exceed 5 cents to successfully enable their deployment on worldwide scale. Of these 5 cents, only 1 or 2 should be used for the manufacturing of the integrated circuit (IC). Another challenge is that the available layout area for the implementation of the IC is in the order of 0.25 mm^2 which, considering current complementary metal–oxide–semiconductor (CMOS) technology, corresponds to a theoretical number of logic gates from 2000 to 4000. Not all the barriers identified in [4] have been removed. Today, the EPC technology is more expensive than what it was originally anticipated—around 10 cents per unit in large quantities. The inclusion of additional features, especially for security purposes, may increase the total end-cost of tags up to 15 cents per unit or more. Although Moore's Law says that the cost of ICs will continue to decrease, cost of analogue devices (i.e., RF front-end of tags) is relatively stable and will remain a constraint [11]. The inclusion of new elements must therefore be clearly justified.

Since EPC tags are powered from the weak energy captured from a reader's electromagnetic waves, their current consumption also needs to be taken into account. This consumption varies according to the operation that is being performed [12] (e.g., responding to a query or writing data into the memory) and other parameters such as the transmission rate, response delay, and memory technology. Most of the operations performed within the modern EPC tags consume about 5–10 μamps—although some special operations, such as write accesses, may consume more. The current consumption of new security primitives must be within this range to allow low-cost tag production. They must also work at the data rate of EPC applications. For example, some supply chain applications demand an average reading speed of about 200 tags/s. This leads to a data transmission rate from tag to reader, of about 640 kbps; and from reader to tag of about 120 kbps. Delays associated to new security mechanisms (e.g., time to perform encryption or random number generation) may also affect the global performance. Delays must hence be taken into account and minimized. We refer the reader to [11] for a more detailed description of aspects that must be considered during the design of new primitives.

Several security proposals aim at including cryptographic primitives on low-cost EPC tags. However, not all the proposals meet the aforementioned constraints or guarantee secure designs. Existing implementations of one-way hash functions, such as MD4, MD5, and SHA-128/SHA-256, exceed cost constraints due to the required number of gates—from 7000 to over 10,000 logic gates according to [13]. The use of cellular automata (CA) theory for the implementation of one-way functions [14] and encryption engines [15]—typically built upon feedback shift registers with a much lower number of gates—has been investigated for the implementation of cryptographic primitives on low-cost RFID tags. However, it has been proved that these implementations are insecure [16,17]. Similarly, the use of linear feedback shift registers and nonlinear feedback shift registers (LFSR & NLFSR) as underlying mechanisms for the implementation of low-cost one-way hash functions and pseudorandom number generators (PRNG)—without appropriate measures that increase the cost—also lead to insecure implementations [18,19]. Light-weight hardware implementations of standard block ciphers to implement one-way hash functions have been discussed. The use of elliptic curve cryptosystems (ECC) [19] for the implementation of primitives for RFID tags has been discussed in [20]. Its use of small key sizes is seen as very promising for providing an adequate level of computational security at a relatively low cost [11]. An ECC implementation for low-cost RFID tags can be found in [21]. In [22], on the other hand, Feldhofer et al. present a 128-bit implementation of the advanced encryption standard (AES) [23] on an IC of about 3500 gates with a current consumption of less than 9 μamps at a frequency of 100 kHz. The encryption of each block of 128 bits requires about 1000 clock cycles. Although, it considerably simplifies previous implementations of the AES, for example, proposals presented in [24,25] that require between 10,000 and over 100,000 gates, respectively, the design is still considered too complex for basic EPC setups [26].

More suitable encryption engine implementations can be found in [27,28]. The first reference presents the implementation of the tiny encryption algorithm (TEA) [29]. It is implemented on an IC of about 3000 gates with a current consumption of about 7 μamps. It fits the timing requirements of basic EPC setups where hundreds of tags must simultaneously be accessed by the same reader. For meeting the constraints, the implementation relies on very simple operators such as XOR, ADD, and SHIFT. The authors of TEA [29] claim that, despite its simplicity and ease of implementation, the complexity of the algorithm is equivalent to data encryption standard (DES) [19]. Variants of the basic algorithm, such as eXtended TEA (XTEA), are however necessary for implementing one-way hash functions. Mace et al. discuss in [28] some of the vulnerabilities of TEA, such as linear and differential cryptanalysis attacks, and present scalable encryption algorithm (SEA). Given the relatively recent invention of these algorithms, their strength is not clear [11].

There are other hardware-based security enhancements for RFID technologies not relying on the implementation of cryptographic primitives. Many signal- and power-based defenses, such as shielding of tags, use of noise and third-party blocker devices have been surveyed in [26]. The use of distance measurements to detect rogue readers has been discussed. In [30], for example, Fishkin et al. propose the inclusion of low-cost circuitry on tags to use the signal-to-noise ratio of readers as a metric for trust. In [31], a similar assumption is used in order to claim that a reader can be authorized to read a tag contents according to its physical distance. The use of trust [32] and trusted computing [33] with similar purposes has also been discussed. For example, Molnar et al. describe in [33] a mechanism consisting of trusted platform modules (TPMs) to enforce privacy policies within the RFID tags. A trusted entity called trusted center (TC) decides whether readers are allowed or not to access tags. Finally, the use of radio fingerprinting [34,35] to detect characteristic properties of transmitted signals and design authentication procedures has been investigated. The authors in [11] consider, however, that this technique is difficult to develop on RFID applications and that the benefits of using it, with respect to performance, cost, and required implementation surface on tags, are unclear. Avoine and Oechslin also debate in [36] the prevention of the traceability via radio fingerprinting.

They conclude that obtaining radio fingerprint of tags is expensive and difficult. The myriad of tags in circulation in future RFID scenarios makes impracticable the individual distinction of them.

Physically unclonable functions (PUFs) and physical obfuscated keys (POKs) are promising for the implementation of new security primitives in low-cost EPC tags. They can be used to handle the authentication threat, as well as the cloning and location tracking threats. Half way between cryptography-based enhancements and physical protection defenses, the ideas behind PUFs and POKs originated in [37] with the conception of optical mechanisms for the construction of physical one-way functions (POWFs). Their use to securely store unique secret keys, in the form of fabrication variations, was proposed as a silicon prototype in [38,39]. These ideas were later improved in [40]. A coating PUF proposed in [41] is implemented with less than 1000 gates. These designs exploit the random variations in delays of wires and logic gates of an IC. For example, the silicon PUF presented in [39] receives input data, as a challenge, and launches a race condition within the IC: two signals propagate along different paths and are compared to determine which one comes first. To decide which signal comes first, a controller, implemented as a latch, produces a binary value. Holcomb et al. [42] propose using the SRAM based on CMOS circuitry to generate physical fingerprints. The key idea is the use of SRAM start-up values as origin of randomness. The use of 256 bytes of Static Random Access Memory (SRAM) can yield 100 bits of true randomness each time that the memory is powered up. While sound in theory, this technique has as important drawback the limitation of memory space of current low-cost tags. The implementation of PUF-based circuits seems to have clear advantages at a cost of less than 1000 logic gates [41]. This technology provides a cost effective and reliable solution that meets the constraints and requirements. Drawbacks, such as the effects of environmental conditions and of power supply voltage [43], must be taken into account. The difficulty of successfully modeling the circuits and their reliability have also raised some concerns. Bolotnyy and Robins [44] address some of these issues. Some attacks on PUF- and POK-based protocols are outlined in [40]. The execution and reinterpretation of existing protocols via new PUF and POK designs—essentially the challenge–response protocols—are outlined in the sequel.

3.4.2 Software Protocols

We review algorithmic solutions and software protocols for handling the threats uncovered in Section 3.3. The solutions rely on the implementation and use of hardware-based primitives discussed in Section 4.1.

Message Authentication Code (MAC)-based security protocols for wireless applications is a typical solution discussed in the literature (e.g., [44–46]). In [45], Takaragi et al. present a very simple MAC-based approach. It uses a static unrewritable 128-bit identifier stored, at manufacturing time, in every tag. This static identifier is not modifiable once the shipment is made. To build up this identifier, the manufacturer uses a unique secret key for each tag and a keyed hash function that accepts as input the secret key and a specific message. All this information (i.e., secret key, hash function, and specific message) is communicated by the manufacturer to the client. By sharing this information among readers and tags, integrity and authenticity of exchanged messages is verified. It therefore reduces the risk of threats to authenticity and integrity by increasing the technical difficulties of performing attacks. However, due to the use of static identifiers embedded in the tags at manufacturing time, the location tracking issue is not solved. Moreover, brute force attacks can break the secrets shared between readers and tags.

The use of public key cryptography and digital signatures is discussed in [47]. The authors address the protection of banknotes embedding the RFID tags. Their approach includes the possibility of deploying cryptographic protocols in RFID applications, but avoids the need to embed cryptographic primitives within the tags. The scheme consists of a public-key cryptosystem used by a central bank aiming to avoid banknote forgery and a law enforcement agency that aims at tracking banknotes. Both authorities, that is, central bank and law enforcement agency, hold an independent pair of public and private keys associated to each banknote. The central bank authority assigns a unique serial number to each banknote. The central bank authority, using its private key, signs the unique serial number. The unique serial number of the banknote and its corresponding digital signature are printed on the banknote as optical data. In addition, the law enforcement agency encrypts with its public key the digital signature, unique serial number, and a random number. The resulting ciphertext is stored into a memory cell of the RFID tag. This memory cell is keyed-protected. The tag only grants write access to this memory cell if it receives an access key derived from the optical data. The random number used to create the ciphertext is also stored into a separated memory cell of the tag. This second memory cell is also keyed-protected. The tag only grants read or write access to this memory cell if it receives an access key derived from the optical data.

Now, a merchant that receives a banknote must verify first the digital signature, printed in the banknote as optical data, using the public key of the central bank. Second, the merchant must also verify the validity of the ciphertext stored in the banknote's tag. To do so, the merchant encrypts the digital signature, serial number, and random number stored in the tag's memory, using the public key of the law enforcement agency and the optical data. If one of these two verification processes fails, the authorities must be warned. To avoid using the same ciphertext on every interaction, Juels and Pappu propose the use of a reencryption process that can be performed by the merchant without the necessity of accessing the private keys of the law enforcement authority. Indeed, based on the algebraic properties of the El Gamal cryptosystem [19], the initial ciphertext can be transformed into a new unlinkable ciphertext only using the public key of the law enforcement authority [26]. This reencryption process is performed outside the tags. Integrity issues of this approach are discussed and fixed in [48]. However, the whole process and requirements for implementing the approach in [47,48] are too complex and expensive for use in EPC supply chain applications.

Mutual authentication protocols among tags and readers are discussed in [49,50]. The work presented by Kinosita et al. in [49] consists of an anonymous ID scheme, in which a tag contains only a pseudonym that is periodically rewritten. Pseudonyms are used instead of real identifiers (e.g., instead of the EPC codes). Similarly, the approach of Juels entitled minimalist cryptography for low-cost RFID tags [50] suggests a very lightweight protocol for mutual authentication between tags and readers based on one-time authenticators. Both solutions rely on the use of pseudonyms and keys stored within tags and back-end servers. Each tag contains a small collection of pseudonyms, according to the available memory of the tag. A throttling process is used to rotate the pseudonyms. Each time the tag is interrogated by a reader, a different pseudonym is used in the response. Authorized readers have access to the complete list of pseudonyms set for each tag and can correlate the responses they receive. Without the knowledge of this list, unauthorized readers are unable to infer any information about the several occurrences of the same tag. The process also forces tags to slow down their data transmissions when queried too frequently, as a defense to potential brute-force attacks. The memory space on current low-cost tags is the main limitation of this approach. Although enhancements can be used to update the list of pseudonyms, communication costs, and integrity threats will remain as main drawbacks.

The use of hash-lock schemes for addressing authentication issues is another possibility. A design can be found in [51]. Weis et al. propose a way to lock tags without storing access keys in them. Only hashes of keys must be known by the tags. Keys must be also stored on back-end servers and be accessible by authorized readers. Most authentication threats are therefore mitigated by locking tags. Cloning and tracking threats are handled by avoiding the use of real identifiers once tags are locked. In [52], Henrici and Müller extend the hash-lock scheme and address some weaknesses in [51] to increase traceability and location resistance. A similar hash-based protocol is presented in [53] in order to deal with those limitations by using time stamps. Other similar hash-based protocols for handling authentication threats can be found in [54–56]. All these protocols rely on synchronized secrets residing in the tags and back-end servers. They require a one-way hash function implemented within the tags. The requirement of reliable hash primitives implemented at the tag level is the main drawback. Workload on back-end servers is also considerably high and can make difficult the deployment in real-world EPC supply chain applications. The Yet Another Trivial RFID Authentication Protocol (YATRAP) protocol presented in [57] reduces the cost of computation by combining precomputed hash-tables for tag verification processes, use of time stamps, and generation of pseudorandom numbers. The protocol is, however, vulnerable to availability attacks when temporal de-synchronizations between tags and readers occur. Some limitations are addressed in [58]. Chatmon et al. define new protocols for anonymous authentication. These improvements notably increase the degree of workload on servers and are highly complex for use in supply chain applications.

3.5 Future Directions for Research

Algorithmic solutions avoiding the execution of on-tag cryptographic processes seems to lead the future of research in RFID security. In this sense, a secret-sharing scheme is presented by Juels et al. in [59] as a defense against the authenticity threats in EPC supply chain applications. Two different models are discussed: dispersion of secrets across space and dispersion of secrets across time. Both models are based on a secret-sharing strategy, where a secret used to encrypt EPCs is split in multiple shares and distributed among multiple parties. In order to obtain the EPC of a tag, a party must collect a minimum number of shares distributed among all the other parties. Authentication is therefore achieved though the dispersion of secrets. The dispersion helps to improve the authentication process between readers and tags, as tags move through a supply chain. Assuming that a given number of shares is necessary for readers to obtain the EPCs assigned to a pallet, for example, a situation where the number of shares obtained by readers is not sufficient to reach the threshold protects the tags from unauthorized scanning (i.e., unauthorized readers that cannot obtain the sufficient number of shares cannot obtain the EPCs either). The approach can be implemented on EPC Gen2 tags without requiring any change to the current tag specification. A limitation is the amount of tag memory space required for storing the shares. However, the shrinking of shares can allow the application of the scheme to current EPC tags. A more important problem is that the location tracking threat is not addressed. Indeed, the shares used in the approach are static. This problem must be solved before deployment of the scheme.

Challenge–response protocols for low-cost EPC tags using physical unclonable functions (PUFs) and POKs have recently gained importance. An approach presented in [60], based on PUFs proposed in [38,39], consists of a challenge–response scheme that probabilistically ensures unique identification of RFID tags. A back-end system must learn challenge–response pairs for each PUF/tag. It then uses these challenges (hundreds of them) at a time, to identify and authenticate tags. Unique identification of tags is only ensured probabilistically. The exposition of tag identifiers to eavesdroppers and lack of randomness in tag responses, make the approach vulnerable to the location tracking threat. Moreover, the great number of challenges that are necessary in the identification process increases the tag response delay and power consumption. Hence, this approach might not meet the constraints and requirements mentioned in Section 4.1. An alternative approach is presented in [61]. Tuyls and Batina discuss an off-line PUF-based mechanism for verifying the authenticity of tags through the PUF technology presented in [41]. Similar to the results presented in [50,62], where readers and tags define *ad hoc* secrets, the PUF-based approach uses the internal physical structure of tags to generate unique keys. A key extraction algorithm from noisy binary data is presented in [61]. The usage of PUF-based keys simplifies the process of verifying tag authenticity. The combination of unique keys generated onboard together with the use of signatures avoid leaking of a single identifier and increases the technical difficulties for an attacker to carry out the location tracking threat. The main drawback is the need of large storage space and reliable searching processes on back-end servers to link readers with PUF/tag identifiers. The use of public key and digital signatures, based on Elliptic Curve Cryptography (ECC), is another important constraint. Following the trend of combining PUFs together with traditional cryptographic primitives and encryption engines, a modification of the tree-based hash protocols proposed in [63] is presented in [64]. Using the notion of POKs introduced in [38] (i.e., application of a fixed hard-wired challenge to the PUF to obtain a unique secret), the authors guarantee the existence of internal keys in basic tree-based hash protocols, now physically obfuscated. They cannot be cloned by unauthorized parties. The use of an AES engine, such as the one presented in [23], is proposed. On the other hand, Bolotnyy and Robins present in [44] a complete set of adapted MAC protocols, based on PUFs, trying to simplify the challenge–response communication scheme of previous proposals and to eliminate requirement of traditional cryptographic primitives. Each tag generates multiple identifiers based on embedded PUFs. Their approach only addresses static identification. It is vulnerable to the location tracking threat identified in Section 3.3. It does not solve the requirement of huge lists of challenge–response pairs for each PUF/tag that must be stored on back-end servers connected to the readers. Indeed, each given pair is of single use to prevent replay attacks.

3.6 Conclusions

At the beginning of this chapter we presented an analysis of threats to the RFID system of the EPC architecture. We identified different groups of threats that we consider relevant for further research. We ranked the eavesdropping, rogue scanning, and tampering threats as critical; and cloning, tracking, and denial of service threats as major. We concluded that they must be handled by appropriate countermeasures. We then surveyed in the sequel practical and theoretical security defenses that can be useful to reduce the risk of the identified threats. We looked at the different defenses from two different research perspectives. On the one hand, we surveyed research on hardware-based defenses that aim at providing additional security primitives on tags such as one-way hash functions, encryption engines, and physically unclonable functions (PUFs). On the other hand, we surveyed research on software protocols that make use of these new on-tag primitives for designing and implementing reliable algorithms for dealing with security and privacy issues. We have seen that the implementation of well-known cryptographic primitives is possible and allows the design of software protocols to reduce the risk of threats ranked as critical or major. The cost and requirements of these proposals are the main difficulties. Indeed, they are too expensive for their deployment in supply chain scenarios based on the EPC technology. We have also surveyed the combination of cryptographic primitives together with the use of PUFs for the design of cost-effective solutions. These solutions present drawbacks, such as the sensitivity of PUFs to physical noise and the difficulty to model and analyze them. They are, however, promising solutions that successfully meet the implementation constraints and requirements for handling the set of threats reported in our work. For the second group, we conclude that the avoidance of on-tag cryptographic processes on current algorithmic solutions seems to lead the future directions of research in RFID security. In this sense, the use of secret-sharing schemes present clear advantages for the management of keys in the design of authentication protocols and to deal with privacy issues. The main drawback is the use of static shares, limiting the use of this approach for addressing the location tracking threat.

Acknowledgments

The authors graciously acknowledge the financial support received from the following organizations: Natural Sciences and Engineering Research Council of Canada (NSERC), Mathematics of Information Technology and Complex Systems (MITACS), Spanish Ministry of Science and Education (TSI2007-65406-C03-03 "E-AEGIS" and CONSOLIDER CSD2007-00004 "ARES" grants), La Caixa (Canada awards), and the IRF project LOCHNESS.

Terminologies

Advanced Encryption Standard (AES)—A block cipher encryption standard, sponsored by the National Institute of Standards and Technology (NIST), for protecting data.

Countermeasure—A defense mechanism designed to mitigate the risk of a threat.

EPC Network—A service-oriented architecture defined by EPCglobal Inc. that proposes the integration of RFID and Internet technologies to enable automatic identification and sharing of item data in supply chain applications.

Electronic Product Code (EPC)—Group of coding schemes defined by EPCglobal Inc.

Elliptic Curve Cryptography (ECC)—A public key cryptosystem based on the algebraic structure of elliptic curves over finite fields.

European Telecommunications Standards Institute (ETSI)—Independent noncommercial organization that produces telecommunications standards to be used in Europe and beyond.

EPC Number—A tag data format compatible with the family of coding schemes proposed by EPCglobal Inc. It typically contains: a header, pointing out the family code that is being used; a manufacturer code; an object class; and a serial number.

EPCGLOBAL Inc.—Joint venture between GS1 (Global Standards One, formerly known as EAN International) and GS1 US™ (formerly the Uniform Code Council, Inc.) created to commercialize the EPC technology.

Linear and Nonlinear Feedback Shift Registers (LFSR & NLFSR)—A digital circuit composed of an n-bit shift register and a feedback function that generates pseudorandom sequences.

Passive Tag—RFID component attached to an item. It contains information about the item. Since it does not have its own power source, it provides the information by backscattering a reader's signal.

Physically Unclonable Function (PUF)—Hardware-based function embedded in a physical structure, that is easy to evaluate but hard to reproduce.

Physical Obfuscated Keys (POK)—Hardware-based function for implementing secrets on digital devices using a physically unclonable function.

Pseudorandom Number Generator (PRNG)—Algorithmic solution to generate deterministic sequences of pseudorandom numbers.

Reader—RFID component that requests and receives information from tagged items.

Tag Identification (TID)—Memory bank or identifier that uniquely identifies an RF tag.

Threat Analysis—Determination and classification, in terms of importance, of threats targeting the security of a system.

Tiny Encryption Algorithm (TEA)—A minimalist block cipher encryption algorithm for protecting data.

Trusted Platform Modules (TPMs)—Hardware-based cryptographic mechanism installed on the motherboard of a digital device (i.e., a personal computer) to enforce security protection and trustworthiness.

Scalable Encryption Algorithm (SEA)—A block cipher encryption algorithm designed to be used on embedded applications such as microcontrollers.

Questions and Sample Answers

1. What is Passive RFID?

    Passive radio frequency identification (RFID) is a wireless communication technology that allows the automatic identification of objects, animals, and persons through radio waves.

2. How a security attack can succeed against EPC network architecture?

    Security attacks can target the different services of the EPC network architecture. They may succeed if weaknesses within the underlying technologies are not handled properly. The exchange of information between EPC tags and readers, for example, is carried out via wireless channels that do not possess basic security attributes such as authenticity, integrity, and availability. This situation allows attackers to misuse the RFID service of an EPC network and perform unauthorized activities such as eavesdropping, rogue scanning, cloning, location tracking, and tampering of data.

3. What is the major challenge for implementing new security features in EPC tags?

    The implementation of new security features in EPC tags faces several challenges. The main one is the cost.

4. What is a threat?

    A threat is the objective of an attacker to violate security properties of a target system, such as authenticity, integrity, and availability.

5. What is an attacker?

    An attacker is an agent that is exploiting a vulnerability of the targeted system to carry out the threat. The exploitation of the vulnerability is defined as the attack.

6. Mention two major areas for research on security countermeasures for RFID technologies.

    Research on security countermeasures for RFID technologies can be divided into two categories: (1) hardware-based security primitives for RFID tags and (2) software protocols using the hardware-based primitives.

7. What does ETSI stand for?

    European Telecommunications Standards Institute (ETSI)

8. When is a threat ranked as minor?

    A threat is ranked as minor when it is unlikely to happen or when its potential impact is low. Minor risk threats typically require no countermeasures.

9. Draw the diagram of Inventory Protocol of a Gen2 Tag.

10. Define: Forward and Backward eavesdropping.
    We define forward eavesdropping as the passive collection of queries and commands sent from readers to tags; and backward eavesdropping as the passive collection of responses sent from tags to readers.

Author's Biography

Joaquin Garcia-Alfaro is a lecturer professor at the Computer Science and Multimedia Studies of the Open University of Catalonia, Spain. He obtained a Bachelor, a Master's, and a PhD (in Computer Science) from Autonomous University of Barcelona (Spain) and TELECOM Bretagne (France) in 2006. From 2007 to 2009, he was postdoctoral fellow at the Computer Science Department of Carleton University, Canada. Since 2009, he also collaborates as an associate researcher at TELECOM Bretagne, France. His research interests include a wide range of network security problems, with an emphasis on the management of security policies, analysis of vulnerabilities, and enforcement of countermeasures.

Michel Barbeau is a professor of computer science. He obtained a Bachelor, a Master's, and a PhD, in computer science, from Université de Sherbrooke, Canada (1985), for undergraduate studies, and Université de Montréal, Canada (1987 and 1991), for graduate studies. From 1991 to 1999, he was a professor at Université de Sherbrooke. During the academic year 1998–1999, he was a visiting researcher at the University of Aizu, Japan. Since 2000, he works at Carleton University, Canada. Wireless communications has been his main research interest. He focuses his efforts on wireless security, vehicular communications, wireless access network management, *ad hoc* networks, and RFID.

Evangelos Kranakis is a professor of computer science. He obtained a BSc (in Mathematics) from the University of Athens (1973) and a PhD (in Mathematical Logic) from the University of Minnesota, Minnesota (1980). From 1980 to 1982 he was at the Mathematics Department of Purdue University, West Lafayette, Indiana, and from 1982 to 1983 at the Mathematisches Institut of the University of Heidelberg, Germany, between 1983 and 1985 he served at the Computer Science Department of Yale University, New Haven, Connecticut, from August to December of 1985 at the Computer Science Department of the Universiteit van Amsterdam, the Netherlands and from 1986 to 1991 at the Centrum voor Wiskunde en Informatica (CWI) in Amsterdam, the Netherlands. Since 1991, he works at Carleton University, Canada. He was director of the School of Computer Science from 1994 to 2000. He received the Carleton Research Achievement award in 2000. He became Carleton University Chancellor's Professor in 2006. He has published in the analysis of algorithms, bioinformatics, communication and data (*ad hoc* and wireless) networks, computational and combinatorial geometry, distributed computing, and network security.

References

1. R. Want, RFID explained: A primer on radio frequency identification technologies, *Synthesis Lectures on Mobile and Pervasive Computing*, Num. 1, 2006, Morgan & Claypool Publishers.
2. G. Roussos, S. Duri, and W. Thompson. RFID meets the internet, *IEEE Internet Computing. Special Issue on RFID*, 13, 105–114, 2009.
3. EPCglobal Overview & Standards, Available from: http://www.epcglobalinc.org/standards/
4. S. E. Sarma, Toward the 5 cent tag, White Paper, November 2001, Auto-ID Center.

5. J. Sounderpandian, R. V. Boppana, S. Chalasani, and A. M. Madni, Models for cost–benefit analysis of RFID implementations in retail stores, *Systems Journal, IEEE*, 1(2), 105–114, 2007.

6. ETSI, Methods and protocols for security; part 1: Threat analysis. ETSI-ts 102 165-1 v4.1.1, 2003.

7. Y. Oren, Remote power analysis of RFID tags, *Cryptology ePrint Archive*, Report 2007/330, IACR, 2007.

8. D. Han, T. Takagi, H. Kim, and K. Chung, New security problem in RFID systems tag killing, *Lecture Notes in Computer Science*, M. L. Gavrilova, O. Gervasi, V. Kumar, C. J. K. Tan, D. Taniar, A. Lagana, Y. Mun, and H. Choo, eds. vol. 3982. Springer, 2006, pp. 375–384.

9. Minime and Mahajivana, RFID Zapper, *22nd Chaos Communication Congress (22C3)*, December 2005.

10. M. Rieback, B. Crispo, and A. Tanenbaum, Is your cat infected with a computer virus?, in *Pervasive Computing and Communications, IEEE*. Pisa, Italy: IEEE Computer Society Press, March 2006, pp. 13–17.

11. P. Cole and D. Ranasinghe, eds. *Networked RFID Systems and Lightweight Cryptography—Raising Barriers to Product Counterfeiting*, 1st ed. Springer, 2008.

12. T. Lohmann, M. Schneider, and C. Ruland, Analysis of power constraints for cryptographic algorithms in mid-cost RFID tags, in *Smart Card Research and Advanced Applications, 7th IFIP WG 8.8/11.2 International Conference (CARDIS 2006), Lecture Notes in Computer Science*, vol. 3928, Berlin/Heidelberg: Springer, April 2006, pp. 278–288.

13. M. Feldhofer and C. Rechberger, A case against currently used hash functions in RFID protocols, *Workshop on RFID Security – RFIDSec 06*, Ecrypt, Graz, Austria, July 2006, pp. 372–381.

14. S. Wolfram, Cryptography with cellular automata, in *Advances in Cryptology, CRYPTO 85, Lecture Notes in Computer Sciences*, vol. 218, New York, NY, USA: Springer-Verlag New York, Inc., 1986, pp. 429–432.

15. S. Sen, C. Shaw, D. R. Chowdhuri, N. Ganguly, and P. P. Chaudhuri, Cellular automata based cryptosystem (CAC), in *4th International Conference on Information and Communications Security (ICICS'02)*. London, UK: Springer-Verlag, 2002, pp. 303–314.

16. P. Bardell, Analysis of cellular automata used as pseudorandom pattern generators, in *International Test Conference*, 1990, Washington DC, pp. 762–768.

17. S. R. Blackburn, S. Murphy, and K. G. Paterson, Comments on theory and applications of cellular automata in cryptography, *IEEE Transactions on Software Engineering*, 23(9), 637–638, 1997.

18. C. Meyer and W. Tuchman, Pseudo-random codes can be cracked, *Electronic Design*, 23, 1972.

19. A. Menezes, P. Van Oorschot, and S. Vanstone, *Handbook of Applied Cryptography*. CRC Press, USA, 1997.

20. J. Wolkerstorfer, Is elliptic-curve cryptography suitable to secure RFID tags? *Workshop on RFID and Lightweight Crypto*, Ecrypt, Graz, July 2005.

21. L. Batina, J. Guajardo, T. Kerins, N. Mentens, P. Tuyls, and I. Verbauwhede, An elliptic curve processor suitable for RFID-tags, *Cryptology ePrint Archive*, Report 2006/227, IACR, 2006.

22. M. Feldhofer, S. Dominikus, and J. Wolkerstorfer, Strong authentication for RFID systems using the AES algorithm, in *Workshop on Cryptographic Hardware and Embedded Systems, CHES 2004, Lecture Notes in Computer Science*, vol. 3156, IACR. Boston, MA, USA: Springer, August 2004, pp. 357–370.

23. J. Daemen and V. Rijmen, *The Design of Rijndael: AES—the Advanced Encryption Standard*. Springer, 2002.

24. S. Mangard, M. Aigner, and S. Dominikus, A highly regular and scalable AES hardware architecture, *IEEE Transactions on Computers*, 52(4), 483–491, 2003.

25. I. Verbauwhede, P. Schaumont, and H. Kuo, Design and performance testing of a 2.29-GB/s Rijndael processor, *IEEE Journal of Solid-State Circuits*, 38(3), 569–572, 2003.

26. A. Juels, RFID security and privacy: A research survey, *IEEE Journal on Selected Areas in Communication*, 24(2), 381–394, 2006.

27. P. Israsena, Securing ubiquitous and low-cost RFID using tiny encryption algorithm, in *International Symposium on Wireless Pervasive Computing, IEEE*. Phuket, Thailand: IEEE Press, January 2006, pp. 1–4.

28. F. Mace, F.-X. Standaert, and J.-J. Quisquater, Asic implementations of the block cipher sea for constrained applications, in *Conference on RFID Security*, Malaga, Spain, July 2007, pp. 103–114.

29. D. J. Wheeler and R. M. Needham, TEA, a tiny encryption algorithm, in *Fast Software Encryption: Second International Workshop (FSE 1994)*, Leuven, Belgium, December, *Lecture Notes in Computer Science*, vol. 1008. Springer, Berlin/Heidelberg, 1995, pp. 363–366.

30. K. Fishkin, S. Roy, and B. Jiang, Some methods for privacy in RFID communication, in *European Workshop on Security in Ad-hoc and Sensor Networks, ESAS 2004, Lecture Notes in Computer Science*, vol. 3313. Heidelberg, Germany: Springer-Verlag, August 2005, pp. 42–53.

31. G. Hancke, Noisy carrier modulation for HF RFID, in *First International EURASIP Workshop on RFID Technology*, Vienna, Austria, September 2007, pp. 63–66.

32. A. Solanas, J. Domingo-Ferrer, A. Martinez-Balleste, and V. Daza, A distributed architecture for scalable private RFID tag identification, *Computer Networks*, 51(9), 2268–2279, 2007.

33. D. Molnar, A. Soppera, and D. Wagner, A scalable, delegatable pseudonym protocol enabling ownership transfer of RFID tags, in B. Preneel and S. Tavares, eds. *Selected Areas in Cryptography, SAC 2005, Lecture Notes in Computer Science*, vol. 3897. Kingston, Canada: Springer, August 2005, pp. 276–290.

34. J. Hall, M. Barbeau, E. and Kranakis, Enhancing intrusion detection in wireless networks using radio frequency fingerprinting, *Communications, Internet, and Information Technology*, 2004, 201–206.

35. J. Hall, Detection of rogue devices in Wireless Networks, PhD dissertation, Carleton University, 2006.

36. G. Avoine and P. Oechslin, RFID traceability: A multilayer problem, in *Financial Cryptography 2005, Lecture Notes in Computer Science*, vol. 3570, IFCA. Roseau, The Commonwealth Of Dominica: Springer-Verlag, February–March 2005, pp. 125–140.

37. R. Pappu, Physical one-way functions, PhD dissertation, MIT, 2001.

38. B. Gassend, Physical random functions, Master's thesis, MIT, 2003.

39. B. Gassend, D. Clarke, M. van Dijk, and S. Devadas, Silicon physical random functions, in *9th ACM Conference on Computer and Communications Security*. New York, NY, USA: ACM, 2002, pp. 148–160.

40. D. Lim, J. Lee, B. Gassend, G. Suh, M. van Dijk, and S. Devadas, Extracting secret keys from integrated circuits, *IEEE Transactions on Very Large Scale Integration (VLSI) Systems*, 13(10), 1200–1205, 2005.

41. B. Skoric and P. Tuyls, *Secret Key Generation from Classical Physics*, Philips Research Book Series, September 2005.

42. D. Holcomb, W. Burleson, and K. Fu, Initial SRAM state as a fingerprint and source of true random numbers for RFID tags, in *Third International Conference on RFID Security*, RFIDSec 2007, Malaga, Spain, 2007, pp. 31–42.

43. D. Ranasinghe, D. Engels, and P. Cole, Security and privacy solutions for low cost RFID Systems, in *2004 Intelligent Sensors, Sensor Networks & Information Processing Conference*, Melbourne, Australia, 2004, pp. 337–342.

44. L. Bolotnyy and G. Robins, Physically unclonable function-based security and privacy in RFID systems, in *International Conference on Pervasive Computing and Communications—PerCom 2007*, IEEE. New York, USA: IEEE Computer Society Press, March 2007, pp. 211–220.

45. K. Takaragi, M. Usami, R. Imura, R. Itsuki, and T. Satoh, An ultra small individual recognition security chip, *IEEE Micro*, 21(6), 43–49, 2001.

46. A. Willig, M. Kubisch, C. Hoene, and A. Wolisz, Measurements of a wireless link in an industrial environment using an IEEE 802.11-compliant physical layer, *IEEE Transactions on Industrial Electronics*, 49(6), 1265–1282, 2002.

47. A. Juels and R. Pappu, Squealing euros: Privacy protection in RFID-enabled banknotes, in R. N. Wright, ed., *Financial Cryptography 2003, Lecture Notes in Computer Science*, vol. 2742, IFCA. Le Gosier, Guadeloupe, French West Indies: Springer, January 2003, pp. 103–121.

48. X. Zhang and B. King, Integrity improvements to an RFID privacy protection protocol for anti-counterfeiting, in J. Zhou, J. Lopez, R. Deng, and F. Bao, eds. *Information Security Conference, ISC 2005, Lecture Notes in Computer Science*, vol. 3650. Singapore: Springer, September 2005, pp. 474–481.

49. S. Kinosita, F. Hoshino, T. Komuro, A. Fujimura, and M. Ohkubo, Non identifiable anonymous-ID scheme for RFID privacy protection, in *Japanese*. English description as part of http://www.autoidlabs.com/whitepaper/KEI-AUTOID-WH004.pdf, 2003.

50. A. Juels, Minimalist cryptography for low-cost RFID tags, in C. Blundo and S. Cimato, eds. *International Conference on Security in Communication Networks, SCN 2004, Lecture Notes in Computer Science*, vol. 3352. Amalfi, Italia: Springer, September 2004, pp. 149–164.

51. S. Weis, S. Sarma, R. Rivest, and D. Engels, Security and privacy aspects of low-cost radio frequency identification systems, in D. Hutter, G. Müller, W. Stephan, and M. Ullmann, eds., *International Conference on Security in Pervasive Computing, SPC 2003, Lecture Notes in Computer Science*, vol. 2802. Boppard, Germany: Springer, March 2003, pp. 454–469.

52. D. Henrici and P. Müller, Hash-based enhancement of location privacy for radio-frequency identification devices using varying identifiers, in *International Workshop on Pervasive Computing and Communication Security—PerSec 2004*, IEEE Computer Society, March 2004, pp. 149–153.

53. G. Avoine and P. Oechslin, A scalable and provably secure hash based RFID protocol, in *International Workshop on Pervasive Computing and Communication Security—PerSec 2005*, IEEE Computer Society Press, March 2005, Kauai Island, Hawaii, pp. 110–114.

54. S.-M. Lee, Y. J. Hwang, D. H. Lee, and J. I. L. Lim, Efficient authentication for low-cost RFID systems, in O. Gervasi, M. Gavrilova, V. Kumar, A. Lagana'a, H. P. Lee, Y. Mun, D. Taniar, and C. J. K. Tan, eds. *International Conference on Computational Science and its Applications, ICCSA 2005, Lecture Notes in Computer Science*, vol. 3480. Singapore: Springer, May 2005, pp. 619–627.

55. E. Y. Choi, S. M. Lee, and D. H. Lee, Efficient RFID authentication protocol for ubiquitous computing environment, in T. Enokido, L. Yan, B. Xiao, D. Kim, Y. Dai, and L. Yang, eds., *International Workshop on Security in Ubiquitous Computing Systems, SECUBIQ 2005, Lecture Notes in Computer Science*, vol. 3823. Nagasaki, Japan: Springer, December 2005, pp. 945–954.

56. S. Lee, T. Asano, and K. Kim, RFID mutual authentication scheme based on synchronized secret information, in *Symposium on Cryptography and Information Security*, Hiroshima, Japan, January 2006.

57. G. Tsudik, YA-TRAP: Yet another trivial RFID authentication protocol, in *International Conference on Pervasive Computing and Communications, PerCom 2006, IEEE*. Pisa, Italy: IEEE Computer Society Press, March 2006, pp. 640–643.

58. C. Chatmon, T. van Le, and M. Burmester, Secure anonymous RFID authentication protocols, Florida State University, Department of Computer Science, Tallahassee, FL, USA, Technical Report TR-060112, 2006.

59. A. Juels, R. Pappu, and B. Parno, Unidirectional key distribution across time and space with applications to RFID security, in *USENIX Security Symposium*. San Jose, CA: USENIX, July 2008, pp. 75–90.

60. D. Ranasinghe, D. Engels, and P. Cole, Low-cost RFID systems: Confronting security and privacy, in *Auto-ID Labs Research Workshop*, Zurich, Switzerland, September 2004, pp. 54–77.

61. P. Tuyls and L. Batina, RFID-tags for anti-counterfeiting, in *Topics in Cryptology, CT-RSA 2006, The Cryptographers' Track at the RSA Conference 2006, Lecture Notes in Computer Science*, USA: Springer, February 2006, pp. 115–131.

62. A. Juels and S. Weis, Authenticating pervasive devices with human protocols, in *Advances in Cryptology, CRYPTO 2005, Lecture Notes in Computer Science*, vol. 3126, IACR. Santa Barbara, CA, USA: Springer, August 2005, pp. 293–308.

63. D. Molnar and D. Wagner, Privacy and security in library RFID: Issues, practices, and architectures, in *Conference on Computer and Communications Security—ACM CCS*, USA: ACM Press, October 2004, pp. 210–219.

64. J. Bringer, H. Chabanne, and T. Icart, Improved privacy of the tree-based hash protocols using physically unclonable function, *Cryptology ePrint Archive*, Report 2007/294, 2007.

Chapter 4

Survey of Anomaly Detection Algorithms
Toward Self-Learning Networks

Tarem Ahmed and Rumana Rahman

Contents

4.1 Introduction

Algorithms for network anomaly detection have traditionally been based primarily on iterative and block-based methods, and a vast array of research has been performed using such approaches [1–4]. Network anomalies nonetheless remain poorly understood in general, and one of the reasons behind this is that Internet Service Providers (ISPs) do not have the tools required to detect anomalies in real time [5]. Most of the existing offline methods require waits of up to hours before alerts occur [6]. By using the common existing means of detection, ISPs typically become aware of major events such as intruders or DoS attacks after the fact, and not while they are in progress. Moreover, anomalies have historically been seen to span a vast range of types and classes, and each class may indicate its presence on raw statistics in a different manner. While it is claimed that there is no universally accepted definition of what constitutes normal behavior, and what precisely creates an anomaly [7], an anomaly is usually said to have occurred when the value of some traffic metric has exhibited a sudden deviation from normal trend [6].

Developing widely applicable definitions or models of normal network behavior and anomalies is thus difficult. Alternative approaches are therefore currently being advocated, where instead of being provided with signatures of the anomalies, the algorithm learns the behavior of normal traffic, and autonomously adapts to shifts in the structure of normality itself [8–11]. Ideally, there should be no parametric model prescribed for normal behavior. The disadvantage of a model is that it imposes limitations on the applicability of an algorithm, and even subtle changes in the nature of network traffic can render the model inappropriate.

This chapter is organized as follows. In Section 4.2, we describe the classic iterative, block-based methods of anomaly detection that are most commonly cited by current researchers of this topic. Section 4.3 discusses some fundamental recursive algorithms that constitute the basis of future directions for research on anomaly detection. Section 4.4 concludes the study.

4.2 Background

Network monitoring may be performed using two major types of approaches: a distributed approach or a centralized approach. Either architecture may be implemented regardless of the means of data collection (wired or wireless).

In the distributed architecture shown in Figure 4.1, the detection algorithms are run locally at each node in the network. Each sensor collects data and each node makes a decision regarding the presence or absence of an anomaly, after running the algorithm locally on the collected data after each time step. Every node then transmits the detection result to other nodes that are directly connected to it. Based on the decision of previous nodes and its own collected data, the new node makes its own decision and again forwards it to other nodes that are connected to it. Eventually, the network converges and every node learns the same detection result and relevant state information. The distributed process of decision making is resistant to the effects of individual node or sensor malfunctioning.

Alternatively, in the centralized architecture shown in Figure 4.2, the detection algorithm is run in a central processor that is connected to all the nodes in the network. Each node collects the readings from their respective sensors after each time step, and sends the raw data to the central processor. The central processor then runs the detection algorithm on the raw data and makes a decision regarding the presence or absence of anomaly. The central processor then forwards the decision to all the nodes for the network to reach convergence. The centralized approach is often

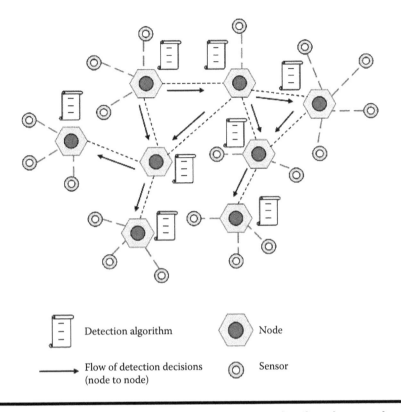

Detection algorithm   Node

Flow of detection decisions   Sensor
(node to node)

Figure 4.1 Monitoring architecture for anomaly detection: a distributed approach.

desirable and necessary to detect anomalies that exhibit themselves through distributed changes in the global measurement vector. The works of Lakhina et al. have shown that traffic volume distributions in large backbone IP networks exhibit a covariance structure, and detecting a break in this structure enables one to unearth a wide range of anomalies [5,12,13]. It is possible to detect a break in the covariance structure only by analyzing the global data at a central repository.

A great deal of research on anomaly detection [14–17] has been based on the insight provided by Lakhina et al. regarding the space occupied by multidimensional time series of network traffic measurements [5,12,13,18,19]. Lakhina et al. showed using data from backbone IP networks that while the full dimensionality of this space is huge, only a few of these dimensions actually contain significant energy. They then used the technique of principal component analysis (PCA) [20,21] to separate the space occupied by a set of traffic metrics into two disjoint subspaces corresponding to normal and anomalous behavior, and then signal an anomaly when the magnitude of the projection onto the residual, anomalous subspace exceeds an associated Q-statistic threshold [22]. The PCA subspace method has been shown to be more effective than more-intuitive exponentially weighted moving average (EWMA) and Fourier approaches [5], which provides another reason behind its popularity in subsequent works.

We begin this section with detailed descriptions of the series of works by Lakhina et al. and then move on to some other classic works on network anomaly detection. It is important to bear in mind that many of the current paradigms of intrusion detection in wireless sensor networks have been developed from [6], and bear similarities to [23], existing algorithms being used in IP networks.

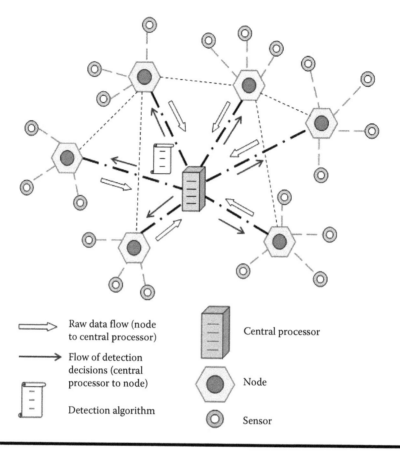

Figure 4.2 Monitoring architecture for anomaly detection: a centralized approach.

Lakhina et al. began by showing the intrinsic low dimensionality of network flows [5]. They study sampled data from the 11 core Abilene backbone routers in the United States, and the 13 core Sprint backbone routers in Europe, with the raw IP packets classified into core origin–destination (OD) flows [24]. Raw source and destination IP (dstIP) addresses are mapped into source and destination backbone routers, by using the BGP forwarding tables according to the methods presented in [25]. Lakhina et al. showed that although the flow space is large ($11 \times 11 = 121$ for Abilene and $13 \times 13 = 169$ for Sprint), most of the energy in the time series of the OD flows may be captured by a very small number (about 5–10) of fundamental flows. They formulate the problem as an eigenvalue problem and define the notion of an *eigenflow*, which is a time series that captures a particular source of temporal variability in the set of (>100) OD flows. Each OD flow is then represented as a weighted sum of *eigenflows*, with the respective eigenvalues providing the weights. They use the technique of PCA to project the high-dimensional raw flow space, into the smaller subspace of 5–10 principal eigenflows. PCA is a coordinate transformation method that maps a given set of data points onto new (orthogonal) axes, called its principal components [20]. It was first proposed in the field of psychology by Hotelling in [21] and is sometimes also referred to as the Hotelling transform. When the data are centered to have zero mean, each principal component has the additional property that it points in the direction of the maximum variance remaining in the data, given the variance already accounted for in the previous components. It is seen that all the flows in the network can be represented remarkably well, in terms of variations with time, by the weighted sum of about five eigenflows. Furthermore, the magnitudes of the eigenvalues go down drastically after the first few (four or five), showing the inherently low dimensionality of the space. Possible reasons for this low dimensionality are the following. First, this will happen if the variation in a small set of dimensions is dominant. That is, if the magnitude of variation among the different dimensions varies greatly, the data may have low dimensionality for this reason alone. Second, if there are common underlying patterns across dimensions, that is, if the dimensions show high correlation. By normalizing each dimension to zero mean and unit variance, it is shown that the correlation across dimensions is possibly the more dominant reason. Lakhina et al. defined eigenflows as three possible types: "d" (deterministic), "s" (spike), and "n" (noise). They provide specific methods to test if a given eigenflow qualifies as any of the three. Eigenflows that satisfy more than one test are categorized as indeterminate. It is seen that almost all eigenflows can be classified as being of exactly one type. This enables the decomposition of each OD flow into its three constituents: deterministic components, spikes, and noise contributions.

Lakhina et al. [12] proposed a method based on PCA to diagnose network-wide volume anomalies from link-level measurement data. A volume anomaly is defined as a sudden (with respect to the time step or bin size used) positive or negative change in an OD flow's traffic. Such an anomaly will propagate through all the links that the said flow traverses. However, it is seen that although the anomaly is easily visible on flow time series for the whole network, the corresponding changes in the traversed links' traffic is hardly discernible. Lakhina et al. proposed an algorithm that does the following: (1) correctly set an interrupt when (the point in time) an anomaly occurs at a link; (2) identify the particular flow (where the total traffic in the link is the sum of the traffic in all flows going through that link) that is the source of the anomaly; and (3) accurately estimate the magnitude (amount of traffic involved) in the anomaly. They begin with separating the space occupied by a set of traffic measurements/metrics into two disjoint subspaces, corresponding to normal and anomalous behavior, respectively. The most significant principal components form the normal subspace, while the remaining forms the anomalous subspace. To construct the subspaces, their algorithm looks for projections onto the principal components that exceed a threshold (such as three standard deviations from the mean), starting from the most significant principal component and proceeding in decreasing order. Once the threshold is encountered, higher axes are assigned to the normal subspace and the lower ones are assigned to the anomalous subspace. Then, to diagnose anomalies, the traffic at a particular link may be resolved into its normal and anomalous components, corresponding to normal and residual traffic, respectively. An anomaly is signaled if the squared prediction error of the residual component exceeds a threshold. The PCA subspace method thus makes use of the spatial correlation (correlation across links) in network measurements. Lakhina et al. tested their algorithm on real anomalous occurrences in Abilene and Sprint Europe datasets, as well as on artificially injected anomalies, and also compared their results with other popular methods of anomaly detection such as EWMA and Fourier approaches. The PCA subspace method was shown to be more effective, and also more useful, as the EWMA and Fourier approaches presume flow time series knowledge for the full network. In contrast, the subspace method operates on link-level data.

In [13], Lakhina et al. used the methods and results of [5,12] to characterize the traffic anomalies into various well-known types. The anomalies are seen to span a remarkably wide spectrum of event types, and nearly all the anomalies detected by the subspace method (and the wide range of possible causes of them) are seen to be of practical interest. Lakhina et al. applied the PCA subspace method of [5,12] to multivariate time series of sampled OD flow traffic from the Abilene network, comprising time series of (1) number of bytes, (2) number of packets, and (3) number of flows. They show that each of these traffic types reveals a different (sometimes non-overlapping) class of anomalies, and thus all three types are important for understanding the reasons behind the anomalies. Reference [13] thus builds on the work in [5,12]. In [5], the authors showed characteristics of byte counts in OD flows; in [13] they investigated packet count and flow counts as well, and also specifically studied each anomaly that is detected. In [12], the subspace method was applied to a different problem-link data. In [13], it is shown that the OD flow data are a much richer source of information than the link-level data examined in [12].

In [18], Lakhina et al. introduced the concept of *entropy* in this context, as a metric for measuring the distribution of traffic features. A *traffic features* is a header field in an IP packet, and the following four main traffic features (header fields) are considered: source IP address (srcIP), dstIP, source port number (srcPort), and destination port number (dstPort). Anomalous events cause changes in the distributions of the header fields in the sample data, and changes in a distribution are best captured by the entropy of the relevant field. The paper shows that analyzing entropies yields two major advances over using volume-based methods: it enables the detection of anomalies that otherwise go undetected; it enables automatic classification of anomalies (into clusters) via unsupervised learning. Analyzing the unusual distributions of traffic features during an anomaly also reveals information about the structure of the anomaly, information that is impossible to obtain from volume metrics alone. Rather than classifying anomalies into types *a priori*, this chapter studies the anomalies from the data. The detection method employed is a multidimensional extension of the subspace method of [5,12]. It expresses each OD flow feature as a sum of normal components and residual (anomalous) components, and signals an anomaly when the energy in the residual components exceeds a threshold. It is seen that using 10 dimensions is sufficient to represent the normal subspace for the entropy flows. This is slightly different from the case with volume metrics, where about five of the principal components contained most of the energy. Lakhina et al. showed that the set of anomalies detected via entropies and volume metrics are, interestingly, largely disjoint. Thus, using entropies complements the use of volume metrics in anomaly detection. The authors also manually inspect each anomaly that is detected, to identify its type. They provide a very thorough and complete discussion of the various types of anomalies that have been detected, and the manner in which each type alters the distribution of the four main header fields in manifesting itself. To identify the type of an anomaly that has been detected, Lakhina et al. used two types of unsupervised clustering algorithms in [18]. They consider the traffic feature anomalies as points in a four-dimensional space and show that each known type of anomaly automatically falls into a specific type of cluster. That is, each type of anomaly naturally appears in an expected position in the four-dimensional space. Lakhina et al. [18] thus demonstrated the utility of using entropies in detecting new anomalies, understanding the structure of anomalies, and in classifying anomalies. This chapter thereby presents a complete diagnosis system that is able to detect a wide range of anomalies with diverse structure, distinguish between different types of anomalies, and finally group together similar anomalies.

In [19], Lakhina et al. demonstrated the power of the subspace methods from [5,12] and the entropies of the four main traffic features introduced in [18] to specifically detect distributed attacks. Distributed denial-of-service (DDOS) attacks are not only prevalent, but also sophisticated. They have been slated by the United States Federal Bureau of Investigations (FBI) as the second largest contributor to all financial losses due to cybercrime [26]. Defending against DDOS attacks is therefore an important problem. Lakhina et al. [18] showed how to detect a wide range of different types of anomalies, using the multidimensional subspace method on time series of entropies. The anomalies studied in [18] included single-source DOS attacks, multi-source DOS attacks (i.e., DDOS attacks), and worm scans. The objective of [19], however, was to specifically evaluate the power of network-wide traffic analysis, via the subspace method, in detecting distributed attacks. Therefore, [19] concentrates on multisource DOS attacks. Reference [19] additionally presents a succinct summary of the technique of decomposing flows into normal and anomalous subspaces by finding temporal patterns that are common to the set of OD flows ([5,12]), and of the extra information provided by the entropy metric ([18]). The results of this paper provide insight into the performance of the multiway subspace method in detecting attacks that are dwarfed in individual flows, and only discernible in multiple, network-wide flows. It is more efficient to counter DDOS attacks at backbones, rather than at edges. However, it is difficult to detect DDOS attacks at backbone networks, as they do not cause significant changes in traffic volumes on individual backbone links. Traditionally, distributed attacks have also been regarded as being fundamentally more difficult to detect than single-source attacks. In contrast, the method presented in [19] is found to work better, the more distributed an attack. This is because the method proposed uses correlations across all network-wide flows simultaneously and does not inspect traffic on individual links in isolation. To specifically study the effectiveness of the multiway subspace method in detecting distributed attacks, the authors manually injected traces of a known DDOS attack on Abilene data. Then, to construct representative distributed attacks, they divided the src IP space in the attack trace into k origin Abilene points of presence (PoP). Here k ranges from 2 to 11, as there are 11 Abilene PoPs. This was done for all the $^{11}C_k$ possible srcIP combinations, and for each of the 11 possible dstIPs, for a total of $^{11}C_k \times 11$ experiments. The entire set of $^{11}C_k \times 11$ experiments was finally repeated at different thinning rates (i.e., selecting one out of every N packets) to evaluate the sensitivity of the detection method to the intensity of the DDOS attack. The value of k is a measure of the degree of distribution of the DoS attack. The results in [19] show that the detection rates are very high even for very low intensity attacks, and that they are also generally relatively higher for larger k. For example, 100% of all DDOS attacks that are split evenly across the 11 Abilene PoPs as origin routers are detected at a thinning rate of 1000, with a 0.999 detection threshold. Thus using the multiway subspace method on traffic feature entropy time series enables the detection of widely distributed attacks, even if the attacks are at a very low intensity. Indeed, a thinning rate of 1000 simulates an intensity of 0.001, or less than 1%.

Apart from PCA, another classic approach to network anomaly detection is the well-known factor analysis method [20]. A thorough comparison of PCA with factor analysis is provided by Tipping and Bishop in [27]. Here, the authors present a probabilistic formulation of standard PCA. They show that probabilistic principal component analysis (PPCA) may be used to obtain the principal components by assuming a latent Gaussian model, in a method similar to factor analysis. Note that standard PCA does not *a priori* assume a model. Tipping and Bishop then develop an expectation-maximization (EM) algorithm [28,29] to iteratively obtain the maximum-likelihood (ML) estimates [30] of the parameters of the assumed Gaussian model. In [31], Tipping and Bishop extend linear PCA into the realms of nonlinearity. Here they analyze a mixture model of probabilistic principal component analyzers and use the EM algorithm to determine the parameters of the mixture. The benefits of using a nonlinear version of PCA is that a greater proportion of the variance may be retained by using fewer components, compared to the linear case. An application of probabilistic PCA is [32], which develops a hierarchical data visualization algorithm. Most visualization algorithms try to find a projection from the original full data space onto a two-dimensional visualization space. For complex high-dimensional datasets, it is unlikely that a single two-dimensional projection can exhibit all the interesting attributes of the original structure. The hierarchical algorithm proceeds by involving multiple two-dimensional visualization spaces, with the following goal: the top-level projection should display the entire dataset and reveal the presence of clusters, while lower-level projections display internal structure within individual clusters, such as the presence of subclusters, which may not be apparent in higher-level projections.

Barford et al. provided general signal analysis methods to detect anomalies, for example, using wavelet filters, in [33]. The wavelet analysis here focuses mainly on aggregated traffic data in network flows. Kim et al. extended this work in [34] by analyzing IP packet header data at an egress router and by studying the correlation among IP addresses and port numbers over multiple timescales with discrete wavelet transforms [6,35].

Soule et al. developed a network anomaly detection scheme based on the Kalman Filter [36] in [37]. In contrast to the PCA approach by Lakhina et al. the Kalman Filter approach predicts the traffic matrix one step into the future. The actual traffic matrix is then estimated based on the new link data, and the difference between the prediction and the actual traffic is used to detect volume anomalies based on different thresholds. Soule et al. also found that the simple generalized likelihood ratio (GLR) test [30] often works better than the wavelet analysis of Barford et al. [33].

We now describe a sample of some of the other well-known methods of anomaly detection. Algorithms for intruder and outlier detection have appeared in a wide range of applications, including jet engine fault detection, radar target detection, mammography, e-commerce, statistical process control, and handwritten digit recognition. It is important to study them, as a great deal of current research is geared toward modifying existing algorithms from other fields to wireless and ad hoc networks. An example is the application of an existing disease propagation model from epidemiology to track viruses and polluted files in peer-to-peer (P2P) networks [38].

Many common outlier detection methods follow statistical approaches, and use the statistical properties of the training data to model the underlying distribution, and then estimate if a test point comes from the same distribution or not. A particular statistical approach may be parametric, or nonparametric. Parametric approaches generally assume that the underlying distribution is a mixture of Gaussians and may be modeled using means and covariance. Many of these approaches are built upon popular algorithms such as the EM algorithm [39], the k-means clustering algorithm [40,41], and hidden Markov models [42].

Nonparametric approaches in general do not make any assumptions regarding the statistical properties of the data. Algorithms in this class include the proposal of Guttormsson et al. for rotor fault detection [43]. Guttormsson et al. first streamlined the data by excluding all points that lie more than three standard deviations away from the mean, and then constructed a surface of various possible topologies around the points. The surface constructed may be a sphere using the Euclidean distance, an ellipse using the Mahalanobis distance [44,45], or a rectangular box. The distance measure is compared to a threshold to flag outliers. This approach is similar to the minimum volume set (MVS) [46] approach that has recently been advocated in identifying the dense normal region in the observation space [6,8,9].

There are a number of techniques that model the density of the training data and then use it for outlier detection. An interesting example is the method by Tax and Duin [47], which assumes a unimodal normal distribution, and then thresholds based on the Mahalanobis distance. Tax and Duin suggested an interesting metric to gauge novelty: the quotient of the distance between the new point and its nearest neighbor in the training set, and the distance of the neighbor with its subsequent nearest neighbor. This method was shown to be effective for distributions with relatively high decay probabilities, and where very little training data exist.

Krishnamurthy et al. first applied the sketch data structure to the heavy-change detection [48], a problem similar to detecting large alpha flows in a network. A sketch is a probabilistic summary data structure based on random projections, and input data are summarized using k-ary sketches. A drawback of the basic k-ary sketch approach is that it is irreversible, and makes it impossible to reconstruct the anomalous keys without querying every IP address. To solve these problems, Schweller et al. developed reversible sketches using an intelligent hashing scheme whereby the keys for the heavy-change observations (i.e., the anomalies) may be recovered in real time [49,50]. Gao et al. extended the earlier work on sketches by developing a high-speed, online intrusion detection algorithm based on two-dimensional sketches [51]. Their algorithm was shown to be able to detect multiple types of attacks simultaneously, and with high accuracy. Li et al. combined the PCA subspace method of Lakhina et al. [5,12,13,18,19] with the sketch data structure [48] to propose an anomaly detection algorithm based on sketch subspaces [17].

Clustering approaches are another popular means of performing anomaly detection. Clustering algorithms [52] typically work by partitioning the data into a number of clusters, and then by evaluating the degree to which each data point belongs to each of the defined clusters. A point is deemed to be an outlier if this degree of membership is below a threshold for all of the defined clusters, and the point is correspondingly determined to belong to none of the clusters. Jiang et al. proposed a two-phase clustering algorithm for novelty detection based on the k-means algorithm and a minimum spanning tree (MST) [53] in [54]. Ensafi et al. suggested sophisticated clustering methods using fuzzy *k*-means and swarm *k*-means algorithms, and apply them to detecting network attacks [55].

As previously mentioned, Lakhina et al. introduced the concept of *entropy* as a metric for measuring the distribution of *traffic features* in [18], and used it to detect distributed anomalies in [19]. There have also been other instances of the use of information theoretic measures [56] in network anomalies detection. Lee and Xiang provide examples of several information theoretic measures that may be used, including entropy and conditional entropy [57]. Gu et al. applied the maximum entropy principle and relative entropy also to detect anomalies that cause gradual changes, as opposed to sudden changes, in network traffic [58].

4.3 Future Directions for Research: Adaptive, Online Algorithms

4.3.1 Foundations

As described in Section 4.2, a great deal of the current research on anomaly and intrusion detection has been built on the offline, PCA approach of Lakhina et al. In [5], Lakhina et al. suggested an online formulation of their proposed PCA-based algorithm. The suggested online extension involved using a sliding window implementation to identify the normal and anomalous subspaces based on a previous block of time. The variation in the structure of multivariate network traffic statistics over time may, however, be non-negligible, as was shown in [9]. Furthermore, the PCA-based detection algorithm is often extremely sensitive to the proper determination of the associated Q-statistic threshold. Ahmed et al. implemented the proposed online version of PCA and observed that although the anomalous and normal subspaces remained relevant over time, using stale measurements to calculate the Q-statistic threshold resulted in an unacceptable number of false positives [9]. Ringberg et al. also pointed out the practical difficulty in tuning the parameters of the PCA method in [59]. The authors conducted a detailed study of the traffic feature time series for the detected anomalies in both Abilene and Geant (Sprint Europe) backbone networks. Their results showed that the false alarm rate is sensitive to small differences in the number of principal components assigned to the normal subspace, and the overall effectiveness of the method is subject to the level of aggregation of the measurements. Thus straightforward extensions to the PCA-based method are not robust, and alternative approaches are necessary for an online application.

In this regard, we advocate the use of recursive, and hence naturally online, algorithms. A family of such algorithms has been built using classic least-squares (LS) estimation techniques [60]. Therefore before progressing onto the online algorithms, we present a review of the fundamental LS estimation techniques themselves.

4.3.2 LS Estimation Techniques

The classical LS problem is the following:

$$\min_{x} \|\mathbf{A}x - \mathbf{b}\|^2 \tag{4.1}$$

where $\mathbf{A} \in R^{m \times n}$, $x \in R^n$, and $b \in R^m$, with $m > n$.

The classical solution to the LS problem is stated in terms of the *normal equations*:

$$\mathbf{A}^T\mathbf{A}x = \mathbf{A}^T\mathbf{b}. \tag{4.2}$$

When a new linear equation $\mathbf{a}^Tx = \beta$ is added to the system, we get the modified LS problem:

$$\min_{x} \|\bar{\mathbf{A}}x - \bar{\mathbf{b}}\|^2, \tag{4.3}$$

where $\bar{\mathbf{A}} = \begin{pmatrix} \mathbf{A} \\ \mathbf{a}^T \end{pmatrix} \in R^{(m+1) \times n}$, $x \in R^n$, and $\bar{\mathbf{b}} = \begin{pmatrix} \mathbf{b} \\ \beta^T \end{pmatrix} \in R^{(m+1)}$.

This process is known as *updating*.

The *Recursive Least-Squares* (RLS) problem is to solve the problem stated by Equation 4.3, without redoing the steps necessary to solve the problem stated by Equation 4.1. Classical RLS algorithms are based on the normal equations [61]. Alternative approaches are based on updating the Cholesky decomposition [62] of the matrix $\mathbf{A}^T\mathbf{A}$ or its inverse:

$$\mathbf{A}^T\mathbf{A} = \mathbf{L}\mathbf{L}^T \tag{4.4}$$

where $\mathbf{L}$ is $n \times n$ lower triangular and $\mathbf{L}^T$ is $n \times n$ upper triangular.

Another approach is to update the QR decomposition [62] of $\mathbf{A}$:

$$\mathbf{A} = \mathbf{Q}\mathbf{R}' \tag{4.5}$$

where $\mathbf{Q}$ is the $m \times m$ orthogonal and $\mathbf{R}' = \begin{pmatrix} \mathbf{R} \\ \mathbf{0} \end{pmatrix}$, with $\mathbf{R}$ being the $n \times n$ upper triangular.

The Cholesky decomposition and QR decomposition methods are related, as

$$\bar{\mathbf{A}}^T\bar{\mathbf{A}} = \mathbf{A}^T\mathbf{A} + \mathbf{a}\mathbf{a}^T = \mathbf{L}\mathbf{L}^T + \mathbf{a}\mathbf{a}^T$$

and

$$\bar{\mathbf{A}}^T\bar{\mathbf{A}} = \mathbf{A}^T\mathbf{A} + \mathbf{a}\mathbf{a}^T = (\mathbf{Q}\mathbf{R})^T(\mathbf{Q}\mathbf{R}) + \mathbf{a}\mathbf{a}^T = \mathbf{R}^T\mathbf{Q}^T\mathbf{Q}\mathbf{R} + \mathbf{a}\mathbf{a}^T = \mathbf{R}^T\mathbf{R} + \mathbf{a}\mathbf{a}^T$$

which implies that

$$\bar{\mathbf{A}}^T\bar{\mathbf{A}} = \mathbf{A}^T\mathbf{A} + \mathbf{a}\mathbf{a}^T = \mathbf{L}\mathbf{L}^T + \mathbf{a}\mathbf{a}^T = \mathbf{R}^T\mathbf{R} + \mathbf{a}\mathbf{a}^T.$$

The parameters of the RLS problem may be time varying, and not time invariant as we had assumed hereto. There are two ways of dealing with time-varying parameters, termed *downdating* and *forgetting*. In turn, there are two ways of forgetting: *exponential forgetting* and *directional forgetting*.

"Downdating" refers to the case when an old linear equation $\mathbf{a}^T x = \beta$ (from sometime in the history of the system and updates) is deleted from the system $\mathbf{A}x = b$. Thus this gives a situation where

$$\bar{\mathbf{R}}^T \bar{\mathbf{R}} = \mathbf{R}^T \mathbf{R} - \mathbf{a}\mathbf{a}^T. \tag{4.6}$$

Downdating is a difficult problem because the singular values of $\bar{\mathbf{R}}$ are less than or equal to the corresponding singular values of $\mathbf{R}$, which means that $\bar{\mathbf{R}}$ may become singular even though $\mathbf{R}$ is nonsingular.

"Forgetting" puts time-dependent weights on the old data, leading to a *weighted least squares* (WLS) problem:

$$\min_{x} \|D(\mathbf{A}x - \mathbf{b})\|^2 \tag{4.7}$$

where $D = \text{diag}(\lambda^{((m-i)/2)})$ and the forgetting factor $\lambda \in (0, 1)$; or equivalently:

$$\min_{x} \sum_{i=1}^{t} \lambda^{t-i} \|\mathbf{A}x - \mathbf{b}\|^2. \tag{4.8}$$

The kind of forgetting just described is "exponential forgetting," in that the data are gradually outdated. This is the common method of forgetting. The method of "directional forgetting" disregards the impact of only *one* linear equation.

The classical linear regression model is

$$y(t) = \theta_1 \cdot \varphi_1(t) + \theta_2 \cdot \varphi_2(t) + \cdots + \theta_n \cdot \varphi_n(t) + \varepsilon(t). \tag{4.9}$$

In a discrete time setting where time keeps incrementing, this leads to the *overdetermined linear system*:

$$\mathbf{Y}_t = \mathbf{\Phi}_t \cdot \mathbf{\Theta} + E_t \tag{4.10}$$

where $\mathbf{Y}_t$ is $t \times 1$ and contains the observed variable y at each time step t, $\mathbf{\Phi}_t$ is $t \times n$ and contains the $1 \times n$ regression vectors φ at each time step, $\mathbf{\Theta}$ contains the weights to be estimated at each time step, and E_t is $t \times 1$ and contains the residual error at each time step.

The LS problem is then

$$\min_{\theta} \|\mathbf{Y}_t - \mathbf{\Phi}_t \theta\|^2 \tag{4.11}$$

yielding the normal equations

$$\mathbf{\Phi}_t^T \mathbf{\Phi}_t \theta = \mathbf{\Phi}_t^T \mathbf{Y}_t. \tag{4.12}$$

If $\mathbf{\Phi}_t^T \mathbf{\Phi}_t$ is nonsingular, then the solution is

$$\tilde{\theta}_t = \left[\mathbf{\Phi}_t^T \mathbf{\Phi}_t \right]^{-1} \mathbf{\Phi}_t^T \mathbf{Y}_t. \tag{4.13}$$

With the assumptions that E_t has $\mathbf{0}$ mean and $\sigma^2 \mathbf{I}$ covariance, the *covariance matrix* of the system is defined as

$$\mathbf{P}_t = \sigma^2 [\mathbf{\Phi}_t^{\mathrm{T}} \mathbf{\Phi}_t]^{-1} \tag{4.14}$$

and the *information matrix* of the system as the inverse of the covariance matrix, $\mathbf{P}_t^{-1}$.

It may be mentioned here that $\min_x \|\mathbf{A}x - b\|^2$ is conventionally referred to as the *LS problem*, while $\min_\theta \|\mathbf{Y}_t - \mathbf{\Phi}_t\theta\|^2$ is conventionally referred to as the *regression problem*. Solutions to the above problems proceed along various avenues, as described below.

4.3.2.1 Algorithms Using Updating of the Covariance Matrix

The conventional RLS algorithm [61] falls into the category of algorithms that rely on updating the covariance matrix $\mathbf{P}_t = \sigma^2 [\mathbf{\Phi}_t^{\mathrm{T}} \mathbf{\Phi}_t]^{-1}$ using the *Matrix Inversion Lemma*. Other classical solutions include Peterka's [63] and Bierman's [64] algorithms. These two algorithms also rely on updating the covariance matrix $\mathbf{P}_t$, except that they work instead with the $\mathbf{LDL}^{\mathrm{T}}$ decomposition of $\mathbf{P}_t$. Recall that $\mathbf{L}$ represents (lower) triangular matrices, while $\mathbf{D}$ represents diagonal matrices.

4.3.2.2 Algorithms Using Orthogonal Transformations

Another class of RLS algorithms uses orthogonal transformations and hyperbolic rotations. The Givens 1 algorithm [60] is the most well known of such methods, and a number of others are described in [65].

4.3.2.3 Algorithms Using Updating of the Information Matrix

Numerical instabilities similar to the case of downdating arise when updating the covariance matrix. Numerical instabilities occur when the eigenvalues of $\mathbf{\Phi}_t$ decay rapidly to zero. Instead of working with the covariance matrix, updating the information matrix

$$\mathbf{P}_t^{-1} = \mathbf{P}_{t-1}^{-1} + \sigma^{-2} \varphi_t \varphi_t^{\mathrm{T}} \tag{4.15}$$

avoids this problem.

The square-root information filter (SRIF) algorithm [64,66] takes this approach, and works on the $\mathbf{LL}^{\mathrm{T}}$ decomposition of $\mathbf{P}_t^{-1}$.

4.3.2.4 Algorithms Using Updating and Downdating of the QR Decompositions

The LINPACK algorithm is an example algorithm that performs both updating and downdating and works on the $\mathbf{R}$ matrix of the QR decompositions of $\mathbf{\Phi}_t$ and $[\mathbf{\Phi}_t \dots \mathbf{Y}_t]$ [67]. The corrected seminormal equations (CSNEs) algorithm is an example algorithm that performs downdating, and also works on the $\mathbf{R}$ matrix of the QR decompositions of $\mathbf{\Phi}_t$ and $[\mathbf{\Phi}_t \dots \mathbf{Y}_t]$ [67]. Both these algorithms require a large amount of data storage, and assume the same $\mathbf{Q}$ matrix for periods of time. Algorithms that modify QR decompositions are thus also called sliding window methods.

Using the $\mathbf{R}$ matrix has the additional drawback that the final equations for the weights $\tilde{\theta}$ involve finding inverses of triangular matrices. Hence, these algorithms are heavy on computation.

By modifying $\mathbf{R}^{-1}$ instead of $\mathbf{R}$ in the QR decompositions of $\mathbf{\Phi}_t$ and $[\mathbf{\Phi}_t \dots \mathbf{Y}_t]$, one may avoid the computation-intensive triangular solutions. Inverse versions of the LINPACK algorithms, known as LINPACK Inverse Updating and LINPACK Inverse Downdating algorithms, as well as a CSNE Inverse Downdating algorithm, are also available [67].

4.3.2.5 Error Analysis

In judging the stability and error propagation characteristics of the classical RLS algorithms, it is important to study the following three primary aspects:

- What bounds can be obtained for the errors that arise in step t?
- How do errors that arise in step t propagate in subsequent steps?
- How do errors that arise in different steps interact and accumulate?

The error in the updating of the covariance matrix may be shown to be roughly bounded as

$$\|\delta \mathbf{P}_{t+1}\|^2 \lesssim \gamma_t^2 \|\delta \mathbf{P}_t\|^2 \tag{4.16}$$

with

$$\gamma_t = \|\mathbf{I} - \mathbf{q}_t \varphi_t^{\mathrm{T}}\| \geqslant 1 \tag{4.17}$$

where $\mathbf{q}_t = \mathbf{P}_t/\sigma^2$ is commonly referred to in RLS literature as the Kalman gain at step t.

The above relation between the accumulated error in step $t + 1$, $\delta \mathbf{P}_{t+1}$, with the accumulated error in step t, $\delta \mathbf{P}_t$, shows that the errors may actually increase! Although this is alarming, it is known that the RLS algorithm works well in practice when the problem is not too ill conditioned. Thus this bound is too pessimistic.

For reference, a problem is said to be well-conditioned, if the condition number (usually denoted by κ) is low, and ill-conditioned, if the condition number is high. The condition number is defined in terms of norms of the **A** matrices. When 2-norms are used, the condition number is given by

$$\kappa = \frac{\max(\mathrm{svd}(\mathbf{A}))}{\min(\mathrm{svd}(\mathbf{A}))}. \tag{4.18}$$

Stability characteristics are dependent on the condition number of the problem, and downdating is in general an ill-conditioned problem.

4.3.2.6 Time-Varying Parameter Estimation

In almost any practical system where recursive LS algorithms are used, the parameters are assumed to vary with time [60]. In some applications the parameters are slowly varying, while in some they are time invariant [8]. It is important to remember that the parameters cannot be allowed to vary too rapidly, lest noise may be interpreted as parameter fluctuations. Contrarily if the parameters were assumed to remain constant over a large period of time, the parameter estimation problem would turn into a tuning process, which is much less demanding.

Slow parameter changes and large parameter changes affect RLS algorithms in different ways. If the parameters change with time, we have to disregard the old observations in the system. There are two ways of dealing with time-varying parameters: *downdating* and *forgetting*. In turn, there are two ways of forgetting: *exponential forgetting* and *directional forgetting*. Exponential forgetting is the common method of forgetting, and this method leads to the gradual discounting of past observations. A problem with exponential forgetting is that although new observations arrive only in the φ-direction, old information is gradually discounted in all directions. In directional forgetting [68], old information is removed in the φ-direction only.

Other alternative types of forgetting have been suggested by Fortescue et al. [69] and by Sanoff and Wellstead [70], where the forgetting factor is also time varying, and a function of the prediction error. Betz and Evans suggest repeatedly resetting the covariance matrix to a large matrix [71].

4.3.2.7 Applications

The RLS algorithm is a popular method for obtaining linear predictors of some data sequence in an online fashion. The RLS algorithm minimizes the mean-squared error over the training data. The algorithm is suitable for learning scenarios as it observes input samples sequentially, one at a time, and does not need to store history of the data. It is thus efficient in terms of computational complexity and storage requirements. Moreover, it satisfies the additional constraint imposed by real-time applications that the computational cost per time step should by bounded by a constant independent of time. Results of applying the various forms of algorithms described above to a variety of fields are easily available. The reader is directed to [61,72] or any other standard book on adaptive filtering. In a separate work, it has already been shown how sampled statistics may be used to infer global characteristics regarding the network traffic flows [73].

4.3.3 Kernel Versions of LS Techniques

Kernel machines are a relatively new class of learning algorithms that use a *kernel mapping* function to produce nonlinear and nonparametric learning algorithms [74,75]. The idea is that a suitable kernel function, when applied to a pair of input data vectors, may be interpreted as an inner product in a high-dimensional Hilbert space known as the *feature space*. This allows inner products in the feature space (inner products of the *feature vectors*) to be computed without explicit knowledge of the feature vectors themselves, by simply evaluating the kernel function

$$k(\mathbf{x}_i, \mathbf{x}_j) = \langle \phi(\mathbf{x}_i), \phi(\mathbf{x}_j) \rangle, \qquad (4.19)$$

where $\mathbf{x}_i$ and $\mathbf{x}_j$ denote the input vectors, while $\phi(.)$ represents the mapping onto the feature space.

Popular kernel functions include the Gaussian kernel with variance σ^2,

$$k(\mathbf{x}_1, \mathbf{x}_2) = \exp\left\{ \frac{\|\mathbf{x}_1 - \mathbf{x}_2\|^2}{2\sigma^2} \right\} \qquad (4.20)$$

and the polynomial kernel of degree p,

$$k(\mathbf{x}_1, \mathbf{x}_2) = (a \langle \mathbf{x}_1, \mathbf{x}_2 \rangle + b)^p. \qquad (4.21)$$

A special case of the polynomial kernel is the linear kernel:

$$k(\mathbf{x}_1, \mathbf{x}_2) = \langle \mathbf{x}_1, \mathbf{x}_2 \rangle. \qquad (4.22)$$

The kernel recursive least-squares (KRLS) algorithm [76] combines the principles of kernel machines [74] and the popular RLS algorithm [61] to provide an efficient and nonparametric approach for performing online data mining. The KRLS algorithm operates on a data sequence of the form $Z_t = \{\mathbf{x}_i, y_i\}_{i=1}^t$ where the input–output pairs $(\mathbf{x}_i, y_i)$ are assumed to be independent, identically distributed (i.i.d.) samples from some distribution $p(Y, \mathbf{X})$. The objective is to obtain the best predictor $\hat{y}_t$ of y_t, given $Z_{t-1} \cup \{\mathbf{x}_t\}$. In conventional RLS, the dimension of the space spanned by the input samples $\{\mathbf{x}_i\}_{i=1}^t$ is constrained by the dimension of the input space. In contrast, KRLS involve a mapping onto a feature space of much higher dimensionality than the input space, and the dimension of the space spanned by $\{\phi(\mathbf{x}_i)\}_{i=1}^t$ has the potential to increase without bounds. At each time step, the dimension will increase unless $\mathbf{x}_t$ satisfies $\phi(\mathbf{x}_t) = \sum_{i=1}^{t-1} a_i \phi(\mathbf{x}_i)$. If the dimension increases, then the new vector provides new information and adds to the predictive power, and thus should be included in the predictor. This leads to the problem that the predictor may require the storage of a large number of input vectors, leading to unreasonable memory and computational requirements. In defining KRLS, Engel et al. address this problem by imposing a minimum threshold on the amount of new information an input vector must provide in order to be added to the predictor [76]. Feature vector $\phi(\mathbf{x}_t)$ is said to be *approximately* linearly dependent on $\{\phi(\mathbf{x}_i)\}_{i=1}^{t-1}$, with approximation threshold ν, if the projection error δ_t satisfies the following criterion:

$$\delta_t = \min_a \left\| \sum_{i=1}^{t-1} a_i \phi(\mathbf{x}_i) - \phi(\mathbf{x}_t) \right\|^2 < \nu. \qquad (4.23)$$

KRLS uses this notion of *approximate linear independence* to obtain a *dictionary* of input vectors $D = \{\tilde{\mathbf{x}}_j\}_{j=1}^m$, where $m < t$, such that $\{\phi(\tilde{\mathbf{x}}_j)\}_{j=1}^m$ approximately spans the feature space. The best predictor $\hat{y}_t$ of y_t in the feature space of the sparse set $\{\phi(\tilde{\mathbf{x}}_j)\}_{j=1}^m$, can then be evaluated:

$$\hat{y}_t = \sum_{j=1}^m \alpha_j \langle \phi(\tilde{\mathbf{x}}_j), \phi(\mathbf{x}_t) \rangle = \sum_{j=1}^m \alpha_j . k(\tilde{\mathbf{x}}_j, \mathbf{x}_t) \qquad (4.24)$$

The weights $\{\alpha_j\}_{j=1}^m$ are learned by KRLS over time through successive minimization of prediction errors in the LS sense.

4.3.4 Adaptive, Online Algorithms Based on LS Techniques

We advocate that truly online and learning methods need to be built on fundamentally recursive algorithms. To that end, we begin our presentation of online algorithms by discussing two recent algorithms that have been developed using KRLS principles.

In [9], Ahmed et al. presented an online, sequential, anomaly detection algorithm that is suitable for use with multivariate data. The proposed Kernel-based online anomaly detection (KOAD) algorithm assumes no model for network traffic or anomalies, and constructs and adapts a *dictionary* of *features* that approximately spans the subspace of normal behavior. The idea is to keep a number of feature vectors that are supposed to be the basis vectors for normal traffic in the space mapped onto by a suitably chosen kernel function. Once a feature vector that represented the current traffic metrics arrives, the algorithm examines the squared error in approximating this new vector as a linear combination of the basis vectors in the dictionary, in the feature space. If the error is less than a predefined lower threshold, the new vector is considered to be in the normal space; if it is larger than an upper threshold, it is thought to significantly deviate from the normal space and hence represent an anomaly; if the error lies between the two thresholds, the algorithm will raise an "orange" alarm, keep tracking the usefulness of the new vector in describing future arrivals for a small interim period, and then take a firm decision on it. At the same time, the obsolete vectors are removed and exponential forgetting is employed to keep the dictionary current. Removal of elements involves dimensionality reduction and is different from the downdating step in standard RLS. These features constitute extensions provided by [9] to the fundamental KRLS algorithm from [76].

The KOAD algorithm has been applied to a camera network using both centralized and distributed monitoring architectures [6], where it was shown to be effective in detective road traffic incidents in an online fashion. Subsequent works has referred to KOAD as one of the primary algorithms for anomaly detection in next-generation communication networks [77], one that is versatile and novel [78]. KOAD also showed the relationship between the space spanned by normal measurement vectors and MVSs [46], an idea that was later cited by [23] as a connection between anomaly detection and intrusion detection.

The kernel estimation-based anomaly detection (KEAD) algorithm was presented in [8], as the next version of the KOAD Algorithm. KEAD infers that any arriving data point is a realization of one of two possible underlying probability distributions, based on whether the data point is normal or anomalous. The algorithm then uses the kernel density estimator (KDE) of the arriving vector with arrivals around it, as the anomaly detection statistic to determine which distribution (normal or anomalous) the given vector most likely arose from. It uses the dictionary of feature vectors, where the dictionary is maintained in the same way as in KOAD, to enable the KDE to be calculated using a sparse representation of the space. KEAD offered the following major improvements. First, supplementary algorithms have been developed to set all major thresholds from the earlier KOAD. Second, the threshold for the KEAD anomaly detection statistic has been mathematically linked to the user-specified false discovery rate using ideas suggested in [79]. Through comparison with existing block-based offline methods, it was demonstrated that KEAD is equally effective but has much faster time-to-detection and lower computational complexity.

4.3.5 Other Online and Adaptive Algorithms

Examples of online algorithms based on parametric approaches include a system based on unsupervised learning of the information source, which incrementally updates the model every time new data arrive [80]. This system was tested on the popular network intrusion database KDD Cup 1999 [81]. Another online technique suggested in the past involved an extension of a method based on deterministic annealing (DA) [82] to detect anomalies in textual data [83].

Yang et al. presented a novelty detection method applied to document classification in [84]. When a new document arrives, their algorithm compares it to all the documents available. If the nearest neighbor in its past has a cosine similarity score below a threshold, then the arriving document is labeled as novel, otherwise it is labeled as old and added to the history. Yang et al. presented an algorithm for detecting novel events in a temporally ordered steam of news stories in [85]. They present an online version where the detection is done in real time as the events occur, or retroactively by using a clustering approach on an accumulated collection. In the online scenario, each arriving documents is marked as either new or old, indicating whether it is the first story describing a novel event or not. A story is represented using a vector of weighted terms.

Brutlag proposes a real-time anomaly detection mechanism that may be incorporated in a network monitoring software in [86]. He presents a dynamic algorithm (as in the parameters of the model keep on updating) to detect aberrant behavior. The primary idea is to define a "violation" as an observation that falls outside an interval (a confidence band), then to define a "failure" (an anomaly) as exceeding a specified number of violations within a specified number of observations (the window length). The model here is an extension of the Holt-Winters forecasting algorithm, which supports incremental model updating via exponential smoothing. This chapter also provides a thorough presentation and explanation of the fundamental Holt-Winters equations and provides interesting methods and guidelines for setting the parameters of the model.

As our discussions so far suggest, much of the other previous work on online network anomaly detection has been based on network traffic models. Hajji uses a Gaussian mixture model in [11] and develops an algorithm based on a stochastic approximation of the EM algorithm [28,29] to obtain estimates of the model parameters. An example of a real-time network anomaly detection method that is not based on an *a priori* model is the time-based inductive learning machine (TIM) of Teng et al. [87]. Their machine constructs a set of rules based on usage patterns. The detection algorithm detects a deviation when the premise of a rule occurs but the conclusion does not follow. This algorithm has subsequently influenced the development of other learning intrusion detection algorithms [88,89].

Kivinen et al. presented an online algorithm applicable to novelty detection that is based on support vector machines (SVM) [74,90] in [91]. Their algorithm minimizes a risk function with the newly arriving data point expressed as a kernel expansion of previous data points. They achieve sparsity by using a power-series approximation in the update equations for the kernel expansion coefficients. Kivinen's algorithm is also dependent on the choice of the loss function, as the derivative of it determines the terms of the coefficient power-series expansion that are retained. The loss function typically used for novelty detection [74] is in turn dependent on a prespecified maximum alarm rate and a prespecified fraction of outliers. Duffield et al. presented an adaptive architecture where anomalies are flagged using flow signatures that are themselves learnt from the data stream [92]. Hero presents the geometric entropy minimization (GEM) algorithm as another application of learning methods to network anomaly detection [10]. GEM is a block-based, nonparametric and adaptive method that does not require the use of a detection threshold. It is based on the minimal covering properties of entropic graphs when constructed from a set of training samples. In a different application but with familiar foundations, Silva and Willett propose a method based on hypergraphs to detect anomalous meetings in a social network [93]. They model the distribution of meetings as a two-component mixture of normal and anomalous events, and then use a variational approximation [94] of the EM algorithm to assess the likelihood of each observation being anomalous.

Extensive reviews of other current approaches to network anomaly detection may be found in [77,95], and in the related work sections of [6,8,9,23,78,96].

4.4 Conclusions

In this chapter we have explained the problem of anomaly network in a network, and then presented possible solutions. We began with a discussion of the classic methods of anomaly detection from history, and then progressed toward the recent trends that involve applying algorithms from the field of machine learning. We advocated the use of recursive algorithms as the basis for developing truly online algorithms. As such, we provided a thorough analysis of one such fundamental recursive algorithms, the RLS algorithm [61]. It is commonly stated that network anomaly detection algorithms commonly fall under three categories: statistical methods, streaming algorithms, and machine learning approaches with a focus on unsupervised learning [77]. We have thoroughly discussed all three types.

All network anomaly detection algorithms must first obtain sampled statistics from the nodes, store the appropriate state information, and then signal anomalies as soon as they occur. A network operator must realize that while some anomalies manifest themselves by creating abrupt changes in some specific link, others may lead to a gradual variation in global network traffic. While some anomalies may be insignificant accident events that are best ignored; others may be caused by an intruder or snooper with malicious intent. The detection algorithm must work on sampled statistics from high-rate links and quickly isolate the appropriate information from the high-speed line cards. The algorithm must be capable of processing long streams of rapidly arriving data with minimal memory requirements and computational power and limited state information. It must be borne in mind that the detection algorithms may be run in a distributed fashion and locally at each low-powered node in an ad hoc wireless, or wireless sensor, network. Therefore a designer of network anomaly detection algorithms must not only obtain receiver operating characteristics (ROC) curves that balance the false alarm or false-discovery rate with the detection rate, but must also ensure low memory, storage, and computational complexities [6,8,9]. First and foremost though, the designer must realize and admit that creating the perfect anomaly detector may always remain a utopian dream [97]!

Terminologies

Anomaly
Approximate linear dependence
Clustering
IP flow space
KEAD
Kernel machines
KRLS
KOAD
LS
PCA
PPCA
RLS

Questions and Sample Answers

1. What is a network anomaly?
   There is no universally accepted, specific definition for an anomaly. An anomaly is understood to be some phenomenon that creates a sudden deviation from the normal operation of a network. It may change some characteristic of the traffic flow at a node, a link, or in the overall network.

2. How have anomalies traditionally been detected?
   Most traditional approaches may be classified into two categories: model based or signature based. In model-based approaches, a mathematical model is used to describe the normal operation of the network, and an anomaly is signaled when an observation cannot be explained by the model within some bounds. In signature-based approaches, the algorithm looks out for observations that fit a database of known anomaly types.

3. What are the recent trends toward anomaly detection?
   Recent efforts are toward the development of detection algorithms that assume no prior model or signature of normal behavior or the anomalies to be guarded against. Instead, data mining

principles are employed to learn the patterns of normal behavior. In addition, the algorithms should be able to raise the alarms in real time.

4. What did Lakhina et al. discover about the structure of multivariate traffic measurements?
   Although the total number of OD flows in large backbone networks may be multidimensional, Lakhina et al. found that only a few fundamental flows are sufficient to explain the time series of the set of flows. These fundamental flows are termed eigenflows, and the respective eigenvalue indicates the fraction of total energy content in each flow.

5. What is the PCA approach to network anomaly detection?
   PCA may be used to separate the space or multivariate measurements into disjoint normal and anomalous subspaces. The magnitude of the projection onto the anomalous subspace may be compared to a threshold (subject to a confidence interval) to signal an anomalous observation.

6. How do statistical parametric approaches to anomaly detection typically work?
   Statistical parametric approaches assume that normal observations arise from some model of an underlying distribution. A mixture of Gaussians is most often used as the model, and some parameter estimation technique, for example, the EM algorithm or variant thereof, is used to estimate the parameters of the model.

7. How do the clustering algorithms for anomaly detection typically work?
   Clustering algorithms partition the data into a number of clusters, and then evaluate the degree to which each observation belongs to a cluster. An observation is deemed anomalous if the degree of membership in every cluster is below a threshold. Variants of the k-means algorithm are popular clustering algorithms.

8. What are the new paradigms of anomaly detection research? Why do we need alternative approaches?
   It has been shown that straightforward extensions of existing offline algorithms to online versions are not robust. Popular algorithms such as the PCA method remain sensitive to parameter and initial conditions. Thus it is better to consider fundamentally recursive and adaptive algorithms as alternative bases for the development of next-generation anomaly detectors.

9. What is the LS problem? What is the RLS problem?
   The classical LS problem is

$$\min_{x} \|\mathbf{A}x - \mathbf{b}\|^2, \tag{4.25}$$

where $\mathbf{A} \in R^{m \times n}$, $x \in R^n$, and $b \in R^m$, with $m > n$.
The classical solution to the LS problem is stated in terms of the *normal equation*

$$\mathbf{A}^\mathrm{T}\mathbf{A}x = \mathbf{A}^\mathrm{T}\mathbf{b}.$$

Now when a new linear equation $\mathbf{a}^\mathrm{T}x = \beta$ is added to the system, the new problem becomes

$$\min_{x} \|\bar{\mathbf{A}}x - \bar{\mathbf{b}}\|^2, \tag{4.26}$$

where

$$\bar{\mathbf{A}} = \begin{pmatrix} \mathbf{A} \\ \mathbf{a}^{\mathsf{T}} \end{pmatrix} \in R^{(m+1) \times n}, \quad x \in R^n, \quad \text{and} \quad \bar{\mathbf{b}} = \begin{pmatrix} \mathbf{b} \\ \beta^{\mathsf{T}} \end{pmatrix} \in R^{(m+1)}.$$

The RLS problem is to solve the problem stated by (3), without redoing the steps necessary to solve the problem stated by (1). This recursive process obviously provides lower computational as well as memory complexity compared to resolving the full system.

10. What are kernel machines?
    Kernel machines are algorithms that involve the so-called *kernel trick:*

$$k(\mathbf{x}_i, \mathbf{x}_j) = \langle \phi(\mathbf{x}_i), \phi(\mathbf{x}_j) \rangle$$

where $\mathbf{x}_i$ and $\mathbf{x}_j$ denote the input vectors, while $\phi(.)$ represents the mapping onto the feature space.
    The above equation means that given two input vectors $\mathbf{x}_i$ and $\mathbf{x}_j$, the inner product of the mappings of these vectors on to a *feature space* of much higher dimension may be evaluated without direct knowledge of the mapping function $\phi(.)$ or the two feature vectors.

11. What is approximate linear dependence?
    Approximate linear dependence is defined in terms of the magnitude of the squared error in representing a new observation as a linear combination of the current dictionary elements.

12. How does the KOAD algorithm work?
    The KOAD algorithm incrementally builds a dictionary of feature vectors that approximately spans the space of normal observations, in the feature space that is mapped onto by a user-chosen kernel function. Once a feature vector that represented the current traffic metrics arrives, the algorithm evaluates this squared error and compares it with two predefined thresholds. If the error is less than the lower threshold, the new vector is considered to be in the normal space; if it is larger than an upper threshold, it is thought to significantly deviate from the normal space and hence represent an anomaly; if the error lies between the two thresholds, the algorithm will raise an "orange" alarm, keep tracking the usefulness of the new vector in describing future arrivals for a small interim period, and then take a firm decision on it.

13. How does the KEAD algorithm work?
    The KEAD algorithm is the next version of the KOAD algorithm. KEAD infers that any arriving data point is a realization of one of two possible underlying probability distributions, based on whether the data point is normal or anomalous. The algorithm then uses the KDE of the arriving vector with arrivals around it, as the anomaly detection statistic to determine which distribution (normal or anomalous) the given vector most likely arose from. It uses the dictionary of feature vectors, where the dictionary is maintained in the same way as in KOAD, to enable the KDE to be calculated using a sparse representation of the space. KEAD offered the following major improvements. First, supplementary algorithms were developed to set all major thresholds from the earlier KOAD. Second, the threshold for the KEAD anomaly detection statistic was mathematically linked to the user-specified false-discovery rate.

14. What are the expectations from the ideal network anomaly detector?
    The ideal anomaly detector must be able to rapidly extract the relevant state information from sampled traffic statistics that are recorded by network monitoring devices in high-speed links. It must then be able to find the signatures of any potential anomaly from the data

itself, and distinguish between measurements that are similar but have arisen from different underlying distributions. The detector must achieve this using low memory and computational complexities, and by balancing the detection accuracy with low false alarm rates and detection delays. It has been argued that creating the ideal anomaly detector may remain a utopian dream.

Author's Biography

Tarem Ahmed received a Bachelor's degree with a double major in physics and economics from Middlebury College, Middlebury, Vermont in 1999. He received a Master's degree in electrical engineering as a Moore Fellow from the University of Pennsylvania, Philadelphia, Pennsylvania in 2000. After working as a research assistant at the Electrical and Computer Engineering department at McGill University, Montreal, Quebec, Canada, he has recently moved to the Department of Electrical and Electronic Engineering at BRAC University, Dhaka, Bangladesh.

Rumana Rahman received a Bachelor's degree in electronics and communication engineering from BRAC University, Dhaka, Bangladesh in 2008. Since graduation, she has joined the Department of Electrical and Electronic Engineering at BRAC University as a junior lecturer. Her research interests lie in wireless broadband communication and networks, and excerpts from her undergraduate final project have recently appeared as the cover story of a leading Bangladeshi technology magazine.

References

1. V. Hodge and J. Austin, A survey of outlier detection methodologies, *Artificial Intelligence Review*, 22(2), 85–126, 2004.
2. M. Markou and S. Singh, Novelty detection: A review–part 1: Statistical approaches, *Signal Process*, 83(12), 2481–2497, 2003.
3. M. Markou and S. Singh, Novelty detection: A review–part 2: Neural network based approaches, *Signal Process*, 83(12), 2499–2521, 2003.
4. A. Lazarevic, L. Ertöz, V. Kumar, A. Ozgur, and J. Srivastava, A comparative study of anomaly detection schemes in network intrusion detection, in *Proceedings of SIAM International Conference on Data Mining*, San Francisco, CA, USA, May 2003.
5. A. Lakhina, K. Papagiannaki, M. Crovella, C. Diot, E. Kolaczyk, and N. Taft, Structural analysis of network traffic flows, in *Proceedings of ACM SIGMETRICS*, New York, NY, USA, June 2004.
6. T. Ahmed, B. Oreshkin, and M. Coates, Machine learning approaches to network anomaly detection, in *Proceedings of USENIX Workshop on Tackling Computer Systems Problems with Machine Learning Techniques (SysML)*, Cambridge, MA, USA, April 2007.
7. L. LaBarre, Management by exception: OSI event generation, reporting, and logging, in *Proceedings of International Symposium Integrated Network Management*, Washington, DC, USA, April 1991.
8. T. Ahmed, Online anomaly detection using KDE, in *Proceedings of IEEE Global Communications Conference (GLOBECOM)*, Honolulu, HI, USA, November 2009.
9. T. Ahmed, M. Coates, and A. Lakhina, Multivariate online anomaly detection using kernel recursive least squares, in *Proceedings of IEEE INFOCOM*, Anchorage, AK, USA, May 2007.
10. A. Hero III, Geometric entropy minimization (GEM) for anomaly detection and localization, in *Proceedings of Neural Information Processing Systems (NIPS)*, Vancouver, BC, Canada, December 2006.
11. H. Hajji, Statistical analysis of network traffic for adaptive faults detection, *IEEE Transactions on Neural Networks*, 16(5), 1053–1063, 2005.
12. A. Lakhina, M. Crovella, and C. Diot, Diagnosing network-wide traffic anomalies, in *Proceedings of ACM SIGCOMM*, Portland, OR, USA, August 2004.

13. A. Lakhina, M. Crovella, and C. Diot, Characterization of network-wide anomalies in traffic flows, in *Proceedings of ACM Internet Measurement Conference (IMC)*, Taormina, Sicily, Italy, February 2004.

14. L. Huang, X. Nguyen, M. Garofalakis, M. Jordan, A. Joseph, and N. Taft, In-network PCA and anomaly detection, in *Advances in Neural Information Processing Systems*, 19th ed., B. Schölkopf, J. Platt, and T. Hoffman, eds, Cambridge, MA, USA: MIT Press, 2007.

15. Y. Huang, N. Feamster, A. Lakhina, and J. Xu, Diagnosing network disruptions with network-wide analysis, in *Proceedings of ACM SIGMETRICS*, San Diego, CA, USA, June 2007.

16. D. Brauckhoff, B. Tellenbach, A. Wagner, M. May, and A. Lakhina, Impact of packet sampling on anomaly detection metrics, in *Proceedings of ACM Internet Measurement Conference*, Rio de Janeiro, Brazil, October 2006.

17. X. Li, F. Bian, M. Crovella, C. Diot, R. Govindan, G. Iannaccone, and A. Lakhina, Detection and identification of network anomalies using sketch subspaces, in *Proceedings of ACM Internet Measurement Conference (IMC)*, Rio de Janeiro, Brazil, October 2006.

18. A. Lakhina, M. Crovella, and C. Diot, Mining anomalies using traffic feature distributions, in *Proceedings of ACM SIGCOMM*, Philadelphia, PA, USA, August 2005.

19. A. Lakhina, M. Crovella, and C. Diot, Detecting distributed attacks using network-wide flow traffic, in *Proceedings of FloCon Analysis Workshop*, Pittsburgh, PA, USA, September 2005.

20. I. Jolliffe, *Principal Component Analysis*, 2nd ed. New York, NY, USA: Springer-Verlag, 2002.

21. H. Hotelling, Analysis of a complex of statistical variable into principal components, *Journal of Educational Psychology*, 24(6), 417–441, 1933.

22. J. Jackson and G. Mudholkar, Control procedures for residuals associated with principal component analysis, *Technometrics*, 21(3), 341–349, 1979.

23. R. Beverly, Statistical learning in network architecture, PhD dissertation, Massachusetts Institute of Technology, Cambridge, MA, USA, June 2008.

24. A. Lakhina, K. Papagiannaki, M. Crovella, C. Diot, E. Kolaczyk, and N. Taft, Analysis of origin destination traffic flows (raw data), Boston University, Boston, MA, USA, Technical Report, BUCS-TR-2003-022, November 2003.

25. A. Feldmann, A. Greenberg, C. Lund, N. Reingold, J. Rexford, and F. True, Deriving traffic demands for operational IP networks: Methodology and experience, *IEEE/ACM Transactions on Networking*, 9(3), 265–279, 2001.

26. L. Gordon, M. Loeb, W. Lucyshyn, and R. Richardson, CSI/FBI computer crime and security survey. 2004. [Online]. Available: http://www.gocsi.com/forms/fbi/csi fbi survey.jhtml.

27. M. Tipping and C. Bishop, Probabilistic principal component analysis, *Journal of the Royal Statistical Society, Series B*, 61(3), 611–622, 1999.

28. G. McLachlan and T. Krishnan, *The EM Algorithm and Extensions*, 2nd ed., Hoboken, NJ, USA: Wiley, 2008.

29. A. Dempster, N. Laird, and D. Rubin, Maximum likelihood from incomplete data via the EM algorithm, *Journal of the Royal Statistical Society, Series B.*, 39(1), 1–38, 1977.

30. A. Papoulis and S. Pillai, *Probability, Random Variables and Stotastic Processes*, 4th ed. New York, NY, USA: McGraw-Hill, 2001.

31. M. Tipping and C. Bishop, Mixtures of probabilistic principal component analysers, *Neural Computation*, 11(2), 443–482, 1999.

32. C. Bishop and M. Tipping, A hierarchical latent variable model for data visualization, *IEEE Transactions on Pattern Analysis and Machine Intelligence*, 20(3), 281–293, 1998.

33. P. Barford, J. Kline, D. Plonka, and A. Ron, A signal analysis of network traffic anomalies, in *Proceedings of Internet Measurement Workshop*, Marseille, France, November 2002.

34. S. Kim and A. Reddy, Statistical techniques for detecting traffic anomalies through packet header data, *IEEE/ACM Transactions on Networking (ToN)*, 16, 562–575, 2008.

35. S. Mallat, *A Wavelet Tour of Signal Processing*, 3rd ed. Burlington, MA, USA: Academic Press, 2008.

36. S. Maybeck, *Stochastic Models, Estimation and Control*. Vol 1. New York, NY, USA: Academic Press, 1979.

37. A. Soule, K. Salamatian, and N. Taft, Combining filtering and statistical methods for anomaly detection, in *Proceedings of ACM Internet Measurement Conference (IMC)*, Berkeley, CA, USA, October 2005.

38. R. Thommes and M. Coates, Epidemiological modelling of peer-to-peer viruses and pollution, in *Proceedings of IEEE INFOCOM*, Barcelona, Spain, 2006.

39. T. Odin and D. Addison, Novelty detection using neural network technology, in *Proceedings of International Congress of Condition Monitoring and Diagnostic Engineering Management (COMADEN)*, Houston, TX, USA, December 2000.

40. L. Tarassenko, A. Nairac, N. Townsend, and P. Cowley, Novelty detection in jet engines, in *Proceedings of IEE Colloquium on Condition Monitoring, Imagery, External Structures and Health*, Birmingham, UK, June 1999.

41. L. Tarassenko, P. Hayton, N. Cerneaz, and M. Brady, Novelty detection for the identification of masses in mammograms, in *Proceedings of International Conference on Artificial Neural Networks*, Cambridge, UK, June 1995.

42. D.-Y. Yeung and Y. Ding, Host-based intrusion detection using dynamic and static behavioral models, *Pattern Recognition*, 36(1), 229–243, 2003.

43. S. Guttormsson, R. Marks, M. El-Sharkawi, and I. Kerszenbaum, Elliptical novelty grouping for on-line short-turn detection of excited running rotors, *IEEE Transactions on Energy Conversion*, 14(1), 16–22, 1999.

44. G. McLachlan, *Discriminant Analysis and Statistical Pattern Recognition*. Wiley-Interscience, USA, 1992.

45. P. Mahalanobis, On the generalised distance in statistics, *Proceedings of National Institute of Science, India*, 2(1), 49–55, 1936.

46. J. Einmal and D. Mason, Generalized quantile processes, *Annals of Statistics*, 20(2), 1062–1078, 1992.

47. D. Tax and R. Duin, Data description in subspaces, in *Proceedings of International Conference on Pattern Recognition (ICPR)*, Barcelona, Spain, September 2000.

48. B. Krishnamurthy, S. Sen, Y. Zhang, and Y. Chen, Sketch-based change detection: Methods, evaluation, and applications, in *Proceedings of ACM SIGCOMM Conference on Internet Measurement*, Miami, FL, USA, October 2003.

49. R. Schweller, Z. Li, Y. C. Y. Gao, A. Gupta, Y. Zhang, P. Dinda, M.-Y. Kao, and G. Memik, Reverse hashing for high-speed network monitoring: Algorithms, evaluation, and applications, in *Proceedings of IEEE INFOCOM*, Barcelona, Spain, April 2006.

50. R. Schweller, A. Gupta, E. Parsons, and Y. Chen, Reversible sketches for efficient and accurate change detection over network data streams, in *Proceedings of ACM Internet Measurement Conference (IMC)*, Taormina, Sicily, Italy, October 2004.

51. Y. Gao, Z. Li, and Y. Chen, A DoD resilient flow-level intrusion detection approach for high-speed networks, in *Proceedings of IEEE International Conference on Distributed Computing Systems (ICDCS)*, Lisbon, Portugal, July 2006.

52. R. Duda, P. Hart, and D. Stork, *Pattern Classification*, 2nd ed. Wiley, USA, 2001.

53. T. Cormen, C. Leiserson, R. Rivest, and C. Stein, *Introduction to Algorithms*, 2nd ed. Cambridge, MA, USA: MIT Press, 2001.

54. M. Jiang, S. Tseng, and C. Su, Two-phase clustering process for outliers detection, *Pattern Recognition Letters*, 22, 691–700, 2001.

55. R. Ensafi, S. Dehghanzadeh, R. Mohammad, and T. Akbarzadeh, Optimizing fuzzy k-means for network anomaly detection using PSO, in *Proceedings of IEEE/ACS International Conference on Computer Systems and Applications*, Doha, Qatar, April 2008.

56. T. Cover and J. Thomas, *Elements of Information Theory*, 2nd ed. New York, NY, USA: Wiley-Interscience, June 2006.

57. W. Lee and D. Xiang, Information-theoretic measures for anomaly detection, in *Proceedings of IEEE Symposium on Security and Privacy*, Oakland, CA, USA, May 2001.

58. Y. Gu, A. McCallum, and D. Towsley, Detecting anomalies in network traffic using maximum entropy estimation, in *Proceedings of ACM Internet Measurement Conference (IMC)*, Berkeley, CA, USA, October 2005.

59. H. Ringberg, A. Soule, J. Rexford, and C. Diot, Sensitivity of PCA for traffic anomaly detection, in *Proceedings of ACM SIGMETRICS*, San Diego, CA, USA, June 2007.

60. A. Sjö, *Updating Techniques in Recursive Least-Squares Estimation*, Lund University, Lund Institute of Technology, Lund, Sweden, Tech. Rep., October 1992.

61. A. Sayed, *Fundamentals of Adaptive Filtering*. Hoboken, NJ, USA: Wiley, 2003.

62. G. Stewart, *Matrix Algorithms: Basic Decompositions*. Philadelphia, PA, USA: Soc. for Industrial and Applied Math., 1998.

63. V. Peterka, Algorithm for LQG self-tuning control based on input-output delta models, in *Proceedings of IFAC Workshop on Adaptive Systems in Control and Signal Processing*, Lund, Sweden, July 1986.

64. G. Bierman, *Factorization Methods for Discrete Sequential Estimation*. New York, NY, USA: Academic Press, April 1977.

65. P. Gill, G. Golub, W. Murray, and M. Saunders, Methods for modifying matrix factorizations, *Mathematics of Computation*, 28(126), 505–535, 1974.

66. P. van Dooren and M. Verhaegen, Numerical aspects of different kalman filter implementations, *IEEE Transactions* on Automatic *Control*, 31(10), 907–917, 1986.

67. Å. Björck, L. Eldén and H. Park, Accurate downdating of least squares solutions, *SIAM Journal on Matrix Analysis and Applications*, 15(2), 549–568, 1994.

68. T. Hägglund, New estimation techniques for adaptive control, PhD dissertation, Lund University, Lund, Sweden, 1983.

69. T. Fortescue, L. Kershenbaum, and B. Ydstie, Implementation of self-tuning regulators with variable forgetting factors, *Automatica*, 17(6), 831–835, 1981.

70. S. Sanoff and P. Wellstead, Extended self-tuning algorithm, *International Journal of Control*, 34(3), 433–455, 1981.

71. R. Betz and R. Evans, New results and applications of adaptive control to classes of nonlinear systems, in *Proceedings of Workshop on Adaptive Control*, Florence, Italy, October 1982.

72. S. Haykin, *Adaptive Filter Theory*, 4th ed. Upper Saddle River, NJ, USA: Prentice-Hall, 2001.

73. N. Duffield, C. Lund, and M. Thorup, Estimating flow distributions from sampled flow statistics, *IEEE/ACM Transactions* on *Networking (TON)*, 13, 933–946, 2005.

74. B. Schölkopf and A. Smola, *Learning with Kernels*. Cambridge, MA, USA: MIT Press, 2001.

75. J. Shawe-Taylor and N. Cristianini, *Kernel Methods for Pattern Analysis*. Cambridge, UK: Cambridge University Press, 2004.

76. Y. Engel, S. Mannor, and R. Meir, The kernel recursive least squares algorithm, *IEEE Transactions* on *Signal Processing*, 52(8), 2275–2285, 2004.

77. M. Thottan, G. Liu, and C. Ji, Anomaly detection approaches for communication networks, in G. Cormode and M. Thottan, eds. *Algorithms for Next Generation Networks*, Springer-Verlag, London, UK, 2010.

78. S. Haggett, Towards a multipurpose neural network approach to novelty detection, PhD dissertation, University of Kent, Canterbury, UK, 2008.

79. C. Scott and E. Kolaczyk, Nonparametric assessment of contamination in multivariate data using minimum volume sets and FDR, University of Michigan, Ann Arbor, MI, USA, Tech. Rep. CSPL-381, April 2007.

80. K. Yamanishi, J.-I. Takeuchi, G. Williams, and P. Milne, On-line unsupervised outlier detection using finite mixtures with discounting learning algorithms, in *Proceedings of ACM SIGKDD International Conference on Knowledge Discovery and Data Mining*, Boston, MA, USA, August 2000.

81. ACM KDD Cup. Annual Data Mining and Knowledge Discovery competition organized by ACM Special Interest Group on Knowledge Discovery and Data Mining. [Online]. Available: http://www.sigkdd.org/kddcup/index.php.

82. K. Rose, Deterministic annealing for clustering, compression, classification, regression, and related optimization problems, in *Proceedings of IEEE*, 86(11), pp. 2210–2239, November 1998.

83. L. Baker, T. Hofmann, A. McCallum, and Y. Yang, A hierarchical probabilistic model for novelty detection in text, Carnegie Mellon University, Pittsburgh, PA, USA, Tech. Rep., 1999.

84. Y. Yang, J. Zhang, J. Carbonell, and C. Jin, Topic-conditioned novelty detection, in *Proceedings of ACM SIGKDD International Conference on Knowledge Discovery and Data Mining*, Edmonton, AB, Canada, July 2002.

85. Y. Yang, T. Pierce, and J. Carbonell, A study on retrospective and online event detection, in *Proceedings of ACM SIGKDD Int. Conf. Research and Development in Information Retrieval*, Melbourne, Australia, August 1998.

86. J. Brutlag, Aberrant behavior detection in time series for network monitoring, in *Proceedings of USENIX System Administrators Conference (LISA)*, New Orleans, LA, USA, December 2000.

87. H. Teng, K. Chen, and S. Lu, Adaptive real-time anomaly detection using inductively generated sequential patterns, in *Proceedings of IEEE Comp. Soc. Symp. Research in Security and Privacy*, Oakland, CA, USA, May 1990.

88. K. Ilgun, R. Kemmerer, and P. Porras, State transition analysis: A rule-based intrusion detection approach, *IEEE Transactions on Software Engineering*, 21(3), 181–199, 1995.

89. T. Lane, Machine learning techniques for the computer security domain of anomaly detection, PhD dissertation, Purdue University, W. Lafayette, IN, USA, August 2000.

90. N. Cristianini and J. Shawe-Taylor, *An Introduction to Support Vector Machines and Other Kernel-based Learning Methods*. Cambridge, UK: Cambridge University Press, 2000.

91. J. Kivinen, A. Smola, and R. Williamson, Online learning with kernels, in *Advances in Neural Information Processing Systems*, 14th ed. Cambridge, MA, USA: MIT Press, 2001.

92. N. Duffield, P. Haffner, B. Krishnamurthy, and H. Ringberg, Rule-based anomaly detection on IP flows, in *Proceedings of IEEE INFOCOM*, Rio de Janeiro, Brazil, April 2009.

93. J. Silva and R. Willett, Detection of anomalous meetings in a social network, in *Proceedings of Annual Conference on Information Sciences and Systems (CISS)*, Princeton, NJ, USA, March 2008.

94. M. Wainwright and M. Jordan, Graphical models, exponential families, and variational inference, *Foundations and Trends in Machine Learning*, 1(1–2), 1–305, 2008.

95. V. Chandola, A. Banerjee, and V. Kumar, Anomaly detection: A survey, *ACM Computing Surveys*, 41(3), Article 15, 2009.

96. H. Ringberg, Privacy-preserving collaborative anomaly detection, PhD dissertation, Princeton University, Princeton, NJ, USA, 2009.

97. S. Venkataraman, J. Caballero, D. Song, A. Blum, and J. Yates, Black-box anomaly detection: is it utopian? in *Proceedings of ACM SIGCOMM Workshop on Hot Topics in Networks (HotNets)*, Irvine, CA, USA, November 2006.

Chapter 5

Reputation- and Trust-Based Systems for Wireless Self-Organizing Networks

Jaydip Sen

Contents

5.1 Introduction

Reputation and trust are two very useful tools that are used to facilitate decision-making in distributed self-organizing networks such as *mobile ad hoc networks* (MANETs) and the Wireless Sensor Networks (WSNs). In simple terms, reputation is the opinion of one entity about another. Essentially, it signifies the trustworthiness of an entity [1]. Trust, on the other hand, is the expectation of one entity about the actions of another [2]. For over the last three decades, formal studies have been done on how reputation and trust can affect decision-making abilities under uncertain situations. However, it is only recently that the concepts of reputation and trust have been adapted to wireless networks, as these concepts can effectively resolve many problems that are otherwise not possible to solve with traditional security and authentication mechanisms.

Two types of wireless networks such as MANETs and WSNs have undergone tremendous technological advances over the last few years. This rapid development brings with it the associated risk of newer threats and challenges and the responsibility of ensuring safety, security, and integrity of information communication over these networks. MANETs are particularly vulnerable to different types of attacks and security threats because of complete autonomy of the member nodes and lack of any centralized infrastructure [3]. Moreover, because every node has a resource constraint, there is an incentive for each node to be programmed to guard its resources by itself. This leads to a manifestation of selfish behavior of every node that is harmful to the network as a whole. WSNs, on the other hand, involve some unique problems due to their usual operations in unattended and hostile areas. Since sensor networks are deployed with thousands of sensors for monitoring even a small area, it becomes imperative to produce sensors at very low costs. This invariably makes it impossible to produce tamper-resistant sensors. It is also very easy for an adversary to physically capture a sensor node (SN) and bypass its limited cryptographic security. The adversary can reprogram the captured node in such a way that it starts causing extreme damage to the system.

These problems can be somewhat resolved by incorporating reputation- and trust-based systems in MANETs and WSNs. The nodes thus make reputation- and trust-guided decisions, for example, in choosing relay nodes for forwarding packets for other nodes, or for accepting location information from beacon nodes (BNs) [1]. This not only provides MANETs and WSNs with the capability of informed decision-making, but also provides them with security against any internal attacks when cryptographic security might have been compromised. The system that discovers, records, and utilizes reputation to form trust, and also uses trust to influence its behavior is referred to as a reputation- and trust-based system. This chapter provides the reader with a complete understanding of reputation- and trust-based systems from the wireless self-organizing networks perspective.

The rest of the chapter is organized as follows: Section 5.2 introduces the concepts of trust and reputation with respect to wireless self-organizing networks. Section 5.3 introduces various features of MANETs and WSNs and discusses different types of misbehaviors that may be exhibited by the nodes in these networks and their effects on the network performance. Section 5.4 discusses various characteristics of trust metric and different classes of reputation- and trust-based systems for wireless self-organizing networks along with their desirable properties. Section 5.5 presents a detailed discussion on various important design issues of reputation- and trust-based systems with regard to the procedures of reputation information collection, processing, modeling, dissemination and final action taken. Section 5.6 makes a critical review of some of the well-known reputation- and trust-based systems for MANETs and WSNs. The strengths and weaknesses of these systems are discussed and their effectiveness is compared with various parameters. Section 5.7 highlights some of the open issues in the domain of reputation and trust from wireless self-organizing networks perspective. Finally, Section 5.8 concludes the chapter.

5.2 Trust-Definition and Concepts

This section presents the trust properties: definitions, classifications, and characteristics. There have been different approaches to define trust. Trust in general is a directional relationship between two entities and plays a major role in building a relationship between nodes in a network. Even though trust has been formalized as a computational model, it still means different things for different research communities.

Gambetta [4] has defined trust as: Trust is a particular level of the subjective probability with which an agent will perform a particular action, both before [we] can monitor such action (or independently of his capacity of ever to be able to monitor it) and in a context in which it affects [our] action. The authors of [5–7] have defined trust as a subjective probability following the definition given by Gambetta. Some authors [2,8,9] have defined trust as a belief in the competence of others, following the definition given by Azzedin and Maheswaran [10]: Trust is the firm belief in the competence (reliability, timeliness, honesty, and integrity) of an entity to act as expected such that this firm belief is not a fixed value associated with the entity but rather it is subject to the entity's behavior and applies only within a specific context at a given time.

Reputation has been defined as [2,5,10,11] an expectation about an agent's behavior as given by Azzedin and Maheswaran [10]: "The reputation of an entity is an expectation of its behavior based on other entities' observations or information about the entity's past behavior within a specific context at a given time."

There are three general types of trusts [12]: basic, general, and situational, which can be applied to any network. The basic trust is based on the previous experience of a node in all situations. If nodes A and B are to communicate with each other, then the basic trust is not the amount of trust node A has on node B; rather, it is the general dispositional trust that the node A has on other nodes [13]. The general trust represents the amount of trust node A has on node B, which is not dependent on a particular situation. The situational trust represents the amount of trust node A has on node B in a particular situation. Situational trust is the most important type of trust in cooperative and self-organizing networks such as MANETs and WSNs. As an example of situational trust in a MANET, a node A may trust another node B that the latter will forward its packets with a reliability of 70%.

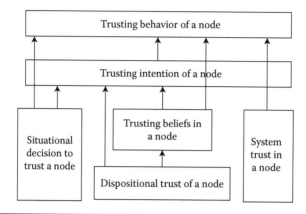

Figure 5.1   Inter-relationship among the trust constructs. (Data from D.H. McKnight and N.L. Chervany, *The meanings of trust*, MIS Research Center, Carlson School of Management, University of Minnesota, 1996.)

There are six trust-constructs of a node in a self-organizing network such as WSN [14]. These constructs are: trusting intention of a node, trusting behavior of a node, trusting beliefs in nodes, system trust in nodes, dispositional trust of a node, and situational decision to trust a node. Figure 5.1 shows the relationship among these constructs as presented in [14].

Trusting intention of a node is the willingness of one node to depend on another node in a specific situation in spite of the knowledge of the risk involved. The trusting intention consists of essential elements such as experience of reliability, evidence of security, and so on. *Trusting behavior* of a node is a voluntary dependence of one node on another node in a specific situation with the existence of risk. The trusting intention of a node supports the trusting behavior. *Trusting beliefs* in nodes is the confidence and belief of one node that the other node is trustworthy in a specific situation, that is, for example, when node *A* believes that node *B* is trustworthy. *System trust* in nodes occurs when nodes believe that proper impersonal structures are in place to encourage successful interaction, such as monitoring and dealing with improper behavior. The system trust heavily depends on the network structure and on the nodes in the network. The *dispositional trust* of a node is the node's general expectation about the trustworthiness of other nodes across different situation, that is, when node *A* is naturally inclined to trust, it has a general trust in other nodes. This is normally the risk a node initially takes when interacting with a new or unknown node. The *situational decision to trust* a node occurs when a node intends to depend on another node in a given situation. For example, if node *B* wants to communicate with node *A*, it should communicate with a trusted third-party management system, which is also trusted by node *A*.

5.3 Trust in Wireless Self-Organizing Networks

This section discusses how the concept of trust can augment security in wireless self-organizing networks such as MANETs and WSNs. In wireless self-organizing networks, there are information asymmetry and opportunism. The nodes in MANETs and WSNs have no way of gathering information about the nodes situated outside their radio range, and the information communication has a great deal of uncertainty associated with it. In systems having asymmetrical designs, some nodes may also be more powerful than the others and may have access to information that others do not have.

The following subsections give a brief background on MANETs and WSNs, the challenges faced in designing reputation- and trust-based systems for these networks, different types of node misbehaviors in such networks, and the effects of these misbehaving nodes on the network performance.

5.3.1 Wireless Self-Organizing Networks

A MANET is a self-configuring system of mobile nodes connected by wireless links. The nodes are free to move randomly that leads to a rapid change in the topology of the network. The network lacks any centralized infrastructure and, therefore, all network activities are carried out by the nodes themselves. Every node acts both as an end-system and as a relay node that forwards packets for other nodes. Since MANETs do not require any fixed infrastructure, they are highly preferred for quickly setting up networks for connecting a set of mobile devices in emergency situations such as rescue operations, disaster relief efforts, or in other military operations. MANETs can either be managed by an organization that enforces access control or they may be open to any participant that is located close enough. The later scenario poses greater security threats. In MANETs, nodes are autonomous and do not have any common interest. It may seem to be advantageous for a node not to cooperate with other nodes in the network and behave selfishly. Hence, the nodes need some sort of incentive and motivation so that they cooperate with each other. The noncooperative behavior of a node may be due to selfish intention, for example to save power, or malicious intention, for example to launch *denial-of-service* attacks.

A WSN is a network of hundreds and thousands of small, low-power, low-cost devices called sensors. The core application of WSNs is to detect and report events. WSNs have found critical applications in military and civilian domain, including robotic landmine detection, battlefield surveillance, environmental monitoring, wildfire detection, and traffic regulation. They have invaluable contributions in life-saving operations, be it the life of a soldier in the battlefield, or a civilian's life in areas of high chances of natural calamities. In WSNs, all the sensors belong to a single group or entity and work toward the same goal, unlike in MANETs. An individual senor has little value of its own unless it works in cooperation with other sensors. Hence, there is an inherent motivation for nodes in WSNs to be cooperative, and so incentive is less of a concern. Since WSNs are often deployed in unattended territories that can often be hostile, they are vulnerable to physical capture by enemies. An obvious solution to this problem is to make the senor nodes tamper proof. However, this makes the SNs prohibitively expensive to manufacture. Since many nodes are often required to cover an area, nodes must be cheap to make use of the network economically feasible. As tamper-proofing the node is not a viable solution, an adversary can modify the sensors in such a way that they start misbehaving and disrupt communication in the network. It may be even possible for the adversary to break the cryptographic security of the captured node and launch attacks from within the network as an insider. Even though cryptography can provide integrity, confidentiality, and authentication, it cannot defend against an insider attack. This necessitates a security mechanism inside a WSN that can cope with insider attacks.

5.3.2 Misbehavior of Nodes

The lack of infrastructure and organizational environment of MANETs and WSNs makes these networks particularly vulnerable to different types of attacks. If the network is not equipped with proper countermeasures, it is possible for a node to gain various advantages by exhibiting malicious behavior such as better service than cooperating nodes, monetary benefits by exploiting incentive measures or trading confidential information, saving power by selfish behavior, preventing someone else from getting proper service, extracting data to get confidential information, and so on. Even if the misbehavior is not intentional, as in the case of a *faulty node*, the effects may be detrimental to the performance of a network. As shown in Figure 5.2, the noncooperative behavior of a node in a MANET is mainly caused by two types of misbehaviors: selfish behavior (e.g., nodes that want to save power, CPU cycles, and memory) and malicious behavior, which is not primarily concerned with power or any other savings but interested in attacking and damaging the network [15]. Karlof and Wagner [16] have identified various types of security threats in a WSN due to malicious nodes and proposed some countermeasures of them. When the misbehavior of a node manifests as selfishness, the system can still cope with it since this misbehavior can always be predicted. A selfish node will always behave in a way that maximizes its benefits, and as such, incentive can be used to ensure that cooperation is always the most beneficial option. However, when the misbehavior manifests as maliciousness, it is difficult for the system to cope with it, since a malicious node always attempts to maximize the damage caused to the system for its own benefit. As such, the only method of dealing with such a node is detection and isolation from the network. Malicious misbehavior in packet forwarding can generally be divided into two types: forwarding misbehavior and routing misbehavior. Some common examples of forwarding misbehavior are packet dropping, modification, fabrication, timing attacks, and silent route change. Packet dropping, modification, and fabrication are self-explanatory. Timing misbehavior is an attack in which a malicious node delays packet forwarding to ensure that the *time-to-live* of the packets are expired, so that it is not immediately understood by other nodes. A silent route change is an attack in which a malicious node forwards a packet through a different route than it was intended to go through. Routing misbehavior may include route salvaging, dropping of error messages, fabrication of error messages, unusually frequent route updates, silent route changes, and sleep deprivation. In route salvaging attack, the malicious node reroutes packets to avoid a broken link, although no error actually has taken place. In silent route change attack, a malicious node tampers with the message header of either control or data packets. In sleep deprivation attack, a malicious node sends an excessive number of packets to another node so as to consume computation and memory resources of the latter. There exist three other types

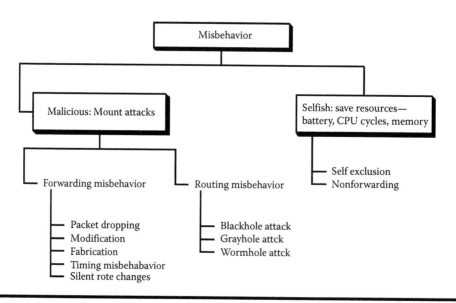

Figure 5.2 Nodes' misbehavior in MANETs and WSNs.

of routing misbehaviors: blackhole, grayhole, and wormhole. A blackhole attack is one in which a malicious node claims to have the shortest path but when asked to forward the packets, it drops them. In a grayhole attack, which is a variation of the blackhole attack, the malicious node selectively drops packets. A wormhole attack, also known as tunneling, is an attack in which the malicious node sends packets from one part of the network to another part of the network, where they are replayed.

The selfish behavior of a node can generally be classified as either self-exclusion or nonforwarding. The self-exclusion misbehavior is one in which a selfish node does not participate when a route discovery protocol is executed. This ensures that the node is excluded from the routing list of other nodes. The selfish node is able to save its power since the node is not required to forward packets for other nodes. A reputation model is an effective way to thwart the intentions of such selfish nodes. Since a node does not forward packets for other nodes in the networks, it is denied any cooperation by other nodes. So, it is in the best interest of a selfish node to be cooperative. On the other hand, the nonforwarding misbehavior is one in which a selfish node fully participates in a route discovery phase but refuses to forward the packets for other nodes at a later time. This selfish behavior of a node is functionally indistinguishable from a malicious packet dropping attack.

Since reputation-based systems can cope up with any type of observable misbehavior, they are useful in protecting a system. Reputation- and trust-based systems enable nodes to make informed decisions on prospective transaction partners. Researchers have steadily been making efforts to successfully model WSNs and MANETs as reputation- and trust-based systems. Adapting reputation- and trust-based systems to WSNs presents greater challenges than MANETs and Peer-to-Peer (P2P) systems due to their energy constraints. CORE [9], Cooperation Of Nodes-Fairness In Dynamic *Ad-hoc* NeTworks (CONFIDANT) [17], RFSN [2], Distributed Reputation-based Beacon Trust System (DRBTS) [1], KeyNote [18], and RT Framework [19] are some of the well-known systems in this area. While RFSN and DRBTS have focused attention on WSNs, the other works have concentrated on MANETs and P2P networks.

5.3.3 Effects of Nodes' Misbehavior

In wireless self-organizing networks, without appropriate countermeasures, the effects of node misbehavior dramatically decrease the performance of a network [15,17,20]. Depending on the proportion of misbehaving nodes and their specific strategies, network throughput may be severely degraded, packet loss in the network may increase appreciably, and the honest nodes in the network may experience denial-of-service attacks. In a theoretical analysis of how much cooperation can help by increasing the probability of a successful forwarding of packets, Lamparter, Plaggemeier, and Westhoff [21] have found that increased cooperation more than proportionately increases the performance in small networks with fairly short routes. Zhang and Lee [22] argue that preventive measures, such as encryption and authentication, may be used in MANETs to reduce the success of intrusion attempts, but cannot completely eliminate them. For example, encryption and authentication cannot defend against compromised mobile nodes, which carry the private keys. No matter what types of intrusion prevention measures are deployed in the network, there are always some weak links that an adversary can exploit to break in. Intrusion detection presents a second wall of defense and it is a necessity in any high-survivability network.

5.4 Reputation- and Trust-Based Systems

Use of reputation- and trust-based systems in Internet, *e*-commerce, and P2P applications has been in place for almost a decade [23–27]. However, it is only recently that efforts have been made to model MANETs and WSNs as reputation- and trust-based systems [15,17,28,29]. This section presents various characteristics of trusts, the goals, properties of reputation systems, and different types of reputation- and trust-based systems.

5.4.1 Trust and its Characteristics

From the perspective of wireless self-configuring networks, Sun et al. [30] have identified some characteristics of trust metric. These characteristics are as follows:

1. Trust is a relationship established between two entities for a specific action. In particular, one entity trusts the other entity to perform an action. The first entity is called the *subject*, and the second is called the *agent*.
2. Trust is a function of uncertainty. In particular, if the subject believes that the agent will perform the action for sure, the subject fully trusts the agent to perform the action and there is no uncertainty; the subject believes that the agent will not perform the action for sure, the subject trusts the agent not to perform the action, and there is no uncertainty either; if the subject does not have any idea of whether the agent will perform the action or not, the subject does not have trust in the agent, In this case, the subject has the highest level of uncertainty.
3. The level of trust can be measured by a continuous real number, referred to as the trust value. The trust value should represent the uncertainty.
4. The subjects may have different trust values with the same agent for the same action. Trust is not necessarily symmetric. The fact that *A* trusts *B* does not necessarily mean that *B* also trusts *A*, where *A* and *B* are two entities.

5.4.2 Reputation Systems: Goals and Properties

The important goals of reputation- and trust-based systems for wireless self-configuring networks have been identified in [25]. These goals are (i) to provide information that allows nodes to distinguish between trustworthy and untrustworthy nodes in the network, (ii) to encourage the nodes in the network to cooperate with each other and become trustworthy, and (iii) to discourage the untrustworthy nodes to participate in the network activities. The authors in [31] have identified two additional goals of reputation- and trust-based systems from the perspective of wireless networks. The first goal is to be able to cope up with any type of observable misbehavior, and the second is to minimize the damage caused by any insider attacks.

To operate effectively and efficiently, reputation- and trust-based systems for wireless communication networks must have three essential properties as identified in [23]. These properties are: (i) the system must have long-lived entities that inspire expectations for future interactions, (ii) the system must be able to capture and distribute feedbacks about current interactions among its components and such information must also be available in future, and (iii) the system must use feedback to guide trust decisions.

5.4.3 Classification of Reputation- and Trust-Based Systems

There are different approaches for classification of various reputation- and trust-based systems. These systems may be categorized from the perspective in which they are initialized, the types of observation they use, the manner in which the observations are accessed, and the way the observed information is distributed in the network [31]. These are discussed in detail subsequently.

Most of the trust- and reputation-based systems are *initialized* in one of the following three ways:

i. All the nodes in the networks are initially assumed to be trustworthy. Every node trusts other nodes in the network. The reputations of the nodes decrease with every bad encounter.
ii. Every node is considered to be untrustworthy in the system bootstrapping stage, and the nodes do not trust each other initially. The reputations of nodes with such system increase with every good encounter.
iii. Every node in the network is considered to be neither trustworthy nor untrustworthy. All nodes start with a neutral reputation value to start with. With every good or bad behavior, the reputation value is increased or decreased, respectively.

On the basis of observations they *use*, reputation- and trust-based systems can be classified into two groups: (i) systems using only first-hand information and (ii) systems using both first-hand and second-hand information. While the systems using the first-hand information rely on the direct observations or experiences encountered by the nodes, the nodes in the systems using second-hand information utilize information provided by the peers in its neighborhood. Most of the current reputation systems use both first- and second-hand information to update the reputation. This allows the systems to make use of more information about the network in computing reputation values. There are systems that use only first-hand information. This makes the systems completely robust against rumor spreading. Observation-based Cooperation Enhancement in *Ad hoc* Networks (OCEAN) [32] and Pathrater [28] are two such systems. In DRBTS [1], certain types of nodes use only second-hand information. In this system, a node does not have any first-hand information to evaluate the trustworthiness of the informers. One way to deal with this reputation is to use a simple majority principle. Reputation systems can broadly be categorized into two types depending on the manner in which different nodes access reputation information in the network. These two types are: (i) symmetric systems and (ii) asymmetric systems. In symmetric reputation systems, all nodes in the network have access to the same level of information, that is, both first-hand and second-hand information. In asymmetric systems, on the other hand, all nodes do not have access to the same amount of information. For example, in DRBTS [1], SNs do not have first-hand information. Thus, in the decision-making process, the SNs are at a disadvantageous situation due to lack of availability of information.

On the basis of the manner in which reputation is *distributed* in the network, reputation systems can be categorized into two groups: (i) centralized and (ii) distributed. In centralized systems, one central entity maintains the reputations of all nodes in the network. This central entity can be a source of security vulnerability and performance bottleneck in the system. Examples of this type are eBay and Yahoo auctions. In distributed systems, each node maintains reputation information of all the nodes about which it is interested. In such systems, maintaining consistency in reputation values maintained in different nodes may be a major challenge. In a distributed system, each node may maintain reputation of the nodes that are within its communication range, or may maintain reputation information of all the nodes in the network. In sensor network applications, every node maintains reputation information only for its neighbors. This reduces the memory overhead for reputation information maintenance. However, for networks with high mobility, maintenance of reputation for as many nodes as possible is a preferred option for every node. This ensures that a node does not get completely alienated if it moves to a new location with a changed neighborhood. This strategy, of course, involves a very large memory overhead.

Irrespective of the type of a reputation- or trust-based system, its objective should effectively be to detect and isolate the misbehaving nodes in the network. It should be self-organized and robust against any insider attacks. The reputation computation and maintenance system should not be vulnerable to manipulation by a malicious attacker. Moreover, it should not involve much memory and communication overhead. All these criteria make designing an effective and efficient reputation system for MANETs and WSNs an extremely challenging task.

5.5 Issues in Reputation Systems for Wireless Communication Networks

This section discusses various issues of reputation- and trust-based systems. Several important design parameters of a reputation- or trust-based system for MANETs and WSNs are discussed in detail, illustrating them with real-world systems whenever appropriate.

5.5.1 Information Gathering

Information gathering is the process in which a node collects information about other nodes it is interested in. This is concerned only with first-hand information. First-hand information is gathered by a node purely on the basis of its observation and experience. However, in CONFIDANT [17], first-hand information is further classified into personal experience and direct observation. Personal experience of a node refers to the information it gathers through one-to-one interaction with its neighbors. Direct observation is the information gathered by a node by observing the interactions among its neighbors. CONFIDANT [17] is currently the only system that makes this distinction.

Most reputation- and trust-based systems make use of a component called Watchdog [28] to monitor their neighborhood and gather information based on the promiscuous mode of observation. Thus, first-hand information is confined to the wireless sensing range of a node. However, the watchdog system is not very effective in situations where directional antennas are deployed and spread spectrum technology is used for wireless communication. This aspect is getting lot of focus in the current research activities on wireless communications.

5.5.2 Information Dissemination

There is an inherent trade-off between the efficiency in using second-hand information and robustness against false ratings. The use of second-hand information gives lot of advantages. First, the reputation of the nodes builds up more quickly due to the ability of the nodes to learn from the mistakes of each other. Secondly, no information in the system goes unused. Finally, over a period of time, a consistent local view stabilizes in the system.

However, sharing information makes the system vulnerable to *false report attacks*. This vulnerability can be somewhat reduced by adopting a strategy of limited information sharing, that is, sharing either only positive information or negative information.

If only positive information is shared, the system is still vulnerable to false praise attacks. With only positive information being shared, the nodes cannot share their bad experiences. This is particularly detrimental since learning from ones own experience in this scenario comes at a very high price. Also, colluding malicious nodes can extend each other's survival time through false praise reports. CORE [9] permits only positive second-hand information, which makes it vulnerable to spurious positive ratings by malicious nodes.

Sharing only negative information protects the system against the false praise attack, but it has its own drawbacks. The nodes cannot share their good experiences. More importantly, malicious nodes can launch bad-mouthing attacks on benign nodes either individually or in collusion with other malicious nodes. CONFIDANT [17] makes use of negative second-hand information in order to proactively isolate misbehaving nodes. This makes the system vulnerable to spurious ratings, and false accusations. Context-aware detection [33] accepts negative second-hand information on the condition that at least four separate sources make such a claim, otherwise the node spreading the information is considered misbehaving. While this distributes the trust associated with the accusation over several nodes and thus distributes the risk, it inadvertently serves as a disincentive to share ratings and warn other nodes by spreading reputation information in the network. It is also not possible to guarantee the availability of four witnesses for an event in a sparsely populated network.

Another way of avoiding the negative consequences of information sharing is not to share any information at all. OCEAN [30] is one such model that builds reputation purely based on the individual observations of the nodes. Although such systems are completely robust against rumor spreading, they have some shortcomings. The time required to build reputation is increased dramatically, and it takes longer duration for reputation to fall, allowing malicious nodes to stay in the system and misuse the system resources.

Systems such as DRBTS [1] and RFSN [2] share both positive and negative information. The negative effects of information sharing, as discussed above, can be mitigated by appropriately incorporating first- and second-hand information into the reputation metric. Using different weighting functions for different information is one efficient technique.

Most of the reputation- and trust-based systems for MANETs and WSNs use one of the three following methods to share information among the nodes: *friends list*, *blacklist*, and *reputation table* (RT). A friends list shares only positive information, a blacklist shares only negative information, while an RT shares both positive and negative information.

Information sharing involves three important issues: (i) dissemination frequency, (ii) dissemination locality, and (iii) dissemination content. These issues are briefly discussed below.

The reputation systems can be of two types on the basis of dissemination frequency they employ: (i) proactive dissemination and (ii) reactive dissemination. In proactive dissemination, nodes communicate reputation information during each dissemination interval. A node publishes the reputation values even if there have been no changes in the stored values in the last dissemination interval. This strategy is more suited to dense network with more activities, as the nodes have to wait till the beginning of the next dissemination interval to publish their reputation information. In reactive dissemination, nodes publish only when there is a predefined amount of change in the reputation values they store or when an event of interest occurs. This method reduces communication overhead in situations where reputations of nodes do not change frequently. However, reactive dissemination may cause congestion in networks with high network activity. In both these types of information dissemination, the communication overhead can be reduced to a large extent by piggybacking the information with other network traffic. In CORE [9], the reputation information is piggybacked on the reply messages and, in DRBTS [1], it is piggybacked on the location information dispatch messages.

Reputation systems may use two types of locality of dissemination of information: (i) local and (ii) global. In local dissemination, the information is published within the neighborhood. It could be either through a local broadcast, multicast, or unicast. In DRBTS [1], the information is published in the neighborhood through a local broadcast. This enables all the BNs to update their RTs accordingly. A reputation system may also choose to unicast or multicast depending on the application domain and security requirements. In global dissemination, the information is propagated to nodes outside the radio range of the node publishing the reputation information. Global dissemination may also use either broadcast, multicast, or unicast technique. For networks with higher node mobility, global dissemination is preferred as it provides nodes with a reasonable understanding of the new locations they are moving to.

Two types of reputation information contents may be disseminated: (i) raw information and (ii) processed information. In case of raw information, the information published by a node is its first-hand information only. It does not reflect the final composite reputation value, as it does not take into consideration the second-hand information of other nodes in the neighborhood. In case of processed information, a node publishes the overall reputation values after computing the composite reputation score.

5.5.3 Redemption and Weighting of Time

An important issue in maintaining and updating reputation is how past and current information are weighted. Different models weight them differently, each with a different rationale. CORE [9] assigns more weight to the past behavior of a node than its current behavior, so that wrong observations or rare behavior changes cannot influence the reputation rating too much. It helps benign nodes that may behave selfishly due to genuinely critical battery conditions. The nodes may also misbehave temporarily due to technical problems such as link failure. CONFIDANT [17] takes the opposite approach—it discounts past behavior by assigning less weight. This ensures that a node cannot leverage on its past good performance and start misbehaving without being punished, and the system becomes more responsive to sudden behavioral changes of nodes. RFSN [2] also gives more weight to recent observations than the past. This forces nodes to be cooperative at all the time. However, there is a problem in adopting the strategy of assigning higher weights to current behavior. In periods of low network activity, a benign node may get penalized. This problem can be resolved by generating network traffic in regions and periods of low network activity using mobile nodes. DRBTS [1] tackles this issue by generating network traffic through BNs when a need arises. Pathrater [28], context-aware detection [33], and OCEAN [32] do not weight ratings according to time.

Ratings are not only weighted to put emphasis on the past or the present, but also to add importance to certain types of observation. CONFIDANT [17] gives more weight to first-hand observations and less to reported second-hand information. CORE [9] also assigns different weights to different types of observations.

Redemption is done in case a node is wrongly identified as a misbehaving node, either because of deceptive observation, spurious ratings, or because of a fault in the reputation system. Redemption is also necessary when a node that was previously isolated from the network because of its misbehavior needs to be allowed to join back, because the cause of its misbehavior has been identified and resolved, for example, a faulty node may have been repaired, a compromised node may have been recaptured by its legitimate user.

CONFIDANT [17] carries out redemption of misbehaving or misclassified nodes by reputation fading, that is, discounting the past behavior even in the absence of testimonials and observations, and periodic re-evaluation, that is, checking from time to time whether the rating of a node is above or below the acceptable threshold. Thus a node that has been isolated from the network because of its misbehavior always gets a chance to rejoin after some time. Since the ratings do not get erased but only discounted, the rating of a previously misbehaving node will still be close to the threshold value and thus the reaction to a current misbehavior will be swift. This will result in faster detection and isolation of that node in case it starts misbehaving. It is thus possible for a node to redeem itself. Since the nodes in the network may differ in their opinion, it is quite likely that a node will not be excluded by all other nodes and thus it can participate partially in the network activities. This will give the node a chance to show good behavior and redeem its reputation value. Even if this is not the case and the suspect node is excluded by everyone, it can redeem itself by means of the reputation fading.

In CORE [9], a node that is isolated because of its misbehavior in the past cannot redeem itself till there is a sufficient number of new nodes arriving in the network that have no past experience with it.

OCEAN [32] relies on a timeout of reputation. The sudden lapse back into the network can pose a problem if several nodes set the timer at roughly the same time. Pathrater [28] and context-aware detection [33] have no provision of redemption.

5.5.4 Weighting of Second-Hand Information

The schemes that use second-hand information have to administer a trust of the witness, that is, the sources of second-hand information, in order to prevent blackmailing attacks. It is thus necessary to use some means of validating the credibility of the reporting node. One method is to use a deviation test as done in [1,11]. If the reporting node passes the deviation test, it is treated as trustworthy and its information is incorporated to update the reputation of the reported node. However, different models choose different strategies for dealing with the second-hand information depending on the application domain and security requirements. For instance, the model presented in [2] uses Dempster–Shafer theory [34] and discounting belief principle [35] to incorporate second-hand information. However, Beta distribution is mostly used in reputation- and trust-based systems. It was first used by Josang and Ismail [27]. Many researchers in the field of security in *ad hoc* networks have used Beta distribution in their analysis. Ganeriwal and Srivastava [2] and Buchegger and Le Boudec [11] are among them. The reason for popularity of Beta distribution is its simplicity as it is indexed by only two parameters.

CONFIDANT [17] assigns weights on the second-hand information according to the trustworthiness of the source and by setting a threshold that had to be exceeded before the second-hand information is taken into account. Second-hand information had to come from more than one trusted source or several partially trusted sources, or any combination thereof, provided that trust times the number of nodes exceeds the trust threshold. This adds a vulnerability to the system where some untrustworthy nodes may eventually be trusted. The notion of trust has been more specifically defined in the enhanced version of CONFIDANT, known as Robust Reputation System (RRS) [11]. In RRS, trust means consistent good performance as a witness, measured as the compatibility between the first- and second-hand information. This dynamic assessment allows the system to keep track of trustworthiness and to react accordingly. If the second-hand information is accepted, it will have a small influence on the reputation rating. More weight is given to the nodes' direct observations.

5.5.5 Spurious Ratings

If second-hand information is used to influence reputation, some nodes may lie and give spurious rating about others. A malicious node may be benefited by falsely accusing an honest node, as this can lead to a denial of service to the latter. A false praise can benefit a colluding malicious node.

Problems related to false accusations are absent in positive reputation systems, since no negative information is maintained [24,36]; however, the disseminated information could be false praise and result in a good reputation for some malicious nodes. Even if the disseminated information is correct, it may not be possible to distinguish between a misbehaving node and a new node that has just joined the network.

If second-hand information is used, an important issue is to decide whether the lying nodes should be punished in the same way as the misbehaving nodes by isolating them from the network services. If nodes are punished for their seemingly inaccurate testimonials, one may end up punishing an honest messenger. This will definitely discourage honest reporting of observed misbehavior. The testimonial accuracy is evaluated according to affinity to the belief of the requesting node along with the overall belief of the network as gathered over time. The accuracy is not measured when compared with the actual true behavior of a node, since the latter is unknown and cannot be proved beyond doubt. Even if it were possible to test a node and obtain a truthful verdict on its nature, a contradicting previous testimonial could still be accurate. Thus, instead of punishing deviating views, it is better to merely reduce their impact on public opinion. Some node is bound to be the first witness of another node's misbehavior, and thus its report will start deviating from the public opinion. Punishing this discovery would be counterproductive, as the goal is precisely to learn about the misbehaving nodes as early as possible.

5.5.6 Identity

The question of identity is of central importance to any reputation systems. Identity may be of three types: *persistent, unique,* and *distinct.* A node cannot easily change its persistent identity. Identity persistence can be achieved by expensive pseudonyms or by a specific security module. Identity persistence is desirable for a reputation system to enable it to gather the behavior history of a node. An identity is unique, if no other node can impersonate the node by using its identity. This can be achieved by cryptographically generated unique identifiers, as proposed by Montenegro and Castelluccia [37]. This property is needed to ensure that the observed behavior was indeed that of the node observed. The requirement of distinct identities is the target of the so-called Sybil attack, as analyzed by Douceur [38], where a node generates several identities for itself to be used at the same time. This property is not of much concern to the reputation system, since those identities that exhibit misbehavior will be excluded while other identities stemming from the same node will remain in the network as long as they behave well. The Sybil attack can, however, influence public opinion, by having its rating considered more than once. A mechanism to defend Sybil attack has been proposed in [39].

5.5.7 Detection

Reputation systems require a tangible object of observation that can be identified as either good or bad. In online auction or trading systems, this is a sale transaction with established and measurable criteria such as delivery or payment delay. In case of reputation systems for MANETs, the analogy of a transaction is not straightforward due to the limited observability and detectability of a mobile node. In order to detect misbehavior, nodes promiscuously overhear the communications of their neighbors. The component used for this type of observation is called Watchdog [28], Monitor [17], or NeighborWatch [32].

The function mostly used to implement the detection component in reputation systems is *passive acknowledgement* [40], where nodes register whether their next hop neighbor on a given route has attempted to forward a packet. Assuming bi-directional links, a node can listen to the transmission of another node that is within its radio range. If within a given time window, a node hears a retransmission of a packet by the next hop neighbor, it has sent packet previously, the behavior is judged to be good. This does not necessarily mean that the packet has been transmitted successfully, since the observing node cannot see what goes on outside its radio range, for example, there could be a collision on the far side of the next hop neighbor.

Several problems with Watchdog have been identified in [28], such as the difficulty in unambiguously detecting that a node does not forward packets in the presence of collisions or in the case of limited transmission power. The watch-dog mechanism in CORE [9] relies on the promiscuous mode of operations of wireless interfaces of the nodes. In addition, the nodes can judge the outcome of a request by rating the end-to-end connection. CONFIDANT [17] uses passive acknowledgement not only to verify whether a node forwards packets, but also as a means to detect if a packet has illegitimately been modified before being forwarded.

5.5.8 Response

Except for Watchdog and Pathrater [28], most of the reputation and trust systems have a punishment component for the misbehaving nodes. The isolation of the misbehaving nodes is done in two steps: these nodes are avoided in routing and then denied cooperation when they request for it. Not using misbehaving nodes for routing but allowing them to use the network resources will only increase the incentive for misbehavior, since it results in power saving due to the decrease in number of packets they have to forward for others.

5.6 Examples of Reputation and Trust-Based Models

In this section, various reputation- and trust-based systems proposed in the literature for MANETs and WSNs are reviewed. For each of the schemes, the working principle is discussed and critically analyzed in terms of its effectiveness and efficiency.

5.6.1 Watchdog and Pathrater

Watchdog and Pathrater components to mitigate the routing misbehavior have been proposed by Marti et al. [28]. They observed increased throughput in MANETs by complementing DSR protocol with a *watchdog* for detection of denied packet forwarding and a *pathrater* for trust management and routing policy, rating every path used. This enables every node to avoid any malicious node on its routing path.

Watchdog determines the misbehavior of a node by copying packets to be forwarded into a buffer and monitoring the behavior of the neighboring nodes with respect to these packets. The watchdog promiscuously snoops to check whether the neighboring nodes forward the packets without modification. If the packets that are snooped match with those in the buffer of the monitor node, they are simply discarded. The packets that stay in the buffer of the monitor node beyond a threshold period of time are flagged as having been dropped or modified. The node responsible for forwarding the packet(s) is then marked as a *suspicious* node. If the number of such failures to forward packets exceeds a predetermined threshold value, the offending node is identified as a *malicious* node. Information about malicious nodes is passed to the *pathrater* component for inclusion in path evaluation.

Pathrater component of a node works to make a rating of all the known nodes in a particular network with respect to their reliabilities. Ratings are made and updated form a particular node's perspective. Nodes start with a neutral rating that is modified over time based on an observed reliable or unreliable behavior during packet routing. Nodes that are observed by the watchdog to have misbehaved are given an immediate rating of -100. The misbehavior of a node is identified on the basis of its packet mishandling and modification activities, whereas unreliability of a node is determined on the basis of its link errors.

From the simulation results, it has been observed that the watchdog and the pathrater are quite effective in routing packets. However, the scheme does not punish malicious nodes that do not cooperate in routing. Rather it relives the malicious nodes of the burden of forwarding for others, while their messages are forwarded in the network by other nodes. In this way, the malicious nodes are encouraged to continue with their misbehavior.

5.6.2 Context-Aware Inference Mechanism

A context-aware inference mechanism has been proposed by Paul and Westhoff [33] in which accusations are related to the context of a unique route discovery process and a stipulated time period. A combination is used that consists of unkeyed hash verification of routing messages and the detection of misbehavior by comparing a cached routing packet to overheard packets. The decision of how to trust nodes in future is based on accusation of others, whereby a number of accusations pointing to a single attack, the approximate knowledge of the topology, and context-aware inference are claimed to enable a node to rate an accused node with certainty. An accusation, however, has to come from several nodes. If a single node makes an accusation, it is itself accused of misbehavior.

5.6.3 Trust-Based Relationship of Nodes in Ad Hoc Networks

Pirzada and McDonald [41] have proposed an approach for building trust relationship between the nodes in an *ad hoc* network. It is assumed that the nodes in the network passively monitor the packets received and forwarded by the other nodes. The receiving and forwarding activities by the nodes are termed as *events*. Events are observed and given a weight, depending on the type of application requiring a trust relationship with other nodes. The weights reflect the significance of the observed events for the corresponding application. The trust values for all events from a node are combined using weights to compute an aggregate trust level for the node. The compound trust values are used as link weights for the computation of routes. Links which connect more trust-worthy nodes will be having smaller weights. A shortest path routing algorithm would compute the most trustworthy paths in a network.

In [42], the authors have presented trust as a measure of uncertainty. Using the theory of entropy, the authors have developed a few techniques to compute trust values from certain observations. In addition, trust models—entropy- and probability-based—are presented to solve the concatenation and multipath trust propagation problems in a MANET.

5.6.4 Trust Aggregation Scheme

Liang and Shi have carried out an extensive work on development of models and evaluation of robustness and security of various aggregation algorithms in an open and untrusted environment [43,44]. They have presented a comprehensive analytical and inference model of trust for aggregation of various ratings received by a node from its neighbors in a WSN. It has been observed that lack of memory space availability is a serious constraint for SNs in storing knowledge in a trust-based framework. The simulation results have shown that it is a computationally more efficient approach to treat the ratings received from different evaluators (i.e., nodes) with equal weights and compute the average to arrive at the final trust value. This approach not only has a very low computational overhead, but also produces very satisfactory result in practice. The authors have also observed that for a trust model to be effective, the most important and critical issue is how it adaptively adjusts the parameters of the model based on the change in the environment.

5.6.5 *Trust Management in* Ad Hoc *Networks*

Yan et al. have proposed a security solution based on trust framework to ensure data protection, secure routing, and other security features in an *ad hoc* network [45]. Mechanisms of logical and computational trust analysis and evaluation are applied on the nodes. Each node evaluates the trust of its peer nodes based on factors such as experience, statistics, data value, intrusion detection results, recommendations from its neighbors and so on.

Ren et al. have presented a technique to establish trust relationships among nodes in an *ad hoc* network [46]. The proposed framework is a probabilistic solution based on a distributed trust model. A secret dealer is introduced only in the system bootstrapping phase to initiate the trust propagation in the network. Shorter and robust trust chains are subsequently developed among the nodes. A fully self-organized trust establishment approach is then adopted to conform to the dynamic membership changes.

In [47], the author has presented the methods of finding paths from a source node to a designated target node in a peer-to-peer computing paradigm. Extending this approach, Zhu et al. [48] provide a practical approach to compute trust in a wireless network by treating individual mobile device as a node of a delegation graph G and mapping a delegation path from a source node S to a target node T into an edge in the corresponding transitive closure of the graph G. From the edges of the transitive closure of the graph G, the trust values of the wireless links are computed. In the proposed trust-based framework, an undirected transitive signature scheme is used within the authenticated transitive graphs.

Davis has presented a trust management scheme based on a structured hierarchical model, which addresses the explicit revocation of certificates [49]. The scheme is robust against false accusation by a malicious node. It uses digital certificates to establish trust. For a node to be trusted, it must possess a valid certificate.

5.6.6 *Trusted Routing Schemes*

The authors in [50] have presented a trusted routing scheme that extends the *Ad hoc* On-demand Distance Vector (AODV) routing protocol [51] to ensure that only trustworthy nodes participate in routing. A new protocol called Trusted Computing *Ad hoc* On-demand Distance Vector (TCAODV) has been proposed to prevent malicious and selfish nodes from misusing network resources. In TCAODV, a public key certificate is used by each node, which is stored within a *trusted root*. The node broadcasts the certificate along with the *hello* messages. The neighbors on receiving the certificate first verify its authenticity by checking the signature of the issuer. If the signature verification is successful, the certificate is stored in the neighbors as the public key of the issuing node. The RREQ packet sent by each node is signed with a sealed signature using integrity metrics from the routing module of the sender. The node that receives the RREQ verifies the signature using the previously received key for the requester node, and determines if the provided measurements are trustworthy. When the destination is not directly reachable by the RREQ, the intermediate node strips off the signature and puts its own signature and integrity measurements. In addition, a per-route symmetric encryption key is established to ensure that only trusted nodes along the path can use the route. Every packet sent along the route is encrypted using the symmetric key. The TCAODV approach has less overhead on the network and can be applied in WSNs and MANETs.

In [52], the authors have presented a scheme for multicast communication in a MANET based on trust metrics. In a multicast MANET, a sender node sends packets to several receiving nodes in a multicast session. Since the membership in a multicast group changes frequently in a MANET, the issue of supporting secure authentication and authorization in a multicast MANET is very critical. The proposed scheme involves a two-step secure authentication method. First, an ergodic continuous Markov chain is used to determine the trust value of each one-hop neighbor. Second, a node with the highest trust value is selected as the Certificate Authority (CA) server. For the sake of reliability, the node with the second highest trust value is selected as the backup CA server. The analytical trust value of each mobile node is found to be very close to that observed in the simulation under various scenarios. The speed of the convergence of the analytical trust value shows that the analytical results are independent of the initial values and the trust classes.

5.6.7 Collaborative Reputation Mechanism in Mobile Ad Hoc Networks

COllaborative REputation mechanism to enforce node cooperation in mobile *ad hoc* networks (CORE) was proposed by Michiardi and Molva to enforce cooperation among nodes in MANETs based on a collaborative monitoring technique [9]. It differentiates between *subjective reputation* (observations), *indirect reputation* (positive reports by others), and *functional reputation* (task-specific behavior), which are suitably weighted to arrive at a *combined reputation* value. The combined reputation value is used to take decisions about cooperation or gradual isolation of a node. Reputation values are obtained by considering the nodes as *requestors* and *providers*, and comparing the expected result to the actually obtained result of a request. Essentially CORE is a distributed, symmetric reputation model that uses both first- and second-hand information for updating reputation. It uses bi-directional communication symmetry and Dynamic Source Routing (DSR) protocol for routing. CORE also assumes wireless interfaces that support promiscuous mode of operation.

In CORE, nodes have been modeled as members of a community who have to contribute on a continuing basis. Otherwise, their reputations degrade, and eventually they are excluded from the network. The reputation is updated with time. More weight is assigned to the past observations than the current observations to ensure that a recent sporadic misbehavior of a node has a minimum influence on the evaluation of its overall reputation value. CORE has two types of protocol entities, a *requestor* and a *provider*.

- *Requester:* it is a network entity that requests for the execution of a function f. A requestor may have one or more providers within its transmission range.
- *Provider:* it is a network entity that can correctly execute the function f.

In CORE, nodes store the reputation values in an RT, with one RT for each function. Each entry in the RT corresponds to a node and consists of four fields: (i) unique ID, (ii) recent subjective reputation, (iii) recent indirect reputation, and (iv) composite reputation for a predefined function.

Each node is also equipped with a watchdog mechanism for promiscuous observation. RTs are updated during the request phase and the reply phase.

The reputation of a node computed from first-hand information is referred to as subjective reputation. It is calculated directly from a node's observation. CORE does not differentiate between interactions and observations for subjective reputation unlike CONFIDANT [17]. The subjective reputation is computed only for the neighbors of the subject node. The subjective reputation is updated only during the request phase. If a provider does not cooperate with a requestor's request, then a negative value is assigned to the rating factor of that observation. This automatically decreases the reputation of the provider. The reputation of a node can take any value between −1 and +1. When a node joins the network for the first time, its reputation is initialized with a value zero.

CORE uses indirect reputation, that is, second-hand information to model MANETs. The impression of one node about another is influenced by other nodes in the network. However, there is a restriction on the type of reputation-information that can be propagated—only positive information exchange is allowed. As discussed earlier, this prevents bad-mouthing attacks on benign nodes. Each reply message includes a list of nodes that cooperated in routing, and thus indirect reputation is updated only during the reply phase.

CORE uses functional reputation to evaluate the trustworthiness of a node with respect to different functions. Functional reputation is computed by combining subjective and indirect reputation for different functions. Different applications may assign different weights to routing and various other functions such as packet forwarding, and so on. The combined reputation value of each node is computed by combining the three types of reputation with suitable weights. The positive reputation values are decremented with time to ensure that nodes cooperate and contribute on a continuous basis. This prevents a node from initially building up a very good reputation by being very cooperative and contributive but start misbehaving after some time.

When a node has to make a decision on whether or not to execute a function for a requestor, it checks the reputation value of the latter. If the reputation value is positive, the function is executed. However, the node is denied any service if its reputation is negative. A misbehaving node with low reputation value can build its reputation by cooperating with other nodes. However, reputation is difficult to build as it gets decreased every time the watchdog detects a noncooperative behavior and also with time to prevent a malicious node from building reputation and then attacking the system resources.

Assignment of more weight to the past reputation in CORE allows a malicious node to misbehave for some time if it has accumulated a high reputation value. False accusation attacks are prevented since only positive information is shared for indirect reputation updates. However, this makes the system vulnerable to false praise attack. The authors argue that a misbehaving node gains no advantage by giving false praise to other unknown entities. This is true only so long as malicious nodes are not colluding. When malicious nodes start collaborating, then they can help prolong the survival time of another node through false praise. However, the effect of false praise is mitigated in CORE to some extent by coupling the information dissemination to reply messages. Moreover, since only positive information is shared, the possibility of retaliation is prevented.

There is an inherent problem in combining the reputation values for various functions into a single global value. This potentially helps a malicious node to hide its misbehavior with respect to certain functions while behaving cooperatively with respect to other functions. The objective of a node to misbehave with respect to a particular function is to save its scarce resources. The node may choose to not cooperate for functions that consume resources such as memory and power and choose to cooperate for functions that do not require these resources much. Nonetheless, functional reputation is a very nice feature of CORE that can be used to exclude nodes from functions for which their reputation value is below the threshold and include them for functions for which they have high reputation values. CORE also ensures that disadvantaged nodes that are inherently selfish due to their critical energy conditions are not excluded from the network using the same criteria as for malicious nodes. Hence, an accurate evaluation of the reputation value is performed that is not affected by sporadic misbehavior. Therefore, CORE minimizes false detection of the misbehavior of a node.

5.6.8 Cooperation of Nodes-Fairness in Dynamic Ad Hoc Networks

CONFIDANT is a security model proposed by Buchegger and Boudec [17] to make misbehavior unattractive in MANETs based on selective altruism and utilitarianism. It is a distributed, symmetric reputation model that uses both first- and second-hand information for computation of reputation values. CONFIDANT uses the DSR protocol for routing and assumes that promiscuous mode of operation is possible. It does not require any tamper-proof hardware, since a malicious node neither knows its reputation values in other nodes nor does it have any access to those entries. The misbehaving nodes are punished by isolating them from accessing the network resources. Moreover, when a node encounters a misbehaving node, it sends a warning message to its trusted members in the network, termed as *friends*. CONFIDANT is based on the principle that reciprocal altruism is beneficial for every ecological system when favors are returned simultaneously because of instant gratification [53]. There may not be any benefit in behaving well if there is a delay in granting a favor and getting back the repayment. As shown in Figure 5.3, each node runs four components in the CONFIDANT protocol: *monitor, trust manager, reputation system,* and *path manager*.

The monitor module in each node passively observes the activities of the nodes within its 1-hop neighborhood. The node can detect any possible deviation made by the next node on the source route. It can also check for any possible content modification of the packets done by its next hop node. The monitor registers these deviations from normal behavior as soon as a bad behavior is

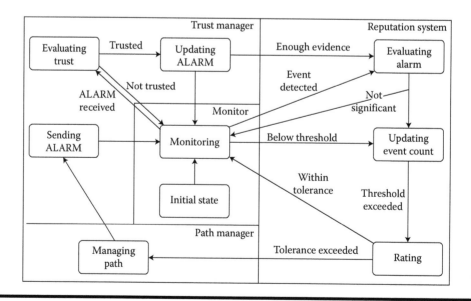

Figure 5.3 Components and state diagram of the CONFIDANT protocol. (Data from S. Buchegger and J.-Y. Le Boudec, *Proceedings of MobiHoc 2002*, Lausanne, CH, June 2002.)

detected, and reports this to the reputation system and the trust manager for evaluation of the new reputation value of the misbehaving node.

The trust manager handles all the incoming and outgoing ALARM messages. Incoming ALARMs can originate from any node. Therefore, the source of an ALARM has to be checked for trustworthiness before triggering a reaction. This decision is made by looking at the trust level of the reporting node. CONFIDANT has provisions for several partially trusted nodes to send ALARMs which will be considered as an ALARM for a single fully trusted node. The outgoing ALARMs are generated by the node itself after having experienced, observed, or received a report of malicious behavior. The recipients of these ALARM messages are called friends, which are maintained in a friends list by each node.

The trust manager consists of three components: *alarm table*, *trust table*, and *friend list*. The alarm table contains information about received alarms, the trust table maintains the trust records of each node to determine the trustworthiness of an incoming alarm, and the friend list contains the list of all nodes to which the node has to send alarms when it detects any malicious activity. The trust manager is also responsible for providing with and receiving routing-related information from other nodes in the network.

The reputation system of every node maintains a table that consists of entries of other nodes and their corresponding reputation values. The reputation rating of a node is updated only when there is sufficient evidence of malicious behavior of that node occurring at least for a threshold number of times. The rating is changed using a function that assigns the highest weight on personal experience, a lesser weight for observations in the neighborhood and an even lesser weight to reported experience. The rationale behind this relative weighting scheme is that nodes trust their own experiences and observations more than those of other nodes. If the computed reputation value of a node falls below a predetermined threshold, the path manager is summoned for further actions.

The Path Manager is the component that is the decision maker. It is responsible for path re-ranking according to the security metric. It deletes paths containing misbehaving nodes and is also responsible for taking necessary actions upon receiving a request for a route from a misbehaving node.

In CONFIDANT, only negative information is exchanged between nodes. The authors argue that it is justified since malicious behavior is an exception and not the normal behavior. However, the exchange of only negative information makes the system vulnerable to false accusation attack on benign nodes by malicious nodes. Unlike CORE, even without collusion, malicious nodes benefit by falsely accusing benign node. With collusion of malicious nodes, this problem may become unmanageable. However, false praise attacks are not possible since no positive information is exchanged. This prevents any possibility of collusion among a set of malicious nodes to prolong their survival time in the network. Since negative information is shared among the nodes, an adversary gets to know his situation and accordingly change his strategy. This may not be desirable. Sharing negative information in the open may also introduce fear of retaliation that may force nodes to conceal their true findings.

In spite of an elegant design of the reputation system, the reputation computation process using experienced, observed, and reported information is not adequately explained in the CONFIDANT mechanism. The nodes that are excluded because of misbehavior are allowed to recover after a certain timeout. This allows a malicious node to re-enter the system and attack repeatedly unless it is permanently denied entry after a certain number of such events. Faulty nodes are treated in the same way as malicious nodes. This may not be always advisable as punishment may make the status of a faulty node even worse. The authors have not provided any reason for differentiating first-hand information as personal experience and direct observation and assigning them different weights.

5.6.9 Observation-Based Cooperation Enhancement in Ad Hoc Networks

OCEAN has been proposed by Bansal and Baker as an extension of the DSR protocol. It consists of a monitoring system and a reputation system [32]. In contrast to other approaches, in OCEAN, nodes rely only on their own observations to avoid vulnerabilities arising out of false accusations and second-hand reputation exchanges. OCEAN categorizes routing misbehavior into two types: *misleading* and *selfish*. If a node has participated in a route discovery process but later on does not forward data packets, it is considered to be misleading as it misleads other nodes to route packets through it. On the other hand, if a node does not even participate in the route discovery, it is considered to be selfish. In order to detect and mitigate the misleading behavior of nodes, after a node forwards a packet to one of its neighboring nodes, it buffers the packet checksum and monitors if the neighbor attempts to forward the packet within a given time. Depending on the activity of the neighboring node, its reputation rating is updated. If the rating falls below a threshold, the neighbor node is added to a faulty list, which is appended to the route request message as a list of nodes to be avoided in routing. All packets originating from the nodes in the avoid list are rejected so that the faulty nodes cannot use network resources. A *timeout* is used to allow faulty nodes to rejoin the network in case they may be wrongly accused or start behaving in a better manner. Each node also has a mechanism of maintaining *chipcount* for each of its neighbors to mitigate selfish behavior. A neighbor node earns chips when it forwards a packet on behalf of the node, and loses chips when it asks the node to forward a packet. If the chipcount of a node falls below a threshold, packets coming from the node are dropped by its neighbors.

5.6.10 Robust Reputation System

Buchegger and Boudec presented an improved version of CONFIDANT called RRS [11]. The RRS introduced a Bayesian framework with Beta distribution for updating reputation. In contrast to CONFIDANT, RSS uses both positive and negative reputation values in the second-hand information. The RRS is robust to false ratings by malicious nodes: accusation or praise.

Every node maintains two metrics: reputation and trust. The reputation metric is used to classify the nodes as either normal or misbehaving, whereas the trust metric is used to classify the nodes as either trustworthy or untrustworthy. The first-hand information is exchanged among the nodes periodically. Whenever second-hand information is received from a node, the information is put under a deviation test. If the incoming reputation information does not deviate too much from the receiving node's opinion, then the information is accepted and integrated with the current reputation value. Since the information sent by the reporting node is supported by the information previously maintained by the receiving node, the reporting node's trust rating is increased. On the other hand, if the reputation report deviates from the record maintained by the receiving node by more than a threshold value, then the reporting node's trust value is decreased. The receiving node also decides whether to integrate the deviating information with its current records, depending on the level of trustworthiness of the reporting node.

In RRS, only fresh information is exchanged. Unlike CORE, RRS gives more weight to the current behavior than the past. This approach is different from the standard Bayesian approach, which gives equal weight to all observations irrespective of their time of occurrence. The authors argue that, if more weight is given to past behavior, then a malicious node can choose to be good initially till it builds a high reputation and trust value and then choose to misbehave. By assigning more weight to current behavior, the malicious node is forced to cooperate on a continuing basis to survive in the network. To accelerate the detection of misbehaving nodes, the authors have utilized selected second-hand information from trusted nodes and the information that has passed the deviation test.

5.6.11 Reputation-Based Framework for High-Integrity Sensor Networks

Ganeriwal and Srivastava have proposed a distributed, symmetric reputation-based framework for high-integrity sensor networks called Reputation-based Framework for Sensor Networks (RFSN) [2]. It classifies the actions of the nodes as cooperative and non-cooperative, and uses both first- and second-hand information for computing the reputation values. The framework employs a beta distribution for reputation representation, updates, and integration. The nodes maintain the reputation and trust values only for their neighboring nodes. RFSN is the first reputation- and trust-based models designed and developed exclusively for sensor networks. RFSN distinguishes between trust and reputation and uses two different metrics for their computation.

As shown in Figure 5.4, the first-hand information from the watchdog mechanism and second-hand information are combined to get the reputation value of a node. The trust level of the node is then computed from its reputation value. On the basis of this computed trust value, the node's strategy for the other node is determined. If the trust value is above a certain threshold, then the strategy is to cooperate with the node otherwise not.

RFSN, like many other systems, employs a watchdog mechanism for collecting first-hand information. The watchdog mechanism consists of different modules, each module monitoring a different function. The higher the number of modules, the greater is the resource requirement on the node. The reputation function is assumed to follow a probability distribution. The authors argue that reputation can only be used to statistically predict the future behavior of the nodes and it cannot be used to deterministically define the action performed by them. The reputation of all nodes that a node i interacts with is maintained in an RT in the node i. The direct reputation, $(R_{ij})_D$, is updated using the direct observations, that is, the output of the watchdog mechanism.

The nodes share their findings with each other. However, only positive information is shared. Higher weights are assigned to second-hand information from nodes that have higher reputation value associated with them. The weight assigned by node i to a second-hand information received from a node k is a function of the reputation of node k as maintained by node i. Like many other reputation and trust-based systems, RFSN uses Beta distribution model for reputation computation.

For each node n_j, a reputation R_{ij} is computed by a neighbor node n_i. The reputation is embodied in the Beta model which has two parameters α_{ij} and β_{ij}. α_{ij} represents the number of

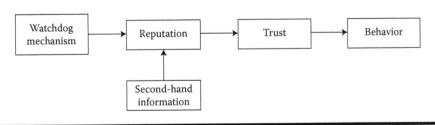

Figure 5.4 Architecture of RFSN system. (Data from S. Ganeriwal and M. Srivastava, *Proceedings of the 2nd ACM Workshop on Security of Ad Hoc and Sensor Networks (SASN '04)*, October 2004, pp. 66–77.)

successful transactions that node n_i had with node n_j, and β_{ij} represents the number of unsuccessful transactions. The reputation of node n_j maintained by node n_i is computed using the following equation:

$$R_{ij} = \text{Beta}(\alpha_{ij} + 1, \beta_{ij} + 1). \tag{5.1}$$

The trust is defined as the expected value of the reputation, shown as

$$T_{ij} = E(R_{ij}) = E(\text{Beta}(\alpha_{ij} + 1, \beta_{ij} + 1)) = \frac{\alpha_{ij} + 1}{\alpha_{ij} + \beta_{ij} + 2}. \tag{5.2}$$

The second-hand information is presented to node n_i by a neighbor node n_k. Node n_i receives the reputation R_{kj} of node n_j from node n_k, in the form of the two parameters α_{kj} and β_{kj}. After receiving this new information, node combines it with its current assessment R_{ij} to obtain a new reputation R_{ij}^{new} as shown in the following equation:

$$R_{ij}^{\text{new}} = \text{Beta}(\alpha_{ij}^{\text{new}}, \beta_{ij}^{\text{new}}), \tag{5.3}$$

where the values of α_{ij}^{new} and β_{ij}^{new} are given by the following equations:

$$\alpha_{ij}^{\text{new}} = \alpha_{ij} + \frac{2\alpha_{ik}\alpha_{kj}}{(\beta_{ik} + 2)(\alpha_{kj} + \beta_{kj} + 2)(2\alpha_{ik})}, \tag{5.4}$$

$$\beta_{ij}^{\text{new}} = \beta_{ij} + \frac{2\alpha_{ik}\beta_{kj}}{(\beta_{ik} + 2)(\alpha_{kj} + \beta_{kj} + 2)(2\alpha_{ik})}. \tag{5.5}$$

RFSN gives more weight to recent observations. This is used for updating reputation value using direct observation. To update the reputation value using second-hand information, Dempster-Shafer theory [34] and belief discounting theory [35] are utilized. The reputation of a reporting node is automatically taken into account in the computation of the reputation of the reported node. This eliminates the need of a separate deviation test. A node with higher reputation gets a higher weight. The trust level of a node is determined using its reputation value. Trust is computed as the statistically expected value of reputation using equation 5.2.

In the final decision-making stage, a node i has to take a decision whether to cooperate with node j. The decision of node i is referred to as its behavior B_{ij} and has a binary value: {*cooperate, don't cooperate*}. Node i uses the value of T_{ij} to take the decision as follows:

$$B_{ij} = \begin{cases} \text{coperate}, & \forall T_{ij} \geqslant B_{ij} \\ \text{don't coperate}, & \forall T_{ij} < B_{ij} \end{cases}. \tag{5.6}$$

The effectiveness of the notion of reputation and trust in RFSN resides in the assumption that the majority of nodes in any neighborhood of a WSN are trustworthy. The trust assessment is used to flush out the bad nodes. RFSN treats misbehaving and faulty nodes the same way. The rationale is that a node that is uncooperative is to be excluded irrespective of the reason of its behavior. The nodes are allowed to exchange only good reputation information and only direct reputation information is propagated. This eliminates the *bad-mouthing attack*. However, it affects the efficiency of the system, as the nodes cannot exchange their bad experiences. The aging factor is also introduced so that differential weights may be assigned to the old and new interactions, higher weight being assigned to the recent experiences.

5.6.12 Distributed Reputation-Based Beacon Trust System

The DRBTS model has been proposed by Srinivasan et al. [1] to solve a special problem in location-beacon sensor networks [1]. DRBTS presents a suite of techniques for detecting and revoking malicious BNs that provide misleading location information in a WSN. It is a distributed security protocol that models a WSN as an undirected graph and makes use of both first-hand and second-hand information. Two types of nodes are considered in the model: BN and SN. The BNs monitor each other and provide information which the SNs may decide to trust using a voting approach. Every BN monitors its one-hop neighborhood for any possible misbehaving BNs and updates the reputation of the neighbor nodes in the respective RTs. BNs use second-hand information for updating the reputation of their neighbors after the second-hand information passes a deviation test. The SNs use the neighbor-RT to determine whether or not to use a given BN's location information based on a simple majority voting scheme. The model is symmetric from the perspective of the BNs but asymmetric from the SNs' perspective. This is because BNs are capable of determining their location, and must pass this information to the SNs. However, without the knowledge of its own locations, an SN has no way of telling if a BN is lying to it. DRBTS enables the SNs to exclude location information from any malicious BN on the fly by using a simple majority principle. This way, DRBTS addresses the malicious misbehavior of any BN.

In DRBTS, information gathering is addressed from two different perspectives: the SN's perspective and the BN's perspective. From a BN's perspective, DRBTS uses a watchdog for neighborhood watch. When an SN sends a broadcast asking for location information, each BN will respond with its location and reputation values for each of its neighbors. The watchdog packet overhears the responses of the neighboring BNs. It then determines its location using the reported location of each BN in turn, and then compares the value against its true location. If the difference is within a certain margin of error, then the corresponding BN is considered benign, and its reputation increases. If the difference is greater than the margin of error, then that BN is considered malicious and its reputation is decreased. From an SN's perspective, there is no first-hand information gathered by it through direct observations. The SNs rely completely on the second-hand information passed to them from nearby BNs during the location request stage. DRBTS also includes a method by which BNs can send out location requests disguised as SNs, in case of low network activity. However, unlike CONFIDANT, DRBTS does not differentiate first-hand information into personal experience and direct observation.

DRBTS also makes use of second-hand information to update the reputation of its neighboring nodes. However, information sharing is only with respect to BNs. SNs do not share any information since they do not collect any first hand observation in their neighborhood. In DRBTS, nodes are allowed to share both positive and negative reputation information. This is allowed to ensure a quick learning time.

Let the BNs i, j, k are one-hop neighbors. An SN is within the range of BN i, but outside the range of BN j and BN k. The SN requests location information and the BN i responds. BN j and BN k listen to this broadcast transmission. BN k, then updates its reputation entry for BN i as follows:

$$R_{ki}^{\text{New}} = \mu_1 \times R_{ki}^{\text{current}} + (1 - \mu_1) \times \tau, \tag{5.7}$$

where $\tau = 1$ if the location was deemed to be truthful and $\tau = 0$ otherwise. μ_1 is a weight factor.

To use second-hand information, assume BN j is reporting about BN k to BN i. Now BN i first performs a deviation test to check if the information provided by BN j is compatible.

$$|R_{ji}^{\text{current}} - R_{ki}^{\text{current}}| \leqslant d. \tag{5.8}$$

If the above test is positive, then information provided is considered to be compatible and the entry R_{ik} is updated as follows:

$$R_{ji}^{\text{new}} = \mu_2 \times R_{ji}^{\text{current}} + (1 - \mu_2) \times R_{ki}^{\text{current}}. \tag{5.9}$$

If the deviation test in equation 5.8 is negative, then j is considered to be lying and its reputation is updated as follows:

$$R_{jk}^{\text{new}} = \mu_3 \times R_{jk}^{\text{current}}. \tag{5.10}$$

Equation 5.10 ensures that the lying nodes are punished so that such misbehavior can be discouraged. In Equation 5.8, d is a threshold deviation value. μ_2 and μ_3 are two weight factors in Equations 5.9 and 5.10 respectively.

Decisions are made from the SN's perspective. An SN, after sending out a location request, waits until a predetermined timeout. A BN has to reply before the timeout with its location information and its reputation ratings for its neighbors. Then, the SN, using the reputation ratings of all the responding BNs, tabulates the number of positive and negative votes for each BN in its range. Finally, when the SN has to compute its location, it considers the location information only from BNs with positive votes greater than negative votes. The remaining location information is discarded.

DRBTS addresses the malicious behavior of beacon nodes. This unique problem that this system solves, though very important to a specific branch of WSNs, is not encountered very frequently. However, the idea can easily be extended to other problem domains.

Table 5.1 compares various trust and reputation mechanisms used in wireless self-organizing networks.

5.7 Open Problems

The research in the field of reputation- and trust-based systems for self-organizing networks such as MANETs and WSNs is still in its incubation phase. There are many open issues that need to be resolved. Some of these open issues are discussed in this section.

The trust-modeling problem is inherently complicated due to uncertainty involved. The only coherent way to deal with uncertainty is by using theory of probability. Even though some of the trust models introduced for WSNs utilize probabilistic solutions coupled with *ad hoc* approaches, none of them produces a complete probabilistic answer to the problem. In [13], the author have presented a Bayesian probabilistic approach for modeling trust and reputation in WSNs, based on sensed continuous data to address security issues and to deal with malicious and unreliable nodes. The proposed mechanism has extended the *beta reputation system* to accommodate continuous sensor data and have utilized a novel Gaussian trust model for building a reputation framework. However, design of a reliable and robust trust framework for self-organizing networks such as MANETs and WSNs is still an open problem.

Another issue is the network bootstrapping problem. Most of the existing reputation- and trust-based systems require appreciable time to build trust among the nodes. Developing an effective

Table 5.1   Comparison of Various Trust and Reputation-Based Mechanisms

Techniques	*Watchdog/ Pathrater*	*CONFIDANT*	*CORE*	*RFSN*	*DRBTS*	*OCEAN*
Architecture	Distributed and cooperative					Standalone
Type of data collection	Reputation	Reputation	Reputation	Reputation	Reputation	Reputation
Data distribution	Negative to source node	Negative to friends	Positive from RREP	Yes	BN to BN BN to SN	No
Observation						
Self to neighbor	Yes	Yes	Yes	Yes	Yes	Yes
Neighbor to neighbor	No	Yes	No	Yes	Yes	Yes
Misbehavior detection						
Selfish-routing	No	Yes	Yes	Yes	Yes	Yes
Selfish-packet forwarding	Yes	Yes	Yes	Yes	Yes	Yes
Malicious-routing	No	Yes	No	Yes	Yes	No
Malicious-packet forwarding	Yes	Yes	No	Yes	Yes	No
Punishment	No	Yes	Yes	Yes	Yes	Yes
Avoid misbehaving node in route discovery	No	No	No	Yes	Yes	Yes

and efficient solution to minimize this latency is a big challenge [31]. While more information availability in the nodes helps in making a reputation- and trust-based system more aware about system-wide events, but it also makes the system vulnerable to *false information attacks*. Moreover, in systems which are based on continuous cooperation among nodes, periods and regions of low network activities pose new challenges. In such systems, *aging* may deteriorate the reputation of honest nodes due to lack of interactions among some nodes.

Another important problem that needs to be addressed is devising a suitable defense against an intelligent adversary strategy. A sophisticated and intelligent adversary may manifest his attack strategy is such a way that it may not be possible for a detection system to catch him. A game theoretic approach may be applied here to investigate the effectiveness of a detection and response system to defend against and counter such attacks.

In some cases, trust-based systems for self-organizing networks may lead to misleading conclusions if only one dimension of trust is used for determining the trustworthiness of a node [13]. A thorough analysis of trust dimensions—the *data trust* and the *communication trust*—must be made, since a trustworthy node from the perspective of data may be untrustworthy from the perspective of communication trust and vice versa. This makes design of a trust framework more challenging, since the computational model must have the ability to integrate the two dimensions of trust.

Designing new algorithms for revocation of trust in the nodes of a self-organizing network is another challenge. The issue of expelling a node from the network due to misbehavior is a decision problem under uncertainty and requires a formal mathematical approach to address. In most of the existing propositions, this problem has been solved using an *ad hoc* approach based on a threshold value.

Some of the existing propositions such as CORE use functional reputation to monitor the behavior of nodes. However, these schemes compute the integrated reputation value of a node from various functional reputation components. This may not be very effective in a real world scenario, since it may allow an adversary to conceal his misbehavior in certain functions while behaving very well for other functions. No research work so far has investigated the possible benefits of using functional reputation values independently. It may be effective for a security system to isolate a node from network resources for a particular function if the node is detected to be misbehaving with respect to that function, rather than judging it with respect to other functions where it may be perfectly well behaving.

Finally, another challenge especially for reputation systems in MANETs is the development of a robust scheme that motivates the nodes to publish their ratings honestly [31]. This is not very easy, as the nodes in a MANET do not belong to the same interest group.

5.8 Conclusion

The existing security measures for wireless self-organizing networks such as MANETs and WSNs are not enough to defend against all possible attacks on these networks. Power constraints and short-range communication between the nodes in these networks make multihop communication an essential feature. Multihop routing requires cooperation between the nodes. As there is no guarantee that all nodes on a routing path are capable of cooperating or willing to cooperate with each other, new mechanisms involving trust and reputation frameworks become essential. Reputation and trust have emerged as two very important tools to facilitate distributed decision-making in cooperative wireless networks. This chapter has provided a detailed understanding of reputation- and trust-based systems from the perspective of wireless self-organizing networks. Many aspects of reputation- and trust-based systems including their goals, properties, initialization process, and classification are discussed. Various important issues of design of such systems and a comprehensive review of some of the research works focusing on adapting reputation- and trust-based systems for MANETs and WSNs are also presented. Finally, some open research challenges in this field are discussed.

Terminologies

Reputation
Trust
Basic
General
Situational
Trusting behavior
Trusting intention
Trusting belief

Situational decision
System trust
Dispositional trust
Wireless self-organizing networks
Mobile *Ad Hoc* Network (MANET)
Wireless Sensor Network (WSN)
Node misbehavior

Questions and Sample Answers

1. What do you mean by "Reputation" and "Trust"?
   In simple terms, reputation is the opinion of one entity about another. Essentially, it signifies the trustworthiness of an entity. Trust, on the other hand, is the expectation of one entity about the actions of another.

2. Define: Trusting intention, Trusting behavior, Trusting beliefs
   Trusting intention of a node is the willingness of one node to depend on another node in a specific situation in spite of the knowledge of the risk involved. The trusting intention consists of essential elements such as experience of reliability, evidence of security, and so on. *Trusting behavior* of a node is a voluntary dependence of one node on another node in a specific situation with the existence of risk. The trusting intention of a node supports trusting behavior. *Trusting beliefs* in nodes is the confidence and belief of one node that the other node is trustworthy in a specific situation, that is, for example, when node A believes that node B is trustworthy.

3. What is a sleep deprivation attack?
   In sleep deprivation attack, a malicious node sends an excessive number of packets to another node so as to consume computation and memory resources of the latter.

4. What are the effects of nodes' misbehavior?
   In wireless self-organizing networks, without appropriate countermeasures, the effects of node misbehavior dramatically decrease the network performance. Depending on the proportion of misbehaving nodes and their specific strategies, network throughput can severely be degraded, packet loss increased, and denial-of-service experienced by honest nodes in the network. In a theoretical analysis of how much cooperation can help by increasing the probability of a successful forwarding of packets, Lamparter, Plaggemeir, and Westhoff have found that increased cooperation more than proportionately increases the performance for small networks with fairly short routes. Zhang and Lee argue that prevention measures such as encryption and authentication can be used in MANETs to reduce the success of intrusion attempts, but cannot completely eliminate them. For example, encryption and authentication cannot defend against compromised mobile nodes, which carry the private keys. No matter what types of intrusion prevention measures are deployed in the network, there are always some weak links that an adversary can exploit to break in. Intrusion detection presents a second wall of defense and it is a necessity in any high-survivability network.

5. What are the characteristics of trust?
   From the perspective of wireless communication networks, Sun et al. [30] have identified some characteristics of trust metric. These characteristics are as follows:

   1. Trust is a relationship established between two entities for a specific action. In particular, one entity trusts the other entity to perform an action. The first entity is called the subject, the second is called the agent.

2. Trust is a function of uncertainty. In particular, if the subject believes that the agent will perform the action for sure, the subject fully trusts the agent to perform the action and there is no uncertainty; if the subject believes that the agent will not perform the action for sure, the subject trusts the agent not to perform the action, and there is no uncertainty either; if the subject does not have any idea of whether the agent will perform the action or not, the subject does not have trust in the agent, In this case, the subject has the highest level of uncertainty.

3. The level of trust can be measured by a continuous real number, referred to as the trust value. Trust value should represent uncertainty.

The subjects may have different trust values with the same agent for the same action. Trust is not necessarily symmetric. The fact that A trusts B does not necessarily mean that B also trust A, where A and B are two entities.

Author's Biography

Jaydip Sen obtained his bachelor of engineering (BE) with honors in electrical engineering with honors from Jadavpur University, Kolkata, India, in 1993, master of technology (MTech) with honors in computer science from Indian Statistical Institute, Kolkata, in 2001, and PhD in network security from Indian Institute of Technology, Kharagpur, India, in 2007. He has 17 years of experience in the filed of networking, communication, and security. He has worked in reputed organizations such as Oil and Natural Gas Corporation Ltd., India, Oracle India Pvt. Ltd., and Akamai Technology Pvt. Ltd. Currently, he is leading the research and development activities in wireless communication in Tata Consultancy Services, Kolkata, India, for the last three years. He has over 16 years of research and development experience. His research areas include security in wired and wireless networks, intrusion detection systems, secure routing protocols in wireless *ad hoc* and sensor networks, secure multicast and broadcast communication in next generation broadband wireless networks, trust- and reputation-based systems, quality of service in multimedia communication in wireless networks and crosslayer optimization-based resource allocation algorithms in next generation wireless networks, sensor networks, and privacy issues in ubiquitous and pervasive communication. He has more than 60 publications in reputed international journals and referred conference proceedings. He has delivered expert talks and keynote lectures in various international conferences and symposia. He is a member of ACM and IEEE and also a working member of IEEE 802.16 group. He is an active member of the security group of IEEE 802.16 standard body and has submitted a number of proposals for the evolving 802.16m standard and ETSI.

References

1. A. Srinivasan, J. Teitelbaum, and J. Wu, DRBTS: Distributed reputation-based Beacon trust system, in *Proceedings of the 2nd IEEE International Symposium on Dependable*, Autonomic and Secure Computing (DASC'06), Indianapolis, USA, pp. 277–283, 2006.
2. S. Ganeriwal and M. Srivastava, Reputation-based framework for high integrity sensor networks, in *Proceedings of the 2nd ACM Workshop on Security of Ad Hoc and Sensor Networks (SASN '04)*, New York, USA, October 2004, pp. 66–77.
3. Y. Hu and A. Perrig, A survey of secure wireless ad hoc routing, *IEEE Security and Privacy*, 2(3), 28–39, 2004.

4. D. Gambetta, Can we trust trust? In *Trust: Making and Breaking Cooperative Relations*, pp. 213–217, Basil Blackwell, Oxford, 1988.
5. A. Abdul-Rahman and S. Hailes, Supporting trust in virtual communities, in *Proceedings of the 33rd Hawaii International Conference on System Sciences*, Maui, Hawaii, Vol. 6, p. 6007, 2000.
6. D. Quercia, S. Hailes, and L. Capra, B-trust: Bayesian trust framework for pervasive computing, in *Proceedings of the 4th International Conference on Trust Management*, Pisa, Italy, Vol. 3986, pp. 298–312, 2006.
7. M. Kinateder, E. Baschny, and K. Rothermel, Towards a generic trust model—comparison of various trust update algorithms, in *Proceedings of the 3rd International Conference on Trust Management*, Rocquencourt, France, Vol. 3477, pp. 177–192, 2005.
8. Z. Liu, A.W. Joy, and R.A. Thompson, A dynamic trust model for mobile ad hoc networks, in *10th IEEE International Workshop on Future Trends of Distributed Computing Systems*, Suzhou, China, pp. 80–85, 2004.
9. P. Michiardi and R. Molva, CORE: A COllaborative REputation mechanism to enforce node cooperation in Mobile Ad Hoc Networks, in *Proceedings of the 6th IFIP Communication and Multimedia Security Conference*, Portoroz, Slovenia, Vol. 228, pp. 107–121, September 2002.
10. F. Azzedin and M. Maheswaran, Evolving and managing trust in grid computing systems, in *Proceedings of the IEEE Canadian Conference on Electrical and Computer Engineering (CCECE'02)*, Winnipeg, Canada, Vol. 3, pp. 1424–1429, 2002.
11. S. Buchegger and J.-Y. Le Boudec, A robust reputation system for peer-to-peer and mobile ad hoc networks, in *Proceedings of P2Pecon 2004*, Harvard University, Cambridge, MA, USA, June 2004.
12. S. Marsh, Formulating trust as a computational concept, Ph.D. Thesis, Department of Computer Science and Mathematics, University of Stirling, 1994.
13. M. Momani, Bayesian methods for modeling and management of trust in wireless sensor networks, Ph.D. Thesis, Faculty of Engineering, University of Technology, Sydney, Australia, July, 2008.
14. D.H. McKnight and N.L. Chervany, *The Meanings of Trust*, Technical Report, MIS Research Center, Carlson School of Management, University of Minnesota, 1996.
15. P. Michiardi and R. Molva, Simulation-based analysis of security exposures in mobile ad hoc networks, in *Proceedings of the European Wireless Conference*, Florence, Italy, pp. 107–121, 2002.
16. C. Karlof and D. Wagner, Secure routing in wireless sensor networks: Attacks and countermeasures, in *Proceedings of the 1st IEEE International Workshop on Sensor Networks Protocols and Applications*, Anchorage, Alaska, USA, pp. 113–127, May 2003.
17. S. Buchegger and J-Y.L. Boudec, Performance analysis of the CONFIDANT protocol (Cooperation Of Nodes-Fairness In Dynamic Ad-hoc NeTworks), in *Proceedings of 3rd ACM International Symposium on Mobile Ad Hoc Networking and Computing (MobiHoc'02)*, Lausanne, Switzerland, pp. 226–236, June 2002.
18. M. Blaze, J. Feigenbaum, J. Ioannidis, and A. Keromytis, *The keynote trust management system version 2*, Internet RFC 2704, 1999.
19. N. Li, J. Mitchell, and W. Winsborough, Design of a role-based trust management framework, in *Proceedings of the IEEE Symposium on Security and Privacy*, Oakland, p. 114, 2002.
20. P. Ning and K. Sun, How to misuse AODV: A case study of insider attacks against mobile ad hoc routing protocols, in *Proceedings of the 4th Annual IEEE Information Assurance Workshop*, West Point, pp. 60–67, June 2003.
21. B. Lamparter, M. Plaggemeier, and D. Westhoff, Estimating the value of co-operation approaches for multi-hop ad hoc networks, *Ad Hoc Networks*, 3 (1), 17–16, 2005.
22. Y. Zhang and W. Lee, Intrusion detection in wireless ad hoc networks, in *Proceedings of the 6th Annual International Conference on Mobile Computing and Networking (MobiCom 2000)*, pp. 275–283, ACM Press, New York, USA, 2000.
23. P. Resnick, R. Zeckhauser, E. Friedman, and K. Kuwabara, Reputation system, *Communications of the ACM*, 43(12), 45–48, 2000.
24. C. Dellarocas, Immunizing online reputation reporting systems against unfair ratings and discriminatory behavior. in *Proceedings of the ACM Conference on Electronic Commerce*, Minneapolis, Minnesota, USA, pp. 150–157, 2000.

25. P. Resnick and R. Zeckhauser, Trust among strangers in Internet transactions: Empirical analysis of eBays's reputation system, *Advances in Applied Microeconomics*, Vol. 11, pp. 127–157, Elsevier Science, Amsterdam, Netherlands, 2002.

26. K. Aberer and Z. Despotovic, Managing trust in a peer-2-peer information system, in *Proceedings of the 10th International Conference on Information and Knowledge Management (CIKM 2001)*, Atlanta, Georgia, pp. 310–317, 2001.

27. A. Josang and R. Ismail, The beta reputaion system, in *Proceedings of the 15th Bled Electronic Commerce Conference*, Bled, Slovenia, pp. 324–327, June 2002.

28. S. Marti, T.J. Giuli, K. Lai, and M. Baker, Mitigating routing misbehavior in mobile ad hoc networks, in *Proceedings of the 6th Annual International Conference on Mobile Computing and Networking (MobiCom 2000)*, Boston, Massachusetts, USA, pp. 255–265, 2000.

29. S. Buchegger and J.-Y. Le Boudec, The effect of rumor spreading in reputation systems in mobile ad hoc networks, in *Proceedings of Wiopt' 03*, Sofia- Antipolis, March 2003.

30. Y.L. Sun, W. Yu, Z. Han, and K.J. Ray Li, Trust modeling and evaluation for ad hoc networks, Technical Report No: 20041017-21, University of Rhode Island, October 2004.

31. A. Srinivasan, J. Teitelbaum, H. Liang, J. Wu, and M. Cardei, Reputation and trust-based systems for ad hoc and sensor networks, A. Boukerche, ed., in *Algorithms and Protocols for Wireless Ad Hoc and Sensor Networks*, Wiley and Sons, 2008.

32. S. Bansal and M. Baker, Observation-based cooperation enforcement in ad hoc networks, Research Report cs.NI/0307012, Stanford University, 2003.

33. K. Paul and D. Westhoff, "Context aware inferencing to rate a selfish node in DSR-based ad hoc networks, in *Proceedings of the IEEE Globecom Conference*, Taipeh, Taiwan, 2002.

34. G. Shafer, *A Mathematical Theory of Evidence*, Princeton University, New Jersey, 1976.

35. A. Josang, A logic for uncertain probabilities, *International Journal of Uncertainty, Fuzziness and Knowledge-Based Systems*, 9(3), 279–311, June 2001.

36. P. Kollock, The production of trust in online markets, E.J. Lawler, M. Macy, S. Thyne, H.A. Walker, eds, *Advances in Group Processes*, Vol. 16, JAI Press, Greenwich, CT, 1999.

37. G. Montenegro and C. Castelluccia, Statistically Unique and Cryptographically Verifiable (SUCV) Identifiers and Addresses, in *Proceedings of the 9th Annual Network and Distributed System Security Symposium (NDSS'02)*, San Diego, California, USA, February 2002.

38. J. R. Douceur, The Sybil attack, in *Proceedings of the 1st International Workshop on Peer-to-Peer Systems*, Cambridge, MA, USA, pp. 251–260, March 2002.

39. J. Newsome, E. Shi, D. Song, and A. Perrig, The Sybil attack in sensor networks: Analysis and defenses, in *Proceedings of the 3rd International Symposium on Information Processing in Sensor Networks*, Berkeley, California, USA, April 2004.

40. J. Jubin and J.D. Tornow, The DARPA packet radio network protocols, *Proceedings of the IEEE Communications and Networks*, 75(1), pp. 21–32, 1987.

41. A. Pirzada and C. McDonald, Establishing trust in pure ad hoc networks, in *Proceedings of the 27th Australian Conference on Computer Science*, Dunedin, New Zealand, 2004, pp. 47–54.

42. Y.L. Sun, W. Yu, Z. Han, and K.J.R. Liu, Information theoretic framework of trust modeling and evaluation for ad hoc networks, *IEEE Journal on Selected Areas in Communication*, 24, 305–317, 2006.

43. Z. Liang and W. Shi, Enforcing cooperative resource sharing in untrusted peer-to-peer environment, *ACM Journal of Mobile Networks and Applications (MONET)*, 10(6), pp. 771–783, 2005.

44. Z. Liang and W. Shi, Analysis of ratings on trust inference in the open environment, Technical Report MIST-TR-2005-002, Department of Computer Science, Wayne State University, February 2005.

45. Z. Yan, P. Zhang, and T. Virtanen, Trust evaluation based security solution in ad hoc networks, in *Proceedings of the 7th Nordic Workshop on Secure IT Systems*, Trondheim, Norway, pp. 37–50, 2003.

46. K. Ren, T. Li, Z. Wan, F. Bao, R.H. Deng, and K. Kim, Highly reliable trust establishment scheme in ad hoc networks, *Computer Networks: The International Journal of Computer and Telecommunications Networking*, 45, 687–699, August 2004.

47. A. Oram, *Peer-to-Peer: Harnessing the Power of Disruptive Technologies*, O'Reilly & Associates, March 2001.

48. H. Zhu, F. Bao, and R.H. Deng, Computing of trust in wireless networks, in *Proceedings of the 60th IEEE Vehicular Technology Conference*, Los Angeles, CA, Vol. 4, pp. 2621–2624, September 2004.

49. C.R. Davis, A localized trust management scheme for ad hoc networks, in *Proceedings of the 3rd International Conference on Networking (ICN'04)*, Guadeloupe, French Caribbean, pp. 671–675, 2004.

50. M. Jarrett and P. Ward, Trusted computing for protecting ad hoc routing, in *Proceedings of the 4th IEEE Annual Communication Networks and Services Research Conference (CNSR'06)*, Moncton, New Brunswick, pp. 61–68, 2006.

51. C. Perkins, E. Belding-Royer, and S. Das, Ad hoc On-demand Distance Vector (AODV) Routing, RFC 3561, July 2003.

52. B.-J. Chang, S.-L. Kuo, Y.-H. Liang, and D.-Y. Wang, Markov chain-based trust model for analyzing trust value in distributed multicasting mobile ad hoc networks, in *Proceedings of the IEEE Asia-Pacific Services Computing Conference*, Yilan, Taiwan, December, 2008, pp. 156–161.

53. R. Dawkins, *The Selfish Gene*, Oxford University Press, 1989.

MOBILE *AD HOC* NETWORK AND VEHICULAR *AD HOC* NETWORK SECURITY

Chapter 6

Security Threats in Mobile *Ad Hoc* Networks

Sevil Şen, John A. Clark, and Juan E. Tapiador

Contents

6.1 Introduction

With the proliferation of cheaper, smaller, and more powerful mobile devices, mobile *ad hoc* networks (MANETs) have become one of the fastest growing areas of research. This new type of self-organizing network combines wireless communication with a high-degree node mobility. Unlike conventional wired networks, they have no fixed infrastructure (base stations, centralized management points and the like). The union of nodes forms an arbitrary topology. This flexibility makes them attractive for many applications such as military applications, where the network topology may change rapidly to reflect a force's operational movements, and disaster recovery operations, where the existing/fixed infrastructure may be nonoperational. The *ad hoc* self-organization also makes them suitable for virtual conferences, where setting up a traditional network infrastructure is a time-consuming high-cost task.

Conventional networks use dedicated nodes to carry out basic functions such as packet forwarding, routing, and network management. In *ad hoc* networks, these are carried out collaboratively by all nodes available. Nodes on MANETs use multihop communication: nodes that are within each other's radio range can communicate directly through wireless links, whereas those that are far apart must rely on intermediate nodes to act as routers to relay messages. Mobile nodes can move, leave, and join the network, and routes need to be updated frequently due to the dynamic network topology. For example, node S can communicate with node D by using the shortest path S-A-B-D as shown in Figure 6.1 (the dashed lines show the direct links between the nodes). If node A moves out of range of node S, it has to find an alternative route to node D (S-C-E-B-D). A variety of new protocols have been developed for finding/updating the routes and generally providing communication between end points (but no proposed protocol has yet been accepted as standard). However, these new routing protocols, based on cooperation between nodes, are vulnerable to new forms of attacks. Unfortunately, many proposed routing protocols for MANETs do not consider security. Moreover, their specific features such as the lack of central points, the dynamic topology, and the existence of highly constrained nodes presents a particular challenge for security.

Much research has been done to counter and detect attacks against the existing MANET routing protocols, including works on secure routing protocols and intrusion detection systems (IDSs). However, for practical reasons, the proposed solutions typically focus on a few particular security vulnerabilities since providing a comprehensive solution is nontrivial. If we are to develop more general solutions, we must first have a comprehensive understanding of possible vulnerabilities and security risks against MANETs. This is the main goal of this chapter. Section 6.2 presents the specific vulnerabilities of MANETs, and deals with the fundamentals of an exemplar routing protocol to help understanding of the attacks given in Section 6.3. An overview of security solutions proposed to prevent and detect attacks on MANETs is presented in Section 6.4. Finally, areas for future research are given.

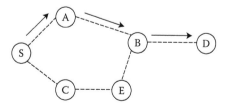

Figure 6.1 Communication between nodes in MANETs.

6.2 Background

The specific features of MANETs present a challenge for security solutions. Many existing security solutions for conventional networks are ineffective and inefficient for many envisaged MANET deployment environments. Consequently, researchers have been working over the last decade on developing new security solutions or changing the current ones to be applicable to MANETs. Since many routing protocols do not consider security, some research focuses on developing secure routing protocols or introducing security extensions to the existing routing protocols. Routing protocols have been proposed to counter selfish activities by forcing the selfish nodes to cooperate. Existing key management mechanisms are usually based on central points where services such as certification authorities (CAs) or key servers (KSs) can be placed. Since MANETs do not have such points, new key management mechanisms have had to be developed to fulfill the requirements. Finally, since prevention techniques are invariably limited in effectiveness, IDSs are generally used to complement other security mechanisms. This applies to MANETs too, and researchers have proposed new IDSs to detect malicious activities on these networks.

If we are to develop more general solutions, we must first have a comprehensive understanding of possible vulnerabilities and security risks against MANETs. They share the vulnerabilities of wired networks, such as eavesdropping, denial of service (DoS), spoofing and the like, which are accentuated by the *ad hoc* context [1]. They also have further vulnerabilities such as those that take advantage of the cooperative nature of routing algorithms. These vulnerabilities of MANETs are summarized in the following section.

6.2.1 Vulnerabilities of MANETs

Wireless links: First of all, the use of wireless links makes the network susceptible to attacks such as eavesdropping and active interference. Unlike wired networks, attackers do not need physical access to the network to carry out these attacks. Furthermore, wireless networks typically have lower bandwidths than wired networks. Attackers can exploit this feature, consuming network bandwidth with ease to prevent normal communication among nodes.

Dynamic topology: MANET nodes can leave and join the network, and move independently. As a result, the network topology can change frequently. It is difficult to differentiate normal behavior of the network from anomaly/malicious behavior in this dynamic environment. For example, a node sending disruptive routing information can be a malicious node, or else simply be using outdated information in good faith. Moreover, mobility of nodes means that we cannot assume nodes, especially critical ones (servers, etc.), are secured in locked cabinets as in wired networks. Nodes with inadequate physical protection may often be at risk of being captured and compromised.

Cooperativeness: Routing algorithms for MANETs usually assume that nodes are cooperative and nonmalicious. As a result, a malicious attacker can easily become an important routing agent and disrupt network operations by disobeying the protocol specifications. For example, a node can pose as a neighbor to other nodes and participate in collective decision-making mechanisms, possibly affecting networking significantly.

Lack of a clear line of defense: MANETs do not have a clear line of defense; attacks can come from all directions [2]. The boundary that separates the inside network from the outside world is not very clear on MANETs. For example, there is no well-defined place where we can deploy our traffic monitoring and access control mechanisms. Whereas all traffic goes through switches, routers, or gateways in wired networks, network information in

MANETs is distributed across nodes that can only see the packets sent and received in their transmission range.

Limited resources: Resource constraints are a further vulnerability. There can be a variety of devices on MANETs, ranging from laptops to handheld devices such as PDAs and mobile phones. These will generally have different computing and storage capacities that can be the focus of new attacks. For example, mobile nodes generally run on battery power. This has led to emergence of innovative attacks targeting this aspect, for example, "Sleep Deprivation Torture" [3]. Furthermore, introduction of more security features into the network increases the computation, communication, and management load [4]. This is a challenge for networks that are already resource constrained.

6.2.2 AODV Routing Protocol

There have been many routing protocols proposed to suit the different needs of MANETs. Unfortunately, most of these routing protocols do not consider security. One of the most popular of them is the AODV routing protocol. In this section, we describe the operation of AODV to understand better the routing attacks explained subsequently. We aim to illustrate the principles of attacks. Other protocols may be susceptible to these or similar attacks, but may also be vulnerable to further protocol-specific attacks. Moreover, the consequences of attacks can have different impacts in different routing protocols (e.g., in proactive vs. reactive routing protocols).

AODV is a reactive routing protocol, discovering routes only when they are needed. "It offers quick adaptation to dynamic link conditions, low processing and memory overhead, low network utilization, and determines unicast routes to destinations within *ad hoc* network" [5]. It is claimed that AODV can handle low, moderate, and relatively high mobile rates, together with a variety of data-traffic loadings [5]. However, it makes no provisions for security.

There are three main types of messages in AODV: route request (RREQ), route reply (RREP), and route error (RERR) messages. When a node wants to communicate with another node in the network and does not have a fresh route to that destination, it starts the route discovery process by broadcasting an RREQ message for the destination node into the network. Intermediate nodes that receive this request either send an RREP to the source node if they have a fresh route to the destination node and the "destination only" flag is not set, or forward the RREQ message to other nodes. A fresh route is a valid route entry whose sequence number is equal to or greater than that contained in the RREQ message. If the request packet has been forwarded by this intermediate node before, it is silently dropped. When the destination node receives an RREQ for itself, it sends back an RREP message on the reverse route. The requesting node and the nodes receiving RREP messages on the route update their routing tables with the new route.

Wireless mobile networks can have frequent link breakages due to the mobility of nodes in the network or simply due to transmission errors. "AODV allows mobile nodes to respond to link breakages and changes in a timely manner" [5]. The methods for a node to control its connectivity to its active next hops on AODV are

- link layer notification using control packets such as link layer acknowledgement messages (e.g., ACK or RTS-CTS);
- passive acknowledgement: notification by listening on the channel to determine if the next node forwards the packet or not; and
- receiving any packet from the next node or sending some request packets to the next node, such as RREQ or ICMP Echo Request, or Hello messages which are periodic control messages sent only to one hop neighbors.

Let us assume that a link breakage to the next hop is detected by the absence of hello messages in the allowed time interval (or with any of the methods above). The routes affected by the link breakage in the routing table are invalidated, and the nodes affected by the link breakage are notified using RERR messages. If the link breakage occurs on an active route, a local repair mechanism can be initiated. In this mechanism, new RREQ messages are broadcasted to the destination by nodes on the existing route which detect the link breakage.

6.3 Attacks on MANET

At the highest level, the security goals of MANETs are not that different from other networks: most typically authentication, confidentiality, integrity, availability, and nonrepudiation. Authentication is the verification of claims about the identity of a source of information. Confidentiality means that only authorized people or systems can read or execute the protected data or programs. It should be noted that the sensitivity of information in MANETs may decay much more rapidly than in other information. For example, yesterday's troop location will typically be less sensitive than today's. Integrity means that the information is not modified or corrupted by unauthorized users or by the environment. Availability refers to the ability of the network to provide services as required. DoS attacks have become one of the most worrying problems for network managers. In a military environment, a successful DoS attack is extremely dangerous, and the engineering of such attacks is a valid modern war-goal. Lastly, nonrepudiation ensures that committed actions cannot be denied. In MANETs, security goals of a system can change in different modes (e.g., peace time, transition to war, and war time of a military network).

The characteristics of MANETs make them susceptible to many new attacks. At the top level, attacks can be classified according to network protocol stacks. Table 6.1 gives a few examples of attacks at each layer. Some attacks could occur in any layer of the network protocol stack, for example, jamming at physical layer, hello flood at network layer, and SYN flood at transport layer—all are DoS attacks. Because new routing protocols introduce new forms of attacks on MANETs, we mainly focus on network layer attacks in this chapter.

6.3.1 Adversary Model

Attackers against a network can be classified into two groups: insider and outsider. Whereas an outsider attacker is not a legitimate user of the network, an insider attacker is an authorized node and a part of the routing mechanism on MANETs. Routing algorithms are typically distributed and cooperative in nature and affect the whole system. Although an insider MANET node can

Table 6.1 Some Attacks on the Protocol Stack

Layer	Attacks
Application layer	Data corruption, viruses and worms
Transport layer	TCP/UDP SYN flood
Network layer	Hello flood, blackhole
Data link layer	Monitoring, traffic analysis
Physical layer	Eavesdropping, active interference

disrupt the network communications intentionally, there might be other reasons for its apparent misbehaviors. A node can be failed, unable to perform its function for some reason, such as running out of battery, or collusions in the network. The threat of failed nodes is particularly serious if they are needed as part of an emergency/secure route [6]. Their failure can even result in partitioning of the network, preventing some nodes from communicating with other nodes in the network. A selfish node can also misbehave to preserve its resources. Selfish nodes avail themselves of the services of the other nodes, but do not reciprocate. In this paper, we mainly concentrate on attacks carried out by malicious nodes that intentionally aim at disrupting the network communication.

We should also consider the misuse goals of attackers. In routing attacks, attackers do not follow the specifications of routing protocols and aim at disrupting the network communication in the following ways:

■ *Route disruption:* modifying existing routes, creating routing loops, and causing the packets to be forwarded along a route that is not optimal, nonexistent, or otherwise erroneous.
■ *Node isolation:* isolating a node or some nodes from communicating with other nodes in the network, partitioning the network, and so on.
■ *Resource consumption:* decreasing network performance, consuming network bandwidth or node resources, and so on.

Ning et al. consider each of these goals in their research that analyses insider attacks against AODV [7]. Achieving these goals depends on the capabilities of the adversary. The following are the main factors affecting the performance of an attack:

Computational power: This clearly affects the ability of an attacker to compromise a network. Such power need not be localized to the attached network—eavesdropped traffic can be relayed back to high-performance super-computing networks for analysis.

Deployment capability: Adversary distribution may range from a single node to a pervasive carpet of smart counterdust, with a consequent variation in attack capabilities [8]. This sort of distinction may affect the ability to eavesdrop, to jam a network effectively, and to escape destruction (e.g., a single powerful jammer can easily be taken out, distributed jamming is harder to extinguish).

Location control: The location of adversary nodes may have a clear impact on what the adversary can do. An adversary may be restricted to placing attack nodes at the geographical boundary of an enemy network (but may otherwise choose the precise locations), may plant specific nodes (e.g., nodes left behind in territory about to be vacated), or may have the ability post facto to create a pervasive carpet of smart dust (where arbitrary degrees of pervasiveness may be achieved).

Mobility: Mobility generally brings an increase in power. (A mobile node can always remain stationary.) On the other hand, mobility may prevent an attacker from continually targeting one specific victim. For example, a node on the move might not receive all falsified routing packets initiated by the attacker. In [9] Sun et al. defined this phenomenon as being a "partial victim." Moreover, they have stated that even if it reduces the damage caused by the attacker, it makes detection more difficult since the symptoms of an attack and those arising due to the dynamic nature of the network are difficult to distinguish. In conclusion, the impact of mobility on detection is a complex matter.

Degree of physical access (including node capture ability and ability to carry out physical deconstruction).

Given the agile nature of MANETs determining an applicable adversary model is difficult. However, systems can be evaluated against a range of representative threat models.

6.3.2 Attacks

We can classify attacks as passive or active.

6.3.2.1 Passive Attacks

In a passive attack, an unauthorized node monitors and aims to find out information about the network. The attackers do not otherwise need to communicate with the network. Hence, they do not disrupt communications or cause any direct damage to the network. However, they can be used to get information for future harmful attacks. Examples of passive attacks are eavesdropping and traffic analysis.

Eavesdropping attacks, also known as disclosure attacks, are passive attacks by external or internal nodes. The attacker can analyze the broadcasted messages to reveal some useful information about the network. Solutions protecting the radio interface from attacks such as eavesdropping (and jamming) attacks have been proposed in the literature (e.g., spread spectrum communication and frequency hopping) [10].

Traffic analysis is not necessarily an entirely passive activity. It is perfectly feasible to engage in protocols, or seek to provoke communication between nodes. Attackers may employ techniques such as RF direction finding, traffic rate analysis, and time-correlation monitoring. For example, by timing analysis, it can be revealed that two packets in and out of an explicit forwarding node at time t and $t + \epsilon$ are likely to be from the same packet flow [11]. Traffic analysis in *ad hoc* networks may reveal

- The existence and location of nodes
- The communication network topology
- The roles played by nodes
- The current sources and destination of communications
- The current location of specific individuals or functions (e.g., if the commander issues a daily briefing at 10 a.m., traffic analysis may reveal a source geographic location)

6.3.2.2 Active Attacks

These attacks cause unauthorized state changes in the network such as DoS, modification of packets, and so on. These attacks are generally launched by users or nodes with authorization to operate within the network. We classify active attacks into four groups: dropping, modification, fabrication, and timing attacks. It should be noted that an attack can be classified into more than one group.

Dropping attacks: Malicious or selfish nodes deliberately drop all packets that are not destined for them. While malicious nodes aim to disrupt the network connection, selfish nodes aim to preserve their resources. Dropping attacks can prevent end-to-end communications between nodes, if the dropping node is at a critical point. It might also reduce the network performance by causing data packets to be retransmitted, new routes to the destination to be discovered, and the like.

Unfortunately, most routing protocols (DSR is an exception [6]) have no mechanism to detect whether data packets have been forwarded or not. However, they can be detected by neighboring nodes through passive acknowledgement or hop-by-hop acknowledgement at the data link layer.

An attacker can choose to drop only some packets to avoid being detected; this is called a selective dropping attack. Besides data packets or route discovery packets, an attacker can also drop route error packets, causing the source node to be unaware of failed links (thus interfering with the discovery of alternative routes to the destination).

Modification attacks: Insider attackers modify packets to disrupt the network. For example, in the sinkhole attack, the attacker tries to attract nearly all traffic from a particular area through a compromised node by making the compromised node attractive to other nodes. It is especially effective in routing protocols that use advertised information such as remaining energy and nearest node to the destination in the route discovery process. A sinkhole attack can be used as a basis for further attacks such as dropping and selective forwarding attacks. A black hole attack is similar to a sinkhole attack that attracts traffic through itself and uses it as the basis for further attacks. The goal is to prevent packets being forwarded to the destination. If the black hole is a virtual node or a node outside the network, it is hard to detect [12].

Fabrication attacks: Here, the attacker forges network packets. In [7], fabrication attacks are classified into "active forge," in which attackers send faked messages without receiving any related message, and "forge reply," in which the attacker sends fake route reply messages in response to related legitimate route request messages.

In the forge reply attack, the attacker forges a Route Reply message after receiving a Route Request message. The reply message contains falsified routing information, showing that the node has a fresh route to the destination node on AODV in order to suppress the real routes to the destination. It causes route disruption by causing messages to be sent to a nonexistent node or putting the attacker itself into the route between the two endpoints of a communication channel if the insider attacker has already had a route to the destination. Figure 6.2 shows an example of a forge reply attack defined in [7]. The best route (with minimum hop) from node S to node D is S-I1-I2-D. Malicious node M forges an RREP message to the source node S through node I1. The message claims to come from the destination node D with higher destination sequence number to suppress the existing route. The faked message results in updating of the route entry to the destination node in the routing tables of node S and I1. Node I1 forwards data packets to the malicious node instead of node I2 since node M seems to have a fresh route to node D, so the new route becomes S-I1-M-I2-D.

Attackers can initiate frequent packets to cause DoS. Example DoS attacks that exploit MANETs' features are sleep deprivation torture attacks, routing table overflow attacks, *ad hoc* flooding attacks, rushing attacks, and the like. The *sleep deprivation torture* attack consumes a node's battery power

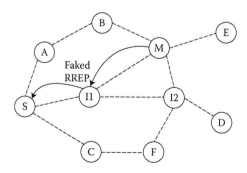

Figure 6.2 A forge reply attack.

and therefore disables the node. It does so by persistently making service requests of one form or another. This attack was introduced by Stajano et al. [3], who emphasized that it is more powerful than the better known DoS attacks such as CPU exhaustion, since most mobile nodes are run on battery power. The *ad hoc flooding attack*, introduced in [13], is another DoS attack against on-demand protocols, in which nodes send Route Request messages when they need a route. The attacker exploits this property of Route Discovery by broadcasting many Route Request messages to a node that is not in the network. Another attack at the Route Discovery phase is the *routing table overflow attack*. Here, the attacker sends a lot of route advertisements for nodes that do not exist. Since proactive protocols update routing information periodically before it is needed, this attack, which results in overflowing the victim nodes' routing tables and preventing new routes from being created, is more effective in proactive protocols than in reactive protocols [14].

Another interesting fabrication attack on MANETs is *the routing cache poisoning attack* [14]. A node can update its table with the routing information in the packets that it hears, even if it is not on the route of the packets. The attacker can make use of this property to poison the routes to a victim node by sending spoofed routing information packets, causing neighboring nodes to update their tables erroneously.

Timing attacks: An attacker attracts other nodes by causing itself to appear closer to those nodes than it really is. DoS attacks, rushing attacks, and hello flood attacks use this technique. *Rushing attacks* [15] occur during the Route Discovery phase. In all existing on-demand protocols, a node needing a route broadcasts Route Request messages and each node forwards only the first arriving Route Request in order to limit the overhead of message flooding. So, if the Route Request forwarded by the attacker arrives first at the destination, routes including the attacker will be discovered instead of valid routes. Rushing attacks can be carried out in many ways: by ignoring delays at message authentication code (MAC) or routing layers, by wormhole attacks, by keeping other nodes' transmission queues full, or by transmitting packets at a higher wireless transmission power [15]. The *hello flood attack* [16] is another attack that makes the adversary attractive for many routes. In some routing protocols, nodes broadcast Hello packets to detect neighboring nodes. These messages are received by all one-hop neighbor nodes, but are not forwarded to further nodes. The attacker broadcasts many Hello packets with large enough transmission power that each node receiving Hello packets assumes the adversary node to be its neighbor. It can be highly effective in both proactive and reactive MANET protocols.

A further significant attack on MANETs is the collaborative *wormhole attack*. Here an attacker receives packets at one point in the network, tunnels them to an attacker at another point in the network, and then replays them into the network from this final point [17]. Packets sent by tunneling forestall packets forwarded by multihop routes as shown in Figure 6.3 and it gives the attacker nodes

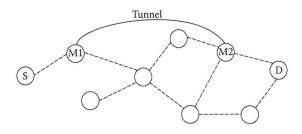

Figure 6.3   Wormhole attack.

an advantage for future attacks. Since the packets sent over tunneling are the same as the packets sent by normal nodes, it is generally difficult to detect wormhole attackers by software-only approaches such as IDS [17]. That is why packet leashes (any information that is added to a packet designed to restrict the packet's maximum allowed transmission distance [17]) have been introduced for preventing wormhole attacks.

6.4 Countermeasures

In general, we prevent compromise where we can (proactive solutions), but seek to detect and deal with it when prevention does not work (reactive solutions). In this section, we discuss proactive and reactive solutions proposed for MANETs.

6.4.1 Prevention Techniques: Secure Routing

Many of the attacks described above could be avoided by including authentication techniques in the routing protocol [18–20]. The main idea here is to guarantee that all nodes wishing to participate in the routing process are authenticated nodes; that is, trusted network elements that will behave according to the protocol rules. Authentication should be enforced during all routing phases, thus preventing unauthorized nodes (including attackers) from participating in the routing and so from launching routing attacks. Authentication can be provided based either on public-key or symmetric cryptography. In the former case, nodes issue digital signatures associated with the routing messages. Signatures can be verified by any other node, providing a secure proof of the identity of the sender. Digital evidence with similar properties can be constructed using secret-key cryptography, such as MACs.

The use of cryptography comes hand-in-hand with an associated problem: the necessity of a mechanism for issuing, exchanging, and revoking keys. Key management in MANETs is generally more difficult than in classical wired networks due to the absence of any infrastructure or central administrative authorities. There is no obvious point(s) where services such as certification authorities (CA) or key servers (KS) can be placed, and the great majority of the solutions proposed so far rely on schemes where the whole key management system is spread out to a subset of the mobile nodes.

Schemes proposed so far are mostly distributed key agreement protocols, such as the classical two-party Diffie–Hellman (DH) scheme [21]. Some works have extended the basic protocol towards n-party versions, in such a way that n-nodes can establish a common key for group communications [22]. Encrypted Key Exchange (EKE) protocols [23] have also been adopted in MANETs. These schemes were proposed with the goal of allowing two parties to generate a long-term common key from a shared password (typically of low entropy and therefore vulnerable to guessing attacks). A common feature of all these approaches (DH, general DH, EKE, etc.) is that some initial values must be shared by all nodes before the protocol can be used. This is generally known as the "bootstrapping" problem and it has received a fair amount of attention in recent years.

The development of public-key infrastructures especially tailored for MANETs has been a hot research topic over the last years. The majority of the solutions rely on a distributed CA based on threshold cryptography [24]. For example, in the scheme proposed in [25], a subset of nodes known as "servers" act collectively as a CA. Each public key belonging to a network node is divided into n shares and distributed among the n servers. A number $k < n$ of servers are required to sign a certificate. Each server generates its partial signature and collects the partial signatures generated by other servers. In global terms, the scheme is robust against any adversary who can compromise no more than $k - 1$ nodes. Mobile Certificate Authority [26] is a similar solution that incorporates a number of criteria such as physical location, computational characteristics, security measures deployed, and so on for choosing which nodes will act as servers.

6.4.2 Intrusion Detection

Since prevention techniques are limited in their effectiveness and new intrusions continually emerge, an IDS is an indispensable part of a security system. An IDS is introduced to detect possible violations of a security policy by monitoring system activities and responding to those that are apparently intrusive. If we detect an attack once it comes into the network, a response can be initiated to prevent or minimize the damage to the system. An IDS also provides information about intrusion techniques, enhancing our understanding of attacks and informing our decisions regarding prevention and mitigation.

Although there are many IDSs for wired networks, they do not find simple application to MANETs. Different characteristics of MANETs make conventional IDSs ineffective and inefficient for this environment. Consequently, researchers have been working recently on developing new IDSs for MANETs, or on modifying current IDSs to be applicable to MANETs.

Next, we give a summary of the different intrusion detection techniques proposed for MANETs. Attacks detected by each technique are identified too.

6.4.2.1 Specification-Based Intrusion Detection

One of the most commonly proposed intrusion detection techniques for MANETs is specification-based intrusion detection, where intrusions are detected as runtime violations of the specifications of routing protocols. This technique has been applied to a variety of routing protocols on MANETs such as AODV [27], optimized link state routing protocol (OLSR) [28,29], and DSR [30]. In [28], each network monitor employs a finite state machine (FSM) to state the specifications of AODV, especially for the route discovery process, and maintains a forwarding table for each monitored node. Each RREP and RREQ message in the range of the network monitor is monitored in a request–reply flow that checks the situations such as if route request packets are forwarded by next node or not, if route reply packets are modified on the path or not, and the like. When a network monitor needs information about previous messages or other nodes that are not in its range, it can ask neighboring network monitors.

In DEMEM [28], a distributed and cooperative IDS is described in which each node is monitored by 1-hop neighbor nodes for the OLSR routing protocol. In addition to 1-hop neighbor monitors, 2-hop neighbors can exchange data to have sufficient evidence about intrusions. The main contribution of DEMEM, as stated by the authors, is to introduce specific IDS messages to help detection.

Specification-based IDSs have generally been used to detect the modification and forge attacks. However, this technique cannot detect attacks that do not violate protocol specifications directly. (Various DoS attacks come in this category.) For that reason, Huang et al. [31] have proposed an IDS that uses a specification-based technique for attacks that violate the specifications of AODV directly and an anomaly-based technique for other kinds of attacks such as DoS. In [29], the authors propose adding a signature analysis tool in the future to detect the DoS attacks that cannot be detected by the extended FSM-based IDS for the OLSR routing protocol.

6.4.2.2 Anomaly-Based Intrusion Detection

This technique profiles the symptoms of normal behaviors of the system, such as usage frequency of commands, CPU usage for programs, and the like. It detects intrusions as anomalies (i.e., deviations from the normal behavior patterns). Various techniques have been applied for anomaly detection, for example, statistical approaches, and artificial intelligence techniques such as data mining and neural networks. The biggest challenge is defining normal behavior. Normal behavior can change over time and IDS systems need to adapt accordingly. That is one of the reasons for false positives—the normal activities that are detected as anomalies by IDS—can be high in anomaly-based detection. On the other hand, it is capable of detecting unknown attacks. This is important in an environment where new attacks and new vulnerabilities of systems are announced constantly.

The first proposed IDS for MANETs uses statistical, anomaly-based detection [2]. In that research, each node has an IDS agent responsible for local detection, and collaborates with neighboring nodes for global detection whenever the evidence is inconclusive and a broader search is needed. SVM Light and RIPPER classifiers are employed on three popular routing protocols (DSR, AODV, DSDV) and compared. The research focuses on two attack types: route logic compromise (e.g., misrouting) and traffic pattern distortion (e.g., dropping, modification, DoS attacks). This is also one of the few approaches to consider mobility data by monitoring node movements using built-in GPS functionality on each node.

Another proposed anomaly-based detection approach for MANETs [9] is zone-based IDS, where the network is divided into zones based on geographic partitioning. The nodes in a zone are grouped into intrazone and interzone nodes (which work as bridges to the other zones). Each node in a zone is responsible for local detection and sending alerts to interzone nodes which make the final decisions. They use a Markov-chain-based local anomaly detection model and evaluate it on route disruption attacks. Link change rate is used to reflect different mobility levels of the system.

Constructing an anomaly-detection model automatically by extracting the correlations among monitored features is proposed in [32]. They introduce simple rules to determine attack types and sometimes attackers. The rules are executed after an anomaly is detected. They are based on statistics such as the number of incoming/outgoing packets on the monitored node and are pre-computed for known attacks. For example, unconditional packet dropping of a node m is formulated as follows [32]:

$$\mathrm{FP}_m(\text{forward percentage}) = \frac{\text{packets actually forwarded}}{\text{packets to be forwarded}}.$$

Blackhole and dropping attacks are used to evaluate the performance of this approach. They observe that MANETs have strong feature correlations in normal behavior patterns. For instance, the correlation packet dropping and route entries updating (while packet dropping is drastically increasing on the network, there is an obvious change in routing updates) are highly correlated.

6.4.2.3 Misuse-Based Intrusion Detection

Misuse-based IDSs compare known attack signatures with current system activities. They are generally preferred by commercial IDSs since they are efficient and have a low false-positive rate. The drawback of this approach is that it cannot detect new attacks. The system is only as strong as its signature database and this needs frequent updating for new attacks.

There has been little research on signatures of new attacks against MANETs, so few misuse-based IDSs have been proposed so far. One of them is based on a stateful misuse detection technique and defines state transition programs for known attacks such as spoofing, dropping, and resource depletion attacks on AODV [33].

In [34], known attacks are formulated as cases for exact/similarity matching on the packet level. However, Snort rules are used as the cases instead of signatures of MANETs' specific attacks. In [35], another misuse-based IDS for MANETs is proposed with descriptions of two attack signatures using FSMs. One of the attacks is a network-level attack against the OLSR routing protocol and the other one is an application-level "stepping stone attack." They evaluated these attacks on a small network. Adding an anomaly detection module to broaden the spectrum of detected attacks has been suggested in both cases [34,35].

A different approach that creates signatures of some known attacks against MANETs automatically is given in [36]. In this research, intrusion detection rules to detect dropping, flooding, and route disruption attacks on AODV are evolved by using evolutionary computation techniques. The performance of evolved programs is demonstrated on simulated networks under varying mobility and traffic levels, and the results are quite promising. In addition, trade-offs between intrusion detection ability of evolved programs (rules) and their energy usage are identified leading to the creation of power-aware programs for such resource-constrained environment in their subsequent research [37].

6.4.2.4 Promiscuous Monitoring-Based Intrusion Detection

Since wireless nodes can overhear traffic in their communication range, promiscuous monitoring is a popular method used to detect misbehavior of nodes such as dropping and modification of packets on MANETs. However, this technique might not detect misbehaving nodes in the presence of ambiguous collisions, or receiver collisions.

The primary work on detecting misbehaving nodes and mitigating their performance effect proposed Watchdog and Pathrater mechanisms on the DSR [38]. Routing control packets in DSR carry all routing path information between nodes on the path. When a node forwards a packet, the Watchdog mechanism of that node monitors the next node to confirm that it also forwards the packet properly. When the number of dropping packets by a node exceeds a threshold, the node is considered as a misbehaving node and a notification is sent to the source node. With the Pathrater, the most reliable path is selected (instead of the shortest path as in DSR) in the presence of misbehaving nodes by using link reliability data and data from the Watchdog.

An approach that uses a reputation mechanism to respond to malicious nodes is given in [39]. Each node is responsible for monitoring the behavior of its next hop neighbors and detecting misbehaving nodes as in [38]. When a misbehaving behavior is detected, the reputation system is called to rate the misbehaving node. The system keeps a local rating list and/or blacklist, which can be exchanged with friend nodes. The rating of a node is based on the times of misbehavior occurrence as in [38]. The rate function also uses weights depending on the source detecting the misbehavior.

In [40], Parker et al. extend the method using promiscuous listening to detect misbehavior in a wide variety of routing protocols (not just DSR). A node listens to all nodes in its transmission range, not just the packets forwarded by the next node as in [38]. It detects dropping and modification attacks that exceed the threshold value in the threshold table for the particular attack class. However, a node moving out of range of the monitoring node before it forwards the packets should be assumed to be carrying out a dropping attack.

There are also a few cooperative approaches proposed to detect misbehaving nodes. In [41], every node counts the packets that it receives and forwards and periodically reports these counts to a coordinator node. Promiscuous monitoring is not used since it depends on the link layer characteristics and the link layer encryption approach [41].

Another approach for detecting dropping attacks on MANETs is presented in [42]. The algorithm only differentiates dropping attacks from the faults due to broken links. Malicious behavior is defined as the dropping of data packets starting at some random time and continuing from that time onwards. The idea behind the algorithm is based on associating the route error messages of the DSR routing protocol with TCP timeouts. In the DSR protocol, a route error control message is sent back to the source node if an intermediate node cannot forward the packet to the next hop. TCP timeout occurs when the sender does not receive an acknowledgement within a specific interval. All route error messages on a per flow basis are collected at the source node. When a TCP timeout occurs at this node, it is controlled if there are any route error messages for this flow within the detection interval or not. If there are, they are associated with a broken link, and otherwise with malicious dropping.

Communication between the IDS agents has also been provided by the use of mobile agents [34,35,43] besides promiscuous monitoring or by exchanging data directly between nodes in the literature.

6.5 Future Directions for Research

Given their flexibility, MANETs are very attractive for military and disaster recovery applications. Moreover, mobile devices are getting smaller, cheaper, more powerful, and more mobile every day. In the future, MANETs will likely be a part of our lives. There has been much research on this promising new networking. Security is one of the hot topics in the area due to new security threats that MANETs have introduced. The threats to MANETs have been examined in many research papers. However, more research needs to be done on identifying new security threats. We believe that with the increase in the use of MANETs, new intrusions are going to emerge continuously.

Since conventional security solutions are not easily applicable to MANETs, new solutions have been proposed for the last decade, which is far fewer than the proposed approaches for conventional networks. None of the proposed systems are necessarily the best solution taking into account different applications which they can have their own requirements and characteristics. They also usually consider few specific attacks and target a specific routing protocol. Furthermore, they emphasize just a few specific MANET features. For instance, the consequences of having limited resources are generally little explored. Some solutions might not be suitable for some nodes, which can have limited computational capabilities and resources. Researchers can develop solutions considering different characteristics of these nodes. Cooperation and communication between nodes is another area that needs to be explored. The proposed network architectures should not introduce new weakness/overheads to the system. To conclude, researchers should focus on developing solutions suitable to MANETs' specific features.

6.6 Conclusions

In this chapter, we have examined the main security issues in MANETs. They have most of the problems of wired networks and many more due to their specific features: dynamic topology, limited resources (e.g., bandwidth, power), lack of central management points. First, we have presented specific vulnerabilities of this new environment. Then, we have surveyed the attacks that exploit these vulnerabilities and the possible proactive and reactive solutions proposed in the literature. Attacks are classified into passive and active attacks at the top level. Since proposed routing protocols on MANETs are insecure, we have mainly focused on active routing attacks that are classified into dropping, modification, fabrication, and timing attacks. Attackers have also been discussed and examined under insider and outsider attackers. Insider attacks are examined on our exemplar routing protocol AODV.

Conventional security techniques are not directly applicable to MANETs due to their very nature. Researchers currently focus on developing new prevention, detection, and response mechanism for MANETs. In this chapter, we summarize secure routing approaches proposed for MANETs. The difficulty of key management on this distributed and cooperative environment is also discussed. Furthermore, we have surveyed IDSs with different detection techniques proposed in the literature. Each approach and technique is presented with attacks they can and cannot detect. To conclude, MANET security is a complex and challenging topic. To propose security solutions well suited to this new environment, we recommend researchers investigate the possible security risks to MANETs most thoroughly.

Terminologies

Intrusion—Any set of actions that attempt to compromise the integrity, confidentiality, or availability of a resource [44].

Intrusion detection system (IDS)—A system to detect the possible violations of a security policy by monitoring system activities and responding to those that are apparently intrusive.

Promiscuous monitoring—The monitoring all packets in a node's transmission range regardless of their destinations in wireless networks.

Denial of service—Attacks that aim to make computer/network resources unavailable to the intended users.

Routing attack—Attacks that seek to manipulate the operation of routing layer and aim to disrupt the routing mechanism of a network.

Mobile agent—Compositions of computer software and data that are able to migrate from one computer to another autonomously and continue its execution on the destination computer [45].

Authentication—The verification of claims about the identity of a source of information.

Cryptography—The study and practice of protecting information by data encoding and transformation techniques.

Key management—The process of generating, distributing, using, exchanging, and updating keys in a cryptography system design.

Questions and Sample Answers

1. Distinguish active and passive modes of attacks on MANETs, and give examples.

   In the passive mode, an attacker node just monitors and aims to find out information about the network. It does not cause any direct damage to the network. Passive attacks can be launched by insider or outsider attackers. Traffic analysis that aims to reveal the location of nodes is an example of a passive attack. On the other hand, in the active mode, attackers cause unauthorized state changes in the network, such as denial of service, modification of packets, and the like. Active attacks are generally launched by insider attackers who have authorization to operate within the network. For example, sink hole is an active attack where the attacker attracts traffic through a victim node by forging falsified routing packets into the network. He changes the operation of the routing mechanism.

2. What new forms of attack are possible in MANETs that do not occur in wired networks?
   The sleep deprivation torture attack is a DoS attack on MANETs. It aims to consume a node's battery power and effectively disables the node since mobile nodes generally run on battery power. The wormhole is another interesting attack where an attacker receives packets at one point in the network and tunnels them to an attacker at another point. There are also other attacks that exploit the vulnerabilities of routing protocols on MANETs. For example, a blackhole attack attracts traffic through itself by advertising falsified routing information.

3. What features of MANETs and MANET routing protocol operation make new attacks possible?
   The use of wireless links makes MANETs particularly susceptible to attacks such as eavesdropping and active interference. The cooperative nature of routing protocols allows an attacker to become a part of the routing mechanism easily and disrupt network communications by disobeying the protocol specifications. Furthermore, resource-constrained nodes can be the target of new attacks such as sleep deprivation torture.

4. For AODV write a concise description of dropping attacks and *ad hoc* flooding attacks.
   In dropping attacks, malicious or selfish nodes deliberately drop all packets which are not destined for them and aim at disrupting network communication. In *ad hoc* flooding attacks, the attacker floods the network with many route request packets and aims to consume the network's resources.

5. What countermeasures have been proposed for dropping attacks and *ad hoc* flooding attacks?
   In the literature, solutions using promiscuous monitoring techniques have usually been proposed to detect dropping attacks. When a node forwards a packet to its next node, it checks whether the next node (in the case it is not the destination node) also forwards the packet or not. There are other approaches such as using active acknowledgment from other nodes or associating route error packets with the lack of TCP acknowledgments. *Ad hoc* flooding is a DoS attack. Anomaly-based intrusion detection techniques are proposed to detect this attack. There is also an attempt to automatically discover intrusion detection rules for this attack. Defining the attack's signature manually is another approach.

6. Why do security solutions for MANETs usually prefer to have a distributed and collaborative approach?
   MANETs do not have any entry points such as routers, gateways, and so on that are typically present in wired networks and can be used to monitor all network traffic that pass through them. A MANET node can see only a portion of a network: the packets it sends or receives, possible together with other packets within its radio range. While some attacks can be detected locally by each node, detection of some attacks (such as network scans, distributed attacks) needs to obtain global data from other nodes in MANETs. For example, routing protocols are usually cooperative in MANETs and attacks against routing protocols can affect many nodes on the network. These attacks can be detected collaboratively by the affected nodes. Moreover, a local response to a malicious node may have very limited effect. A coordinated collaborative response will be much more effective.

7. What attacks on MANETs would not be detectable by autonomous systems running on individual nodes (i.e., with no collaboration)?
   Attacks that have a clear effect on a node can be detected easily by that node. However, some attacks (such as distributed attacks) can be detected only by analyzing the network data. For example, if intrusion detection is carried out locally on the network, network scan can seem

normal to each node. Detecting this attack will probably require distributed and collaborative intrusion detection on MANETs.

8. What is the main difficulty of adapting conventional prevention techniques to MANETs?
   Key management in MANETs is a challenging topic due to the absence of any infrastructure or central administrative authorities. There is no obvious point(s) where services such as CAs or KSs can be placed. So, the great majority of the solutions proposed for MANETs so far rely on schemes where the whole key management system is spread out to a subset of the mobile nodes.

9. What attacks on MANETs are detectable by promiscuous monitoring?
   Techniques that use promiscuous monitoring are usually proposed for detecting misbehaving nodes which carry out attacks such as dropping and misrouting attacks. In this way, the packets sent to a node are monitored to detect if this node forwards the packets properly or not.

10. Identify the attacker goals for selfishness and traffic analysis.
    The main motive for selfish behavior of a node is to preserve its resources. They avail themselves of the services of the other nodes, but do not reciprocate. By traffic analysis, attacker can reveal some information about the network such as the existence and location of nodes, the communications network topology, the roles played by nodes and the like. Then he can use this information to carry out further attacks.

Author's Biography

Sevil Şen is a PhD student at the Department of Computer Science at the University of York. She holds a MSc in computer engineering from Hacettepe University (2005). Her research interests are network security and artificial intelligence.

John A. Clark (MA, MSc, PhD) is professor of Critical Systems. After completing mathematics and then statistics studies at Oxford, he joined the security division of Logica in 1987, working there for five and a half years, first as an evaluator for secure systems and then as a security R&D consultant. He joined York in 1992 and has researched a range of topics including software testing, quantum algorithms and quantum information processing cryptography, security protocols, threat modeling, covert channels, and phishing. His PhD, on the application of metaheuristic search to cryptography, was awarded in 2002. Most recently he has addressed security of mobile *ad hoc* networks and intrusion detection. His major research foci concern the application of nonstandard computational techniques to software engineering and security. He is an author of over 130 research publications.

Juan E. Tapiador is postdoctoral research assistant in the Department of Computer Science at the University of York. He holds a MSc in computer science from the University of Granada (2000), where he obtained the Best Student Academic Award, and a PhD in Computer Science (2004) from the same university. His research interests are in computer security and cryptography.

References

1. Li, Y. and Wei, J. Guidelines on selecting intrusion detection methods in MANET, in *Proceedings of Information Systems Educators Conference*. 2004.

2. Zhang, Y. and Lee, W. Intrusion detection techniques for mobile wireless networks, *Wireless Networks*, 9(5), 545–556.

3. Stajano, F. and Anderson, R. The Resurrecting Duckling: Security issues for ad-hoc wireless networks, in *Proceedings of International Workshop on Security Protocols*, Springer: Berlin/Heidelberg, 1999.

4. Yang, H., Luo, H., Ye, F., Lu, S., and Zhang, L. Security in mobile ad hoc networks: Challenges and solutions, *IEEE Wireless Communications*, 11(1), 38–47, 2004.

5. Perkins, C., Belding-Royer, E., and Das, S. RFC 3561: Ad hoc On-Demand Distance Vector (AODV) Routing, http://www.ietf.org/rfc/rfc3561.txt, 2003.

6. Yau, P.-W. and Mitchell, C.J. Security vulnerabilities in ad hoc networks, in *Proceedings of the 7th International Symposium on Communications Theory and Applications*, pp. 99–104, 2003.

7. Ning, P. and Sun, K. How to misuse AODV: A case study of insider attacks against mobile ad-hoc routing protocols, in *Proceedings of the IEEE Workshop on Information Assurance*, West Point, NY, USA, pp. 60–67, 2003.

8. Chivers, H. and Clark, J.A. Smart dust, friend or foe?—Replacing identity with configuration trust, *Computer Networks*, 46(5), 723–740, 2004.

9. Sun, B., Wu, K., and Pooch, U.W. Zone-based intrusion detection for mobile ad hoc networks, *International Journal of Ad Hoc and Sensor Wireless Networks*, 2(3), 2003.

10. Hubaux, J.-P., Buttyan, L., and Capkun, S. The quest for security in mobile ad hoc networks, *In Proceedings of the 2nd ACM International Symposium on Mobile Ad hoc Networking & Computing*, Long Beach, California, USA, pp. 146–155, 2001.

11. Kong, J., Hong, X., and Gerla, M. A new set of passive routing attacks in mobile ad hoc networks, in IEEE MILCOM, Boston, MA, USA, 2003.

12. Buchegger, S., Tissieres, C., Le Boudec, J.-Y. A test-bed for misbehaviour detection in mobile ad-hoc networks – how much can watchdogs really do?, *Mobile Computing Systems and Applications (WMCSA '04)*, pp. 102–111, 2004.

13. Yi, P., Dai, Z., Zhang, S., and Zhong, Y. A new routing attack in mobile ad hoc networks, *International Journal of Information Technology*, 11(2), pp. 83–94, 2005.

14. Wu, B., Chen, J., Wu, J., and Cardei, M. A survey on attacks and countermeasures in mobile ad hoc networks, *Wireless/Mobile Network Security*, Chapter 5, Springer: US, 2006.

15. Hu, Y.-C., Perrig, A., and Johnson, D.B. Rushing attacks and defence in wireless ad hoc network routing protocols, in *Proceedings of the ACM Workshop on Wireless Security*, 2003.

16. Karlof, C. and Wagner, D. Secure routing in wireless sensor networks: Attacks and countermeasures, *Ad Hoc Networks*, 1(2–3), 293–315, 2003.

17. Hu, Y.-C., Perrig, A., and Johnson, D.B. Packet leashes: A defence against wormhole attacks in wireless ad hoc networks, in *Proceedings of INFOCOM*, San Francisco, USA, 2003.

18. Hu, Y.-C., Perrig, A., and Johnson, D.B. Ariadne: A secure on-demand routing protocol for ad hoc networks, in *Proceedings of the 8th International Conference on Mobile Computing and Networks*, pp. 12–23, 2002.

19. Sanzgiri, K., et al. A secure routing protocol for ad hoc networks, in *Proceedings of the 10th IEEE Conference on Network Protocols*, 2002.

20. Awerbuch, B., et al. An on demand secure routing protocol resilient to Byzantine failures, in *Proceedings of the ACM Workshop on Wireless Security*, Atlanta, GA, USA, 2002.

21. Diffie, W. and Hellman M. New directions in cryptography, *IEEE Transactions on Information Theory*, IT-22(6), 644–654, 1976.

22. Steiner, M., Tsudik, G., and Waidner, M. Diffie-Hellman key distribution extended to group communication, in *Proceedings of the ACM Conference on Computer and Communication Security*, New Delhi, India, pp. 31–37, 1996.

23. Bellovin, S.M. and Merritt, M. Encrypted key exchange: Password-based protocols secure against dictionary attacks, in *IEEE Symposium on Security and Privacy*, pp. 72–84, 1992.

24. Shamir, A. How to share a secret, *Communications of the ACM*, 22(11), 612–613, 1979.

25. Zhou, L. and Haas, Z.J. Securing ad hoc networks, *IEEE Network*, 13(6), 24–30, 1999.

26. Yi, S., Kravets, R. MOCA: Mobile certificate authority for wireless ad hoc networks, in *The 2nd Annual PKI Research Workshop*, 2003.

27. Tseng, C.-Y., Balasubramayan, P., Ko, C., Limprasittiporn, R., Rowe, J., and Levitt, K., A specification-based intrusion detection system for AODV, in *Proceedings of the ACM Workshop on Security in Ad Hoc and Sensor Networks*, 2003.

28. Tseng, C.H., Wang, S.H., Ko, C., and Levitt, K. DEMEM: Distributed evidence driven message exchange intrusion detection model for MANET, *RAID* 2006, *Lecture Notes in Computer Science* 4219, Springer: Berlin/Heidelberg, pp. 249–271, 2006.

29. Orset, J.-M., Alcalde, B., and Cavalli, A. An EFSM-based intrusion detection system for ad hoc networks, in *3rd International Symposium Automated Technology for Verification and Analysis, Lecture Notes in Computer Science* 3707, pp. 400–413, Springer: Berlin/Heidelberg, 2005.

30. Huang, Y. and Lee, W. Attack analysis and detection for ad hoc routing protocols, *RAID* 2004, *Lecture Notes in Computer Science* 3224, pp. 125–145, Springer: Berlin/Heidelberg, 2004.

31. Huang, Y., and Lee, W. A cooperative intrusion detection system for ad hoc networks, in *Proceedings of the 1st ACM Workshop on Security of Ad Hoc and Sensor Networks*, 2003.

32. Vigna, G., Gwalani, S., Srinivasan, K., Belding-Royer, E.M., and Kemmerer, R.A. An intrusion detection tool for aodv-based ad hoc wireless networks, in *Proceedings of the 20th Annual Computer Security Applications Conference*, pp. 16–27, IEEE Computer Society, 2004.

33. Guha, R., Kachirski, O., Schwartz, D.G., Stoecklin, S., and Yilmaz, E., Case-based agents for packet-level intrusion detection in ad hoc networks, in *Proceedings of 17th International Symposium on Computer & Information Sciences*, 2002.

34. Puttini, R.S., Percher, J.-Mr., Me, L., Camp, O., Sousa, Jr. R., Abbas, C.J.B., and Garcia-Villalba, L.J. A modular architecture for distributed IDS in MANET, in *Proceedings of the 2003 International Conference on Computational Science and Its Applications, Lecture Notes in Computer Science* 2669, pp. 91–113, Springer: Berlin/Heidelberg, 2003.

35. Sen, S. and Clark, J.A. A grammatical evolution approach to intrusion detection on mobile ad hoc networks, in *Proceedings of the 2nd ACM Conference on Wireless Network Security*, pp. 95–102, 2009.

36. Sen, S., Clark, J.A., and Tapiador, J.E. Power-aware intrusion detection on mobile ad hoc networks, in *Proceedings of the 1st International Conference on Ad hoc Networks*, 2009.

37. Marti, S., Giuli, T.J., Lai, K., and Baker, M. Mitigating routing misbehaviour in mobile ad hoc networks, in *Proceedings of ACM International Conference on Mobile Computing and Networking*, MOBICOM, pp. 255–265, 2000.

38. Buchegger, S. and Le Boudec, J. Nodes bearing grudges: Towards routing security, fairness, and robustness in mobile ad hoc network, in *Proceedings of Parallel, Distributed and Network-based Processing*, pp. 403–410, 2002.

39. Parker, J., Undercoffer, J., Pinkston, J., and Joshi, A. On intrusion detection and response for mobile ad hoc networks, in *Proceedings of 23rd IEEE International Performance Computing and Communications Conference*, 2004.

40. Anjum, F. and Talpade, R. LiPaD: Lightweight packet drop detection for ad hoc networks. in *Proceedings of IEEE Vehicular Technology Conference*, pp. 1233–1237, 2004.

41. Gavini, S. Detecting packet-dropping faults in mobile ad-hoc networks, Master of Science Thesis, School of Electrical Engineering and Computer Science, Washington State University, 2004.

42. Kachirski, O. and Guha, R. Effective intrusion detection using multiple sensors in wireless ad hoc networks, in *Proceedings of the 36th IEEE International Conference on System Sciences*, 2003.

43. Yi, P., Zhong, Y., and Zhang, S. A novel intrusion detection method for mobile ad hoc networks, in *Proceedings of Advances in Grid Computing* 2005, *Lecture Notes in Computer Science* 3470, pp. 1183–1192, Springer: Berlin/Heidelberg, 2005.

44. Denning, D. An intrusion-detection model, *IEEE Transactions on Software Engineering*, 13(2), 222–232, 1987.

45. Mobile Agent, Available from http://en.wikipedia.org/wiki/Mobile_agent, accessed 01 September 2009.

Chapter 7

Key Management in Mobile *Ad Hoc* Networks

Sudip Misra and Sumit Goswami

Contents

7.1 Introduction

Research advancements on Mobile Ad Hoc Networks (MANETs) have progressed notably in recent years, gradually perfecting itself toward the goal of providing anytime and anywhere networking services. In a MANET, the nodes are mobile and interconnected through wireless interface. They are self-organizing in nature, as they lack centralized routing, server, and administrative infrastructure [1]. Security issues are discussed with reference to services, attacks, and security mechanisms [2]. The services cater for the secure operation of the network, prevention against attacks and the tools and techniques to support the security services. Security mechanism is the process of providing secured services. The basic requirements for a secured networking environment are confidentiality, authentication, integrity, nonrepudiation, and availability. The security mechanisms used for protecting the transmitted data from attacks are encryption, cryptographic hash, and digital signatures. The security attacks can generally be distinguished into two types, as depicted in Figure 7.1.

- *Active attacks:* In active attack, the attacker has access to the transmission channel and the transmission technique, so that he can change the data or transmit his own data in a "camouflaged" manner. Denial of Service (DoS), modification, replay, and insertion are the common active attacks.
- *Passive attacks:* In a passive attack, the attacker can only listen to the network traffic or accumulate data from it, but the data are not altered. Eavesdropping, traffic analysis, and impersonation are general forms of passive attacks.

A MANET should provide a reliable and secure communication mechanism as nodes join or leave the network and their time of association with the network cannot be predicted. The data traffic

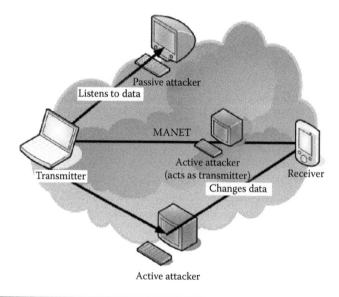

Figure 7.1 Active versus passive attacker.

in the *ad hoc* network travels through multiple hops routed through a vulnerable wireless medium, enhancing the security risk. The *ad hoc* networks generally provide similar security services and mechanism as deployed in wireless local area networks. However, the representative characteristics of an *ad hoc* network render them more vulnerable [3].

7.1.1 MANET: Introduction and Application

The properties of mobility, decentralized nature, infrastructure-less setup, and little or no requirement of planning and ease of deployment, as mentioned earlier, have attracted the attention of researchers. As mentioned earlier, a MANET is a self-organized wireless network with varying topology, characterized by its lack of infrastructure support such as access point or base station [4]. Owing to their spontaneous topological rearrangements, approaching mobile nodes in such networks are continuously associated while the retreating nodes are disassociated with each other. As shown in Figure 7.2, there is no centralized control and nodes cooperate to forward packets over multihop paths [5]. A multihop path is a network route in which a number of intermediate nodes forward the packet to its designated neighbor for communication between the sender and the receiver node. This helps in quick setup of a communication channel between the nodes, making it effective for missions related to defense or disaster management in unconnected and formidable terrain.

A MANET relies on collaborative operations of mobile nodes. As the neighbors of a node may change constantly, trust is an essential feature for cooperation between nodes. A trust establishment feature can enforce security, connectivity, cooperation, and Quality of Service (QoS) in the network. These characteristics of MANET compound the problem of designing a practical and efficient security mechanism. Security and mobility are odds with each other. Security is relatively simpler to enforce by a fixed central authority that is generally in charge of securing the system under consideration [6]. As the MANETs are dynamic and open, nodes leave or join the network at any time and communicate through the publicly accessible electromagnetic spectrum and, thus, open for eavesdropping or injection of fake packets by the adversary. In such scenarios, selfish and malicious nodes can easily creep into the network [7]. The multihop communication depends heavily on the cooperation among nodes and the selfish behavior of a few nodes can affect the speed and reliability of the communication. A node may desire to obtain the services from other nodes but may not participate in routing or selective flooding and thus break the communication link or increase the path length. Various algorithms which generate wave-like data transmission pattern where a receiver node communicates with other nodes in its vicinity to forward the information may fail or become

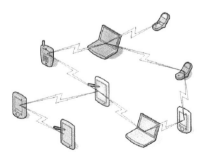

Figure 7.2 Typical architecture of a mobile *ad hoc* network.

less effective with such selfish nodes. It can also lead to the initiation of a polling algorithm to elect another leader if the leader turns out to be selfish even for a small amount of time or the selfish node disrupts communication path to the leader. Generally, leader election algorithms cost heavily on CPU usage, network traffic, and power consumption. Lack of authentication also leads to provision of free-of-charge services to malicious nodes.

The lack of infrastructure makes it difficult to apply conventional security mechanism developed for wired and other wireless networks to MANET directly. As MANET provides promising solution for many applications, security is an essential requirement of MANET. Among all security issues in MANET, key management is a core mechanism to ensure the security of applications and network services in the network. Key management scheme includes key distribution and key revocation.

- *Key distribution:* It is the task of distributing the secret keys to the nodes for secured communication. The keys can be distributed through some secured side channel or a trusted alternate network or delivered physically.
- *Key revocation:* It is the task of securely enlisting and removing the keys, which are known to be compromised. As some nodes of the MANET may be deployed in the hostile region, the security of communication to and from these nodes must be considered. In case the node is captured or compromised, the node should be immediately removed from the network with global knowledge in the MANET about it so as not to trust any communication from it.

Although there are plenty of applications of *ad hoc* networks, still there are plenty to come, as research outputs keep pouring in. The major applications of MANET can broadly be categorized into four categories:

- Defense application
- Rehabilitation and rescue support applications in disasters and calamities
- Civilian applications (environmental, transport, surveillance, etc.)
- Research applications

The core application of *ad hoc* network was, initially, in the military domain to enable mobile battle units to communicate among each other in enemy terrain, which lack friendly infrastructures. The network should be robust as well as amenable to reorganization in case a node is captured or destroyed by an enemy. Thereafter emerged the concept of a "Smart Soldier" whose body and battle gear was fitted with multiple sensors forming the personal area network and these entities communicate among each other and with base station for network-centric operations. An analogous civilian application is the formation of an *ad hoc* network between the personal digital assistant, mobile phone, digital camera, laptop, and so on carried by a person for wireless data transfer with each other or by rescue operators in a disaster affected city as shown in Figure 7.3. It also helps to form a collaborative network among a group of people using their laptops in an isolated place, without the requirement of any other external supporting resource. This property has enhanced the application of the *ad hoc* network in the sites of disasters and natural calamities where the communication network has been damaged. In such scenarios, an *ad hoc* network supports establishment of a communication network even in devastated places. In a disaster situation, it may be safer to send sensor probes in those places that are inaccessible by humans and, thus, act as a force multiplier for the rescue team.

Figure 7.3 Instant formation of *ad hoc* network in a city during disaster recovery operation.

7.1.2 Ad Hoc *Network Security*

MANETs are vulnerable to security attacks as the transmission takes place in the open medium. There is no centralized server, monitoring station, or administrator, and nodes keep on joining and leaving the network. There are communication overheads in detecting optimum routes with power saving and detection of malicious nodes or captured nodes. *Ad hoc* networks have to encrypt the backhaul communication, which represents the critical part of the *ad hoc* network infrastructure [8]. The wired networks are generally secured using firewalls and encryption devices and monitored using proxy, intrusion detection system, and routers. However, such legacy devices are not available within the MANET framework. The intrusion detection techniques used in the wired networks are not effective in the *ad hoc* network, as unknown nodes may join the network at a point of time. The algorithms for asymmetric cryptography have to be modified and customized to make it usable within the MANET. The Public Key Infrastructure (PKI) is difficult to implement with a MANET as they generally lack connectivity with Internet for authentication from the Certifying Authority (CA). The joining of new nodes at a remote location also makes it difficult to distribute its public key to the other nodes in MANET using a secured channel. The security algorithms for *ad hoc* networks are designed to reduce the computational requirement of the *ad hoc* nodes to save on the power. This is in addition to the prime requirement of keeping the dependency of the algorithm on a central node to the minimum. A centralized node is generally required in the security protocols for administrative as well as repository applications. The frequent topological changes in the network design, alive but hibernating nodes and compromised nodes makes it difficult to model the network for simulation and analysis of security protocols for MANET.

The routing service, which supports topological change, can be attacked due to its property of accepting changes. A compromised node or a malicious node, as shown in Figure 7.4, can camouflage itself to be a self-declared leader or server of the network by generating exceptionally high or low traffic and thus misleading the other nodes creating an inaccurate representation of the network and leading to a DoS attack. The density of nodes in an *ad hoc* network is another factor generally ignored in designing security mechanisms. There are a few critical factors such as transmission power level, number of channels used for each node, and the rate of movement of the node which requires to be taken into account while designing the security protocol [9–18]. The nature of attack on an *ad hoc* network for civilian application may be entirely different from the security attack on a tactical battlefield *ad hoc* network.

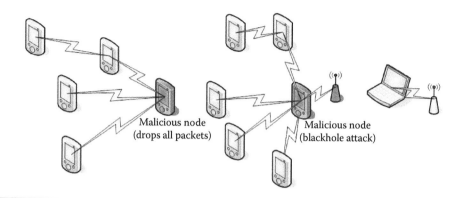

Figure 7.4 Malicious node effecting and *ad hoc* network.

7.1.3 Key Management

Most of the cryptographic mechanisms used in network security use some kind of cryptographic key that is shared between the receiver and the transmitter. The purpose of key management is to generate, distribute, install, update, revoke, and store the associated keys [19]. A trusted third party (TTP) is an entity trusted by the communicating parties to provide the key management service. The TTPs are distinguishable into three categories:

■ *In-Line:* The TTP is known as in-line if all the transmissions are first received by the TTP and then sent to the receiver by the TTP.
■ *Online:* The online TTP receives a copy of the transmitted messages or just a control signal mentioning about the communication being held between the parties, but the effective communication happens directly between the sender and the receiver.
■ *Off-Line:* In case of off-line TTP, it helps to setup the communication between the sender and receiver but is not active at the time of real communication and may not be even available in the network at the time of communication.

The public key cryptography ensures authenticity, confidentiality, integrity, and nonrepudiation. In PKI, if two users wish to communicate, they must exchange their public keys in an authentic manner and this requires the initial distribution of public keys. However, the private key is held only by its owner. It can be generated either by the TTP and transmitted through a secured channel to its owner or the TTP can transmit a seed to the node and the node generates the key using a particular algorithm and the seed received from the TTP. The CA is responsible for key distribution as well as revocation of the key in case it is compromised.

The key distribution involves the services of registration, initialization, certification, and key updation. Registration is the process of verifying the entity for its name, address, and/or email address followed by generation of the certificate by CA for the end user. Thereafter, the end user certificate is verified by the CA's certificate and a public–private key pair is generated for the user. The key has an initial date of validity, which may be updated in future based on its usage.

In PKI, the public keys as well as the Certificate Revocation List (CRL) are hosted by the CA. The list can be hosted on a central server of the CA or distributed across a number of servers for load sharing, redundancy, and faster access. As the list is meant for public viewing, it can also be hosted on a TTP server. The certificates are issued with a period of validity. However, a certificate may be revoked or suspended prior to its expiration. A list of such certificates is maintained in the CRL. The CRL is generally updated whenever a certificate is revoked or suspended. In some cases, the revocation list is updated only periodically with a small time frame for updating. Some CAs prefer updating the revocation list on every suspension or revocation as well as release a revocation list with a certain periodicity, for example, every week or every month. Thus, for effective usage of PKI, the updated CRL should be available with the user to validate each certificate that is not yet expired. The revocation list can be "pulled out" by the user from the central or the distributed server of the CA or a TTP server on a requirement basis. Alternately, the CA can push the CRL update to its users.

There are several reasons [20], which can call for key revocation [21] before a key's expiry. They vary from capture of the node carrying a certificate to the attrition of employees from the organization. Some common reasons for key revocation are described below:

- When the private key of the user or of the issuer (CA) is compromised or is suspected to be compromised.
- Some information in the certificate about the subject or any other information is no longer valid. It may happen that the domain name in the email id of a person or an organization has changed.
- The allotted purpose of the certificate for which it was issued has ceased to exist. For example, a certificate is issued for *e*-governance purpose for a particular policy such as earthquake relief, which is not further required after the cessation of rehabilitation work.
- The signature algorithm used by the CA is broken, in general, or the algorithm of the certified public key is compromised.
- A certificate, being part of the certification path, is revoked. The certificate of the sub-CA might be intact, but that of the CA was compromised, leading to revocation of all the certificates issued by all its sub-CAs.
- Loss or defect of security token, loss of password or personal identification number (PIN). The certificate might be stored in a smart card or USB token and either the owner of the certificate has lost its physical storage or storage is damaged. A password or a PIN protects the token from unauthorized access and that too can be lost or compromised.
- The certified key cannot be utilized any further for its allotted application. It is different from cessation of operation, in that, in cessations a key is no longer required, but in this, the usage changes such as from document signing to server authentication.
- It may happen that a CA does no longer work under its defined policy, for example, it ceases to support a service for certificates.

7.2 Background

PKI is the electronic equivalent of the real world business done using signatures, stamps, transaction logs, and contracts [22,23]. Keeping in view the necessity of secured and trusted electronic communication, various governments and related organizations have taken initiatives for implementation and augmentation of PKI. Rules had been framed to facilitate the enhancement and usage of PKI and create a sense of trust and reliability in electronic transactions. For example, in India, the regulations of PKI have been framed under the Information Technology (IT) Act, 2000. The IT Act 2000 provides the legal sanctity to the digital signatures. The IT Act facilitates the establishment of the Controller of Certifying Authorities (CCA) and bestows to it the responsibility of licensing and regulating the work of the CAs. A CA is a TTP that issues digital signatures to the entity after verifying its identity. The CA is responsible to attest the identity of an entity based on its association with a given key. The CA issues digital signature certificates for electronic authentication of users [24]. Similarly, many other countries have taken the PKI initiative. For example, the European Electronic Signature Standardization Initiative, the European Forum for Electronic Business, the Federal Bridge Certification Authority, the Gatekeeper—Commonwealth strategy for the use of PKI, the Government of Canada PKI, the Infocomm Development Authority Singapore, and the World Internet Security Company [25]. With the establishment of PKI and the generation of trust on it in various electronic transactions, the digital signatures are now accepted as an electronic equivalent of the handwritten signatures. The electronic documents that had been digitally signed are acceptable in lieu of paper documents in India and various other nations.

7.2.1 Security Issues in Wired versus Ad Hoc Network

In the wired networks, the transmission is through a guided medium, but still a number of cryptographic techniques are used to secure the channel from sniffing, traffic analysis, and eavesdropping. The nodes can be physically protected against capture and are generally detectable in case of a malicious attack. On the contrary, the *ad hoc* networks transmit in open and adversaries can listen to the transmission. The nodes in *ad hoc* networks are prone to capture or attacks, as they are mobile. Thus, the data storage and transmission framework in an *ad hoc* network should be resistant to failure of multiple nodes. The *ad hoc* networks also call for a high degree of fault tolerance as it operates in an unpredictable environment and prone to nonmalicious faults such as fading energy level of nodes, fluctuating signal strength with change in distance between nodes and varying transmission, and reception level due to a variety of electromagnetic interference. Wired networks have a continuous connectivity while the *ad hoc* networks suffer from topological variations with time and thus lead to disconnection and reconnection of nodes [26]. The security algorithm should also cater the scenarios of noncooperative node, which is rarely the scenario in a wired network. Thus, a key management framework for *ad hoc* network should support regular disruptions in connectivity, high resilience toward compromised nodes, sniffing, and mobility-associated issues in transmission.

7.2.2 Design Issues

PKI has not become extremely pervasive due to a few problems associated with it. It requires secured and fast management of the keys of the users. As a TTP has to be incorporated into the system, it involves cost for deploying the third-party mediator, thus making it an expensive solution. PKI can also be a security risk in case the private key is compromised. It does not have a separate optimized protocol for implementation over an *ad hoc* network. In case, the private key of an entity is lost or compromised before its expiry, revocation list is a feasible solution to communicate this information to its users. Regular seeking of revocation list leads to flooding, transaction delays, and power consumption.

PKI has an excellent future in MANET if the drawbacks are addressed properly. Digital signatures have been proposed as an effective solution for nonrepudiation in electronic transactions such as *e*-tendering and *e*-filing of income tax returns, and *e*-commerce solutions throughout the world. While using PKI, the following factors arose in our minds, which pull back the implementation of PKI in MANET:

1. It is not optimized for use in a mobile *ad hoc* network environment.

2. There is a time delay between the compromise of the key, its detection, and then finally its entry in the revocation list of the service provider.
3. The confidence of the receiver in the received digitally signed message assuming its authenticity and its negligence in checking for the validity and revocation of the key in the revocation list.
4. There is always some delay in getting revocation list from the service provider as it is constantly updated and is huge in size avoiding its download on a real-time basis.
5. The process of its distribution [27].

There are a plenty of security mechanisms such as transport layer security [28], pretty good privacy [29], terminal access controller access-control system [30], Kerberos [31], which are quite easy to use and are integrated with many existing software and websites. However, PKI is a security mechanism that involves time and money. It also involves extensive computation and, hence, is relatively slow. As the system is already computation intensive, necessary measures should be taken to reduce the communication overhead to make it suitable for use in MANET. The user has to pay for the key and can use it till he/she pays for it. When it is a paid service, the users can demand quality of service. Generally, there are centralized servers of each service provider at favorite data centers leading to the concentration of such servers at a few selective places and leading to extensive communication and computation overheads to the nodes in MANET. There are papers proposing distributed schemes or a CRL holding server in the wired networks as well as MANET [32,33]. It proposes a mechanism to suit the users who frequently change their network locations.

We posed five problems related to the implementation of PKI. The delay in obtaining revocation details in a MANET leads to unwillingness for CRL checks. There are many schemes published in the literature to overcome revocation delays, which we mainly categorized into push and pull mechanisms. When compared on the basis of quality of service parameters such as bandwidth consumption, reliability, and delay, if an algorithm is push based, it will consume more bandwidth in broadcasting, but will be more reliable, while the reverse happens in pull-based techniques. So as to improve on the time delay, the availability of the revocation list, and reduced requirement of communication power, we may either modify the process of revocation or reduce the communication time between the CRL servers and the nodes.

The key revocation techniques are generally analyzed [34] based on the following criteria [35]:

1. *Update cost:* This involves the cost for revocation of certificate. Once a key is revoked, the revocation list has to be updated. Various factors are considered to measure the updating cost such as the bandwidth requirement, the delay in reflection of the certificate in the revocation list, and the operations cost for the same.
2. *Query cost:* Every time a signed message is received, its validity has to be checked from the revocation list. This criterion measures the cost of certificate validity checking. It considers the factors such as bandwidth requirement, operation costs on a CA, and the receiver.
3. *Frequency of forwarding:* This criterion represents one of the security risk of forwarding the revocation list to the users through the network. This is considered for push-based systems. It indicates how frequently the revocation list should be accumulated before being distributed to the users.
4. *Scalability:* As the number of users increases, the network becomes more congested due to increase in the updates and queries. This criterion shows how a scheme performs in large networks and the change in its implementation cost in terms of communication overheads and computational complexity with the increase in the number of nodes in the network. It is measured as the ratio of increased costs over increased size of the communicating nodes.

5. *Implementation complexity:* On the basis of the distribution of servers, underlying algorithms, and revocation scheme, it can be classified as simple or complex.
6. *System risk:* This criterion measures the potential risk that the system can face. On the basis of the level of risk faced by a network, the TTP may be online, in-line or off-line. However, the high-risk networks call for the online TTP, but that it is a security risk for the online key revocation server as it is constantly exposed to the external network and vulnerable to attacks.

7.2.3 Key Management Challenges

PKI was put in place for the wired Internet infrastructure. With the advent of wireless network, in general, and MANET, more specifically, the same continued to be used for wireless without any change in the infrastructure or protocol. Many schemes had been proposed in PKI to reduce the delay in getting response from the server, distributing servers to enhance reliability, and load balancing and implementing TTP servers. Most of them, however, are not appropriate for incorporating in the MANET part of the network, as these solutions are not designed for them. The high mobility of nodes in *ad hoc* networks is also a matter of concern for authentication as well as connectivity, which is not so frequent in the wired world. In order to get a revocation list using the existing scheme, each node, on receipt of every signed message, has to individually contact the central or distributed CA server available on the wired network through the base station or the sink node to retrieve the CRL. Alternately, the CA server can push the CRL and its updates on every key revocation or suspension directly through the base stations to all the nodes in the MANET leading to flooding and other communication overheads. The existing scheme has no separate protocol, which has been optimized for implementation over an *ad hoc* network.

The revocation details are required to reach to the MANET nodes. The key server updates its CRL after receiving any revocation notice and no longer authenticates the revoked certificate and gives out the revocation information. The users of the certificate should be notified with a "push" or a "pull" system [36]. So as to implement a "push" system, a key server could broadcast its latest CRL or its updates periodically to a set of subscribers. The certificate holder has to initiate the "pull" approach by periodically checking from a particular key server from where it obtained the certificate, or before trusting a certificate or decrypting it. Any algorithm, which requires periodic revocation, updates, whether by a push or by a pull access, will lead to delay between the time the certificate is revoked and the time the users get information about it. Push schemes are bandwidth hungry and are bound to fail, if they have to broadcast an entire CRL update every time to a large number of subscribers. It can be tried to reduce the need for revocation by assigning certificates with brief expiration periods, but this approach leads to a system with plenty of overheads. It also calls for frequent distribution of replacement certificates.

7.3 Key Management in MANET

The CRL facilitates managing revoked certificates. CRLs were the initial method to revoke certificates. They used a concept of "black-lists" that indicated all currently valid, that is, nonexpired, but revoked certificates. The CA issued one CRL containing the list of all revoked certificates, thus making these lists quite large for a user to download. The problem is exacerbated, if revocation information needs to be verified online. To shorten the list size, the concept of segmented CRL [37] and delta-CRL, which is just a list of changes to the base CRL, was introduced. In this situation, a complete CRL is issued regularly, but not very frequently to keep the download requirement of the base CRL to the minimum but at the same time provide for a mechanism to keep the system updated between the release of two base CRLs. In between the issues of the base CRLs, delta-CRLs are issued that specify new revocations that have occurred since the release of the last base CRL [38]. The delta-CRL intimates about those certificates whose status has changed between the time the base CRL was issued and the time the delta-CRL was issued. The problem associated with the release of delta-CRLs is the window size which can be variable or constant at an optimized value. In general, if a client last obtained fresh certificate status information at time t and obtains a delta-CRL that references a base CRL that was issued at time $t^* \leqslant t$, then the receiver can use the delta-CRL to update its information without obtaining a new base CRL. The window size of the delta-CRL and the request rate for base CRL are inversely proportional and thus the window size of the delta-CRL has to be highly optimized.

CRL distribution points divide the entire list into logically organized smaller fragments. In order to know the revocation status of a key, only a particular fragment or its neighbor fragments can be downloaded and checked for revocation and thus saves the effort of downloading the entire base CRL. Each certificate has a distribution point associated with it which specifies the address where its revocation information should ideally be stored in case it is compromised. The segmented CRL does not affect the request rate for CRL updates but it reduces the size of the CRL download. It may also be specified that when a certain prespecified number of keys are revoked or the CRL reaches a threshold size, all the certificates are invalidated and reissued. An alternate approach is to associate each key with a predefined small life monitored on the basis of a clock and is revoked after the elapse of a small time frame to prevent its compromise. The time of reissuance of the certificates in such cases can be indicated in the "blacklist" CRL and the certificate issued after the mentioned time only is assumed to be valid.

7.3.1 Hierarchical Model

In the hierarchy-based model of key revocation, there is an intermediate server between the node and the revocation server of the CA. The verifier sends a query message to the intermediate server to obtain the revocation status. Verifiers do not concern about internal storage pattern used to manage certificates in revocation servers. Various types of data structures may be used in different repositories to store the revocation list in an effective manner providing an efficient search and append mechanism. Various techniques in this type of scheme have been reported. Key-Server for Key Distribution [39] introduces a two-level hierarchy consisting of the keyserver level and the enterprise level. The keyserver level contains a set of servers from which enterprise and keyserver certificates can be retrieved. The enterprise level independently maintains the hierarchies of end users [40]. The dual-directional hash chain [41] and hash binary tree [42] are two group-wise key distribution schemes. These are self-healing key distribution schemes effective for secured group communications in Wireless *ad hoc* networks [43,44], as shown in Figure 7.5. The scheme has a feature of periodic rekeying with implicit authentication. The forward and backward secrecy is ensured in the scheme [45] and it has a tolerance for lost rekeying message as it can be regenerated by obtaining the key from another key pair. Online Semi-Trusted Mediator [46] uses a security mediator (SEM) [47]. SEM issues a unique token to a user and only the token holder can sign or decrypt a message using his private key along with the token. Thus, the sender and the receiver must first obtain a message-specific token from the SEM before every usage of his/her private key. This simplifies and centralizes the revocation process as the SEM can be instructed by the administrator about a revoked key and SEM stops issuing tokens against any request from the user of the key or its receiver. This revokes the sender's capability of using the key to encrypt or sign the message and

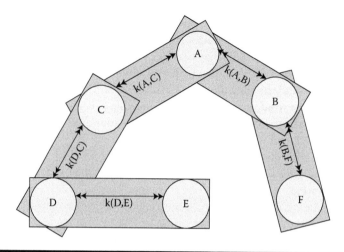

Figure 7.5 Tree-based organization of *ad hoc* nodes and the corresponding keys.

the receiver's capability to decrypt the message. A server also sometimes acts as a group controller in case of a group sharing a key [48].

In hierarchical schemes, the most common internal data structures used by CAs to store the revoked certificates are the sorted trees. The prime advantage of the tree data structure is that the search time is minimal. The idea of a certificate revocation tree is that revocation information is available in the leaves of a binary hash tree, and the root of this hash tree is signed by the CA. The authenticity of the directory information provided by the CA from its leaf node can be reconfirmed by recomputing the root value of the tree from the node of interest and the node values between the leaf node and the root. The calculated root value is affirmed from the signed root value and any alteration in the leaves will lead to alteration of the computed root value and its detection on comparison with the signed root [49].

7.3.2 Threshold Model

Threshold scheme was coined by Shamir [50] in 1979. The data are secured by encrypting it, but to keep the encrypting key secure is an important issue. The most secure way to keep it secure is to keep it at a well-guarded single location, either with central server, haven, or human brain. This has its own demerits attached with it as it is highly unreliable such as failure of the same will result in inability to recover the data. One of the mechanisms is to have the multiple copies of keys stored at different locations. The distribution of keys increases the risk of security breaches manifold. The mechanism, as proposed by Shamir, is that there are k keys and n distribution such that $n = (2k - 1)$. This (k, n) pair is called the threshold scheme. By using this (k, n) scheme, the robustness of the scheme increases as even floor $(n/2) = k - 1$ of the n pieces are destroyed, original key can be recovered. Even if the adversary knows or security breaches are exposed for floor $(n/2) = k - 1$ of the remaining k pieces, the key cannot be regenerated from those. In this scheme, the data are divided into n pieces in a way such that

1. Knowledge of any k or more data pieces makes data computable.
2. Knowledge of $k - 1$ or fewer data pieces gives the incomplete information about the data.

7.3.2.1 Threshold Public Key Management with Partially Distributed Authority

In MANET, the most prevalent Distributed Certificate Authority (DCA) [51] schemes are based upon threshold schemes. The threshold scheme is based upon the PKI in which the public key is known to all users. The TTP (network administrator) takes on the task of distributing the public key certificate *a priori*. It, then, selects the n CA servers on which the partial secret key is shared among them. For (k, n) threshold scheme, there are n servers which has a unique shared secret. Whenever a client wants to communicate a message with others, it sends the message to the servers, the servers with the shared secret generates the partial public key certificates for other nodes in the *ad hoc* mobile network. With the k partial signatures, DCA then computes the complete verifiable signature. This scheme has an advantage over traditional centralized CA scheme in that the nodes have no explicit relationship among each other except that they are part of shared secret key. One cannot deduct the key from others' share. Even if one of the servers is compromised, the attacker cannot compute the original complete certificate. The attacker has to have the information of at least k servers to compute the complete certificate.

The missing servers may also not pose much of the problem till the time the minimum number (threshold servers) of servers is available. This threshold number of servers then calculates the partial signatures, which, in turn, generates the complete certificate. The scheme is not clear about how to decide which n servers will be having the shared keys. The authenticity of the virtual CA is questionable. Further, it is a nonscalable solution. Identity and number of nodes participating in the network is predetermined. It also involves lot of computations and communications that nullify the very need of having this scheme.

7.3.2.2 Distributed CA Based on Threshold Scheme

Kong et al. [52] proposed a scheme based on the threshold scheme [53,54]. In this scheme, instead of having a node to have *a priori* knowledge of its public key, it can acquire it at the time of joining the network. In this case, a number of nodes form the centralized servers. When a new node joins a network and wants to acquire its public key, it gathers partial signatures from its k neighboring nodes. These k neighboring nodes, gives its partial signature which, in turn, is combined by the combinator. Once that is verified, the new joined nodes acquires its public key. In this scheme, the author assumes that all nodes in the network should self-generate their public key in the initialization phase with the help of neighboring nodes and trusted centralized server. Thus, the nodes can leave or join the network by exercising the above mechanism. In this scheme, the attacker can take many trusted signatures and can compromise the network.

7.3.2.3 Public Key Management Based on Identity Threshold

Deng et al. [55] proposed a scheme based on the identity and distributed key generation. Here, the author assumed that the nodes willing to join the network will acquire its identity such as the Internet protocol address through dynamic address allocation. It also assumes that it has the mechanism of acquiring the one hop neighboring nodes addresses and the identities of the other nodes in the network. The entire process of key generation and its usage is supported by four algorithms in the scheme to support the key setup, extraction, encryption, and decryption. The setup algorithm generates the master network public/private key pair. The nodes, thus, generate the public key, which is known to all the nodes and part of the private key is only known to each node. Identity-based authentication ensures availability, confidentiality, and nonrepudiation in addition to authentication between the communicating nodes across the network. The process of authentication is followed by the communication process between nodes. When a node joins the network, it requests for the private key. It acquires the identity during the initialization phase and after that it contacts the neighboring nodes and broadcast its identity to them. To initiate the secure communication, it generates a temporary public key and sends the request message for the private key to the neighboring nodes. The neighboring nodes then collaboratively compute the private key based on the arbitrary string of public key. The private key thus generated by the collaborative nodes is issued to the requesting node which had placed the request using its temporary public key.

7.3.3 Self-Organized Public Key Management

In this scheme, there is no centralized management of the key distribution and it does not require a trusted authority. In this scheme, the user can create, distribute, store, and revoke the keys without the help of the TTP. An externally independent, completely self-organized public key management was proposed by Capkun et al. [56]. Each node is supposed to manage the certificates on its own. The nodes keep the certificate repository with themselves. A node locally generates its own public key and the private key corresponding to this locally generated public key. The user forms the node of a directed graph and the edges form the certificate issued by the user for the nodes. A node gives the public key certificates about other nodes' public key using its own private key. The issuance of certificate is based on the trust relationship between the two nodes. Once the certificate is issued, the nodes perform the certificate exchange. The nodes form a connected graph based on the edges connecting the neighboring nodes. This graph helps the nodes in collecting the certificates from its neighbors. The certificate repository is updated from these collected certificates. The nonupdated certificate repository may also be updated by applying a repository construction algorithm. The authentication process follows by trying to find the directed graph to the destined node. If the path exists, then authentication succeeds, otherwise it fails. Certificates in this proposed scheme are time bound, and the certificate contains the issuance and expiration time.

7.3.4 Mobile Ad Hoc Key Revocation Server Scheme

A new scheme, named Mobile *Ad hoc* Key Revocation Server Scheme (MAKeRS) [57], was proposed to support the movement of *ad hoc* nodes or the receivers of the signatures across a MANET, so as to enable them to remain connected to the strongest server in terms of communication delay and overhead and also a way to ensure the availability of the *ad hoc* Key Revocation Server, even if the server to which it is presently attached goes out of the network for any possible reason. The scheme also supports the movement of MAKeRS in the MANET and provides a mechanism to self-organize the connectivity of the MAKeRS with the base station so as to remain connected to the CA which is accessible through the wired network. The scheme allows the nodes to perform instant and dynamic switchover to another *ad hoc* key revocation server with the Zone of Network Availability (ZoNA) stronger for this node. It supports the mobility of nodes very well.

The MAKeRS scheme is different from the hierarchical server scheme which is the existing scheme as, in the hierarchical scheme, the CRL is available at a central location, which is the server of the CA, and the CRL has to be obtained directly from the CA's server. In the web of trust model, there is no distinction between a CA and the end user [58,59]. In the threshold cryptographic scheme, the CA signing key K is split into n shares, such that the key K is recovered by essentially combining a certain threshold $k < n$. Any k components could combine and generate the threshold for a valid signature, but combination of any number of shares, lesser than the threshold value of k, are unable to do so [60]. A few researchers have even analyzed and simulated the MANET with a CA server within it in a centralized or distributed fashion. Both the web of trust model and the threshold cryptography cannot be used as such in integration with the existing PKI infrastructure.

In the MAKeRS scheme, it has different MAKeRS nodes which will provide CRL to the other nodes in a MANET. Here, the mobile *ad hoc* server, which is nearer to internet connectivity such as the base station, access point, or wired network, will be responsible for downloading CRL from the CA server through its Internet connectivity and this MAKeRS will be responsible for distributing the CRL to other MAKeRS. It refers to the particular MAKeRS which is connected to Internet at a particular point of time as the sink MAKeRS at that instant. A MAKeRS can be mobile, and in such a case, the "pref" field will be used to identify the nearest MAKeRS from an ordinary node. When any node requires a CRL, it will contact the nearest MAKeRS; this can be decided by the distance or number of hops between that node and the MAKeRS node. Knowledge of the available MAKeRS nodes is provided through periodic broadcast by MAKeRS with its ID.

7.3.4.1 Discovery of MAKeRS in a ZoNA

The MAKeRS sends a periodic MAKeRS Query Message (MQM), including the ZoNA options, which can be heard by the receiver node which has just entered the ZoNA, that is, the visiting receiver node. The receiver node, after receiving the MQM and the ZoNA options, performs the local binding update with the new ZoNA. If the list of IDs of MAKeRS is included in the MQM with the ZoNA option, then the new MAKeRSs are detected by the receiver node. The receiver node, then, selects one of the MAKeRS for the authentication of the key revocation. In this way, a considerable amount of time is saved for the authentication between the receiver node and the MAKeRS because the receiver node uses the MAKeRS of the same ZoNA instead of the MAKeRS in some other ZoNA. The receiver node, on moving to the new ZoNA, replaces the ID of the old MAKeRS with the new MAKeRS of the adopted ZoNA for the succeeding key revocation authorization. A detailed schematic diagram to represent this scheme, along with the placement of key revocation servers, is shown in Figure 7.6.

7.3.4.2 Message Format for Neighbor Discovery

There is a requirement of neighbor discovery [61] in the proposed mechanism. The process of neighbor discovery will detect all the available MAKeRS or the overlapping ZoNAs in the communication zone. The format of the MAKeRS option message is given in Figure 7.7, and the description of the fields are given in Table 7.1. When advertising more than one MAKeRS, as many existing MAKeRS options as possible are included in an MQM.

7.3.4.3 Server Discovery Algorithm

Input: MAKeRS option received through MQM.

Output: List of MAKeRS. The nearest MAKeRS/strongest MAKeRS is given a preference value of 15 and the nonavailable/moved-out MAKeRS is set to 0.

Step 1: Scan through all the available MAKeRS options received through MQM.

Step 2: Sort the list of available MAKeRSs in an ascending order of distance, starting with the nearest MAKeRS.

Step 3: Store the ascending list of available MAKeRS in the configuration for key revocation authentication.

Step 4: The receiver node uses this list stored in the configuration for key revocation for revocation authenticator.

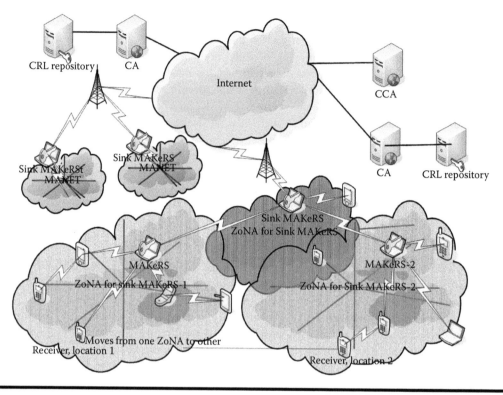

Figure 7.6 Internet is accessible in the MANET only through the MAKeRS nearest to Internet access.

Step 5: If the value of the "Pref" field is set to zero, exclude the MAKeRS entry from the list of MAKeRSs from the configuration for revocation authenticator. The value of zero indicates the failure of the MAKeRS or the large distance to the MAKeRS. This step has to be checked for each MAKeRS.

7.3.4.4 Protocols

The following notations are used for describing the protocols

src: source node
dst: destination node
E_pk: Encryption using public key
D_pr: Decryption using private key
H: Hash function
||: concatenated message

0	8	16	20	29........31
Type	Length	Pref	Distance	Reserved
ID of MAKeRS					

Figure 7.7 MAKeRS option message format.

Table 7.1 MAKeRS Option Message Fields

Field	Description
Type	Message type
Length	Length of the option field
Pref	Preference of a MAKeRS on a scale of 0–15. 15 indicates the highest preference. 0 indicates nonusability
ID of MAKeRS	MAKeRS's ID. It should be a Unique ID
Distance	Number of Hops between MAKeRS and receiver node

Protocol 1: Communication between MAKeRS

```
Makers_src → MAKeRS_dst : E_pk_dst_MAKeRS (send update) // Ask for
update of CRL
MAKeRS_dst → MAKeRS_src : E_pk_src_makers (Update) //dst sends updated CRL
MAKeRS_src : D_pr_src_MAKeRS(E_pr_src_MAKeRS (Update)) // extract update
information
Special Messages like MAKeRS change or MAKeRS failure:
MAKeRS_src→MAKeRS_dst:E_pk_dst_MAKeRS (special message) ||E_pr_src_MAKeRS
(H(special message))
```

Protocol 2: Node to MAKeRS communication for getting CRL list

```
Node → MAKeRS : msg (Request_CRL || MAKeRS ID)   //sends message for getting
                                       //CRL list with MAKeRS ID.
MAKeRS → NODE : E_pk_node(CRL_List) || E_pr_MAKeRS(H(CRL_LIST))
                // this provides list plus authenticate message checking
Node: D_pr_node(CRL_List),
   If H(CRL_LIST) = received H(CRL_List)
   then
      accept
   else
      reject.
```

Protocol 3: Communication between Sink MAKeRS and Server

```
Sink_MAKeRS → Server : E_pk_server (send CRL_List)    //asks for crl list
                                            from server
Server → Sink_MAKeRS : E_pk_sink_MAKeRS(CRL_List) || E_pr_Server
(H(CRL_LIST))
                             //server sends the list to MAKeRS sink node
Sink_MAKeRS : D_pr_Sink_MAKeRS(E_pk_sink_MAKeRS(CRL_List)),
   If H(CRL_LIST) = received H(CRL_List)
   then
      accept
   else
      reject
```

Protocol 4: Communication between nodes

```
Node_src → Node_dst : E_pk_dst_node(message || time stamp) //send message
                                            with time stamp
```

```
Node_dst : D_pr_dst_node (E_pk_dst_node(message || time stamp)
// Destination node gets the message and time stamp. Node records/stores
   the last time stamp so that replayed of messages are discarded/removed.
```

7.4 Future Directions for Research

There has been limited research on key management schemes with consideration for channel capacity and utilization. Various schemes are generally compared based on the CPU utilization and power consumption. However, with the current state of the art, communication capacity is turning to be the bottleneck, rather than the energy consumption or computational power. The number of messages can be equally hampering the performance of the key management algorithm as the size of the message. The future key management schemes should optimally combine the features of bandwidth efficiency, robustness against link failures, and power consumption [62,63].

Most of the key management solutions described in the research papers are theoretical in nature and their efficiency has been proved mathematically or by simulations. Emphasis must be given in implementing these solutions, as there might be various limitations in real world in implementing these solutions. The storage space in the node of a particular MANET such as those in a small sensor may not be even enough to store a key [64]. Research in the field of enhancing battery life, compact batteries with more power, processors with high computation speed and low energy consumption is also demanding. In addition to detection of malicious nodes, it would be highly beneficial to track the malicious node, trace its communication pattern, and ability to detect it on just joining the network. There may be situations when research is initiated to detect malicious node based just on its presence in a MANET's area of communication even if the malicious node is dormant and not communicating at all. Key management algorithms should evolve in such a way that a node leaving a network should not be able to carry a key with itself outside the network, get captured or intentionally pass the key to an adversary which in turn joins back the *ad hoc* network impersonating the friendly node. Process should be evolved to generate a key on the fly based on certain environmental and network parameters at a particular time when a node joins a networks and its destruction in case a node leaves the network. Future research should especially seek techniques for gaining deployment knowledge from transmission pattern as it will support predicting the topological changes and model the behavior of association and dissociation of the nodes. Research in the field of hardware as well as algorithms is also required to make the mobile nodes tamper-proof with least overhead. As the network architecture is dynamic, leading to routing and security issues, decrease in the bootstrapping time required for the network will be a significant contribution to MANET. A scheme should also evolve which provides assured node identification.

7.5 Conclusions

Key management for MANET is a critical issue that has been discussed and solutions to it have been proposed in various papers. Each of the proposed schemes has its own advantage, disadvantage, limitation, operating criteria, design issues, and application. A single key management scheme cannot be marked as better than any other scheme as it cannot fit into various scenarios. The key management schemes are generally dependent on the application scenario for which it is designed. A balance between the usage and the available resources of power and computation determines which key management scheme should be deployed. An *ad hoc* network in a battlefield calls for more security than one used in a group communication used in a classroom; the former may be costly but the later needs to be user-friendly and cheaper. MANET has led to the emergence of many new similar fields of application and research like the Wireless Sensor Network (WSN), Vehicular *Ad hoc* Networks (VANET) in which the key management techniques used in MANETs have to be optimized for usage in these newly emerging networks.

Questions and Sample Answers

1. What are the main differences between a wired network and MANET?

   In a MANET, the nodes are mobile and interconnected through wireless interface. They are self-organizing in nature, as they lack centralized routing, server, and administrative infrastructure. A MANET should provide a reliable and secure communication mechanism as nodes join or leave the network and their time of association with the network cannot be predicted. The data traffic in the *ad hoc* network travels through multiple hops routed through a vulnerable wireless medium, enhancing the security risk. The *ad hoc* networks generally provide similar security services and mechanism as deployed in wireless personal area networks. However, the representative characteristics of an *ad hoc* network render them more vulnerable.

   MANETS have the properties of mobility, decentralized nature, infrastructure-less setup, and little or no requirement of planning and ease of deployment. A MANET is a self-organized wireless network with varying topology, characterized by its lack of infrastructure support such as access point or base station. Due to their spontaneous topological rearrangements, approaching mobile nodes in such networks are continuously associated while the retreating nodes are disassociated with each other. There is no centralized control and nodes cooperate to forward packets over multihop paths. A multihop path is a network route in which a number of intermediate nodes forward the packet to its designated neighbor for communication between the sender and the receiver node. This helps in quick setup of a communication channel between the nodes, making it effective for missions related to defense or disaster management in unconnected and formidable terrain. A MANET relies on collaborative operations of mobile nodes.

2. How does *ad hoc* network security differ from that in a wired network?

   In the wired networks, the transmission is through a guided medium, but still a number of cryptographic techniques are used to secure the channel from sniffing, traffic analysis, and eavesdropping. The nodes can be physically protected against capture and are generally detectable in case of a malicious attack. Contrary to this, the *ad hoc* networks transmit in open and adversaries can listen to the transmission. The nodes in *ad hoc* networks are prone to capture or attacks, as they are mobile. Thus, the data storage and transmission framework in an *ad hoc* network should be resistant to failure of multiple nodes. The *ad hoc* networks also call for a high degree of fault tolerance as it operates in an unpredictable environment and prone to nonmalicious faults like fading energy level of nodes, fluctuating signal strength with change in distance between nodes and varying transmission and reception level due to a variety of electromagnetic interference. Wired networks have a continuous connectivity, whereas the *ad hoc* networks suffer from topological variations with time and thus lead to disconnection and reconnection of nodes. The security algorithm should also cater the scenarios of noncooperative node, which is rarely the scenario in a wired network. Thus, a key management framework for *ad hoc* network should support regular disruptions in connectivity, high resilience toward compromised nodes, sniffing, and mobility-associated issues in transmission.

3. What are the main application scenarios of MANET?

   There are plenty of applications of *ad hoc* networks. The major applications of MANET can be broadly categorized into four categories:

   ■ Defense application
   ■ Rehabilitation and rescue support applications in disasters and calamities
   ■ Civilian applications (environmental, transport, surveillance, etc.)
   ■ Research applications

   The core application of *ad hoc* network was, initially, in the military domain to enable mobile battle units to communicate among each other in enemy terrain, which lack friendly infrastructures. Thereafter emerged the concept of a "Smart Soldier" whose body and battle gear was fitted with multiple sensors forming the personal area network and these entities communicate among each other and with base station for network-centric operations. An analogous civilian application is the formation of an *ad hoc* network between the personal digital assistant, mobile phone, digital camera, laptop, and so on carried by a person for wireless data transfer with each other or by rescue operators in a disaster-affected city. It also helps to form a collaborative network among a group of people using their laptops in an isolated place, without the requirement of any other external supporting resource. This property has enhanced the application of the *ad hoc* network in the sites of disasters and natural calamities where the communication network has been damaged. In such scenarios, an *ad hoc* network supports establishment of a communication network even in devastated places. In a disaster situation, it may be safer to send sensor probes in those places, which are inaccessible by humans and, thus, act as a force multiplier for the rescue team.

4. Why the security protocols of wired network cannot be directly implemented to MANET?

   The lack of infrastructure makes it difficult to apply conventional security mechanism developed for wired and other wireless networks to MANET directly. The wired networks are generally secured using firewalls and encryption devices and monitored using proxy, intrusion detection system, and routers. However, such legacy devices are not available within the MANET framework. The intrusion detection techniques used in the wired networks are not effective in the *ad hoc* network, as unknown nodes may join the network at a point of time. The algorithms for asymmetric cryptography have to be modified and customized to make it usable within MANET. The public key infrastructure is difficult to implement with MANET as they generally lack connectivity with Internet for authentication from the Certifying Authority. The joining of new nodes at a remote location also makes it difficult to distribute its public key to the other nodes in MANET using a secured channel. The security algorithms for *ad hoc* networks are designed to reduce the computational requirement of the *ad hoc* nodes to save on the power. This is in addition to the prime requirement of keeping the dependency of the algorithm on a central node to the minimum. A centralized node is generally required in the security protocols for administrative as well as repository applications. The frequent topological changes in the network design, alive but hibernating nodes and compromised nodes, make it difficult to model the network for simulation and analysis of security protocols for MANET.

   The routing service, which supports topological change, can be attacked due to its property of accepting changes. A compromised node or a malicious node can camouflage itself to be a self-declared leader or server of the network by generating exceptionally high or low traffic and thus misleading the other nodes creating an inaccurate representation of the network and leading to a denial of service attack. The density of nodes in an *ad hoc* network is another factor generally ignored in designing security mechanisms. There are a few critical factors such as transmission power level, number of channels used for each node, and the rate of movement of the node which requires to be taken into account while designing the security protocol. The nature of attack on an *ad hoc* network for civilian application may be entirely different from the security attack on a tactical battlefield *ad hoc* network.

5. State at least five reasons for key revocation?
   Some common reasons for key revocation are described below:
   - When the private key of the user or of the issuer (Certifying Authority [CA]) is compromised or is suspected to be compromised.
   - Some information in the certificate about the subject or any other information is no longer valid. It may happen that the domain name in the email id of a person or an organization has changed.
   - The allotted purpose of the certificate for which it was issued has ceased to exist. For example, a certificate is issued for *e*-governance purpose for a particular policy such as earthquake relief, which is not further required after the cessation of rehabilitation work.
   - The signature algorithm used by the CA is broken, in general, or the algorithm of the certified public key is compromised.
   - A certificate, being part of the certification path, is revoked. The certificate of the sub-CA might be intact, but that of the CA was compromised, leading to revocation of all the certificates issued by all its sub-CAs.
   - Loss or defect of security token, loss of password or personal identification number (PIN). The certificate might be stored in a smart card or USB token and either the owner of the certificate has lost its physical storage or storage is damaged. A password or a PIN protects the token from unauthorized access and that too can be lost or compromised.
   - The certified key cannot be utilized any further for its allotted application. It is different from cessation of operation, in that, in cessations a key is no longer required, but in this, the usage changes such as from document signing to server authentication.
   - It may happen that a CA does no longer work under its defined policy, for example, it ceases to support a service for certificates.

6. What are the advantages of using delta-CRL over CRL?
   To shorten the CRL size, the concept of delta-CRL, which is just a list of changes to the base CRL, was introduced. In this situation, a complete CRL is issued regularly, but not very frequently to keep the download requirement of the base CRL to the minimum but at the same time provide for a mechanism to keep the system updated between the release of two base CRLs. In between the issues of the base CRLs, delta-CRLs are issued that specify new revocations that have occurred since the release of the last base CRL. The delta-CRL intimates about those certificates whose status has changed between the time the base CRL was issued and the time the delta-CRL was issued. The problem associated with the release of delta-CRLs is the window size which can be variable or constant at an optimized value. In general, if a client last obtained fresh certificate status information at time t and obtains a delta-CRL that references a base CRL that was issued at time $t^* \leqslant t$, then the receiver can use the delta-CRL to update its information without obtaining a new base CRL. The window size of the delta-CRL and the request rate for base CRL are inversely proportional and thus the window size of the delta-CRL has to be highly optimized.

7. What are the major criteria for analysis of key revocation techniques?
   The key revocation techniques are generally analyzed based on the following criteria:
   - *Update cost:* This involves the cost for revocation of certificate. Once a key is revoked, the revocation list has to be updated. Various factors are considered to measure the updating

cost such as the bandwidth requirement, the delay in reflection of the certificate in the revocation list, and the operations cost for the same.

■ *Query cost:* Every time a signed message is received, its validity has to be checked from the revocation list. This criterion measures the cost of certificate validity checking. It considers the factors such as bandwidth requirement, operation costs on a CA and the receiver.

■ *Frequency of forwarding:* This criterion represents one of the security risk of forwarding the revocation list to the users through the network. This is considered for push-based systems. It indicates how frequently the revocation list should be accumulated before being distributed to the users.

■ *Scalability:* As the number of users increases, the network becomes more congested due to increase in the updates and queries. This criterion shows how a scheme performs in large networks and the change in its implementation cost in terms of communication overheads and computational complexity with the increase in the number of nodes in the network. It is measured as the ratio of increased costs over increased size of the communicating nodes.

■ *Implementation complexity:* On the basis of the distribution of servers, underlying algorithms, and revocation scheme, it can be classified as simple or complex.

■ *System risk:* This criterion measures the potential risk that the system can face. On the basis of the level of risk faced by a network, the trusted third party may be online, in-line, or off-line. However, the high-risk networks call for the online–trusted third party but that is a security risk for the online key revocation server as it is constantly exposed to the external network and vulnerable to attacks.

8. Differentiate between the push and pull mechanisms of CRL update?

The key server updates its CRL after receiving any revocation notice and no longer authenticates the revoked certificate and gives out the revocation information. The users of the certificate should be notified with a "push" or a "pull" system. So as to implement a "push" system, a key server could broadcast its latest CRL or its updates periodically to a set of subscribers. The certificate holder has to initiate the "pull" approach by periodically checking from a particular key server from where it obtained the certificate, or before trusting a certificate or decrypting it. Any algorithm, which requires periodic revocation, updates, whether by a push or by a pull access, will lead to delay between the time the certificate is revoked and the time the users get information about it. Push schemes are bandwidth hungry and are bound to fail, if they have to broadcast an entire CRL update every time to a large number of subscribers. It can be tried to reduce the need for revocation by assigning certificates with brief expiration periods, but this approach leads to a system with plenty of overheads. It also calls for frequent distribution of replacement certificates.

9. Explain the threshold model of key distribution?

The data are secured by encrypting it, but to keep the encrypting key secure is an important issue. The most secure way to keep it secure is to keep it at a well-guarded single location, either with central server, haven, or human brain. This has its own demerits attached with it as it is highly unreliable such as failure of the same will result in inability to recover the data. One of the mechanisms is to have the multiple copies of keys stored at different locations. The distribution of keys increases the risk of security breaches manifold. The mechanism, as proposed by Shamir, is that there are k keys and n distribution such that $n = (2k - 1)$. This (k, n) pair is called the threshold scheme. By using this (k, n) scheme, the robustness of the scheme increases as even floor$(n/2) = k - 1$ of the n pieces are destroyed, original key can be recovered. Even if the adversary knows or security breaches are exposed for floor $(n/2) = k - 1$ of the remaining k pieces, the key cannot be regenerated from those. In this scheme, data are divided into *n* pieces in a way such that

a. Knowledge of any *k* or more data pieces makes data computable.

b. Knowledge of *k* − 1 or fewer data pieces gives the incomplete information about the data.

10. Discuss the advantages and limitations of a distributed key server?

Distributed key server scheme has an advantage over traditional centralized CA scheme in that the nodes have no explicit relationship among each other except that they are part of shared secret key. One cannot deduct the key from others' share. Even if one of the servers is compromised, the attacker cannot compute the original complete certificate. The attacker has to have the information of at least *k* servers to compute the complete certificate.

The missing servers may also not pose much of the problem till the time the minimum number (threshold servers) of servers is available. This threshold number of servers then calculates the partial signatures which, in turn, generate the complete certificate. The scheme is not clear about how to decide which *n* servers will be having the shared keys. The authenticity of the virtual certifying authority is questionable. Further, it is a nonscalable solution. Identity and number of nodes participating in the network is predetermined. It also involves lot of computations and communications which nullify the very need of having this scheme.

Author's Biography

Dr. Sudip Misra is an assistant professor in the School of Information Technology at the Indian Institute of Technology Kharagpur, India. He held academic positions at Cornell University, Ithaca, New York; Yale University, New Haven, Connecticut; and Ryerson University, Canada. He also worked at Nortel Networks (Canada), Atreus Systems Corporation (Canada), ChartWell, Inc. (Canada), and the Government of Ontario (Canada). He obtained his PhD degree in computer science from Carleton University, Ottawa, Canada. His current research interests include algorithm design and engineering for telecommunication networks.

Dr. Misra is the author/editor of over 90 scholarly research papers and books. He has won five research paper awards in different conferences. He was also the recipient of several academic awards and fellowships such as the (Canadian) Governor General's Academic Gold Medal at Carleton University, the University Outstanding Graduate Student Award in the Doctoral level at Carleton University. In 2008, he was conferred the National Academy of Sciences, India—Swarna Jayanti Puraskar (Golden Jubilee Award). Dr. Misra is the editor-in-chief of two international journals and is an associate editor/editor/editorial board member of around a dozen others published by reputed publishers such as Wiley, Springer, Elsevier, and IET. Dr. Misra is an editor of six books in the areas of wireless *ad hoc* networks, wireless sensor networks, wireless mesh networks, communication networks and distributed systems, network reliability and fault tolerance, and information and coding theory, published by reputed publishers such as Springer and World Scientific. Dr. Misra was also invited to deliver keynote lectures in over a dozen international conferences in the United States, Canada, Europe, Asia, and Africa.

Sumit Goswami holds an MTech degree in computer science and engineering from Indian Institute of Technology Kharagpur, Kharagpur, India, a post graduate diploma in journalism and mass communication and a bachelor's degree in library and information science. He has been working as a scientist with Defence Research and Development Organization (DRDO), Delhi, India, since 2000. His areas of interest include network-centric operations, mobile *ad hoc* and sensor networks, web-hosting security, text mining, and machine learning. He has published 41 papers/chapters in various journals, books, conferences, and seminars. He also chaired a session on the theme

"Computer Architecture" in National Conference on "Emerging Principle and Practice of Computer Science" held at GND Engineering College, Ludhiana, India, in August 2006.

References

1. K. Fokine, Key management in *ad hoc* networks, Available at http://csis.bits-pilani.ac.in/faculty/sundarb/courses/old/spr06/netsec/evals/seminar/readings/refs/adhoc-keymgmt.pdf (accessed on: August 3, 2009).
2. J. Dong, K. Ackermann, and C. Nita-Rotaru, Secure group communication in wireless mesh networks, in: *Proceedings of Ninth IEEE International Symposium on a World of Wireless, Mobile and Multimedia Networks (WOWMOM)*, Newport Beach, CA, 2008, pp. 1–8.
3. W. Stallings, *Cryptography and Network Security: Principles and Practice*, 2nd ed., Prentice-Hall, New Jersey, 1999.
4. C. Davis, A localized trust management scheme for *ad hoc* networks, in *Proceedings of the 3rd International Conference on Networking (ICN'04)*, Guadeloupe, French Caribbean, 2004, pp. 671–675.
5. Y. Dong, H.W. Go, A.F. Sui, V.O. K. Li, L.C.K. Hui, and S.M. Yiu, Providing distributed certificate authority service in mobile *ad hoc* network, in *Proceedings of the First International Conference on Security and Privacy for Emerging Areas in Communications Networks (SECURECOMM)*, Athens, Greece, Sep. 2005, pp. 149–156.
6. S. Capkun, J.-P. Hubaux, and L. Buttyán, Mobility helps security in *ad hoc* networks, in *Proceedings of the 4th ACM International Symposium on Mobile Ad Hoc Networking & Computing*, Annapolis, Maryland, 2003, pp. 46–56.
7. G.F. Marias, K. Papapanagiotou, V. Tsetsos, O. Sekkas, and P. Georgiadis, Integrating a trust framework with a distributed certificate validation scheme for MANETs, *EURASIP Journal on Wireless Communications and Networking*, 2, 1–18, 2006.
8. D.H. Axner, The up side and down side of wireless mesh networks, Available at www.Packethop.com/pdf/business_communications review_january_2006.pdf.
9. P. Krishnamurthy, D. Tipper, and Y. Qian, The interaction of security and survivability in hybrid wireless networks, *WIA 2004, Proceedings of IEEE IPCCC 2004*, Phoenix, AZ, April 14–17, 2004.
10. S. Misra, K.I. Abraham, M.S. Obaidat, and P.V. Krishna, LAID: A learning automata-based scheme for intrusion detection in wireless sensor networks, *Security and Communication Networks*, 2(2), 105–115, 2009.
11. S. Misra, S.K. Dhurandher, A. Rayankula, and D. Agrawal, Using honeynodes for defense against jamming attacks in wireless infrastructure-based networks, *Computers and Electrical Engineering*, 36(2), 367–382.
12. S. Sarkar, B. Kisku, S. Misra, and M.S. Obaidat, Chinese remainder theorem-based RSA-threshold cryptography in mobile *ad hoc* networks using verifiable secret sharing, *Proceedings of the 5th IEEE International Conference on Wireless and Mobile Computing, Networking and Communications (WiMob'09)*, Marrakech, Morocco, October 12–14, 2009.
13. S.K. Dhurandher, S. Misra, M.S. Obaidat, and N. Gupta, An ant colony optimization approach for reputation and quality-of-service-based security in wireless sensor networks, *Security and Communication Networks*, 2(2), pp. 215–224, 2009.
14. S. K. Dhurandher, S. Misra, S. Ahlawat, N. Gupta, and N. Gupta, E2-SCAN: An extended credit strategy-based energy-efficient security scheme in wireless *ad hoc* networks, *IET Communications*, UK, May 2009, pp. 808–819.
15. R. Chandrasekar, M.S. Obaidat, S. Misra, and F. Peña-Mora, A secure data-centric scheme for group-based routing in heterogeneous ad-hoc sensor networks and its simulation analysis, *SIMULATION: Transactions of the Society for Modeling and Simulation International*, 84(2/3), 131–146, 2008.
16. P. Narula, S.K. Dhurandher, S. Misra, and I. Woungang, Security in mobile ad-hoc networks using soft encryption and trust-based multi-path routing, *Computer Communications*, 31(4), 760–769, 2008.
17. S. Misra, S. Roy, M.S. Obaidat, and D. Mohanta, A fuzzy logic-based energy efficient packet loss preventive routing protocol, *Proceedings of the International Symposium on Performance Evaluation of Computer and Telecommunication Systems (SPECTS 2009)*, Istanbul, Turkey, July 13–16, 2009, pp. 185–192.

18. S. Misra, A. Bagchi, R. Bhatt, S. Ghosh, and M.S. Obaidat, Attack graph generation with infused fuzzy clustering, *Proceedings of the International Conference on Security and Cryptology, Part of the International Joint Conference on e-Business and Telecommunications (ICETE 2009)*, Milan, Italy, July 7–10, 2009, pp. 92–98.

19. A. Menezes, P. van Oorschot, and S. Vanstone, *Handbook of Applied Cryptography*, CRC Press, Boca Raton, FL, 1997, ISBN 0849385237.

20. P. Wohlmacher, Digital certificates: A survey of revocation methods, *Proceedings of the ACM Workshops on Multimedia*, CA, 2000, pp. 111–114.

21. P. Kocher, On certificate revocation and validation, in *Proceedings of the Second International Conference on Financial Cryptography*, Anguilla, 1998, pp. 172–177.

22. U. Maurer, New approaches to digital evidence, *Proceedings of the IEEE*, 92 (6), 933–947, 2004.

23. B. Hunter, Simplifying PKI usage through a client–server architecture and dynamic propagation of certificate paths and repository addresses, in *Proceedings of 13th International Workshop on Database and Expert Systems Applications*, France, September 2002, pp. 505–510.

24. The Controller of Certifying Authority, Available from http://www.cca.gov.in. Last accessed April 15, 2009.

25. The World Internet Security Company, Available from http://www1.wisekey.com/products/. Last accessed April 15, 2009.

26. S. Yi and R. Kravets, Composite key management in *ad hoc* networks, in *Proceedings of First Annual International Conference on Mobile and Ubiquitous Systems: Networking and Services (MobiQuitous'04)*, Urbana, USA, August 2004, pp. 52–61.

27. D. Hong and J. Kang, An efficient key distribution scheme with self-healing property, *IEEE Communications Letters*, 9(8), 759–761, 2005.

28. T. Dierks and C. Allen, The TLS protocol, version 1.0, Request for Comments 2246, January 1999.

29. J. Callas, L. Donnerhacke, H. Finney, D. Shaw, and R. Thayer, Open PGP message format, RFC 4880, November 2007.

30. C. Finseth, An access control protocol, sometimes called TACACS, RFC1492, July 1993.

31. B.C. Neuman and T. Ts'o, Kerberos: An authentication service for computer networks, *IEEE Communications*, 32(9), 33–38, 1994.

32. Y.-M. Tseng, A heterogeneous-network aided public-key management scheme for mobile *ad hoc* networks, *International Journal of Network Management*, 17(1), 3–15, 2007.

33. S. Misra, S. Goswami, G.P. Pathak, N. Shah, and I. Woungang, Geographic server distribution model for key revocation, *Telecommunication Systems*, January 2010. [Online]. Available: http://dx.doi.org/10.1007/s11235-009-9254-x.

34. P. Zheng, Tradeoffs in certificate revocation schemes, *ACM SIGCOMM Computer Communication Reviews*, 33(2), 103–112, 2003.

35. M. Myer, Revocation: Options and challenges, in *Proceedings of the Second International Conference on Financial Cryptography*, West Indies, February 23–25, 1998, pp. 165–171.

36. J.K. Millen and R.N. Wright, Certificate revocation the responsible way, in *Proceedings of Computer Security, Dependability and Assurance: From Needs to Solutions (CSDA'98)*, Washington, D.C., 1998, pp. 196–203.

37. D. Cooper, A model of certificate revocation, in *Proceedings of the 15th Annual Computer Security Applications Conference*, Scottsdale, 1999, pp. 256–264.

38. D. Cooper, A more efficient use of delta-CRLs, in *Proceedings of the 2000 IEEE Symposium on Security and Privacy*, CA, 2000, pp. 190–202.

39. P. McDaniel and S. Jamin, A scalable key distribution hierarchy, Technical Report CSE-TR-366-98, EECS, University of Michigan, Ann Arbor, MI, 1998.

40. J.-H. Huang and S. Mishra, Mykil: A highly scalable and efficient key distribution protocol for large group multicast, in *Proceedings of the IEEE 2003 Global Communications Conference (GLOBECOM 2003)*, San Francisco, December 2003.

41. M. Shi, X. (Sherman) Shen, Y. Jiang, and C. Lin, Self-healing group-wise key distribution schemes with time-limited node revocation for wireless sensor networks, *IEEE Wireless Communications*, 14(5), 38–46, 2007.

42. T. Yuan, J. Ma, Y. Zhong, and S. Zhang, Self-healing key distribution with limited group membership property, *First International Conference on Intelligent Networks and Intelligent Systems*, Wuhan, China, 2008, pp. 309–312.

43. I.Z. Berta, L. Buttya, and I. Vajda, A framework for the revocation of unintended digital signatures initiated by malicious terminals, *IEEE Transactions on Dependable and Secure Computing*, 2(3), 268–272, 2005.

44. G. Dini and I.M. Savino, An efficient key revocation protocol for wireless sensor networks, in *Proceedings of the 2006 International Symposium on a World of Wireless, Mobile and Multimedia Networks (WoWMoM'06)*, New York, June 26–29, 2006, pp. 3–5.

45. T. Narten, E. Nordmark, and W. Simpson, Neighbor discovery for IP, version 6, RFC 2461, December 1998.

46. D. Boneh, X. Ding, G. Tsudik, and M. Wong, A method for fast revocation of public key certificates and security capabilities, in *Proceedings of the 10th Conference on USENIX Security Symposium*, Vol. 10, 2001. Washington, D.C., pp. 297–308.

47. B. Libert and J.J. Quisquater, Efficient revocation and threshold pairing based cryptosystems, in *Proceedings of the Twenty-Second Annual Symposium on Principles of Distributed Computing*, Boston, 2003, pp. 163–171.

48. B. Pinkas, Efficient state updates for key management, *Proceedings of the IEEE*, 92(6), 910–917, 2004.

49. M. Naor and K. Nissim, Certificate revocation and certificate update, *IEEE Journal on Selected Areas in Communications*, 18(4), 561–570, 2000.

50. A. Shamir, How to share a secret, *Communications of the ACM*, 22(11), 612–613, 1979.

51. H. Zhou, M.W. Mutka, and L.M. Ni, Multiple-key cryptography-based distributed certificate authority in mobile ad-hoc networks, in *Proceedings of the Global Telecommunications Conference*, 2005. GLOBECOM '05, 2005, St. Louis, Missouri, Vol. 3, 5p.

52. H. Luo, J. Kong, P. Zerfos, S. Lu, and L. Zhang, "Self-securing *ad hoc* wireless networks, in *Proceedings of the IEEE Symposium on Computers and Communications*, Italy, July 2002, pp. 567–574.

53. J. Sen, P.R. Chowdhury, and I. Sengupta, A distributed trust establishment scheme for mobile *ad hoc* networks, in *International Conference on Computing: Theory and Applications*, Kolkata, India, March 5–7, 2007, pp. 51–58.

54. M. Omara, Y. Challalb, and A. Bouabdallahb, Reliable and fully distributed trust model for mobile *ad hoc* networks, *Computers & Security*, 28(3–4), 199–214, 2009.

55. H. Deng and D.P. Agrawal, TIDS: Threshold and identity-based security scheme for wireless *ad hoc* networks, *Ad Hoc Networks*, 2(3), 291–307, 2004.

56. S. Capkun, L. Buttyán, and J.-P. Hubaux, Self-organized public-key management for mobile *ad hoc* networks, *IEEE Transactions on Mobile Computing*, 2(1), 52–64, 2003.

57. S. Misra, S. Goswami, G.P. Pathak, and N. Shah, Efficient detection of public key infrastructure-based revoked keys in mobile *ad hoc* networks, *Wireless Communications and Mobile Computing*, [Online] 1 Oct 2009, Available: http://dx.doi.org/10.1002/wcm.839.

58. P. Zimmermann, *The Official PGP User's Guide*, MIT Press, Cambridge, MA, 1995.

59. G. Arboit, C. Crépeau, C.R. Davis, and M. Maheswaran, A localized certificate revocation scheme for mobile *ad hoc* networks, *Ad Hoc Networks*, 6(1), 17–31, 2008.

60. A. Shamir, How to share a secret, *Communications of the ACM*, 22(11), 612–613, 1979.

61. J. Jeong, K. Lee, J. Park, H. Lee, and H. Kim, The auto configuration of recursive DNS server and the optimization of DNS name resolution in hierarchical mobile IPv6, in *Proceedings of Vehicular Technology Conference (VTC 2003)*, Vol. 5, 2003, Jeju, Korea, pp. 3439–3442.

62. A.M. Hegland, W. Eliwinjum, S.F. Mjolsnes, C. Rong, O. Kure, and P. Spilling , A survey of key management in *ad hoc* networks, *IEEE Communications Surveys & Tutorials*, 4(3), 48–66, 2006.

63. M. Eltoweissy, M. Moharrum, and R. Mukkamala, Dynamic key management in sensor networks, *IEEE Communications Magazine*, 44(4), 122–130, 2006.

64. A. Perrig, R. Szewczyk, V. Wen, D. Culler, and J.D. Tygar, SPINS: Security protocols for sensor networks, *Wireless Networks*, 8(5), 521–534, 2002.

Chapter 8

Combating against Security Attacks against Mobile *Ad Hoc* Networks (MANETs)

Zubair Muhammad Fadlullah, Tarik Taleb, and Marcus Schöller

Contents

8.1 Introduction

Security is a key service for both wired and wireless network communications. In particular, the evolution in the variety and application of *ad hoc* wireless networks has vastly increased the urgency of identifying security threats and countermeasures to thwart these threats. Indeed, the success of *ad hoc* frameworks such as Mobile *Ad Hoc* NETworks (MANETs) relies heavily on the confidence regarding security shown by the relevant users. A MANET is an infrastructure-less network formed by a group of mobile nodes with wireless network interfaces. The mobile hosts dynamically establish paths among one another in order to communicate. In addition to one hop away communication, a mobile node in MANET may also function as a router to relay or forward packets, from a source node to a destination node, over multiple hops. Therefore, the success of MANET communication highly relies on the collaboration of the involved mobile nodes. Such dynamism of MANET-based architectures leads to some inherent weaknesses and a wide variety of attacks target these weaknesses. For instance, by not following the exact specifications of the considered routing protocol in MANET, a malicious node can mount routing attacks to disrupt the routing discovery phase whereby other nodes may not be able to establish a communication-path among themselves. While some attacks may target some specific routing protocols, for example, *Ad Hoc* On Demand Distance Vector (AODV) or DSR, the more sophisticated ones such as blackhole/sinkhole, byzantine, and wormhole attacks lead to serious routing security concerns, addressing which has become one of the hottest topics in MANET research domain.

In this chapter, we explore some of the existing malicious attacks against MANETs and also the techniques to detect them. First, we provide the background that takes an overview on the taxonomy of various attacks against MANETs. Then, we predominantly focus on how the networking and transport layer attacks are carried out against MANETs and how we may deal with such attacks. We also put forward the future research directions and emphasize the need for an intrusion detection system that may be appropriated with the requirements of MANETs and other *ad hoc* networks, and that would be able to detect not a specific type of attacks, but various blends of threats.

8.2 Background: Attack Taxonomy

Broadly speaking, the attacks against MANETs can be categorized into two classes, namely external and internal. In literature, these are synonymous to outsider and insider attacks [1], respectively. While the former is mounted by nodes that do not belong to the target MANET system, the latter is launched from compromised MANET hosts. In contrast to the external attacks, the internal ones have more serious impact on the victim system. This is due to the fact that the internal (i.e., compromised) nodes have knowledge pertaining to valuable information about the network topology and also possess adequate access privileges.

On the basis of the nature of attack interaction, the attacks against MANET may be classified into active and passive attacks. The former consists in replication, modification, or removing information exchanged by other nodes. The active attacks against MANET can lead to congestion, propagation of inaccurate routing information, and possible Denial of Service (DoS) scenario whereby the intended service is prevented from functioning [2–7]. The active attacks are usually launched by either compromised (i.e., malicious) nodes or selfish hosts [5,8] that just drop the received packets for saving their battery resources. The normal operation of the MANET is interrupted by selfish nodes since they do not take part in the routing protocols or forward packets. On the other hand, a compromised node may exploit the routing protocol to broadcast itself as having the shortest communication-path to destination. The selfish hosts (i.e., passive attacks) comprise eavesdropping of information, traffic analysis, and traffic monitoring for building statistical profiles to have an idea about the network operations and possible vulnerability of the target MANETs. The passive attacks are more difficult to detect and counter.

A common active attack is spoofing whereby a compromised node pretends to be a legitimate host. The compromised node usually exploits the lack of authentication in the current MANET protocols [9,10]. As a consequence of spoofing attacks, the other nodes in a MANET get a wrong picture of the network topology and experience network loops or partitioning. Indeed, the lack of authentication in the routing protocols adopted by MANET also leads to fabrication attacks that generate false/erroneous routing messages [11–13].

DoS attacks, in plenty of varieties and guile, remain one of the most common, yet effective, threats against MANETs and other *ad hoc* networks. In a typical DoS attack, an attacker injects a large volume of unnecessary packets into the network in order to consume a substantial amount of network resources. As a consequence, the legitimate MANET nodes compete among one another for the wireless channel and network connections [14,15]. The work in [16] identifies two variations of DoS attacks against MANETs, namely sleep deprivation and routing table overflow attacks, which attempt to deplete the target node's scarce battery power and create routes to nonexisting nodes, respectively. The latter, apart from being a DoS attack, may also be categorized as a comparatively simpler routing attack. More sophisticated routing attacks against MANETs such as wormhole attacks [17,18], Sybil attacks [19], and rushing attacks [20] are more difficult to detect let alone prevent. It should be noted that these attacks take place on the network layer stack. On the transport layer also, MANETs are vulnerable to attacks such as session hijacking and synchronize (SYN) flooding. In addition, attacks are also possible against MANETs in the lower layers. For instance, traffic analysis and monitoring (passive attack), disruption of IEEE 802.11 MAC, and so on may be carried out against MANET-based hosts on the data link layer level. In the physical layer, jamming and other passive threats such as messages interceptions and eavesdropping are known to exist.

However, the focus of this chapter is on the routing attacks against MANETs. To this end, in the remainder of the chapter, we focus on various security attacks against MANETs on the network and on the transport layers.

8.3 Network Layer Attacks against MANETs

The network layer protocols enable the MANET nodes to be connected with one another through hop-by-hop. The intrinsic nature of the MANET routing protocols, thus, ensure the cooperative communication among nodes by enabling them to also act as routers or intermediary devices along the communication path of a source/victim "V" and a destination "D." In literature, different types of attacks against MANET routing protocols have been identified through which a malicious node "M" can absorb network traffic and place itself in between "V" and "D" as shown in Figure 8.1. "M" can then effectively control the network traffic flow from "V" to "D" (and also the other way around) as it becomes a router. "M" may also divert the packets exchanged between "V" and "D" via a nonoptimal or a looped path. This introduces a significant end-to-end delay between "V" and "D." In an even worse scenario, "M" may direct the packets through a nonexisting link. Thus, attacks against the routing protocols in the network layer contribute to a wide range of problems such as the MANET hosts not being able to find any route to destination, face network congestion, and so forth.

In addition, some attacks target specific routing protocols. For instance, if the underlying routing mechanism in Figure 8.1 is Dynamic Source Routing (DSR), then "M" may modify the source route listed in the Route Request (RREQ) and/or Route Reply (RREP) packets, for example, by adding a new node into the route, deleting an existing one from the route, change the sequence of the nodes, and so on. On the other hand, if an AODV is used as the routing protocol, it may happen that "M" advertises a route with a fabricated distance metric that is smaller than the real one. This effectively renders the routing updates from the other MANET nodes invalid. It should also be stressed that "M" does not necessarily perform attacks at the data-forwarding phase only. "$M" may, indeed, launch routing attacks before the routing path has been determined, that is, during the route discovery or the route maintenance phases. These various attacks are described in the remainder of the chapter.

MANET routing discovery phase attacks: Some malicious users willingly do not follow the specifications of the routing protocols used in the target MANET. These attacks usually take place during the routing discovery phase. Examples of these types of threats include routing message flooding (e.g., by exchanging an overwhelming volume of "Hello," "RREQ," and/or "ACK" messages), routing table overflow, routing cache poisoning, and routing loop attacks [21,22]. Indeed, proactive routing algorithms (e.g., Destination-Sequenced Distance-Vector (DSDV) [23] and Optimized Link State Routing (OLSR) [24]) for discovering the routes in MANETs are more prone to these attacks when

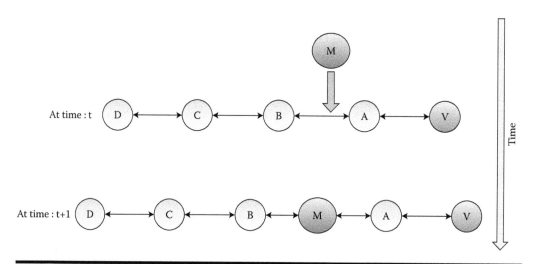

Figure 8.1 Mechanism of a simple routing attack whereby malicious node "M" inserts itself into the MANET topology.

compared with the reactive ones such as DSR [23] and AODV [23]. The reason behind this is that the former attempts to discover the necessary routing information periodically and prior to when such information are required. For instance, a malicious host may overflow a victim node's routing table by transmitting excessive route-advertisements. To this end, the malicious user (e.g., "M" in Figure 8.1) broadcasts routes that may not exist at all in the target MANET topology. Provided that "M" is successful at creating enough nonprevailing routes, a proactive algorithm may be tricked so as not to create additional routes. The proactive routing protocols are vulnerable to routing cache poisoning attacks also whereby "M" exploits the promiscuous mode of updating the routing tables of the MANET nodes. In this case, "M" "poisons" routes to a victim node "V" by broadcasting spoofed packets with source route to "V" via "M" itself. As a consequence, the adjacent nodes, which notice the packets, may add this route to their respective route-caches.

MANET routing maintenance phase attacks: During the route maintenance phase, a number of control messages are exchanged among the participating nodes in the MANET topology. Some of the attacks are mounted during this phase which broadcast spoofed control or signaling messages (e.g., broken link error messages) that trigger reconfiguring or re-establishing the route(s) from a source to a destination. For instance, in order to address the mobility of the nodes within a MANET, routing protocols such as AODV and DSR adopt mechanisms for recovering from broken routes. In such mechanisms, when the destination node and/or other nodes along the path from a source to destination move, the upstream node "U" of the broken link transmits a route error message to each of the other upstream hosts. In addition, "U" also purges this particular route to the destination. A malicious user, "M," may exploit the role of "U" to broadcast false route error messages and prevent the source node (i.e., the victim node in this case) from communicating with the destination.

MANET Data Forwarding Phase attacks: A lot of attacks against MANET routing protocols exploit the information forwarding functionality of the MANET nodes in the network layer [25,26]. These attackers, in these cases, do not disrupt the route discovery and/or maintenance phases. Rather, they willingly disrupt the forwarding of data packets as per the routing table information by a number of means. For instance, a malicious user may drop silently or replay or even modify the inbound packet contents. In addition, the time-sensitive communications may be disrupted by delaying the relaying of data packets to their respective next-hop destinations or simply by injecting and forwarding dummy packets.

The next chapter provides more details pertaining to some of the sophisticated and subtle attacks against MANET routing, and also possible countermeasures against each of these attacks. They include the wormhole, blackhole, Byzantine, rushing, resource consumption, link withholding and spoofing, and replay attacks.

8.3.1 Wormhole Attack

The wormhole attack, one of the most sophisticated and serious threats against MANET routing, comprises a pair of attackers. These two attackers act in collusion to record packets at a particular location in the MANET topology and replay them at another node by using a high-speed private network. Figure 8.2 demonstrates an example scenario of this attack, where "M_1" and "M_2" are the colluding attackers and "V" is the victim node. When "V" broadcasts an RREQ message to find a route to a node "D" (i.e., when "V" and "D" are the source and destination nodes, respectively), the immediate one-hop away neighbors of "V," namely "A" and "F," forward the RREQ message to their respective neighbors "B" and "M_1." However, as "M_1" receives the RREQ from "F," it tunnels the RREQ message to its colluding partner "M_2." The latter then broadcasts the RREQ message to its one-hop away neighboring node "G," through which the RREQ is delivered to the

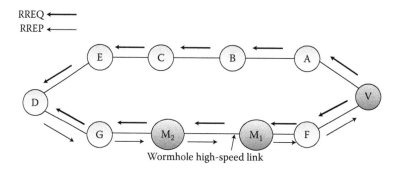

Figure 8.2 A wormhole attack scenario against node V by a pair of colluding attackers M_1 and M_2.

destination node, that is, "D." Due to the high-speed link chosen by the tunnel between "M_1" and "M_2," it takes shorter time for this particular route to deliver the RREQ message to "D" in contrast to that taken over the {V-A-B-C-E-D} path. As a consequence, the route {D-G-F-V} becomes the apparent choice for "D" for issuing a unicast RREP message as a response to the RREQ received from "V." Therefore, "D" ignores the same RREQ that arrives at a later time and, thus, invalidates the legitimate route: {V-A-B-C-E-D}. This forces "V" to select the route {V-F-G-D}, which, indeed, goes through the "M_1" and "M_2," the malicious users, which can tamper with its data packets.

8.3.1.1 Countermeasures against the Wormhole Attack

In order to detect and combat against the wormhole attack, two types of packet leashes were introduced as an effective technique [27], namely temporal and geographical leashes. In the former, every node in the MANET calculates the packet expiration time, t_e, and includes t_e in its packets so that the packet may not travel further than a particular distance, L. When a packet containing this information arrives at a node, the receiver compares the current time with the value of t_e in the packet. With such information, the destination node (e.g., D in Figure 8.2) may then be able to determine, whether an RREQ was tunneled possibly over a high-speed link to serve a malicious purpose. In addition, t_e is authenticated by the involved ends so that it may not be tampered with by malicious nodes such as M_1 and M_2. However, the temporal leash needs all the nodes in the considered MANET topology to be strictly time synchronized with one another. On the other hand, in case of the geographical leash, every node must know two pieces of information, namely its respective position in the MANET it belongs to and the transmission time. This enables the receiver to evaluate neighbor relations by calculating distance between itself and the original source of the packet.

Based on the location information, further solutions evolved to counter the wormhole attacks. For instance, the work in [28] offers protection against wormhole attacks, specifically in MANETs that use OLSR as the routing protocol. This, however, requires an integration of the public-key infrastructure with the time-synchronization between all the nodes. In this scheme, every node, while issuing a HELLO message, inserts its current position and also the current time stamp in the HELLO message. A node that receives a HELLO packet from one of its neighbors can then use the information embedded in the packet to compute the distance between itself and the neighbor. In case the computed distance exceeds the maximum transmission range, the HELLO message is considered to be highly suspicious (i.e., possibly tunneled over a wormhole attack). Interested readers may also refer to additional mechanisms in literature (e.g., SECTOR [29] and directional antenna based detection of wormhole attacks [30]) without the need of clock synchronization among the MANET nodes.

Statistical analysis has also been adopted in literature to detect the wormhole attacks. For example, Qian et al. [31] introduced a statistical analysis over multiple path routing. This scheme computes the relative frequency of every link that is found in all the obtained routes during a single route discovery. The highest relative frequency is then identified as the wormhole link. Although this scheme has low overheads when applied in multipath routing, it does not work well with nonmultipath routing protocols such as AODV.

A potential solution to thwart wormhole attacks is to integrate intrusion detection and/or prevention systems in the MANETs. Other countermeasures have considered, in addition to software-based intrusion detection systems, designing specific hardware and signal-processing techniques. The hypothesis behind such solutions suggests that if the data bits transmitted over some special modulating scheme are known only to the neighboring nodes, they cannot be affected by closed wormholes.

8.3.2 Blackhole Attack

In case of the blackhole attack as shown in Figure 8.3, a malicious node "M" claims to possess an optimum route from a given source, "V," to a destination, "D," and transmits this forged routing information to the other MANET nodes. As a result, the other users are tricked to forward their data packets through the malicious node. For instance, if the target MANET uses AODV as the routing protocol, "M" may generate a false RREP consisting of a nonexisting destination sequence number, which is equal or higher than that in the RREQ from the victim node (i.e., the source node "V"). This implies that the malicious node "M" claims that it possesses a sufficiently fresh route to the destination. This prompts "V" to choose this particular route (i.e., V-A-B-M) for sending data packets through the attacker. Consequently, "M" may willingly delay/drop the data packets or change the contents of the packets. However, "M" runs the risk that its neighboring nodes, for example, "B" and/or "F" may monitor and expose the ongoing attacks. To avoid such

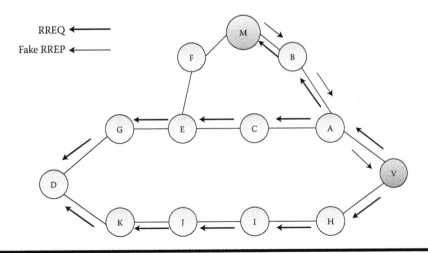

Figure 8.3 A blackhole attack scenario where M generates fake RREP to the RREQ message of the victim source node "S."

detection, "M" may execute more sophisticated versions of the blackhole attack whereby packets are intercepted and rather than dropping every data packet, the intercepted packets are forwarded selectively. Indeed, when "M" makes modifications only to those packets arriving from "V" but not to the ones arriving from "B" or "F," it reduces the chance of detecting this malicious activity by the neighboring nodes.

8.3.2.1 Countermeasures against the Blackhole Attack

In order to defend against possible blackhole attacks, security-aware *ad hoc* routing protocol or SAR was designed [32], which is based on conventional on-demand MANET routing protocols such as AODV and DSR. The SAR employs two techniques as follows. First, a security metric is inserted in the RREQ packet. For achieving a decent trust level (e.g., for avoiding identity theft or identity spoofing), the SAR uses a simple, shared secret to generate a symmetric cryptographic key per trust level using which the packets are encrypted. In other words, a node that belongs to a different trust level is unable to the read the encrypted RREQ or RREP packets. Second, an alternate route discovery mechanism is used. When the other nodes along a route receive the RREQ packets from a particular source, they verify the trust level associated with the security metric information embedded in the packets. Given that the trust level is satisfied, an intermediary node would start processing the packet (i.e., forward it to the next node along the route). If the trust level is not within a prespecified satisfaction level, the intermediary node drops the packet. When the destination node is satisfied with the security attributes or trust levels associated with the overall end-to-end path from the source to destination nodes, it generates an RREP packet with the specific security metric. Otherwise, it notifies the source node that the communication cannot be continued via this route (as it may be already compromised) and thus permits the sender to adjust its security level for finding an alternate route. Indeed, there are chances that a malicious node changes the security metric to a higher or lower level and disrupt the flow of packets. This remains a shortcoming of the SAR approach.

Among other approaches to combat against blackhole attacks, the work in [33] introduces in the routing protocols the use route confirmation request and response, denoted by CREQ and CREP, respectively. In this work, each intermediary node, in addition to sending RREP, sends a CREQ message to its next-hop neighbor toward the direction of the destination node. After receiving the CREQ, the next-hop node searches, in its cache, for a route to the destination. If such a route to the destination is, indeed, available, the next-hop node transmits the CREP message to the source. The source, after receiving this CREP message, checks whether the path in the RREP message is the same as that in the CREP one. If so, it deems the routing information to be correct. However, this approach is not sufficient to counter a pair of nodes working in collusion that attempt to perform blackhole attacks. Because, when the next-hop node is also colluding, it can generate and send forged CREPs containing inaccurate routes. To overcome this issue of colluding nodes, Al-Shurman et al. [34] devised a mechanism that makes the source node wait for RREP messages arriving from more than two nodes. From multiple RREPs, the source node can then evaluate the accuracy pertaining to the path information. This particular approach is also not without its shortcomings, the obvious being the added latency during which the source node must wait for multiple RREP packets to arrive.

A different approach consists in not merely circumventing the blackhole attacks but also detecting them. This approach was inspired by the analysis conducted by Kurosawa et al. [35] that reveals that a malicious user must increase the destination sequence number to such as extent as to convince the source node that the provided path is optimum enough. Following this analysis, a statistical detection scheme was envisioned to discover the anomalies, that is, the increasing differences between the destination sequence numbers of the received RREPs, which would suggest possible blackhole attacks. While such anomaly-based intrusion detection approaches do not produce additional routing traffic or require any modification in the existing routing protocol, they may often be susceptible to high number of false positives and thus disrupt communications.

8.3.3 Byzantine Attack

A byzantine attack comprises either a single compromised node or a group of compromised/colluding nodes in between the route from the source end to the destination node. The compromised node(s) target(s) the MANET by mounting attacks such as creating routing loops and directing data packets via nonoptimal paths that lead to degradation or disruption of the routing services [36].

8.3.3.1 Byzantine Attack Prevention

Crépeau et al. [37] introduced Robust Source Routing (RSR), a secure MANET on-demand routing protocol capable of delivering packets to their respective destinations even in Byzantine attack-like adversarial conditions. RSR, by using Fore-Runner (FR) packets, notifies the intermediate nodes along a route that they are to expect the specified data flows within the given time frames. If an intermediate node has not received any data flow within the expected time, it informs the source node about this event. By this way, the links with selfish and/or active malicious nodes can be identified and isolated.

8.3.4 Rushing Attack

In [20], the authors introduced a new form of routing attack called the rushing attack, which acts as an effective DoS-type threat against all conventional on-demand MANET routing protocols. In fact, even the secure routing protocols [e.g., Secure *Ad hoc* On-Demand Distance Vector (SAODV) and AODV secured with Statistically Unique and Cryptographically Verifiable (SUCV)] were shown to be vulnerable to this particular rushing attack.

In case of a typical on-demand *ad hoc* routing protocol, a node that intends to discover a route to a given destination floods the target network with RREQ packets. In order to keep the impact of the flood as minimal as possible, the nodes in conventional routing protocols forward only the request that arrives first from each Route Discovery. This particular mode of route discovery operation is exploited by the rushing attack. For instance, in case of DSR route discovery, let us refer to Figure 8.4, where "S" and "D" refer to the source and destination nodes, respectively, "M" denotes the rushing attack-node, and "G" and "H" are the one-hop neighbors of "D." If the RREQs for this discovery forwarded by "M" are the first ones to reach "G" and "H" (this is possible in a number of ways as investigated in [20]), then any route discovered by this route discovery operation will include a hop through "M." Simply put, when a neighbor of the target "D," that is, "G" or "H," receives the rushed RREQ from the attacker, it forwards that request alone, and does not forward any further RREQ from this route discovery. Even if nonattacking RREQs from "S" reach "G" and "H" at a later time, those legitimate requests are discarded. As a consequence, "S" fails to discover any useable route or safe route without the involvement of the attacker.

8.3.4.1 Rushing Attack Solution

The authors in [20] proposed Route Discovery Protocol (RAP) that replaces the standard mechanism of the conventional *ad hoc* routing protocols that are inherently vulnerable to rushing attacks. In

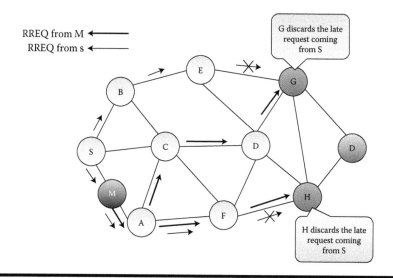

Figure 8.4 A MANET topology where "M" may perform rushing attacks. (Data from Y. Hu, A. Perrig, and D. Johnson, *Proceedings of 2nd ACM Workshop on Wireless Security*, San Diego, CA, September 2003.)

fact, the RAP combines three techniques to prevent the rushing attack, namely a secure neighbor discovery mechanism, a secure route delegation acceptance protocol, and the randomized selection of the RREQ that will be forwarded.

8.3.5 Resource Consumption Attack

The resource consumption or sleep deprivation attack consists in an attacker or a compromised node to consume the resources of the victim node or the target MANET [38]. For example, the aim of the flooding attack is to exhaust resources such as the network bandwidth by forwarding excessive packets to victim nodes, or the computational power and battery life of a victim node by requesting unnecessary route discovery in an excessive volume. A simple example consists of an adversary node targeting AODV routing protocol by transmitting a large number of RREQ packets in a short period of time to a nonexisting destination node. Since no node will respond to these RREQs, these packets will simply flood the MANET, and consume the network bandwidth and deplete the scarce battery power at of the nodes. Furthermore, the work in [39] demonstrates that such a flooding attack can degrade the overall MANET throughput by 84%.

8.3.5.1 Resource Consumption Attack Prevention

Yi et al. [38] envisaged a simple, yet effective, mechanism to prevent the resource consumption attacks, particularly in MANETs that use AODV as the routing protocol. In this mechanism, every node monitors and computes the respective RREQ rates of its neighbors. If the RREQ rate of a neighbor is found to exceed a threshold defined *a priori*, the node blacklists the neighbor and drops further RREQs from that particular neighbor. The main problem of this mechanism is that it is prone to false positives, and may end up blacklisting legitimate nodes. The work in [39], on the other hand, uses a similar anomaly-based detection mechanism which, instead of using a fixed threshold, learns from the statistical analysis of different rates of RREQ packets and computes the threshold on the fly.

8.3.6 Link Withholding and Link-Spoofing Attacks

The names of these attacks are somewhat self-explanatory. In case of a link withholding attack, a malicious node willingly withholds or ignores the requirement to advertise the route to a specific node or a collection of nodes. As a result, the other hosts are unable to find links to communicate with those nodes. In link withholding attack launched against Topology Control (TC) messages in OLSR, Kannhavong et al. [40] show that a malicious node can isolate a particular node and prevent it from communicating with other nodes in the MANET. Their proposed detection technique works based on the hypothesis that if a node receives only a HELLO message from its Multipoint Distribution Relay (MPR), but does not receive any TC message from the MPR, the node evaluates the MPR to be suspicious. The node then switches to another MPR. This approach, however, fails to detect attacks launched by two malicious partners that lie next to one another whereby the first malicious node pretends to advertise a TC message while the second one discards that TC packet.

On the other hand, in link-spoofing attacks, a malicious node advertises forged routes. For instance, an attacker may broadcast a spoofed link with the victim's two-hop away neighbors in an OLSR-based MANET. As a result, the victim chooses the malicious node as its MPR. As MPR, the malicious node can discard the TC messages and other routing traffic from the victim, or modify the data packets arriving from the victim intended for a different destination.

For detecting a link-spoofing attack, the work in [41] envisioned a detection scheme that relies on spatial information obtained from Global Positioning System (GPS) and a time stamp that is encrypted. Each node, in this scheme, advertises to other nodes its GPS coordinates and the time stamp. Thus, it becomes possible to detect possible link-spoofing cases by computing the inter-nodal distances of two given nodes and also to check whether they lie within the maximum transmission range or not. The main problem pertaining to this solution is that every node requires being equipped with GPS, which may not be always possible. Our first case study, provided later in this chapter, will focus on a uniquely crafted scenario for link withholding/spoofing with hints of how to solve such problems.

8.3.7 Replay Attacks

The topology of a MANET frequently changes because of the mobility of its nodes. This dynamic change in the MANET topology means that the current network topology may not prevail even after a few seconds. In replay attacks [42], the malicious nodes record the legitimate control messages (e.g., TC messages in case of OLSR) of other nodes and retransmit them at a later time. As a result, the routing tables of the MANET nodes are updated with old and stale routes. By this way, replay attacks may be exploited for impersonating a particular node or simply disrupting the routing operations of the target MANET.

In order to protect MANETs from replay attacks, the work in [42] employs a solution based on time stamps and asymmetric encryption. The solution simply compares the current time with the time stamp embedded in the received control messages from other nodes. If the time stamp in a received control packet deviates much from the current time, the receiving node considers it to be a possible replay attack and, therefore, the packet is discarded to avoid updating the routing table with stale information. However, this solution still remains vulnerable to wormhole attacks comprising a pair of colluding attackers that employ a high-speed network for replaying messages in a far-away location with rather low latency.

8.4 Transport Layer Attacks against MANET

Transport layer protocols such as Transport Control Protocol (TCP) are used in MANETs for establishing end-to-end connections among MANET nodes and they ensure reliable packets delivery over the end-to-end connections. In addition, similar to the wired communications, flow and congestion control is also possible in MANETs by adopting TCP-like transport protocols. However, due to the intrinsic weakness of TCP, SYN flooding or session hijacking attacks are also possible in the MANET environment. Furthermore, MANETs are attributed with typically higher channel error rates in contrast with their wired counterparts. This augments to TCP-related problems because TCP is unable to differentiate the nature of the loss (e.g., whether the loss is owing to congestion, random error, channel error, or malicious attacks) and as a result, it multiplicatively decreases its congestion window. This eventually affects the performance of the MANET [43] substantially. In the remainder of this section, we take a brief overview of the SYN flooding and session hijacking attacks against MANETs.

8.4.1 SYN Flooding Attack

For two MANET nodes to establish a TCP connection, they need to perform a three-way handshake as shown in Figure 8.5. In the first step, the source node "S" needs to initiate the connection with the destination node "D" by sending a SYN packet along with a sequence number "P." As a response to this, "D" then transmits to "S" a SYN/ACK message, including its own sequence number "Q" and the acknowledgment number "P+1." In the final step, "S" issues an "ACK" message (with ack. number "Q+1") to "D." Thus, "S" and "D" establishes a TCP connection. In case of SYN flooding attack, "S" initiates a large number of TCP connections with the victim node "D." However, "S" spoofs the return address of the SYN packets and thus does not complete step 3 of these TCP

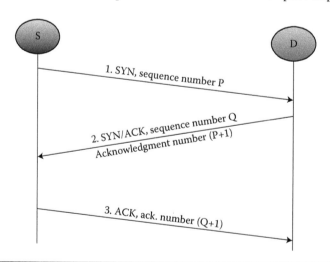

Figure 8.5 TCP three-way handshake mechanism between the source and destination MANET nodes, S and D.

connections (i.e., they are left open in midway). "D," upon receiving the SYN packet from the attacker ("S" in this case), issues immediately the SYN–ACK packets to the spoofed address, which often does not exist in the MANET. As a consequence, "D" awaits reception of ACK packets (in step 3 of the TCP handshake). A large number of these half-opened connections may then overflow the buffer maintained by "D." Such a buffer overflow results in "D" not being able to accept any legitimate request for establishing the TCP connection from other MANET nodes. Although a half-open connection normally should expire within a time-out period, the attacker can exploit this by transmitting SYN packets with spoofed addresses at a rate faster than this time-out value.

8.4.2 Session Hijacking Attack

The attacker in a session hijacking scenario exploits the unprotected session following its initial setup. The attacker forges the IP address of the victim node, computes the sequence number expected by the target, and then launches a DoS attack against the victim. By so doing, the attacker pretends to impersonate the victim node and maintains communicating with the target over the already established TCP session. An example of the session hijacking attack is the TCP–ACK storm problem as depicted in Figure 8.6. Here, nodes "N1" and "N2" have established a TCP connection. An attacker "M" spoofs the IP address of "N2" and injects data into the session of node "N1." Then, "N1" acknowledges the receipt of this information by transmitting an ACK packet to node "N2." As "N2" notices a different sequence number in the received ACK packet from "N1," it reissues its last ACK packet to "N1" in order to resynchronize the TCP session. This process repeats over and over, leading to an ACK storm. Indeed, it is even easier to hijack sessions in a way similar to connection-less transport protocols such as User Datagram Protocol (UDP).

8.5 Case Studies

In this section, we provide two case studies that address attacks on the OLSR protocol. OLSR is a table-driven proactive routing protocol that periodically exchanges messages among the nodes to

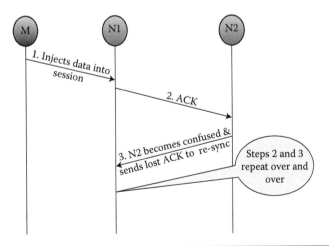

Figure 8.6 TCP ACK storm steps.

maintain the accurate topological information of the considered MANET. Each node can compute the optimal route to a given destination based on the topological information. By employing multipoint relays (MPRs), the OLSR protocol is able to immediately provide optimal routes. Each node chooses a set of its neighbor nodes as MPRs, which are responsible for generating and forwarding the topology information all over the MANET. In the first case study, a collusion attack is presented in which two or more attackers collaborate among one another to launch a routing attack [44]. In the second case study, an effective method is presented for detecting and preventing wormhole attacks against MANETs using OLSR as the routing protocol [45].

8.5.1 A Collusion Attack against OLSR-Based MANETs

The collusion attack demonstrated in [44] is shown in Figure 8.7. In the considered MANET topology depicted in the figure, there are two malicious nodes, namely "M_1" and "M_2." The victim node is denoted by "V." The malicious node "M_1" sends a HELLO message including the address list of the two-hop neighbors of "V." As per the OLSR protocol, "V" then selects "M_1" to be its only MPR. Then, "M_1" selects "M_2" as its only MPR. As a consequence, the TC messages generated by "V" are forwarded only via "M_1." In addition, "M_1" drops the TC message from being forwarded to "B" and "F." Also "M_2" discards the TC message and it is not forwarded to node "F." This collusion attack means that the TC messages from the Victim "V" are not relayed via "M_1" and "M_2" to the remaining nodes (e.g., I, J, F, H, etc.). Therefore, the remaining nodes are unable to construct routes to the victim node "V."

This specific collusion attack may be detected if every node in the considered MANET topology is able to learn the topology setting up to more than two hops. The detection scheme provided by Kannhavong et al. [44] adds in the HELLO message of every node its two-hop neighbors list. This is done to verify if the link information advertised by the one-hop away neighbors are accurate. In case, a node discovers that any of its one-hop neighbors has provided inconsistent routing information, it identifies that neighbor as a malicious one and avoids it.

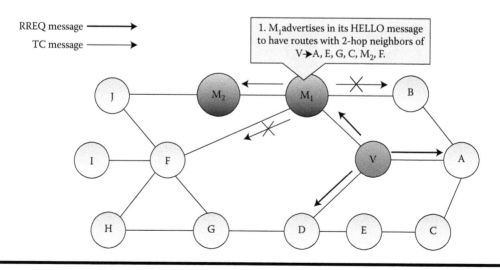

RREQ message ⟶

TC message ⟶

1. M_1 advertises in its HELLO message to have routes with 2-hop neighbors of V→A, E, G, C, M_2, F.

Figure 8.7 **M_1 and M_2 collude with each other to disrupt OLSR routing in the considered MANET.**

8.5.2 Detecting Wormhole Attacks against OLSR Protocols

Nait-Abdesselam et al. [45] envisioned a unique method for detecting and preventing wormhole attacks against OLSR-based MANETs. Their approach consists of a number of steps, namely detecting suspicious links and wormhole verification.

In the first step, the nodes in the MANET attempt to detect links suspected to be part of a wormhole. To this end, every node periodically advertises a HELLO message so that it may discover its own one-hop neighbors. When another node receives this HELLO message, it deems the originator of the message to be its actual neighbor. The wormhole attack, however, may also contribute to such a HELLO message, which is not necessarily originated from just one-hop away from the receiving node. To detect such suspicious links, Nait-Abdesselam et al. define two novel control packets for the OLSR protocol, namely $HELLO_{req}$ and $HELLO_{rep}$. The former is an extension to the original HELLO message used in the conventional OLSR protocol and it can be configured with either of the following two options: (i) it may function as the original message by default, or (ii) it can be used by its originator to request an explicit response from its neighbors. In the second operational mode, the neighbors then respond by issuing $HELLO_{rep}$ to the querying node. The message types of the conventional HELLO, $HELLO_{req}$, and $HELLO_{req}$ are differentiated by employing two unused bits in the original message.

Following each N standard HELLO message transmissions, a MANET node has to transmit a $HELLO_{req}$ message. Here, N reflects the desired security level, that is, if an application requires a high security level and is willing to detect the attackers rapidly, N should be adjusted to a relatively small value. According to the above-mentioned definition of $HELLO_{req}$, this prompts explicit HELLO replies from its neighboring nodes. The originating node waits for the $HELLO_{rep}$ message from its neighbors up to a prespecified time-out value. On the other hand, when a node receives a $HELLO_{req}$ packet, it records the address of the sender and the time left until it is scheduled to send its next HELLO message, denoted by i and Δ_i, respectively. In OLSR, the default value of HELLO message transmission interval is 2. If a receiver is queried by multiple sources, it delays the corresponding replies until it is scheduled to send its normal HELLO message. In addition, it piggybacks the responses to this HELLO message. This is done to avoid flooding the MANET with an excessive number of HELLO replies. Then, for each piggybacked response, the node attaches the recorded address of the sender of the respective $HELLO_{req}$ and the Δ_i values.

Upon receiving a $HELLO_{rep}$ packet, a node needs to verify if the received packet contains information pertaining to any of its pending requests made earlier. If the received $HELLO_{rep}$ has no such information, it is treated as an ordinary HELLO message. Otherwise, the node evaluates if the $HELLO_{rep}$ message came within the scheduled time-out interval. If so, the node considers the route between itself and the node that issued the $HELLO_{rep}$ message to be safe. Therefore, it (i.e., the originator of the $HELLO_{req}$ message) adds the responding node as its neighbor. On the other hand, if the $HELLO_{rep}$ is received by the originator after the expiration of its scheduled time-out value, the originator deems the link between itself and the responder to be suspicious. As a consequence, the originator stops communicating with the suspected node until the wormhole verification process is completed.

The wormhole verification process commences following the detection of suspicious links. This is carried out by the originator of the $HELLO_{req}$ message. The objective of this procedure is to verify if there exists any wormhole tunnel along the route comprising the originator and the other end of the suspicious link. Two more novel messages are annexed in the OLSR protocol for this purpose, namely a probing packet and an acknowledgment to the probing packet, denoted by ACK_{prob}. The originator node transmits a probing packet to each of the suspected nodes. As a response to the probing packet, the other nodes respond to the originator by sending ACK_{prob} packets. For example, a node "A" queries a suspicious node "B" about its wormhole status reputation. The node "B" responds with an ACK_{prob} message that also contains its own opinion about the wormhole status reputation of node "A." In addition, ACK_{prob} packets also contain the processing time information so that an accurate time-out value can be adjusted. For secure exchange of these messages, the transmission of probing and ACK_{prob} packets are encrypted and authenticated. To conclude, if a suspicious link is, indeed, traversing a wormhole tunnel, the originator node compares its evaluation of the reputation of the other endpoint of the suspicious link with the remote node's evaluation of its own reputation status. Experimental results demonstrate that the detection accuracy under this approach depends on the correlation between the number of MANET nodes and the length of a wormhole tunnel. For example, in a small-scale MANET with 15 nodes, the detection accuracy remains over 95% with increasing values of the tunnel length. On the other hand, for larger MANET topologies with 30 up to 50 nodes, the detection accuracy decreases significantly as the tunnel length increases. This is attributed to the fact that if the number of neighboring nodes increases, the malicious nodes are more likely to exploit more neighbors to form longer wormhole tunnels even though the neighboring nodes are faraway from one another. The results also suggest that in OLSR, every node periodically dispatches routing control messages. This increases the overheads in large-scale networks because the traffic increases substantially when these routing control messages are passed through the wormhole tunnels.

8.6 Open Issues and Future Directions of Research

Security is, indeed, a crucial issue in determining the success and wide use of MANET-based applications. As described in the earlier sections, a lot of studies about security threats and vulnerabilities for MANET already exist in literature. Future research should also be focused on reducing the design and deployment cost of the security provisioning so that they would be more suitable for MANET and other *ad hoc* wireless environments such as vehicular networks and wireless sensor networks. In addition, most of the conventional security mechanisms attempt to single out the known threats and deal with them individually. In other words, the countermeasures to a particular type of threat (e.g., the blackhole attack) may not necessarily be sufficient enough to thwart other types of attacks. Therefore, it is important for future researches in this domain to pay more attention toward countering against novel/unknown attacks. Furthermore, in case of cross-layer attacks, the security mechanisms need to be enforced on each layer separately which is cumbersome. This is worth exploring in future to make MANETs more secure and reliable.

By definition, MANETs are self-organized and they are not bound by any infrastructure and/or central authority. This leads to a plethora of open research challenges, including self-organized key management, cooperation incentives, authentication and access control, context-awareness and quality of security services, and so forth. There is also a research scope for integrating security primitives in MANET-based systems. For instance, key management protocols may help enhance the overall security level of a MANET as perceived by its users. Indeed, substantial research may be carried out on designing robust key management systems, trust-based routing protocols, and integrating authentication and encryption in different layers. The major research directions are summarized as follows.

8.6.1 Intrusion Detection and Prevention

In contrast to other wired and wireless networks, MANETs have unique features such as open nature, mobility of the nodes, and dynamic change in the topology. As a result, the conventional intrusion detection and prevention methods from security attacks may not be applicable to MANETs. The intrusion detection theme is of significant importance in discovering a potential attack before it may severely impact the target network and the victim(s). In a nutshell, the intrusion detection techniques that have been so long adopted in older wireless networks are not directly suitable for MANET environments. The future intrusion detection systems for MANETs and other *ad hoc* networks need to be both distributed and cooperative whereby each node should participate in the intrusion detection. That is, the intrusion detection is performed locally at every node and also on a more wide scale where the neighboring nodes share the intrusion detection information and collaborate with one another to trace the suspicious links. The second case study discussed in Section 8.5.2 indicates that researches have already started toward this direction.

Furthermore, upon detecting an intrusion or attack, the adequate response is also required. Intrusion prevention and/or response systems are to evolve to protect MANETs in future so that a wide range of responses may be adopted. The response may involve resetting the communication channels among the nodes, tracing back the compromised nodes, excluding them from the MANET, and so forth. In fact, most of the contemporary work on intrusion preventive and response methods is considered to be the second line of defense. Constructing a reliable trust-based framework for MANET and integrating it with the current preventive methods may be possible in future research work.

8.6.2 Cryptographic Techniques

Cryptographic operations such as encryption and decryption of data packets and control information need to be used wherever applicable to strengthen the security of MANETs. This requires adequate key management schemes. The public key cryptography schemes often depend on the centralized Certificate Authority (CA) entities, and the centralization issue is rather in contrary to the design and functional goals of MANETs. In fact, a number of researches have investigated whether several MANET nodes may be distributed to act as CAs based on a secret-sharing mechanism. However, the mobility of the nodes makes it more difficult to obtain a dynamic reconfiguration of the CAs in MANETs and this issue remains yet to be resolved. Researchers should also decide whether to adopt more efficient distributed trust models such as PGP or employ the simple yet computation-efficient symmetric cryptography for the sake of saving the scarce battery resources of the *ad hoc* wireless nodes. The tradeoff issue pertaining to the level of security and efficiency is definitely going to be an issue in MANET security provisioning.

8.6.3 Resiliency

Due to the fact that many of the attacks against MANETs are unpredictable, a resiliency-oriented security solution is required to be developed in future. This will help the legitimate nodes to recover from a possible network failure as soon as possible. While cryptographic solutions discussed in Section 8.6.2 offer only a subset of solutions, the multifaceted solutions toward a resilient MANET architecture is expected to evolve in future.

8.7 Conclusions

In recent time, mobile *ad hoc* networks have emerged as a promising technology and gained tremendous attention from researchers. Since these networks can rapidly be deployed without the need of any predefined infrastructure, they can easily be applied to various scenarios ranging from emergency operations and disaster relief to military services, vehicular networks, and other sensitive domains. However, their lack of infrastructure and/or central authoritative environment offers plenty of opportunities to malicious nodes for launching a wide array of attacks. Therefore, in order to protect the sensitive information that are exchanged in MANETs, security provisioning is of utmost importance.

One of the main research challenges in MANET security provisioning is that these networks are already resource constrained. For instance, a MANET has scarce battery power, computational/processing ability, and also limited bandwidth. As a consequence, the conventional security schemes for wired and other wireless networks may not be directly applicable to MANETs. This particular issue also makes MANETs much more susceptible to security attacks.

This chapter reviewed the current state of-the-art transport layer and routing attacks. In particular, the counter-measures to circumvent the routing attacks along with their pros and cons have also been delineated. This chapter reveals that while a large body of literature is available in countering many security attacks against MANETs, they usually focus on dealing with only one type of attack. Therefore, we are still far from achieving a perfect solution that may integrate all existing security solutions effectively.

Terminologies

Mobile *Ad Hoc* networks (MANETs)
Security attacks
Routing attacks
OLSR
Blackhole attack
Wormhole attack
Rushing attack
Intrusion detection

Questions and Sample Answers

1. What is the difference between external and insider attacks?
   The attacks against MANETs may be classified into two broad groups: external (outsider) and internal (insider) attacks. In case of an external attack, the attack-node does not belong to the target MANET. On the other hand, insider attacks feature compromised MANET hosts that turn out to be attackers. The internal (i.e., compromised) nodes have knowledge pertaining to valuable information about the network topology and also possess adequate access privileges. As a consequence, the internal attacks have a more serious impact on the victim system.

2. What are selfish nodes in a MANET and how do they lead to security vulnerabilities?
   Selfish hosts in a MANET just willingly drop the received packets for saving their own battery resources. The normal operation of the MANET is interrupted by selfish nodes since they do not take part in the routing protocols or forward packets. This leads to incomplete information of the route from a source to destination via the selfish nodes.

3. In how many ways is it possible to target a MANET on the network layer?
   The network layer attacks against MANETs can be launched in three ways. The first variety of networking attacks target the routing discovery phase. The second type of attack is launched

during the MANET routing maintenance phase. There are also networking attacks during the data forwarding phase.

4. What is the wormhole attack?

 The wormhole attack is considered to be one of the most serious threats against MANETs routing. It consists of two malicious users in collusion. A high-speed private network connects these two attackers. The first malicious node collects packets at a certain point in the MANET topology and replays them at the other node by using the high-speed network. Thus, the packets intended to a particular destination node are dropped.

5. What are the differences between wormhole and byzantine attacks?

 In traditional wormhole attacks, the attacker may trick two nonmalicious nodes to assume that there exists a direct link between them. On the other hand, in case of a Byzantine attack, the wormhole link exists between the compromised nodes and not between the nonmalicious ones. This implies that these end nodes cannot be trusted to follow the routing protocol.

6. What makes the rushing attacks so dangerous?

 Usually, the attacks against MANET target a particular routing protocol (e.g., AODV, OLSR, and so forth). The rushing attack, on the other hand, acts as an effective DoS-type threat against all conventional on-demand MANET routing protocols. In fact, even the secure routing protocols (e.g., SAODV and AODV secured with SUCV) were shown to be vulnerable to this particular rushing attack.

7. How can we prevent rushing attacks?

 The Route Discovery Protocol (RAP), which replaces the standard mechanism of the conventional *ad hoc* routing protocols that are inherently vulnerable to rushing attacks, is used to thwart rushing attacks. The RAP combines three techniques to prevent the rushing attack, namely a secure neighbor discovery mechanism, a secure route delegation acceptance protocol, and the randomized selection of the route request that will be forwarded.

8. Which technique is used to protect MANETs from replay attacks? Is it full proof?

 A time-stamp based asymmetric encryption solution may be employed to protect MANETs from possible replay attacks. This scheme simply compares the current time with the time stamp embedded in the received control messages from other nodes. If the time stamp in a received control packet deviates much from the current time, the receiving node considers it to be a possible replay attack and therefore, the packet is discarded to avoid updating the routing table with stale information. However, this solution is not full proof as it still remains vulnerable to wormhole attacks comprising a pair of colluding attackers that employ a high-speed network for replaying messages in a faraway location with rather low latency.

9. What is the implication of multilayer attacks against MANETs?

 Usually, the individual attacks against MANETs target one of the layers. More sophisticated attacks are expected to evolve to target various layers at the same time. In case of such cross-layer attacks, the security mechanisms need to be enforced on each layer separately which is, indeed, cumbersome.

10. What is the prospect of using key management schemes in enforcing security in MANETs?

 There is a research scope for integrating security primitives in MANET-based systems. For example, key management protocols may assist in enhancing the overall security level of a MANET as perceived by its users. Indeed, substantial research may be carried out on designing robust key management systems, trust-based routing protocols, and integrating authentication and encryption in different layers.

11. Can conventional intrusion detection schemes be used in MANETs?

    In contrast to other wired and wireless networks, MANETs have unique features such as open nature, mobility of the nodes, and dynamic change in the topology. Consequently, the conventional intrusion detection and prevention methods from security attacks may not be directly applicable to MANETs. The intrusion detection theme is of significant importance in discovering a potential attack before it may severely impact the target network and the victim(s).

Author's Biography

Zubair Md. Fadlullah received his MS degree from the Graduate School of Information Sciences (GSIS) at Tohoku University, Japan in March 2008. He obtained his bachelor degree in computer science and information technology from Islamic University of Technology (IUT), Gazipur, Bangladesh, in 2003. Currently, he is working toward a PhD degree at GSIS, Tohoku University. His research interests are in the areas of network security, specifically intrusion detection/prevention, trace-back mechanisms, and quality of protection.

Tarik Taleb is currently working as senior researcher at NEC Europe Ltd., Heidelberg, Germany. Prior to his current position and till March 2009, he worked as assistant professor at the Graduate School of Information Sciences, Tohoku University, Japan. From October 2005 till March 2006, he was working as research fellow with the Intelligent Cosmos Research Institute, Sendai, Japan. He received his B.E. degree in information engineering with distinction, MSc and PhD degrees in information sciences from GSIS, Tohoku University, in 2001, 2003, and 2005, respectively. His research interests lie in the field of architectural enhancements to 3GPP networks (i.e., LTE), mobile multimedia streaming, wireless networking, intervehicular communications, satellite and space communications, congestion control protocols, network management, handoff and mobility management, and network security. His recent research has also focused on on-demand media transmission in multicast environments. Dr. Taleb is on the editorial board of the *IEEE Transactions on Vehicular Technology*, *IEEE Wireless Communications*, *IEEE Communications Surveys & Tutorials*, and a number of Wiley journals. He also serves as Vice Chair of the Satellite and Space Communications Technical Committee of the IEEE Communication Society (ComSoc) (2006–present). He has been on the technical program committee of different IEEE conferences, including Globecom, ICC, and WCNC, and chaired some of their symposia. He is the recipient of the 2009 IEEE ComSoc Asia-Pacific Young Researcher award (June 2009), the 2008 TELECOM System Technology Award from the Telecommunications Advancement Foundation (March 2008), the 2007 Funai Foundation Science Promotion Award (April 2007), the 2006 IEEE Computer Society Japan Chapter Young Author Award (December 2006), the Niwa Yasujirou Memorial Award (February 2005), and the Young Researcher's Encouragement Award from the Japan chapter of the IEEE Vehicular Technology Society (VTS) (October 2003). Dr. Taleb is an IEEE member.

Marcus Schöller received his MSc in computer science (Dipl.-Inform.) from Karlsruhe Universität, Karlsruhe, Germany, in 2001. In 2006, he received his PhD from Karlsruhe Universität. Before joining the Next Generation Networking group at NEC Laboratories, Heidelberg, Germany, he did a postdoctoral year at Lancaster University, Lancaster, United Kingdom. His research interests cover the fields of autonomous communication systems, resilient and survivable network architectures, wireless access technologies, and network security. Marcus has actively contributed to several national and European projects and published in key international conferences and journals.

References

1. C. Douligeris and A. Mitrokosta, DDoS attacks and defense mechanisms: Classification and state-of-the-art, in *Computer Networks: The International Journal of Computer and Telecommunications Networking*, 44(5), 643–666, 2004.
2. L. Zhou and Z. J. Haas, Securing ad hoc networks, *IEEE Network Magazine*, 13(6), 24–30, 1999.
3. A. Mishra, K. Nadkarni, and A. Patcha, Intrusion detection in wireless ad hoc networks, *IEEE Wireless Communications*, 11(1), 48–60, 2004.
4. E. C. H. Ngai, M. R. Lyu, and R. T. Chin, An authentication service against dishonest users in mobile ad hoc networks, *in Proceedings of IEEE Aerospace Conference*, Big Sky, Montana, USA, March 2004.
5. L. Blazevic, et al., Self-organization in mobile ad-hoc networks: The approach of terminodes, *IEEE Communication Magazine*, 39(6), 166–173, 2001.
6. Y. Zhang, W. Lee, and Y. Huang, Intrusion detection techniques for mobile wireless networks, *Wireless Networks Journal (ACM WINET)*, 9(5), 545–556, 2003.
7. W. Zhang, R. Rao, G. Cao, and G. Kesidis, Secure routing in ad hoc networks and a related intrusion detection problem, in *Proceedings of IEEE Military Communications Conference (MILCOM)*, Boston, MA, October 2003.
8. J. Kong, et al., Adaptive security for multi-layer ad-hoc networks, *Special Issue of Wireless Communications and Mobile Computing—Research, Trends and Applications*, 2(5), 533–547, 2002.
9. J. Hubaux, L. Buttyan, and S. Capkun, The quest for security in mobile ad hoc networks, in *Proceedings of 2nd ACM International Symposium on Mobile Ad Hoc Networking and Computing (MobiHoc)*, Long Beach, CA, October 2001.
10. P. Papadimitratos, Z. J. Haas, and E. G. Sirer, Path set selection in mobile ad hoc networks, in *Proceedings of 3rd ACM International Symposium on Mobile Ad Hoc Networking and Computing*, Lausanne, Switzerland, June 2002.
11. C. E. Perkins and E. Belding-Royer, Ad hoc on-demand distance vector (AODV), *Request for Comments (RFC)* 3561, 2003.
12. B. DeCleene, et al., Secure group communications for wireless networks, in *Proceedings of MILCOM'01*, Washington, D.C., October 2001.
13. S. Bo, W. Kui, and U. W. Pooch, Towards adaptive intrusion detection in mobile ad hoc networks, in *Proceedings of IEEE Globecom'04*, Dallas, TX, November 2004.
14. C. Douligeris and A. Mitrokosta, DDoS attacks and defense mechanisms: Classification and state-of-the-art, *Computer Networks: The International Journal of Computer and Telecommunications Networking*, 44(5), 643–666, October 2004.
15. I. Chlamtac, M. Conti, and J. J. Liu, Mobile ad hoc networking: Imperatives and challenges, *Ad Hoc Networks*, 1(1), 13–64, 2003.
16. H. Yang, H. Y. Luo, F. Ye, S. W. Lu, and L. Zhang, Security in mobile ad hoc networks: Challenges and solutions, *IEEE Wireless Communications*, 11(1), 38–47, 2004.
17. Y. Hu, A. Perrig, and D. Johnson, Packet leashes: A defense against wormhole attacks in wireless ad hoc networks, in *Proceedings of IEEE Infocom'03*, San Francisco, CA, March 2003.
18. Y. Hu, A. Perrig, and D. Johnson, Ariadne: A secure on-demand routing protocol for ad hoc networks, *Wireless Networks*, 11(1–2), 21–38, January 2005.
19. J. R. Douceur, The Sybil attack, in *Proceedings of the 1st International Workshop on Peer-to-Peer Systems (IPTPS)*, Cambridge, MA, March 2002.
20. Y. Hu, A. Perrig, and D. Johnson, Rushing attacks and defense in wireless ad hoc network routing protocols, in *Proceedings of 2nd ACM Workshop on Wireless Security*, San Diego, CA, September 2003.
21. W. Lou and Y. Fang, A survey of wireless security in mobile ad hoc networks: Challenges and available solutions, *Ad Hoc Wireless Networks*, X. Chen, X. Huang, and D.-Z. Du, eds. Kluwer Academic Publishers/Springer, Germany, 2003, pp. 319–364.
22. Y. Hu and A. Perrig, A survey of secure wireless ad hoc routing, *IEEE Security and Privacy*, 2(3), 28–39, 2004.
23. C. E. Perkins, *Ad Hoc Networking*, Addison-Wesley, Massachusetts, 2001.
24. T. Clausen and P. Jacquet, Optimized link state routing protocol (OLSR) project, *Hipercom, INRIA*, RFC-3626, 2003.

25. M. Ilyas, *The Handbook of Ad Hoc Wireless Networks.* CRC Press, USA, 2003.

26. K. Ng and W. Seah, Routing security and data confidentiality for mobile ad hoc networks, in *Proceedings of IEEE Vehicular Technology Conference (VTC)*, Jeju, Korea, April 2003.

27. Y.-C. Hu, A. Perrig, and D. B. Johnson, Wormhole attacks in wireless networks, *IEEE JSAC*, 24(2), 370–380, 2006.

28. D. Dhillon, J. Zhu, J. Richards, and T. Randhawa, Implementation and evaluation of an IDS to safeguard OLSR integrity in MANETS, in *Proceedings of IWCMC'06*, Vancouver, BC, Canada, July 2006.

29. S. Capkun, L. Buttyan, and J. Hubaux, Sector: Secure tracking of node encounters in multi-hop wireless networks, in *Proceedings of ACM Workshop on Security of Ad Hoc and Sensor Networks (SASN)*, Washington, USA, October 2003.

30. L. Hu and D. Evans, Using directional antennas to prevent wormhole attacks, in *Proceedings of Networks and Distributed System Security Symposium (NDSS)*, San Diego, CA, February 2004.

31. L. Qian, N. Song, and X. Li, Detecting and locating wormhole attacks in wireless ad hoc networks through statistical analysis of multi-path, in *Proceedings of IEEE Wireless Communication and Networking Conference (WCNC)*, New Orleans, LA, March 2005.

32. S. Yi, P. Naldurg, and R. Kravets, Security-aware ad-hoc routing for wireless networks, Report No. UIUCDCS-R-2002-2290, UIUC 2002.

33. S. Lee, B. Han, and M. Shin, Robust routing in wireless ad hoc networks, in *Proceedings of International Conference Parallel Processing Wksps.*, Vancouver, Canada, August 2002.

34. M. Al-Shurman, S.-M. Yoo, and S. Park, Black hole attack in mobile ad hoc networks, in *Proceedings of ACM Southeast Regional Conference*, Huntsville, AL, April 2004.

35. S. Kurosawa, H. Nakayama, N. Kato, A. Jamalipour, and Y. Nemoto, Detecting blackhole attack on AODV-based mobile ad hoc networks by dynamic learning method, *International Journal of Network Security*, 5(3), 338–346, November 2007.

36. B. Awerbuch, D. Holmer, C. Nita-Rotaru, and H. Rubens, An on-demand secure routing protocol resilient for byzantine failures, in *Proceedings of ACM Workshop on Wireless Security*, Atlanta, GA, September 2002.

37. C. Crépeau, C. R. Davis, and M. Maheswaran, A secure MANET routing protocol with resilience against byzantine behaviours of malicious or sel?sh nodes, in *Proceedings of 21st International Conference on Advanced Information Networking and Applications Workshops*, Niagara Falls, Canada, May 2007.

38. P. Yi, Z. Dai, S. Zhang, and Y. Zhong, A new routing attack in mobile ad hoc networks, International Journal Info. Tech., 11(2), 83–94, 2005.

39. S. Desilva and R. V. Boppana, Mitigating malicious control packet floods in ad hoc networks, in *Proceedings of IEEE Wireless Communication and Networking Conference*, New Orleans, LA, March 2005.

40. B. Kannhavong, H. Nakayama, N. Kato, Y. Nemoto, and A. Jamalipour, Analysis of the node isolation attack against OLSR-based mobile ad hoc networks, in *Proceedings of 7th IEEE International Symposium on Computer Networks (ISCN'06)*, Istanbul, Turkey, June 2006.

41. D. Raffo, C. Adjih, T. Clausen, and P. Mühlethaler, Securing OLSR using node locations, in *Proceedings of European Wireless (EW'05)*, Nicosia, Cyprus, April 2005.

42. C. Adjih, D. Raffo, and P. Muhlethaler, Attacks against OLSR: Distributed key management for security, in *Proceedings of 2nd OLSR Interop/Wksp.*, Palaiseau, France, July 2005.

43. H. Hsieh and R. Sivakumar, Transport over wireless networks, *Handbook of Wireless Networks and Mobile Computing*, I. Stojmenovic, ed., John Wiley and Sons, Inc., New York, 2002.

44. B. Kannhavong, H. Nakayama, N. Kato, A. Jamalipour, and Y. Nemoto, A collusion attack against OLSR-based mobile ad hoc networks, in *Proceedings of IEEE Global Telecommunications Conference (Globecom'06)*, San Francisco, CA, November 2006.

45. F. Nait-Abdesselam, B. Bensaou, and T. Taleb, Detecting and avoiding wormhole attacks in wireless ad hoc networks, *IEEE Communications Magazine*, 46(4), 127–133, 2008.

Chapter 9

Classification of Attacks on Wireless Mobile *Ad Hoc* Networks and Vehicular *Ad Hoc* Networks

A Survey

Vikrant Gokhale, S.K. Ghosh, and Arobinda Gupta

Contents

9.1 Introduction

With larger penetration of mobile devices in the day-to-day life, wireless communication has become an active area of research. Mobile *ad hoc* networks (MANETs) are wireless networks with no infrastructure support and no central control over the nodes in the network. Nodes can join and leave the network dynamically, and the topology of the network can change frequently. Maintaining security in such a volatile and dynamic environment is a challenging task and has become an important area of research. Hubaux et al. [1] brought out various challenges that are faced by MANETs and discussed the need for security in such environments.

Nodes in a MANET can be vulnerable to different types of attacks. These attacks can occur at different layers and can be classified into several ways. Most of the existing works that propose new and secure protocols for MANET have discussed about the issue of classifying the attacks. However, these works have largely remained focused on routing attacks. In [2,3], the authors have discussed the issue of misbehavior of nodes at Medium Access Control (MAC) layer and have attempted the classification of attacks at that layer. Some work has also been done on the attacks on application layer [4,5], but the area of focus has largely remained on sensor networks. It is also important to note that many of the existing works have made some specific and/or implicit assumptions as per their requirements.

In this chapter, we present a survey of attacks on MANETs. The survey attempts to study and classify the different types of attacks that can occur in a MANET and analyses the techniques proposed to mitigate them. Although routing attacks are also considered in detail, the aim of this work is to classify all attacks to give a comprehensive view of possible attacks in a MANET. Table 9.1 gives a brief overview of layers and attacks considered in this work.

The attacks in a MANET can be classified into different ways. One possible way is to classify the attacks as *internal* or *external* attacks. An internal attack is mounted by a node that is part of the system under consideration, whereas an external attack is mounted by a node from outside the system. Internal attacks are sometimes difficult to handle as internal nodes may be more trusted than external nodes, and protecting the network with firewalls may not be helpful. Broadcast attacks

Table 9.1   Schematic of Various Attacks on Individual Layers

Network Layer	Type of Attacks
Application	Malicious code, repudiation
Transport	Session hijacking, SYN flooding
Network	Flooding, blackhole, greyhole, wormhole, link spoofing, link withholding, Byzantine, replay, location disclosure
Data link/MAC	Malicious behavior, selfish behavior, active, passive, internal, external
Physical	Interference, jamming, eavesdropping

or point-to-point attacks are another approach for attack classification because depending on the type of link, the characteristics of attacks vary to a great extent. Another way is to classify the attacks as *active* or *passive*. An active attack modifies the contents of the packets, whereas a passive attack does not. An active attack is usually easier to handle as detecting packet modifications is not difficult. Attacks can also be classified based on the basic mechanism used by the attacker, such as modification, fabrication, impersonation, so on. In this chapter, we first classify the attacks based on the layer of the networking stack in which they occur. Attacks on different layers have different consequences and require different mitigation techniques. Hence, separating the attacks at different layers can be useful. Within each layer, we also classify the attacks based on the above possible classifications. It should also be understood that malicious behavior of a node has also been considered as a type of attack in this work.

The rest of this chapter is organized as follows. In Section 9.2, attacks on physical layer will be considered. In Sections 9.3 through 9.6, the attacks and their mitigation along with discussion and analysis are done for the MAC layer, network layer, transport layer, and application layer, respectively. In Section 9.7, attacks on Vehicular *Ad Hoc* Network (VANET), in particular, have been dealt with owing to peculiar characteristics of it.

9.2 Attacks at the Physical Layer

The attacks on the physical layer are hardware oriented and, although simple to execute, need help from some sort of hardware to come into effect. On the other hand, the attacks on other layers can be thought of as manipulation and use of the existing equipment with only modification to the code which is being used on the equipment. As discussed in [6,7], these attacks are fairly simple to execute and can be launched without having a complete knowledge of the technology.

Some of the attacks observed at this layer are eavesdropping, interference, and jamming attacks [8,9]. Eavesdropping is the intercepting and reading of messages and conversations by unintended receivers. The mobile hosts in MANET share a wireless medium. The majority of wireless communications uses the RF spectrum and broadcast by nature. Signals broadcast over airwaves can easily be intercepted with receivers tuned to the proper frequency. Thus, messages transmitted can be overheard, and fake messages can be injected into the network. Radio signals can be jammed or interfered with, which causes the message to be corrupted or lost. If the attacker has a powerful transmitter, a signal can be generated that will be strong enough to overwhelm the targeted signals and disrupt communications. The most common types of this form of signal jamming are random noise and pulse. Jamming equipment is readily available. In addition, jamming attacks can be mounted from a location remote to the target networks. Navda et al. [10] have proposed frequency hopping as a mitigation technique for jamming attacks. Also, in [11], the authors have proposed a protocol called SPREAD which mitigates the jamming attacks mounted by smart or intelligent jammers. Borisov et al. [12] have discussed how wired equivalent privacy protocol used in IEEE 802.11 MAC can be manipulated and how it is vulnerable to attacks such as message authentication, modification, and spoofing by malicious users with sufficient resources.

Attacks such as physical layer capture [13] have been studied and simulated, but the use of external hardware makes it more difficult to mount such an attack under constantly changing topology. For example, in case of a VANET, the attacker will have to remain in the vicinity of the moving vehicle continuously for disrupting its communication, which is difficult. For jamming attacks, the same constraints remain and though such attacks might of value in stationary topology like sensor networks, it becomes less relevant for dynamic topologies such as VANET.

9.3 Attacks at the MAC Layer

For understanding the attacks on an MAC layer of 802.11, it is necessary to first have a look at the general functioning of the same. Wireless MAC protocols have to coordinate the transmissions of the nodes on the common transmission medium. It also takes care of the transmission within single hop neighbors. Because a token-passing bus MAC protocol is not suitable for controlling a radio channel, the IEEE 802.11 protocol is specifically devoted to wireless local area networks. The IEEE 802.11 MAC protocol uses distributed contention resolution mechanisms for sharing the wireless channel. The IEEE 802.11 working group proposed two algorithms for contention resolution. One is a fully distributed access protocol called distributed coordination function (DCF). The other is a centralized access protocol called point coordination function (PCF). The PCF requires a central decision-maker such as a base station. The DCF uses a carrier sense multiple access/collision avoidance protocol (CSMA/CA) for resolving channel contention among multiple wireless hosts.

Three values for interframe space (IFS) are defined to provide a priority-based access to the radio channel. SIFS is the shortest IFS and is used for Acknowledgment (ACK), Clear to Send (CTS), and poll response frames. DCF Interframe Space (DIFS) is the longest IFS and is used as the minimum delay for asynchronous frames contending for access. PCF Interframe Space (PIFS) is the middle IFS and is used for issuing polls by the centralized controller in the PCF scheme. In case, there is a collision, the sender waits a random unit of time, based on the binary exponential back off algorithm, before retransmitting. In Figure 9.1, nodes N_a and N_c contend to communicate with node N_b. First, node N_a gets access and reserves the channel, and then N_c succeeds and reserves the channel while node N_a has to back off.

In *ad hoc* networks, where mobile nodes communicate with each other through multihop wireless links, the corresponding routing and MAC protocols were designed under the basic assumption that all hosts would obey the protocol specifications and there will be cooperation among the nodes. However, in a dynamic and open environment, such an assumption does not seem to be realistic. A misbehaving node can compromise the network at the physical, MAC, network, or even at the application layer. The MAC layer of 802.11 is potentially an insecure place as no centralized monitoring can be setup in an *ad hoc* environment. The MAC layer misbehavior can be of various types, including increasing unfairness by starving multihop flows [14], increasing delays beyond

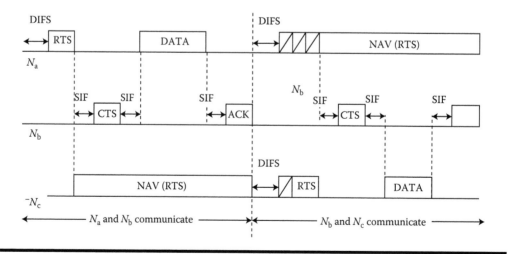

Figure 9.1   Contentions in 802.11.

threshold levels [15], depleting resources or channel capacity and preventing access to channel [16,17]. There is another aspect that needs to be looked into while mitigating these attacks and that is of maintaining anonymity. A diagnosis of wireless packet losses in 802.11 has been studied in [18].

The MAC layer attacks can be classified as to what effect it has on the state of the network as a whole. The effects can be measured in terms of route discovery failure, energy consumption, unfair share of bandwidth, link breakages initiating route discovery, and so on, but this takes into consideration only the effects of attack on macrolevel and finer aspects necessary for control and mitigation of the problem might be overlooked. On the other hand, the attacks can be classified depending on the behavior exhibited by the node while functioning. The misbehavior of a node can be purely in selfish interest or with malicious intents. We next discuss the attacks at the MAC layer in these two categories.

9.3.1 Selfish Misbehavior of Nodes

In *ad hoc* networks, the basic assumption that all nodes are cooperating and are well behaved may not hold well in case of MANET [19]. It should be understood that the attacks under this category are essentially directed to improve self-performance (energy, latency, throughput, etc.) and does not interfere with the operation of the network as a whole. The misbehavior of a node classified as selfish can be attributed to the following factors:

■ conservation of battery power
■ gaining unfair share of bandwidth

Therefore, in the absence of such nodes, the network will function normally, which is not the case otherwise. This is the fundamental difference between the types of misbehavior on which this classification has been made.

Various attacks that can be classified under this category have been studied in the literature. One type of attack is to manipulate the protocol parameters of 802.11 (e.g., shorter DIFS, oversized NAV, back-off manipulation). These types of attack will necessarily give undue advantage to the selfish host in terms of unfair access to channel or bandwidth. Various mitigation techniques have been proposed to overcome such types of attack [15,17,20,21]. Here, the misbehaving host is detected with the help of monitoring the traffic for certain interval of time and if the behavior is found to be inconsistent, then the host is barred from taking part into the transmission.

In other types of attacks, a selfish node may refuse to take part in the forwarding process or drop the packets intentionally in order to conserve the resources. The detection of this type of attacks is again, by using mechanisms, similar to reputation management schemes and heavily depends on monitoring and studying the traffic for certain duration [22]. This type of attacks can be mitigated either by penalizing misbehavior [23,24] or by providing incentives to hosts for proper cooperation [25,26].

One detection scheme for such misbehavior has been proposed by Kyasanur et al. [15], which modifies the 802.11 protocol. However, it works on the assumption that the receiver is trustworthy, which may not be true always in a MANET. To overcome this limitation, Konorski [27] has suggested a modification to the technique proposed by Kyasanur et al. under an assumption that at least one of the parties involved is honest.

Game-theoretic techniques [28–32] have been used to develop protocols which are resilient to misbehavior. The game theoretic approach assumes that all users are selfish and rational. Rational hosts always select a strategy that maximizes their utility (utility is a measure of the benefit obtained by a host). Protocols are designed that reach an equilibrium state, called the "Nash equilibrium," where a selfish host cannot unilaterally gain any advantage over well-behaved hosts.

Most of the protocols using game-theoretic techniques are based on the assumption of "Perfect Information," that is, every host can observe all the actions of other hosts in the network. This assumption is hard to realize in case of MANETs, especially in the context of a wireless network (with fading channels, hidden terminals, etc.). In addition, protocols developed with game-theoretic techniques may not achieve the performance of protocols developed under the assumption that all hosts are well behaved and cooperate with each other (e.g., IEEE 802.11).

9.3.2 Malicious Behavior of Nodes

These attacks are necessarily meant to disrupt the normal operation of the network in terms of network throughput and availability and hence prevent the other legitimate users from communicating.

The attacks observed in these types of scenario are primarily Denial of Service (DoS) attacks. The DoS attacks uses various techniques, namely, back-off interval manipulation [2,20] which exploits the vulnerabilities of 802.11 [17,33], jelly fish attacks [14], and intelligent cheater attacks [2]. In case of intelligent cheater attacks the nodes adhere to the protocol for most of the time and misbehave intermittently.

In the back-off interval manipulation, if the sender is in control of the back-off interval decision as the case is generally, then it can choose smaller and smaller intervals with the intention of launching a DoS attack. This type of behavior can be mitigated using two techniques. In the fit one, the receiver, and not the sender, is in control of adjusting the back-off interval. However, for this scheme to be successful, there is an implicit assumption that one of the communicating parties is not misbehaving. If this condition fails, then this method does not work. The other approach is by establishing a trust-monitoring mechanism that will take the decision of allowing or blocking communication with a particular node.

On the other hand, jelly fish attack is a protocol compliant DoS attack. Here, the misbehaving node adheres to the protocol but stealthily misorder, delay, or periodically drop packets that they are expected to forward. This leads astray end-to-end congestion control protocols. This type of attack is very difficult to detect as the node behaves well most of the time and, therefore, any monitoring mechanism may not be able to revoke the trust level of such nodes. Only prolonged monitoring can help in detecting such misbehavior [14], which is essentially time consuming.

The intelligent cheater attacks are similar to jelly fish attacks where the nodes are behaving for most of the time and misbehave only intermittently. As discussed in the previous paragraph, such attacks are difficult to detect using the approach of trust-monitoring schemes. On the other hand, the damage potential of such attacks is limited because such intelligent nodes make it a point to maintain their trust rating within the threshold [2]. This, in turn, limits their capacity to harm the network.

Various misbehavior detection schemes for MAC layer have been proposed in the literature [2,13,14,16,20,34]. By and large, the most researched problem is that of detecting back-off manipulation. The current literature offers two major approaches to address this problem. The first set of approaches provides solutions based on modification of the MAC layer protocol by making the monitoring stations aware of the back-off value of its neighbor [26,32]. The other approach makes use of misbehavior detection schemes without making any changes in the MAC layer protocol. Here, the set of rules are checked with the observed values, and any change beyond the set limit is construed as misbehavior. However, such schemes fail for attacks such as Jelly fish and intelligent cheater.

9.3.3 *Other Classifications*

For the sake of more clear insight and understanding of the attacks, different classifications can be made depending on the relative positioning of the attacker with respect to the group of nodes under study and whether the attack is a passive or an active attack and so on.

9.3.3.1 *Internal versus External Attacks*

Here, we will be considering a set of nodes on which we want to observe the effects of different attacks. So, if the attacker belongs to this group, the attack can be considered to be internal and vice-versa. The types of attacks discussed earlier under the category of selfish misbehavior mostly fall in the internal type of attacks as notion of being selfish within the group of nodes under consideration is more relevant than a node outside being selfish.

But a node behaving with malicious intents can be an internal as well as an external attacker. The sole intention of the node is to disrupt the network, which can be achieved by being internal to the group of nodes or by being external. For example, the Jelly fish attack can be mounted by two nodes outside the group and still the overall network throughput will effectively get affected.

9.3.3.2 *Active versus Passive Attacks*

The attacks discussed in the above paragraphs can also be classified depending on whether the attack is active or passive. The attacks such as DoS and jamming can clearly be classified as active attacks where the protocol fields are modified. On the contrary, in case of a jelly fish or an intelligent attack, a node does not make any change to the protocol fields but only ensures that the packets are dropped or the routing operation fails. From the nature of attack itself it can be seen that mitigating passive attacks is much more difficult than defending against active attacks where the changes can easily be monitored and action can be initiated.

9.3.4 *Discussion and Analysis*

As can be seen from the above classification, the attacks on MAC layer are mainly defended by forming a matrix of trust using parameters in the protocol fields. Formation of such matrix is necessarily evaluation of accumulated data over a certain period of time and then checking the data for variations above the set threshold. Guang and Assi [2] have discussed attacks where such matrices can be identified for each of the attack mitigation technique.

This approach, however, has an inherent limitation that the newly joined node needs to be trusted initially till such time an opinion has been formed about it. Also, assumptions like authentication and integrity services are running in the network and are effectively working had to be made. So, a secure routing protocol needs to be assumed which itself is another challenging issue in the MANETs [35–37]. Thus, the misbehaving node will have an initial advantage for some time till its credibility is established.

In case of sensor networks where once deployed, for most of the time the topology of the network is stable, such schemes prove to be of value but in volatile environments like VANETs, there might not be sufficient time to establish such rating or matrix for wherein the credibility of the node can be established.

Additionally, if the attack is passive, then the mitigation becomes all the more difficult due to paucity of time and also no visible modification can be detected in the protocol fields. Thus, it can be observed that the approach where metric for evaluation of trustworthiness of node is formed has inherent limitations and can only partially contain or mitigate the misbehavior. It should also be understood that, such MAC layer misbehavior mitigation is subject to the assumptions made like correct working of secure routing protocol.

From the above discussion, it can be seen that providing security against MAC layer misbehavior in MANET cannot be tackled in isolation and at present no full-proof solution for the same exists. A cross layer approach with integration of functionalities from other layers is a necessary requirement for handling this problem.

9.4 Attacks at the Network Layer

Network layer protocols extend connectivity from neighboring 1-hop nodes to all other nodes in a MANET. The connectivity between mobile hosts over a potentially multihop wireless link relies heavily on cooperative reactions among all network nodes. In an ideal situation, all in-between nodes are assumed to be faithful and performing the operations as per the protocol. But, this is a farfetched assumption for the volatile environment such as MANET. As each individual node is responsible for making routing decisions for it, it becomes comparatively easy for a misbehaving node to mount on attack on such a network.

A variety of attacks targeting the network layer have been identified and heavily studied. The basic idea behind the network layer attacks is to absorb network traffic, inject itself into the path between the source and destination, divert and thus control the network traffic flow. By attacking the routing protocols, attackers can achieve these objectives. Simple illustration of the same is shown in Figure 9.2 where attacking malicious node A can inject itself into the routing path between sender S and receiver R.

Many methods are used by the attackers to achieve the ultimate goal of network traffic disruption. The traffic packets could be forwarded to a nonoptimal path, which could introduce a significant delay. In addition, the packets could be forwarded to a nonexistent path and get lost. The attackers can create routing loops and introduce severe network congestion and channel contention in certain areas. Multiple colluding attackers may even prevent a source node from finding any route to the destination, causing the network to partition, which triggers excessive network control traffic, and further intensifies network congestion and performance degradation. Maltz et al. [38] studied the effects of on-demand behavior in routing protocols for multihop wireless *ad hoc* networks.

One way of classifying attacks at the network layer can be based on the phase of routing operation when the attack is carried out. Thus, broad classifications can be made as attacks on routing discovery phase, routing table overflow attacks, route cache poisoning attack, routing maintenance phase attack, data-forwarding phase attack, and so on. But this leaves out many sophisticated attacks such as rushing attack, blackhole attack, and so on.

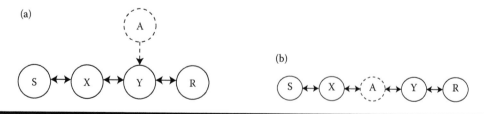

Figure 9.2   Schematic for attack at network layer. (a) Attacking node "A" compromises node "Y" and (b) attacking node "A" injects itself in the routing path between sender "S" and receiver "R."

Most of the attacks in MANET try to utilize the vulnerabilities of the routing protocol being used, that is, either proactive protocols like Optimized Link State Routing (OLSR) or reactive protocols like *Ad hoc* On-Demand Distance Vector (AODV). So, based on the routing protocols and vulnerability being used, the attacks can be classified [39] at network layer. But this also leaves out attacks like rushing attack.

In this chapter, it is intended to list out all attacks on the network layer instead of subclassifying them into categories so as to cover the entire gamut of attacks. The different attacks possible at the network layer are discussed below:

■ *Flooding attack:* The flooding attack [40] aims at exhausting the network resources, such as bandwidth, and consuming a node's resource, such as computational and battery power or to disrupt the routing operation to cause severe degradation in network performance. For example, in AODV protocol, a malicious node can send a large number of Route Requests (RREQs) in a short period to a destination node that does not exist in the network. Because no one will reply to the RREQs, these RREQs will flood the whole network. As a result, all of the node battery power, as well as network bandwidth, will be consumed and could lead to DoS. Boppana and Desilva [41] have shown that a flooding attack can decrease the network throughput by upto 84%.

■ *Blackhole attack:* The blackhole attack has two properties. The node exploits the mobile *ad hoc* routing protocol, such as AODV, to advertise itself as having a valid route to a destination node, even though the route is spurious, with the intention of intercepting packets. Because of this, the route that passes through the attacker gets selected for routing. The attacker consumes the intercepted packets without any forwarding. However, the attacker runs the risk that neighboring nodes will monitor and expose the ongoing attacks. There is a more subtle form of these attacks when an attacker selectively forwards packets. An attacker suppresses or modifies packets originating from some nodes, while leaves the data from the other nodes unaffected, which limits the suspicion of its wrongdoing.

■ *Greyhole attack:* Pirzada and McDonald [42] have discussed a variation in blackhole attack, namely greyhole attack. Here, the node through which the packets are passing does not drop all the packets, but remains selective about it. For example, the node may drop packets with routing information and will route the data packets as per the requirement. Such an attack is difficult to detect as the node misbehaves selectively.

■ *Wormhole attack:* A wormhole attack [43,44] is one of the most sophisticated and severe attacks in MANETs. An attacker records packets at one location in the network and tunnels them into another location. Routing can be disrupted when routing control messages are tunneled. This tunneling between two colluding attackers is referred to as a wormhole [45]. Wormhole attacks are severe threats to MANET routing protocols. For example, when a wormhole attack is used against an on-demand routing protocol such as DSR or AODV, the attack could prevent the discovery of any routes other than through the wormhole. The seriousness of this attack is that it can be launched against all communications that provide authenticity and confidentiality.

■ *Rushing attack:* The tunnel procedure to form a wormhole is used by the colluding attackers. If a fast transmission path (e.g., a dedicated channel shared by attackers) exists between the two ends of the wormhole, the tunneled packets can propagate faster than those through a normal multihop route. This forms the rushing attack [46]. The rushing attack can act as an effective DoS attack against all currently proposed on-demand MANET routing protocols, including protocols that were designed to be secure, such as ARAN and Ariadne [47].

■ *Link withholding attacks:* In this attack, a malicious node does not advertise the information about the links to specific nodes or group of nodes, which may result in losing the links to these nodes. Mounting such attacks is difficult in the dynamic environment of MANET but is more relevant in case of sensor networks.

■ *Link spoofing attack:* In a link spoofing attack, a malicious node advertises fake links with nonneighbors to disrupt the routing operations [48]. For example, in the OLSR protocol, an attacker can advertise a fake link with a target's two-hop neighbors. This causes the target node to select the malicious node to be its multipoint relay. As a multipoint relay node, a malicious node can then manipulate the data or routing traffic, for example, modifying or dropping the routing traffic or performing other types of DoS attacks.

■ *Byzantine attack:* A compromised intermediate node works alone, or a set of compromised intermediate nodes works in collusion and carry out attacks such as creating routing loops, forwarding packets through nonoptimal paths, or selectively dropping packets, which results in disruption or degradation of the routing services.

■ *Colluding misrelay attack:* In this attack, multiple attackers work in collusion to modify or drop routing packets to disrupt the routing operation in a MANET. This attack is difficult to detect using the conventional methods such as *watchdog* and *pathrater* [49]. This can be thought of as a Byzantine attack also. In [50], the authors have described this type of attack and shown that such an attack can drop the network throughput to zero.

■ *Replay attack:* In a MANET, topology frequently changes due to node mobility. This means that the current network topology might not exist in the future. In a replay attack [51], a node records another node's valid control messages and resends them later. This causes other nodes to record their routing table with stale routes. Replay attack can be misused to impersonate a specific node or simply to disturb the routing operation in a MANET.

■ *Location disclosure attack:* An attacker reveals information regarding the location of nodes or the structure of the network. It gathers the node location information, such as a route map, and then plans further attack scenarios. Traffic analysis, one of the subtlest security attacks against MANET, is unsolved. Adversaries try to figure out the identities of communication parties and analyze traffic to learn the network traffic pattern and track changes in the traffic pattern. The leakage of such information is devastating in security-sensitive scenarios [52,53]. This type of attack gives away the anonymity criterion necessary in networks like VANET.

The various types of attacks listed above cover most of the attacks on MANETs observed at the network layer. However, this list is by no means complete as new attacks are being thought of continuously. Mitigation techniques for these attacks have also been studied extensively in the literature.

■ *Flooding attack:* In [54], the authors proposed a simple mechanism to prevent the flooding attack in the AODV protocol. In this approach, each node monitors and calculates the rate of its neighbors' RREQ. If the RREQ rate of any neighbor exceeds the predefined threshold, the node records the ID of this neighbor in a blacklist. Then, the node drops any future RREQs from nodes that are listed in the blacklist. One limitation of this approach is that it cannot prevent against the flooding attack in which the flooding rate is below the threshold. Another drawback of this approach is that if a malicious node impersonates the ID of a legitimate node and broadcasts a large number of RREQs, other nodes might put the ID of this legitimate node on the blacklist by mistake. In [55], the authors proposed an adaptive technique to mitigate the effect of a flooding attack in the AODV protocol. This technique is based on the statistical analysis in detecting malicious RREQ floods and avoiding the forwarding of such packets. Similar to [40], in this approach, each node monitors the RREQ it receives and maintains a count of RREQs received from each sender during the preset time period. The RREQs from a sender whose RREQ rate is above the threshold will be dropped without forwarding. Unlike the method proposed by Yi et al. [40], where the threshold is set to be fixed, this approach determines the threshold based on a statistical analysis of RREQs. The key advantage of this approach is that it can reduce the impact of the attack for varying flooding rates.

■ *Blackhole attack:* Lee et al. [56] have introduced the route confirmation request (CREQ) and route confirmation reply (CREP) to avoid the blackhole attack. Here, the intermediate node not only sends RREPs to the source node but also sends CREQs to its next-hop node toward the destination node. After receiving a CREQ, the next-hop node looks up its cache for a route to the destination. If it has the route, it sends the CREP to the source. Upon receiving the CREP, the source node can confirm the validity of the path by comparing the path in RREP and the one in CREP. If both are matched, the source node judges that the route is correct. The blackhole attack with colluding adversaries cannot be mitigated in this approach, that is, when the next-hop node is a colluding attacker sending CREPs that support the incorrect path. Al-Shurman et al. [57] have proposed a solution that requires a source node to wait until an RREP packet arrives from more than two nodes. Upon receiving multiple RREPs, the source node checks whether there is a shared hop or not. If there is, the source node judges that the route is safe. The main drawback of this solution is that it introduces time delay, because it must wait until multiple RREPs arrive. Kurosawa [58] has analyzed the blackhole attack and showed that a malicious node must increase the destination sequence number sufficiently to convince the source node that the route provided is sufficient enough. On the basis of this analysis, the authors propose a statistical based anomaly detection approach to detect the blackhole attack, based on the differences between the destination sequence numbers of the received RREPs. The key advantage of this approach is that it can detect the attack at low cost without introducing extra routing traffic, and it does not require modification of the existing protocol. However, false positives are the main drawback of this approach due to the nature of anomaly detection.

■ *Greyhole attack:* As described earlier, this attack is carried out by an intelligent node which selectively drops packets but ensures that it will not cross the threshold limit for getting banned from the routing process [42]. Although this type of attack is difficult to detect and mitigate, as an attacker remains within the threshold value of trust, the damage done by such an attack is limited and for volatile topologies like VANET such an attack will not be able to inflict great damage.

■ *Wormhole attack:* Hu et al. [44] have proposed the technique of packet leashes to detect and defend against the wormhole attack. In particular, the authors proposed two types of leashes: temporal leashes and geographical leashes. For the temporal leash approach, each node computes the packet expiration time, t_e, based on the speed of light c and includes the expiration time, t_e, in its packet to prevent the packet from traveling further than a specific distance, L. The receiver of the packet checks whether or not the packet expires by comparing its current time and the t_e in the packet. The authors also proposed TESLA (Timed Efficient Stream Loss-Tolerant Authentication) with Instant Key disclosure (TIK), which is used to authenticate the expiration time that can otherwise be modified by the malicious node. The main drawback of the temporal leash is that it requires all nodes to have tightly synchronized clocks. For the geographical leash, each node must know its own position and have loosely synchronized clocks. In this approach, a sender of a packet includes its current position and the sending time. Therefore, a receiver can judge neighbor relations by computing distance between itself and the sender of the packet. The advantage of geographic leashes over temporal leashes is that the time synchronizations need not to be highly tight. The technique proposed by Raffo [59] offers protection against a wormhole attack in the OLSR protocol. This approach is based on location information and requires the deployment of a public key infrastructure and time-stamp synchronization between all nodes that is similar to the geographic leashes proposed in [44]. In this approach, a sender of a HELLO message includes its current position and current time in its HELLO message. Upon receiving a HELLO message from a neighbor, a node calculates the distance between itself and its neighbor, based on a position provided in the HELLO message. If the distance is more than the maximum transmission range, the node judges that the HELLO message is highly suspicious and might be tunneled by a wormhole attack. Qian et al. [60] have carried out a statistical analysis of multipath, which is an approach to detect the wormhole attack using multipath routing. This approach determines the attack by calculating the relative frequency of each link that appears in all of the obtained routes from one route discovery. In this solution, a link that has the highest relative frequency is identified as the wormhole link. The advantage of this approach is that it introduces a limited overhead when applied in multipath routing. However, it might not work in a nonmultipath routing protocol, such as a pure AODV protocol.

■ *Rushing attack:* Hu et al. [46] have presented this type of attack and have discussed its damage potential. They claim that such attacks cannot be detected by any contemporary mitigating technique and once mounted such attacks make the discovery of route beyond two hops impossible. They have also presented a generic defense technique against such attacks and have presented the simulation results.

■ *Link withholding/message withholding attack:* Kannhavong [61] has shown that by withholding a Topology Control (TC) message in OLSR, a malicious node can isolate a specific node and prevent it from receiving data packets from other nodes. After analyzing and evaluating the impact of this type of attack in detail, the authors proposed a detection technique based on the observation of both a TC message and a HELLO message generated by the Multipoint Relay (MPR) nodes. If a node does not hear a TC message from its MPR node regularly but hears only a HELLO message, a node judges that the MPR node is suspicious and can avoid the attack by selecting one or more additional MPR nodes. Similarly, in [62], Dhillon has proposed an Intrusion Detection System (IDS) to detect TC link and message withholding in the OLSR protocol. In this approach, each node observes whether an MPR node generates a TC message regularly or not. If an MPR node generates a TC message regularly, the node checks whether or not the TC message actually contains itself to detect the attack. The main drawback of these approaches is that they cannot detect the attack that is launched by two colluding consecutive nodes, where the first attacker pretends to advertise a TC message, but the second attacker drops this TC message. Yi et al. proposed a novel intrusion detection method for MANETs in [54].

■ *Link spoofing attack:* To detect a link spoofing attack, Raffo [59] proposed a location information-based detection method using cryptography with a Global Positioning System (GPS) and a time stamp. This approach requires each node to advertise its position obtained by the GPS and the time stamp to enable each node to obtain the location information of the other nodes. This approach detects the link spoofing by calculating the distance between the two nodes that claim to be neighbors and checking the likelihood that the link is based on a maximum transmission range. The main drawback of this approach is that it might not work in a situation where all MANET nodes are not equipped with a GPS. Furthermore, attackers can still advertise false information and make it hard for other nodes to detect the attack. In [61], the authors show that a malicious node that advertises fake links with a target's two-hop neighbors can successfully make the target choose it as the only MPR. Through simulations, the authors show that link spoofing can have a devastating impact on the target node. Then, the authors present a technique to detect the link spoofing attack by adding two-hop information to a HELLO message. In particular, the proposed solution requires each node to advertise its two-hop neighbors to enable each node to learn complete topology up to three hops and detect the inconsistency when the link spoofing attack is launched. The main advantage of this approach is that it can detect the link spoofing attack without using special hardware such as a GPS or requiring time synchronization. One limitation of this approach is that it might not detect link spoofing with nodes further away than three hops.

■ *Byzantine attack:* Authentication of nodes is one of the major approaches against such attacks. Authentication and integrity of packets are generally done using cryptographic techniques such as PKI and so on, which is out of the scope of this chapter. If authenticated nodes, however, mount an attack from within a network, then it is very difficult to detect and mitigate such an attack. Crépeau et al. [63] have presented a secure routing protocol which mitigates such type of attacks.

■ *Colluding mis-relay attack:* A conventional acknowledgment-based approach might detect this type of attack in a MANET, especially in a proactive MANET, but because routing packets destined to all nodes in the network require all nodes to return an ACK, this could lead to a large overhead, which is considered to be inefficient. In [64], the author proposes a method to detect an attack in which multiple malicious nodes attempt to drop packets by requiring each node to tune their transmission power when they forward packets. As an example, the author studies the case where two colluding attackers drop packets. The proposed solution requires each node to increase its transmission power twice to detect such an attack. However, this approach might not detect the attack in which three colluding attackers work in collusion. In general, the main drawback of this approach is that even if we require each node to increase the transmission power by K times, we still cannot detect the attack in which $K + 1$ attackers work in collusion to drop packets. Therefore, further work must be done to counter this type of attack efficiently.

■ *Replay attack:* Adjih et al. [51] have proposed a solution to protect a MANET from a replay attack using a time stamp with the use of an asymmetric key. This solution prevents the replay attack by comparing the current time and time stamp contained in the received message. If the time stamp is too far from the current time, the message is judged to be suspicious and is rejected. Although this solution works well against the replay attack, it is still vulnerable to a wormhole attack where two colluding attackers use a high-speed network to replay messages in a far-away location with almost no delay. This attack will be discussed in the next subsection.

■ *Location disclosure attack:* Location disclosure attacks by nature are passive type of attacks where the sole aim of attacker is to gather only the information about the identity of nodes in the network which can later be utilized in one way or the other. Because of this peculiarity, it is virtually impossible to detect such attacks. In order to guard against such attacks, the technique preferred is the use of pseudoidentities. But this gives rise to another problem of maintaining and updating the list of identities that the node is using. Jian et al. [65] have proposed a location privacy routing protocol for sensor networks, which effectively mitigates such attacks by injecting false packets in the network to minimize the traffic direction information that an adversary can retrieve from eavesdropping. This attack is of more relevance in VANET wherein a driver's location privacy needs to be protected at all times and in [52,53] the authors have proposed protocols which handle this problem.

9.4.1 Other Classifications

Internal or external classification becomes trivial in case of attacks on network layer as all the attacks can be launched by remaining inside a particular set of nodes or by remaining outside the set as well. This is true under the assumption that malicious node can authenticate itself in the network. Liu et al. [66] have argued that cryptographic primitives alone cannot detect such insider attacks and have proposed internal attacker detection scheme for sensor networks. But, still assuming the attacker point of view, most of the attacks on network layer can be mounted irrespective of the position of the attacker and so it may be assumed to be trivial.

In case of MANET, the passive attacks can be of the form of eavesdropping and generating the relative importance of the node in the network topology depending on the traffic. Depending on this information a decision can be reached to physically harm or hijack a node which eventually makes it an active attack. Generally, passive attacks are extremely difficult to detect [42] as there is no change in the protocol fields or behavior of the network. It also becomes more trivial in case of highly dynamic environment like VANET where the topology itself changes very frequently.

On the other hand, as listed out earlier, most of the attacks fall under this category wherein there is less or complete modification to the protocol fields. The impersonation attacks may not completely fall under this category, since there is no apparent modification to the protocol fields. However, in case of impersonation attacks the information itself is coming from an illegitimate node. But, by and large most of the routing attacks can be classified in this category.

The attacks on network layer can also be classified depending on the technique used for launching the attack. Modification, fabrication, and impersonation attacks are briefly discussed in the succeeding paragraphs.

■ *Modification attacks:* Routing protocols for *ad hoc* networks are based on the assumption that intermediate nodes do not maliciously change the protocol fields of messages passed between the nodes. This assumed trust permits malicious nodes to easily generate traffic subversion and DoS attacks. Attacks using modification are generally targeted against the integrity of routing computations and so by modifying the routing information an attacker can cause network traffic to be dropped, redirected to a different destination, or take a longer route to the destination increasing communication delays. An example is for an attacker to send fake routing packets to generate a routing loop, causing packets to pass through nodes in a cycle without getting to their actual destinations, consuming energy and bandwidth. Similarly, by sending forged routing packets to other nodes, all traffic can be diverted to the attacker or to some other node. The idea is to create a *blackhole* by routing all packets to the attacker and then discarding it. *Greyhole* attack discussed earlier is basically a variation in the blackhole attack. A more subtle type of modification attack is the creation of a tunnel (or *wormhole*) in the network between two colluding malicious nodes linked through a private network connection. This exploit allows a node to short-circuit the normal flow of routing messages, creating a virtual vertex cut in the network that is controlled by the two colluding attackers.

■ *Fabrication attacks:* Fabrication attacks are performed by generating false routing messages. These attacks are difficult to identify as they are received as legitimate routing packets. The *rushing attack* is a typical example of malicious attacks using fabrication. This attack is carried out against on-demand routing protocols that hold back duplicate packets at every node. An attacker rapidly spreads routing messages all through the network, suppressing legitimate routing messages when nodes discard them as duplicate copies. Similarly, an attacker can nullify an operational route to a destination by fabricating routing error messages, asserting that a neighbor can no longer be contacted.

■ *Impersonation attacks:* A malicious node can initiate many attacks in a network by masquerading as another node (spoofing). Spoofing occurs when a malicious node misrepresents its identity by altering its MAC or IP address in order to alter the vision of the network topology that a benign node can gather. As an example, a spoofing attack allows the creation of loops in the routing information collected by a node, with the result of partitioning the network. Sybil attacks is a perfect example of such type of attacks [67–69]. Here, the attacking node takes over the identity of another legitimate node to divert or eavesdrop on the traffic which may also lead to a DoS attack.

There are more classifications possible for attacks on MANET, but it does not cover a large number of attacks and for the purpose of clarity they have been discussed here.

9.4.2 Discussion and Analysis

In this section, the majority attacks which take place at the network layer have been identified and their mitigation techniques from the literature have been discussed in brief to get an overview of the attacks. It can be seen that the last classification carried out (i.e., modification, fabrication, and impersonation) covers most of the attacks as they use the method of launching the attack as the criterion.

One of the important attacks discussed in the literature is the rushing attack, which has been discussed earlier. Hu et al. [46] have explained as to why such an attack is feasible and due to its nature how it cannot be detected by any current mitigating technique. For this reason, only the damage potential of this attack is high. The authors have also proposed a generic technique for mitigation that can be used in conjunction with other protocols. Yang et al. [70] have proposed the concept of self-organizing network layer security based on the certifications and tokens which essentially, eventually builds a trust rating. The authors of [71–73] have also discussed about the variations in the trust-based approach and Yang et al. have clearly brought out the limitations and flaws in such trust rating-based schemes. As discussed earlier, rushing attacks can easily be mounted against such security schemes [46]. Lindsay et al. [74] have carried out a comprehensive study of various types of attacks that can be mounted by exploiting the vulnerabilities of trust-based schemes and have also developed some mechanisms against such types of attacks.

Padmanabhan and Simon [75] argue that it is not only important to validate routing updates but also important to ensure robustness of packet forwarding itself, and they also propose a protocol that enables the end hosts or routers to detect and locate the source of routing misbehaviors. Reiterating on the same philosophy, Kefayati et al. [76] have proposed a methodology to improve the end-to-end packet delivery to mitigate blackhole and greyhole attacks.

As can be seen from the literature, most of the attacks at routing layer lead to or culminate in DoS attack [42,46,51]. But that is not the only way in which these attacks affect the system. In any case, the aim of these attacks is to disrupt the normal operation of the network. On the other hand, attacks like location disclosure are truly detrimental not in terms of network throughput but in terms of their outcome where privacy of the node is violated [52]. Zhanz et al. [77] have proposed an on-demand routing protocol for MANETs in which authentication is carried out using a cryptographic concept called pairing and therefore the anonymity of nodes is not compromised. It may be seen that the damage potential of such passive, location disclosure, or identity tracking attacks is not only limited to disrupting the normal operation of network but also depending on the collected network topology an attacker can mount a deadly and accurate attack which might bring down the network without giving any chance for countermeasures. So, more attention needs to be focused on the preventive techniques from such passive attacks. Jakobsson et al. [78] have named such attacks as "stealth attacks" and have argued that the very mechanisms required to defend against all attacks involving propagation of incorrect routing information automatically behoove an attacker attempting to perform a DoS attack, and vice versa.

In [79,80], the authors have proposed a novel approach wherein they claim that contrary to the normal belief that securing MANET becomes a difficult task due to mobility, actually, the mobility helps for creating security associations. Basically, by studying the mobility patterns of the nodes, making security associations becomes easier, which helps in improving security. Such an approach is, however, dependent on lot of other factors [79].

9.5 Attacks at the Transport Layer

The objectives of Transmission Control Protocol (TCP)-like transport layer protocols in MANET include setting up of end-to-end connection, end-to-end Reliable delivery of packets, flow control, congestion control, and clearing of end-to-end connectionand. Similar to TCP protocols in the Internet, the mobile node is vulnerable to the classic Synchronization (SYN) flooding attack or session hijacking attacks. However, a MANET has a higher channel error rate when compared with wired networks. Because TCP does not have any mechanism to distinguish whether a loss was caused by congestion, random error, or malicious attacks, it multiplicatively decreases its congestion window upon experiencing losses, which degrades the network performance significantly. This is more relevant in terms of MANET owing to nonconsistent and frequently changing links.

■ *SYN flooding attack:* The SYN flooding attack is a type of DoS attack. The attacker creates a large number of half-opened TCP connections with a victim node, but never completes the three-way handshake to fully open the connection that is essential for two nodes to communicate using TCP. The three messages exchanged during the handshake allow both nodes to learn that the other is ready to communicate and to agree on initial sequence numbers for the conversation. During the attack, a malicious node sends a large amount of SYN packets to a victim node, spoofing the return addresses of the SYN packets. The SYN–ACK packets are sent out from the victim right after it receives the SYN packets from the attacker and then the victim waits for the response of an ACK packet. Without receiving the ACK packets, the half-open data structure remains in the victim node. If the victim node stores these half-opened connections in a fixed size table while it awaits the acknowledgment of the three-way handshake, all of these pending connections could overflow the buffer, and the victim node would not be able to accept any other legitimate attempts to open a connection. Normally, there is a time-out associated with a pending connection, so the half-open connections will eventually expire and the victim node will recover. However, malicious nodes can simply continue sending packets that request new connections faster than the expiration of pending connections.

■ *Session hijacking:* Session hijacking takes advantage of the fact that most communications are protected (by providing credentials) at session setup, but not thereafter. In the TCP session hijacking attack, the attacker spoofs the victim's IP address, determines the correct sequence number that is expected by the target, and then performs a DoS attack on the victim. Thus the attacker impersonates the victim node and continues the session with the target. The TCP–ACK storm problem could be created when an attacker launches a TCP session hijacking attack. The attacker sends injected session data, and node 1 will acknowledge the receipt of the data by sending an ACK packet to node 2. This packet will not contain a sequence number that node 2 is expecting, so when node 2 receives this packet, it will try to resynchronize the TCP session with node 1 by sending it an ACK packet with the sequence number that it is expecting. The cycle goes on and on, and the ACK packets passing back and forth create an ACK storm. Hijacking a session over User Datagram Protocol (UDP) is the same as over TCP, except that UDP attackers do not have to worry about the overhead of managing sequence numbers and other TCP mechanisms. Since UDP is connectionless, edging into a session without being detected is much easier than the TCP session attacks.

Although these types of attacks have been foreseen at transport layer, no transport layer misbehavior problems have been identified for MANET scenario [81].

9.6 Attacks at the Application Layer

The application layer communication is also vulnerable in terms of security compared with other layers. The application layer contains user data, and it normally supports many protocols such as HTTP, SMTP, TELNET, and FTP, which provide many vulnerabilities and access points for attackers. The application layer attacks are attractive to attackers because the information they seek ultimately resides within the application and it is direct for them to make an impact and reach their goals.

- *Malicious code attacks:* Malicious codes, such as viruses, worms, spywares, and Trojan Horses, can attack both operating systems and user applications [82]. These malicious programs usually can spread themselves through the network and cause the computer system and networks to slow down or even damaged. In MANET, an attacker can produce similar attacks to the mobile system of the *ad hoc* network.
- *Repudiation attacks:* In the network layer, firewalls can be installed to keep packets in or keep packets out. In the transport layer, entire connections can be encrypted, end-to-end. But these solutions do not solve the authentication or nonrepudiation problems in general. Repudiation refers to a denial of participation in all or part of the communication. For example, a selfish person could deny conducting an operation on a credit card purchase, or deny any on-line bank transaction, which is the prototypical repudiation attack on a commercial system. Louridos [83] has discussed this type of attack and proposed some general guidelines to be considered for its mitigation.

For mitigation of attacks at application layer, there is a host of techniques that can be employed. In a network with a firewall installed, the firewall can provide access control, user authentication, packet filtering, and a logging and accounting service. Application layer firewalls can effectively prevent many attacks, and application-specific modules, for example, spyware detection software, have also been developed to guard mission critical services. However, a firewall is mostly restricted to basic access control and is not able to solve all security problems. For example, it is not effective against attacks from insiders. Because of MANET's lack of infrastructure, a firewall is not particularly useful.

In MANET, an IDS can be used as a second line of defense [84]. Intrusion detection can be installed at the network layer, but in the application layer it is not only feasible, but also necessary

[85]. Certain attacks, such as an attack that tries to gain unauthorized access to a service, may seem legitimate to the lower layers, such as the MAC protocols. Some attacks may also be more obvious in the application layer. For instance, the application layer can detect a DoS attack more quickly than the lower layers when a large number of incoming service connections have no actual operations, since low layers need more time to recognize it. In [34,85–88], the authors have discussed and proposed various techniques for intrusion detection in MANET.

It can be seen from the discussion so far that individual layers might not be able to thwart all the attacks being carried out on MANET on their own and some sort of co-operation from other layers of the protocol stack is essential. This technique or approach has been named as cross-layer approach in literature. Stine [89] has discussed in detail why such an approach is necessary particularly for routing solutions in MANET. Similarly, Huang et al. [90] have shown as to how this approach can effectively be used for detecting routing anomalies in *ad hoc* networks. From [43,77,79,91–97], we can say that most of the application layer-oriented solutions in MANET use some or the other form of cross-layering design for mitigation of attacks. Muraleedharan and Osadciw [98] have discussed a similar cross-layer approach, primarily to mitigate the DoS attacks in sensor networks. However, some of the assumptions are not practical for VANET communication which has some special characteristics like volatile topology.

9.7 Attacks on VANET

VANET may be considered as a special case of MANET. It still differs significantly from MANET due to its typical characteristics such as highly dynamic and volatile topology, transient nature of participants, and nonpersistent communication links. With the explosive growth of wireless communication systems, research in vehicular communication and vehicular networking has generated a lot of interest [99]. These characteristics make the typical security arrangements possible in MANET less effective in the context of VANETs. Moreover, the security requirements of VANET are typical and have contradicting requirements in terms of privacy, anonymity, authentication, and nonrepudiation [100]. In [101–104], the authors have discussed in detail the security requirements for vehicular communications and have brought out various vulnerabilities of the vehicular networks. Papadimitratos et al. [105] have proposed a secure architecture for vehicular communication countering the vulnerabilities which also takes into account the privacy issue which is one of the key issues in deployment of VANETs. Stampoulis and Chai [106] have carried out a comprehensive survey of the techniques that have been proposed so far to ensure security and privacy in such networks. Although this survey provides a good insight about the state of security in VANET, it does not dwell on the specifics of various types of attacks and various attack models that can be used for threat assessment.

The IEEE Standard 1609.2 specifies security services for the *Wireless Access in Vehicular Environments* (WAVEs) networking stack and for applications that are intended to run over that stack [107]. Services include encryption using another party's public key and nonanonymous authentication. The safety-critical nature of many *Dedicated Short-Range Communications*/WAVE applications makes it vital that services be specified that can be used to protect messages from attacks such as eavesdropping, spoofing, alteration, and replay. It also takes into account the owner's privacy rights. This means the security services must be designed to respect this right and not leak personal, identifying, or linkable information to unauthorized parties. This standard describes security services for WAVE management messages and application messages, with the exception of vehicle-originating safety messages, to meet these requirements. Although the existence of such a layer automatically guarantees certain authenticity and security to the applications, the exact implementation details and architecture of this layer is beyond the scope of this survey work.

Two broad approaches may be considered while classifying the attacks on VANET. Both these approaches have widely been studied in the literature. In the first approach, the actual attacks such as bogus information attack, disruption of network attack, DoS attack, jamming attack, cheating with identity, speed or positioning information, identity disclosure attack, passive eavesdropping attack, Sybil attacks, and so on, can be considered. VANET being a special case of MANET, all the vulnerabilities of MANET may be considered here also. Specific attacks on VANETs as listed earlier are discussed in brief in succeeding paragraphs, but it should be understood that this is not an exhaustive list and only discusses the major attacks studied in the literature.

- *Sybil attack:* The Sybil attack refers to a malicious node illegitimately taking on multiple identities. In this type of attack, a malicious vehicle creates an illusion of traffic congestion by claiming multiple identities. Not only does this create an illusion, it has the potential to inject false information into the networks via a number of fabricated nonexisting vehicles; it can even launch further DoS attacks by impairing the normal operations of data dissemination protocols [108,109]. Golle et al. [110] have suggested some general techniques such as registration, radio resource testing, and position verification. Leinmuller et al. [111] have proposed various techniques of position verification for mitigating such attacks where numbers of different independent sensors are used to quickly give an estimation of the trustworthiness of other nodes' position claims without using dedicated infrastructure or specialized hardware. In [112,113], the authors have proposed other techniques for position verification using temporal links and have argued about the usefulness of these techniques from security point of view. Harsch et al. [114] have proposed comprehensive position-based routing solution that provides defense against Sybil attacks using cryptographic primitives and also partially mitigates bogus information attack. Xiao et al. [67] have discussed in detail how position verification can only provide limited accuracy and have also proposed enhancements to it. Attacks like cheating with identity [115] or impersonation attack in MANET are same as Sybil attack and, in [68,69], the authors have considered Sybil attacks and its mitigation in sensor networks. Identity authentication of the transmitting node alone cannot prevent Sybil attack [67] as a compromised node with proper credentials may be able to mount such an attack.
- *Jamming attack:* In this type of attack, the attacker needs to be in possession of the proper hardware for mounting the attack. Xu et al. [95] have discussed the possibility of launching such attacks and have proposed various solutions for its mitigation in wireless networks. This attack is generally confined to the physical layer, that is, radio transmission. This attack can be launched by outsiders without any specific information about the network wherein in a particular area, the V2V communication is completely disrupted. But, due to the high dynamism of VANET, mounting such an attack on a larger area consistently would lead to very fast cost escalation and would become impractical.
- *DoS attacks:* The disruption of network operation attack [115] is essentially a DoS attack with broadcast storms flooding the network with unwanted information. Ni et al. [116] have discussed the problem arising due to broadcast storms and Tseng et al. [117] have suggested some techniques for mitigation of this problem. In a DoS attack, the attacker may try to overwhelm a vehicle's network/CPU resources or jam the whole channel of the communication network so that all the critical information cannot be delivered. The main purpose of this attack is to disrupt the functioning of the system. However, such attacks can be detected, and the driver

can be warned that the system is under attack. Moreover, VANET applications need to be designed in a manner such that human beings are not overly dependent on them.

■ *Bogus information attack:* In this case, the attacker disseminates false information in the vehicular network in order to affect the decisions of other drivers. In this also, the attacker can be an outsider or insider and with a robust security layer in place it is difficult for the outsider to mount such an attack. But an insider with malicious intents can provide false authentic information about location or speed information leading to bogus information attack. Golle et al. [109] have discussed the detection and correction of malicious data for VANET under the name of *malicious data attack*. Leinmüller et al. [118] have discussed various position verification techniques that can be used for mitigating the false information attack. In [119,120], the authors have proposed that data aggregation for multihop networks is a vulnerability of this type and also proposed various schemes for overcoming the problem. The nonrepudiation security property is very important for mitigation of such attacks and digital signatures and cryptography are the techniques by means of which this can be ensured.

■ *Identity disclosure attack:* This is a peculiar type of attack in the VANET environment wherein the identity of the vehicle and privacy of the driver may be compromised. Using this information, location profiling may become possible, where attackers gather information on node positions or mobility patterns. Therefore, location privacy is an important issue in *ad hoc* networks. These attacks can be mitigated by techniques like pseudonyms but maintaining such lists and then mapping it back to the owner is in itself a nontrivial task. Calandriello et al. [121] have discussed in detail the pros and cons of this approach and have proposed a method for efficient and robust pseudonyms for VANET. Chapkin et al. [122] have described such an attack based on topology information, thereby nullifying the advantage gained by the use of pseudonyms.

■ *Passive eavesdropping attack:* As brought out earlier, this type of passive eavesdropping attacks are very difficult to detect, can be performed with low effort and cost, and with very low risk of detection of the identity (or whereabouts) of the perpetrator. As such these attacks are particularly dangerous since a small number of malicious parties can disconnect a large network with small effort and minimal risk of tracing. They assume significance in case of VANETs as well, as such monitoring can disclose movement patterns of vehicles and can disclose their identities and pose a risk to the privacy of the driver. Jakobsson et al. [78] have named this type of attack as *stealth attack* and have dealt with this type of attack particularly in case of VANET [123]. They have suggested two approaches to mitigate such type of attacks but no specific methods have been proposed.

■ *Illusion attack:* This is a typical VANET-related attack discussed in [124] wherein the adversary intentionally deceives the sensors on his own car to produce wrong sensor readings which in turn will broadcast false traffic warning messages. This creates an illusion for the other cars about the traffic event and tends to modify the drivers' behavior which is the ultimate aim of any adversary. The authors have correctly brought out that it may not be possible to mitigate such an attack with traditional methods such as trust schemes, message authentication, and message integrity check. Lo and Tsai [125] have proposed a novel *plausibility validation network* model for mitigation of such attacks which takes help of some predetermined rule set and analyses the message contents on these rules to decide the authenticity of the message. They have also provided a set of rules, but no simulation of the same has been carried out to ascertain the results. This attack, although similar in nature to bogus information attack, is fundamentally different owing to the fact that the adversary intentionally tries to modify or spoof sensor output, thereby producing an authenticated message of a bogus event to create an illusion of the same which leads to modification to driver's behavior.

In the second approach, the variations in the threat model, attack model, or adversary model are assumed depending on various underlying assumptions such as availability of underlying anonymous communication, position verification, roadside infrastructure, secure cryptographic techniques and so on. These models are then utilized for carrying out the threat assessment. Papadimitratos et al. [102] have carried out a detailed study of security requirements in the VANET environment and have provided models for the system and the communication, as well as models for the adversaries, and proposed a set of design principles for future security and privacy solutions for vehicular communication systems.

The basic model may be considered depending on the behavior of the node or may be named as behavioral model. The nodes in general will comply with the deployed protocol and may be named as correct. Those who deviate from the protocol can be named as faulty. The faulty behavior may be subclassified as malicious or due to malfunction of some components like sensors. Note that the malfunctioning of a component will eventually look like a malicious behavior. Further, it can be classified as active or passive misbehavior. As brought out earlier, the passive misbehavior detection and mitigation, even in the presence of a security layer specified in IEEE 1609.2, is a nontrivial task. This type of attack has been discussed earlier as a passive eavesdropping attack. The active or passive misbehaviors may take place within the given set of vehicular nodes or from the fixed infrastructure. On the basis of the type of misbehaviors, the attack may be internal or external. Passive insider attacks are the most difficult attacks to detect as there is no parameter from which we can infer that certain node is misbehaving and, moreover, being inside the network, it has the advantage of possessing proper credentials. On the other hand, as it does not affect or change the behavior of any other node, it is harmless from the point of view of immediate danger. Passive outside attacker has the problem of authentication and so it is relatively less harmful compared to a passive insider. Passive identity disclosure attacks may fall in this category. A vehicle behavior analysis to enhance security in VANETs has been presented in [126]. Further, Pease et al. [127] describe a methodology for reaching an agreement in the presence of fault.

An active misbehaving node may control or affect the operation of other nodes while learning information about their behavior. Again, an external active attacker may lack the authenticity due to the security layer of IEEE 1609.2, but it may still affect the operation of legitimate nodes. It may generate data which will force the correct nodes to check it, thereby occupying their resources. This may lead to jamming of communication in correct nodes. This way it can launch DoS attack within its range of communication. An internal active misbehavior is the most dangerous category of attack and has the potential of inflicting immediate harm to the network. Owing to the possession of proper credentials, such an attacker is able to get through the first wall of defense without being detected and this makes the deployment of additional security mechanisms essential. Such an attacker theoretically may mount all the specific attacks discussed earlier, such as bogus information attack, identity disclosure attack, DoS attack, Sybil attack, Illusion attack, and so on. It may modify, forge, replay, omit, delay, fudge, or inject message or transmission in order to attack the network. Such multiple attackers may collude to mount collaborative attacks but this may be limited due to the trust mechanisms in place. It should be understood that such attackers are limited in their ability to inflict damage due to their limited resources, such as memory and computational power. Here, it may be seen that the volatility of VANET environment actually helps in maintaining the security by not allowing adversaries sufficient time to break the security architecture.

The other model that may be considered is Byzantine attacker [128]. Detecting such an adversary may be possible by keeping track of the behavior of the node over a prolonged period of time which might not be feasible in case of VANET. On the other hand, such an attacker may be able to inflict limited degree of damage due to intermittent communication links. The bogus information attack may fall in this category of attacks. It may be inferred that in the presence of a strong security layer this type of adversary may not cause a high degree of damage in a VANET environment.

A more structured approach about the attack model has been proposed in [129]. The authors have carried out a detailed study of the attacks and depending on the security goals envisaged have formed an attack tree and have also used the concept of attack subtrees which can be reused for threat assessment depending on the particular application where it is being used. As a matter of fact, the root of the attack subtree can be considered to be the general attacks listed in the first approach and variations in these attacks then subsequently classified. The major subtrees considered are becoming part of the network, manipulate OBU input and violate privacy. Sybil attack may be considered as a root of the first attack subtree, that is, becoming part of the network and the general mitigation technique that may be considered is the use of robust and efficient cryptographic techniques. Bogus information attack may be considered as the root of the second attack subtree as manipulation of OBU in one way or the other will lead to injection of authenticated false data in the network. Physical security and efficient and accurate position verification techniques may be used to mitigate such type of attacks. Identity disclosure attack may be considered as the root of the third attack subtree, that is, privacy violation. Also, passive eavesdropping may be considered as a part of the privacy violation subtree and pseudonyms may be considered as a mitigation technique for such attacks. If a particular mitigating technique can be used against an attack subtree, then it may be assumed that the variations in the attacks may also be mitigated with some variation in the said technique. Thus, this type of classification may provide some general guidelines for mitigating solutions for VANET security.

The attacks and their mitigation techniques in VANET has been a very active field of research in the past few years, and the survey carried out here largely focuses on the major published work. It only provides an overview of the problem and some mitigation techniques and has included most of the possible attacks mentioned in the literature.

9.8 Conclusion

In this chapter, we have classified the attacks on wireless MANETs with respect to each layer in the protocol stack. The attacks on the physical layer have been discussed in brief owing to its more hardware-oriented nature. In an MAC layer, attacks have been subclassified as selfish and malicious and other classifications such as internal/external and active/passive have also been brought out. The attacks on the network layer and its mitigation have been discussed without any specific classification at first owing to the diversity and reach of such attacks. Attacks on the transport and application layers have been discussed in Sections 9.5 and 9.6 by giving an overview of the types of attacks possible at these layers. Section 9.7 has been devoted to the study of attacks on VANET, considering that it is the current area of research where a lot of activities are being observed. Various attacks and their possible mitigation techniques have been discussed. It can be seen from the discussion that the attacks and their mitigation are a core issue while deploying MANET or VANET. It essentially involves interaction across layers for efficient security measures and this is probably an important area of research in future. While this chapter presented a survey of potential attacks, the next step could be to perform a threat assessment to the functioning of the system from these attacks. Such an analysis could consider both the likelihood of initiating such attacks and its impact on the functioning of the system.

Terminologies

MANET
Wired Equivalent Privacy
IEEE 802.11
Denial of Service (DoS)
Flooding attack
Blackhole attack
Greyhole attack
Wormhole attack
Rushing attack
Link withholding attacks
Link spoofing attack
Byzantine attack
Colluding misrelay attack
Replay attack
Location disclosure attack
Modification attacks
Fabrication attacks
Impersonation attacks
SYN flooding attack
Session hijacking
Malicious code attacks
Repudiation attacks
Intrusion Detection System (IDS)
IEEE 1609.2
Wireless Access in Vehicular Environments (WAVEs)
Sybil attack
Jamming attack
Bogus information attack
Identity disclosure attack
Passive eavesdropping attack
Illusion attack

Questions and Sample Answers

1. What is malicious behavior of nodes?
   Malicious behavior may also be considered as an attack. These attacks are necessarily meant to disrupt the normal operation of the network in terms of network throughput and availability. Thus it may prevent the legitimate users from communicating over the network.

2. What is greyhole attack?
   Greyhole attack is a variation in the blackhole attack. Here, the node through which the packets are passing does not drop all the packets but remains selective about it. For example, the node may drop packets with routing information and will route the data packets as per requirement. Such an attack is difficult to detect as the node misbehaves selectively.

3. Describe session hijacking. Why and how?

Session hijacking takes advantage of the fact that most communications are protected (by providing credentials) at a session setup, but not thereafter. In the TCP session hijacking attack, the attacker spoofs the victim's IP address, determines the correct sequence number that is expected by the target, and then performs a DoS attack on the victim. Thus the attacker impersonates the victim node and continues the session with the target. The TCP–ACK storm problem could be created when an attacker launches a TCP session hijacking attack. The attacker sends injected session data, and node 1 will acknowledge the receipt of the data by sending an ACK packet to node 2. This packet will not contain a sequence number that node 2 is expecting, so when node 2 receives this packet, it will try to resynchronize the TCP session with node 1 by sending it an ACK packet with the sequence number that it is expecting. The cycle goes on and on, and the ACK packets passing back and forth create an ACK storm. Hijacking a session over UDP is the same as over TCP, except that UDP attackers do not have to worry about the overhead of managing sequence numbers and other TCP mechanisms. Since UDP is connectionless, edging into a session without being detected is much easier than the TCP session attacks.

4. Are the attacks against a MANET applicable against VANET?

Yes. As VANET is a special case of MANET, all the vulnerabilities of MANET may be applicable.

5. How can a replay attack be prevented?

A replay attack can be prevented by using time stamping, that is, by comparing the current time and time stamp contained in the received message. If the time stamp is too far from the current time, the message is judged to be suspicious and is rejected.

Author's Biography

Lieutenant commander Vikrant Gokhale is a graduate from Naval College of Engineering, INS Shivaji, Lonavla, in mechanical engineering and is a postgraduate in information technology from Indian Institute of Technology (IIT) Kharagpur, Kharagpur, India. He has a keen interest in networking and has worked on the network security in Vehicular *Ad hoc* Networks. He is a serving officer in Indian Navy and is currently involved in the network administration. He is also associated with highly sophisticated computer-based simulators for three different classes of ships. His areas of interest include virtual networks, cloud computing, and security of *ad hoc* networks.

S.K. Ghosh is presently working as an associate professor in the School of Information Technology, Indian Institute of Technology (IIT) Kharagpur, Kharagpur, India. He has received PhD degree in computer science & engineering from the Department of Computer Science & Engineering, IIT Kharagpur, India. Prior to IIT Kharagpur, he worked for Indian Space Research Organization (ISRO), Department of Space, Government of India, in the field of Satellite Remote Sensing and Geographical Information System. His research interest includes information security and geospatial database.

Arobinda Gupta received his PhD in computer science and engineering from the University of Iowa, Iowa, in 1997. From 1997 to 1999, he was with Microsoft Corp., Redmond, Washington, where he was involved in the design and implementation of distributed directory services for Windows 2000. Since October 1999, he is a faculty member at the Indian Institute of Technology (IIT) Kharagpur, Kharagpur, India, where he is currently an associate professor in the Department of

Computer Science & Engineering and School of Information Technology. His research interests are broadly in the area of design and analysis of distributed algorithms, and in *ad hoc* and sensor networks.

References

1. J. P. Hubaux, L. Buttyán, and S. Capkun, The quest for security in mobile ad hoc networks, in *Proceedings of the 2nd ACM Symposium on Mobile Ad Hoc Networking & Computing (MOBIHOC 2001)*, October 4–5, 2001, pp. 146–155, Long Beach, CA, USA.

2. L. Guang and C. Assi, On the resiliency of mobile ad hoc networks to MAC layer misbehavior, in *Proceedings of the 2nd ACM International Workshop on Performance Evaluation of Wireless Ad Hoc, Sensor, and Ubiquitous Networks*, October 10–13, 2005, Montreal, QC, Canada.

3. A. A. Cardenas, S. Radosavac, and J. S. Baras, Performance comparison of detection schemes for MAC layer misbehavior, in *Proceedings of 26th IEEE International Conference on Computer Communications*, May 6–12, 2007, Anchorage, AK.

4. T. Leinmüller, E. Schoch, F. Kargl, and C. Maihöfer, Improved security in geographic ad hoc routing through autonomous position verification, in *Proceedings of the ACM Workshop on Vehicular Ad Hoc Networks (VANET)*, Los Angeles, USA, 2006, pp. 57–66.

5. Y. Yang, S. Zhu, and G. Cao, Improving sensor network immunity under worm attacks: A software diversity approach, in *Proceedings of the 9th ACM International Symposium on Mobile Ad Hoc Networking and Computing*, May 26–30, 2008, pp. 149–158, Hong Kong SAR, China.

6. S. Ganu, K. Ramachandran, M. Gruteser, I. Seskar, and J. Deng, Methods for restoring MAC layer fairness in IEEE 802.11 networks with physical layer capture, in *Proceedings of the 2nd International Workshop on Multi-Hop Ad Hoc Networks: From Theory to Reality*, May 26, 2006, pp. 7–14, Florence, Italy.

7. E. Bayraktaroglu, C. King, X. Liu, G. Noubir, R. Rajaraman, and B. Thapa, On the performance of IEEE 802.11 under jamming, in *Proceedings of 27th Conference on Computer Communications*, 2008.

8. J. T. Chiang and Y. C. Hu, Dynamic jamming mitigation for wireless broadcast networks, in *Proceedings of the 27th IEEE Conference on Computer Communications (INFOCOM)*, April 13–18, 2008, pp. 1211–1219, Phoenix, AZ, USA.

9. M. Soroushnejad and E. Geraniotis, Probability of capture and rejection of primary multiple access interference in spread spectrum networks. *IEEE Transactions on Communications*, 1991, 39(6), 986–994.

10. V. Navda, A. Bohra, S. Ganguly, and D. Rubenstein, Using channel hopping to increase 802.11 resilience to jamming attacks, in *Proceedings of 26th IEEE International Conference on Computer Communications*, May 6–12, 2007, Anchorage, AK.

11. X. Liu, G. Noubir, R. Sundaram, and S. Tan, SPREAD: Foiling smart jammers using multi-layer agility, in *Proceedings of 26th IEEE International Conference on Computer Communications*, May 6–12, 2007, Anchorage, AK.

12. N. Borisov, I. Goldberg, and D. Wagner, Intercepting mobile communications the insecurity of 802.11, in *Proceedings of the 7th Annual International Conference on Mobile Computing and Networking*, 2001, Rome, Italy.

13. V. Gupta, S. Krishnamurthy, and M. Faloutsous, Denial of service attacks at the mac layer in wireless ad hoc networks, in *IEEE Military Communication Conference (MILCOM)*, October 7–10, 2002, pp. 1118–1123, Anaheim, CA, USA.

14. I. Aad, J. Hubaux, and E. W. Knightly, Denial of service resilience in ad hoc networks, in *Proceedings of the 10th Annual International Conference on Mobile Computing and Networking (MobiCom 2004)*, September 26–October 11, 2004, pp. 202–215, ACM Press, Philadelphia, PA, USA.

15. P. Kyasanur and N. H. Vaidya, Selfish MAC layer misbehavior in wireless networks, *IEEE Transactions on Mobile Computing*, 2005, 4(5).

16. M. Raya, J. P. Hubaux, and I. Aad, Domino: A system to detect greedy behavior in IEEE 802.11 hotspots, in *Proceedings of 2nd International Conference on Mobile Systems, Application and Services*, June 6–9, 2004, pp. 1691–1705, Boston, MA, USA.

17. IEEE802.11 wireless LAN media access control (MAC) and physical layer (PHY) specifications, IEEE, 1999. http://standards.ieee.org/getieee802/802.11.html.

18. S. Rayanchu, A. Mishra, D. Agrawal, S. Saha, and S. Banerjee, Diagnosing wireless packet losses in 802.11: Separating collision from weak signal, in *Proceedings of IEEE INFOCOM Conference*, April 2008, pp. 735–743, Phoenix, AZ, USA.

19. S. Sreepathi, V. Venigalla, and A. Lal, A survey paper on security issues pertaining to ad-hoc networks, http:// www4.ncsu.edu/~sssreepa/Adhoc-networks-Security-Survey.doc.

20. L. Guang, C. Assi, and Y. Ye, DREAM: A system for detection and reaction against MAC layer misbehavior in ad hoc networks, *Computer Communications*, 2007, 30(8).

21. S. Buchegger and J. Le Boudec, Nodes bearing grudges: Towards routing security, fairness, and robustness in mobile ad hoc networks, in *Proceedings of 10th Euromicro Workshop Parallel, Distributed and Network-Based Processing*, January, 2002, Canary Islands, Spain.

22. R. Carruthers and I. Nikolaidis, Certain limitations of reputation–based schemes in mobile environments, in *Proceedings of the 8th ACM International Symposium on Modeling, Analysis and Simulation of Wireless and Mobile Systems*, October 10–13, 2005, Montreal, QC, Canada.

23. S. Buchegger and J. L. Boudec, Performance analysis of the CONFIDANT protocol: Cooperation of nodes—fairness in dynamic ad-hoc NeTworks, in *Proceedings of IEEE/ACM Symposium. Mobile Ad Hoc Networking and Computing*, June 9–11, 2002, Lausanne, Switzerland.

24. L. Buttyán and J.-P. Hubaux, Enforcing service availability in mobile ad-hoc WANs, in *Proceedings of the IEEE/ACM Workshop on Mobile Ad Hoc Networking and Computing (MobiHOC)*, August 11, 2000, pp. 87–96, Boston, MA, USA.

25. L. Buttyán and J.-P. Hubaux, Stimulating cooperation in self-organizing mobile ad hoc networks, *Mobile Networks and Applications*, 2003, 8(5).

26. A. Cardenas, S. Radosavac, and J. S. Baras, Detection and prevention of MAC layer misbehavior for ad hoc networks, in *Proceeding of 11th ACM Conference on Computer and Communications Security*, October 25, 2004, Washington, D.C., USA.

27. J. Konorski, Protection of fairness for multimedia traffic streams in a non-cooperative reless LAN setting, in *Proceedings of 6th International Conference Protocols for Multimedia Systems*, October 17–19, 2001, Netherlands, Vol. 2213.

28. J. Konorski, Multiple access in ad-hoc wireless LANs with non-cooperative stations, in *Proceedings of the 2nd International IFIP-TC6 Networking Conference on Networking Technologies, Services, and Protocols; Performance of Computer and Communication Networks; and Mobile and Wireless Communications*, 2002, Vol. 2345.

29. A. B. MacKenzie and S. B. Wicker, Game theory and the design of self-configuring, *Adaptive Wireless Networks, IEEE Communications Magazine*, 2001, 39(11).

30. A. B. MacKenzie and S. B. Wicker, Stability of multipacket slotted aloha with selfish users and perfect information, in *Proceedings of 22nd Annual Joint Conference of the IEEE Computer and Communications*, March 30–April 3, 2003, pp. 1583–1590, San Francisco, CA, USA, Vol. 3.

31. P. Michiardi and R. Molva, Game theoretic analysis of security in mobile ad hoc networks, Technical Report RR-02-070, Institut Eurecom, April 2002, http://www.eurecom.fr/util/publidownload.en.htm?id=981

32. J. Bellardo and S. Savage, 802.11 denial-of-service attacks: Real vulnerabilities and practical solutions, in *Proceedings of the 12th Conference on USENIX Security Symposium*, June 9–14, 2003, Washington, D.C., USA, Vol. 12.

33. Y. Zhou, D. Wu, and S. M. Nettles, Analyzing and preventing MAC-layer denial of service attacks for stock 802.11 systems, in *Proceedings of the Workshop on BWSA, BROADNETS*, October 25–29, 2004, San Jose, CA, USA.

34. N. Marchang and R. Datta, Collaborative techniques for intrusion detection in MANET, *Ad Hoc Networks*, 2008, 6(4).

35. C. R. Dow, P. J. Lin, S. C. Chen, J. H. Lin, and S. F. Hwang, A study of recent research trends and experimental guidelines in MANET, IEEE 2005, in *Proceedings of the 19th International Conference on Advanced Information Networking and Applications (AINA'05)*.

36. C. H. Lin, W. S. Lai, Y. L. Huang, and M. C. Chou, Secure routing protocol with malicious nodes detection for ad hoc networks, in *Proceedings of the 22nd International Conference on Advanced Information Networking and Applications*, IEEE 2008.
37. J. Veijalainen and A. Visa, Guest editorial security in mobile computing environments, *Mobile Networks and Applications*, 2003, 8(2).
38. D. A. Maltz, J. Broch, J. Jetcheva, and D. B. Johnson, The effects of on-demand behavior in routing protocols for multi-hop wireless ad hoc networks, *IEEE Journal on Selected Areas in Communications (J-SAC), Special Issue on Wireless Ad Hoc Networks*, 1999, 17(8), 1439–1453.
39. B. Kannhavong, H. Nakayama, Y. Nemoto, N. Kato, and A. Jamalipour, A survey of routing attacks in MANET, *Wireless Communications, IEEE*, 2007, 14(5).
40. P. Yi, et al., A new routing attack in mobile ad hoc networks, *Int. J. Info. Tech.*, 2005, 11(2).
41. R. V. Boppana and S. Desilva, Evaluation of a stastical technique to mitigate malicious control packets in ad hoc networks, in *Proceedings of the 2006 International Symposium on World of Wireless, Mobile and Multimedia Networks*, 2006.
42. A. A. Pirzada and C. McDonald, Establishing trust in pure ad-hoc networks, in *Proceedings of the 27th Australasian Conference on Computer Science*, 2004, Vol. 26, pp. 47–54, Dunedin, New Zealand.
43. Y. C. Hu, A. Perrig, and D. B. Johnson, Packet leashes: A defense against wormhole attacks in wireless networks, in *Proceedings of Twenty-Second Annual Joint Conference of the IEEE Computer and Communications*, March 30–April 3, 2003, Vol. 3.
44. Y. C. Hu, A. Perrig, and D. Johnson, Wormhole attacks in wireless networks, *Selected Areas in Communications*, 2006, 24(2).
45. R. Maheshwari, J. Gao, and S. R. Das, Detecting wormhole attacks in wireless networks using connectivity information, in *Proceedings of 26th IEEE International Conference on Computer Communications*, May 6–12, pp. 107–115, Anchorage, AK, USA.
46. Y. C. Hu, A. Perrig, and D. B. Johnson, Rushing attacks and defense in wireless ad hoc network routing protocols, in *Proceedings of the 2nd ACM Workshop on Wireless Security*, September 19, 2003, San Diego, CA, USA.
47. Y. Hu, A. Perrig, D. B. Johnson, and A. Perrig, A secure on-demand routing protocol for adhoc networks, in *Proceedings of the Eighth Annual International Conference on Mobile Computing and Networking (MobiCom 2002)*, September 23–28, 2002, Atlanta, GA, USA.
48. C. Adjih, D. Raffo, and P. Muhlethaler, Attacks against OLSR: Distributed key management for security, in *2nd OLSR Interop/Workshop*, Palaiseau, France, July 28–29, 2005.
49. S. Marti, T. J. Giuli, K. Lai, and M. Baker, Mitigating routing misbehavior in mobile ad hoc networks, in *Proceedings of the 6th Annual International Conference on Mobile Computing and Networking*, 2000, Boston, MA, USA.
50. S. Zhong and F. Wu, On designing collusion-resistant routing schemes for non-cooperative wireless *ad hoc* networks, in *Proceedings of the 13th Annual ACM International Conference on Mobile Computing and Networking*, September 9–14, 2007, Montréal, QC, Canada.
51. C. Adjih, D. Raffo, and P. Muhlethaler, Attacks against OLSR: Distributed key management for security, in *Proceedings of 2nd OLSR Interop/Workshop*, July 28–29, 2005, pp. 1–7, Palaiseau, France.
52. X. Lin, X. Sun, P. Ho, and X. Shen, GSIS: A secure and privacy-preserving protocol for vehicular communications, IEEE November 2007, *IEEE Transactions on Vehicular Technology*, 2007, 56(6).
53. C. Zhang, R. Lu, X. Lin, P. H. Ho, and X. Shen, An efficient identity-based batch verification scheme for vehicular sensor networks, in *Proceedings of The 27th IEEE Conference on Computer Communications*, April 14–18, 2008, Phoenix, AZ, USA.
54. P. Yi, Y. P. Zhong, and S. Zhang, A novel intrusion detection method for mobile ad hoc networks, in *Proceedings of European Grid Conference (EGC 2005)*, February 14–16, 2005, Amsterdam, The Netherlands.
55. S. Desilva and R. V. Boppana, Mitigating malicious control packet floods in ad hoc networks, in *Proceedings of IEEE WCNC'05*, March 13–17, 2005, pp. 2112–2117, New Orleans, LA.
56. S. Lee, B. Han, and M. Shin, Robust routing in wireless ad hoc networks, in *Proceedings of International Conference on Parallel Processing Workshops*, August 18–21, 2002, pp. 73–78, Vancouver, Canada.

57. M. Al-Shurman, S.-M. Yoo, and S. Park, Black hole attack in mobile ad hoc networks, in *Proceedings of the 42nd Annual ACM Southeast Regional Conference*, April 2004, pp. 96–97, ACM Press, Huntsville, AL, USA.

58. S. Kurosawa, H. Nakayama, N. Kat, A. Jamalipour, and Y. Nemoto, Detecting blackhole attack on AODV-based mobile ad hoc networks by dynamic learning method, *International Journal of Network Security*, 2007, 5(3), 338–346.

59. D. Raffo, Securing OLSR using node locations, in *Proceedings of 11th European Wireless Conference 2005—Next Generation Wireless and Mobile Communications and Services*, April 10–13, 2005, Athens, Greece.

60. L. Qian, N. Song, and X. Li, Detecting and locating wormhole attacks in wireless ad hoc networks through statistical analysis of multi-path, in *Proceedings of IEEE Wireless Communications and Networking Conference*, March 13–17, 2005, New Orleans, LA, USA, Vol. 4.

61. B. Kannhavong, Analysis of the node isolation attack against OLSR-based mobile ad hoc network, in *7th International Symposium on Computer Networks*, 2006, Istanbul.

62. D. Dhillon, Implementation & evaluation of an IDS to safeguard OLSR integrity in MANETs, in *Proceedings of the 2006 International Conference on Wireless Communications and Mobile Computing*, July 3–6, 2006, Vancouver, BC, Canada.

63. C. Crépeau, C. R. Davis, and M. Maheswaran, A secure MANET routing protocol with resilience against byzantine behaviours of malicious or selfish nodes, in *Proceedings of the 21st International Conference on Advanced Information Networking and Applications Workshops*, 2007, Vol. 2.

64. Z. Karakehayov, Using REWARD to detect team black-hole attacks in wireless sensor networks, in *Proceedings of Workshop Real-World Wireless Sensor Networks*, June 20–21, 2005, Stockholm, Sweden.

65. Y. Jian, S. Chen, Z. Zhang, and L. Zhang, Protecting receiver-location privacy in wireless sensor networks, in *Proceedings of 26th IEEE International Conference on Computer Communications*, May 6–12, 2007, Anchorage, AK.

66. F. Liu, X. Cheng, and D. Chen, Insider attacker detection in wireless sensor networks, in *Proceedings of 26th IEEE International Conference on Computer Communications*, May 6–12, 2007, Anchorage, AK.

67. B. Xiao, B. Yu, and C. Gao, Detection and localization of sybil nodes in VANETs, in *Proceedings of the 2006 Workshop on Dependability Issues in Wireless Ad Hoc Networks and Sensor Networks*, September 26, 2006, Los Angeles, CA, USA.

68. H. Yu, M. Kaminsky, P. B. Gibbons, and A. Flaxman, SybilGuard defending against sybil attacks via social networks, in *IEEE/ACM Transactions on Networking of 11–15 September*, 2006, San Francisco, CA, USA, Vol. 16, No. 3.

69. R. Muraleedharan, Y. Yan, and L. A. Osadciw, Detecting Sybil attacks in image sensor network using cognitive intelligence, in *Proceedings of the First ACM Workshop on Sensor and Actor Networks*, September 10, 2007, Montréal, QC, Canada.

70. H. Yang, X. Meng, and S. Lu, Self-organized network-layer security in MANET, in *Proceedings of the 1st ACM Workshop on Wireless Security*, September 28, 2002, pp. 11–20, Atlanta, GA, USA.

71. Y. L. Sun, W. Yu, Z. Han, and K. J. R. Liu, Information theoretic framework of trust modeling and evaluation for ad hoc networks, *IEEE, Journal on Selected Areas In Communications*, 2006, 24(2).

72. G. F. Marias, K. Papapanagiotou, V. Tsetsos, O. Sekkas, and P. Georgiadi, Integrating a trust framework with a distributed certificate validation scheme for MANETs, *EURASIP Journal on Wireless Communications and Networking*, 2006, 2006(2).

73. W. Yu, Y. Sun, and K. J. R. Liu, HADOF defense against routing disruptions in MANET, in *Proceedings of 24th Annual Joint Conference of the IEEE Computer and Communications Societies*, March 13–17, 2005, pp. 1251–1261, Vol. 2.

74. Y. Lindsay, Sun, Z. Han, W. Yu, and K. J. Ray Liu, Attacks on trust evaluation in distributed networks, in *Proceedings of 40th Annual Conference on Information Sciences and Systems*, March 22–24, 2006, pp. 1461–1466.

75. V. N. Padmanabhan and D. R. Simon, Secure traceroute to detect faulty or malicious routing, *ACM SIGCOMM Computer Communications Review*, 2003, 33(1), 77–82.

76. M. Kefayati, H. R. Rabiee, S. G. Miremadi, and A. Khonsari, Misbehavior resilient multipath data transmission in MANET, October 30, 2006, Alexandria, VA, USA.

77. Y. Zhanz, W. Liu, and W. Lou, Anonymous communications in mobile ad hoc networks, in *Proceedings of 24th Annual Joint Conference of the IEEE Computer and Communications Societies*, March 13–17, 2005, Vol. 3.

78. M. Jakobsson, S. Wetzel, and B. Y. Stealth, Attacks on ad hoc wireless networks, in *Proceedings of IEEE Vehicular Technology Conference*, October 4–9, 2003, Orlando, FL, USA.

79. S. Capkun, J. P. Hubaux, and L. Buttyan, Mobility helps security in ad hoc networks, in *Proceedings of the 4th ACM International Symposium on Mobile Ad Hoc Networking & Computing*, June 9–14, 2003, pp. 46–56, Annapolis, MD, USA.

80. F. Li and J. Wu, Mobility reduces uncertainty in MANETs, in *Proceedings of 26th IEEE International Conference on Computer Communications*, May 6–12, 2007, Anchorage, AK.

81. G. Athanasiou, L. Tassiulas, and G. S. Yovanof, Overcoming misbehavior in mobile ad hoc networks: An overview, *ACM Crossroads 11.4: Mobile and Wireless Networking*, 2005.

82. M. Walfish, M. Vutukuru, H. Balakrishnan, D. Karger, and S. Shenker, DDoS defense by offense, in *Proceedings of the SIGCOMM 2006*, September 11–15, 2006, Pisa, Italy.

83. P. Louridas, Some guidelines for non-repudiation protocols, *ACM SIGCOMM Computer Communication Review*, 2000, 30(5).

84. P. Kamat, A. Baliga, and W. Trappe, An identity based security framework for VANETs, in *Proceedings of the 3rd International Workshop on Vehicular Ad Hoc Networks*, September 29, 2006, Los Angeles, CA, USA.

85. D. Subhadrabandhu, S. Sarkar, and F. Anjum, Statistical framework for intrusion detection in ad hoc networks, in *Proceedings of 25th IEEE International Conference on Computer Communications*, April 2006, Barcelona, Spain.

86. Y. Huang and W. Lee, A cooperative intrusion detection system for ad hoc networks, *Proceedings of the 1st ACM Workshop Security of Ad Hoc and Sensor Networks*, 2003, Fairfax, VA, USA, pp. 135–147.

87. Y. Zhang and W. Lee, Intrusion detection in wireless ad-hoc networks, in *Proceedings of the 6th Annual International Conference on Mobile Computing and Networking*, 2000, Boston, MA, USA.

88. Y. Zhang, W. Lee, and Y. Huang, Intrusion detection techniques for mobile wireless networks, *Wireless Networks*, 2003, 9(3).

89. J. A. Stine, Cross-layer design of MANETs: The only option, in *Proceedings of Military Communications Conference*, October 23–25, 2006, Washington, D.C., USA.

90. Y. Huang, W. Fan, W. Lee, and P. S. Yu, Cross-feature analysis for detecting ad-hoc routing anomalies, in *Proceedings of the 23rd International Conference on Distributed Computing Systems*, May 19–22, 2003, pp. 478–487.

91. M. Spohn and J. J. Garcia-Luna-Aceves, Neighborhood aware source routing, in *Proceedings of the 2nd ACM International Symposium on Mobile Ad Hoc Networking & Computing*, 2001, Long Beach, CA, USA.

92. M. Conti, R. D. Pietro, and L. V. Mancini, A randomized, efficient, and distributed protocol for the detection of node replication attacks in wireless sensor networks, in *Proceedings of the 8th ACM International Symposium on Mobile Ad Hoc Networking and Computing*, September 9–14, 2007, Montréal, QC, Canada.

93. W. He, Y. Huang, K. Nahrstedt, and W. C. Lee, Alert propagation in mobile ad hoc networks, *ACM SIGMOBILE Mobile Computing and Communications Review*, 2008, 12(1).

94. H. Yang, F. Ye, Y. Yuan, S. Lu, and W. Arbaugh, Toward resilient security in wireless sensor networks, in *Proceedings of the 6th ACM International Symposium on Mobile Ad Hoc Networking and Computing*, May 9–14, 2005, pp. 34–45, Urbana-Champaign, IL, USA.

95. W. Xu, W. Trappe, Y. Zhang, and T. Wood, The feasibility of launching and detecting jamming attacks in wireless networks, in *Proceedings of the 6th ACM International Symposium on Mobile Ad Hoc Networking and Computing*, May 9–14, 2005, pp. 46–57, Urbana-Champaign, IL, USA.

96. P. Ning and K. Sun, How to misuse AODV: A case study of insider attacks against mobile ad-hoc routing protocols, in *Proceedings of Information Assurance Workshop*, 2003, IEEE Systems, Man and Cybernetics Society, June 3, 2003.

97. S. Zhong, M. Jadliwala, S. Upadhyaya, and C. Qiao, Towards a theory of robust localization against malicious beacon nodes, in *Proceedings of the 27th IEEE International Conference on Computer Communication*, April 15–17, 2008, IEEE Communication Society, Phoenix, AZ, USA.

98. R. Muraleedharan and L. A. Osadciw, Security: Cross layer protocol in wireless sensor network, in *Proceedings of 25th IEEE International Conference on Computer Communications*, April 23–29, 2006, Barcelona, Spain.

99. T. Ernst, The information technology era of the vehicular industry, *ACM SIGCOMM Computer Communication Review*, 2006, 36(2).

100. A. Aijaz, B. Bochow, F. Dötzer, A. Festag, M. Gerlach, R. Kroh, and T. Leinmüller, Attacks on inter-vehicle communication systems—an analysis, in *Proceedings of 3rd International Workshop on Intelligent Transportation*, March 14–15, 2006, Hamburg, Germany.

101. M. Raya and J. Hubaux, The security of VANETs, in *Proceedings of the 2nd ACM International Workshop on Vehicular Ad Hoc Networks*, September 2, 2005, Cologne, Germany.

102. B. Parno and A. Perrig, Challenges in securing vehicular networks, http://sparrow.ece.cmu.edu/~adrian/projects/cars.pdf, accessed on July 31, 2008.

103. P. Papadimitratos, V. Gligor, and J. P. Hubaux, Securing vehicular communications—assumptions, requirements, and principles, in *Proceedings of the Workshop on Embedded Security in Cars (ESCAR)*, November 14–15, 2006, Berlin, Germany.

104. M. Raya, P. Papadimitratos, and J. P. Hubaux, Securing vehicular communications, *IEEE Wireless Communications*, 2006, 13(5).

105. P. Papadimitratos, L. Buttyan, J. P. Hubaux, F. Kargl, A. Kung, and M. Raya, Architecture for secure and private vehicular communications, in *Proceedings of 7th International Conference on ITS*, 6–8 June 2007, Sophia Antipolis.

106. A. Stampoulis and Z. Chai, A survey of security in vehicular networks, http://zoo.cs.yale.edu/~ams257/projects/wireless-survey.pdf, accessed on July 31, 2008.

107. IEEE Vehicular Technology Society, IEEE trial-use standard for wireless access in vehicular environments—security services for applications and management messages, IEEE Std 1609.2™, 2006.

108. R. M. Yadumurthy, A. Chimalakonda, M. Sadashivaiah, and R. Makanaboyina, Reliable MAC broadcast protocol in directional and omni directional transmissions for VANETs, in *Proceedings of the 2nd ACM International Workshop on Vehicular Ad Hoc Networks*, September 2, 2005, Cologne, Germany.

109. H. Wu, R. Fujimoto, R. Guensler, and M. Hunter, MDDV: A mobility-centric data dissemination algorithm for vehicular networks, in *Proceedings of the 1st ACM International Workshop on Vehicular Ad Hoc Networks*, October 1, 2004, Philadelphia, PA, USA.

110. P. Golle, D. Greene, and J. Staddon, Detecting and correcting malicious data in VANETs, in *Proceedings of the 1st ACM International Workshop on Vehicular Ad Hoc Networks*, October 1, 2004, Philadelphia, PA, USA.

111. T. Leinmuller, C. Maihofer, E. Schoch, and F. Kargl, Improved security in geographic ad hoc routing through autonomous position verification, in *Proceedings of the 3rd International Workshop on Vehicular Ad Hoc Networks*, September 29, 2006, Los Angeles, CA, USA.

112. G. Yan, G. Choudhary, M. C. Weigle, and S. Olariu, Providing VANET security through active position detection, in *Computer Communications*, September 10, 2007, Montréal, QC, Canada.

113. N. Patwari and S. K. Kasera, Robust location distinction using temporal link signatures, in *Proceedings of the 13th Annual ACM International Conference on Mobile Computing and Networking*, September 9–14, 2007, Montréal, QC, Canada.

114. C. Harsch, A. Festag, and P. Papadimitratos, Secure position-based routing for VANETs, in *Proceedings of Vehicular Technology Conference*, September 30–October 3, 2007, Baltimore, MD, USA.

115. M. Raya and J. P. Hubaux, Security aspects of inter-vehicle communications, in *Proceedings of 5th Swiss Transport Research Conference*, March 9–11, 2005, Monte Verità, Ascona, Switzerland.

116. S. Ni, Y. C. Tseng, Y. S. Chen, and J. Sheu, The broadcast storm problem in a mobile ad hoc network, in *Proceedings of the 5th Annual ACM/IEEE International Conference on Mobile Computing and Networking*, August 15–19, 1999, pp. 151–162, Seattle, Washington D.C., USA.

117. Y. Tseng, S. Y. Ni, and E. Y. Shih, Adaptive approaches to relieving broadcast storms in a wireless multihop mobile ad hoc network, in *International Conference on Distributed Computing Systems*, April 16–19, 2001, Phoenix, AZ, USA.

118. T. Leinmüller, E. Schoch, and F. Kargl, Position verification approaches for vehicular ad hoc networks, *IEEE Wireless Communications*, 2006, 13(5), 16–21.

119. M. Raya, A. Aziz, and J. Hubaux, Efficient secure aggregation in VANETs, in *Proceedings of the 3rd International Workshop on Vehicular Ad Hoc Networks*, September 29, 2006, Los Angeles, CA, USA.

120. F. Picconi, N. Ravi, M. Gruteser, and L. Iftode, Probabilistic validation of aggregated data in VANET, in *VANET'06*, September 29, 2006, Los Angeles, CA, USA, Vol. 31, No. 12.

121. G. Calandriello, P. Papadimitratos, J. Hubaux, and A. Lioy, Efficient and robust pseudonymous authentication in VANET, in *Proceedings of the 4th ACM International Workshop on Vehicular Ad Hoc Networks*, September 10, 2007, Montréal, QC, Canada.

122. S. Chapkin, B. Bako, F. Kargl, and E. Schoch, Location tracking attack in ad hoc networks based on topology information, in *Proceedings of IEEE International Conference on Mobile Adhoc and Sensor Systems (MASS-2006)*, October 9–12, 2006, Vancouver, Canada.

123. M. Jakobsson, X. Wang, and S. Wetzel, Stealth attacks in vehicular technologies, in *Vehicular Technology Conference*, September 26–29, 2004, Los Angeles, CA, USA.

124. A. Aijaz, B. Bochow, F. Dotzer, A. Festag, M. Gerlach, R. Kroh, and T. Leinmuller, Attacks on inter vehicle communication systems—an analysis. s.l. Technical report. The Network on Wheels Project, 2005 http://www.network-on-wheels.de/Documents.html.

125. N.-W. Lo and H.-C. Tsai, Illusion attack on VANET applications–a message plausibility problem, in *Proceedings of IEEE Global Communications Conference, GLOBECOM-2007 Workshops*, November 26–30, 2007, Washington, D.C., USA.

126. R. K. Schmidt, T. Leinmüller, E. Schoch, A. Held, and G. Schäfer, Vehicle behavior analysis to enhance security in VANETs, in *Proceedings of 4th Workshop on Vehicle to Vehicle Communications (V2VCOM 2008)*, June 3, 2008, Eindhoven, the Netherlands.

127. M. Pease, R. Shostak, and L. Lamport, Reaching agreement in the presence of faults, *Journal of the ACM*, 1980, 27(2), 228–234.

128. S., Bruce, Attack trees: Modeling security threats. *Dr. Dobbs Journal*, 1995, 3(5).

129. L. Blazevic, L. Buttayan, S. Capkun, S. Giordano, J. P. Hubaux, and J. Y. Le Boudec, Self organization in mobile ad hoc networks: The approach of Terminodes, *IEEE Communications Magazine*, 2001, 39(6), 166–174.

130. W. Stallings, *Wireless Communication and Networks*, 2nd ed., 2004, Prentice Hall.

Chapter 10

Security in Vehicular
Ad Hoc Networks

Vikas Singh Yadav, Sudip Misra, and Mozaffar Afaque

Contents

10.1 Introduction

With the increasing number of vehicles on the streets, an increasing population of vehicle manufacturers are looking for value-added services for providing their customers with increased safety and information. Toward this goal, Vehicular Communication (VC) is likely to play a major role. VC involves the use of short-range radios in each vehicle, which would allow various vehicles to communicate with each other and with road-side infrastructure. These vehicles would then form an instantiation of *ad hoc* networks in vehicles, popularly known as Vehicular *Ad Hoc* Networks (VANETs). VANETs are envisaged to provide safety-related information, traffic management, and infotainment services. These are the major areas in which applications are likely to develop and find commercial deployment. The first two, that is, safety and traffic management, require real-time information, and this conveyed information can affect life or death decisions. Without security, a VANET system is vulnerable to a number of attacks such as propagation of false warning messages and suppression of actual warning messages, thereby causing accidents. This makes security a factor of paramount importance in building such networks.

VANETs are of prime importance, as they are likely to be among the first commercial application of *ad hoc* network technology. Vehicles will act as nodes that are capable of forming self-organizing networks with no earlier knowledge of each other. The potential of VANET technology is high with a range of applications being deployed in aid of consumers, commercial establishments such as toll plazas, entertainment companies as well as law enforcement authorities. However, without securing these networks, they would lend themselves to blatant abuse, leading to major problems and immense damage to life and property. The implementation of security has to be accomplished, keeping in mind the conflicting requirements of personal users—car manufactures as well as law enforcement authorities.

This chapter focuses on providing an overview of the security of VANETs. We will begin by covering the various basic parameters such as vulnerabilities, challenges, adversaries and types of attackers specific to VANETs. We will then focus on security architecture of VANETs, followed by various proposed solutions by researchers for VANET security. Finally, we will cover the latest research challenges, including providing some thoughts for future directions of research.

10.2 Vehicular Networks: An Overview

VANETs constitute all types of *ad hoc* networks formed by the use of short-range radios installed in private (personal consumer) and public (public transport and law enforcement authorities) vehicles. Therefore, the first requirement of VANETs is to have each vehicle equipped with short-range radios for communication. The other components of a VANET node include those for providing detailed position information, road-side infrastructure units (RSUs), and central authorities responsible for identity management and registration. Communication in these networks involve both Vehicle-to-Vehicle (V–V) and Vehicle-to-Infrastructure (V–I) communications. Vehicles communicate with one another when they are within their transmission ranges. Vehicles will also communicate with road-side infrastructure, wherever they are present. The road-side infrastructure is spread regularly or sporadically depending on the region and extent of deployment. This infrastructure, though desirable to be present always in range, may in practice only be present sporadically, considering the high cost of development. Initially, it is most likely to be present at intersections and region borders. They serve the vehicles by providing information on safety, traffic conditions and infotainment, and information from central authorities.

Vehicular networks need a frequency band to operate and toward this most countries have allocated spectrum specifically for VC. In 1999, the U.S. Federal Communications Commission (FCC) allocated a block of spectrum in the 5.850–5.925 GHz band for applications primarily intended for VC [1]. Similar bands exist in Europe and Japan. The emerging de facto standard for VC is the Dedicated Short-Range Communications (DSRCs) [2]. DSRC is based on IEEE 802.11 technology and proceeds toward standardization under the name of IEEE 802.11p [3]. DSRC is quite attractive due to the large bandwidth and the possibility of using multiple channels. The IEEE standards propose employing multiple 10 MHz channels, each capable of carrying 27 Mbps of data for VC. Up to seven channels are available in the 5.9 GHz bands and one channel is supposed to be dedicated for safety applications [1]. The remaining channels are intended to be used for content distribution and delivery.

Major applications of VANETs include providing safety information, traffic management, toll services, location-based services, and infotainment. One of the major applications of VANET include providing safety-related information to avoid collisions, reducing pile up of vehicles after an accident, and offering warnings related to the state of roads and intersections. Affixed with the safety-related information are the liability-related messages, which would determine which vehicles are present at the site of the accident and later help in fixing responsibility for the accident. VANETs can be used to prevent collisions between vehicles by providing information to the driver about whether the vehicle ahead is braking, if the speed is too high or the distance to other vehicles or objects is getting too close. Eight safety applications based on deliberations between government agencies and private industry have been identified in [4], which are traffic signal violation warnings, curve speed warnings, emergency electronic brake lights, pre-crash warnings, cooperative forward collision warnings, left-turn assistance, lane change warning, and stop-sign movement assistance.

Another attractive application is for traffic management, where it is ensured that the vehicles choose the shortest route to a destination, avoid busy and congested areas and also enable traffic diversions in case of traffic jams or accidents. VANETs also have the potential to make various toll services easier to implement by enabling online toll collection as well as to provide information to drivers on cheapest routes between a source and a destination. Location-based services are already available through GPS, but the same can be enhanced by real-time data from road-side infrastructure and other vehicles, which periodically disseminate the information. Finally, various infotainment services that include access to Internet, music, and advertisements are likely to be provided with the help of VANETs.

A number of organizations and industry consortiums are involved in developing standards for VANETs. For example, the IEEE is involved in standards development related to the physical, medium access and security issues as well as in defining higher layer services and interfaces for intelligent transportation. By the end of 2006, the IEEE P1609 standards for wireless access in vehicular environments (WAVE) had specified the application layer and message formats for operation in the 5.9 GHz DSRC communications. The IEEE 802.11p standard [5], which is a modification of the popular IEEE 802.11 (Wi-Fi) standard, looks at issues related to the highly dynamic environment and the extremely short time durations, during which communications must be completed due to the high speed of the communicating vehicles. Several consortiums with industry and/or public participation are also working on furthering the development and deployment of vehicular networks. Some examples include the Car-2-Car Consortium [6], which has, as one of its primary objectives, the creation and establishment of an open and interoperable standard for V2V communications in Europe using Wi-Fi-like components. Some communication protocols are being developed by the Network-on-Wheels (NOW) group [7], which is associated with the Car-2-Car Consortium [6]. Ford and General Motors created a Crash Avoidance Metrics Partnership (CAMP) and with the National Highway Transportation Safety Administration, this partnership is working on projects such as enhanced digital maps for safety, driver workload metrics, and forward crash warning requirements [8]. Other VC projects being implemented include the Berkeley PATH [9] and the Fleetnet [10] projects in the United States and Germany, respectively.

10.3 Background: Need for Security

As in any major public network, VANETs, when deployed without considering the security requirements, lend themselves vulnerable to a host of attacks. The danger involved in possible road accidents and loss of life further impress upon the need for fail-proof security for VANETs. For example, safety-related applications need a high level of security, as a single vehicle sending out false warnings can disrupt the traffic of a whole highway. Similarly, in traffic information service, a greedy driver can send out multiple false messages under different identities to divert traffic to ensure a smooth drive for him. This can be resolved using authentication, which raises a fresh concern of privacy, as vehicles can be tracked and monitored using the identities sent out by them along with their messages. However, consumers would not like their private information being shared over a public network. Balancing these conflicting requirements is essential while implementing security in VANETs. Therefore, security and privacy are prerequisites for deployment of VANETs and striking an adequate balance between the two will be the key in their successful deployment. Figure 10.1 shows an example of secure VC.

A number of research efforts are on in the field of VANET security. Prominent among them are The US Vehicle Safety Communication Consortium (VSCC), which promotes and produces the DSRC standards for VC, part of which is the IEEE P1609.2/D2 draft standard [3]. VANET security is partially considered in the European Global System for Telematics (GST) [11], German

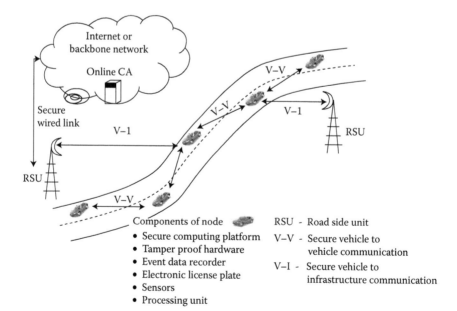

Figure 10.1   Secure vehicular communication.

Network on Wheels (NoW) [7] project and security workgroup of the Car-2-Car Communication Consortium (C2C-CC). SEcure VEhicular COMmunications (SEVECOM) [12] is a three-year European funded project in which universities, car manufactures, and car equipment suppliers are collaborating on the design of a baseline architecture which is practical and also provides a level of protection sought by users and legislators.

10.3.1 Security Requirements

A comprehensive discussion of basic requirements for a VC system is provided in [13]. Other related works which discuss the security requirements [14,15]. Based on the above works we outline the following major security requirements for a VANET:

- *Authentication* is a major requirement for VANETs. This is simply because it ensures that various messages are sent by actual nodes and not by a node representing multiple identities or a node impersonating as someone else. Sybil attacks are also avoided if authentication is assured, as a malicious node cannot send messages from nonexistent nodes. This attack can be used by greedy drivers to divert traffic from their routes by simulating a congested road by sending false messages. Authentication, however, raises privacy concerns, as a basic authentication scheme of attaching the identity of the sender with the message would allow tracking of vehicles. It, therefore, is absolutely essential to authenticate that a sending vehicle has a certain property that provides authentication as per the application. For example, in location-based services this property could be that a vehicle is in a particular location from where it claims to be.
- *Message integrity* is important, as it needs to ensure that the message is not modified in transit. This, coupled with authentication, assures VANET nodes that the messages they receive are not false.
- *Message nonrepudiation* is required so that sender cannot deny having sent that message. This, however, further exacerbates the identity management issue. Only specific authorities should be allowed to identify a vehicle from the authenticated messages it sends.
- *Entity authentication* is a property that enables a receiver to ensure that the sender generated a message and is still active in the network. This is required to ensure that a particular message was generated by a sender within a small time interval just before the receipt of the message at the receiver.
- *Access control* is required to ensure that all nodes function according to the roles and privileges authorized to them in the network. Toward access control, authorization specifies what each node can do in the network and what messages can be generated by it.
- *Message confidentiality*, though strictly not very essential, in a VANET, can still be utilized when certain nodes want to communicate with each other in private. Such a case can arise when law enforcement vehicles communicate with each other for disseminating private information regarding suspected location of criminals or speed check points.
- *Privacy* is important to ensure that the user information is not leaked or distributed to parties not authorized to access such information. Third parties should also not be able to track vehicle movements as it is a violation of personal privacy. Therefore, a certain degree of anonymity should be available for messages and transactions of vehicles. However, in liability-related cases, specified authorities should be able to trace the user identities to determine the responsibilities. Location privacy is also important so that no one should be able to learn the past or future locations of vehicles.

- *Availability* is essential to enable VANET services to be operational in the presence of faults and attacks. This implies that VANETs should be resilient to denial of service (DoS) attacks by having alternate means of communication and redundant infrastructure.
- *Real-time guarantees* is essential in a VANET, as many safety-related applications depend on strict time guarantees. This can be built into protocols to ensure that the time sensitivity of safety-related applications such as collision avoidance is met.

10.3.2 Challenges

For implementing VANET security, it is essential to understand the unique challenges faced in such networks. A good discussion of challenges in securing VANETs is given in [16]. The major challenges are outlined below:

- *Tradeoff between authentication and privacy:* To ensure that certain nodes do not impersonate another node, it is essential to authenticate all message transmissions. However, this leads to identification of vehicles from the messages they send. This can enable tracking of vehicles, which most consumers would not like to enable in their systems. Privacy is a major issue in a VANET, because cars are highly personal devices. This has to be balanced with the need for establishing accountability and liability of vehicles and their drivers. This requires an authentication system to be designed that enables messages to be anonymous for general nodes but also enables identification by central authorities in liability-related cases like accidents.
- *High mobility:* VANETs are characterized by highly mobile nodes which will result in frequent changes in topology and brief connectivity between the nodes. In such situations, VANET protocols cannot be handshake based. Most of the communications are between nodes that have never interacted before and will probably not interact again in future. This characteristic rules out learning- or reputation-based schemes where nodes learn about each others behavior.
- *Scale of network:* VANETs are likely to be among the largest *ad hoc* networks, requiring scalable solutions for an adequate availability and a sufficient performance. This aspect rules out having prestored information about other nodes or distribution of centralized information to all nodes. Also, security and privacy policies will differ from region to region owing to the worldwide deployment of this network. Coordination of such a network will be difficult and would require specific relationships between various regions.
- *Real-time guarantees:* The major VANET applications are safety related for collision avoidance, hazard warning, and accident warning information. These applications require strict deadlines for message delivery. Any security protocol implemented for VANETs would need to take this into consideration and have low processing and message overheads.
- *Incentives:* For effective deployment of VANET technology, it is imperative to offer incentives to the involved parties for them to adopt the system. With security the cost and the complexity of this system would further increase. It, therefore, becomes imperative to offer all concerned the correct incentives to adopt this technology and the security being implemented.
- *Location awareness:* For most VANET applications to be truly effective, certain location-based service is essential. This increases the reliance of the VANET system on GPS or other specific location-based instruments. Any error in these is likely to reflect in the VANET applications.

10.3.3 Adversaries

Before developing a VANET security system, it is imperative that we understand what type of adversaries would target the system and type of attacks they are capable of launching against the system. In this section, we discuss the probable adversaries and attacks they can launch.

The attackers can be divided into the following general categories:

- *Selfish drivers:* Even though majority of the drivers in a system would be honest and adhere to the rules, it is natural that some drivers would try to gain specific advantages from the system. In such a situation, the driver may send false information to divert traffic and gain a free path on his route. This is the most common form of attacker, but can easily be put off with a basic authentication system and fear of law enforcement authorities if he believes that there is high probability of getting caught.
- *Eavesdroppers:* These adversaries would like to collect information about drivers and use this to understand drivers' behaviors and traffic pattern. Also, commercial firms can use this to offer content in infotainment services, when the customer or driver has no interest in getting such services as in mobile networks. Moreover, drivers would not like their personal information to be divulged to third parties.
- *Teenage hackers:* These adversaries try and hack into any major system that gets deployed publicly. They try to find bugs in the software and cause traffic disruptions just for fun.
- *Insiders:* These adversaries include persons working in car companies and installing the VC system. They are capable of loading malicious software in cars that could cause immense damage. Also, if manufacturers are entrusted with the responsibility of key distribution, then an insider may create keys acceptable to all users for his cars, that is, compromise private keys of vehicles in a Public Key Infrastructure (PKI) setting.
- *Malicious attackers:* These attackers could be criminals or terrorists having access to more sophisticated tools and hardware than normal attackers. Criminals may have specific targets for financial gains or would like to carry out personal harm to rivals. Terrorists can use sophisticated technology to disrupt vehicular traffic to cause maximum damage when using bombs or launching gun attacks. These are the most dangerous of attackers and specific measures need to be taken to guard the system against such attacks.

A good classification of attackers is given in [14] and an adversarial model for a VANET system is presented in [13]. They classify attackers as Insider versus Outsider, Malicious versus Rational, and Active versus Passive. An Insider is a part of the network and has the requisite certificate to authenticate him into the network. A malicious attacker can become a part of the network by extracting the cryptographic keys from the hardware unit of a vehicle. Outsiders do not possess such credentials and can only launch limited attacks against the VANET system. A malicious attacker generally wants to wreak havoc and disrupt the system whereas a rational attacker is generally a greedy driver who wants to gain advantage for oneself. Finally, active attacker is an active member of the network and can launch attacks by injecting false messages, forge, alter and replay messages. Passive attackers can only snoop and gain information which they can use to track vehicles but not affect their behavior in any way.

10.3.4 Attacks

VANETs are susceptible to various types of attacks. They vary according to the situation, attacker's intent, and the amount of damage and scope. These are discussed in detail in [16,17]. We give a brief outline of the major attacks possible.

- *Denial of service:* This can be done by channel jamming, that is, block access to a communication channel by high power transmission on the communication channel or by injection of dummy messages. Channel jamming can be done relatively easily and can be very effective in disrupting a communication network. Also, an attacker could inject a large number of dummy messages in the network to flood the network and not allow critical safety messages to reach the desired recipients.

■ *Impersonation:* An attacker may take on someone else's identity and gain certain advantages or cause damage to other vehicles. A vehicle may impersonate an emergency vehicle to gain access to highway or a vehicle may send out another person's identity in an accident case to escape liability.

■ *Message falsification:* An attacker can send false messages in a VANET network such as false hazard warnings to divert traffic from a route for freeing up resources for it.

■ *Message alteration:* This form of attack can be done where either an individual entry in a message can be altered or full message contents can be altered. For example, a hazard warning can be changed to a no-hazard message to cause traffic accidents.

■ *Message delay and suppression:* In case of accidents, certain vehicles may delay or suppress safety messages. Owing to the real-time nature of such information, this may cause immense damage. By selectively dropping messages or delaying their transmission, critical information may not reach vehicles in time.

■ *Privacy violation:* In order to authenticate messages to prevent impersonation, a simple mechanism is to associate the identity of vehicles with the messages they send using asymmetric key cryptography. However, this lends itself to people being able to identify the sender of the message. Thus, vehicles can be tracked and anyone can identify a vehicle's owner. This raises some serious privacy issues as in all applications like safety, traffic management and toll access. The messages would reveal the driver's identity, his location, his actions and preferences. Consumers would not like to adopt a technology that violates their privacy.

■ *Replay attacks:* Vehicle can easily eavesdrop and log messages of other vehicles and replay them later to gain access to specific resources like toll services or to send false alarms. Therefore, mechanisms to ensure replay attacks are not possibly need to be built into the VANET security architecture.

■ *Hardware tampering:* Attackers can also tamper with the security hardware of a vehicle to steal identities as well as extract cryptographic keys. Therefore, specific mechanism like tamper-proof hardware needs to be implemented to ensure such attacks cannot easily be accomplished.

■ *Sensors tampering:* Another easy attack is to fool around with the vehicle's hardware sensors. If the main system is tamper proof, it is easy to fool the vehicle's sensors with wrong information by simulating false conditions. Examples include tampering with the GPS system and temperature sensors.

10.3.5 VANET Properties Supporting Security

VANET systems have certain properties such as high mobility and immense scale, which make them a unique type of *ad hoc* network. Some of its unique properties support security while others hinder security. We discuss the positive aspects in this section.

■ *High processing power and adequate power supply:* A major aspect of VANET is that unlike nodes in other *ad hoc* networks, the VANET nodes are vehicles which have their own power in the form of batteries and can have high computing power. This means that unlike a majority of the *ad hoc* networks, they do not need power-efficient protocols. Also, sufficient computing power allows the nodes to run complex cryptographic calculations. In the latter case, however, a limiting factor is the fleeting encounter of nodes with each other. This is not the case, however, when vehicles communicate with road-side infrastructure.

■ *Known time and position:* As most vehicles in VANETs are expected to be equipped with GPS receivers, the location of a node with time would be available. This would simplify implementation of various security protocols.

■ *Limited physical access:* Access to a VANET node is limited to its driver or authorized personnel as a physical locking mechanism is present in every vehicle. This highly aids the physical security of VANET nodes.

■ *Periodic maintenance and inspection:* In most cases, cars receive periodic maintenance, which can be used for regular checks and updates of firmware and software. In case public key cryptography is implemented, it can also be used for updating certificates and keys, along with provision of fresh Certificate Revocation Lists (CRLs).

■ *Central registration:* Another advantage of the VANETs is that, unlike other *ad hoc* networks, all the nodes (i.e., vehicles) are registered with a central authority and already have a unique identity in the form of a license plate. There is an existing infrastructure that maintains records of all vehicles. This existing setup could be leveraged to enhance security of VANETs by setting up a vehicular PKI and make the registration authorities act as Certification Authorities (CAs). However, this would also require a change in their working setup that will require effort both in terms of time and money.

■ *Honest majority:* We can safely assume in a VANET network that majority of nodes will be honest and law abiding as the existing setup will continue with the same set of drivers on the vehicles. This makes easier to detect and isolate malicious nodes with polling and voting mechanism helping to implement the same. Another reason for less number of malicious nodes is that few people like modifying or assembling their own car unlike PCs where tinkering is quite prevalent.

■ *Existing law enforcement infrastructure:* Unlike other *ad hoc* networks here, there is an existing agency that can catch and apprehend wrong doers. This will serve as a major deterrent to attackers. Although it would require additional training on part of law enforcement officers to adapt themselves to this technological network, it is still a major deterrent.

■ *Nodes limited to a certain physical region:* Vehicles will be limited to roads in a VANET network and as most roads are mapped it will be easy to pinpoint node locations. This makes node locations geographically limited and their movements trackable. Both these aspects help in implementation of protocols such as geographic routing.

10.4 Security in VANETs

Implementing security in VANETs has unique challenges as discussed in Section 10.3.2. The case is more complicated due to differing requirements of different applications of VANETs. Security for secure dissemination of safety information requires a different approach than that required for traffic management applications. However, we can balance the differing requirements and still create baseline architecture for implementing security in VANETs. Before we discuss the various specific security issues in VANETs, we discuss in this section a base architecture based on common elements of various proposals presented over the years. One of the first such attempts was made in [18], where security architecture is presented using an AAA (Authorization, Authorization and Accounting) framework using tamper-proof security hardware, Vehicular PKI and use of road-side infrastructure. This is further refined in [13], where additional elements are added to address more issues. We will discuss this architecture as it appears the most deployable with an existing framework of registration authorities being utilized to act as CAs.

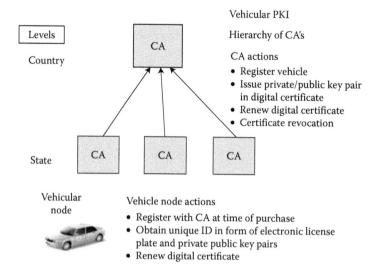

Figure 10.2   VANET security architecture.

Figure 10.2 shows a typical VANET security architecture. All proposed security architecture for VANETs fundamentally deals with the following aspects: authorities, a trusted component, identity management, privacy protection, pseudonym management and secure communication. We discuss these and other specific mechanisms, including detection of false data and faulty and malicious nodes, secure aggregation of data and secure location verification. We discuss these aspects serially in the following sections.

The basic architecture consists of *Network nodes* that can be either *Vehicles* or *Road-side Infrastructure* and existing *Registration Authorities* for vehicle registration and record maintenance. These nodes will be installed with required *sensors* for gaining information, *processing units* for processing the collected or received information, and *communication system* for disseminating information to and receiving information from other nodes. This will enable a basic network system that can easily be manipulated and subverted by adversaries both within and outside the network. To implement security in this system, we need additional infrastructure and mechanism in place. The aspects to be addressed are the security requirements discussed in Section 10.3.1.

A secure system, besides the basic network nodes, will consist of a *Vehicular PKI, a Secure computing platform and various security mechanisms.* Secure mechanisms comprise Identity management using Electronic License Plates with certified public and private keys attached to the owner, Authentication and Integrity using Digital Signatures, Privacy using Pseudonyms, Pseudonym handling and Certification Revocation mechanisms. A Vehicular PKI will consist of the national- and state-level registration authorities acting as CAs which will issue certified public/private key pairs to vehicles. A Secure Computing platform on a vehicle will consist of tamper-resistant hardware and firmware. Its job is to store cryptographic material (private keys) and a trusted (tamper proof) clock. Digital Signatures will provide the required authentication and integrity along with nonrepudiation using timestamps. Privacy is introduced by using Pseudonyms in the form of additional set of public/private keys that are given to the user. These keys are used for a short period of time and changed frequently. These keys do not contain identity related information but can be traced back to the owner in liability related cases with the help of central authorities. The aim in using pseudonyms is to ensure that a vehicle cannot be tracked and a message cannot be attributed to its sender by other vehicles. Pseudonyms handling includes the time to change pseudonyms and when to change them. These depend on the level of anonymity required by a vehicle. This also depends on the type of application the VANET system is providing. Finally, when a vehicle becomes faulty or is detected as an illegitimate or malicious vehicle, Certificate Revocation mechanisms are required to revoke both long-term certificates and set of Pseudonyms currently being used by the vehicle. We will cover all these aspects in greater detail in the following sections after detailing the other research initiatives for developing a secure architecture for VANETs.

The security architecture developed by the VSCC and subsequently submitted to IEEE P1609.2 can be seen as the only approach for a security architecture in vehicular networks that is under standardization so far [19]. It defines a public-key-infrastructure (PKI)-based approach for securing messages sent in a vehicle-to-vehicle and vehicle-to-infrastructure fashion. The standard, however, does not address privacy issues, multihop communication, and how the network can be protected against malicious-certified nodes.

The Daimler Chrysler group also published security architecture in [20] in the form of a layered structure with multiple views of the system. The security architecture of the system discussed in this paper contains the Vehicle Manufacturer and the Registration Authority for registration of nodes and assigning node identifiers, the Inspection site for test and certification of nodes, an Escrow entity with authority to identify and revoke certification of nodes and finally the communication infrastructure consisting of communication systems, processing and databases necessary to carry out online testing, pseudonym provision for nodes and infrastructure based data assessment and intrusion handling.

10.5 Central Authorities and Vehicular PKI

The security system will require a Vehicular PKI with the existing authorities responsible for registration of vehicles acting as CAs responsible for issue of certified public/private key pairs to vehicles. Every region will have a separate CA. Each vehicle will be issued with an Electronic License Plate which will be a unique identity similar to a registration number issued to vehicles presently with additional certified public/private keys in a digital certificate. Cross certification will be required when a vehicle goes to different regions, that is, different state or different country. This will require that each vehicle stores number of trusted root CA certificates containing their public keys.

Each owner will register his vehicle with its own region's CA while purchasing the vehicle and get a pair of private and public keys in a digital certificate. Fresh key pairs will be issued when the previous certificate expires or is revoked. The digital certificate contains the node's identity as well as various parameters like cryptographic algorithm used as well as the certificate lifetime. Each vehicle gets one long-term identity for identification from the CA. However, the need to maintain privacy introduces the concept of short-term public/private key pairs which have no direct link to the identity of the vehicle. The task of issuing these keys can also be entrusted to the CAs. In some proposals, a different entity is entrusted with this task. Certificate revocation is also a responsibility of the CAs as and when vehicles are taken out of service or are declared as malicious nodes for illegal activity.

Each vehicle will be issued with a digital certificate which will contain its public key digitally signed by the CA. While implementing a PKI mechanism, it is not possible for a node to store public keys of all other nodes even it is for one region alone. Therefore, after digitally signing a message, each node will send his digital certificate (containing his public key) with each message.

The public key in the digital certificate can be verified by extracting it from the digital certificate and verifying it with the public key of the root CAs stored in each node. This can be refreshed and updated during periodic inspections or secure transfers from road-side infrastructure.

10.6 Secure Computing Platform: Hardware and Software for a Secure VANET Node

VANETs require nodes to be smart intelligent vehicles. As described in [21], a vehicle is smart if it is equipped with recording, processing, positioning, and location capabilities and if it can run wireless security protocols. Already, modern cars possess processors that control vehicular functions ranging from fuel injection to automatic climate control. A smart car, as detailed in [21], has event data recorder (EDR), which keeps a record of all major actions involving the car, a GPS receiver for providing real-time location of the car, a front end radar for detecting obstacles and short distance radar for use of parking.

To implement security in a smart car, we need to have specific hardware to be installed in a car. As it is intended to have a vehicular PKI for identification of vehicles and message authentication and integrity, we need a secure place to store private keys issued to each vehicle. This would be a Secure Computing platform which will be a Tamper-Proof Device (TPD) consisting of tamper-proof hardware and firmware. This should have both processing and storage facility. It should store the private keys issued to each vehicle, perform cryptographic operations using the private key, and have a trusted clock. It should have its own battery for maintaining this clock. A trusted clock is necessary to ensure that the clock time is not changed and the vehicle forced to produce messages for future time which can be captured and later used in some malicious manner. This clock can be synchronized with the clocks at nodes installed as a part of road-side infrastructure, that is, RSUs. Tamper-proof hardware implies that if the module is manipulated with an attempt to extract the secret keys the hardware would delete the private keys to ensure that they are not revealed. Also, the secure computing platform performs the cryptographic operation involving the private key inside the tamper-proof hardware to ensure that the private key never leaves the module and therefore can never be revealed to any attacker. The tamper-proof hardware will also store the short-term private keys that are generated as a part of a public/private key pair as pseudonyms to maintain privacy of the vehicle.

Along with this we need a logging mechanism that can record all information related to a vehicle's movement and actions during accident cases. This would be done by an EDR which is similar to a Black Box for an aeroplane and store critical information such as speed and position for reconstructing what happened during an accident. Such a device is already present in high-end cars and heavy vehicles like trucks.

10.7 Implementing Message Authentication and Integrity Using Digital Signatures

Digital Signatures are going to be the basic mechanism for implementing message authentication and data integrity. This is considered to be the most viable solution in VANETs for the above requirements. Each node will digitally sign the message before transmission and the receiver will verify the same before accepting the message as valid. The method to digitally sign a message includes using a hash function to produce a hash of the message (much smaller in size) and then encrypt the hash with the private key. The message is transmitted along with its digital signature and the receiver on reception verifies the digital signature by decrypting it to get the hash of the original message. The received message is then hashed using the same hashing function and the two hashes compared with each other. If they are the same, then it is confirmed that the message has not been altered in transit as also the message is from the sender as no one else could have encrypted it with its private key. To protect against replay attacks, the timestamp from the trusted clock can be added to the message. This will ensure that the message is fresh and cannot be replayed later. To reduce the security overhead, the common approach as suggested in [14] is to use Elliptical Curve Cryptography (ECC). It is shown that the message overhead is of the order of a few megabytes and the critical overhead is signature verification time as it has to verify multiple messages simultaneously while signing only one message. This can be reduced by verifying only messages where the content is relevant and if a signer is verified once it need not verify its messages again especially in traffic congestion scenarios where number of messages are redundant.

Keeping in mind the need to maintain privacy of the vehicle, the authors in [14] propose using anonymous key pairs that change very frequently according to driving speed. These pseudonyms are issued in a set to be used for a short period of time. These keys do not reveal the identity of the node. For the purpose of liability identification, a mapping is maintained between the long-term identity and short-term anonymous keys at the CA. The node will use these short-term keys once for a short period of time and then switch to another key pair.

10.8 How is Privacy and Identity Management being Handled

The issue of privacy of each vehicle and desire of drivers not be tracked at all times, coupled with the need of law enforcement authorities to ensure liability, makes it imperative to have a scheme that balances both of these conflicting requirements. Most reasonable among the solutions proposed seems to be the use of pseudonyms as short-term public keys with no identity information linked with these keys. Pseudonymous authentication requires that each node is equipped with multiple credentials in the form of certified public keys that do not reveal the node identity. Each node, be it a vehicle or RSU, will have a long-term identity in the form of an electronic license plate issued at the time of registration. This identity will be coupled with a private/public key pair and attributes of the node. Along with this, the node will also obtain over a secure communication channel a set of pseudonyms for use after authenticating itself using its long-term public/private key pair. These pseudonyms will have no information of the identity of the vehicle. The private keys of each pseudonym will be stored in the tamper-proof hardware. Moreover, a vehicle can only have one pseudonym active at one time, thereby preventing a vehicle from sending multiple messages at one time and pretend to act as multiple vehicles.

As the number of short-term keys required is large, it is necessary that fresh sets are taken at regular intervals. In [14], it is suggested this be done, initially, by the certifying authority while registering the vehicle and subsequently at the yearly inspection of a vehicle. However, this requires a large set of keys to be stored as also the need for ensuring timely inspections. Moreover, certificate revocation will become a difficult in this case. Other option is that of an online CA as suggested in [16]. Each vehicle will have to be in touch with the online CA regularly, if not all times for ensuring that it does not run out of pseudonyms and gets a new set as soon as the previous one is about to finish. However, this scheme requires a permanent online CA at all times. Another problem is that as these pseudonyms will carry the identity of the CA for liability and signing purposes it is imperative that a vehicle moving into another region obtains a fresh set, after getting certified, for that region. This is necessary, as it would be trivial to track a vehicle outside its parent region as it may be the only vehicle in the new region with its CA's signed pseudonyms. Scalability is a major issue with the solution of having an online CA. Initially, the vehicles can obtain pseudonyms from RSUs located regularly in specific areas like toll highways where VANET services are offered. This, however, suffers from the problem of maintaining data about all these vehicles at different places for liability-related cases. Initial deployment of VANETs is still therefore a cost-intensive aspect, for which better approaches need to be devised.

As a vehicle's location information included in messages can still be used to track the vehicle, it becomes imperative for better schemes to avoid such tracking. Therefore, in [22], a scheme is proposed where vehicles change pseudonyms in special regions called mix zones [23] that are not monitored by attackers. A mix zone is a region where a number of vehicles mix together and as they change pseudonyms inside the mix zone it becomes difficult for an adversary to track a vehicle when it exits the mix zone among the multiple vehicles exiting at nearly the same time. However, such mix zones may not always be available or it may not be possible for a vehicle to detect that it is in a mix zone. Therefore, the authors in [24] suggest a proposal to create a mix zone. These zones are created using encrypted communication between the nodes. However, such zones will have a smaller size and will accommodate lesser number of vehicles.

10.9 Certificate Revocation

With the large number of nodes in a VANET environment, it is natural that some vehicles will get faulty and there will be malicious nodes looking to gain an unfavorable advantage or trying to causing harm to other vehicles. When these vehicles are detected, it is necessary to revoke their identity and evict them from the network to prevent them from causing further harm. Another case could be where the CRLs would be created by the CA and distributed periodically to all nodes. The method of distribution has to be catered to the specific needs of the VANET environment.

Certificate Revocation and CRL distribution are discussed in [25] with different protocols suggested depending on whether adequate RSUs are available or not. Their first protocol called RCCRL (Revocation using Compressed CRL) utilizes Compressed CRLs made using Bloom filters which is a form of lossy compression [26]. These Compressed CRLs are distributed to all users to disseminate information on revoked keys. This scheme is used when specific keys have to be revoked of each vehicle. However, there are cases where all the keys of a vehicle have to be revoked. In this case, a better option is send a message to the TPD to revoke all certificates stored in it. The authors propose another protocol called RTPD (Revocation of TPD) in which the CA sends a message to TPD to delete all the keys and cryptographic functions stored in it. The message contains the vehicles identity and timestamp and is encrypted with its public key. The message can be decrypted by the vehicle's TPD only which affectively avoids any blocking by the vehicle itself as it does not know the contents of the message till it is decrypted. Such a message can be routed to the vehicle using the nearest RSU or if sufficient infrastructure is not present then it can be disseminated via low-speed broadcast radio service such as FM radio.

10.10 Secure Aggregation of Data

Safety-related applications are the major application for VANETs. They are likely to be the major incentive for deployment and commercialization of VANET technology. In all safety-related applications, there is a lot of redundant data which are relayed. When a vehicle approaches an accident site or a hazardous area not fit for driving, it gets a lot of similar messages from vehicles near the location. This information has to be relayed to other vehicles behind it. The most basic approach of forwarding all messages is not effective as it leads to information overload as a large number of similar messages are forwarded. It is better to find a way to aggregate the information in similar fields and then forward the data. However, the question is how the nodes to which the aggregated data are forwarded will trust the aggregated data and authenticate various data from multiple senders. The authentication and integrity of aggregated data have to be handled differently depending on whether the messages are exactly similar or messages in which few fields have different values but similar type of data that can be replaced with a group value which though not exactly the same but is near to it.

A basic proposal was made in [27], which dealt with the case where the individual messages are identical. In this proposal, the vehicles form location-based groups as most safety messages are destined for vehicles in a particular location. Based on the location, a leader is elected who is located in the centre of the region. This leader now communicates with the other groups on behalf of its group. Therefore, the number of messages is drastically reduced. However, the issue remains as to how can the data aggregated by the leader be trusted. The authors present three schemes for secure data aggregation: Combined Signatures, Overlapping Groups, and Dynamic Group Key Creation.

The first scheme tries to combine signatures of various members of the group. The most basic way to combine signatures is Concatenated Signatures where a node just appends its signatures to the existing signatures and rebroadcasts the message. This method has the same security overhead as it has the same number of signatures. However, the network overhead is reduced. This is because the messages are aggregated at the source and then forwarded. The second method is to use a concept of Onion Signatures in which a node, instead of appending its signature, re-signs the received signature and forwards the message with the previous signature and the new signature. This reduces the overhead to only the message and two signatures. In this case, an invalid signature at any node renders the so-called Onion Signature invalid. The computation overhead is increased as the signatures have to be verified one by one. To strike a balance between the above two schemes, a hybrid scheme is proposed which uses several onion signatures each of given depth with the number and depth depending on whether communication or computation overhead is more important.

In Overlapping Groups scheme, each location-based group has its own symmetric key and communicates securely using this key. The nodes, which belong to more than one group, have the symmetric key of both the groups. This scheme suffers from the lack of nonrepudiation within the group. However, this can be countered by the assumption that majority of VANET nodes are honest. This scheme has less communication and computational overhead, but also requires secure position verification for formation of location-based groups.

The third scheme called Dynamic Group Key Creation assumes the existence of sporadic road-side infrastructure and an online CA. In this scheme, dynamic groups are created based on vehicles sharing the same driving pattern. The group leader then requests the CA for a group asymmetric public/private key pair. This is disseminated to all the group members using the agreed upon symmetric key. Any group member can send a message on behalf of the group using this key pair. The nonrepudiation property is preserved by the CA assigning all members of the group a unique ID.

10.11 Detection of Malicious Data and Secure Position Verification

As discussed in Section 10.3.4, it is possible that a vehicle's sensors may be tampered with and thereby the vehicle will send false data involuntarily. An example would be if a vehicle's ice sensors are fooled by inserting them into ice or covering them with ice, thereby convincing the vehicle to send out false ice hazard warnings. It is, therefore, imperative not only to authenticate message senders and detect false nodes but false messages as well. Most application in VANET area requires accurate positions of vehicles for effective running. Therefore, having accurate position data of all nodes is one of the most essential tasks. Accurate positions of all nodes is also essential for geographic routing where messages are forwarded to intermediate nodes for transmission to the destination data based on its location. We, therefore, consider here the various proposals for secure position verification and then a generic proposal for all types of data.

The first proposal for secure position verification was given in [21], using an approach known as *verifiable multilateration*. It works on the premise that adequate base stations are present and do not require vehicles to have GPS devices for positioning. In this approach, the base stations build a trustworthy network and four of them collaborate to fix the position of the vehicle by measuring one by one the time between sending a challenge to the vehicle and receiving its answer. In this approach, the trick is that a node cannot advance the time of the reply, thereby fixing its near distance. It also cannot delay its reply, as that would need it to advance the time for reply to another base station which is not possible. The second proposal in [28] is one in which there are specific sensors which use threshold values and other checks to determine that the position information it receives from other nodes is correct. In this proposal, all nodes have means to determine their own position through GPS and they do not rely on the existence of base stations. Each node sends its location information using beacon messages and the receiving node verifies this information using specified checks. The first is *Acceptance Range Threshold*, in which a node uses its communication range to define a maximum threshold value of distance of the node whose position is being verified. This limit can help detect wrong position being advertised by a node. The second check is the *Mobility Grade Threshold* in which it is assumed that there is a maximum speed for a node. This is used to check the subsequent beacon messages from a node to detect if any false change of position is being sent as there is limit on the node's movement because of its last known position. The third check is *Maximum Density Threshold* where keeping the physical dimensions of vehicles in mind it is checked whether a particular region has more vehicles than it can possibly accommodate. The fourth check is a map-based one where it is checked whether a node's advertised location is a correct location for a vehicle on the map. This basically checks that a vehicle's location is confined to roads and another building or structure does not exist at that location. Finally, it proposes that a node overhears other node's claims of its position at different times and its packet forwarding to detect if it is forging its location. This approach can be used with other approaches as it is not reliable on its own.

Another proposal in [29] outlines a more complete approach which tackles the problem of detection and correction of malicious data. It works on the concept of having each vehicle maintaining a model of VANET system based on the physical information collected from messages it receives. This model contains all the information which the node has of the network. This model is checked for consistency based on various checks similar to the one outlined in the previous proposal. For this model, the authors use a heuristic named adversarial parsimony which assumes that attacks involving a few malicious nodes are more likely than an attack involving large number of nodes. Once a new message is received, its information is added to the model. If the model becomes inconsistent, then a minimal set of malicious vehicles and messages are searched which if removed make the model consistent. The authors have showed that this model is effective in handling various attacks. However, this model needs to be developed further, including the minimal set search algorithm for finding the set of nodes and messages whose removal make the model consistent again.

10.12 Future Directions for Research

Although significant research has been done in the area of VANET security, both by academic researchers as well as by industry groups, there are still certain key areas that require focus to ensure effective deployment of VANETs. We discuss below some of the relevant areas that require attention in this section.

- *Effective trade-off between liability and privacy:* Even though the use of pseudonyms is seen as a good measure of ensuring the privacy of a consumer without losing liability in accident-related cases, the overhead of road-side infrastructure and a Vehicular PKI is immense. Moreover, a large-scale deployment is necessary for this scheme to work. In a truly *ad hoc* network, without road-side infrastructure, this scheme will not be successful. It is, therefore, essential to research and find simpler and more effective schemes for ensuring privacy with liability.
- *Cost-effective tamper-proof modules:* It has been discussed earlier in this chapter that for effective implementation of security in nodes, it is essential to have tamper-proof hardware. However, presently, the cost of such hardware is high, thereby increasing the implementation cost. It is, therefore, necessary to find security schemes which rely on better methods of managing security keys or cheaper tamper-proof hardware needs to be developed.
- *Integration with and usage of other communication systems such as GSM, CDMA, and WiFi:* The majority of efforts toward VANET research have been to set up a new communication network for VANETs. However, with a number of existing communication networks setup, it needs to be seen as to how these can be leveraged to be used for VANET communication. Moreover, their security features can be utilized to set up a secure VANET. Therefore, more search effort needs to be focused on using the available infrastructure for VANET applications and security implementation.
- *Secure localization–vulnerability of GPS:* GPS has been subjected to a series of attacks such as spoofing and signal jamming. Although attempts have been made to correct this problem, no definitive solution is available yet. It is, therefore, necessary to search for better schemes for finding the location of vehicles or improving the security of the existing technologies.

10.13 Conclusions

The need for safer driving conditions and better traffic management has helped development of smart cars and VANET technology. The potential of VANET applications is immense, considering the large amount of vehicles on the road. However, most of the VANET applications such as safety messaging and hazard warning have stringent time requirements and malfunctioning systems and malicious attackers can cause loss of life and injury due to accidents. It is, therefore, imperative to develop a strong security system for VANET. Toward this aim, adequate research needs to be done to develop effective security mechanisms. In the last few years, substantial research efforts have been conducted by academic researchers and in collaboration with industry and government agencies. In this chapter, we covered the major VANET security research developments which have taken place. However, more work needs to be done in this field. Important research areas have been highlighted to develop and deploy effective systems. VANET technology has the ability to transform the way vehicles travel from one place to another and offer a whole gamut of services from safety messaging to infotainment. It is an interesting research area and we are sure that this field will see some more exciting developments in the next few years.

Terminologies

VANETs—VANETs or Vehicular *Ad hoc* Networks are communication networks formed by vehicles equipped with wireless radios for exchange and dissemination of information for various applications.

Road-side infrastructure units (RSU)—These are communication base stations located near roads and highways with which vehicles communicate for getting various information and are connected with the backbone network by static wired or broadband wireless links.

Liability—This is property of a VANET network that specifies and fixes the responsibility of actions of various vehicles can be fixed in the event of an accident happening between various vehicles.

Privacy—This property of a VANET network ensures that the identity of a node is not revealed in the messages it sends which may enable tracking or recording of its actions and locations. This cannot be fully adhered to owing to the adherence of liability property and therefore the node identity needs to be revealed in accident cases to the proper authorities.

Pseudonyms—These are short-term identities assigned to the VANET node that can be used only once. These are used to maintain the privacy of VANET nodes. They keep on changing and the VANET node periodically gets fresh Pseudonyms from a central authority.

Mix zone—These are areas in VANET network region which is not in the surveillance range of attackers and therefore are suitable for a VANET node to change their pseudonym or short-term identity to prevent tracking. As multiple vehicles exit the mix zone at the same time, it is difficult for the attacker to keep track of a vehicle while it changes its pseudonym inside the mix zone.

Verifiable multilateration—This is the form of position verification to obtain the correct position of a VANET node. Here, four base stations issue challenges to the node which replies to each one by one. Each base station fixes the nodes distance by measuring the time to receive the reply and collaborate to fix its location.

Secure aggregation—This is a property of VANETs wherein a VANET node is able to verify the aggregated data sent to it from another node. This is done using digital signatures in various ways.

Questions and Sample Answers

1. What is the need for security in a Vehicular *Ad Hoc* Network (VANET)?
   As in any major public network, VANETs, when deployed without considering the security requirements, lend themselves vulnerable to a host of attacks. The danger involved in possible road accidents and loss of life further impress upon the need for fail-proof security for VANETs. For example, safety-related applications need a high level of security, as a single vehicle sending out false warnings can disrupt the traffic of a whole highway. Similarly, in traffic information service, a greedy driver can send out multiple false messages under different identities to divert traffic to ensure a smooth drive for him. This can be resolved using authentication, which raises a fresh concern of privacy, as vehicles can be tracked and monitored using the identities sent out by them along with their messages. However, consumers would not like their private information being shared over a public network. Balancing these conflicting requirements is essential while implementing security in VANETs. Therefore, security and privacy are prerequisites for deployment of VANETs and striking an adequate balance between the two will be the key in their successful deployment.

2. What are the basic security requirements for VANETs?

The major security requirements for a VANET are as under:

- *Authentication* is a major requirement for VANET. This is simply because it ensures that various messages are sent by actual nodes and not by a node representing multiple identities or a node impersonating as someone else. This attack can be used by greedy drivers to divert traffic from their routes by simulating a congested road by sending false messages.

- *Message integrity* is important, as it needs to ensure that the message is not modified in transit. This, coupled with Authentication, assures VANET nodes that the messages they receive are not false.

- *Message nonrepudiation* is required so that sender cannot deny having sent that message. This, however, further exacerbates the identity management issue. Only specific authorities should be allowed to identify a vehicle from the authenticated messages it sends.

- *Entity authentication* is a property which enables a receiver to ensure that the sender generated a message and is still active in the network. This is required to ensure that a particular message was generated by a sender within a small time interval just before the receipt of the message at the receiver.

- *Access control* is required to ensure that all nodes function according to the roles and privileges authorized to them in the network. Toward access control, Authorization specifies what each node can do in the network and what messages can be generated by it.

- *Message confidentiality*, though strictly not very essential, in a VANET, can still be utilized when certain nodes want to communicate with each other in private.

- *Privacy* is important to ensure that user information is not leaked or distributed to parties not authorized to access such information. Third parties should also not be able to track vehicle movements as it is a violation of personal privacy. Therefore, a certain degree of anonymity should be available for messages and transactions of vehicles. However, in liability-related cases, specified authorities should be able to trace the user identities to determine responsibilities.

- *Availability* is essential to enable VANET services to be operational in the presence of faults and attacks. This implies that VANETs should be resilient to DoS attacks by having alternate means of communication and redundant infrastructure.

- *Real-time guarantees* is essential in a VANET, as many safety-related applications depend on strict time guarantees. This can be built into protocols to ensure that the time sensitivity of safety-related applications such as collision avoidance is met.

3. What are the major challenges in implementing security in VANETs?

For implementing VANET security, it is essential to understand the unique challenges faced in such networks. The major challenges are outlined below:

- *Tradeoff between authentication and privacy:* To ensure that certain nodes do not impersonate another node, it is essential to authenticate all message transmissions. However, this leads to identification of vehicles from the messages they send. Privacy is a major issue in a VANET, because cars are highly personal devices. This has to be balanced with the need for establishing accountability and liability of vehicles and their drivers. This requires an authentication system to be designed that enables messages to be anonymous for general nodes but also enables identification by central authorities in liability-related cases like accidents.

- *High mobility:* VANETs are characterized by highly mobile nodes which will result in frequent changes in topology and brief connectivity between the nodes. In such situations, VANET protocols cannot be handshake based.

- *Scale of network:* VANETs are likely to be among the largest *ad hoc* networks, requiring scalable solutions for adequate availability and sufficient performance. This aspect rules out having prestored information about other nodes or distribution of centralized information to all nodes.

- *Real-time guarantees:* The major VANET applications are safety related for collision avoidance, hazard warning and accident warning information. Any security protocol implemented for VANETs would need to take this into consideration and have low processing and message overheads.

- *Incentives:* For effective deployment of VANET technology, it is imperative to offer incentives to the involved parties for them to adopt the system. It is imperative to offer all concerned the correct incentives to adopt this technology and the security being implemented.

- *Location awareness:* For most VANET applications to be truly effective, certain location-based service is essential. This increases the reliance of the VANET system on GPS or other specific location-based instruments. Any error in these is likely to reflect in the VANET applications.

4. List the various possible attacks on VANETs.

   VANETs are susceptible to various kinds of attacks. They vary according to the situation, attacker's intent, amount of damage and scope. A brief outline of the major attacks possible is given below:

   - *Denial of service:* This can be done by channel jamming, that is, block access to a communication channel by high power transmission on the communication channel or by injection of dummy messages.

   - *Impersonation:* An attacker may take on someone else's identity and gain certain advantages or cause damage to other vehicles. A vehicle may impersonate an emergency vehicle to gain access to highway or a vehicle may send out another's person's identity in an accident case to escape liability.

   - *Message falsification:* An attacker can send false messages in a VANET network such as false hazard warnings to divert traffic from a route for freeing up resources for it.

   - *Message alteration:* This form of attack can be done where either an individual entry in a message can be altered or full message contents can be altered. For example, a hazard warning can be changed to a no-hazard message to cause traffic accidents.

   - *Message delay and suppression:* In case of accidents, certain vehicles may delay or suppress safety messages. Due to the real-time nature of such information, this may cause immense damage. By selectively dropping messages or delaying their transmission, critical information may not reach vehicles in time.

   - *Privacy violation:* Authentication raises some serious privacy issues as in all applications like safety, traffic management and toll access; the messages would reveal the driver's identity, his location, his actions and preferences. Consumers would not like to adopt a technology which violates their privacy.

   - *Replay attacks:* Vehicle can easily eavesdrop and log messages of other vehicles and replay them later to gain access to specific resources like toll services or to send false alarms. Therefore, mechanisms to ensure replay attacks are not possibly need to be built into the VANET security architecture.

   - *Hardware tampering:* Attackers can also tamper with the security hardware of a vehicle to steal identities as well as extract cryptographic keys. Therefore, specific mechanism like tamper-proof hardware needs to be implemented to ensure such attacks cannot easily be accomplished.

■ *Sensors tampering:* Another easy attack is to fool around with the vehicle's hardware sensors. If the main system is tamper proof, it is easy to fool the vehicle's sensors with wrong information by simulating false conditions.

5. List the various types of adversaries which may attack a vehicular network.

Before developing a VANET security system, it is imperative that we understand what type of adversaries would target the system and type of attacks they are capable of launching against the system. There are a number of probable adversaries and attacks they can launch. The attackers can be divided into the following general categories:

■ *Selfish drivers:* Even though majority of the drivers in a system would be honest and adhere to the rules, it is natural that some drivers would try to gain specific advantages from the system.

■ *Eavesdroppers:* These adversaries would like to collect information about drivers and use this to understand driver behaviors and traffic pattern.

■ *Teenage hackers:* These adversaries try and hack into any major system which gets deployed publicly. They try to find bugs in the software and cause traffic disruptions just for fun.

■ *Insiders:* These adversaries include persons working in car companies and installing the VC system. They are capable of loading malicious software in cars that could cause immense damage.

■ *Malicious attackers:* These attackers could be criminals or terrorists having access to more sophisticated tools and hardware than normal attackers. Criminals may have specific targets for financial gains or would like to carry out personal harm to rivals. Terrorists can use sophisticated technology to disrupt vehicular traffic to cause maximum damage when using bombs or launching gun attacks.

6. Which are the major industry and research bodies working in the field of VANET security?

A number of organizations and industry consortiums are involved in developing standards for VANETs. For example, the IEEE is involved in standards development related to the physical, medium access and security issues as well as in defining higher layer services and interfaces for intelligent transportation. By the end of 2006, the IEEE P1609 standards for WAVE had specified the application layer and message formats for operation in the 5.9 GHz DSRC communications. The IEEE 802.11p standard, which is a modification of the popular IEEE 802.11 (Wi-Fi) standard, looks at issues related to the highly dynamic environment and the extremely short time durations, during which communications must be completed due to the high speed of the communicating vehicles. Several consortiums with industry and/or public participation are also working on furthering the development and deployment of vehicular networks. Some examples include the Car-2-Car Consortium, which has as one of its primary objectives, the creation and establishment of an open and interoperable standard for V2V communications in Europe using Wi-Fi-like components. Some communication protocols are being developed by the NOW group, which is associated with the Car-2-Car Consortium. Ford and General Motors created a CAMP and with the National Highway Transportation Safety Administration, this partnership is working on projects such as enhanced digital maps for safety, driver workload metrics, and forward crash warning requirements. Other VC projects being implemented include the Berkeley PATH and the Fleetnet projects in the United States and Germany, respectively.

7. What is the PKI structure required for VANETs and how the existing registration authorities can be utilized for implementing this structure?

The security system in a Vehicular Network will require a Vehicular PKI with existing authorities responsible for registration of vehicles acting as CAs responsible for issue of certified public/private key pairs to vehicles. Every region will have a separate CA. Each vehicle will be issued with an Electronic License Plate which will be a unique identity similar to a Registration number issued to vehicles presently with additional certified public/private keys in a digital certificate. Cross certification will be required when a vehicle goes to different regions, that is, different state or different country. This will require that each vehicle stores number of trusted root CA certificates containing their public keys.

Each owner will register his vehicle with its own region's CA while purchasing the vehicle and get a pair of private and public keys in a digital certificate. Fresh key pairs will be issued when the previous certificate expires or is revoked. The digital certificate contains the node's identity as well as various parameters like cryptographic algorithm used as well as the certificate lifetime. Each vehicle gets one long-term identity for identification from the CA. However, the need to maintain privacy introduces the concept of short-term public/private key pairs which have no direct link to the identity of the vehicle. The task of issuing these keys can also be entrusted to the CAs. In some proposals, a different entity is entrusted with this task. Certificate revocation is also a responsibility of the CAs as and when vehicles are taken out of service or are declared as malicious nodes for illegal activity.

Each vehicle will be issued with a digital certificate which will contain its public key digitally signed by the CA. While implementing a PKI mechanism, it is not possible for a node to store public keys of all other nodes even it is for one region alone. Therefore, after digitally signing a message each node will send his digital certificate (containing his public key) with each message. The public key in the digital certificate can be verified by extracting it from the digital certificate and verifying it with the public key of the root CAs stored in each node. This can be refreshed and updated during periodic inspections or secure transfers from road-side infrastructure.

Author's Biography

Vikas Singh Yadav holds a bachelors degree in telecommunications and information technology from Jawaharlal Nehru University (JNU), New Delhi, India. His research interests include network security, wireless *ad hoc* networks and delay tolerant networks. He is presently pursuing his master of technology (MTech) in computer science and engineering from Indian Institute of Technology (IIT) Kharagpur, Kharagpur, India. He is working on a project on routing in delay tolerant networks for his MTech Thesis. He has written various articles on networking and security for a number of magazines and journals.

Sudip Misra is an assistant professor in the school of information technology at the Indian Institute of Technology (IIT) Kharagpur, Kharagpur, India. Prior to he held academic positions in Cornell University (New York), Yale University (Connecticut), and Ryerson University (Canada). He also worked at Nortel Networks (Canada), Atreus Systems Corporation (Canada), ChartWell, Inc. (Canada), and the Government of Ontario (Canada). He received his PhD in computer science from Carleton University, Ottawa, Canada. His current research interests include algorithm design and engineering for telecommunication networks. Dr. Misra is the author/editor of over 90 scholarly research papers and books. He has won five research paper awards in different conferences. He was also the recipient of several academic awards and fellowships such as the (Canadian) Governor General's Academic Gold Medal at Carleton University, the University Outstanding Graduate Student Award in the Doctoral level at Carleton University. In 2008, he was conferred the National Academy of Sciences, India—Swarna Jayanti Puraskar (Golden Jubilee Award). Dr. Misra is the Editor-in-Chief of two international journals and is an associate editor/editor/editorial board member of around a dozen others published by reputed publishers such as Wiley, Springer, Elsevier, and IET. Dr. Misra is an editor of six books in the areas of wireless *ad hoc* networks, wireless sensor networks, wireless mesh networks, communication networks and distributed systems, network reliability and fault tolerance, and information and coding theory, published by reputed publishers such as Springer and World Scientific. Dr. Misra was also invited to deliver keynote lectures in over a dozen international conferences in the United States, Canada, Europe, Asia, and Africa.

Mozaffar Afaque holds a bachelors degree in computer engineering from Zakir Husain College of Engineering and Technology, Aligarh Muslim University, Aligarh, India. His research interests include *ad-hoc* networks, GIS and distributed systems. He is presently pursuing his master of technology (MTech) in computer science and engineering from Indian Institute of Technology (IIT) Kharagpur, Kharagpur, India. He is working on a project on misbehavior detection in vehicular networks for his MTech Thesis.

References

1. FCC. FCC allocates spectrum in 5.9 GHz range for intelligent transportation system uses. October 1999, http://www.fcc.gov/Bureaus/Engineering_Technology/News_Releases/1999/nret9006.htm
2. DSRC: Designated Short Range Communications. http://grouper.ieee.org/groups/scc32/dsrc/index.html
3. IEEE P1609.2/D2—Draft Standard for Wireless Access in Vehicular Environments—Security Services for Applications and Management Messages, November, 2005.
4. C. L. Robinson et al., Efficient coordination and transmission of data for cooperative vehicular safety applications, in *Proceedings of ACM Workshop on Vehicular Ad Hoc Networks (VANET)*, Marina del Rey Marriott, Los Angeles, CA, USA, September 29, 2006.
5. The IEEE 802.11 Task Group to define enhancements to 802.11 required to support Intelligent Transportation Systems. http://grouper.ieee.org/groups/802/11/Reports/tgp_update.htm
6. C2C CC: Car-to-Car Communication Consortium. http://www.car-to-car.org/
7. NoW: Network on Wheels. http://www.network-on-wheels.de/
8. Crash Avoidance Metrics Partnership. http://www.campivi.com/
9. PATH: California Partners for Advanced Transit and Highways. http://www.path.berkeley.edu/
10. FleetNet: Internet on the Road. http://www.et2.tuharburg.de/fleetnet/english/vision.html
11. GST: A Global System for Telematics enabling on-line safety services. http://www.gstproject.org/
12. SEVECOM: Secure Vehicular Communications. http://www.sevecom.org
13. P. Papadimitratos et al., Secure vehicular communications: Design and architecture, *IEEE Communications Magazine*, 2008, 46(11), 100–109.
14. M. Raya and J.-P. Hubaux, The security of vehicular ad hoc networks. In: *Workshop Security in Ad hoc and Sensor Networks (SASN)*, Hilton Alexandria Mark Center, Alexandria, VA, USA, November 7, 2005.
15. F. Kargl, Z. Ma, and E. Schoch, Security engineering for VANETs, in *Proceedings of 4th Wksp. Embedded Sec. in Cars*, Berlin, Germany, November 2006, pp. 15–22.
16. B. Parno and A. Perrig, Challenges in securing vehicular networks. In: Workshop Hot Topics in Networks (HotNets–IV), 2005.

17. A. Aijaz, B. Bochow, F. Dotzer, A. Festag, M. Gerlach, R. Kroh, and T. Leinmuller, Attacks on inter vehicle communication systems–an analysis. In: *3rd International Workshop on Intelligent Transportation*, WIT, 2006.

18. M. Raya, P. Papadimitratos, and J.-P. Hubaux, Securing vehicular communications. *IEEE Wireless Communications Magazine*, Special Issue on Inter-Vehicular Communications, 2006, 13(5), 8–15.

19. IEEE P1609.2 Version 1—Standard for wireless access in vehicular environments: Security services for applications and management messages, in development, 2006.

20. M. Gerlach, A. Festag, T. Leinmuller, G. Goldacker, and C. Harsch, Security architecture for vehicular communication, in *5th International Workshop on Intelligent Transportation (WIT)*, Hamburg, Germany, March 2007.

21. J.-P. Hubaux, S. Capkun, and J. Luo, The security and privacy of smart vehicles, *IEEE Security and Privacy Mag.*, 2004, 2(3), 49–55.

22. L. Buttyan, T. Holczer, and I. Vajda. On the effectiveness of changing pseudonyms to provide location privacy in VANETs, in *Proceedings of ESAS*, Sidney Sussex College, Cambridge, UK, July 2–3, 2007.

23. A. R. Beresford and F. Stajano, Mix-zones: User privacy in location-aware services, in *Proceedings of PerSec*, Orlando, FL, USA, March 14, 2004.

24. J. Freudiger et al., Mix-zones for location privacy in vehicular networks, in *Proceedings of 1st International Workshop Wireless Networking for Intelligent Transportation Systems (Win-ITS)*, Vancouver, BC, Canada, August 2007.

25. P. Papadimitratos, G. Mezzour, and J.-P. Hubaux, Certificate revocation list distribution in vehicular communication systems, *ACM VANET*, San Francisco, CA, 2008.

26. B. H. Bloom, Space/time trade-offs in hash coding with allowable errors. *Commun. ACM*, 1970, 13(7), 422–426.

27. M. Raya, A. Aziz, and J.-P. Hubaux, Efficient secure aggregation in VANETs, in *ACM Workshop on Vehicular Ad hoc Networks (VANET)*, Marina del Rey Marriott, Los Angeles, CA, USA, September 29, 2006.

28. T. Leinmueller, C. Maihoefer, E. Schoch, and F. Kargl. *Improved Security in Geographic Ad-hoc Routing through Autonomous Position Verification*, 2006, ACM Press, New York.

29. P. Golle, D. Greene, and J. Staddon, Detecting and correcting malicious data in VANETs, in *Workshop On Vehicle Ad hoc Networks (VANET)*, Loews Philadelphia Hotel, Philadelphia, PA, USA, October 1, 2004.

Chapter 11

Toward a Robust Trust Model for Ensuring Security and Privacy in VANETs

Une Thoing Rosi and Syed Ishtiaque Ahmed

Contents

11.1 Introduction

With the evolution of smart mobile devices, the systems involving in interaction and cooperation with different similar/nonsimilar devices are being ubiquitous. These devices are autonomous and self-organizing, and depend on communication, cooperation, and collaboration with other devices/peers for effective operation. This communication and cooperation with other peers raises the issue—Trust; to whom it may trust with how much confidence—to which devices it will allow access to own resources and from which devices it will take services—whose information will it believe. The peer with which a device communicate and cooperate must be trustworthy, otherwise they can make harm. In this circumstance, evaluating trustworthiness of a peer in distributed systems came into focus of research at the end of twentieth century. At that time AT&T implemented trust management systems at their projects PolicyMaker (1996) [1], Keynote (1998) [2,3], and REFEREE (1997). Trust management is integrated with several components of network management such as risk management, access control, and authentication. In Europe, Abdul-Rahman and Hailes developed first (in 2000) a reputation-based trust management system for virtual communities. In 2003, Kamvar et al. [4] proposed an Eigentrust algorithm for reputation management in P2P systems which is still most cited (according to Google scholar 1224 times cited up to July 2009) and most popular algorithm for reputation management. Trust management systems now became focal attention to researcher community with the increasing popularity of autonomously distributed and mobile systems. Some popular autonomous systems are MANET [5], Wireless Mesh Network [6,7], and P2P networks [8,9]. A lot of research works are now continuing and many are published every year in these fields. Like every autonomous networks, newly emerging Vehicular *Ad Hoc* Networks (VANETs) also require trust management systems for risk management, access control, and authentication. Researcher communities are working toward trust management systems for VANETs since the beginning of research on VANETs. In 2002, Buchegger and Le Boudec proposed the CONFIDANT protocol [10], which is pioneer in reputation-based trust management system for Mobile *Ad Hoc* Networks (MANETs).

VANETs, which are a form of MANETs, have some distinctive properties compared with MANETs which influence in the attempts of construction of trust models for VANETs. Highly dynamic network, frequently changing topology, and ephemeral nature of the network connectivity are the key nature of these networks. As no vehicle can establish long-term connectivity with any other vehicle, the trust metric to a vehicle is valid for a short period. Another characteristic of these networks is interaction with high number of peers within a very short period. A vehicle may get connected and disconnected with thousands of other vehicles depending on the traffic, speed, and location it is crossing over. Some important services of VANETs are safety warnings, traffic updates, route suggestions, and so on. Time freshness and location relevance as well as trustworthiness of these services are very important for taking right decision at the right time; otherwise, it may get into accident, may get into congestion, or follow a wrong route.

To date, researchers have proposed many trust models for autonomous/distributed systems (P2P, WMN, MANET, etc.); all of them established trust frameworks with the peer entity or network node—the reputation, past behavior, and tendency of a node based on experience, observation, and reports/opinion from other neighboring nodes. However, in case of VANETs, node identities are largely irrelevant as a node is getting connected and disconnected with thousands of other nodes within few minutes and network connection is not stable for a long period; rather, time freshness of data containing security warnings, traffic updates with their location relevance, and authenticity are much important. In summary, authenticity of data/information/message is much important than identity, behavior history, reputation of particular nodes supplying the information. Of course, we can expect an authentic message from a well-reputed node, but a more convincing idea is to take account of information/opinion of hundred other vehicles with which one is getting connected to evaluate authenticity of an information. In this way, we must also consider each individual vehicle's trustworthiness based on their credentials and security status as some vehicle may become faulty (mechanically), and some may intentionally spread misleading messages. Raya et al. [11] focused this issue by coining the term data-centric trust and established data-centric trust notion to realize this framework. In this chapter, we extended the data-centric trust framework for more pragmatic application in the realization of VANETs' trust framework.

As this chapter is organized, Section 11.2 details background of trust models for distributed systems in the sequence of their chronology. In Section 11.3, notions and preliminaries of trust model and VANETs are discussed. Section 11.4 discusses the state-of-the-art of the VANET trust model, emphasizing mainly on data-centric trust notion for ephemeral networks. Section 11.5 discusses some challenges for implementing trust model for VANETs.

11.2 Background/Related Works

"Increased flexibility through programmability" is the current trend that is having big influence in most research works of computer science. In every networked system, researchers are working to eliminate network ossification by replacing manual/hard-coded part with some flexible small programming command. Traditional authorization systems did not support filtering of user's to allow access of its resources to other users, allowing unwanted/malicious peers easy access to resources and consequently leading to inefficiency to the systems. The trust management approach has come to cover the inadequacy of traditional authorization mechanisms, allowing access only to the trustworthy peers. "Trust management" was first coined by Matt Blaze [12] while he was a researcher at AT&T labs at 1996. At that time, AT&T implemented trust management models in their projects PolicyMaker (1996), KeyNote (1998) [13] (RFC 2704 [14]), and REFEREE (1997) [15]. Since then trust management approach has become a focal point of research, and researchers are trying to integrate trust management systems in every type of autonomous, self-organizing, and distributed systems.

Trust management models are categorized into two different approaches. They are: (1) Certificate based, and (2) Reputation based. In *Certificate-based* models, a trusted third party is responsible for providing certificates to every transaction in the network. Every network node can verify the authenticity of the certificate and hence transaction using the third party. This approach is prone to single point failure. However, more than one third-party server can be used to improve the condition. In *Reputation-based* models, history of behavior of individuals are observed directly or indirectly, and kept into note. An individual's reputation is the sum of its history of previous behavior. This approach is slow evolving with time, and reputation information is propagated from individual to individual using gossip-based protocols. This approach is also based on query and answer about individual's behavior among peers. In summary, this approach does prediction about individual's behavior based on history of its previous behavior. However, nowadays, a hybrid of both reputation- and certificate-based approaches is available. Reputation-based trust management systems became very popular among researcher communities of its applicability of distributed environments and avoiding single point failure. In 2000, Abdul-Rahman et al. [15] implemented Reputation-based trust model for virtual communities, which introduced concepts of keeping database of recorded experience and gossip-based protocols to exchange recommendation. Aberer et al. [16], in 2001, proposed another reputation-based trust model, which introduced negative traits of individual (complaints) as metric for reputation value. CONFIDANT [18], published in 2002, a project at EPFL [19], is another reputation-based trust model that introduced features like detection and isolation of misbehaving nodes, sharing of warning data and neighbor watch. CONFIDANT dynamically monitors its neighbor to update neighbor's reputation value with time. In case CONFIDANT finds any misbehaving node, it takes action in terms of its own routing and forwarding, and informs other neighbor nodes by sending ALARM messages.

The most cited paper in reputation management system is Eigentrust [20], published in WWW 2003, and developed by then Stanford student Sep Kamvar. In July 2009, the paper was cited 1224 time as per Google Scholar. Eigentrust classified threat scenarios in P2P as Malicious Individual, Malicious Collective, Camouflaged Collective, and Malicious Spies, and proposed robust solution to isolate them. Eigentrust implemented reputation-based trust management, and it evolves slowly with time.

STRUDEL [21] provide trust framework to detect free-riders in the Coalition Peering Domain in Mesh Network, and consequently isolate the free-riders with time. STRUDEL uses Bayesian model to assess the trustworthiness of individual peers. B-Trust [22] is a distributed trust framework that evolves trust metric based on Bayesian formulation while supporting anonymity and resistance from Sybil attack. MobiRate [23] is a reputation-based trust model with gossip-based protocol that uses tamper-evident hash chain to protect the reputation record from forgery. MobiRate is applicable in content-sharing process of mobile devices.

Table 11.1 depicts some projects realizing trust models along with their working domain, publication venue, and institutions involved.

For more study on related works, a detailed summary of trust models for *ad hoc* networks is available at [24]. Although a little bit outdated, it represents a good summary. An analysis of trust models based on their characteristics is also available at [25].

11.3 Trust Model for VANET Preliminaries

11.3.1 Characteristics of Trust Models and its Metrics

Trust models and their evaluation metrics have some notable characteristics. Characteristics of trust can be listed under the following headings:

Context dependence: Trust metric may differ based on the context over which trust is evaluated. Specific contexts based on which trust value will be evaluated must be defined by the authority in the protocol before bootstrapping/initiating the protocol. For example, contexts of VANETs will include authenticity of message, service specific authenticity and so on.

Subjective: Trust value about an entity by other entities widely differs by entity to entity. Entity $x1$ trust entity y does not mean that entity $x2$ trusts y. Entity $x1$ may trust y with value 0.9 while entity $x2$ may trust y with value 0.3 in the scale between 0 and 1.

Asymmetric: Entity x trusts y does not mean that entity y trusts x. Each entity has independence in the evaluation of others.

Table 11.1 Some Projects Realizing Trust models

Project Name/People, Year	Implementation Domain/Features	Publication	Institution
PolicyMaker (1996), KeyNote (1998)	Public key infra-structure, certificate based	RFC 2704, LNCS 1550 Springer-Verlag	AT&T Labs
REFEREE (1997)	Web applications	Computer Networks and ISDN Systems Volume 29, Issue 8–13 (September 1997)	AT&T Labs
Abdul-Rahman et al. (2000)	Reputation based trust model for virtual communities	33rd Hawaii Inter-national Conference on System Sciences	UCL, UK
CONFIDANT (2002)	Dynamically moni-tor neighbor's, uses ALARM messages to report misbehaving nodes	SIGMOBILE, 2002	EPFL, Switzerland
Eigentrust (2003)	P2P, reputation based	WWW, 2003	Stanford university
B-trust (2006)	Bayesian framework with support of anonymity and sybil attack	Springer i-Trust 2006	UCL, UK
STRUDEL (2006)	Mesh network, coalition peering domain	SAC, 2006	University college London
Data-centric Trust (2008)	Ephemeral networks (VANET)	INFOCOM 2008	EPFL, Switzerland Carnegie Mellon univ.
Mobirate (2008)	Mobile devices, reputation based	Ubicomp 2008	University college London, UK
QuanTM (2009)	Combines trust and reputation manage-ment for policy eval-uation	Eurosec 2009	Univ. of Pennsylvania, USA

Measurable belief: Trust metric of a node in a network is measured within a certain numeric scale. In most trust models, this takes a real number between 0 and 1 where 1 means highest trust value while 0 means lowest trust value.

Evolves with time: As trust values reflect the actions of an entity over passage of time, it is varied with time. Honest actions increase trust value, whereas dishonesty decreases trust value.

Reflexive: Trust metrics have reflexive property. An entity may trust itself with a certain metric or confidence value.

Trust management model's characteristics can be listed under following headings:

Centralized versus decentralized trust model: In centralized models, a globally trusted entity calculates trust values for every node in the system. There are two problems associated with this approach. One is single point failure, and the other is that different users may have different opinions about the same target which is suppressed in this scheme. However, centralized approaches cannot be implemented in autonomous and *ad hoc* networks.

In decentralized trust management models, each user calculates trust values around it with its own policy. This is a bottom-up approach, and most widely implemented and used as a part of Pretty Good Privacy (PGP) [26] for public key certification.

Proactive versus reactive computation: In proactive computational model, reputation/trust values of network nodes are computed periodically and keep into record. This approach uses more bandwidth and space for maintaining trust relationship as reputation value exchange in network is done frequently and stored locally.

11.3.2 Objectives of VANETs

In general, the objectives of deployment of VANETs are to enhance the safety and efficiency of transportation systems. VANETs will provide different services among which some will save life through preventing accidents, some will bring luxury in trip, and some services will support commercial transactions. Here some prominent services of VANETs are listed below:

Traffic updates: VANETs will provide localized traffic updates. Driver's can get fresh traffic information of a distant location. This information may help the drivers to take decision to avoid congestion before reaching the congested place.

Route suggestion: VANETs will provide comparative analysis of possible routes toward a destination. This information may help drivers choosing right route.

Emergency warning signals: VANETs will help propagate warning messages when a car ahead abruptly brakes.

Commercial purpose: Private organizations may advertise their services and goods using VANETs. For example, a hotel may advertise its service using an RSU (Road-side Unit) before it. Client's can book any service from distant using the VANETs. For example a client may book a hotel room before reaching the hotel, using VANETs.

Environmental warning signal: VANETs may provide warnings about environmental hazards. For example, ice on the pavement, road damage, site construction, and so on.

11.3.3 Components and Key Characteristics of VANETs

In order to achieve the objectives of VANETs and realize these as a successful technical deployment in mass people, vehicles and infrastructures should be equipped with cutting-edge devices, and a sound protocol should be developed to achieve maximum output from these networks. VANETs are composed of the following components, and have the following characteristics:

1. Each vehicle constructing VANETs are equipped with an onboard processing unit to process data and wireless modules to build the network.
2. Infrastructure is constructed using RSUs that are also equipped with a processing unit and a wireless module to build the network.

3. There are authorities (public/private) who control the participation of the network through providing certificate and credentials. In addition to certification, authorities may posses other administrative powers such as identity management, key revocation, and so on. These authorities are trusted to nodes participating in VANETs. City or state transportation authorities will serve as public authority.
4. A subset of RSUs serves as gateway to and from the authority.
5. Each vehicle is equipped with a clock and a positioning system (GPS). These devices allow the vehicles to include time and location information with each outgoing messages. Time and location is important to maintain time freshness and location relevance of the messages.
6. Each vehicle v_k is equipped with a pair of private/public cryptographic keys ($P_r k / P_u k$), and certificate issued by authority X, $Cert_x(P_u k)$.
7. Authentication of message source is done via digital certificates (e.g., X.509). Authentication helps prevention from Sybil attacks. By source authentication, the identity/type of the node that sent the message can be identified. From identity, trustworthiness of a node can be measured by finding previous history of that node.
8. Local broadcasts (1 hop) and Geocast (flooding in a fixed area) are mainly used in VANETS. However, uni- and multicast are still possible but inefficient due to high nobility of network nodes.
9. Each vehicle frequently and periodically transmits packets. These packets include velocity, coordinates, time, position of own, and network events, safety messages, and messages/reports from others.

11.3.4 Adversary Model for VANETs

Vehicles that do not comply with or deviate from the implemented network protocol are adversaries. In other words, adversaries are the network entities that can make harm by sending wrong messages, jamming networks and in many other ways. To protect the network from adversaries', proper identification of them and prepare for adequate defense are necessary.

Who Are They?

In broad sense adversaries are classified into two:

1. Those intentionally deviate from protocol are called Attackers.
2. Those due to fault, unintentionally deviate from protocol are Faulty nodes.

Both Attackers and Faulty nodes can cause damage to the network.
  Some researchers classify [27,28] under more specific headings:

Insider versus outsider: Insiders are authenticated members of the network that can legitimately communicate with other members. An insider possesses a certified public key from the authority. Outsiders are the network members who does not possess a certified public key, or whose public key has been revoked by the authority.

Malicious versus rational: Malicious attackers seek no personal benefits from the attacks and aim to harm the members or functionality of the network. Hence, malicious attackers may employ any means disregarding corresponding costs and consequences. On the contrary, rational attackers seek personal benefit and hence are more predictable in terms of attack means and attack target.

Active versus passive: Active attackers can generate packets or signals, whereas passive attackers eavesdrop through the network channel.

In [23], adversaries are classified as three types: independent, collusive, and random.

Independent: Some independent nodes can be mechanically faulty, or intentionally attack the network for its own advantage.

Collusive: A group of nodes may collaborate to cause attack, spread wrong messages and jam the network.

Random: Some nodes may be more intelligent; they sometime behave well, and sometime behave malicious.

What They Do:

Adversaries can do many types of harm in the network. Some of its activities are as follows:

1. They can replay false message/report throughout the network.
2. They can jam communication through spreading spam.
3. They can modify a message and replay wrong information throughout the network. However, modification of a message can be detected by digital certificates and hence outsiders cannot modify a message.
4. They can inject faulty data and reports in the network.
5. They can control the inputs to benign nodes and induce them to generate faulty reports.

11.3.5 Salient Features of Trust Metrics in Distributed Systems and VANETs

Unlike connected and static networks (e.g., Internet), VANETs and many other distributed systems are rapidly changing. VANETs network topologies are not stable. If a vehicle is now connected with 50 other vehicles in a network, 5 min later it can be connected with 50 totally different other vehicles. This ephemeral nature of network connectivity leads to the following features:

Locality: Trust information exchanged locally and valid within that locality for VANETs. Change of locality may lead to different trust metrics.

Staleness of data with time: Validity of the trust metrics decay exponentially with time [24]. As network is changing frequently, time freshness of the information is much important.

Incompleteness and uncertainty of data: Evidences are gathered from the local peers; so they can be incomplete, and certainty of the information cannot be guaranteed with full confidence.

Ephemeral nature of the network: As discussed above, network connectivity in VANETs are changing in every minute. So, a network may last for minutes, and hence VANETs are ephemeral.

Rapid network partitioning and merging: Partition and merging of network happens frequently with the frequency of change of topology.

11.4 State of the Art: Data Centric Trust Management Model

11.4.1 Preliminaries

In data-centric trust management model, as proposed at Raya et al. [11] (INFOCOM 2008), trust in data (e.g., reported event) is derived/evaluated from multiple pieces of evidence (e.g., reports from multiple vehicles). Each piece of evidence is weighted according to some well-established rules which take into account various trust metrics, time freshness of report, location relevance, and so on. Each reported evidence and their respective weights, as calculated by certain rule, serve as input to the decision logic module. The output of the decision logic module is the level of trust on these data.

Decision logic can be implemented using several techniques, for example, voting, Bayesian learning, Dempster–Shafer theory. In [11], the authors provided a detailed comparison of the several decision logics which include voting, Bayesian inference, and Dempster–Shafer theory of evidence. Bayesian approach works with *prior knowledge*, whereas Dempster–Shafer works with *uncertainty* about the data.

11.4.2 Framework

In data trust model, there is an authority responsible for assigning identities and credentials (certified public keys) for all the network nodes (vehicles, RSUs). The authority can revoke the credentials, if a node does any misconduct. Each vehicle is equipped with devices that are able to perceive some predefined basic events, and report the events with outgoing messages. Example of basic events can be "ice on the road," "road blocked," "traffic jam ahead," and so on. A composite event is formed from two or more basic events happening simultaneously. An authority may define a set of basic events $\Omega = \{\alpha_1, \alpha_2, \ldots, \alpha_I\}$, that can be perceived by on-vehicle device and reported via outgoing messages.

$\Theta = \{\theta_1, \theta_2, \theta_3, \ldots, \theta_N\}$ is a set of type of nodes (vehicle/RSU). Each node participating in VANETs belongs to a class, and default trust level to that node depends on the class to which it belongs to. For example, all police cars belong to a class, and private vehicles belong to another class. Definitely, police vehicles have higher default trust level than public vehicles, given that all sensing and network devices of police vehicles are working properly. Class of a vehicle reporting a message is determined by verification of credentials. This classification of vehicles/RSUs is logical in the sense that some nodes highly protected against attack, and well equipped; so they exhibit less faulty behavior, and hence considered more trustworthy, and have a higher class.

11.4.3 Dynamic Factors

Besides default trust values of data, there are some dynamic factors that can alter the trust value of the data. They are listed as follows:

1. A node can become faulty (mechanical) or compromised by attackers, and in this case the credentials of the node should be revoked by the authority.
2. Proximity to both time and/or geographic location is another dynamic trust metric to consider. The closer the reporter is to the location to the event, the more likely to have accurate information on the event. Similarly, the more recent and closer to the event occurrence time a report is generated, the more true information is expressed.

While implementing dynamic factors (see Figure 11.1) one must use digital signatures to ensure that location and time information cannot be modified within the report.

11.4.4 Decision Logic

In data-centric trust management model, a vehicle computes combined trust of an event based on the reports it receives from different distinct vehicles. In this section, different model of decision

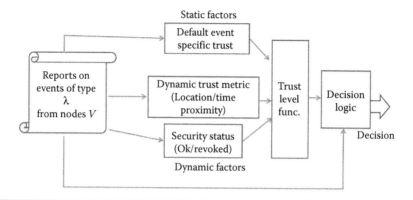

Figure 11.1 Architecture.

logics to compute the data trust are discussed. Prominent decision logics are:

1. Majority voting
2. Most trusted report (MTR)
3. Weighted voting (WV)
4. Bayesian inference (BI)
5. Dempster–Shafer theory (DST)

11.4.4.1 Majority Voting

In this scheme, majority wins [29]. Combined trust level to event Ai is defined by

$$d_i = \frac{1}{K} \sum_{k=1}^{K} F(e_k^i).$$

Where K is the total number of vehicles from whom it gets messages at that time. Here, $F(e_k^i) = 1$, if node v_k reports the event α_i; or $F(e_k^i) = 0$, otherwise. The value of d_i is a real number between 0 and 1. The value of d_i closer to 1 represents a more trustworthy report.

11.4.4.2 Most Trusted Report

The MTR decision logic outputs a trust level equal to the maximum value of trust levels about the report. The equation for combined trust level corresponding to event α_i is defined by

$$d_i = \max_k(e_k^i)$$

for example, an event is reported by a private vehicle and a police vehicle. Trust level of the reports of the above two vehicles is 0.5 and 0.85, respectively. Then the combined trust level of the event is set to 0.85. The point of using MTR is to show the effect of isolated high trust values (data/entities) on the system.

11.4.4.3 Weighted Voting

The WV approach sums up all the votes supporting an event with each vote weighted by the corresponding trust level of the entity supporting the event. The combined trust level according to this approach is

$$d_i = \frac{1}{K} \sum_{k=1}^{K} F(e_k^i) * W_k$$

Like majority voting scheme, here $F(e_k^i) = 1$, if node v_k reports the event α_i; or $F(e_k^i) = 0$, otherwise. The values of W_k are the respective weight of the event reported by a particular node, and take a real value between 0 and 1.

Since the above three decision logics do not provide formalisms for handling unions and intersections of events, decisions on composite events are harder to do using the above three techniques. However, Bayesian inference and Dempster–Shafer theory provide formalisms for handling unions and intersections of events.

11.4.4.4 Bayesian Inference

Bayesian inference (BI) [30] for trust establishment is one of the most frequently used decision logics for trust management model. In BI, evidence or observations are used to update or to newly infer the trust level of a report/entity. The combined trust level corresponding to event α_i is expressed in terms of the prior probability $P[\alpha_i]$ using the Bayes' theorem as follows:

$$P[\alpha i|e] = \frac{P[\alpha_i] \prod_{k=1}^{K} P[e_k^j|\alpha_i]}{\sum_{h=1}^{I} \left(P[\alpha_h] \prod_{k=1}^{K} P[e_k^j|\alpha_h] \right)}$$

Here, $P[e_k^i|\alpha_i]$, which is equal to the trust level of a report, is the probability that report by node v_k confirms event α_i, given α_i is happened. Computation of trust level for composite events that are unions or intersections of basic events are done following the rules of probability theory.

In BI, for an event α_I and for a node v_k, a node either confirms or refutes an event. For example, if a node confirms the presence of an event with probability p, then it refutes the absent of the event with probability $1 - p$. But, in most of the cases, a node may be uncertain about occurrence of an event that cannot be formulated with BI. Dempster–Shafer theory in this regard provides uncertainty notions to represent uncertainty.

A good mathematical framework for Bayesian framework-based trust evaluation is available at [23].

11.4.4.5 Dempster–Shafer Theory

In BI, nodes either confirm or refute an event. However, there are cases where a node can give no information or be uncertain about status an event. The DST model handles the scenario when there is no information available. Lack of knowledge/uncertainty about an event is not necessarily the refutal of the event. In the DST model [31], if there are two conflicting events, uncertainty of one of them can be considered as supporting evidence for the other. The level of uncertainty is bounded between supporting and nonrefuting evidence. Lower bound of uncertainty is *belief* which is a supporting evidence, and upper bound of uncertainty is *plausibility* which is a nonrefuting evidence. For example, an uncertainty level of p is p degree of belief in the event and 0 degree of belief in its absence. *Belief* value corresponding to an (composite/basic) event α_i and provided by report k is computed as

$$\text{bel}_k(\alpha_i) = \sum_{q:\alpha_q \cap \alpha_i} m_k(\alpha_q)$$

In this case, α_I is composed of basic events α_q. $m_k(\alpha_q)$ is basic belief assignment corresponding to the event α_q. If α_I is a basic event, then $\text{bel}_k(\alpha_i) = m_k(\alpha_i)$.

The plausibility value corresponding to event α_I is the sum of all evidence that does not refute α_I and is computed as

$$\text{pls}_k(\alpha_i) = \sum_{r:\alpha_r \cap \alpha_i \neq \phi} m_k(\alpha_r)$$

Both belief and plausibility are related by

$$\text{pls}(\alpha_i) = 1 - \text{bel}(\bar{\alpha}_I)$$

The combined trust level corresponding to event α_I is the belief corresponding to α_i

$$d_i = \text{bel}(\alpha_i) = m(\alpha_i) = \oplus_{k=1}^{K} m_k(\alpha_i)$$

Pieces of evidence can be combined using Dempster's rule for combination:

$$m_1(\alpha_i) \oplus m_2(\alpha_i) = \frac{\sum_{q,r:\alpha_q \cap \alpha_r = \alpha_i} m_1(\alpha_q) m_2(\alpha_r)}{1 - \sum_{q,r:\alpha_q \cap \alpha_r \neq \phi} m_1(\alpha_q) m_2(\alpha_r)}$$

11.4.4.6 Application of Decision Logic

A pragmatic implementation of decision logic will need hybridization of the above logics. In most cases, a reporting node (here vehicle) may be uncertain about an event. For example, a vehicle may not know about an accident happened behind it due to sensing problem; in this case, it will be uncertain about the event. The Bayesian approach does not support uncertainty. It takes only yes/no about happening of an event as input. In this case, DST is much preferable than Bayesian approach for its support of uncertainty of an event. However, a hybrid of the above approaches may yield even pragmatic and good solution for implementing decision logic in real-life events. More about decision logics are available at the literatures [32–34].

11.5 Challenges and Questions

In What Layer should Trust Framework be Implemented?

An important issue to think over is the layer at which the trust protocol will operate. To get a feasible solution, one must analyze the issues like services required by the trust protocol, services offered by trust protocol, relationship to other security components, system architecture, and so on. In case of VANETs, different classes of services along with their priority must be taken into account.

Separation of Policy from Detail Implementation

To support "Increased flexibility through programmability"-trend, policy language must be separate from the difference in the implementation framework. This separation of policy language will also depend on the layer in which the trust management system is implemented. In VANETs, centralized authority may be a city or state transport authority. Transport authority's policy of service may vary time to time based on the response from public users. Frequent change of policy may be needed. For ensuring flexibility, policy languages must be isolated from detail implementation.

11.6 Future Directions

Data-centric trust model need more enhancement for its practical application in the VANETs. In future, there will have a need for aggregation and trade-off among decision logics for practical use of these in VANETs. Moreover, architecture of the trust models may evolve with time, keeping the need of successful implementation of VANETs. In this regard, deployment of VANETs using test beds with trust architecture integrated in it, and performance evaluation of the system is utmost necessary. A generic "trust management language" is of need to implement the trust models. Collaboration among vehicle manufacturer, researchers, and computer scientists may lead to a successful implementation of VANETs.

11.7 Conclusion

In this chapter, we gathered the state-of-the-art of trust models for VANETs. Developing trust management language for realizing trust models is a challenging task. In near future, we shall work on the development of trust management language and will implement it using test beds.

Terminologies

Dempster–Shafer Theory—DST is a mathematical theory of evidence that allows us to combine evidence from different sources and arrive at a degree of belief that takes into account all the available evidence.

Trust Model—Trust model is an entity relationship model in which two groups are involved. *Trustor*—the subject that trusts a target entity—and *Trustee*—the entity that is trusted.

Bayesian Inference—In statistics, Bayesian inference is an inference method in which evidence or observations are used to update or to newly infer the probability that a hypothesis that may be true.

Questions and Sample Answers

1. Who and when first coined the term "trust management"? Name two pioneering trust management languages.
   Matt Blaze, 1996. PolicyMaker and KeyNote.

2. Which factors make trust management for VANETs distinct from MANETs?
Ephemeral network connectivity, huge amount of information flow, unstable network, irrelevance of node identity.

3. When does first data-centric trust notion came into focus of researchers?
Data-centric trust notion first come into focus in researchers by the work of researchers of EPFL, Switzerland, and later their publication on it.

4. What are the differences between data-centric and entity-centric trust notion?
In data-centric trust, trust evaluation is primarily based on the reliability of the data; identities of the individual nodes producing the data have less influence in determining the trust value. In entity-centric trust notion, trust evaluation is primarily based on the history, past behavior, and reputation of the individual nodes/entity producing the data.

5. Are unicast and multicast applicable for VANETs?
Unicast and multicast may be applicable in VANETs; however, due to high rate of node mobility and ephemeral nature of network connectivity, the success rate would be very low. Rather, broadcast is preferable in VANETs.

6. Give a detailed classification of the adversary models of VANETs.
In broad sense adversaries are classified into two:
1. Those intentionally deviate from protocol are called attackers.
2. Those due to fault, unintentionally deviate from protocol are faulty nodes.
Both attackers and faulty nodes can cause damage to the network. Some researchers [27,28] classify under more specific headings:
Insider versus outsider: Insiders are authenticated members of the network that can legitimately communicate with other members. An insider possesses a certified public key from the authority. Outsiders are the network members who does not possess a certified public key, or whose public key has been revoked by the authority.
Malicious versus rational: Malicious attackers seek no personal benefits from the attacks and aim to harm the members or functionality of the network. Hence, malicious attackers may employ any means disregarding corresponding costs and consequences. On the contrary, rational attackers seek personal benefit and hence are more predictable in terms of attack means and attack target.
Active versus passive: Active attackers can generate packets or signals, whereas passive attackers eavesdrop through the network channel.
In [23], adversaries are classified as three types: independent, collusive, and random.
Independent: Some independent node can mechanically be faulty, or intentionally attack the network for its own advantage.
Collusive: A group of nodes together with may cause attack, inform wrong, and jam the network.
Random: Some nodes may be more intelligent; they sometime behave well, and sometime behave malicious.

7. What is trust model? What characteristics does it possess?
Trust models and their evaluation metrics have some notable characteristics. Characteristics of trust can be listed under the following headings:
Context dependence: Trust metric may differ based on the context over which trust is evaluated. Specific contexts based on which trust value will be evaluated must be defined by the

authority in the protocol before bootstrapping/initiating the protocol. For example, contexts of VANETs will include authenticity of message, service specific authenticity, and so on.

Subjective: Trust value about an entity by other entities widely differs by entity to entity. Entity $x1$ trust entity y does not mean that entity $x2$ trusts y. Entity $x1$ may trust y with value 0.9, whereas entity $x2$ may trust y with value 0.3, in the scale between 0 and 1.

Asymmetric: Entity x trusts y does not mean that entity y trusts x. Each entity has independence in the evaluation of others.

Measurable belief: Trust metric of a node in a network is measured within a certain numeric scale. In most trust models, this takes a real number between 0 and 1 where 1 means highest trust value while 0 means lowest trust value.

Evolves with time: As trust values reflect the actions of an entity over passage of time, it is varied with time. Honest actions increase the trust value whereas dishonesty decreases the trust value.

Reflexive: Trust metrics have reflexive property. An entity may trust itself with a certain metric or confidence value.

Trust management model's characteristics can be listed under following headings:

Centralized versus decentralized trust model: In centralized models, a globally trusted entity calculates the trust values for every node in the system. There are two problems associated with this approach. One is single point failure, and the other is that different users may have different opinions about the same target which is suppressed in this scheme. However, centralized approaches cannot be implemented in autonomous and *ad hoc* networks.

In decentralized trust management models, each user calculates trust values around it with its own policy. This is a bottom-up approach, and most widely implemented and used as a part of PGP [26] for public key certification.

Proactive versus reactive computation: In a proactive computational model, reputation/trust values of network nodes are computed periodically and kept into a record. This approach uses more bandwidth and space for maintaining trust relationship as reputation value exchange in network is done frequently and stored locally.

8. What are the salient features of VANETs?

In order to achieve the objectives of VANETs and realize these as a successful technical deployment in mass people, vehicles and infrastructures should be equipped with cutting-edge devices, and a sound protocol should be developed to achieve maximum output from these networks. VANETs are composed of the following components and have the following characteristics.

1. Each vehicle constructing VANETs are equipped with an onboard processing unit to process data and wireless modules to build the network.
2. Infrastructure is constructed using road-side units (RSUs), which are also equipped with a processing unit and a wireless module to build the network.
3. There are authorities (public/private) who control the participation of the network through providing certificate and credentials. In addition to certification, the authorities may posses other administrative powers such as identity management, key revocation, and so on. These authorities are trusted to nodes participating in VANETs. City or state transportation authorities will serve as public authority.
4. A subset of RSUs serves as gateway to and from the authority.
5. Each vehicle is equipped with a clock and a positioning system (GPS). These devices allow the vehicles to include time and location information with each outgoing messages.

Time and location is important to maintain time freshness and location relevance of the messages.

6. Each vehicle v_k is equipped with a pair of private/public cryptographic keys (P_rk/P_uk), and a certificate issued by an authority X, $\text{Cert}_x(P_uk)$.

7. Authentication of message source is done via digital certificates (e.g., X.509). Authentication helps prevention from Sybil attacks. By source authentication, the identity/type of the node that sent the message can be identified. From identity, trustworthiness of a node can be measured by finding previous history of that node.

8. Local broadcasts (1 hop) and Geocast (flooding in a fixed area) are mainly used in VANETs. However, uni- and multicast are still possible but inefficient due to high nobility of network nodes.

9. Each vehicle frequently and periodically transmits packets. These packets include velocity, coordinates, time, position of own, and network events, safety messages and messages/reports from others.

9. By drawing a schematic diagram, describe a state-of-the-art of data-centric trust model.

10. Name prominent decision logics applicable for VANETs. Which logic would be the most suitable for VANETs?

Prominent decision logics are

1. Majority voting
2. Most trusted report (MTR)
3. Weighted voting (WV)
4. Bayesian inference (BI)
5. Dempster–Shafer theory (DST)

A pragmatic implementation of decision logic will need hybridization of above logics. In most cases, a reporting node (here vehicle) may be uncertain about an event. For example, a vehicle may not know about an accident happened behind it due to sensing problem; in this case, it will be uncertain about the event. The Bayesian approach does not support uncertainty. It takes only yes/no about happening of an event as input. In this case, DST is much preferable than the Bayesian approach for its support of uncertainty of an event. However, a hybrid of the above approaches may yield even pragmatic and good solution for implementing decision logic in real-life events.

Author's Biography

Une Thoing Rosi received BSc engineering degree in computer science and engineering from Bangladesh University of Engineering & Technology (BUET), Dhaka, Bangladesh, in 2009. He has been serving as a faculty of CSE at United International University (UIU), Dhaka, Bangladesh since February 2009. He has published numerous articles on VANET/ITS in reputed (ACM/IEEE sponsored) international conferences. He also served as a program committee of renowned international conferences, and also reviewed articles of international journals and conferences. His research interest includes distributed systems and information retrieval, in general. He is the recipient of Erasmus Mundas scholarship by EU, Dean's award, and University Scholarships from BUET.

Syed Ishtiaque Ahmed received BSc engineering degree in computer science and engineering from Bangladesh University of Engineering & Technology (BUET), Dhaka, Bangladesh, in 2009. He has been serving as a faculty of the CSE department at BUET since October 2009. Prior to BUET, he served as faculty of United International University (UIU) from February 2009 to September 2009. He has published a book on computational geometry and numerous articles on computational geometry and networking in reputed international conferences and journals. He also served as a program committee of renowned international conferences, and also reviewed articles of international conferences. His research interest includes computational geometry and networking, in general. He is the recipient of Dean's award and University Merit Scholarships from BUET.

References

1. Blaze, M., Feigenbaum, J., and Lacy, J. Decentralized trust management, In *Proceedings of IEEE Symposium on Security and Privacy*, Oakland, CA, May 1996, pp. 164–173.
2. Blaze, M., Feigenbaum, J., and Keromytis, A. KeyNote: Trust management for public-key infrastructures. In *Proceedings of the 6th International Workshop on Security Protocols, volume 1550 of Lecture Notes in Computer Science*, Cambridge, UK, April 1998, Springer, pp. 59–63.
3. Blaze, M., Feigenbaum, J., Ioannidis, J., and Keromytis, A. The KeyNote Trust Management System, Version 2. RFC-2704. IETF, September 1999. Available from http://www.faqs.org/rfcs/rfc2704.html
4. Sepandar, D. K., Schlosser, M. T., and Molina, H. G. The eigentrust algorithm for reputation management in P2P networks. In *Proceedings of the 12th International World Wide Web Conference*, Budapest, Hungary, May 20–24, 2003, pp. 640–651.
5. IETF MANET working group, available from http://www.ietf.org/html.charters/manet-charter.html
6. Aguayo, D., Bicket, J., Biswas, S., Judd, G., and Morris, R. Link-level measurements from an 802.11b mesh network, *ACM SIGCOMM*, Portland, OR, 2004.
7. Draves, R., Padhye, J., and Zill, B. Routing in multi-radio, multi-hop wireless mesh networks. In *Proceedings of Mobicom*, Philadelphia, PA, 2004.
8. Gnutella Protocol Specification. Available from http://www.limewire.com/developer/gnutella_protocol_0.4.pdf
9. Stoica, I., Morris, R., Liben-Nowell, D., Karger, D. R., Frans Kaashoek, M., Dabek, F. and Balakrishnan, H. Chord: A scalable peer-to-peer lookup protocol for internet applications. In *Proceedings of SIGCOMM, 2001*, San Diego, CA, 2001.
10. Buchegger, S. and Le Boudec, J. Performance analysis of the CONFIDANT protocol: Cooperation of nodes—fairness in dynamic ad-hoc networks, in *Proceedings of IEEE/ACM Symposium on Mobile Ad Hoc Networking and Computing (MobiHOC)*, Lausanne, CH, June 2002.
11. Raya, M., Papadimitratos, P., Gligor, V. D., and Hubaux, J.-P. On data-centric trust establishment in ephemeral ad hoc networks, in *Proceedings of INFOCOM*, Phoenix, AZ, 2008.
12. http://www.crypto.com/papers/
13. Blaze, M., Feigenbaum, J., and Keromytis, A. D. KeyNote: Trust management for public-key infrastructures (Position Paper), *Lecture Notes in Computer Science; Vol. 1550 archive*, in *Proceedings of the 6th International Workshop on Security Protocols*.
14. Chu, Y.-H., Feigenbaum, J., LaMacchia, B., Resnick, P., and Strauss, M. REFEREE: Trust management for Web applications. *Computer Networks and ISDN Systems Archive* 1997; 29(8–13): 953–964.
15. Abdul-Rahman, A. and Hailes, S. Supporting trust in virtual communities, in *Proceedings of the 33rd Hawaii International Conference on System Sciences*, Wailea Maui, Hawaii, 2000.
16. Aberer, K. and Despotovic, Z. Managing trust in a peer-2-peer information system, in *Proceedings of the 10th ACM International Conference on Information and Knowledge Management*, Atlanta, November 2001.
17. Buchegger, S. and Le Boudec, J. Performance analysis of the CONFIDANT protocol: Cooperation of nodes—fairness in dynamic ad-hoc networks, in *Proceedings of IEEE/ACM Symposium on Mobile Ad Hoc Networking and Computing (MobiHOC)*, Lausanne, CH, June 2002.
18. http://icapeople.epfl.ch/sbuchegg/confidant.html

19. Sepandar, D., Kamvar, Schlosser, M. T., and Garcia-Molina, H. The eigen trust algorithm for reputation management in P2P networks, in *Proceedings of WWW2003*, Budapest, Hungary.

20. Quercia, D., Lad, M., Hailes, S., Capra, L., and Bhatti, S. STRUDEL: Supporting trust in the dynamic establishment of peering coalitions, in *Proceedings of the 2006 ACM Symposium on Applied Computing*, Dijon, France.

21. Quercia, D., Hailes, S., and Capra, L. B-Trust: Bayesian trust framework for pervasive computing. In *Lecture Notes in Computer Science*, Volume 3986/2006, ISBN 978-3-540-34295-3.

22. Quercia, D., Hailes, S., and Capra, L. MobiRate: Making mobile raters stick to their word, in *Ubi-Comp'08: Proceedings of the 10th International Conference on Ubiquitous Computing*, Seoul, South Korea.

23. Baras, J. T. and Baras, J. S. Trust evaluation in anarchy: A case study on autonomous networks, in *Proceedings of INFOCOM 2006. 25th IEEE International Conference on Computer Communications*, Barcelona, Catalunya, Spain.

24. Theodorakopoulos, G. Distributed trust evaluation in ad hoc networks. MS Thesis, University of Maryland.

25. Theodorakopoulos, G. and Baras, J. S. On trust models and trust evaluation metrics for ad hoc networks. *IEEE Journal on Selected Areas in Communication* 2006, 24, 2.

26. Zimmermann, P. R. *The Official PGP User's Guide*. Cambridge, MA: MIT Press, 1995.

27. Parno, B. and Perrig, A. Challenges in securing vehicular networks, in *Workshop on Hot Topics in Networks (HotNets-IV)*, College Park, Maryland, USA, 2005.

28. Raya, M. and Hubaux, J-P. The security of vehicular ad hoc networks, in *Proceedings of SASN'05*, Alexandria, VA, USA.

29. Ostermaier, B., Dotzer, F., and Strassberger, M. Enhancing the security of local danger warnings in VANETs—A simulative analysis of voting schemes, in *Proceedings of the Second International Conference on Availability, Reliability and Security*, Vienna, Austria, pp. 422–431.

30. Pearl, J. *Probabilistic Reasoning in Intelligent Systems: Networks of Plausible Inference*. Morgan Kaufmann, 1988, ISBN: 1558604790, 9781558604797.

31. Shafer, G. *A Mathematical Theory of Evidence*, Princeton University Press, 1976, ISBN13: 978-0-691-10042-5.

32. Josang, A. An algebra for assessing trust in certification chains. In *Proceedings of NDSS'99, Network and Distributed Systems Security Symposium*, The Internet Society, San Diego, 1999.

33. Lindsay, Y., Wei Yu, S., Han, Z., and Ray Liu, K. J. Information theoretic framework of trust modeling and evaluation for ad hoc networks. *IEEE Journal of Selected Areas of Networks* 2006, 24, 305–317.

34. Theodorakopoulos, G. and Baras, J. S. On trust models and trust evaluation metrics for ad hoc networks. *IEE Journal of Selected Areas of Networks*, 2006, 24, 318–328.

Chapter 12

Sybil Attack in VANETs
Detection and Prevention

Jyoti Grover, M.S. Gaur, and V. Laxmi

Contents

12.1 Introduction

Vehicular *Ad-Hoc* Network (VANET) is a specific type of Mobile *Ad-Hoc* Network (MANET) that provides communication between (1) nearby vehicles and (2) vehicles and nearby roadside equipments [1–3]. VANETs are one way to implement Intelligent Transportation System (ITS), a technique for imparting information and communication technology [4,5] to transport infrastructure and vehicles. It is based on IEEE 802.11p standard [6] for Wireless Access for Vehicular Environment (WAVE). These networks have no fixed infrastructure, and they rely on the themselves for implementing any network functionality [5]. A VANET is a decentralized network as every node performs the functions of both host and router. The main benefit of VANET communication is enhancement of passenger safety by exchanging warning messages between vehicles. VANETs differ from MANETs in high mobility of nodes, large scale of networks, geographically constrained topology, and frequent network fragmentation. Most of the research on VANET is focused on Medium Access Control (MAC) layer and the network layer [1,3]. VANETS aim to build applications such as collision avoidance, route changing, and so on. Security of vehicular networks is still largely an explored area.

VANET, being a wireless network, inherits all the security threats that a wireless system has to deal with [7]. VANET security is critical because a poorly designed VANET is vulnerable to network attacks, and this can compromise the safety of drivers. A security system should ensure that transmission comes from a trusted source and is not a tampered en-route by other sources. It should also strike a balance with privacy because implementing security and privacy together in a system is contradictory. There are various types of possible attacks on VANETs. It is imperative that VANET security should be capable of handling every type of attack.

VANET security is different from that of wireless and wired networks because of its unique characteristics of mobility constraints, infrastructure-less framework, and short duration of link between nodes. In a wired network, infrastructure has components for specific functions, for example, routers decide the route to destination while network hosts send and receive messages. Security implementation is relatively easy as networks need to be physically tampered for eavesdropping. Wireless networks use infrared or radio frequency signals to communicate among devices. These networks can be either (a) infrastructure based or (b) infrastructure-less. Infrastructure-based wireless networks are based on Public Switched Telephone Network (PSTN) switches, MSCs, base stations, and mobile hosts. In *ad-hoc* networks, a type of infrastructure-less wireless networks, nodes perform all operations such as routing, packet forwarding, and network management, and so on. The existing security solutions use traditional digital signature [8,9] and certificates using Public Key Infrastructure (PKI).

In VANETs, primary focus of security is on safety-related applications. Nonsafety applications have less stringent security requirements. There is no prior trust relationship between the nodes* of VANETs because of its infrastructure-less nature. Any node can join and leave the network at anytime without informing other nodes in vicinity. Cooperative security schemes are more efficient in VANETs as node misbehavior can be detected through collaboration between the number of nodes by assuming that majority of nodes are honest.

---

* The terms node and vehicle are used synonymously in VANETs.

In this tutorial, we are focusing on one important attack named Sybil attack on VANETs. Sybil attack can occur in every scenario where there is no centralized unit controlling all the entities in the network. As wireless communication is more prone to security threats, Sybil attack leaves its impact on all wireless networks.

The rest of chapter is organized as follows: VANET architecture is presented in Section 12.2 followed by a summary of attacks in Section 12.3. Sybil attack is described in detail in Section 12.4. Trust establishment techniques that are at the core of VANET security are described in Section 12.5. Both infrastructure-based and dynamic trust establishment techniques are presented, highlighting their strengths and weaknesses. Section 12.6 reviews detection techniques for Sybil attack. Open research issues in VANETs are briefly discussed in Section 12.7 followed by concluding remarks in Section 12.8.

12.2 VANET Architecture

VANET architecture employs two types of communication devices: (1) On-board Units (OBUs) and (2) Road-side Units (RSUs). As name suggests, OBU is installed in a vehicle and RSUs are placed on roadside. Each OBU consists of an Event Data Recorder (EDR), Global Positioning System (GPS) receiver, computing platform, and a radar. GPS receiver provides information about geographic location, speed, direction of movement, and acceleration of a node at specified time intervals. EDR device records the transmitted and received messages [3,5]. Information stored in EDR can assist in recreation of an accident/emergency situation for subsequent analysis after the occurrence of an event. The computing device is used to take appropriate actions in response to messages received from other nodes. Radar is used for detecting obstacles near the vehicle. Each vehicle also has an omni-directional antenna that the OBU uses to access a wireless channel. An RSU is similar to an OBU in that it has an antenna, computing device, transceiver, and sensors. It is a stationary device mounted on roadside. An RSU may be installed at road intersections or embedded in traffic-light for traffic control. It can be deployed for commercial use also. For example, a restaurant can use an RSU for advertisement of its presence. An RSU may use either directional antenna or omni-directional antenna depending on the type of application. Figure 12.1 shows a typical VANET architecture.

VANET is not a pure *ad-hoc* network as an infrastructure in the form of RSUs may exist in some parts of the network. Sometimes, on highway, there may not be any infrastructure. VANETs

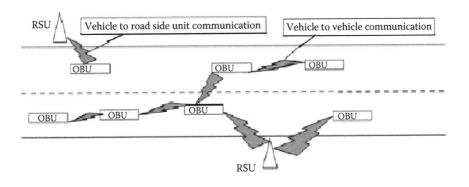

Figure 12.1 Architecture of VANET.

Sender ID	Sender position	Speed	Time	Direction	Acceleration	Warning message

Figure 12.2 Contents of safety message.

support two types of communications: (1) Vehicle-to-Vehicle and (2) Vehicle-to-RSU. A V2R communication enables vehicular safety applications, including collision warning as well as other ITS applications such as local traffic information for routing and high-speed tolling.

All the nodes know their own position and motion details. At periodic intervals, they exchange this information with neighboring nodes. Each vehicle stores information about itself and neighboring vehicles in a local database. Records of this database are periodically broadcasted to neighboring nodes and roadside equipments. These broadcasted messages assist in updating the information. Figure 12.2 illustrates the structure of a typical broadcasted message.

12.3 Attacks on Vehicular Networks

Before designing any security solution for VANETs [10,11], we should know different types of security threats, their capabilities, and the types of attackers also.

12.3.1 Classification of Attackers

Attackers can be classified according to scope, nature, and behavior of attacks [12,13]. Some types of attackers are discussed in following paragraph:

1. Some attackers eavesdrop only on the wireless channel to collect traffic information which may be passed onto other attackers. As these attackers do not participate in the communication process of the network, they are called *passive* attackers. On the other hand, some attackers either generate packets containing wrong information or do not forward the received packets. These are called *active* attackers.
2. Attacker may be an authentic member of a VANET having authentic public keys and access to other members of the network. Such attackers are called *insider*. Outside attackers (*outsider*) are intruders and they can launch attacks of less diversity.
3. Some attackers are not personally benefited from the attack. Their aim is to harm other members of the network or disrupt the functionality of a VANET. These attackers are *malicious*. On the other hand, *rational* attacker seeks personal benefit and is more predictable in terms of type and target of the attack.
4. *Local* attacker launches an attack with a limited scope, that is, an attack is restricted to a particular area. An attack can be *extended*, where an attacker can control several entities distributed across the network.

12.3.2 Types of Attacks

Owing to the large number of autonomous network members and the presence of human factor, misbehavior of nodes in future vehicular networks cannot be ruled out. Several types of attacks [13,14] have been identified and classified on the basis of layers used by the attacker. At the physical and link layer, an attacker can disturb the network system by overloading the communication channel with useless messages. An attacker can inject false messages or rebroadcast an old message also. Some attackers can tamper with an OBU or destroy an RSU. At network layer, an attacker can insert false routing messages or overload the system with routing information. Privacy of drivers can be disclosed by revealing and tracking the position of drivers. Some of these attacks are briefly explained subsequently.

1. *Bogus information:* In this case, attackers are insiders, rational, and active. They can send wrong information in the network so that it can affect the behavior of other drivers. For example, an adversary can inject wrong information about a nonexistent traffic jam or an accident diverting vehicles to other routes and freeing a route for itself.
2. *Cheating with sensor information:* This attack is launched by an attacker who is insider, rational, and active. He uses this attack to alter the perceived position, speed, and direction of other nodes in order to escape liability in case of any mishap.
3. *ID disclosure:* An attacker is insider, passive, and malicious. It can monitor trajectories of a target vehicle and can use this information for determining the ID of a vehicle.
4. *Denial of service (DoS):* Attacker is malicious, active, and local in this case. Attacker may want to bring down the network by sending unnecessary messages on the channel. Example of this attack includes channel jamming and injection of dummy messages.
5. *Replaying and dropping packets:* An attacker may drop legitimate packets. For example, an attacker can drop all the alert messages meant for warning vehicles proceeding toward the accident location. Similarly, an attacker can replay the packets after that event has been occurred to create the illusion of accident.
6. *Hidden vehicle:* This type of attack is possible in a scenario where vehicles smartly try to reduce the congestion on the wireless channel. For example, a vehicle has sent a warning message to its neighbors and it is awaiting a response. After receiving a response, the vehicle realizes that its neighbor is in a better position to forward the warning message and stops sending this message to other nodes. This is because it assumes that its neighbor will forward the message to other nodes. If this neighbor node is an attacker, it can be fatal for the system.
7. *Worm hole attack:* It is challenging to detect and prevent this attack. A malicious node can record packets at one location in the network and tunnel them to other location through a private network shared with malicious nodes. Severity of the attack increases if the malicious node sends only control messages through the tunnel and not data packets.
8. *Sybil attack:* In this attack, a vehicle forges the identities of multiple vehicles. These identities can be used to play any type of attack in the system. These false identities also create an illusion that there are additional vehicles on the road. Consequence of this attack is that every type of attack can be played after spoofing the positions or identities of other nodes in the network.

12.3.3 Security Requirements

A security system should ensure that any life critical information can be neither inserted nor modified by any attacker [10,15,16]. Most of the security mechanisms result in a significant overhead, thereby reducing the capabilities of the system in terms of latency and channel capacity. A security system used for VANET should satisfy the following requirements:

1. *Time delivery constraints:* Safety messages are time sensitive as delays in an emergency/accident should be as low as possible. Most safety applications require a latency time of less than 100 milliseconds. The security system should display all the warning messages to the driver before it is too late to react to the warning.
2. *Accurate location:* A system should display a warning message at the right location, that is, a driver should receive a warning message before he passes over the geographic position where a critical event has taken place.
3. *Privacy:* In wireless communication, an information is sent via broadcast channels so that anybody can receive it. This information contains privacy data such as location, speed, and sensor data of vehicle. Complete identity of any vehicle transmitting/forwarding message should not be revealed or else other vehicles can spoof this identity to break-in security measures. Data need to be de-linked from the driver's identity.
4. *Message integrity:* Messages must be protected from any alteration. Integrity means ensuring that a message is not tampered during transmission. It is not concerned with the origin of the packet.
5. *Accountability:* The sender of the message should not be able to deny that he has sent the message. This feature allows the security system to identify and ensure appropriate actions against the entity engaged in suspicious activity.
6. *Authenticity:* The security system must ensure that packets are generated by a trusted source. A network should be able to identify and discard any unauthenticated data. Privacy and anonymity are also important, but authenticity is an essential requirement.
7. *Access control:* Access to services provided by the infrastructure node is determined by local policies. Authorization is part of an access control and determines what type of service can be provided/availed by a network node. The system must have the capability to reject messages from known compromised units.

It needs to be noted that there is a contradiction in implementation of security and privacy in VANETs. To ensure accountability, messages need to be uniquely signed. But unique signatures will allow the signer to be tracked revealing its true identity.

12.4 Sybil Attack

Sybil attack, first discussed by Douseur [17], is a serious threat as it impairs the functionality of VANETs. In this attack, an attacker node sends messages with multiple identities to other nodes in the network. The attacker simulates several nodes in the network. The node spoofing the identities of other nodes is called malicious node/Sybil attacker, and the nodes whose identities are spoofed are called Sybil nodes. Almost every other attack can be launched in a network in the presence of Sybil attack. One possibility could be an illusion of a traffic jam or accident so that other vehicles change their routing path or leave the road for the benefit of the attacker.

Sybil attacker can also inject false information in the networks via some fabricated nonexisting nodes. For example, in the case of an accident on a highway, the first vehicle observing the accident is sending change route/deceleration warning message to all the following vehicles. Receivers may forward this message to warn followers, if any. This forwarding process can be disrupted by Sybil vehicles by not forwarding the warning message. This may put the life of passengers in danger.

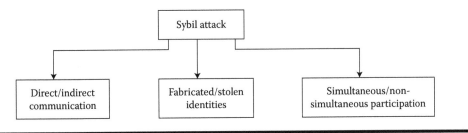

Figure 12.3 Different forms of Sybil attack.

Sybil attacks can be classified into three categories based on the type of communication, identity, and their participation in the network [18]. These categories are illustrated in Figure 12.3 and briefly discussed in following paragraphs.

1. *Communication category:* When an honest node sends a radio message to a Sybil node, one of the malicious nodes listens to the message. In the same way, messages sent from Sybil nodes are actually sent from one of the malicious devices. Communication to/from Sybil nodes can be direct or indirect. In a *direct* mode, all the Sybil nodes created by malicious node communicate with legitimate nodes. In an *indirect* communication, legitimate nodes reach the Sybil nodes through a malicious node.
2. *Identity category:* In a Sybil attack, an attacker creates a new Sybil identity. This identity can be a random 32-bit integer (*fabricated* identity) or an attacker can spoof legitimate identity of one of its neighbors (*stolen* identity).
3. *Participation category:* Multiple Sybil identities created by malicious nodes can *simultaneously* participate in the attack or the attacker can present these Sybil identities *one by one*. A particular identity may leave or join the network many times, that is, one identity is used at a time. The number of identities the attacker uses is equal to or less than the number of physical identities.

An attack through multiple Sybil nodes can adversely affect proper functioning of any network. Some of these functions where vulnerabilities can be introduced are

1. *Data aggregation:* Through multiple identities, a malicious node may contribute to the aggregation many times and can alter the result of data aggregation. If we are calculating the average number of packets dropped in the network, the packets dropped by Sybil nodes will be added in total. In this case, network performance will appear to be reduced significantly.
2. *Fair resource allocation:* Sybil nodes can also impact the fairness of resource allocation if resources are allocated per node. In the presence of Sybil identities, a malicious node receives a larger share of any resource. This result in DoSs to legitimate nodes as their share of resources is reduced leading to a DoS attack.
3. *Routing:* Sybil attacks are effective against functioning of routing protocols in VANETs. In multipath routing, disjoint paths are used. The presence of Sybil identities of a malicious node on these paths can impair routing. Geographic routing is also vulnerable as a malicious node can appear in more than one place at once.
4. *Voting:* Sybil attack can update the output of voting scheme incorrectly. If the attacker creates enough Sybil nodes that participate in determining a misbehaving node, a legitimate and well-behaved node can be expelled from the network.

5. *Misbehavior detection:* An attacker can bypass a mechanism to detect a malicious node by spreading the blame through the Sybil nodes. If the detection mechanism uses multiple observations to locate a malicious node, the attacker can still escape detection by using different nodes at different times. If some Sybil nodes are detected and expelled from the network for malicious behavior, the attacker uses other identities.

12.5 Trust Establishment

For prevention and detection of Sybil attacks, trust must be established among the participating nodes. It is a major challenge as a receiving node needs to ensure authenticity and trust-ability of the received messages before reacting to them. It is assumed that each node in a VANET is equipped with a trust system to take trust decisions. There are two options for trust establishment: (1) based on static infrastructure and (2) on dynamic establishment of trust in a self-organized manner. Trust based on static infrastructure is more efficient than dynamic infrastructure. The only concern is unavailability of fixed infrastructure in some locations. If all the nodes establish trust with other nodes in a VANET, the probability of occurrence of a Sybil attack can be reduced.

12.5.1 Infrastructure-Based Trust Establishment

In this section, we are discussing various approaches for establishing infrastructure-based trust. A good reference is [19].

1. *Digital signature and certificate-based system:* This is the most popular technique for trust establishment [12,20]. Safety messages are not meant to be confidential so they do not need any privacy. As a result, safety messages require authentication but do not require any encryption. A set of Public/Private key pairs is assigned to each vehicle to sign each message digitally and authenticate itself to receivers. Each message sent in the network contains a digital signature and corresponding certificate for the purpose of authentication and integrity. PKI is mostly used as a self-trust management technique, owing to issue of liability, it cannot be implemented. A centralized authority is required to issue digital certificates.

   Every vehicle is registered with a national/regional authority and is allocated a unique identifier called Electronic License Plate (ELP). This electronic identification is used for tracking of vehicles. In a PKI solution, a safety message is signed with the private key of a vehicle and includes certificates issued by Central Authority (CA) as follows. Let V is the sending vehicle, "*" stands for all the receivers, M is the message, T is the time-stamp to ensure the validity of the message, $PrKv$ is private key, and Cert_v is the public key certificate of V.

$$V \to^*: \langle M,\ \text{Sign}_{PrKv}\{M \mid T\},\ \text{Cert}_v \rangle$$

   For ensuring privacy, a vehicle has to store a large key/certificate set and the keys need to be changed after an interval of time for cryptographic security [12]. All the secret information (Public/Private key pair) is stored in a Tamper-proof Device (TPD) to prevent duplication and modification by an unauthorized vehicle. This device offers physical protection of keys residing in it and ensures that they cannot be modified or read by a malicious outsider. It is also responsible for signing all outgoing messages. Access of this device is limited to authorized people.

In PKI solution, the resulting message can become three times the size of the original message. TPD should be designed so that it can store cryptographic material and perform cryptographic operations in an efficient manner. The cost of TPD should be minimal.

2. *Pseudonyms:* In a certificate-based system, identity of a vehicle can be revealed when it is interacting with other nodes through its public key [20,21]. Privacy preservation using pseudonyms is proposed in [22]. This would not establish anonymity but protect privacy. Public keys or ELPs, in case of VANETs, need to be changed at periodic intervals. This change is performed by some CAs which also grants pseudonym. Association between pseudonyms and real-world entities is known to CA only. This solution is difficult to implement because of high mobility of nodes and dynamic nature of VANETs. Binding of these pseudonyms with a particular vehicle at a particular time requires an accurate synchronization. Revocation and reuse of these pseudonyms is another issue as numbers of vehicles on roads are increasing day-by-day.

3. *Group signatures:* A group signature scheme [20,23] provides both security and privacy in a VANET. This scheme allows a vehicle to sign the message on behalf of a group. A single group public key is used and it does not reveal the identity of the signer. It is not possible to verify if two signatures are issued by the same group members because each member of the group is assigned a unique private key. Private key of group members is used for generating signatures with group public key. A vehicle outside this group can verify that the message is generated by this group but cannot detect which node has generated this message. A node designated as group manager is required to resolve the signatures of individual nodes. The group manager uses his secret key and given signature to determine the identity of the group member who generated the message.

    A vehicle that does not have group manager's secret key cannot determine the identity of a group member. This ensures that members of the group are anonymous within the group and also with other group members. In addition, no outsider can issue signatures. Only the group members can sign correctly. A group signature scheme consists of the following procedures:

    i. *Set-up:* This protocol is used for interaction between a designated group manager and members of the group to decide group public key, private key of group manager, and all the members of the group.

    ii. *Sign:* In this protocol, a group member signs a message m with its private key p and returns a signature s.

    iii. *Verify:* It is used for verification of a valid signature. It requires a message m, signature s and group's public key Y.

    iv. *Open:* This returns the identity of the member who signed the message. This requires a signature s and group manager's private key.

    This technique is quite promising for VANETs since it does not require a permanent online connection to infrastructure. It works well in a dynamic environment, privacy can be established, and verification process is fast. Special attention should be given when vehicles leave one group and join another group. In this case, cell dimension should be less than the diameter of transmission range of vehicles. There is a need for efficient group management, key certification, and key revocation techniques because of the dynamic nature of groups. As numbers of vehicles are increasing day-by-day, some types of hierarchical structure are needed for efficient group management.

4. *Pair-wise keys:* If two nodes want to communicate with each other for a long time, they have to stay in the range of each other. For one-to-one communications of such nature, pair-wise keys are used. Symmetric keys are more efficient in terms of time and space overhead, than asymmetric keys [12,20]. Main challenge is distribution of key pairs. It is very difficult to

preload pair-wise shared keys to vehicles because of large-scale and dynamic nature of VANETs. One method for establishing pair-wise key makes use of PKI and digital certificates. Vehicle A encrypts the message comprising of identity of B, time-stamp T, and session key K with B's public key $PuKB$. This message is also signed with $PrKA$ private key of A. Subsequent message exchange can use this session key, and integrity of message can be verified through Hashed Message Authentication Code (HMAC) with key K.

$$A \rightarrow B : \langle E_{PuKB}\{B|K|T\}, \text{ Sign}_{PrKA}\{B|K|T\}\rangle$$
$$A \rightarrow B : \langle m, \text{ HMAC}_K(m)\rangle$$

This scheme does not scale well as number of digital signatures increase with growing number of vehicles in a VANET. For a fewer numbers of vehicles, this scheme is not justifiable because of lack of congestion on the wireless channel. For critical safety applications, symmetric session keys are not used as nonrepudiation property cannot be established.

5. *Threshold cryptography:* The concept of threshold cryptography is discussed in [19,24]. In this technique, no centralized trust system is required. In (n, t) threshold cryptography, a system shares a secret between n nodes so that any t nodes can rearrange the secret. This system is robust because any malicious node has to attack at least t nodes to obtain the secret. The only hurdle for the implementation of this solution is to decide an appropriate value of n and t. This technique will not work if there are less than t nodes at any time as some nodes are leaving the network.

12.5.2 Dynamic Trust Establishment

Trust establishment techniques [19,25] should adapt to the dynamic environment of a VANET. All the techniques discussed in Section 12.5.1 fail to adjust with changes in the VANET environment. Trust decisions must be made autonomously because fixed security infrastructure is not guaranteed at all times in a VANET. Decisions must be based on the partial information collected for a short time from unknown nodes. Self-organized trust establishment is required because of nonavailability of infrastructure and shared global knowledge among the participating nodes. Such a mechanism builds trust on the basis of mutual communication: (1) between the vehicles (direct method) and (2) exchange of trust information of other nodes (indirect method) or combination of both direct and indirect techniques.

12.5.2.1 History-Based Trust Establishment

In this technique, the history of behavior of nodes is stored in every vehicle. The following phases are required for creating history.

1. *Monitor:* This monitoring unit runs in background and monitors a neighbor by observing its routing protocol behavior.
2. *Trust manager:* If this behavior is consistent for a specified period of time, the node is considered trusted otherwise it is labeled as a malicious node. Trust manager stores trust value of the nodes. If the trust value falls below a threshold value, the path manager is involved.
3. *Path manager:* It isolates malicious nodes by ignoring sending/receiving packets to/from these nodes.

This is a high-level modular design but in real-world it is difficult to deploy this solution as it is difficult to distinguish between misbehavior of nodes and errors due to fast changes in topology. Storage of history requires more space and computing power.

12.5.2.2 Self-Certified Pseudonym-Based Trust Establishment

This approach [21] is used to provide privacy and Sybil-freeness without requiring any continuous availability of centralized authority. Users can compute pseudonyms from their cryptographic identities themselves. If initial identity domain is Sybil free, this Sybil freeness can be propagated to other identity domains even without continuous involvement of Trusted Third Party (TTP). CA will be needed only for initial setup of Sybil-free domain. Initially user acquires membership certificate by enrolling with CA. User can create a self-certified pseudonym per identity domain by using this membership certificate. These pseudonyms are valid for the domains for which they were issued. Pseudonyms for different domains are unlinkable. Sybil attacker has to identify the relationship between two pseudonyms generated for different identity domains. Attacker can eavesdrop on the wireless channel and find if any pseudonyms belong to the same user. This technique can help to prevent Sybil nodes in a VANET.

12.5.3 Analysis of Trust Establishment Approaches

For achieving authenticity and integrity, the message should be signed before sending and it should be verified at the time of receiving. In VANETs, public key signatures are desirable because broadcasted applications are dominant and multiple targets may be within the range. Digital signatures provide authenticity and integrity of messages. It also maintains sender's accountability for their messages. In addition to time-stamp, every message should also include sequence numbers to prevent replay attack through replaying of old messages at different time and/or places.

Unique signatures can disclose the identities of the sender especially when safety applications are broadcasting a safety message at high frequency. In such circumstances, a sender can easily be tracked and its movements can be monitored. It is important to maintain the privacy of vehicles in a VANET. Safety messages must not contain any data that identify the vehicle or allow recipients to link messages. For example, if a vehicle is receiving the same packet from different locations and at different time, then it can link that the message is coming from same node and track the movement of this node by calculating the distance and time between the transmissions of two messages. On the other hand, this knowledge of same origin of messages received through different paths and/or times may be used by vehicle to drop extra packets.

Transactional applications can use encryption of sent data. The encryption must be semantically secure so that it should produce two unrelated cipher texts if same message is encrypted twice. Encryption helps in implementing anonymity in a VANET; a vehicle can be tracked by its unique MAC address, Internet Protocol (IP) address, digital signatures and certificates, and account information for transactional applications.

In schemes involving *digital signatures* and *public key certificates* for safety messages, multiple digital certificates and public/private key pairs can be issued for each vehicle. According to an analysis by Raya and Hubaux [12], a vehicle should change its anonymous key after every one minute to avoid being tracked. If we assume that an average driver uses his vehicle 2 hours per day, the number of required keys per year is ~43,800 and shall consume ~21 Mbytes of storage space.

Random MAC address can be used to ensure anonymity in a VANET. MAC address is a unique identifier attached with every node. It is used for uniquely identifying each host and allows each message to be sent to a specific host. It is sufficient that MAC addresses of all the neighbors should be different. So for implementing anonymity, random MAC addresses can be used to avoid associating a vehicle to a particular MAC address. Maximum immediate communication range in VANETs is 1000 meters. With a very large address space (2^{46}) and small groups, it is very rare that two vehicles will be assigned same MAC address at the same time. MAC addresses can be changed very frequently to avoid being tracked.

Short lifetime of IP addresses owing to mobility of vehicular nodes can be used advantageously for anonymity. When vehicles move from one RSU communication zone to another RSU range, this address is changed. This change reduces the probability of being tracked.

Most befitting technique for VANET security implementation is to use dynamic trust establishment. But it is difficult to implement because of dynamic characteristics of VANETs. Self-organized trust establishment mechanism can build trust based on the mutual communication between the vehicles. Overhead of information exchange, storage, and analysis is high. Efficient algorithms for dynamic trust establishment should be designed to deal with VANET characteristics.

12.6 Detection of Sybil Attack

In literature, different techniques [26] are proposed for detection of Sybil attack in VANETs. Douceur has shown in [17] that Sybil attacks are always possible in the absence of any logical centralized authority. As there is no centralized entity in VANETs, detection of Sybil attacks is very difficult. Some constraints such as validating all entities simultaneously by all nodes and strict coordination among entities are necessary for detection of a Sybil attack. Some techniques are described below.

12.6.1 Resource Testing

This technique, proposed by Douceur [17], can be used to detect Sybil attack discussed in [18,21, 25–27]. It is assumed that every physical entity is equipped with limited computational resources. A typical puzzle is given to all the nodes in the network for testing computational resources. If resources of a single node are used to simulate multiple entities, any particular entity will be resource constrained in computation, storage, and bandwidth. This approach is not suitable as an attacker may have more computational resources when compared with honest nodes. Yet another problem is that this technique may create network congestion because more number of requests/replies are used for identification of nodes. Radio resource testing can also be used for detecting Sybil nodes. It is based on the assumption that any node has only one radio so any radio cannot send and receive more than one channel at a time. This technique also fails because the attacker can use multiple radio devices simultaneously.

12.6.2 Public Key Cryptography

Security issue of Sybil attacks can be solved by using public key cryptography and authentication mechanism as described in [12,24]. In this security solution, signatures are combined with digital certificates provided by TTP and asymmetric cryptography is used. Certificates are issued by CA and there is a hierarchy of these CAs. For each region, there is one CA. These CAs communicate with each other through secure channel and keep track of issued certificates used by every signed message. Figure 12.4 illustrates the underlying concept. In this method, vehicles send signed information

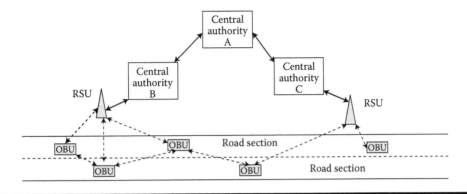

Figure 12.4 Hierarchy of central authority.

and verification is dependent on receivers. This technique can prevent Sybil attacks as only messages with valid certificates are considered and invalid messages are ignored. The only requirement is that each node should be assigned one certificate at a time. For privacy implementation, these certificates are changed from time-to-time. But in VANETs, it is difficult to deploy PKI as there is no guarantee of the presence of infrastructure. It is very complex, consumes large memory, and time consuming as well. To be part of a VANET, each vehicle stores the following cryptographic information.

a. An ELP issued by the government. These identities should be unique and cryptographically verifiable that is, a certificate is attached to all identities issued by CA. ELP should be changed when the owner moves to different region.

b. Anonymous key (Public/Private) pairs used to preserve privacy in VANETs. This key pair is authenticated by the CA, but does not contain any relationship with the identity of a vehicle. For maintaining privacy, a set of anonymous key pairs are used.

12.6.3 Passive Detection through Single Observer

This detection method does not require any special hardware [25]. In this detection, a single vehicle monitors network traffic passively and requires only a small amount of memory to record its observations. Identities of nodes (their MAC/IP addresses, public keys) are stored and profiles of nodes that send traffic together are built. Some affinity function is used to detect a Sybil attack. As multiple identities of single Sybil node are bound to a single physical node, the attacker and Sybil nodes will move simultaneously in an attack, while honest nodes are free to move at will. As the nodes move geographically on specific portion of road, these Sybil identities will appear or disappear simultaneously. If an attacker is using single channel radio, multiple Sybil identities will transmit serially whereas multiple independent identities will transmit in parallel. This simultaneous movement of nodes and mode of transmission (serial or parallel) stored in a profile can detect the Sybil identities. This solution does not work well as nodes moving in group and exchanging messages can falsely be detected as a Sybil attack. If a single attacker is using multiple radios simultaneously, detecting a Sybil attack is very difficult. As the network becomes dense, that is, number of nodes per area increase, detection of Sybil nodes will be very difficult and false-positive rate will increase significantly. False-positive rate is the rate at which more number of non-Sybil identities (honest nodes) is falsely identified as Sybil identities.

In passive detection, any node that wants to detect a Sybil node records the identities of all the nodes it overhears broadcasting within its range over some interval of time. This observation period depends on the mobility of nodes and the area to be covered. Only identities and position of nodes at regular interval of time are stored. After the observation period, it is very easy to correlate the nodes. Correlation can be carried out by these steps:

1. If two nodes i and j are observed together in a profile, then affinity behavior between these nodes are calculated in terms of T_{ij} (time interval in which nodes i and j observed together), L_{ij} (time interval in which either i or j were alone), and N (number of intervals in observation period).

$$A_{ij} = (T_{ij} - 2L_{ij}) \frac{T_{ij} + L_{ij}}{N}$$

2. After computation of affinity function between each pair of nodes, the observer constructs a graph with identities as vertices and affinity values as weight of edges. Only those edges are included whose weights are greater than a threshold value based on many parameters such as observation period, total number of nodes, mobility of nodes, and so on.
3. Depth First Search (DFS) is run over each vertex to find connected components, each component representing a Sybil attacker. There may be several connected components, but the largest one is chosen as a Sybil attacker, the underlying assumption is that there is one Sybil attacker per network.

Many parameters need to be determined for proper functioning of passive detection of a Sybil node. These include sampling rate, length of this sample period, threshold value, changing parameters in response to change in network conditions such as size, topography, mobility of nodes, simulation time, and so on. Computational power is another constraint. If the mobility of nodes in network is very high, observation period should be long. Affinity value A_{ij} should be so long that each identity of an attacking node must transmit enough packets within given observation period to participate in the network activity. In smaller topographies, it is very difficult to differentiate Sybil identities from real nodes because there is insufficient mixing of nodes. As the topography size increases to an optimal value, the number of meaningful observations increases and false-positive rate decreases significantly. Beyond this optimal value, the number of observations that a single node can make is reduced and as nodes are spread apart, accuracy of identifying the Sybil identities decreases.

12.6.4 Passive Detection through Multiple Observers

A single observer is limited in the area that can be monitored. In sparse networks, accuracy of Sybil node detection by single observer decreases as nodes cannot hear its neighbors because of low transmission range of nodes of VANETs. So, multiple trusted nodes [25] are required to share their observation of traffic (in different area and at different time) with each other to improve the Sybil node detection rate. It is assumed that these multiple observers can trust each other. Number of exchanges between these trusted observers within the observation period should be decided for detection of maximal set of Sybil identities in the whole network. There should be efficient connectivity between these trusted observers also. These passive observers can be roadside base stations or leaders of geographical groups in the simulation area. For multiple observers, affinities can be calculated as:

$$A_{ij} = (T_{ij} - 2L_{ij})w_{ij}$$

$$w_{ij} = \begin{cases} (T_{ij} + L_{ij})/N & \text{if } (T_{ij} + L_{ij}) < N \\ 1 & \text{if } (T_{ij} + L_{ij}) \geq N \end{cases}$$

where G is total number of nodes sharing observations with each other, $T_{ij}(n) = $ number of intervals in which nodes i and j were observed together by node n. $L_{ij}(n) = $ number of intervals in which either i or j were observed alone by node n. N is total number of intervals in observation period.

$$T_{ij} = \sum T_{ij}(n); \quad n \in G$$

$$L_{ij} = \sum L_{ij}(n); \quad n \in G$$

It is assumed that all the nodes (including attackers) are using a single radio channel. Sybil identities are different from group of honest nodes in two ways. First, a physical position of all the Sybil identities is same. Secondly, Sybil identities will try to send the messages serially on the channel so that they can send all the messages without any loss. Sybil attacker smartly synchronizes all the messages that are to be sent on the network. On the other hand, the honest nodes moving in the group can transmit the messages in parallel because all the nodes are autonomous. So, by analyzing whether nodes are accessing channel serially or in parallel, Sybil identities can be identified. It is very important in the case of multiobserver Sybil attack detection that all the observers are trusted. If attacker has spoofed the identity of anyone of the observer in the group, this detection technique will not work properly. So, building trust between these multiple observers is the key to success of this detection method.

12.6.5 Sybil Node Detection by Propagation Model

Sybil attack can also be detected by using a propagation model as described in [27]. This is an active detection technique. In this technique, the received signal power from a sending node is matched with its claimed position. By using the propagation model, received signal power can be used to calculate the position of the node. If both the positions (calculated and claimed) do not match, this may be a Sybil node. This technique is unsuitable for detection of a Sybil attack as a malicious node can use the same propagation model to compute the transmission signal strength required to fool detection system in estimating the next position of the node. The above-technique is vulnerable to fabricated measurements by Sybil nodes. VANET infrastructure like roadside base stations, high mobility and traffic pattern in conjunction with statistical analysis of the physical position of node can be used to derive the corresponding trajectory.

Every node in a VANET periodically broadcasts beacon message containing its identity (ID), position (obtained from GPS). Such a node is called a *claimer*. This is in addition to periodic exchanges of neighbor table, signal strength measurement, beacon number, and signed certificate issued by a centralized authority. Every node stores complete network information in its table. Neighboring nodes performing measurement of signal strength of a claimer are termed *witness*. Received signal strength of beacon message is compared with that estimated by propagation model derived from observations. The nodes verifying the positions of beacon messages are called *verifier*.

Verifying interval may be longer than the beacon interval. If estimated position of claimer is far removed from claimed position, it is regarded as a suspect node.

Accuracy of signal strength measurement is limited. It cannot be concluded that a Sybil attack is happening even after identifying the suspected node(s). All the neighboring nodes need to be verified at the same time. Only when physical positions of more than two nodes are same, it is a Sybil attack. Claimer's physical position can be localized through measured signal strength of all witnesses. If some of these witnesses are malicious, they can localize the position incorrectly.

Specific traffic patterns and base station supports in vehicular environment can be used for removing Sybil witness. It is assumed that base stations are sparsely distributed along the roads. Base stations can issue a position certification containing location and time-stamp for each vehicle passing nearby. This certificate can be used to verify that a vehicle was near the base station at any particular time. For example, if two vehicles coming from opposite sides cross each other at some point on the road, it can be ensured that they are coming from opposite directions through exchange of their position certifications. It is assumed that witness nodes for a claimer need to be from opposite traffic flow. In such a scenario, it can be ensured that the witnesses coming from opposite sides are physical rather than Sybil vehicle as Sybil vehicles coming from opposite sides will fail the verification of base station certification. The remaining vehicles coming from opposite sides are actually honest nodes. If these two assumptions work well, witness will consist of only physical vehicles and Sybil vehicles will be excluded. But, in real-world VANETs, it is not necessary that infrastructure is always present. It may happen that some places do not have any fixed base station support.

Signal strength approach has a limited accuracy. Small-scale attacks cannot be detected. It is very difficult for a malicious vehicle to change signal strength distribution. Any change in signal strength will, therefore, be detected by a receiver. If each vehicle is given limited space, malicious vehicles can fabricate only few Sybil nodes. More realistic radio propagation model is required to support high mobility of nodes in VANETs.

12.6.6 Active Detection by Position Verification

This technique [28] uses front and rear radars in each vehicle for detection of obstacles in the front and rear of any vehicle. Position-based cells [29] are the basis of this approach. Area is divided into equal-sized location-based cells. Each vehicle in the cell can directly communicate with every other vehicle in the cell. Each vehicle broadcasts a beacon message (GPS position, identity, etc.). After receiving this beacon message, the receiver matches its GPS position with the position calculated by the radar. If these two positions match, the message is accepted and sender is labeled an honest node; otherwise, it is concluded that the sender is a Sybil node.

1. *Network model:* A GPS can determine its own position. It is also used for time synchronization between different units of a VANET. It is assumed that the radar is omni-directional. Neighboring nodes can be found by exchange of beacon messages with one hop neighbors. In order to verify the position information of neighbors detected by radar with its GPS coordinates, nodes in the VANETs should be divided into different cells that can be either dynamic or position based. Dynamic cells [29] are flexible but inefficient. Vehicles map their position to these cells. GPS coordinates of the cell are compared with the center position and diameter of cell to get their host cell. Position-based cells can be used to build a communication network. This can be understood by taking a simple example. Assume that a cell is located at every 150 meter on road. This cell's radius is 75 meters, and all the vehicles inside this cell can directly communicate with other nodes of the cells without any routing algorithm.

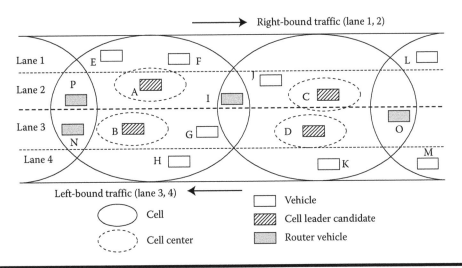

Figure 12.5 Representation of cell leader and cell router.

Transmission range of radar is same as the diameter of cell so that radar can directly detect all the vehicles in the cell.

Figure 12.5 depicts the cells on road. Rounded circles near to cell centers and vehicles in the overlapped area behave as routers. There are four vehicles A, B, C, and D in the center of the cell and four vehicles E, F, G, and H in the overlap area. The size of this intersection area is decided in advance at the time of creation of cells. This overlap size depends on the size of the cell and road conditions. The need of routers depends on transmission range of radar and overlap size. On a highway, overlapping area may be large to contain more vehicles as routers.

If the size of cell is very large, overlap may only be limited to a small portion of the cell. If the transmission range can directly reach the cell leader directly, there is no need of a large overlapped area. Each vehicle can decide the width of an overlapped region between two cells by consulting the loaded digital map partitioned into different cells. Each vehicle decides its own position by determining its position from GPS and preloaded digital map. Within each cell, there is a unique cell leader for each direction. A vehicle close to the cell center is chosen as the cell leader. A score-based system can be used to determine the position of the cell leader. Each vehicle gets a score depending on the distance from the cell. The vehicle with highest score becomes the cell leader and announces its new designation. If more than one vehicle gets the same score, the vehicle with a lower ID gets precedence.

Every vehicle periodically sends position information to cell leader. This position information is also received by all cell members. The position of other vehicles in their direct line of sight is also included to prevent collisions. Next transmission time is decided by measuring the position of vehicle with respect to the cell leader. Neighboring vehicles do not broadcast their coordinates if they agree with transmitting vehicle. All members of the cell aggregate this information to build the traffic view. If a new vehicle enters the system, it waits for some minimum time to collect information transmitted by other members of the cell and learn the cell leader's ID. It also detects its neighbors by activating its radar. If there is no response on address of cell leader even after the end of waiting time, the new vehicle takes over the role of a cell leader.

2. *Local detection of malicious node:* Security at global level can be achieved by observing history of vehicles movements and verifying the consistency of positions. Any member of the cell can verify GPS coordinates of a node by comparing them readings from its radar. If the broadcasted GPS coordinates match with those determined by the radar, this message is accepted or else it is a malicious vehicle position [30]. A cell leader periodically broadcasts the traffic view within the cell. This view can also be ascertained by all the vehicles in this cell as radar range is not less than the diameter of the cell or it can be constructed through message exchanges taking place in the cell. If there is a mismatch, a cell leader is malicious and it can be detected assuming most of the vehicles in the cell are trustable.

3. *Global detection of malicious node:* This is done through a verification process of data transmitted by a suspect vehicle. Record verification can be either reactive or proactive. Any vehicle in the cell or cell leader itself can initiate the verification process. Reactive verification is done when a vehicle is in dispute. Proactive means that a vehicle can randomly pick a record and request verification. Usually, a cell leader initiates the verification process but any other vehicle can initiate if the cell leader fails to do so. Every vehicle in the cell knows exact positions of all the vehicles in the cell. A node can start verification process by sending query about a specified vehicle to its neighbors. After receiving the response from neighbors and computing their positions, the requester comes to an agreement about the entire position of the neighbor. Now radar-detected data and data received from trusted neighbors are compared. If similarity value is above the threshold value, then these data are accepted, otherwise they are dropped.

In global security implementation, a vehicle without history is not trustable. History table is used to verify the position based on the movement consistency. If there is any inconsistency, the particular record is more likely to be picked for verification. If a node is found malicious [16,30] after verification, they are evicted from VANETs. It is worth noting that messages broadcasted by a cell leader are picked up by approaching traffic rather than the one that has already passed that location. A vehicle in nearby cell has the highest probability to verify malicious vehicles positions. Within each cell, local security is maintained by detecting and removing Sybil nodes. Globally, Sybil nodes are detected and removed by using history of nodes within the cells.

This technique does not ensure that an attack detected is a Sybil attack. The Sybil detection requires another step of checking whether identities are also spoofed along with positions. Errors in this technique arise from imprecision of GPS and radar systems.

12.6.7 Sensor-Based Position Verification

These techniques use multiple sensors rather than using fixed infrastructure to detect malicious behavior of the nodes in the network [31]. Sensor data are used for verification of position information determined by GPS receivers and detection of false position information. A trust value is calculated to decide if a node is trustworthy. These sensors work collectively for building trust values. This Sybil detection approach is active in nature. Some of the techniques are as follows:

1. *Acceptance range threshold (ART):* Depending on the channel radio used in the network, a maximum acceptance range threshold is decided. All the beacon messages coming from nodes claiming to be at distance larger than ART are discarded.

2. *Mobility grade threshold (MGT):* It is assumed that nodes can move only with some well-defined maximum speed. Beacon message also contains time-stamps. Average speed of a node

is computed from consecutive beacon messages it broadcasts. If this speed exceeds MGT, messages from this node are ignored.

3. *Maximum density threshold (MDT):* This threshold is based on the assumption that only a limited number of nodes can reside in a certain area. If the number of nodes exceeds this MDT, further position beacons from this area are rejected. This threshold can be used to prevent Sybil attack through creation of large number of virtual nodes and thwart attempts to take control of the whole network.

4. *Map-based verification:* Many vehicles use navigation systems where street maps are accessible by position verification system. In this verification, the system can check whether the vehicles are physically present on the streets or not.

The above-techniques use autonomous sensors for verification of position-faking nodes. Cooperative sensors can also be used for communicating and exchanging information in order to detect position-faking nodes. This technique can be used only when autonomous sensors indicate that position faking exists in the network but cannot detect malicious node-faking position. In this case, sensors cooperate with each other by sharing their results with other sensors. For example, nodes proactively (periodically) exchange their neighbor tables and position verification can be done on a demand basis (reactively).

12.6.8 Analysis of Sybil Detection Solutions

Trusted certification technique is the only technique with the potential of preventing Sybil attacks. Trusted certification relies on a centralized authority that ensures that each entity is assigned only one identity. However, for privacy maintenance, these identities are changed from time to time. This technique is expensive and is not scalable, that is, performance degrades in large-scale networks. Certifying authority also needs to ensure that lost identities are revoked. In situations where no infrastructure is present, trust establishment is dynamic. Resource testing technique is also ineffective because of the amount of time needed and requirement of other verifying devices. The assumption of limited computational power per vehicle is not realistic as an attacker can have more computational resources when compared with honest nodes.

Positions-based techniques, employing GPS, radar communication, or sensors, are limited by tolerance and imprecision of these devices. Propagation model-based techniques suffer from imprecision of devices. By varying transmitted power, an attacker can easily defeat such models. In other approaches, nodes are passively keeping an eye on other nodes in VANETs and creating their profiles. Such solutions will work only if majority of nodes are honest. If majority of nodes are malicious, honest nodes have to pay off the penalty. The nodes cooperating in detecting the Sybil attack in a VANET should have their clocks synchronized, mechanism to establish trust and, then, trusted nodes may be able to establish their relative positions. They will estimate the distance from each sender by comparing the time. In this way, a Sybil attacker node can be detected.

Using VANET features such as traffic pattern and mobility constraint, Sybil nodes can be identified if there is a fixed infrastructure comprising of fixed base stations on the sides of road. These base stations issue time-stamp-based certificates to all the vehicles passing them. Spoofing multiple locations, as in Sybil attack, are difficult in such a system. All solutions are inefficient and lack in some aspect for a complete VANET security solution. Once the Sybil node is detected by any technique, it is evicted from a VANET [30].

12.7 Future Directions for Research

Sybil attack in VANETs is initiated mainly by spoofing the identities of nodes. In VANETs, safety messages are sent to nodes in a specific area where the nodes can directly be affected from critical situations. Hence, geographic routing based on the position of nodes can be useful in VANETs. Position information of vehicles plays an important role in the detection of a Sybil attack and localization of an attacker. Position verification technique should work well even in the presence of some malicious nodes. In spite of these facts, all the proposed methods have some limitations. Dynamic trust establishment is the main approach for detecting a Sybil attack as it does not require a centralized entity. VANETs should support some mechanism for mutual authentication of nodes and access control. Dynamic certification approach for large-scale networks is one of the unexplored areas in VANETs. The main goal is to efficiently use VANET features such as mobility and RSU availability for detection of a Sybil attack. It is an open issue to detect all the Sybil nodes when more than one Sybil attackers exist. If the number of Sybil attackers increases, they may overtake the entire system. Attackers can target the trusted nodes having long trust history. Modeling radio propagation in a VANET environment with high mobility feature is another open area.

12.8 Conclusion

VANETs are more prone to security threats when compared with fixed wired networks [11,12,15]. The broadcast nature of the wireless channels, the absence of a fixed infrastructure, the dynamic network topology, and the self-organizing characteristic of the network increase the vulnerabilities of a VANET. Security has to be taken into account at deployment of link layer and network layer protocols in a VANET.

Theoretically, a centralized trust establishment technique (i.e., popularly used in fixed wired/wireless networks) can be used for maintaining trust in VANETs, but practically dynamic trust establishment is a better solution. It is difficult to implement dynamic trust because of the very nature of a VANET. Self-organized trust establishment mechanisms through the mutual communication between the vehicles are another possibility. Efficient algorithms adapting to VANET characteristics is a prerequisite for dynamic trust establishment. Yet another issue is that protecting privacy while maintaining security in VANETs is a difficult task.

In this study, we have presented Sybil attacks, in which one adversary entity controls many different identities. We also discussed challenges posed by this attack and proposed an analysis of various solutions of detection of Sybil attacks. Sybil attack is a problem for peer-to-peer networks, mobile networks, and reputation systems. Different solutions can detect, limit, or prevent attacks in different scenarios. In this chapter, we limited our discussions about the applicability of Sybil attack in VANETs only. The current efforts of research community are to support VANETs with practical security solutions, which can cope with the challenging environment of VANETs. A robust security infrastructure has to consider all aspects of a Sybil attack and should be easy to integrate with the existing system in a cost-effective manner.

Terminologies

VANET—Vehicular *Ad hoc* Network
MANET—Mobile *Ad hoc* Network

ITS—Intelligent Transportation system
GPS—Global Positioning System
RSU—Road-side Unit
OBU—On-board Unit
TPD—Temper Proof Device
ELP—Electronic License Plate
CA—Central Authority
MAC—Medium Access Control
IP—Internet Protocol
HMAC—Hashed Message Authentication Code
PKI—Public Key Infrastructure

Questions and Sample Answers

1. What is the role of trust establishment to detect Sybil attack in VANETs? Describe various trust establishment methods.
   Detection of Sybil attack is based upon trust, so trust establishment plays large role in detection of Sybil attack. Profile of all the nodes are stored by some trusted nodes and behavior of nodes are analyzed. If the behavior deviates from normal behavior, then it is detected as a malicious node. Trusted nodes are decided cooperatively by other nodes in the network. Parameters for selection of trusted nodes are total time spent in the network, their behavior history, and so on. If some suspicious nodes become part of this set of trusted nodes, then most of the malicious/Sybil nodes would not be detected. Two techniques are used for trust establishment, static and dynamic. Static trust establishment approach is not suitable in VANET scenario because of no assurance of presence of infrastructure always. Dynamic approach is suitable for VANETs as topology of the network is continuously changing.

2. Explain how self-certified Sybil-free pseudonyms prevent Sybil nodes in VANETs.
   Self-certified Sybil-free pseudonyms are used to provide privacy-friendly Sybil freeness without requiring continuous availability of trusted centralized authority. CA is required only at the time of bootstrapping a Sybil-free domain. Identity domains are constructed at the initial setup of the network topology when CA is available. Each user has unique pseudonym in every identity domain. Sybil freeness can be propagated to other identity domains as the nodes propagate to other domains. All the nodes acquire a membership certificate from CA. Each user can create self-certified pseudonym per identity domain. Pseudonyms issued for different domains are mutually unlinkable and even they cannot be linked to their membership certificate by the CA. If some nodes have the same pseudonyms, then automatically it is detected. Hence, Sybil-free pseudonyms are useful to detect and prevent Sybil nodes in VANETs.

3. Imagine the Sybil attacker spoofs the random positions or correct positions of nodes. Will there be any difference in performance of VANETs for this implementation?
   If the Sybil attacker is spoofing random positions, that is, it is spoofing the ID of some other node and not using the actual position of this node, then the total generated packets and dropped packets will be more when compared with ideal case. On other side, if a Sybil attacker will spoof the ID of a node along with its actual position, then the number of generated and dropped packets will decrease. The reason behind is that every node has some restriction on a number of transmitted and received packets.

4. Why a traditional PKI-based technique is inefficient for preventing Sybil attacks in VANETs? PKI-based certificates contain only key information and do not contain corresponding vehicle's unique physical information. This technique is vulnerable to Sybil attack because a malicious vehicle can use stolen valid key pair and certificate. It takes very long time to establish Vehicular Public Key Infrastructure (VPKI). Centralized key management and certificate authorization is not realistic in VANET environment because of number of vehicles with different manufacturers and regions exist at the same time. The main hurdle in implementing the VPKI technique is key distribution and certificate management, including issuing, storing, and revocation. It is very time-consuming also. It is very easy to track and collect vehicles behavior because of the use of long-term key pairs and certificates.

5. Assume a VANET scenario in a simulator (NCTUNS or NS-2) where 10 nodes are moving on a road segment. Create three Sybil nodes as part of this VANET and incorporate them in the network. Analyze the impact of these Sybil nodes on the performance of VANET in terms of (a) number of packets transmitted but not received, (b) throughput of transmitted packets among various entities, (c) delay in transmission packets, (d) repeat the same experiment with varying number of Sybil nodes from 3 to 5.

   Ten mobile nodes can be created in any network simulator, and path of movement of these nodes can be specified by traffic simulator. Incorporate some routing protocol in this network where each node is maintaining a record of its neighbors. For creating three Sybil nodes, ID of these nodes can be copied from the IDs of its neighbors. It differs from one technique to other depending on the way how vehicular nodes are created. Values of different parameters can be recorded in ideal case where no Sybil nodes exist. On the other hand, these parameters can be analyzed in the presence of a Sybil attack. Definitely, an average number of generated packets and dropped packets will increase and delay will be more in a Sybil attack scenario. As the number of Sybil nodes increase, the generated and dropped packets also increase.

6. Consider certain number of nodes on one road segment in a VANET. In this VANET, few nodes are transmitting packets independently to each other. One of the possibilities for a Sybil attack could be the transmission of simultaneous packets when the nodes are moving. Are there any other symptoms for confirmation of the Sybil attack?

   A Sybil attack can be confirmed by verifying the physical position of the transmitting nodes. If physical position of all the Sybil identities is same, then the Sybil attack is confirmed. Sybil identities try to send packets serially on the channel so that packets can be delivered without any loss, while honest nodes moving in the group can transmit the messages in parallel because all the nodes are autonomous.

7. Assume a VANET consists of 50 nodes. Out of 50 nodes, a group of 10 trusted nodes are sharing their observation for detecting Sybil nodes. The time interval at which a node observes another node is 5 minutes. Simulate this situation of network condition for a time period of 60 minutes. Comments on the role of observation period in a Sybil attack detection.

   There will be 12 time intervals required for the above case. In question, the observation period is set to 60 minutes. If numbers of observing intervals increase in the observation period, then more number of false Sybil identities will be found and by decreasing the number of intervals, very few false Sybil identities will be found. There is a symmetric relationship between the observation period and the number of observing intervals. This can be proved by applying various observation periods and time interval in affinity function equation given in Section 12.6.3.

8. What is the main benefit and drawback of detecting a Sybil attack by propagation model technique?
Detecting Sybil attack by using suitable propagation model has many advantages. It can estimate the physical position of Sybil node and its corresponding trajectory also. It is very easy to find all the Sybil nodes originating from the same physical malicious node. It does not require any special hardware rather it is based on statistical analysis of signal strength by which a sender is transmitting the message. It gives good results in large-scale VANETs because limited memory space is occupied by each vehicle. So there is a limit of creating Sybil identities. Owing to the limited accuracy of this approach, it is not suitable for small-scale VANETs. This is the main drawback of this technique.

9. Assume that a Sybil attack detection algorithm does not need the support of other vehicular nodes. The only requirement is that each node of this VANET should store authorized infrastructure's signature, which is gathered during movement of nodes within the infrastructure. The signature is in the format of <infrastructure$_i$, time>. Explain how this detection algorithm detects a Sybil attack efficiently than other methods.
Suppose in this scenario, there are five vehicles and five RSUs.

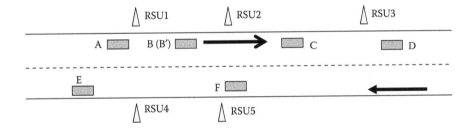

A table of authorized RSUs' signatures can be maintained which is collected by vehicular nodes in motion.

Time	Node A	Node B	Node B$'$	Node C	Node D
T0	None	RSU1	RSU1	RSU2	RSU3
T1	RSU1	RSU2	RSU2	RSU3	None
T2	RSU2	RSU3	RSU3	RSU5	None

If we observe this table, then it will be analyzed that nodes B and B$'$ are having the same signatures which cannot be possible because at no point of time two vehicles will have the same signatures. It is very easy to detect that B and B$'$ are physically same, that is, a Sybil node.

10. What is the limitation of techniques described in Section 12.6.6? Suggest a modification for dealing with this limitation.

The technique described in Section 12.6.6 verifies only the position of a receiver. If the claimed position of a sending node does not match with the position detected by a receiver radar, then it is detected as malicious node. It is not ensured that the detected attack is a Sybil attack. One more step is required to confirm whether identities are also spoofed along with the positions. Accuracy of this approach is limited because in VANETs vehicles are moving with high mobility. There is also imprecision of GPS and radar systems in this technique. We should combine this technique with some other detection techniques like multiple observer approach. After detecting the malicious node by position verification approach, it should be detected by a set of trusted nodes. If majority of trusted nodes agree to a point, only then it should be concluded as a Sybil node.

Author's Biography

Jyoti Grover received her BE in computer science and engineering from MD University, Rohtak, India, in 2002, and MTech degree in computer science and engineering from GJ University, Hisar, India, in 2004. She is currently pursuing her PhD in computer engineering from Malaviya National Institute of Technology, Jaipur, India.

Dr. M.S. Gaur is a professor of computer engineering, Malaviya National Institute of Technology, Jaipur, India. He received his PhD in electronic and computer science from University of Southampton, United Kingdom, in 2004. His current research focuses on network security, distributed systems, network on chip, and simulation. He has published 5 papers in referred journals and 21 papers in international conferences. He is member of review and program committees of various conferences.

Dr. V. Laxmi received her PhD in electronic and computer science from University of Southampton, United Kingdom, in 2003. She is currently a reader at the Department of Computer Engineering, Malaviya National Institute of Technology, Jaipur, India. Her current research interests include cryptographic system, biometrics, and image analysis. She has published 6 papers in referred international journals and 32 papers in referred international conferences.

References

1. J. Guo and N. Balon, *Vehicular Ad Hoc Network and Dedicated Short Range Communications*, 2006. University of Michigan-Dearborn. Available at: http://www.nathanbalon.com/project/cis95
2. H. Fusslur, M. Transier, S. Schnaufer, and W. Effelsberg, Vehicular ad hoc network: From vision to reality and back, *The Fourth IEEE/IFIP Annual Conference on Wireless On demand Network Systems and Services*, Vol. 4, 80–83, 2007.
3. A. Boukerche, *Algorithms and Protocols of Wireless and Mobile Ad hoc Networks Book*, 1st edition, John Wiley & Sons, Inc., 2009.
4. S. S. Manvi and M. S. Kakkasageri, Issues in mobile ad hoc networks for vehicular communications, *IETE Technical Review* 25(2), 59–72, 2008.
5. M. L. Sichitiu and M. Kihl, Inter-vehicle communication systems: A survey, *Communications Surveys & Tutorials, IEEE*, 10(2), 88–105, Second Quarter 2008.

6. D. Jiang and L. Delgrossi, IEEE 802.11p: Towards an international standard for wireless access in vehicular environments. In *IEEE Vehicular Technology Conference, VTC Spring 2008*, Singapore, pp. 2036–2040.
7. Y. Qian and N. Moayeri, Design secure and application-oriented VANETs. In *Proceedings of IEEE VTC'2008- Spring*, Singapore, May 11–14, 2008.
8. M. El Zarki, S. Mehrotra, G. Tsudik, and N. Venkatasubramanian, Security issues in a future vehicular network. In *Proceedings of Euro Wireless 2002*, Florence, Italy, February 2002.
9. K. Plößl, T. Nowey, and C. Mletzko, Towards a security architecture for vehicular ad hoc networks. In *Proceedings of the First International Conference on Availability, Reliability and Security (ARES)*, April 2006.
10. T. Leinmuller, E. Schoch, and C. Maihofer, Security requirements and solutions concepts in vehicular ad hoc networks. In *Proceeding of Fourth Annual Conference on Wireless on Demand Network Systems and Services*, 2007.
11. P. Papadimitratos, V. Gligor, and J.-P. Hubaux, Securing vehicular communications—assumptions, requirements, and principles. In *Proceedings of the Workshop on Embedded Security on Cars (ESCAR) 2006*, November 2006.
12. M. Raya and J.-P. Hubaux, Securing vehicular ad hoc networks. *Journal of Computer Security*, 15(1), 39–68, 2007.
13. A. Aijaz, B. Bochow, F. Dtzer, A. Festag, M. Gerlach, R. Kroh, and T. Leinmuller, Attacks on inter-vehicle communication systems—an analysis. In *Proceedings of the 3rd international Workshop on Intelligent Transportation (WIT)*, March 2006.
14. M. Raya, P. Papadimitratos, and J.-P. Hubaux, Secure vehicular communications, *IEEE Wireless Communications Magazine*, 13(5), 8–15, 2006.
15. P. Papadimitratos, L. Buttyan, J.-P. Hubaux, F. Kargl, A. Kung, and M. Raya, Architecture for secure and private vehicular communications. In *Proceedings of the 7th International Conference on ITS Telecommunications*, June 2007.
16. P. Golle, D. Greene, and J. Staddon, Detecting and correcting malicious data in VANETs. In *VANET '04: Proceedings of the First ACM Workshop on Vehicular Ad Hoc Networks*, 2004, pp. 29–37, ACM Press.
17. J. R. Douceur, The Sybil attack. In *Proceedings of the International Workshop on Peer to Peer Systems*, March 2002, pp. 251–260.
18. J. Newsome, E. Shi, D. Song, and A. Perrig, The Sybil attack in sensor networks: Analysis and defences. In *Proceedings of International Symposium on Information Processing in Sensor Networks*, April 2004, pp. 259–268.
19. P. Wex, J. Breuer, A. Held, and T. Leinmüller, Trust issues for vehicular ad hoc networks. In *67th IEEE Vehicular Technology Conference (VTC2008-Spring)*, Marina Bay, Singapore, 2008.
20. A. Khalili, J. Katz, and W. Arbaugh, Toward secure key distribution in truly ad-hoc networks. In *Proceedings of the IEEE Workshop on Security and Assurance in Ad hoc Networks, in Conjunction with the 2003 International Symposium on Applications and the Internet*, Orlando, FL, January 28, 2003.
21. L. A. Martucci, M. Kohlweiss, C. Anderson, A. Panchenko, Self-certified Sybil-free pseudonyms. In *WiSec'08: Proceedings of the First ACM Conference on Wireless Network Security*, New York, NY, USA, ACM Press, 2008, pp. 154–159.
22. J.-P. Hubaux, S. Čapkun, and J. Luo, The security and privacy of smart vehicles, *IEEE Security and Privacy*, 4(3), 49–55, 2004.
23. X. Sun, X. Lin, and P. Ho, Secure vehicular communications based on group signature and ID-based signature scheme. In *Proceedings of the IEEE International Conference on Communications*, 2007.
24. A. Khalili, J. Katz, and W. A. Arbaugh, Towards secure key distribution in truly ad-hoc networks. In *Proceedings of IEEE Workshop on security and Assurance in Ad-Hoc Networks*, 2003.
25. C. Piro, C. Shields, and B. N. Levine, Detecting the Sybil attack in mobile ad hoc network. In *Proceedings of the International Conference on Security and Privacy in Communication Networks*, 2006, pp. 1–11.
26. B. N. Levine, C. Shields, and N. B. Margolin, A survey of solutions to the Sybil attack. *Tech report 2006-052*, University of Massachusetts Amherst, Amherst, MA (October 2006).

27. B. Xiao, B. Yu, and C. Gao, Detection and localization of Sybil nodes in VANETs. In *Proceedings of the Workshop on Dependability Issues in Wireless Ad Hoc Networks and Sensor Networks (DIWANS '06)*, Los Angeles, CA, USA, pp. 1–8, September 2006.

28. G. Yan, S. Olariu, and M. C. Weigle, Providing VANET security through active position detection, *Computer Communications*, 31(12), 2883–2897, 2008.

29. M. Raya, A. Aziz, J.-P. Hubaux, Efficient secure aggregation in VANETs. In *Proceedings of the ACM Workshop on Vehicular Ad Hoc Networks (VANET)*, Los Angeles, CA, 2006, pp. 67–75.

30. M. Raya, P. Papadimitratos, I. Aad, D. Jungels, and J.-P. Hubaux, Eviction of misbehaving and faulty nodes in vehicular networks, *IEEE Journal on Selected Areas in Communications*, Special Issue on Vehicular Networks, 2007.

31. T. Leinmüller, E. Schoch, and F. Kargl, Position verification approaches for vehicular ad hoc networks, *IEEE Wireless Communications Magazine*, 13(5), 16–21, 2006.

WIRELESS SENSOR NETWORK SECURITY

Chapter 13

Key Management Schemes of Wireless Sensor Networks
A Survey

Syed Muhammad Khaliq-ur-Rahman Raazi,
Zeeshan Pervez, and Sungyoung Lee

Contents

13.1 Introduction

Key management is the most important aspect of security in any type of network. Other aspects of security such as authentication and privacy also depend upon key management. Concept of key in computer science and wireless sensor networks is same as that in our daily lives. Basically, key is a secret that is known to relevant parties only just like in a household, where only insiders have the house's keys. Relevant parties use keys to conceal secret information that they need to exchange with each other. Apart from that, keys are also used to identify parties which have permission to access certain information.

In real-life scenario, physical keys are used to lock a room, a bag, or a cupboard. Only authorized persons are given copies of the key, which can open the lock. In this way, things inside lock are kept confidential from other users and only authenticated users can access inside the lock. The same concept is applied in the scenario of computer science and computer networks but in a different way.

In computer network, data and information, which are not something physical, need to be secured. They are a stream of bits stored in electronic devices and exchanged between two or more parties through a wire or a wireless medium. In wireless medium, it is impossible to prevent unauthorized parties from eavesdropping and impersonating authorized parties. In wired medium, it is nearly impossible. In order to keep these streams of bits confidential from an outsider, sender performs certain mathematical operations on these bits and sends them to the receiver through a wired or wireless medium. Upon receiving this information, the receiver performs some mathematical operations, which are reverse of the mathematical operations performed by the sender, to produce the original stream of bits. However, just performing some mathematical operations cannot secure important information. If an adversary knows the mathematical operations, it can crack confidential information very easily. Even if the adversary does not know the mathematical operations used for decryption in certain scenario, it can guess them by applying already-known mathematical operations of cryptography. If a pair of communicating nodes does not employ an already-known mathematical operation, then they have to agree upon new mathematical operations and their reverse every time they communicate, which further increases the complexity of the problem. In order to simplify this problem, random keys are used.

Keys are fixed length streams of random bits, which are known only to the authorized parties. Sender encrypts data/information in the key, that is, performs mathematical operations on data/information and key collectively. This produces a stream of bits that does not reveal any information about the original stream of bits. Only authorized parties can decrypt or come to know original data/information. This allows us to use those mathematical functions that are proven to be secure and have inverse operations, which can reveal original information with the help of key. In order to have keys readily available for every communication, keys need to be managed securely and efficiently. Key management of any computer network depends upon its characteristics, limitations, and applications.

Wireless sensor networks consist of a large number of low-cost, resource-constrained sensor nodes. Each node has to sense certain phenomena and forward its readings toward a central node, which is also called the base station. Figure 13.1 shows an example of a wireless sensor network. Base

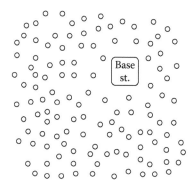

Figure 13.1 A generic wireless sensor network.

station uses sensed information according to the application requirements. Wireless sensor network applications include habitat monitoring, military surveillance, border monitoring, and health care.

Sensor nodes have low memory, computation, communication, and energy capabilities. One has to be very careful about resource consumption while proposing a solution for any problem in this domain. The same thing holds for key management also. Apart from making sensor networks as secure as possible, key management should introduce as less overhead as possible. Normally, this is a trade-off, which also depends upon the application scenario. In the next section, we will shed some light on the background of key management and wireless sensor networks so that the readers can understand the motivations behind different key management solutions proposed so far. After that, we will discuss the key management solutions proposed for wireless sensor networks so far. However, before discussing key management solutions, we will shed some light on the possible security threats or attacks that can take place in wireless sensor networks. Also, we will also discuss future research directions in key management before we draw conclusions from this chapter.

13.2 Background

Key management has been an important research area since the start of computer networks. Before that, research was mainly focused on the security of computing devices. Before the start of computer networks, it was emphasized that computers must have security programs that secure the computer itself and the peripheral devices that were used to transfer information from one point to another.

With the start of computer networking and later on its commercialization, importance of key management grew substantially. Information was shared among different computers through a wired or wireless network and it was not possible to monitor all the links. Also, types of attacks on confidential information grew substantially. All these circumstances invoked research for better security especially key management.

With growth of computer networks, information was also shared among different computers, who had not interacted with each other before. In order to secure such communications, Diffie–Hellman key exchange algorithm was proposed [1]. In Diffie–Hellman key exchange, two parties agree on two large prime numbers g and p and choose one random value each. For example, Alice chooses a random value x and Bob chooses a random value y. Then they compute a secret key in the way shown in Figure 13.2. While this protocol provided a secure way for two unknown parties to communicate, it did not have any authentication mechanism.

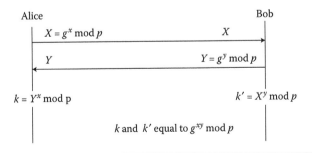

Figure 13.2 Diffie–Hellman key exchange algorithm.

1. Randomly select two large prime numbers "p" and "q".
2. Compute $N = p \cdot q$ and $\phi(N) = (p - 1)(q - 1)$.
3. Select any encryption key "e" such that $1 < e < \phi(N)$ and $\gcd(e, \phi(N)) = 1$.
4. Solve the following equation to find decryption key "d" :-

   $e \cdot d = 1 \bmod \phi(N)$

   where $0 \leq d \leq N$
5. Publish the public encryption key: $KU = \{e, N\}$.
6. Keep the secret private decryption key: $KR = \{d, p, q\}$.

Figure 13.3 Computation of public and private key pairs in RSA algorithm.

Soon after the Diffie–Hellman protocol, the concept of Public Key Cryptography was introduced by Rivest et al. [2]. Their protocol is normally referred to as Rivest, Shamir, and Adleman (RSA) protocol In RSA algorithm, every computer computes its public and private keys, as shown in Figure 13.3. Anything encrypted using a public key can be decrypted using the corresponding private key. By using this protocol, any two parties can communicate with each other using their public–private key pairs.

Although both RSA and Diffie–Hellman secure communication between two parties, who do not know each other, they do not have any mechanism to authenticate the other party. In early 1990s, when internet was growing rapidly, Kerberos [3] matured. Kerberos is a network authentication protocol based on secret key cryptography. Kerberos is based on the concept of trusted third party, which authenticates the communicating parties and provides keys or communication. Using Kerberos, a communicating party can be sure about the authenticity of the other communicating party. Apart from keeping information confidential, Kerberos also protects against the replay of data packets sent from sender to receiver. Apart from these advantages, Kerberos also has some drawbacks. In Kerberos, the trusted third party is a single point of failure. Also, clocks of communicating parties must be synchronized with the trusted third party for this protocol to work.

With the growth of internet in the era of 1990s, many applications, which used the internet, were developed. These applications used some predefined protocols such as Simple Mail Transfer Protocol (SMTP) for e-mail, File Transfer Protocol (FTP) for file sharing, and Hypertext Transfer Protocol (HTTP) for sharing information over the internet. In the design of internet, all application level protocols depend upon transport layer protocols for communication. Transport layer security (TLS) [4], which encapsulated and secured the transport layer of the internet, was proposed in late 1990s. TLS uses concepts trusted third party and public key cryptography to establish secure connections between the communicating parties.

With the evolution of the internet and growth in the use of personal electronic devices, electronic hardware also evolved rapidly. Size of computation and communication hardware such as processor, memory, and antenna kept shrinking along with its cost. This goes on until the current day. New models, with more capabilities and reduced size, are introduced in market, previous ones become obsolete and their cost falls to earth very quickly. Reduced size and reduced cost of hardware led to the use of computing devices in monitoring certain activity, phenomena, or biometric from human body. This led to the evolution of wireless sensor networks and opened a lot of new research challenges.

Wireless sensor networks consist of low-cost computing devices that can also sense their environment and forward sensed data to a nearby node. Sensors can be used to sense temperature, air pressure, salinity, moisture, movement, biometric, or any other phenomena. Exact number of sensor nodes used in an application depends upon the nature of application. Some applications, like border monitoring or military surveillance applications, require hundreds or may be thousands of sensor nodes. If we need to support such a huge number of sensors in a single wireless sensor network, we need to keep the cost of a single node to a minimum.

Reducing the cost of a sensor node has a direct effect on its computation power, communication power, and memory. Also, sensor nodes have limited battery life. In certain circumstances, like battlefields, it is impossible to recharge their battery. Limited battery life, lesser memory, low computation power, and small range of communication capability of a sensor node makes wireless sensor network a special type of computer network. Wireless sensor networks differ from the internet not only because of the limited capabilities of sensor nodes but also because of *ad hoc* and data-centric nature of wireless sensor networks. In wireless sensor networks, all sensed data are directed toward a central computing device called the base station. Also, old nodes may die down, new nodes may join the network and nodes may change their position during normal network operation of wireless sensor networks.

Research in network security matured with the spread of internet. Many security mechanisms were proposed even after Kerberos and TLS. However, these security mechanisms were best suited for the internet and not for wireless sensor networks. Main reason why these traditional security mechanisms were not viable for wireless sensor networks was that they were too heavy for simple sensor devices. Even these days, either sensor devices do not have enough computation power and/or memory to handle traditional security mechanisms or these mechanisms cause a lot of energy drainage from sensor nodes.

Inapplicability of traditional security mechanisms in wireless sensor networks opened a new research area in network security. A number of researchers have proposed novel and interesting ideas for key management in wireless sensor networks. Some ideas apply to simple sensor networks and some ideas apply to clustered sensor networks [5,6]. (As wireless sensor networks evolved, some researchers proposed clustering techniques to increase efficiency of wireless sensor networks.) In our discussion, we will also identify whether a scheme is applicable to clustered sensor networks or simple sensor networks or both. However, before proceeding with the discussion of key management schemes, it is better to discuss the types of attacks that can occur in wireless sensor networks. Discussion of attacks against security will make it easy for us to identify strengths and weaknesses of various key management schemes.

13.3 Security Threats in Wireless Sensor Networks

Main goal of a key management scheme is to ensure confidentiality of information. Also, keys can be helpful in authenticating legitimate nodes. An adversary may try to crack secret key and extract confidential information from the messages exchanged between communicating nodes. If keys are used for authentication purposes, adversary may try to act as a legitimate node and try to extract confidential information from other nodes. While trying to crack a secret key, adversaries try to learn message patterns and guess the secret key. Also, they try to save some encrypted messages, which they can replay later on. In order to prevent adversaries from guessing secret keys, it is important to refresh keys at appropriate time intervals. Time intervals depend upon frequency of communication and frequency of key usage.

Apart from trying to crack secret information, adversary can also harm a sensor network in several other ways. It can try to jam wireless signals of a sensor network. Also, it can try to create noises and disrupt communication. In other words, adversary can carry out denial-of-service (DoS)

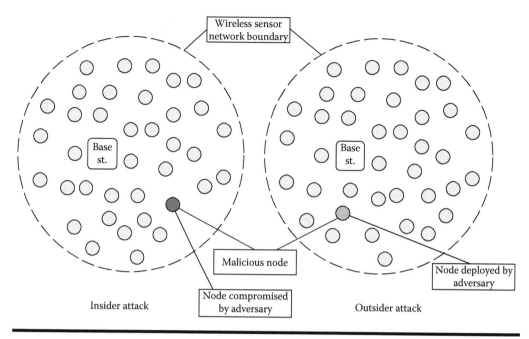

Figure 13.4   Insider and outsider attack scenario.

attacks. Apart from that an adversary can try to drain sensor nodes' energy by initiating bogus messages or replaying old messages. Although many of such attacks, like signal jamming, cannot be handled by key management schemes, we think it is important to list them at least. Readers can refer to the security mechanisms at physical layer to find remedy of jamming attacks. Zia and Zomaya [7] classify all security attacks in wireless sensor networks into four classes: interruption, interception, modification, and fabrication. Interruption is when a communication link in interrupted. Interception takes place when a sensor node or its data are compromised. In modification, adversary gains access and tampers with the data. Finally, fabrication takes place when an adversary injects false data into the network. References [8–10] are also very helpful resources for studying about security attacks that can take place in wireless sensor networks.

Broadly speaking, we can classify attacks on wireless sensor networks into two categories: outsider attacks and insider attacks. In an outsider attack, adversary is not a part of the network. For example, jamming attack is carried out on physical layer of the sensor nodes by a node, which is not part of the sensor network so it is classified as an outside attack. In an insider attack, an insider node (a node from within the network) is compromised through node tampering or through a weakness in its system software. Figure 13.4 elaborates the scenario of insider and outsider attacks. Now, we will discuss each attack one by one in a bit detail. We will also identify whether an attack is a type of insider attack or an outside attack and in which category it falls according to [7].

Passive information gathering and message corruption: Passive information gathering and message corruption are the simplest attacks that can take place in wireless sensor networks. If information is not encrypted, an adversary can listen to the communication passively. Passive information gathering can be classified as interception carried out by an outsider node. In message corruption, adversary modifies contents of a message before it gets to the receiver. All key management schemes designed for wireless sensor networks can provide protection against these attacks.

Node compromise: An adversary gets control of a node by compromising it through a weakness in its system software. After compromise, adversary gains access to data, information, and cryptographic keys stored in the node. Adversary can also cause the node to malfunction and generate inaccurate data. This attack is carried out by an outsider node but later on, adversary uses the compromised node to carry out an insider attack. Adversary can interrupt communication, intercept messages, modify data packets transferred from one node to another, and fabricate messages.

Node tampering: An adversary tampers with a sensor node physically and gains access to data, information, and cryptographic keys stored on it. Just like node compromise, adversary can also cause the tampered node to malfunction. Classification of this type of attack is same as that of node compromise.

False node: Malicious node is introduced in the network by an adversary. This malicious node tries to inject malicious data and attract other nodes to send data to it. For example, it can advertise shortest route to the base station so that other nodes route their packets through it in order to save energy.

Node outage: Adversary removes a node from the sensor network or a node's energy is exhausted. Typically, these types of attacks are not dealt with the key management schemes. Rather, there are other security mechanisms that help in resolving this issue and finding other more appropriate routes to the base station.

Traffic analysis: An adversary can analyze the communication patterns of a sensor network and cause harm to the network. For example, if all packets are routed through a single node, it can try to compromise that node first. This is the reason why cluster head nodes should have more security as compared to other nodes in clustered sensor networks. Apart from that, traffic analysis attacks also highlight the need for refreshing keys at regular intervals.

Acknowledgement spoofing: This type of attack targets routing algorithms of wireless sensor networks. In this case, an adversary can spoof link layer acknowledgement after overhearing packets. Suppose there are two nodes A and B. Node A wants to send some data to the base station through node B but node B is dead. A compromised or outsider node E overhears the initial message sent by node A and spoofs an acknowledgement to node A at link layer. Based on spoofed acknowledgement, node A starts forwarding its packets to the base station through node E. After that, node E drops some or all packets forwarded to it by node A. This attack is classified under interruption and takes place if the forwarding node is not authenticated.

Spoofed, altered, or replayed routing information: An attacker can use compromised nodes or outside malicious nodes to play with routing information in such a way that it creates routing loops, attracts or repels network traffic, alters source routes, or generates false error messages. Apart from other hazards, this type of attack cause large network delays and also drain the battery power of sensor nodes very quickly.

Selective forwarding: In this case, an adversary compromises an insider node or uses an outsider malicious node to create a black hole in the target sensor network. The malicious node deliberately drops data packets in order to disrupt working of the target sensor network.

Sinkhole attacks: In this case, adversary tries to attract network traffic toward a malicious node. After that, the malicious node can carry out selective forwarding on the traffic. Sinkhole and selective forwarding attacks are most effective if the compromised node or the outsider malicious node is near the base station.

Sybil attacks: In these types of attacks, a malicious node presents multiple identities in the sensor network. In doing so, it can either steal other nodes' identities or it can try to fabricate new identities itself. Basically, sybil attacks reduce effectiveness of fault tolerant schemes such as distributed storage.

Also, sybil attacks can affect routing algorithms. For example, it can cause a routing algorithm to determine two disjoint paths, which are not disjoint in reality.

Wormhole attacks: Wormhole attack is carried out using two distant malicious nodes, which can communicate with each other, through an out-of-band communication channel, which is invisible to the underlying sensor network. One of the malicious nodes is placed near the base station and the other one is placed near the sensor nodes, which generate data. Using this low latency link the malicious node, which is placed near the data generating sensor nodes, convinces data generating nodes that it is just one or two hops away from the base station. This can cause sinkhole in the network. Also, this can create routing confusion especially in malicious node's neighbors, who might think that the other malicious node, near to the base station, is their neighbor.

Hello flood attacks: In this case, an adversary sends a HELLO packet itself or replays a routing protocol's HELLO packet with more signal strength. As a result, each of the other sensor nodes thinks that the malicious node is its neighbor. Then the malicious node can advertise a low latency link creating a wormhole. Also, sensor nodes waste their energies in responding to HELLO floods.

DoS attacks: An adversary can carry out DoS attack by disrupting communication between sensor nodes. Typically, DoS attacks occur at the physical layer of wireless sensor networks. Radio Jamming, which we have already discussed, is a classical example of a DoS attack.

While going through the types of attacks in wireless sensor networks, a very important thing to note is that most attacks involve either a malicious outsider node or a compromised insider node. Therefore, most of the attacks can be avoided by having highly effective node authentication and compromised node eviction scheme. Strength of both node authentication and compromised node eviction depend upon the underlying key management scheme. This highlights the importance of key management in wireless sensor network security once again. In the next section, we will discuss various key management solutions for wireless sensor networks proposed so far in the literature.

13.4 Key Management

After going through the previous section, we learn the importance of key management in wireless sensor networks. Also, we learn what is expected from a key management scheme that is designed for wireless sensor networks. Whenever we think of the points that should be kept in mind while designing a key management scheme for wireless sensor networks, resource constraints (processing and memory capabilities) and energy constraint of the sensor nodes always come first. Otherwise, traditional key management schemes, which we have already discussed in the section of background, are very useful. It was due to the constraints of sensor nodes that always lightweight key management schemes proposed for wireless sensor networks. However, maintaining required level of security in wireless sensor networks is also very important. Now we will discuss various key management schemes for wireless sensor networks proposed in the literature so far. Xiao et al. [11] presented a very useful survey of key management schemes for wireless sensor networks. We will start from most simple key management solutions for wireless sensor networks and then discuss more complex ones later on. Also, we will include some of the useful findings of [11].

According to [11], a key management scheme, which is designed for wireless sensor networks, should support certain characteristics. Strength of any key management scheme for wireless sensor networks depends on how many of those characteristics are present in any scheme. When we talk about the required level of security, it means that any key management scheme should provide authenticity, confidentiality, integrity, scalability, and flexibility. Apart from authenticity and confidentiality, maintaining integrity of a secret key is also very important. With integrity, we mean that the adversary should not be able to forge a key or change it altogether. Since the number of nodes may vary in wireless sensor networks and in some cases it may increase substantially, key management scheme should be scalable to cater for this scenario. Finally, wireless sensor networks are dynamic in nature. Old nodes, which run out of energy, die down with time and new ones can be added at any time. Key management scheme for wireless sensor networks should be flexible enough to cater for such scenarios.

Apart from the required level of security, key management schemes designed for wireless sensor networks should also cater for constraints related to sensor nodes. Apart from limited bandwidth, memory and computation capabilities, sensor nodes do not have any prior knowledge regarding their deployment. Limited transmission range and limited battery life also add to the constraints. Limited battery life is the primary reason why asymmetric key management strategies are not considered suitable for wireless sensor networks. Asymmetric key management schemes perform intense mathematical calculations, drains a lot of energy from sensor nodes. Since sensor nodes can only transmit up to short distances, some sensor network data collection techniques employ in-networking processing [12,13]. In in-network processing, all nodes send their data to a few nodes, which aggregate messages and transmit only processed information toward the command node. In order to avoid unnecessary communication, some schemes in wireless sensor networks require nodes to overhear messages from other nodes [14,15]. It may not be possible for a key management scheme to have all the above characteristics. Also, it is very difficult to design a key management scheme that is optimal for all topologies of sensor networks and their applications. Nonetheless, application developers can choose suitable key management schemes according to their application requirements.

Single group key for a network: It is by far the simplest key management scheme used for wireless sensor networks. In this case, a single key is loaded into every sensor node before deployment. All sensor nodes communicate using that single key. This scheme is very lightweight in terms of memory, computation, and communication requirements. It is also flexible and scalable but at the same time, it is also very vulnerable. If a single key is used for a long time, chances of cryptanalytic attack on the key gets higher and it is easier for an adversary to compromise the key. In this scenario, if a node is compromised or the key is revealed in some other way whole network is compromised. There is no way we can refresh the key or revoke the compromised sensor node from network and retain rest of the network.

Pair-wise key establishment: Establishing pair-wise key between every pair of nodes in a sensor network is the most secure key management scheme for wireless sensor networks. Every node is preloaded with a key for communication with every other node. For instance, if there are n nodes in a network, every node will have $n - 1$ keys stored in its memory. This scheme is possibly the most secure for wireless sensor networks. Pair-wise key establishment not only provides confidentiality and authenticity in a network but also provides efficient mechanism for revocation of a compromised sensor node in the network. However, this scheme is not at all efficient in terms of scalability and memory requirements. If the number n becomes too large as in many applications of wireless sensor networks, this scheme becomes impractical. Also, communication between every pair of sensor nodes is not necessary in wireless sensor networks.

Random pair-wise key establishment: Chan et al. [16] argue that all nodes in a sensor network need not share pair-wise keys. In their approach, two nodes share a pair-wise key with some probability p and p must be chosen carefully in order to keep the network connectivity up to a desired level. Also, node revocation does not need to involve the base station. A node's status in the network depends upon the consensus among nodes, with which it communicates. If a certain number of nodes, with which it communicates, say that node A is compromised, all of them will terminate their communication with node A. Although this scheme works well for small networks, it does not scale well enough if network size becomes too large.

Trusted key distribution center (KDC): In this approach, we mitigate the drawbacks of pair-wise key management by storing all pair-wise keys in a KDC. This KDC can be the base station or a cluster head node in clustered sensor networks. Although this approach is secure and resilient against node capture and node replication, this approach is also not scalable. This is because every pair of nodes has to obtain keys from the trusted base station for every session. Apart from the communication overhead introduced in this approach, links around the base station may become overloaded. If trusted KDC is a sensor node, for example, a cluster head node in clustered sensor networks, then its memory requirements increase manifold. Also, it must have far better energy and communication capabilities. In addition to all that, the trusted KDC becomes a single point of failure especially if it is one of the sensor nodes.

Random key predistribution scheme: In wireless sensor networks, it is not necessary that keys are established among every pair of sensor nodes. For a wireless sensor network to work, it is important that every sensor node gets sufficient bandwidth and neighboring nodes, who can relay its messages to the base station through various paths. For example, if node A has 15 nodes in its neighborhood, it can establish pair-wise keys with only 4 of them and those 4 neighboring nodes can provide node A distinct routes to the base station, then node A does not need to establish pair-wise keys with rest of the 11 nodes. Random key predistribution scheme was proposed by [17]. In the first phase of their scheme, a key ring of K keys and their identifiers is stored in the memory of each node prior to deployment. Every pair of nodes shares a key with some probability. In discovery phase, every node broadcasts its key identifiers and challenges to find those nodes, with which it shares a key. If some keys are left unused after the discovery phase, they can be used to establish keys between nodes which do not share a common key. For example, node A shares a key x with node B and node B shares another key y with node C while nodes A and C do not share a key. If node B has a key z, which it does not share with any node, it can send key z to both nodes A and C so that they can communicate with each other using z. In this scheme, there are group keys that are shared between the base station and all other nodes. In order to revoke a compromised sensor node, the base station compiles the list of keys known to the compromised node, uses a group key to sign the list and broadcasts it into the network using another one. Upon receiving the list, all nodes delete the keys, which are known to be the compromised node, from their memory. Apart from the fact that shortest path to the base station might not be established in this scheme, another drawback is that node revocation might cause many other links, which use one of the deleted keys, to break.

Q-composite random key predistribution scheme: Q-composite random key predistribution scheme, which was an improvement to the random key predistribution scheme, was proposed by [16]. In Q-composite scheme, two sensor nodes must share at least q number of keys if they want to establish a link between themselves. In this way, two linked nodes will have other keys for communication if one of the keys is compromised. In this case, size of the random key pool need to be reduced to maintain the probability that two nodes share q common keys. This poses another security problem: adversary will need to compromise only a few sensor nodes to compromise most of the keys.

Multi-path key reinforcement scheme: In basic random key predistribution scheme, multiple nodes may share more than one key. In this case, if one node is compromised, there is a chance that links between other noncompromised sensor nodes may also be compromised. In order to solve this problem, Chan et al. [16] proposed that keys that are used for communication on links between other noncompromised nodes should be refreshed but not through already established link. For this purpose, they use multiple disjointed paths between two nodes. If nodes A and B share a common key k, and they have h disjointed paths between them, node A generates h random values and sends each one of them to B through a separate disjointed path. Then both nodes A and B compute a key k' using key k and all h random values. Even if a node in a path is compromised, adversary will not know k' and k can be refreshed through k'. In order to keep the chances of eavesdropping to a minimum, size of disjointed paths should be kept small. Apart from increased network communication, this scheme also increases the computation overhead of sensor nodes by requiring them to generate random values, which require a lot of energy.

Polynomial pool-based key predistribution: In polynomial pool-based key predistribution scheme [18], a setup server generates one t-degree polynomial for each sensor node. These polynomials hold the property $f(x, y) = f(y, x)$. For example, if node i receives a polynomial $f(i, y)$ and node j receives a polynomial $f(j, y)$, they can compute a common key using identity of the other node. This scheme is scalable. However, the whole network is compromised, if t nodes are compromised. Memory requirement of this scheme is directly proportional to the value of t.

Grid-based key predistribution: This approach is similar to the polynomial-based key predistribution approach. In this approach, a matrix is stored in each node's memory. If two nodes i and j want to establish a pair-wise key for communication, they must have a common row or column in the matrix. If none of the rows or columns matches, then they must find alternate path to each other in path key establishment stage. This scheme offers greater probability of key establishment as compared to the random pair-wise key establishment scheme. This scheme reduces communication and computation overhead but increases the storage overhead.

Public key cryptography in wireless sensor networks: In previous sections, we discussed that like other traditional key management schemes, public key cryptography cannot be used in wireless sensor networks due to highly sophisticated computations involved in it. Contrary to this point of view, many researchers argue that the use of public key cryptography on wireless sensor networks cannot be ruled out completely especially the Elliptic Curve Cryptography (ECC), which has been used in wireless sensor networks recently [19–21]. Also, public key schemes have been used on 8-bit processors [19]. ECC can provide same level of security as that of RSA with much smaller key. According to [19], 160-bit ECC key has the same level of security as that of 1024-bit RSA. Also, the difference in the number of bits is not constant because 224-bit ECC has the same level of security as that of 2048-bit RSA key. ECC-based public keys have been used in TinyOS [20], an operating system developed specifically for wireless sensor networks.

Some schemes provide hybrid approach for key management, that is, they mix both symmetric and asymmetric key management approaches for providing security in wireless sensor networks [22,23]. LSec [24] also uses hybrid approach for key management in wireless sensor networks. In the first phase, they perform authentication and authorization using symmetric keys. In the second phase, keys are distributed using random secrets. This is performed using asymmetric cryptography.

SHELL: SHELL [25] is a location-aware combinatorial key management scheme designed for clustered sensor networks. We will discuss SHELL and the rest of the schemes in a bit more detail because they are state-of-the-art solution of key management so far in the literature. SHELL assumes large-scale sensor networks with cluster sizes of the order of hundreds of nodes. SHELL uses a small number of keys to manage large sensor networks using combinatory. SHELL employs EBS system of matrices [26] to use small number of keys for large networks. In addition to using small number of keys for large networks, SHELL also gets rid of single point of failure by using neighboring cluster heads for key management. SHELL targets sensor networks that are hierarchical, that is, a cluster head node manages a large number of sensor nodes and a base station manages multiple cluster head nodes. In other words, this scheme is suitable for networks, which support in-network processing of information. Also, this scheme supports overhearing of messages as one key is known to a large number of nodes.

SHELL assumes that the cluster head nodes can broadcast messages to all the sensor nodes in its cluster. Also, the cluster head node can reach all nodes in its own cluster. However, if a cluster head node wants to communicate with some node, which is not in its cluster, it has to go through the neighboring cluster head node. In short, cluster head nodes have more communication, computation, storage and power capabilities as compared to other sensor nodes. Base station or the command node has minimal involvement in this key management protocol. Another important assumption taken in SHELL is that two compromised nodes cannot come to know the location of each other, that is, they cannot launch a coordinated attack. Also, two compromised nodes cannot communicate through an out-of-band communication channel. Lastly, an attacker does not know memory contents of a sensor node before deployment. EBS system of matrices is used by SHELL, another state-of-the-art key management scheme; we think it is important to discuss EBS system of matrices briefly.

Table 13.1 shows an example of EBS matrix. Size of an EBS matrix depends upon the number of nodes and the number of keys used to manage those nodes. In EBS matrix, the number of columns equals the number of nodes in a cluster. The number of rows equals the number of keys used to manage those nodes. Total number of keys in a network is $k + m$. Out of these $k + m$ keys, every sensor node knows a distinct set of k keys, that is, set of keys known to one of the sensor nodes cannot be exactly identical to the set known to some other sensor node.

If a node is compromised, set of m keys, which are not known to the compromised node, are used to refresh the k keys known to the compromised node. Suppose that in Table 13.1, Node N_1 is compromised. Set of k keys known to N_1 is K_1, K_2 and K_4. If the managing node generates new values of K_1, K_2 and K_4, encrypts each one of them in K_3 and K_5 separately, and broadcasts in the cluster, all of the nodes will be able to decrypt the message except the node N_1. Number of nodes that can be supported by $k + m$ keys can be expressed by the formula:

$$\text{Number\_of\_nodes} = \frac{(k + m)!}{k!m!}$$

It is very important to note that the number of nodes supported by $k + m$ keys grows exponentially with the values of k and m. Values of k and m can be adjusted according to the network and its security requirements. Higher value of m results in higher security but with increased overhead.

Initially, each sensor node is preloaded with a discovery key Ksg and two other keys KS_{CH} and KS_{Key}. Ksg is recomputed with one-way hashing function, such as SHA1 [13] or MD5 [3], stored

Table 13.1 Example of an EBS Matrix

	N_0	N_1	N_2	N_3	N_4	N_5	N_6	N_7	N_8	N_9
K_1	1	1	1	1	1	1	0	0	0	0
K_2	1	1	1	0	0	0	1	1	1	0
K_3	1	0	0	1	1	0	1	1	0	1
K_4	0	1	0	1	0	1	1	0	1	1
K_5	0	0	1	0	1	1	0	1	1	1

in the node. The one-way hashing function is only known to the sensor node and the command node. Ksg is used to recover the network if the cluster head node is compromised. Both KS_{CH} and KS_{Key} are used for initial key distribution. In a cluster head node, the key Kgc, which is used for communication between the cluster head node and the command node, is preloaded along with the key Ksg of all nodes that lie in its cluster. Gateways can also communicate between themselves using another type of key provided by the command node. Command node generates the key, used for communication between the cluster head node, and renews them at regular intervals.

In SHELL, cluster head node is responsible for the formation of EBS matrix and generation of communication keys of its own cluster. Also, it is responsible for refreshment of its cluster's data keys. On request, the cluster head nodes generates administrative key of other clusters. In addition to that, the cluster head node is responsible for detecting and evicting compromised sensor nodes in its cluster. Every node is authenticated by the command node right after the initial deployment. After the initial deployment, gateways form their EBS matrices first. Each EBS matrix, along with the list of sensors in that cluster, is shared between the gateways and the command node. For each cluster, more than one neighboring cluster head nodes are designated by the command node for managing the administrative keys. For example, if there are 12 keys used in a cluster, command node can designate 4 neighboring cluster head nodes to manage 3 keys each for that cluster. The cluster head node shares the relevant portions of EBS matrix with each of the neighboring cluster head node.

Command node shares the key KS_{CH} of each sensor node with its cluster head. It also shares key KS_{Key} of every sensor node with the relevant neighboring cluster head nodes, that is, the one's responsible for managing any of the administrative keys that will be known to the sensor node. For key distribution, each relevant neighboring gateway generates one message per individual administrative key in its cluster for each sensor node. The message is first encrypted with the KS_{Key} of the node and then the administrative key of the sensor node's gateway. Gateway decrypts the message, encrypts it with KS_{CH} of the node and sends it to the sensor node. In order to share communication keys, cluster head nodes generate them and send them to their neighboring cluster head nodes. Neighboring clusters then send them to sensor nodes in the same way as they send the administrative keys.

If a cluster head node is compromised, either a new cluster head node is deployed or its sensors are redistributed among other cluster head nodes. The new gateway makes a new EBS matrix and repeats the process of initial deployment and initial key distribution. If a sensor node is compromised, keys known to the compromised node are changed with the method mentioned above in the description of EBS matrices. Advantages of SHELL are that it is highly scalable and resilient against node capture attacks. Also, it has very effective node authentication mechanisms. However, it is susceptible to collusion attacks. Collusion attack takes place when two or more compromised nodes collaborate with each other to attack a sensor network. In the same paper, they have also proposed mechanisms to prevent compromised nodes from collusion by assigning the keys strategically.

MUQAMI+: MUQAMI+ [27], which is an extended version of MUQAMI [28], is also an EBS-based key management scheme for clustered sensor networks. Just like SHELL, MUQAMI+ is highly scalable, resilient against node capture and has effective node authentication mechanisms. Apart from that, it does not have single point of failure in a cluster or in a network. However, its mechanism of avoiding single point of failure is different and more efficient than that of SHELL scheme. Instead of relying on the neighboring cluster head nodes for key management, responsibility of key management is distributed among few key-generating nodes within a cluster. This reduces communication, computation, storage overhead, and energy consumption of the sensor nodes in the network. A big advantage of this scheme is that it is very flexible, that is, it allows the responsibility of being cluster head node and being a key-generating node to be shifted seamlessly from one node to another. Therefore, if this scheme is employed in a sensor network, responsibilities can be transferred among nodes according to their capabilities and energy levels.

In MUQAMI+, first the cluster head nodes are deployed. After that, sensor nodes are deployed, which report to the nearest cluster head nodes with the help of a preloaded key. Cluster head node authenticates every sensor node from the command node before adding it to the cluster. Then the command node sends initial values of administrative keys to all sensor nodes through the cluster head node in such a way that the cluster head node does not get to know the key values. If the cluster head node comes to know about the administrative keys, it will become single point of failure for the cluster. However, the cluster head node does get to know about the key identities known to a particular node and key-generating responsibilities assigned to a node. The cluster head node builds EBS matrix based on this information. If a key needs to be refreshed, the cluster head node sends a message to the key generating node and the key-generating node refreshes the key.

If a sensor node is compromised, the cluster head asks the key-generating nodes, which generate the k keys known to the compromised node, to send new values of administrative keys encrypted in the old one, to the cluster head node using pair-wise keys between the key-generating node and the cluster head node. Cluster head node forwards these k values to the other m key generating nodes, which broadcast them after encrypting in their administrative key values. In case a cluster head or a key-generating node is compromised, its responsibility is shifted to some other node in the cluster.

One problem with this scheme is that some sensor nodes need to generate keys, which increases computation overhead substantially. Authors of MUQAMI+ propose to solve this issue using one-way hashing functions. In one-way hashing functions, we cannot compute the previous value using the current value. In MUQAMI+, one-way hashing functions are stored in sensor nodes. The command node sends initial and final values to the key-generating nodes. The key-generating node computes all intermediate values using the first value. Last value is used to confirm the end of a key chain. After this, the key-generating node stores the key chain in its memory and uses it in a reverse manner, that is, last value first. In this way, an adversary cannot compute next value of a key even if it knows the current value. However, this solution incurs storage overhead and a small communication overhead. Effectively, we can say that there is a tradeoff between computation overhead and storage and communication overhead.

LEAP+: Localized Encryption and Authentication Protocol (LEAP+) [10] is also a state-of-the-art solution for key management in wireless sensor networks. Its initial version was proposed as LEAP [29]. Later on, it was proposed as LEAP+ with some extensions. Although it can be used in both homogeneous and heterogeneous (clustered) sensor networks, it is more suitable for homogeneous sensor networks. This scheme is highly scalable and resistant to collusion attacks. Also, compromised node revocation is very simple and sensor node compromise does not affect other parts of the network.

In LEAP+, every sensor node uses a pseudo-random function to compute keys. The pseudo-random function uses node identities and some preloaded key values to compute keys. When sensor nodes are deployed, they compute their individual keys, which they share with only the command node. After that, they exchange their identities with their neighboring nodes and compute pair-wise keys with their neighbors. In order to broadcast some message, they need a key that is known to all the neighboring nodes. They compute this key for broadcasting purposes and send it to all the neighboring nodes individually. Finally, a global key, which is managed by the command node, is used for broadcasting in the whole network.

If a sensor node is compromised, all of its neighboring nodes delete pair-wise keys shared with the compromised node. After that, every neighboring node computes new value of its key, which is used for broadcasting purposes, and sends it to rest of the neighbors individually. Global key is also refreshed in the end. Apart from the increased computation overhead of LEAP+, another drawback of this scheme is that it assumes the network is safe during some initial time period. LEAP+ also has effective mechanisms for authenticated broadcast.

13.5 Future Directions of Research

Until now, research related to key management in wireless sensor networks has been very generic. All the key management schemes, proposed so far, are designed from broad perspective, that is, they are either designed for clustered sensor networks or homogeneous sensor networks. With the passage of time, applications and usage of wireless sensor networks have increased. Still, wireless sensor networks are being employed in newer application areas. For example, wireless sensor networks have been applied to healthcare scenarios, providing ubiquitous healthcare for patients. Another new idea is to attach sensors to the devices worn by patients, so that their health can be monitored in real time. Temperature, blood pressure sensors can be embedded in wrist watches, necklaces, so on. Apart from that, wireless sensor networks have been applied to houses and apartments resulting in smart homes.

With the introduction of sensor networks in new application areas, the characteristics of sensor networks will change according to application scenarios. For example, a small body area network need not be scalable. Also, if it is under constant human observation, adversary cannot tamper a sensor node. (If a conscious patient is wearing a sensor node, it is under constant human observation.) A sensor node in a smart home can have constant source of power and need not use battery. In short, requirements and characteristics of different application areas, employing wireless sensor networks, will differ and would require separate mechanisms for key management. Therefore, future directions of research are that need for those key management schemes will arise that should be specific to different application areas, for example, healthcare applications and smart home applications.

13.6 Conclusions

We have discussed the security requirements of key management and possible security threats in wireless sensor networks in detail. Also, we have discussed the constraints of sensor nodes and wireless sensor networks. In addition to that, various key management schemes, which were designed according to the requirements and characteristics of wireless sensor networks, were discussed. We learn that we do not have absolute criteria to rate a key management scheme better than all other schemes. There are different types of wireless sensor networks and different application areas, in which they are used. One key management scheme can be more efficient in one application area or one topology while another scheme can be more efficient in some other application area or some other topology. In future, the number of application areas is expected to increase.

Terminologies

Adversary—Any party, person, or a device which tries to reveal secret information to unauthorized users or tries to disrupt the network operation or causes harm to the network in any other way.

Cryptography—Mechanisms that are used to hide secret information from unauthorized users. Keys are used to encrypt (conceal) and decrypt (reveal) the secret information.

Symmetric key cryptography—Cryptography mechanism in which the same key is used for both encryption and decryption.

Asymmetric key cryptography—Cryptography mechanisms in which different keys are used for encryption and decryption.

Cryptanalytic attacks—Cryptanalytic attacks are attempts made by an adversary to crack cryptographic information such as secret keys shared between communicating parties.

Denial-of-service—Denial-of-Service, abbreviated as DoS, is a type of attack on wireless sensor network. In a DoS attack, the adversary tries to disrupt the normal network operation by not allowing the sensor nodes to properly communicate with each other.

Single point of failure—This is a term used to express a scenario, in which an attack at a single place can bring down the whole system or a unit. In case of wireless sensor networks, single point of failure can be a node, whose compromise can compromise the whole network or a whole cluster.

Questions and Sample Answers

1. Why is it thought that it is not viable to use asymmetric or public key cryptography in wireless sensor networks?

    Normally, processors are required to perform complex mathematical computations if asymmetric or public key cryptography is used. Wireless sensor networks are resource-constrained devices having limited battery power, processing capability, and memory capacity. If we use public key cryptography in wireless sensor networks, it uses up a lot of memory and takes a lot of time on sensor nodes to execute. In trying to perform tough mathematical calculations, sensor nodes lose their battery power very quickly. Therefore, it is thought that public key cryptography is not viable for wireless sensor networks.

2. Can RSA and Diffie–Hellman algorithms perform authentication? Why?

    RSA and Diffie Hellman algorithms are not capable of authenticating the other party. The reason is that both algorithms were designed to facilitate communication between two parties, which do not know each other in advance. In Diffie–Hellman key exchange, a node agrees on a secure secret key with an unknown party through an insecure medium. In RSA, public key of a node is published. Anyone can send messages to the node using its public key. The node uses its private key to decrypt the message.

3. What is the concept of a trusted third party?

    If two parties do not know each other in advance and they want to communicate with each other, then there should be a mechanism, using which they can authenticate each other. Concept of trusted third party is all parties trust a central server and register themselves on it. Whenever two parties try to establish a connection, they authenticate each other from the trusted central server, which is the third party.

4. What are the main goals of a key management scheme?

    Main goal of a key management scheme is to maintain confidentiality of secret information and help in authenticating legitimate parties or nodes in a network. Apart from that, a key management scheme for wireless sensor networks should also be able to deal with the issue of node compromise.

5. Why is it important to refresh secret keys after some time interval? How do we determine the time interval?

   Adversaries always try to crack or guess the keys used to conceal important information by launching cryptanalytic attacks. If same key is used to conceal secret information for a long time period, an adversary may become successful in guessing the secret key. Therefore, it is important to refresh secret keys at regular time intervals. These time intervals depend upon the frequency of information exchange and the time required by an adversary to find success in cracking a secret key.

6. What is the difference between an outsider attack and an insider attack? Elaborate with examples.

   In an outsider attack, the adversary is not part of the network. On the other hand, the adversary is part of the network in case of an insider attack. Example of an outsider attack is a malicious sensor node placed within the sensor network. The malicious node manages to get it authorized and listens to secret information or injects false information in the network. Example of an insider attack is that a legitimate sensor node from the network is compromised through software or through node tampering and then the compromised node listens to secret information or injects false information in the network.

7. What are the two most important requirements of a key management scheme designed for wireless sensor networks? Why?

   The two most important requirements of a key management scheme designed for wireless sensor networks are: (1) It should have highly effective sensor node authentication mechanisms; (2) It should also have effective mechanisms to deal with sensor node compromise. These are the two main requirements because most of the attacks on security of wireless sensor networks involve either an unauthorized outsider node or a compromised insider node.

8. What attacks on wireless sensor networks cannot be handled by a key management scheme? Give an example.

   Attacks that are carried out on physical layer of wireless sensor networks cannot be handled by a key management scheme. Example of such an attack is jamming attack. This is a type of denial-of-service attack that is launched by disrupting radio communication, through which sensor nodes communicate with each other.

9. Why do we require other key management schemes when we have pair-wise key distribution scheme, which has all security features required by a wireless sensor network?

   Although pair-wise key distribution scheme provides high level of security, it is impractical to use it in wireless sensor networks because it is not scalable and its storage overhead is too high. Also, it is not required that every sensor node in a wireless sensor network shares a key with every other sensor node in the network.

10. How does an EBS-based key management scheme manage a large number of sensor nodes using a small number of keys?

    When using EBS matrix for key management, each sensor node must have a distinct key combination stored in it, that is, no other node knows the same set of keys. EBS matrix has a property that as the number of keys grow linearly, the number of available distinct key combinations grow exponentially. This allows the management of large number of sensor nodes with a small number of keys.

11. How does SHELL avoid a single point of failure in a cluster of wireless sensor networks? SHELL scheme avoids single point of failure in a wireless sensor network by allocating the responsibility of key management to cluster head nodes of the neighboring clusters while cluster head node of the subject cluster does not get to know those keys. When this technique is used for key management, there is no single node in the network whose compromise can result in the compromise of the whole cluster or network.

12. How does MUQAMI+ avoid single point of failure after bringing key management responsibility within the subject cluster? Does not it add to the burden to sensor nodes? When the responsibility of key management is brought within the subject cluster, it is divided among a few key-generating nodes, which manage one key each. Also, cluster head node of the subject cluster does not get to know the keys managed by key-generating nodes. This way, compromise of a single sensor node cannot result in the compromise of entire cluster or entire network. It does not add significant burden to the sensor nodes because a very small number of nodes are required to manage keys. Also, MUQAMI+ allows the responsibility of key management to be shifted from one node to another seamlessly.

13. Why don't we consider the command node a single point of failure? In a wireless sensor network, command node is the node which receives all the data from the network. Normally, it is not a sensor node. Rather, it is a computer or a laptop class device, which has more capabilities than a simple sensor node. In some application scenarios of wireless sensor networks, there can be more than one base station collecting data from the network.

Author's Biography

Syed Muhammad Khaliq-ur-Rahman Raazi received the BS degree in computer software engineering from National University of Sciences and Technology (NUST), Rawalpindi, Pakistan, in 2002. He got his MS degree from Lahore University of Management Sciences (LUMS), Lahore, Pakistan, in 2006. He also has more than two years of industrial experience as software engineer. Currently, he is a PhD candidate in the department of computer engineering, Kyung Hee University (Global Campus), South Korea. His research interests include key management, security, sensor networks, and cloud computing.

Zeeshan Pervez is a PhD student at Kyung Hee University (Global Campus) South Korea, in Ubiquitous Computing Lab, and holds a master's degree in information technology form National University of Sciences and Technology (NUST), Rawalpindi, Pakistan, in 2006. His area of interest includes cloud computing security, intellectual property rights management, and service-oriented architecture.

Sungyoung Lee received his BS from Korea University, Seoul, Korea. He got his MS and PhD degrees in computer science from Illinois Institute of Technology (IIT), Chicago, Illinois, in 1987 and 1991, respectively. He has been a professor in the Department of computer engineering, Kyung Hee University, South Korea, since 1993. He is a founding director of the Ubiquitous Computing Laboratory, and has been affiliated with a director of Neo Medical Ubiquitous— Life Care Information Technology Research Center, Kyung Hee University since 2006. Before joining Kyung Hee University, he was an assistant professor in the department of computer Science,

Governors State University, Illinois, from 1992 to 1993. His current research focuses on ubiquitous computing and applications, context-aware middleware, sensor operating systems, real-time systems and embedded systems. He is a member of the ACM and IEEE.

References

1. Diffie, W. and Hellman, M., New direction in cryptography. *IEEE Transactions on Information Theory*, 1976; IT-22: 644–654 (Invited Paper).
2. Rivest, R. L., Shamir, A., and Adleman, L., A method for obtaining digital signatures and public-key cryptosystems. *Communications of the ACM* 1978; 21(2): 120–126. http://doi.acm.org/ 10.1145/359340.359342.
3. Kohl, J. and Neuman, C., The kerberos network authentication service (v5), *The Internet Engineering Task Force (IETF) website (www.ietf.org)*, Fremont, CA, 1993.
4. Dierks, T. and Allen, C., The TLS protocol version 1.0, *The Internet Engineering Task Force (IETF) website (www.ietf.org)*, Fremont, CA, 1999.
5. Gupta, G. and Younis, M., Load-balanced clustering of wireless sensor networks, in *IEEE International Conference on Communications*, 2003. ICC'03, vol. 3, pp. 1848–1852 vol. 3, May 11–15, 2003. http://ieeexplore.ieee.org/stamp/stamp.jsp?arnumber=1203919&isnumber=27115.
6. Younis, O. and Fahmy, S., Heed: A hybrid, energy-efficient, distributed clustering approach for ad hoc sensor networks. *IEEE Transactions on Mobile Computing* 2004; 3(4): 366–379. http://dx. doi.org/10.1109/TMC.2004.41. Student Member-Ossama Younis and Member- Sonia Fahmy.
7. Zia, T. and Zomaya, A., Security issues in wireless sensor networks, in *Proceedings of the International Conference on Systems and Networks Communication (October 29–November 03, 2006). ICSNC*. IEEE Computer Society, Washington, DC, p. 40, 2006. http://dx.doi.org/10.1109/ICSNC.2006.66.
8. Karlof, C. and Wagner, D., Secure routing in wireless sensor networks: Attacks and countermeasures, in *Proceedings of the First IEEE. 2003 IEEE International Workshop on Sensor Network Protocols and Applications, 2003*, Anchorage, AK, USA, pp. 113–127, 11 May 2003, http://ieeexplore.ieee.org/stamp/stamp. jsp?arnumber=1203362&isnumber=27104.
9. Roosta, T., Shieh, S. W., and Shankar Sastry, S., Taxonomy of security attacks in sensor networks and countermeasures, in *The First IEEE International Conference on System Integration and Reliability Improvements*, Hanoi, Vietnam, December 2006.
10. Zhu, S., Setia, S., and Jajodia, S., LEAP+: Efficient security mechanisms for large-scale distributed sensor networks. ACM Transactions on Sensor Networks, 2006; 2(4): 500–528. http://doi.acm.org/ 10.1145/1218556.1218559.
11. Xiao, Y., Rayi, V. K., Sun, B., Du, X., Hu, F. and Galloway, M. A survey of key management schemes in wireless sensor networks. *Computer Communications* 2007; 30(11–12): 2314–2341. http://dx.doi.org/10.1016/j.comcom.2007.04.009.
12. Intanagonwiwat, C., Govindan, R., and Estrin, D., Directed diffusion: A scalable and robust communication paradigm for sensor networks, in *Proceedings of the 6th Annual international Conference on Mobile Computing and Networking (Boston, Massachusetts, United States, August 06–11, 2000)*, MobiCom '00. ACM Press, New York, NY, pp. 56–67, 2000. http://doi.acm.org/10.1145/345910.345920.
13. Karlof, C., Sastry, N., and Wagner, D. TinySec: A link layer security architecture for wireless sensor networks, in *Proceedings of the 2nd International Conference on Embedded Networked Sensor Systems (Baltimore, MD, USA, November 03–05, 2004), SenSys '04*. ACM Press, New York, NY, pp. 162–175, 2004. http://doi.acm.org/10.1145/1031495.1031515.
14. Karlof, C., Li, Y., and Polastre, J., Arrive: An architecture for robust routing in volatile environments. *Tech. Rep. CSD-03-1233*, University of California at Berkeley.
15. Madden, S., Szewczyk, R., Franklin, M. J., and Culler, D., Supporting aggregate queries over ad-hoc wireless sensor networks, in *Proceedings of the Fourth IEEE Workshop on Mobile Computing Systems and Applications (June 20–21, 2002). WMCSA*. IEEE Computer Society, Washington, DC, p. 49, 2002.

16. Chan, H., Perrig, A., and Song, D. Random key predistribution schemes for sensor networks, in *Proceedings of the 2003 IEEE Symposium on Security and Privacy (May 11–14, 2003)*. SP. IEEE Computer Society, Washington, DC, pp. 197–213, 2003.

17. Eschenauer, L. and Gligor, V. D., A key-management scheme for distributed sensor networks, in *Proceedings of the 9th ACM Conference on Computer and Communications Security (Washington, DC, USA, November 18–22, 2002)*, V. Atluri, Ed. CCS '02. ACM Press, New York, NY, pp. 41–47, 2002. http://doi.acm.org/10.1145/586110.586117.

18. Liu, D. and Ning, P., Establishing pairwise keys in distributed sensor networks, in *Proceedings of the 10th ACM Conference on Computer and Communications Security (Washington D.C., USA, October 27–30, 2003)*, CCS '03. ACM Press, New York, NY, pp. 52–61, 2003. http://doi.acm.org/10.1145/948109.948119.

19. Gura, N., Patel, A., Wander, A., Eberle, H., and Shantz, S. C., Comparing elliptic curve cryptography and RSA on 8-bit CPUs, in *Cryptographic Hardware and Embedded Systems—CHES 2004* (2004), Lecture Notes in Computer Science, pp. 119–132, 2004, Berlin/Heidelberg: Springer. http://www.springerlink.com/content/87aejjlhqn6fuxpy=0pt.

20. Malan, D. J., Welsh, M., and Smith, M. D., A public-key infrastructure for key distribution in TinyOS based on elliptic curve cryptography, in *2004 First Annual IEEE Communications Society Conference on Sensor and Ad Hoc Communications and Networks, 2004. IEEE SECON 2004*, pp. 71–80, 4–7 Oct. 2004, http://ieeexplore.ieee.org/stamp/stamp.jsp?arnumber=1381904&isnumber=30129.

21. Wander, A. S., Gura, N., Eberle, H., Gupta, V., and Shantz, S. C., Energy analysis of public-key cryptography for wireless sensor networks, in *Proceedings of the Third IEEE international Conference on Pervasive Computing and Communications (March 08–12, 2005)*. PERCOM. IEEE Computer Society, Washington, DC, pp. 324–328, 2005. http://dx.doi.org/10.1109/PERCOM.2005.18.

22. Huang, Q., Cukier, J., Kobayashi, H., Liu, B., and Zhang, J., Fast authenticated key establishment protocols for self-organizing sensor networks, in *Proceedings of the 2nd ACM international Conference on Wireless Sensor Networks and Applications (San Diego, CA, USA, September 19–19, 2003)*. WSNA '03. ACM Press, New York, NY, pp. 141–150, 2003. http://doi.acm.org/10.1145/941350.941371.

23. Kotzanikolaou, P., Magkos, E., Vergados, D., and Stefanidakis, M., Secure and practical key establishment for distributed sensor networks, in *Security and Communication Networks*, Wiley InterScience, NJ, USA, 2009. http://dx.doi.org/10.1002/sec.102.

24. Shaikh, R. A., Lee, S., Khan, M. A. U., and Song, Y. J., LSec: Lightweight security protocol for distributed wireless sensor network, in *11th IFIP International Conference on Personal Wireless Communications, LNCS 4217*, (pp. 367–377), Albacete, Spain: Springer-Verlag, 2006.

25. Ghumman, K., Location-aware combinatorial key management scheme for clustered sensor networks. *IEEE Trans. Parallel Distrib. Syst.* 2006; 17(8): 865–882. http://dx.doi.org/10.1109/TPDS.2006.106.

26. Eltoweissy, M., Heydari, H., Morales, L., and Sadborough, H., Combinatorial optimization of group key management. *Journal of Network and Systems Management*, 2006; 12(1): 33–50.

27. Raazi, S. M. K., Lee, H., Lee, S., and Lee, Y., MUQAMI+: A scalable and locally distributed key management scheme for clustered sensor networks. *Annals of Telecommunications* 2010; 65(1–2): 101–116.

28. Raazi, S. M. K., Khan, A. M., Khan, F. I., Lee, S., Song, Y., and Lee, Y., MUQAMI: A locally distributed key management scheme for clustered sensor networks, in *Trust Management, Volume 238/2007 of IFIP International Federation for Information Processing*, pp. 333–348, Boston: Springer, 2007. 10.1007/978-0-387-73655-6_22.

29. Zhu, S., Setia, S., and Jajodia, S., LEAP: Efficient security mechanisms for large-scale distributed sensor networks, in *Proceedings of the 10th ACM Conference on Computer and Communications Security (Washington D.C., USA, October 27–30, 2003)*. CCS '03. ACM, New York, NY, pp. 62–72, 2003. http://doi.acm.org/10.1145/948109.948120.

Chapter 14

Key Management Techniques for Wireless Sensor Networks
*Practical and Theoretical Considerations**

Effie Makri and Yannis C. Stamatiou

Contents

* Parts of this chapter were reprinted from E. Makri and Y. Stamatiou, Deterministic and randomized key pre-distribution schemes for mobile ad-hoc networks: Foundations and example constructions, in Zhen Jiang, Yi Pan, Ed. *From Problem Toward Solution: Wireless Sensor Networks Security*, Chapter 11, pp. 211–233, USA: Nova Science Publishers, Inc., 2009. With permission.

14.1 Introduction and Background

A wireless *mobile ad-hoc network* (or MANET for short) is a collection of wireless nodes that form a temporary network dynamically on an "as-needed" basis. This is done without the use of any pre-existing infrastructure.

A *distributed sensor network* (DSN) is a mobile *ad hoc* network which consists of nodes of limited computation and communication abilities [1]. These types of networks are mainly used for military purposes for monitoring and collecting information from hostile environments. The types of sensors include acoustic, seismic, and magnetic.

Wireless links both in MANETs and DSNs are open to both passive and active attacks, and nodes might roam in hostile environments where they are susceptible to capture and tampering. Therefore, security is of prime importance in these types of networks, more so than in networks with a fixed infrastructure. Security mechanisms need to be adaptable to the changes in the network topology due to roaming nodes and frequent changes in node membership. They also need to be scalable to handle increasing number of nodes.

Cryptographic schemes protect information traversing through the network, but in order for these schemes to be successful, they are dependent on key management. Key management is the process of establishing cryptographic keys between a sender and a receiver. Cryptosystems are often attacked through their key management infrastructure.

Traditional key management schemes (e.g., based on Trusted Third Parties) are not suited to MANETs or DSNs due to the absence of a fixed network structure as well as the different nature of the security threats that beset wireless communications. The purpose of this chapter is to examine some of the most prominent key management techniques that have been proposed for MANETs and DSNs.

Although network security, as a field, has coexisted with conventional networks for decades now, the "diffusive" nature of wireless information transmission as well as the lack of structure in mobile *ad hoc* networks introduces new threats and security demands for such networks. More specifically, we can identify the following special security requirements [2]:

Increased probability of the occurrence of a security violation event: The open nature of wireless transmissions renders them particularly vulnerable to eavesdropping attacks. Given the fact that at any time instance there are numerous active wireless connections in a mobile network, it is almost certain that at least one security violation will occur given a sufficiently large timing span.

Mobility and service ubiquity: Since users of mobile networks move constantly, there is no fixed communication infrastructure on which services can be reliably (and quickly) delivered. This creates difficulties in establishing connections between distant users (as well as service access points) since there may be situations where such a connection may be routed through other mobile stations. Such a need for multihop connectivity creates reliability as well as security problems.

Changes in network characteristics: Transmission errors as well as link and node failures create additional difficulties in establishing reliable network connections.

Dynamic user management: Mobile users may join and leave a network in a dynamic, unpredictable fashion. In addition, there is a need for management of malicious network users by revoking their security credentials (e.g., revoking digital certificates, key rings etc.) as well as management of new users (e.g., creating, dynamically, certificates, and new key rings).

Unpredictable network size variability: Since mobile networks can grow in an unpredictable fashion, security protocols should be able to scale well with network size.

All these requirements, not usually existent in the design and operation of conventional networks, constitute real challenges in the design of security protocols for mobile ad hoc networks. In what follows, we will briefly survey works that attempt to address these requirements and provide viable solutions for the security of such networks.

14.1.1 Single Network-Wide Key

This is the most simple key establishment scheme, whereby a single key is preloaded onto all the nodes of the WSN. Once the WSN is deployed, the key is used by every node in the network to encrypt and decrypt messages. This scheme provides many advantages, such as minor memory, computation, and broadcast demands, since only a single key is stored on the nodes' memory and nodes do not undertake the execution of key discovery or key exchange. On the other hand, this scheme's main drawback is that the capture of a single node renders the whole network as compromised.

14.1.2 Pairwise Key Establishment Scheme

On the other end of the spectrum to the Single Network-Wide Key scheme is the Pairwise Key Establishment technique. This is one of the most secure schemes in that it offers node-to-node authentication and resilience to node capture, among others, but is lacking in efficiency due to the large number of keys required to satisfy the scheme. Specifically, if the WSN consists of N nodes, then each node would be assigned a unique pairwise key with all nodes of the network, resulting in an additional overhead required for each node to establish $N-1$ unique keys that are stored in the memory of each sensor node. It is evident here that an increase in the size of the WSN is prohibitive, since an increase in the number of sensor nodes results in an increase in the number of keys needed to be stored in the memory.

The authors of [3] proposed an alternative to the pairwise scheme, namely the random pairwise keys scheme. In the random scheme, supposing that p is the probability that the WSN is almost completely connected (according to the Erdős–Rényi random graph model [4]), then it is not necessary to install $N-1$ unique keys on the memory of each sensor node, but np. Using pairwise keys here offers node-to-node authentication properties when the identity of the node on the sensor is stored in addition to the shared key.

Key revocation in this scheme may be performed in a distributed manner in that the nodes of the network may broadcast "public votes" against misbehaving nodes. Each node will have pre-knowledge of a threshold t which if exceeded in public votes for a specific node, and then all communication is broken off with that node.

14.1.3 Random Key Predistribution

14.1.3.1 The Basic Scheme

The Basic Random Key Predistribution scheme of [5] consists of three phases, namely (i) key predistribution, (ii) shared-key discovery, and (iii) path-key establishment.

i. *Key predistribution:* The first of the three phases comprises of five off-line steps in order to equip sensor nodes with keys. First of all, a large pool of P keys is generated along with the corresponding key identifiers. Secondly, k keys are randomly drawn out of the key pool (without replacement), and these along with their corresponding key identifier, constitute the key ring of each sensor node. Next, the key rings are loaded onto the memory of each sensor, and the key identifiers of a key ring along with the associated identifier of the node are saved onto a trusted controller node. Finally, for each node, the ith controller node is loaded with the key shared with that node—this means that the ith controller nodes know exactly which keys are on sensor x and shares a common key with that sensor (see Figure 14.1).

ii. *Shared key discovery:* The second phase of the basic scheme takes place during the actual initialization of the WSNs in the operational environment. Here, every node discovers its neighbors in wireless communication range with which it shares keys.

This is done by each node broadcasting the key identifiers on their key ring, i.e. each node broadcasts a list along $\left[\alpha, E_{k_i}(\alpha)\right]$, $i = 1, 2, \ldots, k$ with α being a challenge. The decryption of with a recipient's appropriate key will reveal α (Figure 14.2).

At the end of this phase, the topology of the WSN is established, whereby links exist between the two sensor nodes only if they have a key in common.

iii. *Path-key establishment:* After the Shared Key Discovery phase, various nodes in the WSN may be left without any links to other nodes. In this case, a path-key is assigned to a selected pair of sensor nodes that are in wireless communication range, but do have a key in common (Figure 14.3).

In Figure 14.4, sensor z generates a key between sensor x and sensor y which is used to encrypt message m, and becomes a Key Distribution Centre.

Owing to the limited communication range observed by sensor nodes, certain questions are raised regarding the connectivity of a DSN. The authors of [3] bring forth these questions regarding

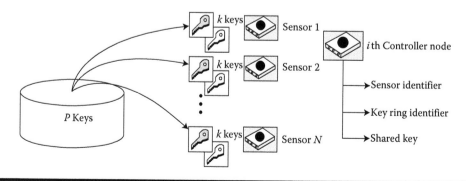

Figure 14.1   Key predistribution phase.

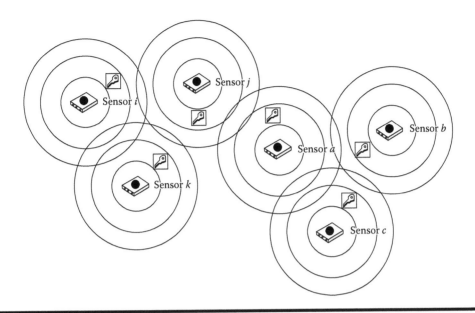

Figure 14.2 Shared key discovery.

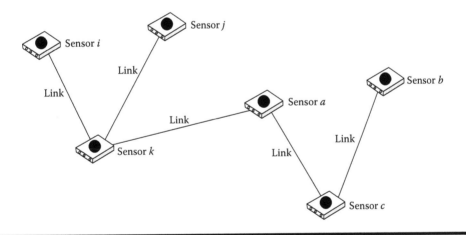

Figure 14.3 Path-key establishment.

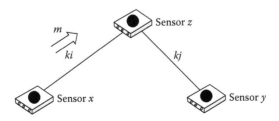

Figure 14.4 Path-key establishment.

the degree* of a node as well as the number of keys k on a sensor key ring, and the number of keys generated in the key pool P.

The first question queries on the degree d of a sensor node in order for a DSN of n nodes to be fully connected, and the second question inquires on required number of k keys installed on the key ring of a sensor node, as well as the number of keys in the pool P, given the degree and the communication constraints of the sensor. Taking into consideration the memory constraints of the sensor node that limits k, which then, would be the size P of the pool of keys?

Using random graph techniques (see [6] for more on these techniques), the authors determine the value of p and d given the value n, the number of nodes in the DSN. For a desired probability of graph connectivity P_c, we have

$$P_c = \lim_{n \to \infty} P_r[G(n, p) \text{ is connected}] = e^{e^{-c}} \quad \text{with} \quad p = \frac{\ln(n)}{n} + \frac{c}{n}, \quad c \in R.$$

Exercise 1 Prove this fact (Hint: Consult [6]).

Erdös and Rényi [6,7] showed that, for monotone properties, there exists a value of p such that the property moves from a "nonexistent" to "certainly true" in a very large random graph. The function defining p is called the threshold function of a property. Taking into account the wireless connectivity constraints of the sensor nodes, neighborhoods are limited to n' nodes, where n' is significantly less than n. This means that the probability p of sharing a key between any two nodes in a neighborhood of size n' is given by $p' = (d/(n' - 1))$, which is significantly greater than p. The value of P, which is the size of the key pool, from which the k keys of a key ring are randomly chosen, is not limited by the memory constraints of the sensor nodes, as is k. For a given p' that maintains DSN connectivity with an expected node degree d, it holds $p' = 1 - \left(((P - k)!)^2 / ((P - 2k)! P!)\right)$. However, if a node has been compromised, the keys on that node's key ring need to be revoked. In this case, the corresponding controller node broadcasts a signed revocation message, containing the list of the k key identifiers.

In the Basic Scheme, the size of the key pool P is chosen such that two random sets of selected keys of size k will have at least one key in common with a probability of p. Furthermore, the key discovery phase in effect determines the connectivity of the WSN, and thus if p is the probability that a shared key exists between two sensor nodes, and N is the size of the network, then $d = p(N - 1)$ is the expected degree of a node, i.e. the average number of links between a node in the network and its neighbors. The value of d should be determined so as to have a WSN that is almost certainly connected, when viewed as a communication network. Furthermore, since sensor nodes have limited communication range capabilities, if N' is the number of sensor nodes within a specific wireless communication range, and given d, then the values of k and P should be selected so as to have a successful key setup phase between neighboring nodes (see [3] for more on these considerations).

14.1.3.2 The q-Composite Random Key Predistribution Scheme

The q-composite Random Key Predistribution Scheme [3] is based on the basic scheme but instead of requiring two neighboring nodes to possess one common key between them, the q-composite

---

* Degree of a sensor node is considered the average number of edges connecting that node with its graph neighbors and is given by $d = p * (n - 1)$, where p is the probability that a shared key exists between two sensor nodes, and n is the number of nodes in the DSN.

scheme requires the existence of q common keys ($q > 1$), which results in increased network resilience against node capture.

The phases need to accomplish secure communication between the nodes of the WSN are identical to that of the basic scheme with the only difference on the number of common keys that are required. Once key discovery is complete, if the actual number of keys shared between neighboring nodes is q', where $q' \geq q$, then a new communication key K is generated as a hash of all the shared keys: $K = \text{hash}(k_1 \parallel k_2 \parallel \cdots \parallel k_{q'})$. The key setup is not performed between nodes that have less than q keys in common.

Since the required number of common keys increases, the pool size P needs to be decreased, introducing the caveat of allowing an attacker to gain a larger sample of the key pool by compromising a fewer number of nodes.

Comparing the basic and the q-composite scheme, the q-composite scheme offers better resilience against node capture for a small number of compromised nodes, as compared with the basic scheme. On the other hand, when the number of nodes compromised is large, larger fractions of the network are revealed to the adversary.

Given the probability c of full-network connectivity and the expected number of neighbors to each node n', the expected degree d of each node is calculated using:

$$d = \left(\frac{n - 1}{n} \right) (\ln(n) - \ln(-\ln(c)))$$

and then the desired probability that any two nodes can perform the key setup p is calculated as $p = d/n'$.

All that remains is the calculation of the random key pool size P, which should not be too large as to decrease the probability p that two nodes share q keys, nor too small so as to compromise the security of the network. This means that the largest possible value for P is such that if k is the number of keys a node can hold on its key ring, then any two random samples from the pool of size k have at least q elements in common with a probability of p. An analysis of the scheme follows below.

Any given node has $\binom{P}{k}$ ways of picking its k keys from the key pool of size P, and any two nodes have $\binom{P}{k^2}$ ways to pick k keys each. If the two nodes have i keys in common, there are $\binom{P}{i}$ ways to pick these, and once they have been picked then there are $2(k - i)$ distinct keys left in the key rings to be picked from the remaining $P - i$ keys in the pool, and there are $\binom{P-i}{2(k-i)}$ ways to do this. These keys must be partitioned between the two nodes, and the number of ways to do this is $\binom{2(k-i)}{k-i}$. Therefore, the total number of ways to choose two key rings with i keys in common from a key pool of size P is $\binom{P}{i} \binom{P-i}{2(k-i)} \binom{2(k-i)}{k-i}$, giving us:

$$p(i) = \frac{\binom{P}{i} \binom{P - i}{2(k - i)} \binom{2(k - i)}{k - i}}{\binom{P}{k^2}}.$$

If p_{connect} is the probability of any two nodes sharing sufficient keys to form a secure connection, then $p_{\text{connect}} = 1 - (\text{probability that the two nodes share insufficient keys to form a connection})$.

That is, $p_{connect} = p(0) + p(1) + \cdots + p(q-1)$. Thus, for a minimum key overlap q, and the minimum connection probability p, the key pool size P is chosen such that $p_{connect} > p$.

This scheme shows that for a small-scale attack, the network is more resilient, showing that for $q = 2$, the amount of additional communications compromised when 50 nodes are captured is 4.74% as opposed to 9.52% in the scheme proposed in Ref. [8]. On the other hand, when a large number of nodes are compromised, the q-composite scheme reveals a larger portion of the network to an adversary. This has the advantage due to the fact that it is much more difficult to detect large-scale attacks as supposed to small scale ones, which are also less expensive to mount.

The *Multipath Key Reinforcement* method presented by [3] and first explored by [9] trades off some communication overhead for more resilient security of an established link key. This method is recommended to be used in conjunction with the Basic Scheme, but not advised to be used with the q-Composite Scheme. Going back to the Basic Scheme, once the Shared Key Discovery phase is complete, pairs of sensor nodes share keys. Such keys can also be on the key ring of other nodes, the capture of which jeopardizes the secure links in the network.

In more detail, now, suppose two nodes share a common key. This key is likely to exist on the key ring of other nodes as well. In the case that any of the nodes are captured, the link between the original two nodes is under risk, therefore an update of the common key is required. The key update can be performed over multiple independent paths, rather than directly between these two nodes, if the first node has sufficient routing information so as to have knowledge of the various paths to the second node (paths are link-disjoint, that is, they do not share any network link). If there are j such paths, the first node generates j random values $k_1, k_2, \ldots, k_j$, whereby each value has the same length as the original key shared by the two nodes. These j values are routed along the different paths to the second node. Upon receipt of these values, the second node computes the new key by XOR-ing (bit-wise) all these values with the original key k in order to form the new key $k' : k' = k \oplus k_1 \oplus k_2 \oplus \cdots \oplus k_j$. In this way, a potential eavesdropper would have to capture all nodes along the j paths in order to recover the value k', given that it already has acquired the original key k. Thus, this scheme provides improved resilience to node captures but at the expense of increased communication overhead (routing of the random values to destination nodes) and memory requirements locally to each node (in order to store path information).

The multipath key reinforcement scheme is comparable to the q-composite scheme in that it magnifies the difficulty for an adversary to compromise a link by compelling the adversary to posses multiple keys in order to be able to eavesdrop on a link. These benefits though are not without trade-offs. In the case of the q-composite scheme, there is a smaller size for the original pool of keys, and in the case of the multipath key reinforcement, the trade-off is the increased network overhead.

14.1.4 Deterministic Key Distribution

Deterministic (as opposed to randomized) key distribution schemes have been proposed by many authors [1,10,11]. WSNs can be considered in terms of the physical layer and the network layer, the former represented by a random geometric graph, whereby the nodes of the graph are the sensor nodes within wireless communication range of each other. The network layer, on the other hand, is represented by a graph such that adjacent nodes are one which share a secret key—called a network graph—and is determined by the key predistribution scheme independently of the sensor node distribution, hence being random graphs.

In the deterministic key predistribution schemes, the assignment of keys is deterministic, whereby regular graphs are used as network graphs. However, working with properties of combinatorial key sets can be tricky and some slippery points can be overlooked. It is very instructive to see the discovery of a fault in the work of [12] by the authors of [13].

14.1.5 Combinatorial Key Predistribution

14.1.5.1 Set-Based Constructions for Key Predistribution

In theory, the key sets of nodes could be chosen at random from within a universe set X and the nodes can be equipped with a suitable key selection protocol for establishing a shared key for their secure communication. There are nevertheless some sets that may give rise to some undesirable characteristics. For instance, it is not advisable to have two network nodes a, b with key sets A and B, respectively, such that $A \subseteq B$. One reason is that if node b is ever compromised by an adversary, it will be possible for the adversary to eavesdrop on any communication involving node a. Thus, there will be a need to cancel the validity of all of node a's keys leaving node a with no valid keys from which to choose. Another problem which may arise is that if for three key sets A, B, C, the condition $A \cap B \subseteq C$ holds, then the node with key set C, although not containing, perhaps, either of the two key sets A, B, it may, nevertheless, contain their intersection. In what follows, we will give some sufficient conditions for ensuring that these problems do not arise. In this section we will describe deterministic key predistribution schemes based on a set system with special combinatorial properties [14].

Definition 1 *Let S be a family of sets and $A \in S$. Then A is called S-free if $A \nsubseteq \cup_{S_i \in S} S_i$.*

By limiting our set systems so as to be regular (i.e., all have the same number of elements), we avoid the first problem outlined above.

Claim 1 If F is a regular set family, then no member of F can contain some other member.

Lemma 1

Let A be a set of a set family F and $N(A) = \{A_1, \ldots, A_d\}$ its $d \geq 1$ neighbors. Then the following hold:

1. *If $\sum_{j=1}^{d} |A \cap A_i| < A$, then A is $N(A)$-free.*
2. *If A is $N(A)$-free, then for each l-element subset $i_1, \ldots, i_l$ of $\{1, \ldots, d\}, 1 \leq l \leq d$, it holds that*

$$\sum_{j \in \{i_1, \ldots, i_l\}} |A \cup A_j| - \sum_{i, j \in \{i_1, \ldots, i_l\} i < j} |A \cap A_i \cap A_j| < A.$$

Proof. We will first prove Statement 1 of Lemma 1. Assume, toward a contradiction, that $\sum_{j=1}^{d} |A \cap A_i| < A$, yet A is not $N(A)$-free. Then there exist a subset of its neighbors whose union contains A, that is, there exist indices $i_1, \ldots, i_l \in \{1, \ldots, d\}$ such that $A \subseteq A_{i_1} \cup \ldots \cup A_{i_l}$. Then, from this relation, it follows that $|A| = |A \cap (A_1 \cup \ldots \cup A_{i_l})| \leq \sum_{j=1}^{l} |A \cap A_{i_j}|$, which contradicts our assumption that $\sum_{j=1}^{d} |A \cap A_i| < A$.

We will now prove Statement 2 of Lemma 1. Since A is $N(A)$-free, for any subset $\{i_1, \ldots, i_l\}$ of $\{1, \ldots, d\}$, A cannot be a subset of $A_{i_1} \cup \ldots \cup A_{i_l}$. Thus

$$A \supset A \cap (A_1 \cup \ldots \cup A_d) \Rightarrow |A| > |A \cap (A_1 \cup \ldots \cup A_d)|.$$

Using the inclusion–exclusion principle up to the second term (see, for instance, [15]), from the last equation, we obtain the following:

$$|A \cap (A_{i_1} \cup \ldots \cup A_{i_l})| > \sum_{j \in \{i_1, \ldots, i_l\}} |A \cap A_j| - \sum_{i,j \in \{i_1, \ldots, i_l\} i < j} |(A \cap A_i) \cap (A \cap A_j)|$$

$$= \sum_{j \in \{i_1, \ldots, i_l\}} |A \cap A_j| - \sum_{i,j \in \{i_1, \ldots, i_l\} i < j} |(A \cap A_i \cap A_j)|$$

and the required inequality follows. ■

Exercise 2 In the final step of the proof of Lemma 1, give the details of the application of the Principle of Inclusion–Exclusion.

Lemma 2

Let A be a set of r-regular set system F defined on a universe set X of cardinality n. Let $N(A) = \{A_1, \ldots, A_d\}$ be the set of the $d \geq 1$ neighbors of A. Then for any two of its neighbors A_{i_1}, A_{i_2}, if $|A \cap A_{i_1}| = |A_{i_1} \cap A_{i_2}| = s$ and $3r - 2s > n$ then $A \cap A_{i_1} \not\subseteq A_{i_2}$.

Proof. Assume, toward a contradiction, that $A \cap A_{i_1} \subseteq A_{i_2}$. Then $A \cap A_{i_1} \subseteq A_{i_1} \cap A_{i_2}$. Since $|A \cap A_{i_1}| = |A_{i_1} \cap A_{i_2}|$, we have that $A \cap A_{i_1} = A_{i_1} \cap A_{i_2}$ and, thus, all three sets have the same pairwise intersection which, also, equals the intersection of all three taken together:

$$|A \cap A_{i_1}| = |A_{i_1} \cap A_{i_2}| = |A \cap A_{i_2}| = |A \cap A_{i_1} \cap A_{i_2}|.$$

Since, also,

$$|A \cup A_{i_1} \cup A_{i_2}| = |A| + |A_{i_1}| + |A_{i_2}| - |A \cap A_{i_1}| - |A \cap A_{i_2}| - |A_{i_1} \cap A_{i_2}| + |A \cap A_{i_1} \cap A_{i_2}|$$

it follows that $|A \cup A_{i_1} \cup A_{i_2}| = 3r - 2s$, which contradicts the assumption $3r - 2s > n$. ■

Lemma 3

Let A be a set of a r-regular set system F defined on a universe set X of cardinality n and $N(A) = \{A_1, \ldots, A_d\}$ the set of the $d \geq 1$ neighbors of A. Then for any three of its neighbors $A_{i_1}, A_{i_2}, A_{i_3}$, if their pairwise intersections are of cardinality s and $2r + |A_{i_2} \cup A_{i_3}| - |A \cap (A_{i_2} \cup A_{i_3})| - |A_{i_1} \cap (A_{i_2} \cup A_{i_3})| > n$. Then it holds that $A \cap A_{i_1} \not\subseteq A_{i_2} \cup A_{i_3}$.

Proof. Let $R = A_{i_2} \cup A_{i_3}$. Assume, toward a contradiction, that $A \cap A_{i_1} \subseteq A_{i_2} \cup A_{i_3}$. Then

$$|A \cup A_{i_1} \cup R| = |A| + |A_{i_1}| + |R| - |A \cap A_{i_1}| - |A \cap R| - |A_{i_1} \cap R| + |A \cap A_{i_1} \cap R|$$

since $A \cap A_{i_1} \subseteq A_{i_2} \cup A_{i_3} = R, A \cap A_{i_1} = A \cap A_{i_1} \cap R$. Thus

$$\begin{aligned}|A \cup A_{i_1} \cup R| &= |A| + |A_{i_1}| + |R| - |A \cap R| - |A_{i_1} \cap R| \\ &= 2r + |A_{i_2} \cup A_{i_3}| - |A \cap R| - |A_{i_1} \cap R|\end{aligned}$$

which is a contradiction since $2r + |A_{i_2} \cup A_{i_3}| - |A \cap R| - |A_{i_1} \cap R| > n$. ■

Note that since $|A_{i_2} \cup A_{i_3}| - |A \cap R| - |A_{i_1} \cap R| < r - 2s$, if the condition of Lemma 3 holds implies (as expected) the condition of Lemma 2.

Corollary 1

Let A be a set of r-regular set system F defined on a universe set X of cardinality n. Let $N(A) = \{A_1, \ldots, A_d\}$ be the set of the $d \geq 1$ neighbors of A. Then for any three of its neighbors $A_{i_1}, A_{i_2}, A_{i_3}$, if their pairwise intersections are of cardinality s and $3r - 4s > n$, then $A \cap A_{i_1} \nsubseteq A_{i_2} \cup A_{i_3}$.

Proof. In the condition of Lemma 3, we can replace the quantity $|A_{i_2} \cup A_{i_3}| - |A \cap R| - |A_{i_1} \cap R|$ with the smaller quantity $r - 2s$. ■

14.1.5.2 Constructions Based on Hadamard Matrices

A simple set system construction method that one can use relies on the well-known Hadamard matrices whose definition is the following [16,17]:

Definition 2 *A Hadamard Matrix of order n is an $n \times n$ matrix whose elements are either $+1$ or -1, whereby the rows are pairwise orthogonal (that is, their pairwise inner product is 0) and $HH' = nI$ where H' is the transpose of H and I is the identity matrix.*

Hadamard matrices of orders 1 and 2 exist, but every other Hadamard matrix is of order n, where n is a multiple of 4.

One type of Hadamard matrices are the square Sylvester construction, which are based on the fundamental matrix $H_1 = \begin{bmatrix} +1 & +1 \\ +1 & -1 \end{bmatrix}$. Higher order matrices are constructed by recursion, for example,

$$H_2 = \begin{bmatrix} H_1 & H_1 \\ H_1 & -H_1 \end{bmatrix} = \begin{bmatrix} +1 & +1 & +1 & +1 \\ +1 & -1 & +1 & -1 \\ +1 & +1 & -1 & -1 \\ +1 & -1 & -1 & +1 \end{bmatrix}$$

and

$$H_3 = \begin{bmatrix} H_2 & H_2 \\ H_2 & -H_2 \end{bmatrix} = \begin{bmatrix} +H_1 & +H_1 & +H_1 & +H_1 \\ +H_1 & -H_1 & +H_1 & -H_1 \\ +H_1 & +H_1 & -H_1 & -H_1 \\ +H_1 & -H_1 & -H_1 & +H_1 \end{bmatrix}.$$

Exercise 3 Compute, explicitly, the matrix H_3.

By deleting the first row and first column, we obtain a set system in which all sets have cardinality $(n/2) - 1$ and pairwise intersections of size $(n/4) - 1$ on a universe X of size $n - 1$. By choosing any subset of these rows that have -1 in the same column and by considering them as sets on the universe set X minus the element that correspond to this column, we obtain a set system that satisfies the condition of Lemma 2:

$$3r - 2s = 3\left(\frac{n}{2} - 1\right) - 2\left(\frac{n}{4} - 1\right) = n - 1 > n - 2 = |X'|.$$

We will, now, define a class of matrices, the *intersection constrained* matrices, which can be used in building set systems with parameters that satisfy the condition of Lemma 2 or the condition of Lemma 3.

Definition 3 *The set $H^{n,k}_{r,s,c(r,s)}$ is defined to consist of all $k \times n$ matrices representing k sets defined over a universe set of size n, such that for any two of its rows (sets)R_i and R_j (interpreted as sets), the following three conditions hold:*

- $|R_i| \geq r$
- $|R_i \cap R_j| \leq s$
- The condition $c(r,s)$ holds

For instance, the following is a matrix that belongs in $H^{n,k}_{r,s,c(r,s)}$ with $n = 10$, $k = 4$, $r = 4$, $s = 1$ and $c \equiv (3r - s) - n > 0$ (condition of Lemma 2):

$$\begin{bmatrix} -1 & -1 & -1 & +1 & +1 & -1 & +1 & -1 & -1 & +1 \\ -1 & -1 & +1 & -1 & -1 & +1 & -1 & +1 & -1 & +1 \\ -1 & +1 & -1 & -1 & -1 & -1 & +1 & +1 & +1 & -1 \\ +1 & -1 & -1 & -1 & +1 & +1 & -1 & -1 & +1 & -1 \end{bmatrix}$$

We would like to construct arbitrarily large matrices that satisfy a given condition c. As in the case of Hadamard matrices, we will join smaller matrices into larger ones trying either to preserve or enforce the condition c. We will use the symbolism H^{i*} to denote the repetition of H by i times, side by side. We will prove the following lemma for the case of $c = 3r - 2s$ (condition of Lemma 2).

Lemma 4

Let H, H' be arbitrary $k \times n$ matrices with elements ± 1. Assume that for H' the maximum intersection size of its rows is s''. Consider the following matrix:

$$\begin{bmatrix} H & H'^{i*} \\ f(H) & H^{i*} \end{bmatrix}$$

for some function f defined on ± 1 matrices so as not to modify their dimensions. Let s' be the maximum intersection size for rows of the matrix $\begin{bmatrix} H \\ f(H) \end{bmatrix}$. Then if $s' > s''$, r is the minimum number of $+1$ in rows and $i > (2s' - 2s'')/(3r - n - 2)$, $\begin{bmatrix} H & H'^{i*} \\ f(H) & H^{i*} \end{bmatrix} \in H^{in,2k}_{ir,s'+(i-1)s'',c(ir,s'+(i-1)s'')}$.

Consider, for instance, the matrix

$$H = \begin{bmatrix} -1 & -1 & -1 & +1 & +1 & -1 & +1 & -1 & -1 & +1 \\ -1 & -1 & +1 & -1 & -1 & +1 & -1 & +1 & -1 & +1 \\ -1 & +1 & -1 & -1 & -1 & -1 & +1 & +1 & +1 & -1 \\ +1 & -1 & -1 & -1 & +1 & +1 & -1 & -1 & +1 & -1 \end{bmatrix}$$

and

$$f(H) = -H = \begin{bmatrix} +1 & +1 & +1 & -1 & -1 & +1 & -1 & +1 & +1 & -1 \\ +1 & +1 & -1 & +1 & +1 & -1 & +1 & -1 & +1 & -1 \\ +1 & -1 & +1 & +1 & +1 & +1 & -1 & -1 & -1 & +1 \\ -1 & +1 & +1 & +1 & -1 & -1 & +1 & +1 & -1 & +1 \end{bmatrix}$$

Then $s'' = 1$ and $s' = 3$. Let, also, $H' = H$. Then $\begin{bmatrix} H & H^{i*} \\ f(H) & H^{i*} \end{bmatrix} \in H_{8i,2+i,c(8i,2+i)}^{10i,8}$ for $i > 2$. Now the condition of Lemma 3 is too stringent to create among the sets of a set system. Moreover, since the coefficient of r is smaller than the coefficient of s, the construction of Lemma 4 is not applicable. We can, however, create probabilistically sets for which the property $A \cap A_{i_1} \not\subseteq A_{i_2} \cup A_{i_3}$ holds with high probability.

Let us consider a $k \times n$ matrix H whose rows correspond to sets and columns to universe set values. We will change from the ± 1 notation to the notation of 0/1 for convenience. Fix $p, 0 < p < 1$, and set each entry $H(i,j)$ of the matrix, independently of all other positions, to the value 1 with probability p and to the value 0 with probability $1 - p$. Take, now, any two row pairs $\{i_1, i_2\}$ and $\{i_3, i_4\}$ of the resulting matrix. Then the following is easily seen to hold:

Lemma 5

$A_{i_1} \cap A_{i_2} \not\subseteq A_{i_3} \cup A_{i_4}$ iff $\exists j : H(i_3, j) + H(i_4, j) - H(i_1, j) \cdot H(i_2, j) < 0$.

The positions $H(i_3, j), H(i_4, j), H(i_1, j)$ and $H(i_2, j)$ are random variables which assume the value 1 with probability p and the value 0 with probability $1 - p$. Thus

$$Pr[\exists j : H(i_3, j) + H(i_4, j) - H(i_1, j) \cdot H(i_2, j) < 0]$$
$$= 1 - Pr[\forall j : H(i_3, j) + H(i_4, j) - H(i_1, j) \cdot H(i_2, j) \geq 0]$$
$$= 1 - [1 - (p(1 - p))^2]^n.$$

The number of possible row pairs is equal to $\binom{k}{2}\binom{k-2}{2}$. Let Y_i be the indicator random variable corresponding to the ith such row pair such that

$$Y_i = \begin{cases} 1 & \text{if the } i\text{th set does not have the property} \\ 0 & \text{if the 0th set has the property} \end{cases}$$

Then the random variable Y, defined as $Y = Y_1 + Y_2 + \cdots + Y_{\binom{k}{2}\binom{k-2}{2}}$, counts the number of row pairs for which the property does not hold. Using Markov's inequality, $Pr[Y > 0] \leq E[Y]$ [6], the probability that there is at least one such row pair can be bounded above by $\binom{k}{2}\binom{k-2}{2}[1 - (p(1-p))^2]^n$.

Exercise 4 Prove this bound. ■

From this bound, we see that, keeping k constant and increasing n, we can make the probability of the appearance of four sets $A_{i_1}, A_{i_2}, A_{i_3}$ and A_{i_4} such that $A_{i_1} \cap A_{i_2} \not\subseteq A_{i_3} \cup A_{i_4}$ does not hold arbitrarily close to 0.

We can now generalize with the following:

Theorem 1

Let H be a $k \times n$ matrix with entries 1 or 0 chosen at random with probability p and $1 - p$, respectively. Then the probability that at least one set of l indices $i_1, i_2, \ldots, i_l$ exist for which the following holds:

$$A_{i_1} \cap A_{i_2} \not\subseteq \cup_{i \neq i_1, i_2} A_i$$

is bounded from above by $\binom{k}{2}\binom{k-2}{l}[1 - (p(1-p))^2]^n$. Then for any $\varepsilon > 0$, we can compute a value for n for which this probability is smaller than ε by solving the inequality

$$\binom{k}{2}\binom{k-2}{l}[1 - (p(1-p))^2]^n < \varepsilon$$

for n:

$$n > \frac{1}{\ln([1 - (p(1-p))^2])} \cdot \ln\left(\frac{\varepsilon}{\binom{k}{2}\binom{k-2}{2}}\right).$$ ■

Exercise 5 Prove Theorem 1.

14.2 Advanced Concepts for Key Management and Trust in WSNs

In this section, we discuss some considerations with regard to *formally* definable properties that hold, almost certainly, in the limit in randomly growing combinatorial structures that model shapeless computing systems (e.g., dynamic ambient intelligence networks and WSNs). These properties are statements about the key sets of the nodes or about their interconnections, that aid or hinder (depending on the network structure) the formation of proper agreed upon key sets. Our treatment is, rather, theoretical but can be applied to WSNs because such structures are, usually, massive and their interconnection unpredictable and, thus, can be thought of as comprising a "random graph structure." The approach is based on the results that establish the limit behavior of predicates written in the *first* and *second order* logic. Our central viewpoint is that dynamic, global computing systems are not amenable to a static, completely formal definition of their properties (in our case, properties about their key sets). We, rather, maintain that the study of these properties should be of a *statistical, asymptotic* nature, considered in the limit as the number of the network's components and interconnections grow according to some predetermined growth rates. Thus, our main goal is to define "good quality properties" as *emerging properties* appearing when a set of predicates (expressed within some logic formalism) hold, asymptotically, almost certainly in random communication structures.

The proposed approach requires, first, that one adopts a random graph model that describes as accurately as possible the target dynamic system (WSN, for instance). Then a number of properties that model "quality of key sets" are stated using, for instance, first-order logic or some second-order logic fragment. In addition, conditions are established under which these properties appear (or do not appear) in the limit, as the system grows [18].

14.2.1 *Random Graph Models*

As we discussed above, the departure point of our approach is that dynamic, boundary-transcending computing systems (like WSNs) are not amenable to a static consideration of quality of key sets. Thus, our main goal is to define quality as an emerging statement among entities of the system, which appears when a set of predicates hold, asymptotically, almost certainly in random communication structures that model computing systems and the interaction between constituent devices.

One of the most well-studied and most intuitively appealing formalism for studying *emergent properties* is the *graph*. This quality metric model can be used to evaluate key set quality assertions in a distributed information system. Generally, graphs can be used to represent the following statements: key set of node A intersects with key set of node B, or, key set of node A is a proper subset of key set of node B. Then one may analyze the graph and prove, for instance, that the intersection of no two key sets belongs totally to the key set of another key set. However, things get complicated if very large network graphs are considered that evolve in an unpredictable way, such as the WWW society (see Bollobás [5] for a thorough treatment of threshold phenomena in relation to random graph properties).

In this section we will refer to the basic random graph models that are currently used to model entities and relations among them as graphs: nodes represent entities and edges among entities represent relations among key sets (e.g., intersection). But why random? Randomness in the graph model has been studied extensively and many rigorous results exist for proving that evolving graphs have a number of interesting, emerging, global properties. But this is a matter of convenience in proving things about big structures, such as the dynamic networks and its key set properties. Actually, randomness is a way to model the unpredictability of how the network structure changes by the addition (and deletion) of huge numbers of links (communication links or key set relationships in our case) on a daily basis. Since unpredictability without any previous knowledge about possible biases permits the "full randomness" assumption, random graphs may uncover many interesting properties of the network graph.

We will assume from now on, for simplicity, that the key set relationships are symmetric and no weights (that is, strength or importance estimates) exist for these relationships. The basic definitions can be extended, but we will refrain from doing so in order to exemplify the basic techniques. In what follows, by n, we will denote the number of network nodes and by Ω the set of all possible $\binom{n}{2}$ edges between these nodes [6].

- Model $G_{n,m}$: select the m edges of G by selecting them uniformly at random, independently of one another from Ω.
- Model $G_{n,p}$: include each edge of Ω in G independently of the others and with probability p.

- Model $G_{n,R_0,d}$: generate n points in some d-dimensional metric space uniformly at random and draw an edge between two points only if their distance is at most R_0.
- Model $G_{k,m,p}$: each node i of the k available creates a set S_i by selecting uniformly at random each of the available m objects with probability p. Then an edge is formed between two nodes i, j only if $S_i \cap S_j \neq \emptyset$. This is the random intersection graph model.

14.2.2 A Randomized Scheme Based on the Fixed Radius Model

14.2.2.1 Random Points in Euclidean Spaces

With regard to the distribution of the distance [19] between points chosen uniformly at random to lie within a Euclidean sphere, the following was proved in [10]: the probability density function and cumulative distribution function for the distance x between two random points within a d-dimensional Euclidean ball of radius R are given, respectively, by the following equations:

$$P_d(s) = \frac{s^{d-1} \int_{s/2}^{R} (R^2 - x^2)^{(d-1/2)}}{(1/2d) B((d/2) + (1/2), (1/2)) R^{2d}}$$

$$D_d(x) = \int_0^s P_d(s)\, ds = \left(\frac{x}{R}\right)^d - \frac{B_\alpha((1/2), (d/2) + (1/2))}{B((d/2) + (1/2), (1/2))} \left(\frac{x}{R}\right)^d$$

$$+ 2^d \frac{B_\alpha((d/2) + (1/2), (d/2) + (1/2))}{B((1/2), (d/2) + (1/2))}$$

with $0 \leq x \leq 2R$, $\alpha = 1/4(x/R)^2$. The function $B_\alpha(x, y)$ is the *incomplete beta function*:

$$B_\alpha(x, y) = \int_0^\alpha t^{x-1}(1 - t)^{y-1}\, dt$$

while $B(x, y)$ is the beta function which is equal to $B_\alpha(x, y)$ for $\alpha = 1$. For more on these function, see, for example, [8].

For the two-dimensional case, which is of interest for *surface* sensor networks, we have the following: For the circle (the two-dimensional Euclidean ball) of radius R it holds that

$$P_2(s) = \frac{2s}{R^2} - \frac{s^2}{\pi R^4}\sqrt{4r^2 - s^2} - \frac{4s}{\pi R^2} \arcsin \frac{s}{2R}, \quad \text{and}$$

$$D_2(x) = 4\left(\frac{x}{2R}\right)^2 + \frac{1}{2\pi}\left(\frac{x}{2R}\right)\left(4 - 4\left(\frac{x}{2R}\right)^2\right)^{3/2} - \frac{3}{\pi}\left(\frac{x}{2R}\right)\sqrt{4 - 4\left(\frac{x}{2R}\right)^2}$$

$$+ \frac{2}{\pi} \arcsin \frac{x}{2R} - \frac{8}{\pi}\left(\frac{x}{2R}\right)^2 \arcsin \frac{x}{2R}.$$

Exercise 6 Prove the above derivations.

We will now study the threshold behavior of the fixed radius random graph model with regard to properties expressible in the first-order language of graphs (see Section 2.3 for the details).

For the two-dimensional sphere (circle) the probability that $A_{s,t}$ fails for $G_{n,R_0,d}$ is bounded from above as follows:

$$Pr[A_{s,t} \text{ fails in } G_{n,R_0,2}] \leq \binom{n}{s+t} \left[1 - D_2(R_0)^s(1 - D_2(R_0))^t\right]^{n-(s+t)}. \quad (14.1)$$

If $\sigma = (R_0/2R) = c$ is a constant, $0 < c < 1$, then the right-hand side of Equation 14.1 tends to 0. If $\sigma = (R_0/2R) = f(n) = \omega(1/\sqrt{n})$, then the right-hand side of Equation 14.1 also tends to 0.

Proof. From Equation 14.1, it follows that

$$Pr[A_{s,t} \text{ fails in } G_{n,R_0,2}] \leq \binom{n}{s+t} \exp\left[-D_2(R_0)^s(1 - D_2(R_0))^t(n - (s + t))\right]. \quad (14.2)$$

Our goal is to find a condition on c such that the right-hand side of Equation 14.2 tends to 0. Then $Pr[A_{s,t}$ fails in $G(n, R, 2)]$ tends to 0 and, consequently, $Pr[A_{s,t}$ holds in $G_{n,R_0,2}]$ tends to 1, establishing the fact that any first-order property holds, asymptotically, in $G_{n,R_0,2}$ with probability 1 or 0.

Let σ be a constant c, $0 < c < 1$. Then $D_2(R_0)$ is a constant too. Thus, the exponential factor of the right-hand side of Equation 14.2

$$\exp\left[-D_2(R_0)^s(1 - D_2(R_0))^t(n - (s + t))\right] \quad (14.3)$$

tends to 0, for fixed s, t and n tending to infinity. Therefore, the probability $Pr[A_{s,t}$ fails in $G_{n,R_0,2}]$ also tends to 0.

Let, now, $\sigma = f(n) < 1$, a function of n tending to 0. Then using power series analysis around 0, we obtain that

$$D_2(R_0) = 4\sigma^2 + \frac{1}{2\pi}\sigma(4 - 4\sigma)^{3/2} - \frac{3}{\pi}\sigma\sqrt{4 - 4\sigma^2} + \frac{2}{\pi}\arcsin\sigma - \frac{8}{\pi}\sigma^2 \arcsin\sigma$$

$$= 4\sigma^2 - \frac{32}{3\pi}\sigma^3 + \frac{16}{15\pi}\sigma^5 + O(\sigma^6). \quad (14.4)$$

The term $D_2(R_0)^s(1 - D_2(R_0))^t$ in the exponent in Equation 14.3 can be approximated as follows:

$$D_2(R_0)^s(1 - D_2(R_0))^t = 4s\sigma^2 - \frac{32s}{3\pi}\sigma^3 - \left[16st + 8s(s - 1)\right]\sigma^4$$

$$+ \left[\frac{256st}{3\pi} + \frac{16s}{15\pi} + \frac{128s(s - 1)}{3\pi}\right]\sigma^5 + O(\sigma^6) \quad (14.5)$$

with s, t constants. Then, from Equations 14.4 and 14.5, it follows that if $\sigma = f(n) = \omega(1/\sqrt{n})$, then the right-hand side of Equation 14.3 tends to 0, for any s, t, completing the proof.

Exercise 7 Prove Equation 14.5.

The generalization, now, follows readily: Let $\sigma = (R_0/2R) = c$ be a constant, $0 < c < 1$. Then for any first-order property A, then $Pr[G_{n,R_0,d}$ has $A]$ tends to 1 or 0. If $\sigma = (R_0/2R) = f(n) = \omega(1/\sqrt[d]{n})$, then $Pr[G_{n,R_0,d}$ has $A]$ tends to 1 or 0 too. Although the property of forming a connected graph cannot be described in the first-order theory of graphs, in [20] it is shown that for slightly larger values of σ, the network is almost certainly connected. More specifically, we need only to increase the threshold probability (in the two-dimensional case) from $1/\sqrt{n}$ to $\sqrt{\log(n)}/\sqrt{n}$ to, also, ascertain connectivity in the resulting graph. See, also, [21] for proofs of threshold behavior of the geometric random graph model.

14.2.2.2 A Key Predistribution Scheme Based on Random Points on Circular Disks

On the basis of these considerations, we will propose a key management scheme that does not rely on predistribution but rather creates (and destroys) key sets dynamically for each node taking into account its physical position so as to form an interdependence between the key sets of physically nearby nodes and, thus, help these node to reach agreement on the keys that will be used for their communication.

For the details now, assume we have n nodes randomly distributed within a circle of radius R. We first fix a value C which, for each node, will define a circle centered at the node within which candidate keys will be considered. The radius C also models the communication range of each of the n nodes, meaning that their communication devices have sufficient power to transmit only within distance C away from the node (circle of radius C).

Assuming that the nodes are placed uniformly at random within the area of radius R, a fixed radius random graph with n nodes is formed so as to include edges between nodes only if their distance is at most $2C$ (that is, their communication ranges–circles intersect). We also assume that each of the nodes knows its coordinates (e.g., Cartesian coordinate on the plane).

We now consider a discretization, a lattice, of the area with radius R which is known to the nodes. Thus, each of the nodes will occupy a point of the lattice. We are interested in estimating the number of lattice points lying within a radius C from a given node (we disregard minor discretization discrepancies since, asymptotically, they do not affect the estimates we will use). This is actually a problem, known as the *Gauss circle problem*, which asks for an estimate for the number of points within distance C of a given lattice point. This estimate is given by $N(C) = \pi C^2 + E(C)$, with $E(C) \leq 2\sqrt{2}\pi C$. See [22] for these values and very informative illustrations related to the Gauss circle problem as well as the more formal exposition of [23]. The combined (x, y) coordinates of these lattice points (which can easily be systematically produced by each node with only information its current position on the lattice) can form a set of keys to be used while interacting for establishing secure communication with its nearby neighbors within distance $2C$ (which are actually within the physical communication reach).

Regarding the relationship between R, the range within which the nodes are moving, C, the transmission range, and n, the number of nodes, we can draw some useful conclusions from the consideration of Theorem 3. In our context, $R_0 = 2C$. Thus, $\sigma = C/R$. Let $C = C(n)$ and $R = R(n)$ be functions of n tending to infinity, with $C(n) = o(R(n))$. The assumption of $R(n)$ and $C(n)$ tending to infinity reflects the fact that as more nodes appear within a range, we should allow them to move in a wider area and, also, increase their communication range. The assumption $C(n) = o(R(n))$ reflects the fact that we should not force the nodes to increase the communication range too much, compared with the region within which they move, since the power dissipation will be excessive while, in addition, problems will appear with nodes eavesdropping on the communication of other nodes. Thus, $\sigma = (C(n)/R(n)) \to 0$ and, according to Theorem 3, the extension property holds with probability approaching 1 as the number of nodes increases. This means that all properties expressible in the first-order language of graphs hold (asymptotically with n) either with probability 1 or 0. What we need to do next is to define *good* properties with regard to the chosen key sets that can be expressed in this graph language.

An example of such a property is *there is no triangle*. Having a triangle in the fixed radius random graph model means that for a pair of nodes which are sharing keys, there is another node that shares key with both nodes, a thing which might cause problems since it reduces the candidate keys which can possibly used for secure communication. This is because the two nodes should avoid the selection of keys that are also shared with the third node of the triangle.

Other good properties could be the following:

- For any node v, its key set A_v is not a subset of the key set of any other node. Note that his property holds for the fixed radius random graph model and the key management scheme we introduced above.
- For any node v, its key set A_v cannot be a subset of the union of the key sets of at least l other nodes. Although this property cannot, possibly, be expressible in the first-order language of graphs, it nevertheless can be approximated by a property that is expressible: *no l nodes of the graph are adjacent, simultaneously, to any given node.*

14.2.3 First-Order Language of Graphs

We are interested in discovering conditions under which a random graph model displays a 0–1 behavior for certain properties that can also be relevant to key set security issues. By a "0–1 property" we mean a property that either holds with probability tending to 1 or with probability tending to 0, in the limit as the random graph grows. In this section we will be focused on properties expressible in the *first-order language* of graphs. This language can be used to describe some useful (and naturally occurring in applications) properties of random graphs under a certain random graph model, using elements of the first-order logic [18,24]. The alphabet of the first-order language of graphs consists of the following elements [25]:

- Infinite number of variable symbols, for example, x,w,y ... which represent graph vertices.
- The binary relations == (equality between graph vertices) and: (adjacency of graph vertices) which can relate only variable symbols, for example, x:y means that the graph vertices represented by the variable symbols x, y are adjacent.
- Universal, $\exists$, and existential, $\forall$, quantifiers (applied only to *singletons* of variable symbols).
- The Boolean connectives used in propositional logic, that is, $\vee, \wedge, \neg, \Rightarrow$.

An example of graph property expressible in the first-order language of graphs is the existence of a triangle: $\exists x \exists y \exists w (x : y) \wedge (y : w) \wedge (w : x)$. Another property is that the diameter of the graph is at most 2 (can easily be written for any fixed value k instead of 2): $\forall x \forall y [x = y \vee x : y \vee \exists w (x : w \wedge w : y)]$. However, other equally important graph properties, like connectivity, cannot be expressed in this language.

Exercise 8 Try to state connectivity of a graph in the first-order language.

We will now define the important *extension statement* in natural language, although it clearly can be written using the first-order language of graphs (see Spencer [26] for the details):

Definition 4 (Extension statement $A_{s,t}$) *The extension statement $A_{s,t}$, for given values of s,t, states that for all distinct $x_1, x_2, \ldots, x_s$ and $y_1, y_2, \ldots, y_t$ there exists distinct z adjacent to all x_i s but no y_j.*

The importance of the extension statement $A_{r,s}$ lies in the following theorem. When applied to the first-order language of graphs

---

Theorem 2

Let G to be a random graph with n nodes and $A_{r,s}$ to be an extension statement, then if $A_{r,s}$ for all r,s $\lim_{n\to\infty} Pr[G \text{ has } A_{r,s}] = 1$, then for every statement A written in the first-order language of graphs either $\lim_{n\to\infty} Pr[G \text{ has } A] = 0$ or $\lim_{n\to\infty} Pr[G \text{ has } A] = 1$.

The connection between 0 and 1 laws in random graphs and first-order logic was first noted by Fagin in the seminal paper [27]. ■

14.2.4 Second-Order Language of Graphs

Although the extension property can be used in order to settle the existence of 0–1 behavior for all properties expressible in the first-order language of graphs in any random graph model, things change dramatically when properties are considered that are expressed in the *second*-order language of graphs.

The second-order language of graphs is defined exactly as the first-order language (see Section 2.3), except that it allows quantification over subsets of graph vertices (predicates) instead of single vertices. An example of such a property follows [15].

Definition 5 (Separator) *Let $F = \{F_1, F_2, \ldots, F_m\}$ be a family of subsets of some set X. A separator for F is a pair (S,T) of disjoint subsets of X such that each member of F is disjoint from either S or from T. The size of the separator is $\min(|S|, |T|)$.*

In the context of our problem, this property may be interpreted as follows. Let us assume that $|F_i| = 2$, modeling an edge of a graph. Thus, the sets F_i model a graph's links between pairs of nodes. With this constraint, the separator property says that in a graph there exist two disjoint sets of nodes S and T such that any set of two adjacent (that is, communicating) nodes is disjoint from either S or T. In other words, it is not possible to have one node belonging to one of the two disjoint sets S and T and the other node belonging to the other. This might mean that no two communicating nodes are authenticated by two different authentication bodies (the two disjoint sets of nodes). Thus, the two nodes can communicate with each other more since they are not authenticated by two disjoint (that is, unrelated) authentication bodies. Each of the two disjoint sets may form, for instance, Certification Authority (CA) providing authentication services.

In order to cast the separator property into the language of graphs, we set X to be a set of vertices and the subsets F_i to be of cardinality 2 so as to represent graph edges. Then the separator property can be written in the framework of the second-order language of graphs as follows:

$$\exists S \exists T \forall x \forall y [\neg(Sx \wedge Tx) \wedge (Axy \rightarrow \neg(Sx \wedge Ty \vee Sy \wedge Tx))].$$

Let us define another property:

Definition 6 (Key set representatives) *A graph G has the key set representative property if there exists a set of vertices such that any vertex in the graph is an adjacent with at least one of these vertices.*

A formal definition using a second-order logic is the following: $\exists S \forall x \exists y [Axy \wedge Sy]$.

The extension statement cannot, unfortunately, be used in order to examine whether (and under which conditions on the random graph model parameters) the separator property or the key set representative property is a 0–1 property since these properties cannot be written in the first-order language of graphs. However, in 1987, Kolaitis and Vardi initiated in [28] a research project in order to characterize fragments of the second-order logic that display 0–1 behavior (that is, they have a 0–1 law). (The interested reader may consult the review paper [29] by the same authors.) Without delving into the details, one of the important conclusions reached at by this project is that there are second-order fragments that do not have a 0–1 behavior while other second-order fragments do.

Let Σ_1^1 denote the existential second-order logic (that is, formulas contain only existential quantification over second-order variables, that is sets). Let FO denote the first-order logic formalism and L be any fragment of FO. Then a $\Sigma_1^1(L)$ sentence over a vocabulary R is an expression of the form $\exists S \phi(R, S)$, where S is a set of relation variables and $\phi(R, S)$ is a first-order sentence on vocabulary (R,S). In general, 0–1 behavior is not displayed by Σ_1^1 [29]. Thus, in order to discover fragments of Σ_1^1 that do have such a behavior, a restriction is imposed on the first-order part (that is, the sentence ϕ written in L) of the sentences considered. This restriction refers to the pattern of quantifiers that appear in the first-order sentence ϕ. Some restricted first-order logics that have been studied in connection to Σ_1^1 are the following:

1. The *Bernays–Schönfinkel class*, which is the set of all first-order sentences with quantifier prefixes of the form $\exists^* \forall^*$ (that is, the existential quantifiers precede the universal quantifiers).
2. The *Ackermann class*, which is defined as the collection of first order sentences of the form $\exists^* \forall \exists^*$ (that is, the quantification prefix contains only one universal quantifier).
3. The *Gödel class*, which is defined as the collection of first order sentences of the form $\exists^* \forall \forall \exists^*$ (that is, the prefix contains two consecutive universal quantifiers).

The separator property defined earlier belongs to the second order fragment Σ_1^1(Gödel) since it contains (in the first order part) two consecutive universal quantifiers. On the other hand, the key set representative property belongs to the second order fragment $\Sigma_1^1(Ackermann)$ since it contains a single universal quantifier. The key set representatives property can be proved to be a 0–1 property since the second order logic fragment $\Sigma_1^1(Ackermann)$ has a 0–1 behavior in general [29]. This means that, asymptotically, it holds with either probability 0 or 1 depending on the random graph model parameters. On the other hand, the separator property is not guaranteed to be a 0–1 property since the Σ_1^1(Gödel) second-order logic fragment does not display a 0–1 behavior in general [29,30]. See, also, the series of papers [31–33] for interesting results pertaining to 0–1 laws of fragments of second-order logic.

14.2.5 Undecidable Probabilities

Thus, sentences (properties) that can be written in fragments of second-order logic that have a 0–1 behavior (e.g., $\Sigma_1^1(Ackermann)$) are 0–1 properties. However, some second-order logic fragments allow the construction of sentences that have no limiting probability and, thus, are not 0–1 properties.

Theorem 3 (Trachtenbrot–Vaught Theorem [34])

There is no decision procedure that separates those first-order statements S that hold for some finite graph from those S that hold for no finite graph.

With regard to random graphs now which, as we show, in conjunction with the first- and second-order language of graphs, can be used to express, formally, complex relationships that can be related to key sets, we have the following result [7]:

Theorem 4

There is no decision procedure that separates those first-order statements S that hold almost always for the random graph $G_{n,p}$ from those for which $\neg S$ holds almost always.

This theorem is targeted to $G_{n,p}$ random graphs, with $p = n^{\alpha}$, α being a rational number between 0 and 1. In summary, for any first-order statement A about a finite graph, a first-order statement A^* is given that holds almost always in $G_{n,p}$, if A holds for some finite graph, while it never holds, if A holds for *no* finite graph. Now, if a formal procedure (algorithm) existed for deciding such statements for the $G_{n,p}$ model, then relationship between A and A^* would allow using the procedure to separate those first-order statements A that hold for some finite graph from the statements that hold for no finite graph, contradicting the Trachtenbrot–Vaught theorem.

More specifically, let us consider the following statement S: There is no isolated vertex in the graph, which can be written as $\forall y \exists z (y : z)$. Let S^* be the corresponding statement, for the random graph $G_{n,p}$ with $p = n^{-2/5}$ [7]:

$$\exists x_1 \exists x_2 \exists x_3 \exists x_4 \left[\forall y \text{MEM}(y; x_1, x_2, x_3, x_4) \implies \exists z \text{MEM}(z; x_1, x_2, x_3, x_4) \wedge ADJ(y, z) \right]$$

and ADJ being the following first-order language predicates:

$$\text{MEM}(y; x_1, x_2, x_3, x_4) \Leftrightarrow \exists z [(z : x_1) \wedge (z : x_2) \wedge (z : x_3) \wedge (z : x_4) \wedge (z : y)]$$
$$ADJ(u, v) \Leftrightarrow \text{MEM}(u; x_1, x_2, x_3, x_4) \wedge \text{MEM}(v; x_1, x_2, x_3, x_4) \wedge \exists t \text{MEM}(t; x_1, x_2, u, v).$$

Then:

$$\lim_{n \to \infty} Pr[G_{n,p} \text{ has } S^*] = \begin{cases} 0 & \text{if } S \text{ holds for no finite graph,} \\ 1 & \text{if } S \text{ holds for some finite graph.} \end{cases}$$

Thus, a decision procedure that could differentiate between statements that hold almost always in $G_{n,p}$ and the statements whose negation holds almost always would provide a decision procedure to differentiate between those statements S that hold for *some* finite graph and those that hold for no finite graph, contradicting the Trachtenbrot–Vaught theorem. ■

The morale of this discussion is that it may not even possible to mechanically analyze whether a given state of affairs (e.g., an assertion about key set) or its negative, within the world of discourse, is expected to almost certainly appear. Thus, it may be the case that one may have to observe the target world for sufficiently much time in order to be able to make a safe prediction about the state of affairs that will finally prevail in the limit (see [35] for a discussion on the possibility of founding general properties of randomly evolving structures on formalism alone).

14.2.6 Set Systems Based on Special Polynomials

14.2.6.1 Some Definitions

We will adopt the notation of [36]. Let q be a positive integer and $L \subset Z$. We say that $r \in L (\text{mod } q)$ if there exists $l \in L$ such that $r \equiv l(\text{mod } q)$. Otherwise, we say that $r \notin L(\text{mod } q)$. Let X be a *universe* set with $|X| = n$. A *set system* F on the universe set X is a collection of nonempty subsets of X. Also, by X_k we will denote the collection of all subsets of X with k elements (see, also, [26,37] for more on set systems with special intersection properties).

Definition 7 *A set system F is called r-regular if all its members have cardinality $r \geq 1$.*

Definition 8 *A set system F is called L-avoiding* mod q *if $\forall E \in F, |E| \notin L(\text{mod } q)$.*

Definition 9 *A set system F is called L-intersecting* mod q *if $\forall E, F \in F : E \neq F, |E \cap F| \in L(\text{mod } q)$.*

14.2.6.2 The BBR Polynomials

One way of constructing set families with interesting properties is through the use of the following theorem:

Theorem 5 (Frankl [38])

Let $g(x)$ be a polynomial of the form $g(x) = \sum_{i=0}^{d} b_i x_i$, where the b_i are nonnegative integers. Choose $q \in Z$ and $L \subset Z$, and suppose F is a set system over X which is L-intersecting mod q. *Then there is a set system G on a universe of size $g(|X|)$, with $|G| = |F|$, which is g(L)-intersecting* mod q. *If we further have that, for all sets $E \in F, g(|E|) \notin g(L)(\text{mod } q)$, then G is also g(L)-avoiding* mod q. ■

Theorem 6 (Barrington, Beigel, and Rudich)

Let $P_1, \ldots, P_r$ be r distinct prime numbers, with $r \geq 1$. Let t be an integer of the form $t = \prod_j p_j^{e_j}$, and let q be a positive integer divisible by $\prod_j p_j$. Then there exists a polynomial $Q_{q,t}(x)$ such that:

1. $Q_{q,t}(x) = \sum b_i x_i$, where $0 \leq b_i \leq q$
2. $Q_{q,t}(x) \equiv 0(\text{mod } q)$ if and only if $x \equiv 0(\text{mod } t)$

3. $\deg Q_{q,t}(x) \leq \max_j p_j^{e_j}$

4. $Q_{q,t}(x)$ takes only 2^r values $(\bmod\, q)$

On the basis of these theorems, we can start with some randomly chosen regular set family (whose members intersect in a random fashion) and construct another family with controllable intersection sizes. The general scheme is as follows:

- Choose a universe set X of size n.
- Select k random sets, $S_1, \ldots, S_k$, of cardinality l from within the set X_l.
- Form a $k \times k$ matrix M whose entry (i, j) equal $|S_i \cap S_j|$.
- Construct the set family $S_{1'}, \ldots, S_{k'}$ using a polynomial as defined in Theorem 3 and applying Theorem 2 with this polynomial.
- Construct a $k \times k$ matrix M' whose entry (i,j) equals $|S_{i'} \cap S_{j'}| = Q_{q,t}(|S_i \cap S_j|)$.
- For each value h of the possible 2^r values of $Q_{q,t}$ compute $M_{h'}$ such that $M_{h'}(i,j) = 1$ if and only of $M'(i,j) = h \bmod q$. Otherwise, $M_{h'}(i,j) = 0$.
- For each $M_{h'}$, any maximal (no need to be maximum) clique constitutes a family of sets whose pairwise intersections are equivalent $\bmod\, q$.

Although finding maximum cliques is a computationally intractable problem, finding a maximal clique can be done efficiently. One way to do so is the following: we start from any graph node and consider it a clique of size 1. Then we produce larger cliques by merging cliques of smaller numbers of nodes. Two cliques C_1 and C_2 can be merged if each node belonging to clique C_1 is adjacent to each node of clique C_2. This is a linear time algorithm that can be implemented efficiently based on efficient implementations of Union-Find algorithms. From each such maximal clique, we can single out several regular set families, one for each intersection value.

There is another interesting property that a regular family has. Let us first give two definitions, that of a *sunflower* and that of a Δ-*system:*

Definition 10 *A sunflower with k petals and a core Y is a collection of sets $S_1, \ldots, S_k$ such that $S_i \cap S_j = Y$ for all $i \neq j$. The sets S_i-Y are called the petals and should be nonempty.*

Definition 11 *A set family $F = \{S_1, \ldots, S_k\}$ is called a weak Δ-system, if there is some λ such that $|S_i \cap S_j| = \lambda$, whenever $i \neq j$.*

The following was proved in 1973 by Deza:

Theorem 7

Let F be an r-uniform weak Δ-system. If $|F| \geq r^2 - r + 2$ then F is a sunflower.

Exercise 9 Prove Theorem.

Thus, if the singled out cliques (which contain sets whose pairwise intersections are equal) contain more than $r^2 - r + 2$ members, then all the sets participating in it form a sunflower, that is, they share a set of keys. This may have applications in establishing key sets which, when necessary, can lead the possessing nodes to mutual agreement on a common subset of they keys (the core of the sunflower).

14.3 Conclusions

In this chapter, we have attempted to describe a number of techniques that are applicable to the key management problem for mobile ad hoc and sensor networks. Our focus was on the mathematical techniques and the way in which they can be applied to this problem. The main objective was to demonstrate that there is a rich theory that researchers can employ for the development of robust key management schemes. We saw, for instance, that a number of proposed schemes rely on the principles of public key cryptography and build on ideas already exploited in Public Key Infrastructures for conventional networks. Some other schemes resort to the use of specially constructed (either deterministically or probabilistically) set systems (combinatorial designs). Judging from the variety of methodologies as well as richness of obtained results found in the reviewed papers, we believe that wireless network security has reached a level of maturity that allows the formation of design principles guided by theoretical foundations, much like the principles that have existed for decades now for conventional networks. One characteristic of mobile networks, however, which distinguishes them from the conventional ones, has yet to be explored more fully. This is the lack of structure and how it affects the way attacks may spread in the network as well as efforts for recovery. Solving the key management problem is one facet of the problem. However, it is still not clear how to confront massive attacks in mobile networks, given that no structure exists and, thus, no network information can be exploited to locate attack points as well as attack spread patterns. This issue is clearer in conventional networks where numerous models of virus/disease spread as well as predator/prey interaction can be exploited to model attack and defense using specially designed random walks. The very lack of structure in mobile networks necessitates a reconsideration of all such models and techniques in order to be applicable to these networks too.

In this chapter we have, also, reviewed a number of formalisms with respect to their expressive and deductive power when describing trust/security-related properties of large combinatorial structures, with emphasis on properties related to key sets of network nodes. Our view is that key set properties can be reduced to a number of predicates that appear as a limiting behavior in systems under certain conditions. These systems are modeled within the formalism of a random graph model according to the context of the target system. Then the properties can be written formally using the first- and second-order language of graphs. If the properties can be written in the first-order language of graphs, then one can use the extension statements in order to establish the conditions under which the model displays 0–1 behavior and, thus, all the properties hold asymptotically with either probability 0 or 1. On the other hand (and, perhaps, more interestingly) if a property cannot be written in the first-order language of graphs, then one may try to see if it can be defined within the vocabulary of a second-order logic fragment that has 0–1 behavior. Otherwise, the question of whether the property holds almost certainly or not remains open and needs the application of a more difficult to apply methodology as the one used for proving that the Kernel property is not a 0–1 property [31]. Our view is that in order to study the key set quality within the realm of dynamically changing complex computing systems, one has to resort the discovery of formally definable key set properties (that are apt for the application at hand—for example, the separator property) and see what happens when the system grows [24].

Terminologies

Ad-hoc network
BBR Polynomials
Combinatorial Key Pre-Distribution
Deterministic Key Distribution
First-order language of graphs
Hadamard Matrix
Key management problem
Key Predistribution
Pair-Wise Key Establishment Scheme
Path-Key Establishment
q-Composite Random Key Predistribution Scheme
Random graph models
Random Key Pre-distribution
Random points in Euclidean spaces
Second-order language of graphs
Security requirements
S-free set family
Shared Key Discovery
Single Network-Wide Key
Undecidable probabilities

Questions and Sample Answers

1. What is MANET?
   MANET is a wireless mobile ad hoc network, which is a collection of wireless nodes that form a temporary network dynamically on an "as-needed" basis. This is done without the use of any pre-existing infrastructure.

2. What is a DSN?
   Distributed sensor network (DSN) is a mobile ad hoc network, which consists of nodes of limited computation and communication abilities. These types of networks are mainly used for military purposes for monitoring and collecting information from hostile environments. Types of sensors include acoustic, seismic, and magnetic.

3. What are the security challenges for wireless mobile ad-hoc networks and wireless sensor networks?
   ■ Increased probability of the occurrence of a security violation event
   ■ Mobility and service ubiquity
   ■ Changes in network characteristics
   ■ Dynamic user management
   ■ Unpredictable network size variability

4. Why use of pair-wise key is prohibitive?
   If the WSN consists of N nodes, then each node would be assigned a unique pairwise key with all nodes of the network, resulting in an additional overhead required for each node to establish $N - 1$ unique keys that are stored in the memory of each sensor node.

5. Why mobility and service ubiquity pose security challenges in MANET?
   Since users of mobile networks move constantly, there is no fixed communication infrastructure on which services can reliably be (and quickly) delivered. This creates difficulties in establishing connections between distant users (as well as service access points) since there may be situations where such a connection may be routed through other mobile stations. Such a need for multihop connectivity creates reliability as well as security problems.

6. What is key predistribution?
   Key predistribution is the storing of secret keys in the sensor node memories before deploying them over the target area of deployment.

7. What is a path-key? Why is it needed?
   After the Shared Key Discovery phase, various nodes in the WSN may be left without any links to other nodes. In this case, a path-key is assigned to a selected pair of sensor nodes that are in wireless communication range, but do have a key in common.
   If a sensor z generates a key between sensors x and y which is used to encrypt message m, and becomes a Key Distribution Centre.

8. Write the basic idea of q-composite Random Key Predistribution Scheme?
   The q-composite Random Key Predistribution Scheme is a based on the basic scheme but instead of requiring two neighboring nodes to possess one common key between them, the q-composite scheme requires the existence of q commons keys ($q > 1$), which results in an increased network resilience against node capture.

9. Define Key set representative.
   A graph G has the key set representative property if there exists a set of vertices such that any vertex in the graph is an adjacent with at least one of these vertices.

10. What is Trachtenbrot–Vaught theorem?
    There is no decision procedure that separates those first-order statements S that hold for some finite graph from those S that hold for no finite graph.

Author's Biography

Effie Makri holds a degree in informatics engineering from the Technological Educational Institute of Athens (1994) and an MSc degree in electronic engineering from the University of Dublin, Trinity College (1996). She is currently doing PhD studies in the mathematics department of the University of the Aegean on Objective Number Theory and its applications to key agreement protocols for wireless *ad-hoc* sensor networks. Her research interests include optimization of broadband ATM networks, airborne radar clutter, wireless LANs, middleware and object technologies, wireless ad-hoc and sensor networks and recently development of semantically enhanced knowledge management systems using ontological models. Effie Makri has worked on a number of EU R&D projects in the field of security, telecommunications, location-based services, and next generation networks.

Yannis C. Stamatiou holds a degree of Computer Engineering & Informatics, from the University of Patras (1990) and a PhD on "Theory and Applications of the Constraint Satisfaction Problem: Distributed Environment-Parallel and Randomized Algorithms–Nonmonotonic Reasoning." He also holds a certificate of postgraduate studies from the Greek Open University in the field of

"Educational Systems for Open and Distant Learning" and he was a postdoctoral fellow at the computer science department of Carleton University, Ottawa, Canada (1999). His scientific interests lie in the field of security and cryptography. He has participated as a technical manager in many European R&D projects related to security and has served as an expert evaluator in three European calls for proposals within FP7. At present, he is an assistant professor at the University of Ioannina, Department of Mathematics, Greece, and consultant on ICT security and cryptography at the Research Academic Computer Technology Institute (RACTI), Greece.

References

1. J. Lee and D. R. Stinson, Deterministic key pre-distribution schemes for distributed sensor networks, in *Proceedings of ACM Symposium Applied Computing*, Santa Fe, NM, 2005.
2. J. Kong, P. Zerfos, H. Luo, S. Lu, and L. Zhang, Providing robust and ubiquitous security support for mobile ad hoc networks, in *Proceedings of the 9th International Conference on Network Protocols (ICNP)*, Riverside, California, November 2001.
3. H. Chan, A. Perrig, and D. Song, Random key predistribution schemes for sensor networks, in *Proceedings of the 2003 IEEE Symposium on Security and Privacy*, Oakland, CA, May 11–14, 2003, pp. 197–213.
4. P. Erdős and A. Rényi, On random graphs. I. Publicationes Mathematicae (Debrecen), 6, 290–297, 1959.
5. L. Eschenauer and V. D. Gligor, A key-management scheme for distributed sensor networks, in *Proceedings of the 9th ACM Conference on Computer and Communications Security*, Washington, DC, November 2002, pp. 41–47.
6. B. Bollobás, *Random Graphs*, Second Edition, UK: Cambridge University Press, 2001.
7. P. Dolan, *Undecidable Statements and Random Graphs, Annals of Mathematics and Artificial Intelligence*, 6(1–3), 17–25, 1992.
8. M. Abramowitz and I. E. Stegun, eds., *Handbook of Mathematical Functions*, Washington, DC: US Department of Commerce, National Bureau of Standards, 1972.
9. R. Anderson and A. Perrig, Key infection: Smart trust for smart dust. Unpublished Manuscript, November 2001.
10. J. Lee and D. R. Stinson, A combinatorial approach to key predistribution for distributed sensor networks, in *IEEE Wireless Communications and Networking Conference*, New Orleans, LA, 2(13–17), 1200–1205, 2005.
11. D. R. Stinson and R. Wei, Generalized cover-free families, *Discrete Mathematics*, 279, 463–477, 2004.
12. A. C.-F. Chan, Distributed symmetric key management for mobile Ad hoc networks, in *Proceedings of the IEEE INFOCOM'04*, Hong Kong, March 2004.
13. J. Wu and R. Wei, Comments on Distributed Symmetric Key Management for Mobile Ad hoc Network from INFOCOM 2004, manuscript.
14. E. Makri and Y. C. Stamatiou, Deterministic key pre-distribution schemes for mobile ad-hoc networks based on set systems with limited intersection sizes, in *2nd IEEE International Workshop on Wireless and Sensor Networks Security* (WSNS'06), Vancouver, Canada.
15. S. Jukna, *Extremal Combinatorics with Applications in Computer Science*, Germany: Springer, 2001.
16. C. C. Gumas, A century old, the fast Hadamard transform proves useful in digital communications, *Personal Engineering & Instrumentation News*, 6(6), 1184–1238, 1978.
17. A. Hedayat and W. D. Wallis, Hadamard matrices and their applications, *The Annals of Statistics*, 6(6), 1184–1238, 1978.
18. V. Liagkou, E. Makri, P. G. Spirakis, and Y. C. Stamatiou, The threshold behaviour of the fixed radius random graph model and applications to the key management problem of sensor networks, in *ALGO-SENSORS*, pp. 130–139, 2006.
19. S.-J. Tu and E. Fischbach, Random distance distribution for spherical objects: general theory and applications to physics, *Journal of Physics. A, Mathematical and General*, 35, 6557–6570, 2002.

20. P. Gupta and P. R. Kumar, Critical power for asymptotic connectivity, in *Proceedings of Conference on Decision and Control*, Tampa, USA, 1998.
21. A. Goel, S. Rai, and B. Krishnamachari, Monotone properties of random geometric graphs have sharp thresholds, Manuscript.
22. E. W. Weisstein, Circle lattice points, From MathWorld-A Wolfram Web Resource. http://mathworld.wolfram.com/CircleLatticePoints.html
23. G. E. Andrews, *Number Theory*, USA: W.B Saunders Company, 1971. Also by Dover Publications, 1994.
24. V. Liagkou, E. Makri, P. G. Spirakis, and Y. C. Stamatiou, Trust in global computing systems as a limit property emerging from short range random interactions, *The Second International Conference on Availability Reliability and Security (ARES 2007)*, Vienna, Austria, 741–748, 2007.
25. J. Spencer, *The Strange Logic of Random Graphs*, USA: Springer, 2001.
26. L. Babai, P. Frankl, S. Kutin, and D. Štefankoviě, Set systems with restricted intersections modulo prime powers, *Journal of Combinatorial Theory Series A*, 95, 39–73, 2001.
27. R. Fagin, Probabilities on finite models, *Journal of Symbolic Logic*, 41(1), 50–58, 1976.
28. P. G. Kolaitis and M. Y. Vardi, The decision problem for the probabilities of higher-order properties STOC, *Association for Computing Machinery*, USA, 425–435, 1987.
29. P. G. Kolaitis and M. Y. Vardi, 0–1 Laws for fragments of existential second-order logic: a survey, in *MFCS*, Bratislava, Slovak Republic, pp. 84–98, 2000.
30. P. G. Kolaitis and M. Y. Vardi, 0–1 Laws and decision problems for fragments of second-order logic. *Information Computation*, 87(1/2), 301–337 (1990).
31. J.-M. Le Bars, Fragments of existential second-order logic without 0–1 Laws, in *Proceedings of 13th IEEE Symposium on Logic in Computer Science*, Indianapolis, IN, pp. 525–536, 1998.
32. J.-M. Le Bars, Counterexamples of the 0–1 law for fragments of existential second-order logic: an overview, *Bulletin of Symbolic Logic*, 6(1), 67–82, 2000.
33. J.-M. Le Bars, The 0–1 law fails for monadic existential second-order logic on undirected graphs, *Information Processing Letters*, 77(1), 43–48, 2001.
34. B. A. Trachtenbrot, Impossibility of an algorithm for the decision problem on finite classes, *Doklady Akad. Nauk. S.S.R.* 70, 569–572, 1950.
35. V. Liagkou, P. Spirakis, and Y. C. Stamatiou, Can formalism alone provide an answer to the quest of a viable definition of trust in the WWW society?, in *Proceedings of 3rd International Conference on e-Democracy*, Greece, 2009.
36. D. A. M. Barrington, R. Beigel, and S. Rudich, Representing Boolean functions as polynomials modulo composite numbers, *Computational Complexity*, 4, 367–382, 1994.
37. S. Kutin, Constructing large set systems with given intersection sizes modulo composite integers, *Combinatorics, Probability and Computing*, 11, 476–486, 2002.
38. P. Frankl, Constructing finite sets with given intersections, Combinatorial Mathematics (Marseille-Luminy, 1981), North-Holland, Amsterdam, *Annals of Discrete Mathematics*, 17, 289–291, 1981.
39. S. A. Çamtepe and B. Yener, Key distribution mechanisms for Wireless Sensor Networks: a Survey, *Technical Report TR-05–07*, Rensselaer Polytechnic Institute, Computer Science Department.
40. E. Makri and Y. Stamatiou, Deterministic and randomized key pre-distribution schemes for mobile ad-hoc networks: Foundations and example constructions, in Zhen Jiang, Yi Pan, Ed. *From Problem Toward Solution: Wireless Sensor Networks Security*, Chapter 11, pp. 211–233, USA: Nova Science Publishers, Inc., 2009.

Chapter 15

Bio-Inspired Intrusion Detection for Wireless Sensor Networks

Swapna Ghanekar, Nancy Alrajei, and Fatma Mili

Contents

15.1 Introduction

All networked systems are inherently vulnerable. Their proper functioning depends on being able to communicate between entities in a reliable (with no loss of data quality) and secure fashion (no unauthorized entities gaining access to the information being communicated). The communication links that are the foundation of such systems present general risks on both fronts: they can be error and fault prone, especially when they are wireless; they are a relatively easy target for snooping, especially when they are wireless. Wireless sensor networks (WSNs) have all the characteristics of general wireless networked systems. They also have additional characteristics; some make them less vulnerable and others make them more vulnerable.

A WSN typically consists of one or more base stations and hundreds or even thousands of sensor nodes [1]. A typical set up is illustrated in Figure 15.1. The base station is the main storage and processing center. Requests (queries) emanate from the base station and are broadcasted to the sensor nodes, telling them what data to sense (collect), how frequently, and what to do with it. Typical uses of sensor networks include environmental monitoring applications where the nodes are spread over an area of interest and are requested (by the base station) to collect and send information of interest. This information may include temperature, humidity, wind velocity, concentration of chemicals, and so forth. We briefly discuss the specifics of sensor networks that are relevant to security. Some are robustness characteristics, others are vulnerability characteristics.

15.1.1 Security Robustness Characteristics of Sensor Networks

No data to protect: By nature of sensor networks, a single sensor node carries any critical information at any point in time. Each sensor node is nothing but a memory-less witness to its environment.

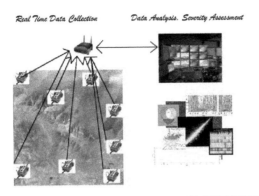

Figure 15.1 Typical architecture of a wireless sensor network.

It senses data when requested to do so and generally transmits it to other nodes along its routing path. It may store a limited number of values as requested by the base station or based on its stored program. Similarly, it receives data that are raw or aggregated and forwards it along the routing path, again based on instructions from the base station or from its stored program. Because of this, there is very little value to data possessed by any single node or being transmitted between any pair of nodes. As a result, there is very little benefit to capturing individual nodes to extract their data or to snooping on communication channels trying to listen to the content of the exchanges.

No node or no data are irreplaceable: In traditional networks, the data exchanged between nodes are not only valuable, but also unique in some way. Data transmitted by one node are not interchangeable with data transmitted by any other node. With sensor networks, a very high spatial and temporal redundancy makes the network very resilient to the failure of some of the nodes or some of the communication channels. For example, consider a network that is collecting and sending temperature information every 100 seconds. Because temperature is continuous over time and space, the values of neighboring nodes are highly correlated and the values at a given node over a short period of time are also highly correlated. As a result, the loss of some of the nodes or the incidence of errors in some readings and some communications can easily be overcome and corrected. A malicious attacker will not see any benefits to injecting false values into the network or disabling a small number of nodes or communication channels. Most of such attacks will result in little or no disruption to the correct working of the network. This redundancy is not specific to temperature alone. Most sensor networks monitor natural phenomena; these phenomena are continuous; their variations are known beforehand; and it is relatively easy to detect and dismiss faulty values.

While the above two characteristics make WSNs resilient, others make them easy targets.

15.1.2 Security Vulnerability Characteristics of Sensor Networks

Architecture is open: In order to be truly effective, WSNs must be flexible and self-configuring. In other words, the set of nodes that are part of the network must be allowed to change over time, allowing nodes whose power is depleted or who have been compromised to "leave" the network and allowing the addition of new nodes as needed. Similarly, the physical and logical relationship between the nodes must allow for various patterns (e.g., spaced along a perimeter fence) or in random arrangements (e.g., scattered over an area from an aircraft). To be effective, the network must automatically adapt to these changing conditions, adding, and dropping nodes as necessary. This flexibility makes it relatively easy for intruder nodes to insert themselves in the network.

Communication channel is open: Sensor nodes employ a broadcast scheme for relaying messages. This means that any entity within listening range can get the details of messages sent. These messages can also be altered and sent back to the network. This openness makes it very easy for malicious nodes to detect communications, change them, and broadcast them to the rest of the network.

Power is a bottleneck: As mentioned above, individual nodes are very simple entities. They are able to sense specific parameters; they can receive messages from and send messages to other nodes or to the base stations; they have a limited storage and processing capacity; and they have a limited, battery-delivered power. These same characteristics that do not make them a worthwhile target also make them a very easy one. In order to disable a node, it suffices to overwhelm it with requests and force it to communicate with a high frequency. While it may be of little value and of little consequence to disable one node, a concerted effort to disable many nodes by keeping them busy answering malicious requests rather than performing real services and eventually leading them to completely deplete their power and die can cause a major disruption in the whole network. Causing a large number of nodes to either die out or to be unavailable for useful functionality is an instance of a Denial of Service (DoS) attack.

Detection requires resources: DoS attacks are relatively common in networked systems, and there are a number of approaches developed to detect them and neutralize them. Unfortunately, most of these approaches rely on extensive memory to store a trace of past attacks and some level of processing power to identify patterns in these attacks and match them with ongoing activities. Because of the limited power, storage, and computation capabilities of sensor nodes, most of these approaches would consist of a high or impossible overhead on the nodes, and their cost would very easily outweigh their benefits.

As stated above, the characteristics of WSNs constitute a mix of challenges and opportunities. As such, traditional approaches to security that tend to be centralized, computationally extensive, and based in secrecy, firewalls, and closed architectures are not a good fit for WSNs. In some respects, WSNs have as much in common with biological multi-organism systems than they do with artificial computer network systems. For this reason, many researchers turned to naturally occurring immunity processes for inspiration.

We will provide a brief overview of the security risks that networked systems are subjected to and some sample approaches in Section 15.2. In Section 15.3, we discuss naturally occurring and artificial immune systems. In Section 15.4, we present a sample of bio-inspired security systems in sensor networks. We summarize and conclude in Section 15.5.

15.2 Background

Site security is a standard component of any networked system. The host of the system establishes their security policies spelling out authorities and access rights. The security system is the set of hardware and software components (e.g., firewalls, routers) devoted to enforcing these policies. Every such system includes a prevention mechanism complemented with detection and handling mechanism. The prevention is the "front door" (firewall) that locks the system in. Because no front door is 100% secure, there is a need for a detection mechanism that can recognize nonlegitimate activities and a handling mechanism for reacting to such activities. This is illustrated in Figure 15.2.

In this chapter, we are primarily interested in the Intrusion Detection [2] aspect of security.

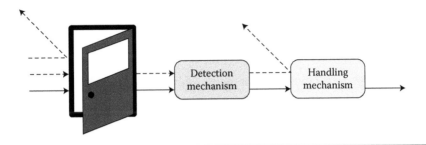

Figure 15.2 Prevention, detection, and handling of intruders.

The function of an intrusion detection system (IDS) is to identify actors within the system that are intruders. What constitutes an intruder is obviously context and policy dependent. Whereas a firewall filters out any entity that is not able to prove its credentials; the IDS system must identify entities that have misrepresented themselves in some way and have been let in within the walls of the system. The task of the IDS is then to detect them based on other features, typically *behavioral features*. To understand what such features are used, we focus on the types of malicious activities they might engage in, or the types of attacks that they may unleash with a focus on the attacks of interest to us, that is, DoS attacks.

15.2.1 Types of DoS Attacks that an IDS Must Deter

DoS attacks encompass any behavior that would impede the functioning of the network; they are defined in contrast, for example, with privacy violation attacks which access data without disrupting (at least not immediately or directly) the function of the network. There are two major strategies in DoS attacks: compromising the function of the network by injecting inaccurate information and compromising the availability of the network (DoS) attacks [3–5]. A key difference is that the compromising of the functionality alone may go undetected for a long time, thus lead to a network doing something "wrong", whereas the compromising of the availability will also make the system nonfunctional but in a visible and massive way. The following list gives some common DoS attacks.

Attacks that target the compromising of the functionality include:

Selective forwarding: An intruder node or a legitimate node that is compromised disrupts the functioning of the network by failing to perform its function. Typically, almost every node in a network is both a generator of information and a relay of information. In selective forwarding, the malicious node becomes an unreliable relay. For this, it may deliberately drop all or some of the information that it is requested to relay to other nodes. The malicious node may continue passing along *some* of the packets to maintain a less suspicious appearance. Also, the node may be selective in which packets it forwards and which packets it drops.

Misdirection: A malicious node may divert the messages it receives by forwarding them to the wrong nodes. This requires the message to go through more hops (a longer path) before it reaches its destination, thus using more energy. This type of attack is not really applicable to WSNs where messages are often broadcasted for all to hear.

Falsified routing information: Routing information can be spoofed, altered, or replayed. In a WSN, nodes can overhear what their neighbors are sending, so a malicious node can obtain a legitimate message and rebroadcast it with modified routing information. This gives the malicious node the ability to damage the network. For example, it could cut-off portions of the network or create routing loops.

Attacks that target the compromising of the availability include:

Sinkholes/black holes: A malicious node creates a sinkhole or black hole in the network by advertising as many nodes as possible that it is along a low-cost route. This allows the malicious node to become the center of communication, putting it in a better position to discard or inject information.

Wormholes: With this attack, two or more malicious nodes collude to subvert the network. These nodes employ an additional communication system that allows them to communicate directly with each other, often over a farther distance than good nodes can communicate. This makes some attacks, such as misdirection, more effective.

Hello flood: A malicious node may take advantage of the fact that many sensor network protocols have a neighbor detection phase. With this attack, the malicious node falsely announces itself as a neighbor to many nodes. Variations in this seemingly innocuous activity can be paralyzing by the amount of activity it generates, and, in the case of WSNs, it can quickly deplete the power of most of the nodes within communication range of the malicious node.

Understanding the nature of these attacks is an important context for discussing the different approaches that have been used in detecting them.

15.2.2 Elements of a DoS Intruder Detection System

In order to detect DoS attacks, systems need to know the characteristics of the behavior of intruder nodes in a way that allows them to distinguish them from innocent nodes; they need to have a mechanism by which they can monitor nodes in real time and classify their behavior as legitimate or suspicious. For example, if a node receives a message from a neighbor that is addressed to some specific node, say X, it should retransmit the message either to node X if it is within transmission range, or to another neighbor that is along the path to X. Furthermore, the node should only retransmit the message the minimum number of times required for the next hop to receive the message. There are a number of metrics that can be collected about individual nodes or about the whole network. Metrics that qualify the activity of a specific node include:

Activity frequency: The time elapsed between two consecutive messages. The level of activity of malicious nodes tends to distinguish them from normal nodes.

Retransmission delay: This is the time elapsed between when a message is received and when it is transmitted (assuming the recipient is not the final destination).

Transmission span: The number of nodes that hear the messages transmitted by a given node.

Integrity: This indicates whether the content of a message received is the same as the contents of the same message when retransmitted.

Repetition rate: Typically, unless there is a transmission error, every message needs to be sent only once. When the number of times that the same message is sent by the same node increases, this may be a sign of suspicious behavior.

Metrics that qualify the activity of the network: Many of the metrics described above for individual nodes can be generalized to measure the level of suspicious activity within a network or within a sub-network. Additional metrics are global reflections of the network rather than of a specific node. These include:

Jamming frequency: This indicates the percentage of messages that collide and need to be resent.

When a network collects these metrics, triggers can be used to flag nodes as suspicious and worthy of further inspection. Examples of such rules include [6]:

Hyperactivity: A node that has an exceptionally *high activity frequency* (i.e., sends messages at an exceptionally high frequency) may be suspicious.

Procrastination: A node with repeated *high transmission delay* may be suspicious of selected forwarding.

Screaming: A node with a frequent *high transmission span* may actually be flooding the networking by sending a message everywhere rather than toward a specific destination.

Stuttering: A node with a *high repetition rate* may be flooding the network, resending the same message multiple times.

Similarly, when there is a *high jamming frequency* in the network, this may be a signal that too many messages are being sent indiscriminately.

The monitoring and detection of these patterns can be performed in multiple ways:

- Each node monitors its own activity. This allows nodes to recognize when they have been compromised, assuming the intrusion does not disable this process.
- Each node watches its neighbors. This can be very effective but is also sure to result in a very high overhead on the network.
- Specific nodes are elected as watchdogs [7]. There has to be a sufficient number of nodes to cover the whole network. The watching function can be assigned as a permanent function or it can be a rotation among all the nodes using some rotation schedule and policy.

15.2.3 Approaches to DoS Intruder Detection

We discuss a representative sample of approaches used for intruder detection in WSNs.

15.2.3.1 IDS Based on Deviation from Normal

Onat and Miri [8] present an IDS for WSNs that analyzes deviations from normal behavior to identify intruders successfully.

Scope: In this system, two types of intrusion are detected, namely node impersonation and resource depletion.

Assumptions: This approach relies on a *static uniform architecture*. In particular, it assumes that:

- All nodes have the same make and the same physical characteristics, including power.
- The topology never changes. No node enters or leaves the network after it starts operating. This precludes nodes dying because their power has been depleted or nodes being added to ensure full coverage.

The approach also relies on *full context awareness*.

- Each node is assumed to know the identity of every node within communication range from it.
- Each node can identify each of the other nodes.

Who does the monitoring: Every node monitors every one of its neighbors. Each node is able to raise an alarm about any of its neighbors that it deems suspicious.

Statistics collected: The information collected about each node consists of a log. The log is a sliding window of a fixed size tracking all communications the node has had within the time window. From the log, different statistics are computed. They include transmission span (measured as the average received power) and activity frequency (measured as the average number of packets received per unit of time).

Anomaly detection rules used:

Impersonation: A node is deemed suspicious if it does not exactly imitate its target, that is the received power of its messages suggests that the node is different.

The rationale behind this rule is that an intruder node would require more power than a regular node in order to deplete the power of its neighbors. The residual power of each node is computed from its activity; if the residual power is higher than the expected, then the node must be of a different make.

Resource depletion (Hyperactivity): A node that has a *high Activity Frequency* is suspicious.

Results: Each of the statistics—average received power and average number of packets received per unit of time—were tested via simulation. The simulation modeled a wireless channel with received power decreasing with distance and with included random variation. The network underwent an initial training period during which all nodes behaved normally. After training, intruder nodes increased either power or transmission rate.

For received power, three things were tested: the probability of a false alarm versus the length of the intrusion buffer, the received power versus the detection probability for varying buffer lengths, and the received power versus time required for detection for varying buffer lengths. For all buffer lengths, as power increased, detection probability increased and the time required for detection decreased. Reducing buffer length enhanced this result with the smallest buffer length producing the highest detection probability and the smallest detection time; however, the false alarm rate also increased with decreasing buffer length.

For packet rate, the following three things were tested: the probability of a false alarm versus the length of the intrusion buffer, packet rate versus the detection probability for varying threshold values (the ratio between current packet rate and historic packet rate required to raise an alarm), and packet rate versus time required for detection for varying threshold values. Unlike received power, packet rate did not change the probability of detection or the time required for detection. A lower threshold increased the detection probability and reduced the time required, but also produced more false alarms.

15.2.3.2 Selective Forwarding Attack Detection Scheme

Yu and Xiao [9] present a method for detecting intruders in WSNs.

Scope: This method is specifically for detecting intruders that are carrying out a selective forwarding attack.

Assumptions:

■ There is a small initial deployment phase during which nodes cannot be compromised. Nodes can obtain their location and loosely synchronize with the base station during this phase.
■ Malicious nodes try to maintain a less suspicious appearance by only dropping some of the packets they receive.
■ The malicious node is considered a legitimate part of the network via its possession of the appropriate keys.
■ The nodes carry out some preexisting routing protocol.

Who does the monitoring: Nodes that are upstream and downstream of the intruding node. Here, upstream refers to nodes that are between the source and the intruder, whereas downstream refers to the nodes that are between the intruder and the base station.

Statistics collected: With this method, some set of nodes between the source and the base station, for example, every third node, is required to send an acknowledgement back toward the source node when it receives information to pass to the base station. Each node counts the number of acknowledgement messages it receives. In addition, each node examines the packet id for each message to ensure that it is receiving continuous ids.

Anomaly detection rules used:

Lethargy: A node that is not sending expected information is suspicious.

Some predictable number of nodes is required to send acknowledgement messages. If the proper number of acknowledgements is not received within the expected time, a node may suspect its immediate downstream neighbor.

Nodes may suspect their upstream neighbor if they receive a discontinuous packet id. This indicates that a message may have been lost.

Dishonesty: A node is suspicious if it provides information that is different than what was expected.

Results: This method was simulated to evaluate its performance. The simulation included the effects of transmission loss and a retransmission mechanism that would allow a message to be retransmitted up to five times. The method was evaluated in terms of the undetected rate (the portion of attacks that were successful) and overhead (the ratio of system overhead for a network using this method vs. one that does not). For a low channel error rate, the number of undetected attacks is low. This holds true even as the percentage of packets that malicious nodes drop is increased from 10% to 30%. In terms of overhead, channel error rate and the number of compromised nodes had little effect. Instead, a better indicator of relative overhead is the percentage of packets dropped by the intruders.

15.2.4 Summary Anomaly-Based IDS Systems for WSNs

In summary, IDS systems based on anomaly detection behavior are based on the continuous monitoring of the nodes. This monitoring consists of computing metrics based on the behavior of the nodes over sliding time windows. The vector of metrics collected is then compared with the profile of a normal node. Deviations from the norm are considered suspicious. Although these approaches can be very effective in capturing behavior and identifying deviations, it has the following drawbacks:

- The monitoring entities are based on one of two approaches, anomaly detection, or misuse detection. In anomaly detection, a profile of *normal* behavior is created. Any deviation from this normal behavior is considered suspicious. The difficulty with this is to capture the profile in a way that is sufficiently abstract that does not get caught in irrelevant details but also sufficiently detailed so as to capture all features relevant to normalcy. Perfection is never achieved in these profiles; the process is a trial and error approximation. In misuse detection, a catalog is created of *known misuses*. A signature then is created for each of these misuses. Monitoring entities watch to identify occurrences whose signature matches one of the previously identified misuses. This approach suffers from the same problems as the anomaly detection approach. In addition, it runs the risk of being constantly behind the curve; always fighting the previous wars. A judicious attacker can easily deflect detection by coming up with new attacks that do not match the previously known ones.
- Whether the system uses anomaly detection or misuse detection, it must construct a profile, for example, in the form of a set of rules. To be effective, these profiles are rarely as simple as a single-term condition such as, for example, "Activity Frequency greater that 0.7." The profiles involve multiple terms computed over time.
- Due to the potential complexity of the profiles, and thus of comparing current situations with these profiles, and the fact that monitoring entities need to collect large amounts of data about the communications of the monitored entities in order to match them with these profiles, there is a high overhead involved in terms of listening to communications, storing them, synthesizing them, comparing them with stored profiles, and acting on them as deemed necessary.
- When a single entity has the authority to flag other entities are suspicious, the process is both error prone and may be subject to intrusion. A compromised node may maliciously flag innocuous nodes. In order to protect against abuse of authority, the flagging power of multiple nodes must overlap and their opinions must be combined in order to determine suspicious nodes with a higher accuracy. This magnifies even further the overhead associated with monitoring.

In light of all of these complexities, difficulties, and continuous vulnerabilities of networked systems, researchers looked for inspiration from the existing resilient systems that seem to obtain and maintain immunity with relatively low cost. Nature provides us with a number of examples that have been perfected over millions of years. Multi-cell organisms are complex systems able to perform complex functions even though they are composed with a large number of very simple cells. This is not unlike sensor networks that are composed by a large number of simple sensor nodes.

15.3 Natural and Artificial Immune Systems, General Principles

In his essay about system management titled "Promise you a Rose Garden," Burgess [10] decries what he calls the "great monitoring myth" that equates system management with monitoring and detection. His point is that, on the one hand, the continuous monitoring is a huge distraction;

on the other hand, by the time anything is detected, it is generally too late. This has led Burgess and many other researchers to look for a different paradigm to improve the safety and security of networked systems. One source of inspiration has been biological multicell organisms with their robust, adaptable, and very efficient immune systems. The following subsections provide a description of natural immune system concepts and describe various artificial immune systems derived from those concepts.

15.3.1 Natural Immune Systems

In nature, all living organisms have survived as a species thanks to an elaborate immune system. Immune systems across species share some common characteristics. For example, they are all layered systems with multiple levels of protection so that should one fail, subsequent levels will still be able to provide protection. The different layers of the natural immune systems can be characterized into four different classes: the physical layer, the innate layer, the adaptive layer, and the Danger theory layer staked as shown in Figure 15.3.

Physical barriers: This layer is the *wall* surrounding the organism. It can also be thought of as the main door allowing the filtering of legitimate visitors from intruders. The physical layer is generally a combination of mechanical (e.g., the shell on an egg or the skin on a human), chemical (e.g., stomach acid), or biological (e.g., bacteria in the intestinal tract) barriers. These barriers prevent many intruders from entering an organism (Figure 15.4).

Innate immunity: This layer is the *starting capital* of all members of a species. It is the set of defenses that they are all born with. It is also referred to as the "nonspecific" immune system. This layer is the same in an organism from when it is born to when it dies. It handles intruders that have passed through the physical barriers. The major functions of the innate system are to: (1) attract immune cells to an infected area, (2) mark intruders for removal from the system, (3) remove intruders from the system, and (4) present antigen to the adaptive immune system. An important component of this system is phagocytes, which are cells that find and consume intruders. The most efficient type of phagocyte is called a macrophage. Phagocytes roam through the organism looking for specific intruders (Figure 15.5).

Adaptive (acquired) immunity: Whereas the innate system is generic and identical for all members of a species, the adaptive immunity system, as its name implies, is specific to an instance's history of exposure. It reflects both the environment and context of the instance as well as its track record

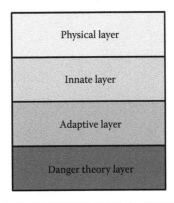

Figure 15.3 Multi-layered immune systems.

Figure 15.4 Physical layers in biological systems.

in fighting off whatever it was exposed to. Whereas all instances of a species have identical innate immunity, generally every instance has a unique immunity signature. For this, the adaptive immunity is also referred to as the "specific" immune system. This layer of the immune system grows and changes over time based in the environment. It looks for particular intruders and provides memory so that the system can mount a quick defense against intruders it is already familiar with. This layer randomly creates detectors that do not match "self" using a process called negative selection. When a detector has made a match, it is improved upon through clonal selection. The basic underlying process of the adaptive immune system is the process by which organisms can be classified. Enzymes react with other organic molecules based on their three-dimensional, special configuration—rather

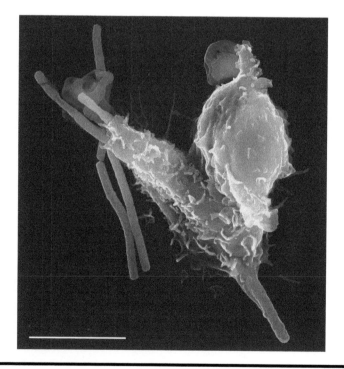

Figure 15.5 Innate immune system.

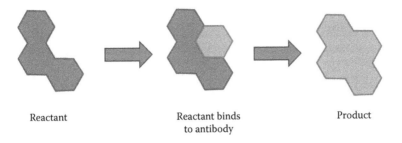

Reactant   Reactant binds   Product
                to antibody

Figure 15.6   Process of attaching to molecules.

than their exact chemical formula. When a "reactant" produced by the immune system matches a molecule of a certain shape, it attaches to it to create a product that can then be processed (Figures 15.6 and 15.7).

The following list describes some of the key concepts of the adaptive immune system:

■ *Antigen:* The chemical signature that identifies all cells—"self" and "non-self"—within an organism. The presence of "non-self" antigen triggers an immune response.
■ *Antibody:* The detectors used within the organism to identify "non-self". These detectors chemically bind with "non-self" antigen. The antibodies do not necessarily bind perfectly with an antigen; they may bind "well enough."
■ *B cells:* Randomly generated detector cells that produce antibody and compose adaptive immune system memory. These cells are one of the key components of the adaptive immune system. They try to match foreign cells within an organism. If they have made a successful match, they replicate and become memory cells waiting for any similar intruders.
■ *Negative selection:* The process by which antibody matching "self" is deselected. This process occurs before B cells are sent to detect foreign cells.
■ *Clonal selection:* The process whereby mature B cells (successful matches) replicate quickly and with slight modification in order to find a better match with the antigen. The original B cell and most of its descendants die during the process of removing the intruders. However, some B cells remain and serve as memory.

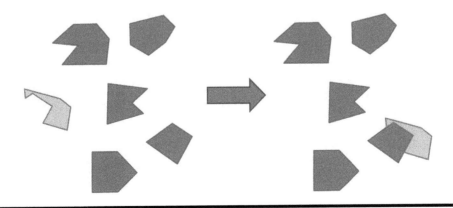

Figure 15.7   The antibody (light gray) matches an antigen.

Table 15.1 Important Differences Between Innate and Adaptive Immunity

Innate Immunity	Adaptive Immunity
Non-specific	Specific
Immediate maximal response	Delayed maximal response
No memory	Maintains memory
In most life forms	Only in jawed vertebrates
Constant from birth	Continually changing

See Table 15.1 for a summary of some important differences between the innate and adaptive immune systems.

Danger theory: This concept is based on the idea that the immune system must be doing something more than just "self"/"non-self" discrimination. The idea that every organism that is endogenous is automatically innocuous and every organism that is exogenous is automatically pathogenic is too simplistic and inaccurate. In reality, endogenous cells can be pathogenic, as is illustrated by cancerous cells, for example. Similarly, exogenous organisms can be innocuous and even often beneficial and necessary as is the case in many symbiotic relationships with beneficial bacteria. Danger theory stipulates that a defense system must be set up to identify danger. Danger can take the form of any signal that something bad is taking place. For example, cells that die of a natural death and cells that die prematurely send off different signals. The presence of a large number of cells sending a premature death signal (e.g., inflammation) is an indicator of danger. The danger layer is not meant to replace any of the other layers, but to complement them. When B cells (antibody) detect exogenous antigens, they become alert, and look for danger signals with a heightened vigilance. If a danger signal is detected, they are almost sure that the exogenous antigens are pathogenic and need to be neutralized. Also, the B cells that detected this pathogen need to be enforced. This is done by lengthening their life span and creating more clones for them.

This can be thought of with a two-signal model. The first signal occurs when antibody matches antigen. The second signal occurs with co-stimulation (verification that the antigen really is an intruder). With this model, B cells that receive both signals will be activated; cells receiving only the first signal will die; and cells receiving only the second signal will do nothing.

15.3.2 Artificial Immune Systems

Computer-based protection systems that are heavily influenced by natural immune systems are called Artificial Immune Systems (AIS). The label AIS has relatively loosely been used because there are no strict requirements as to what qualifies as an AIS and what does not. The rise in popularity in AIS systems and systems that claim the label AIS have led researchers to identify a set of properties of Natural Immune Systems that are critical to their success and good performance. In [11], the authors list the following set of characteristics:

◾ *Pattern recognition and self-identity:* Natural immune systems are very successful at distinguishing between entities that are part of "self" and entities that are "non-self."

■ *Uniqueness:* Each instance of each organism has its own immune system that is particular to that system based on the environment of that system. In other words, imparting organisms with innate immunity is not sufficient.

■ *Diversity, disposability, and robustness:* Immune systems offer a layered approach to finding and eliminating "non-self." No particular layer is responsible for the success of the entire system. Furthermore, no single cell of any layer is responsible for the success of the entire system. The latter point implies that we cannot relegate the monitoring and further actions to some bottleneck entity. Any entity participating in the monitoring and detection must have other entities duplicate its effort.

■ *Autonomy and distributivity:* There is no central control commanding the system components. The immune system is a self-driven system where each participating entity "knows" what to do, is able to make autonomous decisions, and does not depend on directions from other entities. The success of the system depends on having each entity (or most entities) performing their tasks correctly.

■ *Dynamically changing coverage:* Since the total number of possibly anomalous patterns is very large compared with the number of detectors available, the discovery and detection of intruders is a statistical process. To maximize its coverage and the likelihood of its success, immune systems must employ a changing set of detectors. This is in part what makes immune systems unique to individuals.

■ *Noise tolerance:* Noise and uncertainty are part of natural processes. Immune systems must be unfazed by either. A detector designed to match a specific pattern must also match slight variations in the pattern. If these variations prove to be innocuous, adjustments to the process can be taken later. Thus, in general, a "good enough" match starts the immune response.

■ *Resilience:* Even when under severe resource constraints, a natural immune system can continue to provide some basic level of immune coverage. Thus immune system processes must be on or off; as much as possible, they must always be on to some degree determine by the resources available among other things.

■ *Learning and memory:* The specific immune system is adaptable. It is continuously adapting its behavior to its experience, that is, intruders it has detected, detectors that proved successful, false alarms, and so forth.

15.4 Representative Sample of AIS for Sensor Networks

In the following sections, we discuss a sample of approaches published in the literature establishing AIS systems for sensor networks. In particular, we show examples of implementations of each of the layers. For each of the approaches described, we will discuss the following aspects:

1. The general motivation and focus of the approach.
2. The scope of the approach in terms of the four layers defined. For those approaches that focus on the last two layers, we will discuss in some detail the following issues:
   ■ Which layers of the immune system is being emulated
   ■ The way in which the system defines (learns) self
   ■ How the process of generating antigens is implemented
   ■ The matching process
   ■ What ensues a successful match

3. General assessment of the approach in terms of the criteria listed in Section 15.3.2.
4. Experimental results.

15.4.1 Sample Implementation of the Innate Immune System

15.4.1.1 Background

In their paper titled Native Artificial Immune System, the authors [12] discuss their design and implementation of a bio-inspired system to secure computer networks.

15.4.1.2 Immunity Layer Simulated

The focus of their approach is on the innate immune system.

15.4.1.3 Characterization of Intruders (Self vs. Non-self)

This method is based on the observation that DoS attacks are often preceded by an information gathering phase. This phase results in the creation of unexpected processes. By noticing and eliminating these processes, the system can be made more secure.

The system begins with an initial training phase. This helps identify which processes are normal. These processes compose the list of innate components that are accepted as "self," which determines how the system starts at "birth." For a web server example, innate processes include system processes and web surfing processes.

15.4.1.4 Identifying Intruders and Ensuing Processes

When an intruder is using the server, it will do other things, such as open a terminal or compile code. These processes should be identified as "non-self." In keeping with the innate immunity analogy, the system contains macrophages that take the form of processes that poll the server. When a macrophage has identified a "non-self" process, it will kill that process.

15.4.1.5 General Assessment and Simulation Results

Because the approach is closely tailored to the processes that are being sought, one would expect this approach to have 100% accuracy, no false positives or false negatives. This is our interpretation, given that the authors did not elaborate on their simulation and results.

15.4.2 Example Implementation of an Adaptive Immune System: Immunity-Based Intrusion Detection for WSNs

15.4.2.1 Background, Motivation

Liu and Yu presented [13] an immunity-based system for intrusion detection in WSNs. They start with four assumptions: nodes are not added after the initial deployment, there is a flat routing structure, malicious nodes are just like normal nodes except that they are also carrying out some DoS attack, and there is a sufficiently long training period.

This algorithm is applied to TinyOS beaconing. Nodes communicate with the sink by passing packets in a hop-by-hop manner. Nodes may overhear what their neighbors send; they may not overhear information from nodes that are outside their communication range. The algorithm is composed of four steps: self-acquisition, detector generation, detection, and clonal selection.

15.4.2.2 Immunity Layer Simulated

This system focuses on the adaptive layer. The antibodies are generated continuously to reflect the nodes' history and experience.

15.4.2.3 Characterization of Intruders (Self vs. Non-self)

The first step, self-acquisition, is the training phase. Nodes listen in on neighbor communication. For each message heard, the node keeps track of next hop neighbor, hop count to the sink, and loss rate. This information is coded into an antigen string of ones and zeros and represents "self." The idea behind collecting this information is that the DoS attacks presented—jamming, sinkhole, wormhole, and blackhole—each affect at least one of these fields.

After the initial training period, the detector-generation step allows each node to create a set of antibody for detecting "non-self." Random antibodies are created, also as a string of ones and zeros. However, the string is treated as a set of three shorter strings—one for each feature—to decrease the computation time. During this step, antibodies undergo negative selection. All antibodies that have some number of continuous matching bits as any antigen are deselected. Each antibody in the detector set has some finite life.

15.4.2.4 Identifying Intruders and Ensuing Processes

The third step is the actual detection. As it overhears its neighbors' communication, each node creates antigen strings. As each string is created, each antibody checks to see if it matches the antigen. At the end of its life, each antibody determines whether or not it has made some threshold number of matches. If the antibody has not made the required number of matches, it dies and is replaced; otherwise, it raises an alarm and proceeds to step four.

The final step is clonal selection. During this step, successful antibodies are stored for system memory. Each time an antibody reaches this phase, its lifetime is increased and its threshold for success is decreased. This method introduces a costimulation mechanism; an inspector may signal the network to add a false positive into the group of "self" antigens.

15.4.2.5 General Assessment and Simulation Results

This method was simulated with five different attacks—routing loops, jamming, sinkholes, wormholes, and black holes. The proposed system identified all the attacks. However, it was subject to a number of false positives—20%, 92.3%, <10%, <10%, and 63.2%, respectively, for the aforementioned attacks. The simulation also tested the effect of adding the costimulation mechanism. This mechanism was able to reduce the number of false positives significantly.

15.4.3 Example Implementation of Danger Theory: An Artificial Immune System Approach with Secondary Response

15.4.3.1 Background, Motivation

Sarafijanovic and Le Boudec presented [14] an AIS approach that learns from experience to defend against new attacks that use concepts from the adaptive immune system and danger theory. This method is evaluated in the context of detecting routing misbehavior when using Dynamic Source Routing (DSR). The analogy of this method with natural immune systems is mapped as shown in Table 15.2.

Table 15.2   Mapping Between Immunity and Algorithm Concepts

Natural Immune System	Approach From [14]
Body	The entire network
Self	Normally behaving nodes
Non-self	Misbehaving nodes
Antigen	A pattern based on observation of DSR
Antibody	A randomly selected pattern of similar format to the antigen

15.4.3.2 Layer Simulated

This method focuses on the complementary implementation of the adaptive layer and the danger theory layer.

15.4.3.3 Characterization of Intruders (Self vs. Non-self)

This method creates antigen by observing the network and counting events. For the DSR example, the authors give the list of protocol events listed in Table 15.3. The network observations of protocol events are combined into genes, which are the features used to look for anomalous behavior. The authors give the following genes for their example:

■ Gene1 = \#E in sequence
■ Gene2 = \#(E*(A or B)) in sequence
■ Gene3 = \#H in sequence
■ Gene4 = \#(H*D) in sequence

Table 15.3   Protocol Events

Label	Protocol Event
A	RREQ sent
B	RREP sent
C	RERR sent
D	DATA sent and IP is not of monitored node
E	RREQ received
F	RREP received
G	RERR received
H	DATA received and IP destination address is not of the monitored node

Source: Data from Sarafijanovic, S. and Le Boudec, J.Y. *IEEE Transactions on Neural Networks,* 2005, 16: 1076–1087.

After information is gathered to form genes, an antigen is created by coding the information as a string of ones and zeros. The information for each gene is put into a bin based on its value. The bit corresponding to its bin is set to one and all other bits for the gene are set to zero. For example, if a gene may range in value from 1 to 10 and there are five bits to represent each gene, then values one and two will correspond to the gene 00001, three and four will correspond to 00010, five and six will correspond to 00100, and so on. Notice that the antigen has exactly one bit set for each gene.

Once an initial set of "self" is determined, antibodies are generated. Each antibody takes the same form as the antigens. However, the antibody is randomly selected and may have any number of bits set for each gene. An antibody and an antigen match if every "1" in the antigen corresponds to a "1" in the antibody. This algorithm uses negative selection to remove all bad antibodies.

15.4.3.4 Identification of Intruders

After the initial antigen and antibody have been created, the system moves to the detection phase. During this phase, the set of antibodies are used to detect against antigen that are created from new observations. To eliminate false positives, it is not enough for an antibody to match an antigen. If an antibody matches an antigen, suspicious behavior is said to be "detected." However, a node is not classified as "misbehaving" until some number of detections has occurred against it.

Antibodies that have made at least one detection are candidates for clonal selection. This will occur only for antibodies that receive the danger signal. For this algorithm, signal 1 is given to some top percentage of antibodies, where the antibodies are ranked by the number of detections they make. The antibodies that received signal 1 each create one new detector, which is a mutation of the original antibody. The clones undergo negative selection. The creation and negative selection processes are repeated until each antibody has a clone that is not negatively selected. This survival of the negative selection process is signal 2.

15.4.3.5 Assessment and Simulation Results

This method was simulated both with and without clonal selections. The simulation implemented a selective forwarding attack and was evaluated in terms of time required to classify an intruder, true positives (the percentage of misbehaving nodes that were successfully identified), and false positives (the percentage of nodes that were falsely accused of misbehaving). Both with and without clonal selections, this method was able to successfully identify all intruders while maintaining a low false-positive rate. The time required for classification was lower with clonal selection than that without, and the clonal selection trials produced a much improved secondary response time.

15.4.4 Adaptive Immunity for WSNs

15.4.4.1 Motivation

Here we present an ongoing research that seeks to create an adaptive immune system for WSNs. This system is adaptive because it learns from its environment by maintaining memory of attacks it has encountered and it periodically refreshes its detectors so that it will discover new attacks.

15.4.4.2 Determining "Self"

This simulation begins with an initial training phase that allows the nodes to determine "self." The simulation iterates through the first data set, epoch by epoch. Each node collects sensor readings, and (every 10th epoch) the nodes use those readings to calculate four features of the data: first derivative, second derivative, minimum, and maximum. The value calculated for each feature is used to encode a gene, which is a 1-by-n matrix with n representing the selected gene length.

A single 1 is placed into the gene array depending on the calculated value. For example, if we have a gene array of length 5 and we are looking at minimum and it varies between 0 and 10, then each place in the array represents a range of 2. The following five array elements represent the ranges: [0, 2], (2, 4], (4, 6], (6, 8], and (8, 10]. If we have just calculated the minimum to be 1.2, then a 1 gets placed in the first element of the array as shown in the example gene in Figure 15.8. All other elements get set to zero. The genes for all the features are combined into an m-by-n matrix, with m representing the number of features. See Figure 15.9 for an example antigen.

Note that only one 1 is present in each row of an antigen matrix. All antigen created during this phase collectively becomes the list of acceptable patterns, that is, "self."

With the initial set of "self" antigen created, each node randomly creates antibodies, which are also encoded as a matrix of 1's and 0's. However, any number of 1's is permitted in each row of the antibody.

Negative selection is used to deselect any antibody that matches any antigen from our first data set. All antibodies that match any "self" antigen are re-created. A candidate antibody matches antigen if it has a 1 in every location that an antigen has a 1. For example, the antibody in Figure 15.10a does not match the example antigen in Figure 15.9 because it does not have a 1 in row 2, column 2. The example antibody in Figure 15.10b, however, does not match the example antigen; it would be selected for the initial detector set. As each antibody is selected as a detector, it is given a randomly selected lifespan.

Next, the simulation iterates through the second set of data and creates antigen as before. Each of these new antigens is checked against the set of antibody of each of its neighbors. Each node that has an antibody that matches a neighbor's antigen marks its neighbor as suspicious. Antibodies that have made a match are given an increased lifespan to simulate immunological memory.

15.4.4.3 Simulation Examples

The data used for the simulation come from the Intel Berkeley Research Lab [15]. Fifty-four sensor nodes were distributed through the lab to collect data between February 28 and April 5, 2004. The data contain the following fields: date, time, epoch, moteid, temperature, humidity, light, and voltage. The simulation extracts two sets of data. The first is used for determining good behavior whereas the second is used to look for any anomalous behavior. Each of these data sets contains one day's worth of data. The second data set is altered to deliberately introduce incorrect data.

For the simulation examples, let data set A be data from February 28, 2004, and data set A' be data from the same day with values for node 7 modified to be incorrect. Also, let data set B' be data from February 29, 2004, that has had values for node 7 modified to be incorrect.

The examples presented here use temperature or humidity data. For each of these two data distributions, data set A is first compared with data set A' and then with data set B'. Figures 15.11 through 15.14 give the number of times each node was flagged as suspicious (match) of found acceptable (no match). For both temperature and humidity data, the desired result is for node 7 to have 0 for "not matched" and for all other nodes to have 0 for "matched."

When data set A is used to detect on A' for temperature data, node 7 matches every time and no other node matches. This is expected because set A is being checked against itself except when looking at node 7.

1	0	0	0	0

Figure 15.8 Example gene.

1	0	0	0	0
0	1	0	0	0
0	1	0	0	0
0	0	0	1	0

Figure 15.9 Example antigen.

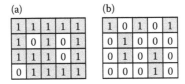

Figure 15.10 Example antibodies: (a) example antibody 1; (b) example antibody 2.

mote id	1	2	3	4	5	6	7	8	9	10	11	12	13	14	15	16	17	18	19	20	21	22	23	24	25	26	27
match	0	0	0	0	0	0	10	0	0	0	0	0	0	0	0	0	0	0	0	0	0	0	0	0	0	0	0
no match	6	4	8	8	0	4	0	8	8	8	8	4	4	6	2	6	6	10	6	10	10	12	10	10	8	8	4

mote id	28	29	30	31	32	33	34	35	36	37	38	39	40	41	42	43	44	45	46	47	48	49	50	51	52	53	54
match	0	0	0	0	0	0	0	0	0	0	0	0	0	0	0	0	0	0	0	0	0	0	0	0	0	0	0
no match	10	6	4	10	4	4	4	6	6	6	8	4	8	8	8	6	8	12	10	10	12	4	6	8	6	6	6

Figure 15.11 Temperature, data A used to detect data A′.

mote id	1	2	3	4	5	6	7	8	9	10	11	12	13	14	15	16	17	18	19	20	21	22	23	24	25	26	27
match	0	0	0	0	0	0	12	0	0	0	0	1	0	0	0	0	0	0	0	0	0	1	0	1	0	0	0
no match	8	8	8	10	0	6	0	8	8	8	8	3	4	6	4	6	6	12	6	12	12	11	10	9	10	10	4

mote id	28	29	30	31	32	33	34	35	36	37	38	39	40	41	42	43	44	45	46	47	48	49	50	51	52	53	54
match	0	0	0	0	0	0	0	0	0	0	0	0	0	0	0	0	0	2	2	2	0	0	0	0	0	2	0
no match	10	6	4	10	4	4	6	12	10	10	10	6	8	8	8	8	8	10	10	10	12	8	6	8	4	2	6

Figure 15.12 Temperature, data A used to detect data B′.

mote id	1	2	3	4	5	6	7	8	9	10	11	12	13	14	15	16	17	18	19	20	21	22	23	24	25	26	27
match	0	0	0	0	0	0	10	0	0	0	0	0	0	0	0	0	0	0	0	0	0	0	0	0	0	0	0
no match	6	4	8	8	0	4	0	8	8	8	8	4	4	6	2	6	6	10	6	10	10	12	10	10	8	8	4

mote id	28	29	30	31	32	33	34	35	36	37	38	39	40	41	42	43	44	45	46	47	48	49	50	51	52	53	54
match	0	0	0	0	0	0	0	0	0	0	0	0	0	0	0	0	0	0	0	0	0	0	0	0	0	0	0
no match	10	6	4	10	4	4	4	6	6	6	8	4	8	8	8	6	8	12	10	10	12	4	6	8	6	6	6

Figure 15.13 Humidity, data A used to detect data A′.

With temperature data A and B′, the simulation checked for a match 428 times. Out of this, a match was found 23 times, 11 more times than expected. This corresponds to an unexpected results rate of 3%. The unexpected results indicate either patterns that are falsely identified (patterns that do not indicate fault that were missing from the initial data set) or anomalies in the data (time periods where nodes were doing faulty or incorrect things).

For humidity data, using data set A to detect on A′ also shows a match every time node 7 is checked and shows no matches for any other nodes.

With humidity data A and B′, the simulation checked for a match 428 times. The 12 matches corresponding to node 7 were expected, but the other 24 matches were not expected. This corresponds to an unexpected results rate of 6%. Again, this may have been caused by missing patterns in data set A or by fault or incorrect data in data set B′.

15.5 Future Directions for Research

The increasing interest in bio-inspired methods and systems is not accidental. As software systems increase in complexity, the traditional tools and approaches are reaching their limits. A look to nature reveals that very successful, resilient, fault-tolerant systems exist in nature. These systems are not designed in a top-down fashion through decomposition, but in a bottom-up fashion relying on very simple units. From the interaction of these simple units *emerges* a behavior that, when seen from a more abstract point of view, is complex, efficient, and resilient.

Research on emergent behavior is taking place at different layers:

1. *Modeling:* Creating programs that mimic biological system in their approach to performing a function or addressing a problem. The implementations such as the one described for

mote id	1	2	3	4	5	6	7	8	9	10	11	12	13	14	15	16	17	18	19	20	21	22	23	24	25	26	27
match	0	0	0	0	0	0	12	0	0	0	0	0	0	0	0	2	0	0	0	2	0	2	0	1	2	2	2
no match	8	8	8	10	0	6	0	8	8	8	8	4	4	6	4	4	6	12	6	10	12	10	10	9	8	8	2

mote id	28	29	30	31	32	33	34	35	36	37	38	39	40	41	42	43	44	45	46	47	48	49	50	51	52	53	54
match	1	1	0	0	0	0	0	1	0	0	0	0	0	0	1	0	1	0	0	0	0	0	2	0	0	4	0
no match	9	5	4	10	4	4	6	11	10	10	10	6	8	8	7	8	7	12	12	12	12	8	4	8	4	0	6

Figure 15.14 Humidity, data A used to detect data B′.

Immunity-Based Intrusion Detection for WSNs falls under this category. Creating such programs requires first that we have at our disposal a good understanding and detailed model of a similar functionality in a natural system (e.g., auto-immune systems), second that we are able to map the natural systems to the context we are interested in, and third that we are able to choose the right parameters to bypass millions of years of natural selection and zoom in into the right combination of parameters. In the case of immunity in sensor networks, the first element is present. Natural sciences provide us with a sufficiently well-understood and well-documented model of the different components of immune systems. We expect that progress in science will continue and that the results of that progress can easily be incorporated in programs created to reflect this progress. The mapping between the natural systems and the artificial representation is often the step that presents the most challenges. For example, natural immune systems have an unmistakable means for identifying self. In most contexts, mimicking this behavior requires some creativity or trial and error in determining necessary and sufficient characteristics of self. For sensor networks, some implementations rely on manufacturer's identifiers; others rely on keys known only by self; others rely on patterns of behavior that need to be learned and that may error prone. Finally, every model created has a number of parameters that then need to be "guessed" and fine-tuned.

2. *Architectural design:* This is the level above modeling. In architectural design, the goal is also to create a system that uses a natural system as an inspiration and exhibits a type of emergent behavior. In contrast with modeling where the focus is on getting the model up and running within a specific context (e.g., DoS intruder detection in sensor networks organized as a hierarchy of clusters), architectural design is a higher level approach identifying the various components of a class of models that share a common functionality. For example, the architecture may be designed for intruder detection in sensor networks regardless of the type of intrusion and regardless of the organization used in the sensor network. The architecture identifies the different components with their functionality and their constraints but does not specify how the components are implemented. Work on architectures generally follows work on models. Traditionally, it is only after we have developed some models, created enough proofs of concept, and proofs of efficiency that the community turns toward working at a more generic level.

3. *Framework and theory:* One of the challenges with emergent behavior is that it removes us from the realm of predictability. When we write a program, we no longer have a certainty about what happens when. Instead, we develop the individual components with their simple behavior and rely on the interaction between them. As the traditional methods of program verification are no longer applicable, alternative theories need to be developed to help us reason about the system we create. Alternative theories are currently being developed for emergent behavior. Chemical Organization Theory is one such example. The theory consists of an algebraic chemistry with molecular species (the units in the system) and reaction rules that dictate how the contact between some elements generates new elements. The theory defines some good properties that a theory must possess, such as closure (all elements generated are from the molecular species of the theory) and self-maintenance (the rules preserve the existence of a subclass of the elements). Chemical organization theory is but only one of these theories. Others include organic computing theory, chaos theory, and complexity theory.

To summarize, the field of bio-inspired computing is an exciting and promising field. The results seen and published so far establish the feasibility of the approach. The practice remains an art, yet a science is slowly but surely developing. There is room for contribution at the implementation of new models in specific applications, in the development of general architectures that can be used as a basis for many implementations, and there is still a lot to learn in terms of theoretical tools that can be used to reason about these systems, establish properties of their behavior.

15.6 Conclusions

Although many applications of sensor networks present very little risk in terms of security, some high level of protection is a prerequisite for other applications to be viable. The distributed nature of sensor networks and the redundancy in their functionality and their structures present both challenges and opportunities in terms of protection from harm. In this chapter we have focused on the most probable type of attacks: DoS attacks. We have also focused one type of defense: Intruder detection. We discussed a variety of approaches traditionally used in intruder detection and have shown their shortcomings with respect to WSNs. These shortcomings led to looking for alternate strategies, mainly by seeking inspiration from natural systems. Natural immune systems have been developed and refined over millions of years. They are generally multilayered starting from a protective wall, preventing intruders from penetrating the system. When this wall fails, innate and adaptive processes are in charge of identifying and neutralizing the intruder. Innate systems are hereditary and are hardwired and common to all individuals; whereas adaptive systems encompass a random component and reflect the exposure and history of the system. We have discussed different research efforts to emulate the innate and adaptive systems in sensor networks. Also, because intruders in nature are characterized as much by their effect as they are by their belonging to self, danger theory is added to further distinguish between harmless (or even beneficial) and harmful entities.

There is a growing interest in this field. This is reflected in part by the increasing number of journals, conferences, and other venues devoted to it. For example, the following journals have had a special issue on artificial immune systems recently: *Applied Soft Computing, Biosystems, Evolutionary Computation, Evolutionary Intelligence, Genetic Programming and Evolvable Machines, International Journal on Unconventional Computing, IEEE Transactions on Evolutionary Computation, IEEE Transactions on Systems, Man and Cybernetics, Natural Computing,* and *Physica D.* In terms of conferences, the following are exclusively or in part focused on AIS systems in sensor networks: BIC-TA (http://www.bic-ta.org/), Bio-Inspired Computing: Theories and Applications, BIONETICS (http://www.bionetics.org/), Bio-Inspired Models of Network, Information, and Computing Systems, and BLISS (http://www.see.ed.ac.uk/bliss09/), Bio-inspired, Learning, and Intelligent Systems for Security. All of this scholarly activity is bound to generate new ideas and bring about interesting results.

Terminologies

WSN—Wireless sensor networks
IDS—Intrusion detection systems
AIS—Artificial immune systems
Bio Inspired
Natural immune systems

Questions and Sample Answers

1. What characteristics of the nodes make the sensor network less vulnerable to attack than traditional networks?
   The information provided by a single node may not mean much without the context of information from other nodes. Additionally, no node carries unique data; similar data can be found in neighboring nodes.

2. What characteristics of a sensor network make it more vulnerable to attack than a traditional network?
   A sensor network is very open. It has an open, possibly changing topology and it relies on a broadcast communication medium. Additionally, sensor nodes are very resource constrained. This means that the nodes cannot easily utilize a resource-intensive detection algorithm, and an attacker will have an easier time carrying out an attack.

3. What are some of the differences between security approaches for wireless sensor networks and traditional networks?
   Traditional approaches may be centralized, computationally expensive, and based on a closed architecture. Wireless sensor networks, on the other hand, need approaches that are decentralized, resource-light, and open/adaptable.

4. What are the components of a network security system and what is the purpose of each of these components?
   The components of a network security system are a prevention mechanism, detection mechanism, and handling mechanism. The prevention mechanism acts as a "front door" to keep intruders out. The detection mechanism analyzes everything that comes in through the "front door" and identifies intruders. The handling mechanism removes any intruders from the system.

5. What is a denial of service attack?
   A denial of service attack is a malicious attempt to stop the network from operating as it should.

6. List some of the denial of service attacks that stop the network from functioning.
   Denial of service attacks that stop the network from functioning include selective forwarding, misdirection, and falsified routing information.

7. List some of the denial of service attacks that stop the network from being available.
   Denial of service attacks that stop the network from being available include sinkholes/black holes, wormholes, and hello flood.

8. What are some of the metrics that can be used to distinguish a misbehaving node?
   Some metrics that can be used to distinguish a misbehaving node include activity frequency, retransmission delay, transmission span, integrity, and repetition rate.

9. What triggering rules can be derived from the metrics used to distinguish a misbehaving node?
   High activity frequency, high transmission delay, high transmission span, and high repetition rate correspond to hyperactivity, procrastination, screaming, and stuttering rules, respectively.

10. How intrusion detection can be carried out in a sensor network?
    Intrusion detection can be carried out by a node monitoring itself, by all nodes monitoring their neighbors, or by specific watchdog nodes.

11. What is the purpose of each of the layers in an immune system?

The physical layer blocks out intruders, preventing them from being able to invade the organism. Innate immunity provides a nonspecific response to intruders, killing known intruders. Adaptive immunity provides a specific response to intruders, learning over time what is "non-self" and providing immunological memory. Danger theory adds to the self/non-self determination.

12. What is the purpose of antigen and antibody in the adaptive immune system?

Antigen is the chemical signature identifying all cells—"self" and "non-self." The presence of "non-self" antigen triggers an immune response. Antibodies act as detectors by binding with "non-self" antigen.

13. What is negative selection and clonal selection?

Negative selection is the process by which suitable detectors are selected. All antibodies that bind with "self" are eliminated so that the set of detectors will find only "non-self." Clonal selection allows mature B cells to replicate quickly and with slight modification so that a more effective match can be made and so that there is a memory of the attack.

Author's Biography

Swapna Ghanekar received a BSE in aerospace engineering from the University of Michigan, Ann Arbor and an MS in computer science and engineering from Oakland University, California. She is currently working on a PhD at Oakland University. Her research interests include intrusion detection in wireless sensor networks.

Nancy Alrajei received her bachelor's degree from Palestine Polytechnic University, Palestine in Information Systems. After that, she joined Oakland University and completed a master's degree in information systems engineering. She is currently a fourth-year PhD student at Oakland University. Her area of research is intrusion detection in sensor networks.

Fatma Mili received her PhD in computer science from Universite Pierre et Marie Curie. She has been at Oakland University since with sabbatical and invited positions at Universite Laval, Canada, University of Buenos Aires, Argentina, Ecole des Arts et Metiers, France, and Ecole Nationale des Sciences Informatiques, Tunisia. Her research interests are in optimization, formal modeling, and distributed computing.

References

1. Lewis, F. L. *Wireless Sensor Networks. Smart Environments.* New York: John Wiley & Sons, 2004.
2. Krontiris, I., Benenson, Z., Dimitriou, T., Freiling, F. C. and Dimitriou, T. Cooperative intrusion detection in wireless sensor networks. *Wireless Sensor Networks,* LNCS, Vol. 5432. Berlin/Heidelberg: Springer, 2009.
3. Zhou, Y., Fang, Y. and Zhang, Y. Securing wireless sensor networks: A survey. 3, 3rd Quarter 2008, *IEEE Communications Surveys & Tutorials,* 10, pp. 6–28.
4. Karlof, C. and Wagner, D. Secure routing in wireless sensor networks: Attacks and countermeasures, in *First IEEE International Workshop on Sensor Network Protocols and Applications,* Anchorage, AK, 2002, pp. 113–127.

5. Wood, A. and Stankovic, J. A. A Taxonomy for Denial-of-Service Attacks in Wireless Sensor Networks. s.l., in *Handbook of Sensor Networks: Compact Wireless and Wired Sensing Systems*, CRC Press, Boca Raton, FL, 2004.

6. da Silva, A. P. R., Martins, M. H. T., Rocha B. P. S., Loureiro, A. A. F., Ruiz, L. B. and Wong, H. C. Decentralized intrusion detection for wireless sensor networks, in *Q2SWinet '05: Proceedings of the 1st ACM International Workshop on Quality of Service & Security in Wireless and Mobile Networks*, New York, ACM Press, pp. 16–23, 2005.

7. Roman, R., Zhou, J. and Lopez, J. Applying intrusion detection systems to wireless sensor networks, in *IEEE, 2006. IEEE Consumer Communications & Networking Conference (CCNC 2006)*, Las Vegas, pp. 640–644.

8. Onat, I. and Miri, A. An intrusion detection system for wireless sensor networks. s.l.: IEEE, August 2005, in *IEEE International Conference on Wireless And Mobile Computing, Networking, and Communications (WiMob '05)*, vol. 3, pp. 253–259.

9. Yu, B. and Bin, X. Detecting selective forwarding attacks in wireless sensor networks, in *Parallel and Distributed Processing Symposium*, 2006, IPDPS 2006, 20th International, April 25–29, 2006.

10. Burgess, M. *Promise You A Rose Garden: An Essay About System Management*, January 1, 2007. http://research.iu.hio.no/papers/rosegarden.pdf.

11. de Castro, L. N. and Timmis, J. *Artificial Immune Systems: A New Computational Intelligence Approach*. London: Springer, 2002.

12. Pagnoni, A. and Visconti, A. An innate immune system for the protection of computer networks. Dublin: s.n., 2005, in *Proceedings of the 4th International Symposium on Information and Communication Technologies (WISICT '05)*, Cape Town, South Africa, pp. 63–68.

13. Liu, Y. and Yu, F. Immunity-based intrusion detection for wireless sensor networks, in *IEEE International Joint Conference on Neural Networks (IJCNN 2008)*, Hong Kong, China, June 2008, pp. 439–444.

14. Sarafijanovic, S. and Le Boudec, J. Y. An artificial immune system approach with secondary response for misbehavior detection in mobile ad hoc networks. *IEEE Transactions on Neural Networks*, 2005, 16: 1076–1087.

15. Madden, S. *Intel Lab Data*. http://db.csail.mit.edu/labdata/labdata.html.

16. Matsumaru, N. and Dittrich, P. Organization-oriented chemical programming for the organic design of distributed computing systems, in *Proceedings of the 1st International Conference on Bio Inspired Models of Network, Information and Computing Systems*, Cavalese, Italy, ACM Press, 2006, Vol. 275.

17. Schöler, T. and Müller-Schloer, C. First steps towards organic computing systems: Monitoring an adaptive protocol stack with a fuzzy classifier system, in *Proceedings of the 2nd Conference on Computing Frontiers*, Ischia, Italy, s.n., 2005, pp. 10–20.

18. Sharma, K. Designing knowledge based systems as complex adaptive systems, in *Proceeding of the 2008 Conference on Artificial General Intelligence 2008: Proceedings of the First AGI Conference*, vol. 171, s.l.: IOS Press, Memphis, TN, 2008, pp. 424–428.

Chapter 16

Biological Inspired Autonomously Secure Mechanism for Wireless Sensor Networks

Kashif Saleem, Norsheila Fisal, Sharifah Hafizah Syed Ariffin, Sharifah Kamilah Syed Yusof, and Rozeha A. Rashid

Contents

16.1 Introduction

Wireless communication plays an important role in these days in the sector of telecommunication and has huge importance for future research. There has been an exponential growth in wireless communication due to the development of different devices and applications. In addition, there is an explosive increase in integration and convergence of different heterogonous wireless networks to ensure effective and efficient communication. These technologies primarily includes Wireless Wide Area Networks (WWANs), Wireless Local Area Networks (WLANs), Wireless Personal Area Networks (WPANs), and the Internet. The cellular networks can be classified under the WWAN, Bluetooth, and Ultrawide Bands classified as WPANs, and finally the WLANs and High-Performance Radio Local Area Networks (HiperLANs) belongs to the WLAN class.

Researchers at the University of California, Berkeley, recently develop a new approach in wireless system design: one that involves low-cost embedded devices that can be implemented for a variety of applications [1]. These small and low-cost sensor nodes became technically and economically feasible [2]. The sensor node is a miniaturized device which is equipped with sensors such as temperature, humidity, light, sound, and so on. Nevertheless, due to the extremely small architecture as shown in Figure 16.1, the sensors lacks in storage space, energy supply, and communication width. For

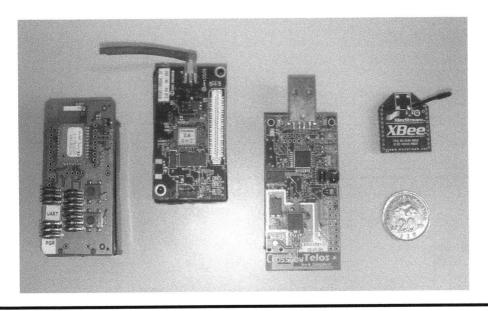

Figure 16.1   Sensinode, Micaz, Telosb, and XBee (from right to left).

example, a sensor typically has 8–120 kB of code memory and 512–4096 bytes of data memory. The transmission bandwidth ranges from 10 to 115 kbps.

The sensor nodes are programmed to work in a self-organized way. Owing to their autonomous capability, the sensor nodes can transfer the sensed data node-by-node to the destination known as sink node. The sink node is also called base station. The amount of base stations (such as laptop, personal digital assistant (PDA) gateway to other networks, and so on) in the deployed network depends on the application requirements. Enormous numbers of these disposable sensor nodes come up with a wireless sensor network (WSN) as shown in Figure 16.2.

16.1.1 IEEE 802.15.4

One of the most famous initiatives consolidating the possible deployment of WSN systems was IEEE 802.15.4, which specifies a physical (PHY) and a medium access control (MAC) layer dedicated for low-rate wireless personal area network (LR-WPAN). The main motivation of IEEE 802.15.4 is to develop a dedicated standard, and not to rely on the existing technologies such as Bluetooth or WLAN, and to ensure low-complexity energy-efficient implementations. IEEE 802.15.4 offers simple energy efficient and inexpensive solution to a wide variety of applications in WSNs. It supports simple one hop star network and multihop peer-to-peer network [3] as shown in Figure 16.3. Wireless links under IEEE 802.15.4 can operate in three license-free industrial scientific medical frequency bands. These accommodate over the air data rates of 250 kb/s in the 2.4 GHz band, 40 kb/s in the 915 MHz band, and 20 kb/s in the 868 MHz. In total, 27 channels are allocated in 802.15.4, with 16 channels in the 2.4 GHz band, 10 channels in the 915 MHz band, and 1 channel in the 868 MHz band [4].

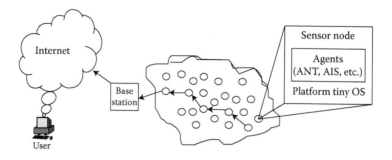

Figure 16.2 Wireless sensor network.

Figure 16.3 Multihop communication.

16.1.2 Types of Applications

The WSN serves an extremely valuable position in sensing and monitoring systems. The monitoring systems contain military and civil applications such as target field imaging, intrusion detection, weather monitoring, security and tactical surveillance, distributed computing, detecting ambient conditions such as temperature, movement, sound, light, or the presence of certain objects, inventory control, and disaster management as illustrated in Figure 16.4. In these applications, the random deployment of wireless sensor nodes can be through some air support or can be planted manually as given in Figure 16.2. For example, a number of sensors can be dropped in a disaster management

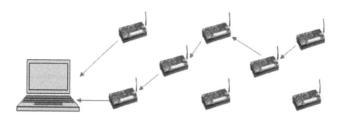

Figure 16.4 Sensor board.

application. The deployed sensor nodes through their autonomous communication assist in rescue operations by locating survivors, identifying risky areas, and making the rescue team more aware of the complete situation.

16.1.3 Resource Constraint

In sensing and monitoring systems, new gadgets and software advancements are very frequently available to the end-user. This development subsequently increases the complexity of the network. Also some of infrastructure less WSNs deployment area is out of human reach. Since the nodes in a network can serve as routers and hosts, they can forward packets on behalf of the other nodes and run user applications [5]. The resources in a sensor node are so limited that every possible means of reducing the usage of these resources are aggressively required. In essence, sensor networks will provide the end user with intelligence and a better understanding of the environment. WSN demands self-organized communication, which means the network can easily manage itself according to the changes in its environment. Furthermore, because of resource constraint and vulnerabilities of wireless communication, it is easier to suffer all kinds of attacks, if the sensor nodes are deployed in the unprotected/hostile environment [6]. Some of these attacks include signal jamming, eavesdropping, tempering, spoofing, resource exhaustion, altered routing information, selective forwarding, sinkhole attacks, Sybil attacks, wormhole attacks, and flooding attacks [7].

Since many sensor networks will be deployed in critical applications, security is essential. Unfortunately, security may be the most difficult problem to solve in WSNs. Widespread acceptance and adoption of these protocols in real-world WSNs would not be possible until their security aspects have thoroughly been investigated. However, security in these nature-inspired routing protocols is still an open issue [8]. Most self-organized communication and coordination solutions do not address security, so it is easy for an adversary to exploit those implemented solutions on a given WSN.

Owing to the wireless nature of WSNs, an adversary can deploy his own node that can take many actions to create a denial of service attack. Some of these are simply broadcasting at high energy, advertising that it is the fastest path to everywhere and simply throwing away packets that arrive, or sending wake-up calls to neighbors to exhaust their power. Note that when an adversary deploys a node to cause denial of service, it is the self-organizing and positive characteristic of WSNs that opens the system to various security breaches. Protocol solutions for media access control, routing, congestion control, and others all attempt to operate with minimum overhead and cost. This also subjects them to security problems. For example, a good solution for large-scale sensor networks is to give routing priority to packets passing through a node rather than admitting new packets. This helps prevent long delays for packets that have to traverse a large part of the sensor network.

Basically, the research challenges for security in sensor networks are vast and difficult. Lightweight schemes are required. Solutions must exploit the nature of the sensor network, possibly related to issues such as most data are valid only for a short time [7,9,10]. Therefore, lightweight security will be effective as individual nodes may possess little knowledge by themselves. Protecting the data aggregation function may also be possible. In summary, new ideas on the fundamental limits for security in these systems are needed. The new mechanism can maintain the features of WSNs such as multihop routing and dynamically environmental changes in a complete autonomous mode.

16.1.4 Self-Organization

In order to address autonomous capability for multihop WSNs, it has been visualized that self-organized network application can understand the network operational objectives. Additionally, probabilistic methods that provide scalability and preventability can be found in nature and adapted to technology. Researchers anticipate self-organization methods as the general solutions to the depicted communication issues in WSNs and sensor and actor networks (SANETs). Centralized management and optimized control will be replaced by methodologies that focus on local knowledge about the environment and adequate decision-making processes. Similar problems are known and well-studied in nature. Therefore, such biological solutions should be adapted to enhance the communication in *ad hoc* networks and WSNs [11].

It is observed that various biological principles are capable of overcoming the above adaptability issue. The area of bio-inspired network engineering which has the most well-known approaches is swarm intelligence (ANT Colony, Particle swarm), Artificial Immune System (AIS), and intercellular information exchange (Molecular biology) [8,12–14]. AIS has shown brilliant results for misbehavior detection in WSNs. The principles of AIS help in designing and implementing the security framework.

AIS learns the normal behavior of the system and then monitors the system for occurrences of abnormal patterns. The most interesting working behavior is the self-optimization and learning process. Two immune responses were identified. The primary one is to launch a response to invading pathogens leading to an unspecific response (using leukocytes). In contrast, the secondary immune response remembers past encounters, that is, it represents the immunologic memory. It allows a faster response the second time around, showing a very specific response (using B cells and T cells). The system, therefore, has the ability to detect previously unknown attacks. Therefore, it seems obvious to apply the same mechanisms for self-organization and self-healing operations in computer networks [11]. The authors of [8] proposed misbehavior detection in nature-inspired MANET protocol, BeeAdHoc. iNet proposed in [15] detects and eliminates the antigens (e.g., viruses) from the BiSNET/e enabled networks. However, self-healable and complete security is still an open issue. Widespread acceptance and adaptation of these protocols in real-world wireless networks would not be possible until their security aspects have thoroughly been investigated.

16.1.5 Objectives

This chapter proposes a biological, inspired autonomously secure mechanism. The security mechanism is developed to secure the WSN from the most common network layer attacks such as sink hole attack, select forward, black hole, message alter attack, hello attack, and worm hole. On network layer, the optimal route is discovered through Ant Colony Optimization (ACO) as described in [16]. Furthermore, a self-optimized routing protocol is enriched with a self-security mechanism. The preliminary report on this work may be seen in [17]. Security methods are inspired by immune system (IS) to perform an autonomous protection over WSN. The IS of mammals builds the basis for research on the AIS. The scope of AIS is widespread. There are applications for fault and anomaly detection, data mining (machine learning, pattern recognition), agent-based systems, control, and robotics [11]. Whereas variety of different techniques have been applied to error detection. Dynamic clonal selection algorithm (DynamiCS) [18,19] has been acquired from AIS, in order to adapt to a continuously changing environment.

The above-mentioned techniques will be accomplished by assigning each procedure to the group of agents. The agents such as search agent, data agent, and so on will work in a decentralized way to collect data and/or detect an event on individual nodes. Once collected, it will be transferred securely to the required destination/destinations through multihop communication. While moving from one node to another, the agent checks characteristics of next node with the preinitialized table. Over autonomous security is based on packet-receiving rate, packet-dropping rate, packet mismatch rate, and packet sending power metrics. Relying on these parameters, certain alarm will be generated by the agent or agents. The actions will be applied according to the type of alarm triggered. Eventually, the autonomous security mechanism will come up with the architecture that can prevent the WSN from the regular network layer attacks.

16.1.6 Organization of the Chapter

The next section reviews the related research a bit about optimum route discovery through ACO and in detail on the security of network routing protocol using various approaches, including the security through AIS. Section 16.3 describes the way to implement this autonomously secure routing mechanism. Section 16.4 shows the work and results obtained through the work done yet. The future directions for research and conclusion section are stated under Section 16.5.

16.2 Background and Related Research

16.2.1 Overview of Ant Routing in WSNs

Ant colony algorithms were first proposed by Dorigo et al. as a multiagent approach to difficult combinatorial optimization problems such as the traveling salesman problem (TSP) and the quadratic assignment problem (QAP), and later introduced the ACO meta-heuristic [20].

ACO algorithms are a class of constructive meta-heuristic algorithms that mimic the cooperative behavior of real ants to achieve complex computations and have been proven to be very efficient to many different discrete optimization problems. Many theoretical analyses related to ACO show that this optimization can converge to the global optima with nonzero probability in the solution space [21] and their performance has greatly matched many well-studied stochastic optimization algorithms, for example, genetic algorithm, pattern search, GPASP, and annealing simulations [20].

Singh et al. [22] have given an on-line ACO algorithm using AntNet techniques for MSDC that has been formalized to be a typically Minimum Steiner Tree problems. They also have proposed an improved algorithm by adding another type of ants, random ants, just like the newspaper deliverers, whose main task is to dissipate information gathered at the nodes among other neighboring nodes. Practically, simulation results also show that their algorithms are significantly better than address-centric routing. In these proposed algorithms, the forward ants normally spend a long time. There is a bug of dead lock in their algorithms. In their improved algorithm, a large amount of random ants are needed.

Zhang et al. [23] proposed three ant-routing algorithms for sensor networks. The sensor-driven and cost-aware ant routing (SC) algorithm is energy efficient but suffers from a low success rate. The flooded forward ant routing algorithm has shorter time delays; however, the algorithm creates a significant amount of traffic. Despite high success rate shown by the FP algorithm, it is not energy efficient.

Adaptive ant-based dynamic routing (ADR) algorithm using a novel variation of reinforcement learning (RL) was proposed by Lu et al. [24]. The authors used a delay parameter in the queues to estimate RL factor. In [25], the author proposes a novel approach for WSN routing operations. Through this approach, the network life time is maintained to a maximum period, while discovering the shortest paths from the source nodes to the base node using an evolutionary optimization technique. The research has also been implemented on microchip PIC® series hardware, called PIC12F683.

In [26], the authors propose two adaptive routing algorithms based on ant colony algorithm, the AR algorithm and the improved adaptive routing (IAR) algorithm. To check the suitability of ADR algorithm in the case of sensor networks, they modified the ADR algorithm (removing the queue parameters) and used their RL concept and named it the AR algorithm. The AR algorithm did not result in optimum solution. In IAR algorithm by adding a coefficient, the cost between the neighbor node and the destination node, they further improve the AR algorithm. In [27], the authors propose E&D ANTS based on Energy*Delay metrics for routing operations. Their main goal is to maintain network lifetime in maximum and propagation delay in minimum using a novel variation of an RL. E&D ANTS results were evaluated with AntNet and AntChain schemes. In [28], the authors propose a mechanism which maintains the network life time, while discovering the shortest paths from the source nodes to the base node using swarm intelligence-based optimization. In their proposed algorithm, the two parameters such as battery level and shortest path are considered. The authors also claimed that they have implemented their routing protocol on a proposed router chip and have tested on Proteus simulation platform.

16.2.2 Comparison of the Most Recent ANT-Based Routing in WSNs

SC and the approach discussed in [25] depend on the energy metric, whereas FF based on delay. IA and IAR are the modification of ADR, which used a delay parameter in the queues to estimate the RL factor. In FP, they combine the forward ant and data ant to enhance the success rate. E&D ANT based on energy*delay metrics for routing operations. In our proposed BIOSARP, the best values of velocity, packet reception rate (PRR), and the remaining power mechanism [29] are used to select forwarding node, because velocity alone does not provide the information about quality of link. The best link quality usually provides low packet loss and energy efficient [30]. Another novel feature of BIOSARP is it utilizes the remaining power parameter to select the forwarding candidate node. The remaining power assists the source node or intermediate node to distribute the forwarding load to all the available forwarding candidates and hence avoid the routing holes problem. BIOSARP is enhanced with any-cast forwarding scheme to route the data toward the best and nearest destination (Table 16.1).

16.2.3 Security in WSNs

The attractive features of WSNs involved many researchers to work on various issues. In WSNs, the routing strategies are getting more attention and the other hand security issues are not taken under consideration up to the required level. It is imperative that the security concerns be addressed from the beginning of the system design [31].

Many WSN routing protocols are quite simple, and for this reason are sometimes even more susceptible to attacks against general *ad-hoc* routing protocols. In WSNs, an adversary can either deploy his own node or compromise some nodes. The compromised node can take many actions to create a network layer attacks. Most network layer attacks against sensor networks fall into one of the following categories: spoofed, altered, or replayed routing information; selective forwarding; sinkhole attacks; Sybil attacks; wormholes; HELLO flood attacks and acknowledgement spoofing [7]. In the descriptions below, attacks that are based on manipulated sensor data are divided into two classes: that include attacks that try to manipulate user data directly and attacks that try to influence the underlying routing topology.

Table 16.1 Comparison of the Most Recent ANT-based Protocols

Title of the Mechanism	Velocity or Deadline	Remaining Power	Link Quality	Types of Forwarding
SC		Energy efficient		Multicast (one-to-many) and converge-cast (many-to-one)
FF			Network layer estimation	Multicast and converge-cast
FP	Forward agent + data agent		Network layer estimation	Multicast and converge-cast
AR	Heuristic correction factor			Broadcast, unicast and multicast
IAR	Heuristic correction factor			Geographic routing
Okdem	Path costing	Energy level		Broadcast, unicast and multicast
E&D ANTS		Energy	Network layer estimation	Broadcast, unicast and multicast

16.2.3.1 Spoofed, Altered, or Replayed Routing Information

The most direct attack against a routing protocol is to target the routing information that changed between nodes. By spoofing, altering, or replaying routing information, adversaries may be able to create routing loops, attract or repel network traffic, extend or shorten source routes, generate false error messages, partition the network, increase end-to-end latency, and so on [7,32].

16.2.3.2 Selective Forwarding

In a selective forwarding attack, malicious nodes may refuse to forward certain messages and simply drop them [7]. The neighboring nodes will conclude that the current route has failed and they decided to seek another route. Selective forwarding attacks are typically most effective when the attacker is explicitly included on the path of a data flow. However, it is conceivable that an adversary overhearing a flow passing through neighboring nodes might be able to emulate selective forwarding by jamming or causing a collision on each forwarded packet of interest. Thus, an adversary who is launching a selective forwarding attack will likely follow the path of least resistance and attempt to include herself on the actual path of the data flow. In the next two sections, we discuss sinkhole attack and the Sybil attack, two mechanisms by which an adversary can efficiently include herself on the path of the targeted data flow.

16.2.3.3 Sinkhole Attacks

In a sinkhole attack [7,32], the adversary goal is to attract almost all the traffic from a particular area through a compromised node, creating a metaphorical sinkhole with the adversary at the center.

Because nodes on or near the path that packets follow have many opportunities to tamper with application data, sinkhole attacks can enable many other attacks (selective forwarding, for example). Sinkhole attacks typically work by making a compromised node look especially attractive to the surrounding nodes with respect to the routing algorithm. For instance, an adversary could spoof or replay an advertisement for an extremely high-quality route to a base station. Some protocols might actually try to verify the quality of route with end-to-end acknowledgements, containing reliability or latency information. In this case, a laptop-class adversary with a powerful transmitter can actually provide a high-quality route by transmitting with enough power to reach the base station in a single hop, or by using a wormhole attack discussed in the following section. Owing to either the real or imagined high-quality route through the compromised node, it is probably each neighboring node of the adversary will forward packets destined for a base station through the adversary, and propagate the attractiveness of the route to its neighbors. One motivation for mounting a sinkhole attack is that it makes selective forwarding trivial. By ensuring that all traffic in the targeted area flows through a compromised node, an adversary can selectively suppress or modify packets originating from any node in the area.

16.2.3.4 Sybil Attacks

In a Sybil attack [33], a single node presents multiple identities to other nodes in the network. The Sybil attack can significantly reduce the effectiveness of fault-tolerant schemes such as distributed storage [34], dispersity [35], multipath routing [36], and topology maintenance [37,38]. Douceur [39] showed that, without a logically centralized authority, Sybil attacks are always possible except under extreme and unrealistic assumptions of resource parity and coordination among entities. Sybil attacks also pose a significant threat to geographic routing protocols. Location aware routing often requires nodes to exchange coordinate information with their neighbors to efficiently route geographically addressed packets. It is only reasonable to expect a node to accept a single set of coordinates from each of its neighbors, but by using the Sybil attack an adversary can be in more than one place at once.

16.2.3.5 Wormholes

In the wormhole attack [40,41], an adversary tunnels messages received in one part of the network over a low latency link and replays them in a different part. The simplest instance of this attack is a single node located between the two other nodes forwarding messages between the two of them. An adversary located close to a base station may be able to disturb routing by creating a well-placed wormhole. An adversary could convince nodes who would normally be multiple hops from a base station that they are only one or two hops away via the wormhole. This can create a sinkhole: since the adversary on the other side of the wormhole can artificially provide a high-quality route to the base station, potentially all traffic in the surrounding area will be drawn through her if alternate routes are significantly less attractive.

16.2.3.6 HELLO Flood Attack

Many protocols require nodes to broadcast HELLO packets to announce themselves to their neighbors, and a node receiving such a packet may assume that it is within the radio range of the sender [7]. This assumption may be false; a laptop-class attacker which is broadcasting routing or other information with large enough transmission power could convince every node in the network that the adversary is its neighbor. For example, an adversary who is advertising a very high-quality route to the base station to every node in the network could cause a large number of nodes to attempt to use this route, but those nodes sufficiently far away from the adversary would be sending packets into oblivion. The network is left in a state of confusion. A node realizing the link to the adversary is false could be left with few options: all its neighbors might be attempting to forward packets to the adversary as well. Protocols that depend on localized information exchange between neighboring nodes are also subject to this attack. An adversary does not necessarily need to be able to construct legitimate traffic to use the HELLO flood attack. This can simply re-broadcast overhead packets with enough power to be received by every node in the network. HELLO floods can also be thought of as one-way, broadcast wormholes.

It is interesting to note that flooding usually used to denote propagation of a message to every node in the network over a multihop topology. In contrast, despite its name, the HELLO flood attack uses a single hop broadcast to transmit a message to a large number of receivers.

16.2.3.7 Acknowledgment Spoofing

Several sensor network routing algorithms rely on implicit or explicit link layer acknowledgments. Owing to the inherent broadcast medium, an adversary can spoof link layer acknowledgments for overhead packets addressed to neighboring nodes [7]. Goals include convincing the sender that a weak link is strong or that a dead or disabled node is alive. For example, a routing protocol may select the next hop in a path using link reliability. Artificially reinforcing a weak or dead link is a clever way of manipulating such a scheme. Since packets sent along weak or dead links are lost, an adversary can effectively mount a selective forwarding attack using acknowledgement spoofing by encouraging the target node to transmit packets on those links.

Because of resource constraint and vulnerabilities of wireless communication, it is easier to suffer all types of attacks if the sensor nodes are deployed in the unprotected/hostile environment. These attacks involve signal jamming and eavesdropping, tempering, spoofing, resource exhaustion, altered or replayed routing information, selective forwarding, sinkhole attacks, Sybil attacks, wormhole attacks, flooding attacks, and so on [32]. Many papers have proposed prevention countermeasures of these attacks and the majority of them are based on encryption and authentication. However, these prevention measures in WSNs can reduce intrusion, to some extent, but cannot eliminate them at all. A simple example is that these two measures take no effect on these attacks caused by these compromised nodes with legal keys. In this case, Intrusion Detection System (IDS) can work as second secure defense of WSNs to further reduce attacks and insulate attackers.

In traditional networks, traffic and computation are typically monitored and analyzed for anomalies at various concentration points. Though, this is often expensive in terms of memory and energy consumption of a network, as well as its inherently limited bandwidth. WSNs require a solution that is distributed and inexpensive in terms of communication, energy, and memory requirements. Therefore, these traditional techniques of IDS must be modified or new techniques must be developed to make intrusion detection work effectively in WSN.

16.2.4 Overview of IDS-Based Security

IDS in traditional network has widely been proposed and applied. Siriaj et al. [42] presented a model of decision engine for intelligent IDS. This decision engine uses Fuzzy Cognitive Maps and fuzzy rule-bases for causal knowledge acquisition and to support the causal knowledge reasoning process. Harmer et al. [43] presented the AIS architecture for computer security.

Albers and Camp [44] proposed a type of general intrusion detection architecture based on the implementation of a local IDS (LIDS) at each node.

Kachirski and Guha [45] proposed a multi-sensor IDS based on mobile agent technology. They divided the mobile agents into three types of agents: monitoring agent, decision agent, and action agent.

In WSNs, Alpcan et al. [46] and Agah et al. [47] proposed to adopt game theory for decision and analysis in intrusion detection of WSNs. They investigate the basic decision and analysis processes involved in information security and intrusion detection, as well as a possible usage of game theory for developing a formal decision and control framework. Generic model for distributed IDSs was introduced by defining a network of sensors, and propose two simple, flexible, and easy-to-implement schemes utilizing both cooperative and noncooperative game theoretic concepts.

ACO is also utilized for intrusion detection in [48,49]. The authors have introduced the concept of tabu list, where for every session the list would like to store the pheromone trace or path that is prone to attack.

In [50], the ACO-based intrusion feature selection algorithm is proposed. The FDR is taken in as the heuristic information for ACO. The authors have adopted the Least Square-based SVM estimation to avoid training of a large number of SVM classifier. The results have been demonstrated, by which they have show the detected attacks as probe, dos, and U2R&R2L intrusions.

16.2.5 Overview of AIS-Based Security

AISs are adaptive systems, inspired by theoretical immunology and observed immune functions, principles, and models, which are applied to problem solving. Adaptability in the IS ensues from features such as learning and memory that endow the IS with the ability to fight a large variety of invaders [19]. The application of AIS to fault tolerance was initially motivated by Avizienis, who described the analogy between the IS and fault tolerance [51]. Since then, several approaches have been proposed in literature that have applied AIS to problems related to both software and hardware fault tolerance [19].

In [6], the authors have proposed a novel IDS (SAID) to be suitable for deploying in WSNs. SAID with three-logic-layer architecture adopt the merits of LIDS and distributive and cooperative IDS and is self-adaptive for intrusion detection of resource-constraint WSNs. SAID can actively trigger agent evolution to more effectively prevent intrusion when WSN suffers unknown attacks. For distributive cooperation attacks, these distributive mobile monitor agents will cooperatively collect abnormal information of network to help a correct intrusion decision. Knowledge base is deployed base station where the complex algorithm (e.g., genetic algorithm) for agent evolution can be computed and intrusion rules can be stored.

In [52], the authors proposed a new group-based intrusion detection scheme which is a detection-based technique. In this scheme, the authors partition the sensor nodes in a network into a number of groups. The nodes in a group have the same sensing capability and are physically close to each other. And the proposed intrusion detection algorithm is scheduled to run for each group. The authors use data released from the Intel Berkeley Research Lab. Through experiments they have shown that their scheme can achieve a lower false alarm rate and a higher detection accuracy rate with less power consumption.

In [19], the authors detail the investigations undertaken to develop an immune-inspired adaptable error detection (AED) technique for ATMs. The proposed framework for AED consists of two levels of error detection. One level of the framework is local to a single ATM, while the other is a network-wide AED. In the given architecture, each ATM hosts a local AED, while the network-wide

AED is implemented within the central management system. The implementation undertaken in this work was limited to the local AED. An AIS algorithm was found to possess these characteristics, and was evaluated by using relevant criteria that include: (1) classification performance of the algorithm in discriminating normal behaviors from potential failure behaviors and (2) the measurement of the time interval between detection and the actual system failure. Exposed results demonstrate that the described AED technique could detect an incipient system failure approximately 12 hours for one data set, and 2 hours for a second data set.

In [53], the authors published a system named AISEC which was capable of classifying emails as interesting or noninteresting and removed uninteresting mail from a user's inbox. Furthermore, the system was shown to be capable of continuous learning; following changes in a user's interest, the system could adapt to the new interests. The results were published from a single set of 2268 emails of which 32.7% were classified as uninteresting and the remainder were interesting. The results were compared to the results from the performance on the same data set with a native Bayesian system; although performance was similar overall, AISEC showed improved performance during certain periods of time. It was postulated that this was due to the ability of AISEC to adapt to changes in the data. However, this hypothesis was never explicitly tested by examining the data in detail or by testing specific scenarios in which emails were known to change in content.

In [54], the authors have re-evaluated the AISEC with the different set of test emails and have also extended the method discussed in ref. [53]. While extending, the authors have specifically addressed certain objectives. Initially, they investigate the sensitivity of the algorithm parameters to different data sets. Afterwards, they test the ability of the algorithm to adapt to changing interests, by setting up a number of test scenarios in which the users interest in emails from a particular source changes from interesting to uninteresting (and vice versa) over a period of time. Subsequently, AISEC has been modified, while improving the speed of algorithm at which it adapts and the overall accuracy of the classification algorithm. In addition, they changed the AISEC in .NET as an add-in for Microsoft Outlook and named as AISEC-Outlook. When the user supplies negative feedback from misclassification of an item in the Inbox, the email is now added to the repository of B-Cells responsible for classifying mails. On the basis of this feedback, AISEC Outlook rewards the B Cells.

In [55], Lee and Suzuki proposed and evaluated a decentralized self-healing mechanism that detects and recovers from wormhole attacks. Upon detecting a wormhole attack, the proposed mechanism, called SWAT[a], isolates wormhole nodes from the network by eliminating links connected to them, and recovers the routing structure distorted by the wormhole nodes. SWAT is designed as a decentralized in network detection mechanism that uses network connectivity information only. Finally, they have implemented SWAT on MICA2 and TelosB motes and acquire the result under tinyviz. The simulation results shown in this paper yields 100% wormhole attack detection, 0% false detection, 100% wormhole node isolation, and 0% false isolation in dense networks. Through SWAT, they have come to tackle only wormhole attacks on WSNs.

Since the characteristics of WSNs (e.g., resource constrains of sensor nodes, *Ad Hoc* mechanism, the sensor node may be static after deployment, and so on), most IDS for internet network or mobile *Ad Hoc* network cannot be applied in WSN well. Therefore, the researches in IDS of WSN are still at the beginning [6]. The operation of BeeHive algorithm requires an initialization phase (30 seconds) even before the AIS learning could start. It is followed by the learning (50 seconds) and protection phases to, respectively, learn the BeeHive normal behavior and detect the routing attacks [8]. The existing security methods are not applicable for self-organize routing in WSNs because the execution time for one-hop is high and WSNs have density deployment where hundreds of nodes need more time to process security mechanism.

16.2.6 Overview of Keying-Based Security

Su et al. [56] proposed two approaches to improve the security of clustering-based sensor networks: authentication-based intrusion prevention and energy-saving intrusion. The proposed authentication-based intrusion prevention is enhanced from µTESLA, which uses one-time key chains. Therefore, each CH needs to be loosely time synchronized with its member nodes. All sensor nodes are loosely time synchronized. The synchronization in WSNs is very hard and mostly because of huge number of nodes it is impossible to have an accurate synchronization.

Owing to the factor of initialization phase, the WSN need security mechanism to be in operation before the network deployment. As stated in [57], Node cloning attacks can be mounted only during deployment since a cloned node cannot initiate the protocol with success; it can successfully be connected only by acting as a responding node. Recent progress in implementation of elliptic curve cryptography (ECC) on sensors proves Public Key Cryptography (PKC) is now feasible for resource constrained sensors [58]. Given the efficient low-layer primitive in place, the high-layer PKC-based security scheme design in sensor networks, however, is not straightforward due to the special hardware characteristics and requirements of sensor networks. Therefore, the performance of PKC-based security schemes is still not well investigated.

In [59], a scheme has been proposed which explores the superimposed s-disjunct code for a timely clone attack detection. A fingerprint can easily be encoded with a very short bit stream, which results in small message overhead. Their mechanism can identify cloned sensors with high detection accuracy at the expense of a very low communication/computation/storage overhead. Given scheme conducts fingerprint verification locally (via neighboring nodes) and globally (via the base station) for each message broadcasted by any node; therefore, clone attackers can be detected in real-time.

In [60], the authors presented the security enhancement that uses the encryption and decryption with authentication of the packet header to supplement secure packet transfer. SRTLD solves the problem of producing real random number problem using random generator function encrypted with mathematical function. The output of random function is used to encrypt specific header fields in the packet such as source, destination addresses, and packet ID. Moreover, the data authenticity problem is solved in SRTLD using an authentication procedure applied after decryption. In this mechanism, they assume that each sensor node is static, aware of its location, and the sink is a trusted computing base.

The mechanism such as Pairwise Key Establishment (PKE) based on transitory master keys as discussed in [57] is particularly useful for the purpose. LEAP++ consists of system setup, PKE, and authentication key disclosure. Security in natural inspired routing protocols is still an open issue [8]. Widespread acceptance and adoption of these protocols in real-world wireless networks would not be possible until their security aspects have thoroughly been investigated.

16.2.7 Comparison of the Most Common Secure Routing Protocols in WSNs

The MAC and physical layers are based on IEEE 802.15.4, which is designed for low rate communication such as WSNs. The review of literature concludes that the data security and routing design in WSNs are not easy works due to the numerous constrains in WSN such as memory storage, power limitation, and unreliable wireless communication. The aforementioned limitations should be considered when security based on routing is designed.

The security measures as we have discussed above have still a lot of holes and could not tackle the most common WSN routing attacks. In front of the currently running security protocols, BIOSARP will protect WSN from the most common attacks yet stated [7,61]. Our autonomously secure routing mechanism BIOSARP will make security decisions based on PRR, packet dropping rate, packet mismatch rate, and packet sending power metrics. As further the security enhancement based on human nerve barrier system will also helps to cut down the feedbacks. This barrier system will help to differentiate between a good and malicious node to be or not to be a member of the running sensor network. Other security attacks and intrusions are handled by the Artificial Immune misbehavior/IDS. Finally, BIOSARP routing scheme has three type of security: built-in security due to random selection for next hop, authentication-based security, and intrusion detection. Hence, it is more difficult for an adversary to attack and intercept a message. Table 16.2 summarizes the most common secure routing protocol in WSNs.

Table 16.2 Comparison of the Most Recent Security Based Protocols

Title of the Protocol	Tackled Attacks	Limitations
[56] by Su	Bogus routing information Hello floods Black hole Select forward	Sensor nodes cannot move and new sensor nodes cannot be added after the pairwise keys are established
BeeAIS	Non-self antigens	Assuming no attack in first 30 seconds
SAID	Worm hole Sybil	Assumption of no attack while initialization
[52] by Li	Fabric information attack Select forward Sink hole Hello attack Worm hole attack	Group-based intrusion mechanism and only internal detection
[59]	Clone attacks	Have less overheads but tackling only clone attacks
LEAP++	Clone attacks DoS attacks Worm hole attacks	Assuming NCC is willing invest large enough time Assume all nodes deployed at a time are programmed to start neighbor discovery at some delay time after being airdropped
STRLD	HELLO flood Selective forwarding	Each sensor node is static and aware of its location Sink is a trusted computing base

16.3 Methodology

16.3.1 System Design

System design deals mainly with the development of state machine diagrams, the routing management, neighborhood management, energy management, and security management, as shown in Figure 16.5.

16.3.2 Routing Management

Routing management will be dependent mostly on forwarding metrics calculation. While establishing the forwarding procedure, the routing management will look for the next best node toward the destination through the routing table, available at every node. By acquiring the optimal route from the routing table, routing management will finalize the forwarding process. Otherwise, if it could not find next node toward destination, then routing management will call the process of neighbor discovery under neighborhood management, as shown in Figure 16.5.

16.3.3 Neighbor Management

The neighborhood management will then search for the best neighboring node by calling method. Calling method take place through broadcasting hello massages. The node that broadcasts will

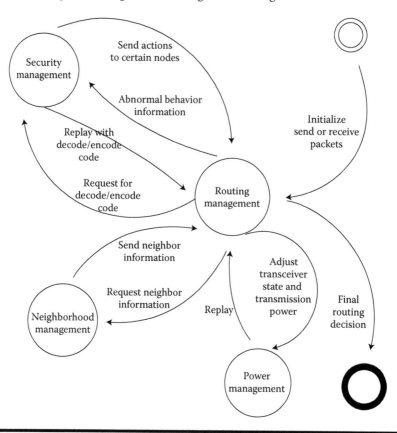

Figure 16.5 System diagram.

receive replies from the neighboring nodes along with their characteristics. On the base of these replays, it provides the final solution back to the routing management state. According to this solution, the routing management state will update the routing table on a current node.

16.3.4 Power Management

The key role of power management state is to check the remaining power and inform accordingly to the higher state. The power management state can also adjust the power for the transceiver according to the environmental conditions. Under this state, the energy parameter is imported from the physical layer into the network layer. In wireless sensor node, there are five levels of power transmission. At the time of forwarding, the first level is utilized; however, if node is out of reach, then the power level is increased in stages. Helping neighborhood management state in the energy aware route discovery and power level management is controlled by the power management state.

16.3.5 Forwarding Criteria

Inside routing management, the forwarding metrics calculation takes place, as shown is Figure 16.6. The forwarding metrics as given in Table 16.3 are calculated to get the optimal route decision toward the destination. If the error occurred while processing this state, it will be controlled by routing problem handler as elaborated in Figure 16.6. The error can be like required neighbor not present or the best neighboring node is lost or the required parameter is not there. Otherwise, if there is no error while forwarding calculation, then the anycast state will be called for forwarding the required packets.

Common functions under neighbor management state are neighbor table maintenance, neighbor discovery, insert new neighbor, neighbor replacement, and so on, as exposed in Figure 16.7. The routing table which is the most important thing while performing routing is maintained under this stage. If the best node toward the destination could not be found, the child state neighbor discovery is initiated. The explored new nodes will be checked with the old records by neighbor replacement process. While inserting a new record, the routing table space is first checked by neighbor table

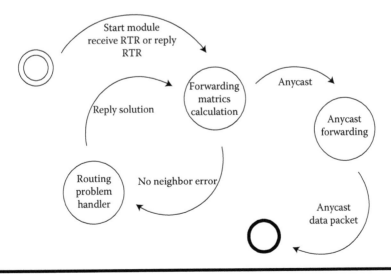

Figure 16.6 Routing management.

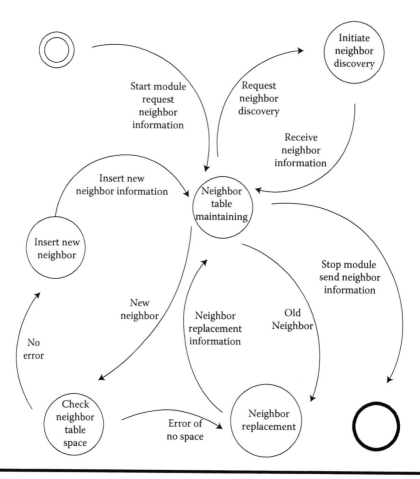

Figure 16.7 Neighbor management.

space state according to the wireless sensor node memory. Finally, the new record is inserted in routing table through insert new neighbor state.

While performing the routing management, the forward agents and search agents are also checking the characteristics of next node through the parameters, as given in Table 16.4. With this check the abnormality over the network is detected. The alarm based on the type of abnormality is then generated by particular agents. The alarm message is then handled by the security management

Table 16.3   Routing Metrics

	Link Quality	*Energy*	*Velocity*
Node 1	α^1	β^1	V^1
Node 2	α^2	β^2	V^2
.	.	.	.
.	.	.	.
.	.	.	.
Node *n*	α^n	β^n	V^n

Table 16.4   Security Metrics

	Packet Receiving Rate	Packet Dropping Rate	Packet Mismatch Rate	Packet Sending Power
Tackled Attacks	Sink Hole Attack [52]	Select Forward, Black Hole [52]	Message Alter Attack [52]	Hello Attack, Worm Hole [52,59]
Node 1	α^1	ρ^1	υ^1	ζ^1
Node 2	α^2	ρ^2	υ^2	ζ^2
.	.	.	.	.
.	.	.	.	.
.	.	.	.	.
Node n	α^n	ρ^n	υ^n	ζ^n

module. Under security module decision, defense agents are working. Decision agent takes the alarm message and makes the decision accordingly through Equation 16.7.

According to the information from the decision agents, the defense agents can visit the neighbor nodes of the attack node and take appropriate actions. These actions will include (1) requiring the neighbor nodes to low the priority or refuse relaying the packages from these attack nodes; (2) telling the sending node another routing path in order to circumvent the attacker; or (3) repairing the attacked node by renewing the encryption keys of nodes. By these ways, they can isolate the attack nodes successfully. The defense agent can also copy themselves to surround the intruders when it arrive nearby the adversary and suicide after an appropriate period in order to reduce the communication overhead of WSN. It is very necessary when the WSN suffers the attacks, especially for resource exhaustion attacks [6].

16.3.6 Optimal Route Discovery

BIOSARP system is mainly based on route section. The optimal route discovery is tackled by ACO [62]. Routing decision will be achieved through a probabilistic decision rule, as shown in the following equation [26]:

$$P'_{\text{id}} = \frac{P_{\text{id}} + \beta x C_i}{1 + \beta x(|N| - 1)} \qquad (16.1)$$

where P'_{id} is the normalized sum of the probabilistic entry, P_{id}, of the routing table with a correction factor C_i and the coefficient β. The value of β in Equation 16.2 weights the desirability of the correction factor C_i with respect to the probability values (P_{id}) stored in the routing table. The coefficient β has a value between 0 and 1. C_i is the cost from the current node k to the neighbor node i, which is calculated using the following equation [26]:

$$C_i = 1 - \frac{D_{k,i}}{\sum_{j=i}^{|N|} D_{k,i}} \qquad (16.2)$$

where $D_{k,i}$ distance node k and node i.

The decision will depend on our used metrics as, velocity, the remaining power and link quality mechanism as given in Table 16.3.

16.3.7 Determination of Packet Velocity

Velocity factor has been integrated to support real-time traffic over WSNs. The velocity factor depends on end-to-end delay from source to destination node. The maximum packet velocity (V) between a pair of nodes is calculated using the following equation [63]:

$$V = \frac{d(S, N)}{\text{Delay}(S, N)} \quad (16.3)$$

where $d(S, N)$ is the one-hop distance between source node S and destination node N. The total delay to one hop neighbor (N) from the source (S) can be estimated using the following equation [29]:

$$\text{Delay}(S, N) = T_c + T_t + T_p + T_q + T_b + T_s = \frac{\text{Round\_trip\_time}}{2} \quad (16.4)$$

where T_c is the time it takes for S to obtain the wireless channel with carrier sense delay and backoff delay. T_t is the time to transmit the packet that is determined by channel bandwidth, packet length, and the adopted coding scheme. T_p is the propagation delay that can be determined by the signal propagation speed and the distance between S and N. In sensor networks, the distances between sensor nodes are normally very small, and the propagation delay can normally be ignored. T_q is the processing delay that depends on the network data processing algorithms to process the packet before forwarding it to the next hop. T_b is the queuing delay, which depends on the traffic load. In a heavy traffic case, queuing delay becomes a dominant factor. T_s is sleep delay which is caused by nodes periodic sleeping. When S gets a packet to transmit, it must wait until N wakes up. Equation 16.3 shows that the delay between the two pair of nodes varies since the T_c and T_b delays differ for all nodes.

16.3.8 Determination of Link Quality

The link quality of the wireless medium determines the performance of the WSN. In designing BIOSARP, the link quality is considered in order to improve the delivery ratio and energy efficiency. It should be noted that the link quality is measured based on PRR to reflect the diverse link qualities within the transmission range. PRR is determined by the following equation [29].

$$\text{PRR} = \left[1 - \left(\frac{8}{15} \right) \left(\frac{1}{16} \right) \sum_{j=2}^{16} (-1)^j \binom{16}{j} \exp\left(20\,\text{SNR} \left(\frac{1}{j} - 1 \right) \right) \right]^m \quad (16.5)$$

SNR is calculated as [29]:

$$\text{SNR} = P_t - PL(d) - S_r \quad (16.6)$$

where P_t is the transmitted power (in dBm) and S_r is the sensitivity of a receiver (in dBm).

16.3.9 Security Management

ACO will be further enhancing with an additional security management module, as shown in Figure 16.5. This module is based on AIS to self-secure and self-heal the network from the foreign bodies or attacks. AIS make its decision by sensing or detecting the intrusion based on additional parameters/metrics: packet dropping rate, packet mismatch rate, and packet sending power as given in Table 16.4. AIS will do the detection for intruders that will be further increased by key management system.

Monitoring will be performed depending on the parameters as given in Table 16.4. When the intrusion alarm generates by the monitoring agents (search and forwarding agents), the security management makes the decision through decision agents. The decision agents are similar with B-Cell in IS. All types of decision agents is distributive, mobile, cooperative, and redundant. Thus, these decision agents can cooperate effectively to make a correct decision for distribute attacks. The major objective of decision agents is to detect the existence of non-self patterns within a potentially large set of the existing non-self patterns. The matching criteria depend on the following equation [6]:

$$\text{match}(f, \varepsilon, I, D) = \begin{cases} \text{malicious,} & f(1, \alpha) \geq 1- \\ \text{benign,} & \text{otherwise} \end{cases} \tag{16.7}$$

where I is the input string, D the matching string of a decision agent, f the matching function, and ε is the matching threshold.

To simplify the pattern matching, the statistical matching rules are adopted [6]. The correlation coefficient produces a number between -1 and 1 that relates how similar the two input sequences are. It is defined as [6]:

$$X, Y, \varepsilon\{0\ldots255\}^N, \quad N = \frac{l}{8}, \quad \rho = \frac{\sum_{i=1}^{n}(X_i - \bar{X})(Y_i - \bar{Y})}{\sqrt{\sum_{i=1}^{n}(X_i - \bar{X})^2 \sum_{i=1}^{n}(Y_i - \bar{Y})^2}} \tag{16.8}$$

Inside security architecture (Figure 16.8), the agents will perform the actions accordingly as described in Figure 16.9 through state diagrams. Monitoring agents (forward and search agents) will perform the checking process based on the parameters as given in Table 16.4. The security management implements certain actions through defense agents. The defense agents act somewhat similar as the antibody that is secreted by lymphocyte. Their function modules involve the self-copy, isolation, and suicide.

Moreover, in this routing algorithm, we enhanced the ant routing algorithm with new idea taken from the structure of the universe to minimize energy consumption. As in the universe, every galaxy has one source (sun) to do broadcast and stars (destinations) to receive. While acquiring, we

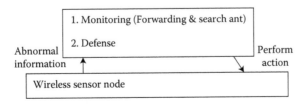

Figure 16.8 Two layer security architecture for intrusion detection.

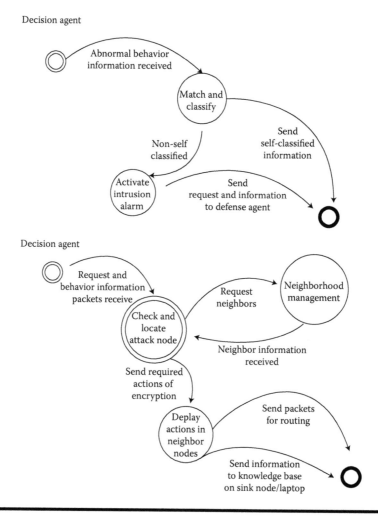

Figure 16.9 Decision and defense agent.

change the broadcasting to be for a certain time limit according to WSN environmental conditions, as shown in Figure 16.10. Modifications will help WSN in the means of saving energy in the initialization phase. Additionally, the network has been secured by random encryption system as explained in [60]. This is inspired by human immune barrier system, as illustrated in Figure 16.10.

16.4 Simulation

16.4.1 Simulation Tools

The scenario was simulated using network simulator 2 (NS2) [64]. The BIOSARP is implemented under NS2, the code is written in C++ and OTcl programming language. Twenty-five wireless sensor nodes were deployed as shown in Figure 16.11. Each link is bidirectional and the weighting value of the link depends on the power consumption (in nJ/bit), ant's moving time delay (ms), and PRR. After the source nodes produce a quantity of artificial ants, the destination nodes are randomly chosen by an average probability. When one packet passes through a node at a certain

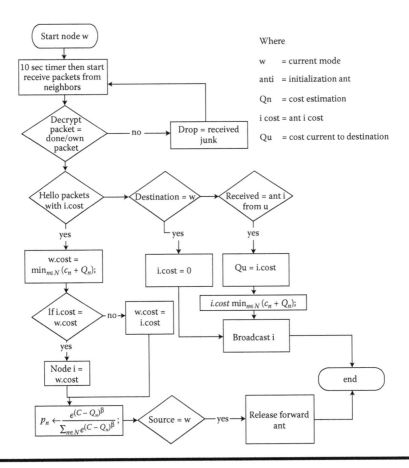

Figure 16.10 Enhanced initialization mechanism with encryption system.

speed, the node takes the first step to gather all the ant agents into buffer storage and then selects the optimal path from its routing table to transfer packets.

16.4.2 Graphical Animation of the Network

During the animation produced by an animator (nam), as shown in Figure 16.11, we can examine the output of the network. The cbr traffic is produced first from node 2 to node 6, then the Poisson traffic from node 8 to node 23. Each node contains a table with the pheromone value. As an example, pheromone table at node 0 is shown in Table 16.5. The pheromone table at each node contains the pheromone value for the next node toward the required destination. Although the network is online, the routing table is directly built up through pheromone table exponential transformation. In this way, all the ants disperse in as many paths as possible to achieve the balance of the load.

16.4.3 Network Model and Performance Parameters

Twenty-five wireless sensor nodes were deployed onto 50×50 meter^2 grid, as shown in Figure 16.11. A fixed size of one packet is considered in our simulation. The experimental parameters used to configure the system according to WSNs are listed in Table 16.6. In order to avoid cycles and the routing table's freezing, we need to initialize $\tau 0$ as shown in [65].

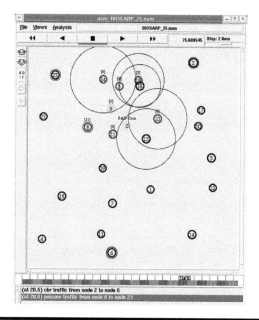

Figure 16.11   Graphical representation of network topology.

Table 16.5   Pheromone Table at Node 20

Dest	Next	pH Value
0	3	0.043210
0	17	0.429941
0	10	0.526850

Table 16.6   System Parameters

Parameters	Values
Propagation Model	Two Ray Ground
phyType	Phy/WirelessPhy/802_15_4
macType	Mac/802_15_4
CSThresh_	8.54570e–07 (15 meters)
RXThresh_	8.54570e–07 (15 meters)
Frequency	2.4e+9
Traffic	CBR, POISSON
Packet size	70

In this case, ant agents can adjust to the more efficient path when the network traffic loads change and the congestion fades away. Simulation methods for the AntNet were attempted in [66] where the parameters $(c, a, a', \varepsilon, h, t)$ were set to $(2, 10, 9, 0.25, 0.04, 0.5)$.

16.5 Results

The results presented in this paper, through the logical implementation of WSNs, are the real-time graphical network topology and graphs by the accumulated results. Graphical representation is through the network nam under NS2, as shown in Figure 16.11.

Secondly, the results presented the graphs generated by the trace graph 2.05 [67]. These graphs are depending on the results extracted from a trace file produced under NS2. Through parameters adjustment under trace graph 2.05, we get the network information accordingly as given in Figure 16.12.

16.5.1 Performance Analysis

The given BIOSARP is compared with routing protocols such as AODV and DSR routing protocol. The simulation evaluates the performance of every protocol in terms of generated packets throughput and dropping packets throughput over WSN. Figure 16.13 shows the graph generated by using BIOSARP. In the beginning, the throughput of generating packets over WSN is less. However, when BIOSARP build up the route knowledge in a short while over the network, the throughput of generating packets is enhanced whereas the packets dropping is minimized.

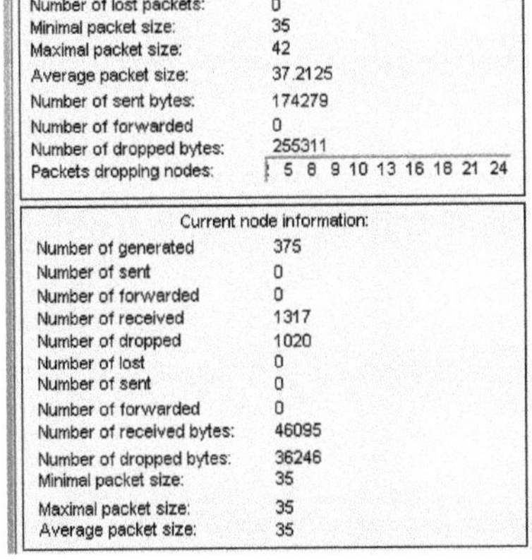

Figure 16.12   Network information.

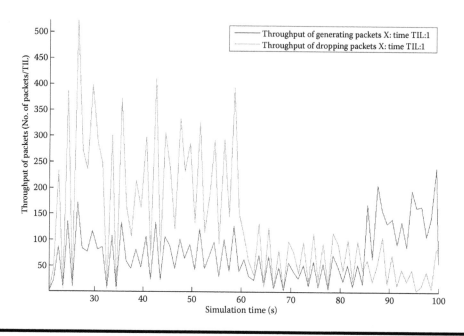

Figure 16.13 Throughput with BIOSARP.

Figure 16.14 shows the results accumulated by AODV routing protocol over WSNs. The throughput of generated packets is very less and dropping is very high. Even with the passage of time the ratio cannot be maintained, shown in Figure 16.14. Figure 16.15 shows the DSR outputs. In

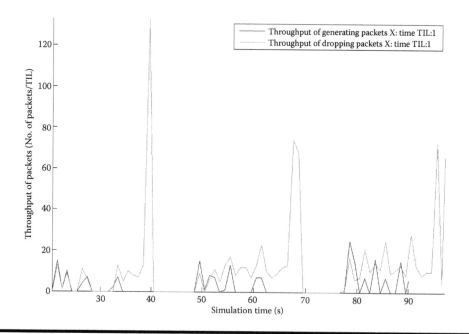

Figure 16.14 Throughput with AODV.

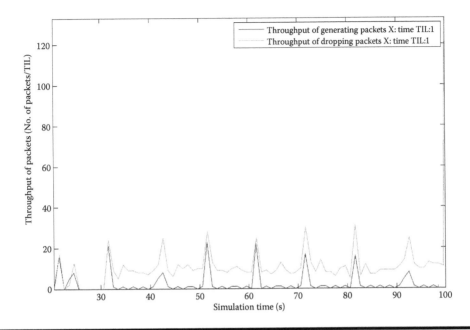

Figure 16.15 Throughput with DSR.

the beginning, the DSR maintains the throughput but as the simulation goes on we find an increase in the throughput of dropping packets. Finally, the throughput of generated packets stops and dropping packets goes to extreme. With the help of these results, we can check the different states of network, like the sleep/wake state. By checking these states we can also improve our parameters for BIOSARP to minimize the packet dropping while maximizing data throughput rate for real-time traffic transfer.

16.6 Future Directions for Research

BIOSARP routing protocol in WSNs provides a channel for further enhancement toward future applications such as coexistence with wireless network, monitoring applications for indoor and outdoor application, and so on. However, further work can be carried out to enhance the performance of the proposed routing protocol. The suggestions for future works are as follows:

- Testing the architecture of AIS as described in this chapter in the real test bed according to different applications/scenarios.
- To develop an artificial intelligent algorithms using fuzzy logic to exclude the repeating sensory data before packets are sent to the sink in the next hop neighbors.
- Other ant colony variants such as RL, Max–Min Ant System can be considered.
- To investigate the performance of TCP/IP data traffic in the BIOSARP routing protocol for real-time multimedia applications.
- To develop Wireless Multimedia Sensor Networks (WMSNs) for a large class of applications such as multimedia surveillance networks, target tracking, environmental monitoring, and traffic management systems. WMSN can be established using the BIOSARP routing protocol.

16.7 Conclusion

A biological inspired secure autonomous routing mechanism named as BIOSARP is proposed for WSNs. The routing decision is based on ACO and the self-security mechanism on AIS application. The decision for getting WSN secure depends on our use metrics PRR, packet dropping rate, packet mismatch rate, and packet sending power. The proposed mechanism will successfully detect the non-self antigens (most common known attacks) as sink hole attack, select forward, black hole, message alter attack, hello attack, worm hole. We also find that the proposed system will provide security at no additional control or energy costs to the system. Our proposal demonstrates that AIS-based security has the potential to offer significantly higher performance in WSN due to its significantly less energy, control, and computational cost. The efficient utilization of these resources is a key challenge in WSNs. While enhancing we will also improve this AIS principle by adopting a special feature from human nerve structure called barrier system.

Acknowledgment

The authors wish to express their sincere appreciation, sincerest gratitude to Ministry of Higher Education Malaysia for their full support and Research Management Center (RMC), Universiti Teknologi Malaysia (UTM) for their contribution. Special thanks to researchers in Telematic Research Group (TRG), UTM.

Terminologies

WSN—Wireless Sensor Network
ACO—Ant Colony Optimization
AIS—Artificial Immune System
BIOSARP—BIOlogical-inspired self-organized Secure Autonomous Routing Protocol
DynamiCS—Dynamic Clonal Selection
IS—Immune System
PKC—Public Key Cryptography
PKE—Pairwise Key Establishment
PRR—Packet Reception Rate
TSP—Traveling Salesman Problem

Questions and Sample Answers

1. What is Wireless Sensor Network?
   Wireless sensor network (WSN) is a distributed, autonomous network of communication and sensory nodes that collaborate to detect and relay information on phenomena under investigation to external coordinators.

2. Describe the resources limitations in Wireless Sensor node.
   The wireless sensor node is a miniaturized device typically has 8–120 kB of code memory and 512–4096 bytes of data memory. The transmission bandwidth ranges from 10 to 115 kbps.

3. Illustrate the applications which can benefit from WSNs?

The military and civil applications can benefit from WSN such as target field imaging, intrusion detection, weather monitoring, security and tactical surveillance, distributed computing, detecting ambient conditions such as temperature, movement, sound, light, or the presence of certain objects, inventory control, and disaster management.

4. Why self-organization?

   The determinism and the controllability of the overall system are reduced by self-organization, which also helps to overcome all scalability problems.

5. Give the principles found in nature which can overcome the adaptability issue found in WSNs.

   The principles that can overcome the adaptability issue found in WSNs are ANT Colony and Artificial Immune System.

6. What are the common network layer attacks found in WSNs?

   Most common network layer attacks against sensor networks are spoofed, altered, or replayed routing information; selective forwarding; sinkhole attacks; Sybil attacks; wormholes; HELLO flood attacks; and acknowledgment spoofing.

7. Explain the key factors of AIS.

   AIS learns the normal behavior of the system and then monitors the system for occurrences of abnormal patterns. The most interesting working behavior is the self-optimization and learning process.

8. How does AIS work to provide security over WSNs?

   The agents such as search agent, data agent, and so on will work in a decentralized way to collect data and/or detect an event on individual nodes. While moving from one node to another, the agent checks characteristics of next node with the preinitialized table. The table contains the packet receiving rate, packet dropping rate, packet mismatch rate, and packet sending power of neighboring nodes. Relying on these parameters certain alarm will be generated by the agent or agents. The alarm messages will be then processed by decision agents. The decision agent based on AIS mechanism comes up with the security decision. According to this decision, the actions will be applied by defense agents.

9. Explain the way by which AIS can make decision.

   When the intrusion alarm generates by the monitoring agents (search and forwarding agents) based on the parameters as given in the table, the security management makes the decision through decision agents.

   The decision agents are similar with B-cell in immune system. All types of decision agents are distributive, mobile, cooperative, and redundant. Thus, these decision agents can cooperate effectively to make a correct decision for distributing attacks. The major objective of decision agents is to detect the existence of non-self patterns within a potentially large set of existing non-self patterns. The matching criteria depend on the equation given below.

$$\text{match}(f, \varepsilon, I, D) = \begin{cases} \text{malicious,} & f(1, \alpha) \geq 1 - \varepsilon \\ \text{benign} & \text{otherwise} \end{cases}$$

   where I is the input string, D the matching string of a decision agent, f the matching function, and ε is the matching threshold.

10. Describe the barrier system? How can WSN benefits from the barrier system?

	Security Metrics			
	Packet Receiving Rate	*Packet Dropping Rate*	*Packet Mismatch Rate*	*Packet Sending Power*
Tackled Attacks	*Sink Hole Attack*	*Select Forward, Black Hole*	*Message Alter Attack*	*Hello Attack, Worm Hole*
Node 1	α^1	ρ^1	ν^1	ζ^1
Node 2	α^2	ρ^2	ν^2	ζ^2
.	.	.	.	.
.	.	.	.	.
.	.	.	.	.
.	.	.	.	.
Node n	α^n	ρ^n	ν^n	ζ^n

The central nervous system protects itself with a special barrier to prevent harmful substances passing from the blood into the brain. The blood–brain barrier consists of blood vessel walls which at this interface are especially strong. Blood vessel wall is built by endothelial cells, which form especially strong connections. This ensures that pathogens and immune cells floating in the blood do not access the brain and the spinal cord, thus controlling border traffic. Usually, inflammatory cells are not able to penetrate this barrier unless they camouflage themselves and can crack the required access code.

WSN can have benefit from this blood–brain barrier in the initialization phase while ACO is building up knowledge over the network.

11. Suppose the WSN is under attack by sinkhole, selective forwarding, message alter and hello message attacks. How can ACO and AIS works together to defend and secure WSN?

   Monitoring agents are forward ants and search ants performing the checks while moving from one node to another. When intrusion alarm is generated by monitoring agents, the decision agent activated under security module and serves the alarm messages. The decision agent detects the existence of non-self nodes or patterns based on the matching criteria. If the detection of non-self patterns is true, then security management implements certain actions through defense agents. The defense agents act somewhat similar as the antibody that is secreted by lymphocyte. Their function modules involve the self-copy, isolation, and suicide.

Author's Biography

Kashif Saleem received his BSc in computer science from Allama Iqbal Open University, Islamabad, Pakistan, in 2002. Postgraduate diploma in computer technology & communication from Government College University, Lahore, Pakistan, in 2004. Master of engineering degree in electrical (electronics & telecommunication) from University Technology Malaysia (UTM), Malaysia in 2007. Currently, he is pursuing his PhD in electrical engineering at the Faculty of Electrical Engineering, University Technology Malaysia (UTM), Malaysia, under the supervision of Prof. Dr. Norsheila Fisal.

Norsheila Fisal received her BSc in electronic communication from the University of Salford, Manchester, United Kingdom, in 1984. She received her MSc degree in telecommunication technology, and PhD degree in data communication from the University of Aston, Birmingham, United Kingdom, in 1986 and 1993, respectively. Currently, she is the professor with the Faculty of Electrical Engineering, University Technology Malaysia and Director of Telematic Research Group (TRG) Laboratory.

Sharifah Hafizah Syed Ariffin received her BEng (Hons), electronic and communication engineering, from London Metropolitan University, London, United Kingdom, and her MEE by research (mobility management in wireless telecommunication) from University Technology Malaysia (UTM), Malaysia, in 1997 and 2001, respectively. Obtained her PhD in telecommunication (accelerated simulation) from Queen Mary University, London, United Kingdom, in 2006. She is a senior lecturer at UTM and her research interests include network modeling and performance, accelerated simulation, self similar traffic and priority scheduling.

Sharifah Kamilah Syed Yusof received her BSc (Cum Laude) (electrical engineering) from George Washington University, in 1988. Obtained her MEE and PhD in electrical engineering from University Technology Malaysia (UTM), Malaysia, in 1994 and 2006, respectively. Her current research areas are OFDM-based system, software-defined radio, and cognitive radio.

Rozeha A. Rashid received her BSc in electrical engineering from the University of Michigan, Ann Arbor and her MEE in Telecommunication from University Technology Malaysia (UTM), Malaysia in 1989 and 1993, respectively. She is a senior lecturer at UTM and her research interests include wireless communication, sensor network, and cognitive radio.

References

1. A. Cerpa, J. L. Wong, L. Kuang, M. Potkonjak, and D. Estrin, Statistical model of lossy links in wireless sensor networks, in *ACM/IEEE IPSN*, Los Angeles, USA, 2005.
2. J. N. Al-Karak and A. E. Kamal, Routing techniques in wireless sensor networks: A survey, *Wireless Communications IEEE Journal*, 11, 6–28, 2004.
3. M. Chen, V. C. M. Leung, S. Mao, and Y. Yuan, Directional geographical routing for real-time video communications in wireless sensor networks, *Computer Communications Journal*, 30, 3368–3383, 2007.
4. K. Romer and F. Mattern, The design space of wireless sensor networks, *IEEE Wireless Communications Journal*, 11, 54–61, 2004.
5. M. Frodigh, P. Johansson, and P. Larsson, Wireless ad hoc networking—The art of networking without a network, *Ericsson Review* 4, 248–263, 2000.
6. J. Ma, S. Zhang, Y. Zhong, and X. Tong, SAID: A self-adaptive intrusion detection system in wireless sensor networks, *Information Security Applications*, LNCS, Springer, Berlin/Heidelberg. 4298, pp. 60–73, 2007.
7. C. Karlof and D. Wagner, Secure routing in wireless sensor networks: Attacks and countermeasures, *Ad Hoc Networks*, 1, 293–315, 2003.
8. N. Mazhar and M. Farooq, BeeAIS: Artificial immune system security for nature inspired, MANET routing protocol, BeeAdHoc, Springer-Verlag, Berlin Heidelberg, LNCS 4628, pp. 370–381, 2007.
9. E. Felemban, C. G. Lee, E. Ekici, R. Boder, and S. Vural, Probabilistic QOS guarantee in reliability and timeliness domains in wireless sensor networks, in *24th Annual Joint Conference of the IEEE Computer and Communications Societies*, *IEEE Proceedings*, Miami, FL, USA, 2005, pp. 2646–2657.

10. A. D. Wood and J. A. Stankovic, Denial of service in sensor networks, *Computer*, 35(10), 54–62, 2002.

11. F. Dressler, Benefits of bio-inspired technologies for networked embedded systems: An overview, in *Dagstuhl Seminar Proceedings 06031, Organic Computing–Controlled Emergence*, Dagstuhl, Southwest Germany, 2006.

12. S. Balasubramaniam, D. Botvich, W. Donnelly, M. Foghluh, and J. Strassner, Biologically inspired self-governance and self-organisation for autonomic networks, in *Proceedings of the 1st International Conference on Bio Inspired Models of Network, Information and Computing Systems*. vol. 275, Cavalese, Italy: ACM, p. 30, 2006.

13. S. Balasubramaniam, W. Donnelly, D. Botvich, N. Agoulmine, and J. Strassner, Towards integrating principles of molecular biology for autonomic network management, in *Hewlett Packard University Association (HPOVUA) Conference*, Nice, France, 2006.

14. P. Boonma and J. Suzuki, MONSOON: A coevolutionary multiobjective adaptation framework for dynamic wireless sensor networks, in *Proceedings of the 41st Hawaii International Conference on System Sciences (HICSS) Big Island*, HI, 2008.

15. C. Lee and J. Suzuki, Autonomic network applications designed after immunological self-regulatory adaptation, in *International Conference on Integration of Knowledge Intensive Multi-Agent Systems*, IEEE, Ed., Waltham, MA: KIMAS, 2007.

16. K. Saleem, N. Fisal, S. Hafizah, S. Kamilah, and R. A. Rashid, Ant based self-organized routing protocol for wireless sensor networks, *International Journal of Communication Networks and Information Security*, 2, 42–46, 2009.

17. K. Saleem, N. Fisal, M. S. Abdullah, A. B. Zulkarmwan, S. Hafizah, and S. Kamilah, Proposed nature inspired self-organized secure autonomous mechanism for WSNs, in *Asian Conference on Intelligent Information and Database Systems*, Quang Binh University, Dong Hoi City, Quang Binh Province, Vietnam, pp. 277–282, 2009.

18. P. J. Bentley, Towards an artificial immune system for network intrusion detection: An investigation of dynamic clonal selection, in *Evolutionary Computation, CEC '02. Proceedings of the 2002 Congress*. Vol. 2, Honolulu, HI, IEEE Computer Society, 2002.

19. R. de Lemos, J. Timmis, M. Ayara, and S. Forrest, Immune-inspired adaptable error detection for automated teller machines, *IEEE Transactions on Systems, Man, and Cybernetics, Part C: Applications and Reviews*, 37, 873–886, 2007.

20. G. Chen, T.-D. Guo, W.-G. Yang, and T. Zhao, An improved ant-based routing protocol in wireless sensor networks, in *Collaborative Computing: International Conference on Networking, Applications and Worksharing, 2006*. CollaborateCom 2006, New York, NY, pp. 1–7, November 2006.

21. T. Stuetzle and M. Dorigo, A short convergence proof for a class of ACO algorithms, *IEEE Transactions on Evolutionary Computation*, 6, 358–365, 2002.

22. G. Singh, S. Das, S. Gosavi, and S. Pujar, Ant Colony Algorithms for Steiner Trees: An application to Routing in Sensor Networks, in L. N. de Castro, F. J. von Zuben, Eds., *Recent Developments in Biologically Inspired Computing*, Idea Group Publishing, pp. 181–206, 2004.

23. Y. Zhang, L. D. Kuhn, and M. P. J. Fromherz, Improvements on ant routing for sensor networks, in M. Dorigo et al. eds *ANTS 2004*, LNCS 3172, Springer, Berlin Heidelberg, pp. 154–165, 2004.

24. Y. Lu, G. Zhao, and F. Su, Adaptive ant-based dynamic routing algorithm, in *Proceedings of the 5th World Congress on Intelligent Control and Automation*, Hangzhuo, China, pp. 2694–2697, June 2004.

25. S. Okdem and D. Karaboga, Routing in wireless sensor networks using ant colony optimization, in *Proceedings of the First NASA/ESA Conference on Adaptive Hardware and Systems (AHS'06)*, Istanbul, 2006.

26. R. G. Aghaei, M. A. Rahman, W. Gueaieb, and A. E. Saddik, Ant colony-based reinforcement learning algorithm for routing in wireless sensor networks, in *Instrumentation and Measurement Technology Conference—IMTC Warsaw*, Poland, IEEE, 2007.

27. Y.-F. Wen, Y.-Q. Chen, and M. Pan, Adaptive ant-based routing in wireless sensor networks using Energy*Delay metrics, *Journal of Zhejiang University SCIENCE A* vol. 9, 531–538, 2008.

28. S. Okdem and D. Karaboga, Routing in wireless sensor networks using an ant colony optimization (ACO) router chip, *Sensors*, 9, 909–921, 2009.

29. A. Ali, L. A. Latiff, M. A. Sarijari, and N. Fisal, Real-time routing in wireless sensor networks, in *The 28th International Conference on Distributed Computing Systems Workshops*, Beijing, China, 2008.

30. J. Zhao and R. Govindan, Understanding packet delivery performance in dense wireless sensor networks, in *Proceedings of the 1st International Conference on Embedded Networked Sensor Systems*, Los Angeles, USA, 2003.

31. J. P. Walters, Z. Liang, W. Shi, and V. Chaudhary, Wireless sensor network security: A survey, in Y. Xiao, ed., *Security in Distributed, Grid, mobile, and Pervasive Computing*, CRC Press, USA, pp. 368–403, 2007.

32. A. K. Pathan, H. W. Lee, and C. S. Hong, Security in wireless sensor networks: Issues and challenges, in *Proceedings of 8th IEEE ICACT 2006*, Phoenix Park, Korea, pp. 1043–1048, 2006.

33. C. Intanagonwiwat, R. Govindan, and D. Estrin, Directed diffusion: A scalable and robust communication paradigm for sensor networks, in *Proceedings of ACM MobiCom '00*, Boston, MA, 2000.

34. M. Castro and B. Liskov, Practical byzantine fault tolerance, in *OSDI: Symposium on Operating Systems Design and Implementation*, New Orleans, LA, 1999.

35. A. Banerjea, A taxonomy of dispersity routing schemes for fault tolerant real-time channels, in *Proceedings of ECMAST*, Louvian-la-Neuve, Belgium, pp. 129–148, 1996.

36. K. Ishida, Y. Kakuda, and T. Kikuno, A routing protocol for finding two node-disjoint paths in computer networks, in *International Conference on Network Protocols*, pp. 340–347, November 1992.

37. Y. Xu, J. Heidemann, and D. Estrin, Geography-informed energy conservation for ad hoc routing, in *Proceedings of the Seventh Annual ACM/IEEE International Conference on Mobile Computing and Networking*, Rome, Italy, 2001.

38. C. Benjie, J. Kyle, B. Hari, and M. Robert, Span: An energy-efficient coordination algorithm for topology maintenance in ad hoc wireless networks, *ACM Wireless Networks Journal*, 8, 481–494, 2002.

39. J. R. Douceur, The Sybil attack, in *Proceedings of IPTPS'02*, MIT Faculty Club, Cambridge, MA, 2002.

40. Y.-C. Hu, A. Perrig, and D. B. Johnson, Packet leashes: A defense against wormhole attacks in wireless networks, *IEEE Infocom, Twenty-Second Annual Joint Conference of the IEEE Computer and Communications*, IEEE Societies, San Francisco, California, USA, pp. 1976–1986, 2003.

41. Y.-C. Hu, A. Perrig, and D. B. Johnson, Wormhole detection in wireless ad hoc networks, *Technical Report*, Department of Computer Science, Rice University, 2002.

42. A. Siraj, R. B. Vaughn, and S. M. Bridges, Intrusion sensor data fusion in an intelligent intrusion detection system architecture, in *Proceedings of the 37th Annual Hawaii International Conference on System Sciences (HICSS'04)*, Hawaii, 2004.

43. P. K. Harmer, P. D. Williams, G. H. Gunsch, and G. B. Lamont, An artificial immune system architecture for computer security applications, *IEEE Transactions on Evolutionary Computation*, 6, 252–280, 2002.

44. P. Albers and O. Camp, Security in ad hoc networks: A general intrusion detection architecture enhancing trust based approaches, in *First International Workshop on Wireless Information System, 4th International Conference on Enterprise Information System*, Universidad de Castilla-La Mancha Ciudad Real, Spain, 2002.

45. O. Kachirski and R. Guha, Elective intrusion detection using multiple sensors in wireless ad hoc networks, in *Proceedings of the 36th Annual Hawaii International Conference on System Sciences (HICSS'03)*, Hawaii, USA, p. 57.1, January 2003.

46. T. Alpcan and T. Basar, A game theoretic approach to decision and analysis in network intrusion detection, in *Proceedings 42nd IEEE Conference on Decision and Control*, The Hyatt Regency Resort & Spa. Maui, Hawaii, USA, pp. 2595–2600, 2003.

47. A. Agah, S. K. Das, and K. Basu, A game theory based approach for security in wireless sensor networks, in *IEEE International Conference on Performance, Computing, and Communications*, Phoenix, Arizona, pp. 259–263, 2004.

48. S. Banerjee, C. Groşan, A. Abraham, and P. K. Mahanti, Intrusion detection on sensor networks using emotional ants, *International Journal of Applied Science and Computations, USA*, 12, 152–173, 2005.

49. S. Banerjee, C. Grosan, and A. Abraham, IDEAS: Intrusion detection based on emotional ants for sensors, in *Proceedings of the 2005 5th International Conference on Intelligent Systems Design and Applications (ISDA'05)*, Wroclaw, Poland, 2005.

50. H.-H. Gao, H.-H. Yang, and X.-Y. Wang, Ant colony optimization based network intrusion feature selection and detection, in *Proceedings of 2005 International Conference on Machine Learning and Cybernetics*, Guangzhou, China, pp. 3871–3875, 2005.

51. A. Avizienis, Toward systematic design of fault-tolerant systems, *Computer*, 30, 51–58, 1997.

52. G. Li, J. He, and Y. Fu, A distributed intrusion detection scheme for wireless sensor networks, in *The 28th International Conference on Distributed Computing Systems Workshops*, Beijing, China, 2008, pp. 309–314.

53. A. Secker, A. A. Freitas, and J. Timmis, AISEC: An artificial immune system for e-mail classification, in *The 2003 Congress on Evolutionary Computation, 2003. CEC '03*, Canberra, Australia, pp. 131–138. 2003.

54. N. Prattipati and E. Hart, Evaluation and extension of the AISEC email classification system, *Artificial Immune Systems*, 5132, 154–165, 2008.

55. C. Lee and J. Suzuki, SWAT: A decentralized self-healing mechanism for wormhole attacks in wireless sensor networks, in Y. Xiao, H. Chen and F. Li eds. *Handbook on Sensor Networks*. World Scientific Publishing Co., 2010.

56. C.-C. Su, K.-M. Chang, M.-F. Horng, and Y.-H. Kuo, The new intrusion prevention and detection approaches for clustering-based sensor networks, in *Wireless Communications and Networking Conference*, New Orleans, USA, pp. 1927–1932, 2005.

57. C. H. Lim, LEAP++: A robust key establishment scheme for wireless sensor networks, in *The 28th International Conference on Distributed Computing Systems Workshops*, Beijing, China, 2008.

58. H. Wang, B. Sheng, C. C. Tan, and Q. Li, Comparing symmetric-key and public-key based security schemes in sensor networks: A case study of user access control, in *The 28th International Conference on Distributed Computing Systems*, Beijing, China, 2008.

59. K. Xing and F. Liu, Real-time detection of clone attacks in wireless sensor networks, in *The 28th International Conference on Distributed Computing Systems*, Beijing, China, 2008.

60. A. Ali and N. Fisal, Security enhancement for real-time routing protocol in wireless sensor networks, in *5th IFIP International Conference on Wireless and Optical Communications Networks. WOCN'08*, pp. 1–5, 2008.

61. X. Du, Y. Xiao, H.-H. Chen, and Q. Wu, Secure cell relay routing protocol for sensor networks: Research articles, *Wireless Communications and Mobile Computing, Special Issue: Wireless Network Security*, 6, 375–391, 2006.

62. S. Lange and M. Middendorf, Design aspects of multi-level reconfigurable architectures, *Journal of Signal Processing Systems*, 51, 23–37, 2008.

63. A. Ali and N. Fisal, A real-time routing protocol with load distribution in wireless sensor networks, *Computer Communications*, 31, 3190–3203, 2008.

64. S. Mccanne, S. Floyd, and K. Fall, Network simulator 2 (NS-2) version 2.28, http://www-nrg.ee.lbl.gov/ns/, http://www.isi.edu/nsnam/ns.

65. B. Barin and R. Sosa, A new approach for AntNet routing, in *Ninth International Conference on Computer Communications and Networks, 2000. Proceedings*, Las Vegas, NV, USA, pp. 303–308, 2000.

66. G. D. Caro, F. Ducatelle, and L. M. Gambardella, AntHocNet: An adaptive nature-inspired algorithm for routing in mobile ad hoc networks, *European Transactions on Telecommunications*, 16, pp. 443–455, 2005.

67. J. Malek, Trace graph–Network Simulator NS-2 trace files analyser, http://www.tracegraph.com/2003.

Chapter 17

Controlled Link Establishment Attack on Key Pre-Distribution Schemes for Distributed Sensor Networks and Countermeasures

Thanh Dai Tran and Johnson I. Agbinya

Contents

17.1 Introduction

Significant advances in digital circuitry, wireless communications, battery technology, and micro-electromechanical systems (MEMS) have paved the way for the emergence of low-cost, battery-powered, multifunctional, and network-enabled tiny sensor nodes (nodes for short) that have limited energy resources, computation, memory, and communication capacities. These nodes consist of four basic parts: a sensing unit, a data-processing unit, a communication unit, and a power unit [1]. They are capable of capturing various physical phenomena, such as temperature, pressure, humidity, light, object motion, soil composition, noise level, presence of a certain object, and so on, as well as mapping such physical characteristics of their immediate surroundings to quantitative measurements [2].

Typically, a DSN often comprises hundreds to thousands of such nodes linked locally by a low-bandwidth wireless communication, controlled and managed by one or few powerful control nodes (often called base stations or sinks). Usually, nodes are deployed densely in a designated area by a single authority such as the government or military unit and then, automatically form a network. Their deployments can be in a random fashion (e.g., scattered from an airplane) or planted manually (e.g., fire alarm sensor in a building) without any infrastructure support. Once deployed, nodes are virtually static over most of the network lifetime. A base station or sink may be a static or mobile node functioning as a gateway to other networks or data centers via high-bandwidth communication links (either leased lines or broadband wireless links). DSNs have created new paradigms for wide-ranging applications in both military and civilian operations such as battlefield surveillance, enemy tracking, military facility monitoring, agriculture and environmental monitoring, civil engineering, medical care, smart spaces, and scientific explorations.

In terms of data dissemination, DSNs can involve single- or multihop communications. The former type may take place between the base station and neighboring nodes. Meanwhile, the latter type occurs between two nodes which are multihop away and involves a sequence of hops through a series of pairwise adjacent nodes. In such communication, each node communicates directly only to its neighboring nodes. Messages to a geographically distant node will be relayed by the sender node's neighbors and the neighbors' neighbors, and so forth.

In many circumstances, especially in military applications, security for DSNs is of paramount concern to users due to a number of design challenges, including limited resources, adversarial environments, unattended deployments and operations, wireless broadcast nature of communication, multihop communication, lack of infrastructure, and no tamper-resistant hardware [3]. These design challenges may lead to many malicious attacks. For instance, an attacker can plant his own nodes in uncontrolled environments to actively join the network. He can replace, compromise, or physically damage the existing nodes. Because of the openness of wireless medium, he can silently listen to radio channels to capture data, security credentials, or to collect enough information to derive the credentials. Moreover, he can make corrupt use of multihop communication to insert, modify, replay, block or delete messages to mislead sensing applications. He can also use wireless devices with various capabilities to impersonate some of the network nodes, play man-in-the-middle, hijack a session, and jam a part of or whole network [4].

Under the above circumstances, security services such as encryption and authentication are key enablers for the success of those applications. In a typical DSN, node-to-node communication is the most common communication pattern [5–7]. Therefore, the provision of these services boils down to establishing pairwise keys among communicating nodes.

Generally speaking, approaches to the Pairwise Key Establishment (PKE) in general network environments can roughly be classified into three categories: centralized key schemes, public key schemes, and key predistribution schemes [8]. In DSNs, due to the existence of base stations, the centralized key schemes such as Security Protocols for Sensor Networks (SPINS) [5] can be used. In this scheme, the base station acts as a Key Distribution Center (KDC) to generate, regenerate, and distribute pairwise keys. However, with only one managing entity, the KDC could become a single point of failure. The entire network security will be affected if there is a problem with the KDC. During the time when the KDC is not working, the network becomes vulnerable as keys are not generated, regenerated, distributed, and updated. Furthermore, the network can be to large to be managed by a single entity, thus affecting network scalability [9]. In addition, this centralized approach could incur a large amount of communication overhead as two neighbors might be required to do handshakes through the KDC at a distant place [10].

Public key schemes such as the Diffie–Hellman [11] or RSA [12] are widely regarded as impractical to be employed in DSNs because of their code size, data size, processing time, and power consumption. For example, RSA is computationally intensive and usually execute thousands or even millions of multiplication instructions to perform a single security operation. As pointed out in [13,14], public key algorithms such as RSA require on the order of tens of seconds and up to minutes to perform encryption and decryption operations in constrained wireless devices or take a microprocessor thousands of nano-joules to do a simple multiply function with a 128-bit result. This leads to vulnerability to DoS attacks. On the contrary, symmetric key algorithms consume much less computational energy than their public key counterparts. For instance, Carman et al. [14] report that on the MC68328 DragonBall processor, the energy consumption for a 1024-bit AES encryption operation is much lower than that for a 1024-bit RSA encryption operation; that is, about 0.104 versus 42 mJ.

Recent studies [15–17] have demonstrated that it is feasible to apply Elliptic Curve Cryptography (ECC) [18,19] to PKE in DSNs. The pre-eminence of ECC is that it seems to offer equal security for a much smaller key size, thus reducing processing and communication overheads. While public key cryptography may be possible in sensor nodes, the public key operations are still prohibitively expensive to be employed on such a large scale as DSNs.

In the last few years, research has witnessed a considerable body of work [7,8,20–25] on developing pairwise key predistribution schemes (PKPSs) which allow preloaded keys or keying materials (hereafter referred to as *cryptographic secrets*) to be used directly [20–23] or indirectly via dynamic pairwise key generation in the PKE process [7,8,24,25]. The development of these schemes follows the trend that the later schemes enhance or/and extend security or/and performance properties of, or introduce additional ones to the earlier schemes. In this trend, the most developed schemes exhibit a number of appealing properties including resilience to large scale of node compromises, guaranteed key establishment, direct key establishment, resilience to dynamic topology, and efficiency. Owing to these salient properties, the PKPSs have been regarded as the most preferable and extensively adopted alternative to be used in DSNs.

Nonetheless, the practical applicability of the PKPSs is undermined by a controlled link establishment attack whose goal is to gain partial or even full control of DSNs via illegitimate link establishment. This attack has severely ruinous impacts on many applications that require collaborative efforts of nodes such as data aggregation mechanisms, routing protocols, distributed voting schemes, and misbehavior detection systems, and so on. This chapter examines the attack and recent advances in countermeasures against it. The discussion begins with some background on the PKPSs for DSNs, immediately followed by a main section on the attack itself. The description of the countermeasures then follows. Finally, we conclude with possible future research directions on this issue.

17.2 Background on PKPSs for DSNs

In DSNs, most of the PKPSs make use of the fact that DSNs virtually operate under administration of a single central authority to ease key deployment tasks. Accordingly, before deployment, nodes are predistributed with cryptographic secrets. After those nodes are deployed into a designated terrain, they perform several rounds of communications to agree on pairwise keys based on the preloaded secrets. The key challenge is how to devise an efficient method for predistributing the cryptographic secrets to nodes prior to deployment such that an optimal compromise between conflicting requirements of optimal resource usage, network scalability, high shared-key connectivity, and network resilience could be reached. In this section, we first present the abstract concept and expected properties of the PKPSs for DSNs and then review their state of the art.

17.2.1 Mathematical Model

At the highest level of abstraction, a DSN is modeled as a set of N nodes $\Gamma = \{S_i | i = \overline{1, N}\}$ most of which have the same hardware configuration and software functionalities. For this network, a key sharing graph $G(V, E)$ is defined where V is the set of vertices representing all the nodes and E is the set of edges connecting nodes in the network. For any two nodes in V, there exists an edge between them if and only if they can find out a shared secret key between themselves finally.

The main idea of all PKPSs is that a set $\Psi = \{\kappa_i | i = \overline{1, P}\}$ of the cryptographic secrets is generated. Each node S_i is assigned a subset $\Lambda_i = \{\kappa_{ij} | i = \overline{1, N}, \ j = \overline{1, k}\}$ derived from the set Ψ prior to node deployment. After deployment, any pair of neighboring nodes, for example, S_i and S_j, use the cryptographic secrets in $\Lambda_i \cap \Lambda_j$ to establish pairwise keys for securing communication links. The set Ψ and $\Lambda_i s$ are generated in such a way that the graph $G(V, E)$ is connected eventually.

17.2.2 Expected Properties of PKPSs

Owing to the number of design challenges mentioned above, the expected properties of PKPSs for DSNs are very different from and much more challenging than those targeted at general network environments. Some of these properties are conflicting requirements and thus might not be possible to be integrated into a single solution. These properties summarized from the literature [7,8,21] are as follows:

- *Low energy consumption:* Since nodes are powered by batteries with limited power, PKPSs should have low communication and computation costs.
- *Low cost:* Since nodes are expected to be inexpensive, the associated hardware costs should be low.
- *Low memory usage:* Since nodes have very limited memory, the memory requirements of PKPSs should be low.

- *Distributed manner:* Owing to the lack of *a priori* knowledge about the postdeployment network topology and fixed infrastructure, PKPSs should be designed to operate in a distributed fashion.
- *Resilience to large number of node compromises:* In hostile environments, the attacker can manipulate the system through capturing and compromising some nodes. The memory of captured nodes can be read, erased, or tampered with. Therefore, the attack would know cryptographic secrets of compromised nodes. This property guarantees that the large number of disclosed cryptographic secrets due to the memory manipulation does not lead to the disclosure of secrets held by noncompromised nodes.
- *Guaranteed key establishment:* PKPSs should guarantee that any two nodes can establish a pairwise key whenever needed.
- *Direct key establishment:* PKPSs should allow two nodes that can communicate (directly or indirectly) with each other to establish a pairwise key without revealing secrets to or obtaining secrets from any third parties (e.g., a central on-line server or other helper nodes). The involvement of third parties is highly undesirable because third parties may have been compromised, they may not be available, and more messages have to be exchanged among involved nodes.
- *Node-to-node identity authentication:* Nodes are able to verify the identities of the nodes with whom they are communicating. The attacker is unable to impersonate the identity of any node if that node has not already been compromised.
- *Resilience to dynamic topology:* PKPSs should work even if one or both nodes are mobile. In some applications, a mobile sink may perform some tasks in a sensor network, which require secure communication between the mobile sink and sensor nodes.

17.2.3 State-of-the-Art of PKPSs

Researchers' efforts to address the PKE problem in DSNs began with the pioneering work [20], namely the *basic scheme* proposed by Eschenauer and Gligor. This scheme is aimed at tackling the two insufficient solutions offered by traditional key predistribution: either a single network-wide key or a set of $N-1$ separate keys, each being pairwise privately shared with another node, must be installed in every node. The single network-wide key solution is inadequate since the compromise of a single node may lead to the compromise of the entire network due to the impossibility of key revocation and the lack of tamper-resistant hardware. In contrast, the pairwise private sharing of keys between any pair of nodes offers absolute resiliency since node compromise does not lead to the disclosure of information about keys shared between any pair of noncompromised nodes. Unfortunately, when N is very large, this solution requires a very large amount of memory per node which is impractical for nodes due to their extremely limited memory storage. Furthermore, this solution does not allow new nodes to be added with ease to a deployed network because the existing nodes do not have the new nodes' keys.

In the basic scheme, each node is predistributed with a key ring, randomly chosen from a large key pool such that any pair of nodes can share at least one key with a certain probability. After deployment, two neighbors can discover a shared key directly or negotiate a pairwise key indirectly through a secure path which consists of a number of pairs of nodes having a direct shared key. In fact, the basic scheme relies on the random graph theory [26] to determine the key sharing probability p between any pair of nodes. A random graph $G(N, p)$ is a graph of N nodes for which the probability that a link exists between two nodes is p. The graph is fully disconnected if $p = 0$ or fully connected if $p = 1$. There is a transition from the disconnected graph to the fully connected graph when p increases. The scheme exploits this property by setting p larger than a certain value such that the entire network is almost connected. In this setting, the size of the key pool and the size of the key ring are taken into account to achieve such a property.

Chan et al. [21] further extended the basic scheme into two mutually exclusive variants *q-composite keys scheme* and *multipath key enforcement*. The goals of these variants are both to strengthen the security of a link key against small-scale node capture attack. This is done by either increasing the amount of key overlap of q common keys ($q > 1$) instead of only one as required in the basic scheme for the link key establishment or establishing the link key through multiple paths.

One severe security problem with the basic scheme is the lack of node-to-node authentication because of the possibility of the reuse of the same key by multiple nodes. This problem is a strong motivation for two pieces of work [21,27] which use node identity information in the PKE to avoid the lack of authentication. In the random-pairwise keys scheme [21], each node identity is paired with a set of randomly chosen distinct node IDs and a pairwise key is generated for each pair of nodes. The key is stored in both nodes' key rings, along with the ID of the other node that also holds the key. Later, the node IDs are used to discover shared keys among neighboring nodes. In the co-operative protocol [27], the node identity information is used to derive key rings for nodes.

Another salient thread of PKPSs represented by [8,24,25] manages to offer both the resiliency to the small-scale node compromise and node-to-node authentication. Du et al. [8] developed a multiple-space key predistribution scheme that combines the idea of k-secure property of Blom's scheme [28] with the idea of global key pool in the basic scheme. Similarly, Liu et al. [25] and Tran and Hong [24] presented the polynomial pool-based scheme and matrix pool-based scheme that rely on Blundo et al.'s scheme [29] and Matsumoto–Imai's scheme [30], respectively, to generate the global key pool. In these schemes, each key is associated with the IDs of nodes sharing it. By this way, the node-to-node authentication property can be achieved through a challenge–response approach. Specifically, a verifier node can send an encrypted random number or a challenge to a suspected node. The suspected node can prove its ID by returning the plain-text challenge to the verifier node.

Despite the improved security features, the aforementioned schemes do not guarantee direct key establishment, resiliency to large-scale node compromise, and tolerance to dynamic network topology. Therefore, Zhang et al. [7] proposed a random perturbation-based scheme to address those limitations. In fact, this scheme is a variant of the basic polynomial-based scheme [29] in which the scheme does not give each node an original share of a symmetric polynomial but the perturbed share, which is the sum of the original share and a perturbation polynomial introduced by the authors themselves. The benefit of this addition is twofold. On one hand, two nodes can still establish a key using the perturbed shares. On the other hand, the adding introduces prohibitively high complexity to the attacker in order to break the symmetric polynomial even if he/she has compromised a large number of nodes. This means that any number of compromised colluding nodes have negligible probability to break the pairwise key established by a pair of noncompromised nodes.

Approaching the problems of the resource constraints and the node compromise in a different way, many researchers proposed to exploit location and deployment information in the PKE [22,31–33]. Taking the proposal in [22], for example, it takes advantage of deployment knowledge of group-based deployment model together with the idea of key pool in the basic scheme. The deployment knowledge is defined as the knowledge about the nodes that are likely to be the neighbors of each node. This knowledge can be modeled using nonuniform probability density functions, such as normal Gaussian distribution. Making good use of the knowledge, the authors proposed to divide the global key pool into different key pools. Each of them is corresponding to one deployment group. The goal of setting up the key pools is to allow the nearby key pools that are corresponding to deployment groups of neighboring locations to share more keys, while pools far away from each other share fewer keys or no keys at all. By doing like this, each node needs only to carry a fraction of the keys required by the other key predistribution scheme [20,21] while attaining the same level of shared-key connectivity. Furthermore, this reduction in key storage substantially improves network's resilience against node capture.

Due to the fact that the knowledge about location and deployment is not always available, the above schemes may not be applicable in many application scenarios. To avoid this problem, Anjum [34] proposed a location aware approach for the PKE. This approach, location-dependent key (LDK) management, is aimed at improving network resiliency against node compromise by diversifying network-wide pre-distributed keys on each key ring of a sensor node into location-specific keys. This scheme has two noticeable advantages. First, it confines the impact of node compromise to the vicinity of compromised nodes only. This means that compromise of a node in a location affects the communications only around that location. Second, the scheme guarantees that sensor nodes in different locations have different keys without resort to any knowledge about the deployment of sensor nodes. However, several significant disadvantages make the scheme impractical. First, the assumption of the presence of anchors is not always reasonable. Moreover, the requirement of the tamper-resistant features from anchors adds more cost to the scheme. Secondly, in order to support incremental node addition, either the anchors are required to transmit beacons periodically or newly deployed sensor nodes have to send signals to the anchors. These are very problematic requirements since the former results in quick energy depletion of the anchors and the latter facilitates the energy depletion attack on the anchors. Thirdly, the transmission pattern of the anchors make easier for attackers to locate the anchors for compromise. Finally, this scheme assumes that the new nodes can sustain node compromise in its initialization phase, which may not be true in some circumstances.

In some applications, DSNs are deployed to operate for a long period of time. In those cases, the average lifetime of each battery-powered node is much shorter than the overall network lifetime. Consequently, the periodical multiphase deployment of nodes is needed to maintain network connectivity. Motivated by this problem and the node compromise problem, Castelluccia and Spognardi [35] proposed *RoK*. This scheme supports the multiphase deployment by allowing nodes deployed at different times to establish secure links. It exhibits the self-healing feature under the temporarily active node compromise and the limited and constant ratio of compromised links under constantly active node compromise.

17.3 Controlled Link Establishment Attack

Although the PKPSs, presented in Section 17.2, exhibit many features suitable for DSNs, they are still vulnerable to two malicious attacks, including node replication attack (NRA) and key-swapping collusion attack (KSCA). They are grouped into a new broader attack termed *controlled link establishment attack* (CLEA). The reason is that their eventual attack goals are identical, that is to say, gaining partial or even full control of DSNs via illegitimate link establishment. This section is to detail these attacks together with their impacts.

17.3.1 Node Replication Attack

The NRA was first mentioned in the research work proposed by Chan et al. [21]. The pioneering research on the NRA was conducted by Parno et al. [36] followed by a series of research efforts [37–39] to thwart the attack. In a different approach, Fu et al. [40] focused on analysis, characterization, and discussion of the relationship among the replicas, sensor networks, and resiliency of various PKPSs again the NRA. Their study aims at providing practical insights into the design of more secure and efficient key establishment schemes rather than a countermeasure against the attack.

In fact, to launch the NRA, the attacker has to execute four steps: *compromising nodes, generating replicas, inserting replicas, establishing controlled links*. First, the attacker manages to stealthily infiltrate or break into the network to capture a limited number of legitimate nodes. Having captured these nodes, the attacker can exploit the unshielded nature of the nodes to extract their cryptographic secrets [41]. Secondly, having obtained the secrets, the attacker could then duplicate the compromised nodes by loading the obtained secrets onto multiple generic nodes. Thirdly, the attacker inserts the replicas back into the original network. He can realize this in several ways such as physical installation of each node or random scattering. Finally, the replicas attempt to establish pairwise keys with legitimate nodes. The detailed process of establishing pairwise keys varies according to the implementation of PKPSs, which is already discussed in Section 17.2. The links secured by the established pairwise keys are called controlled links because they are under the control of the replicas. In addition, these replicas are allowed to communicate and collaborate with each other under the attacker's control. Figure 17.1 illustrates the NRA with 10 replicas in a network of 50 nodes.

Once successfully mounted, the NRA can cause substantially disastrous impacts on DSNs. Indeed, attackers can carry out a wide variety of malicious attacks on various mechanisms and protocols using their control of replicas and controlled links. Several potential attacks are described roughly as follows:

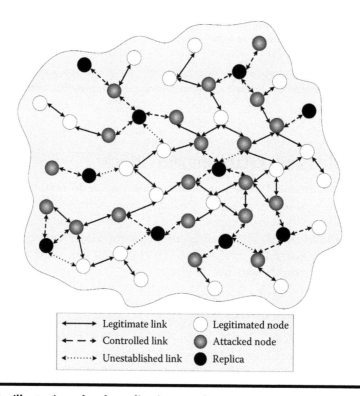

←——→ Legitimate link	◯ Legitimated node
←– –→ Controlled link	⬤ Attacked node
◄·······► Unestablished link	⬤ Replica

Figure 17.1 An illustration of node replication attack.

■ *Routing attacks:* As pointed out in [4], making illicit use of controlled links, replicas can poison or spoil routing tables of their neighbors by spoofing, altering, or replaying routing information. In consequence, the attacker might be able to create routing loops, attract or repel network traffic, lengthen or shorten source routes, generate false error messages, partition the network, increase end-to-end latency, and counteract redundant/multipath routing, and so on.

■ *Data aggregation attack:* Many sensor applications [2,42–44] emphasize the importance of data aggregation or in-network processing in eliminating data redundancy, minimizing the number of transmissions, and thus saving network resources such as energy, bandwidth to prolong network lifetime. However, by using the replication attack, the attacker having replicas reporting incorrect sensor readings might be able to significantly and badly skew the computed aggregates network-wide.

■ *Distributed voting attack:* Owing to the lack of infrastructure support and being distributed by nature in most sensor applications, many sensor network protocols and mechanisms make use of a distributed voting scheme whose voters are sensor nodes themselves to resolve tasks that require collaborative decision-making. Nevertheless, this also offers a good chance of a successful NRA. The attacker via replicas can cast biased votes in an overwhelming number to gain the right of determining the outcome of any decision-making process. For instance, in reputation-based schemes [21], the attacker can revoke legitimate nodes by claiming that the nodes are misbehaving. In contrast, the attacker can also use votes to protect the replicas from being detected.

■ *Resource allocation attack:* In some scheduling algorithms such as MAC protocols, network resources are allocated soundly based only on proper operations of sensor nodes. In such cases, the replicas can be ordered not to follow the algorithms to obtain an unfair share of any resources and, thus, putting legitimate nodes at resource starvation.

17.3.2 Key-Swapping Collusion Attack

This attack was first studied by Moore in [45]. The study is developed based on observations on shortcomings of PKPSs exemplified by the Chan et al.'s random pairwise keys scheme [21]. These shortcomings arise from the fact that while the illegal usage of compromised cryptographic secrets is feasible network-wide via the existing routing mechanisms or out-of-band channels, it is unlikely to be detected as the consequence of the lack of collaboration among locally communicating nodes. The enabling assumption is that DSNs are deployed in hostile environments where nodes are subject to node compromise but cannot afford expensive tamper-proof hardware. As a consequence, the attacker can compromise a set of nodes and then manipulate their memories and program codes. The manipulated nodes referred to as attacker-controlled nodes are then injected back into the networks and under the attacker's control. Thereafter, attacker-controlled nodes start colluding with each other by swapping their cryptographic secrets to establish illegitimate links with their neighbors for the purpose of seizing nodes' communication channels. As indicated in [45], only a small percentage of network compromise, approximately 5%, can yield control over half of the valid communication channels. Therefore, a colluding minority can launch the similar attacks as the NRA does on routing, data aggregation, distributed voting, resource allocation, and so on.

However, methods of performing key swapping have not been examined in Moore's work as well as in the literature. The following are to present three approaches to the key swapping: short distance, long distance, and mixed distance.

17.3.2.1 Short-Distance Collusion Attack

Such an attack is the most cost-effective way to be launched where it is applicable. It works provided that the following conditions are satisfied:

- Attacker-controlled nodes receive no attacker's communication support after having been replaced into the network. This means that the attacker neither installs any extra communication hardware onto attacker-controlled nodes nor deploys additional communication infrastructure into the sensor field.
- Each of attacker-controlled nodes is able to be reachable by the others via colluding tunnels established based on the existing routing mechanisms. For example, attacker-controlled nodes can make use of the existing routing infrastructure to discover each other initially. Thereafter, they exchange necessary information such as a list of closest attacker-controlled nodes to establish colluding tunnels. These tunnels are then used to swap or update attacker-controlled nodes' key rings.
- Every attacker-controlled node can be reachable by at least one of the others via a short distance. Ideally, the distance should be at most four-hop long since within such a distance attacker-controlled nodes can swap their keys in a timely manner without consuming too many resources and easily escape from being detected by wormhole attack detection mechanisms such as geographical leashes and temporal leashes [46].

For the sake of presentation, let us assume a short-distance collusion attack scenario of eight attacker-controlled nodes numbered from 1 to 8 where, for each node, it is always possible to find another node such that the distance between them is not more than four hops away as illustrated in Figure 17.2. Take the collusion among three attacker-controlled nodes 1, 2, and 3, for example; although node 3 has four nodes in its neighborhood, it can only establish a pairwise key with node 11. Furthermore, according to a PKPS, the cryptographic secrets of nodes 1 and 2 can be used to establish pairwise keys with two nodes 9 and 10, respectively. When no collusion occurs, the

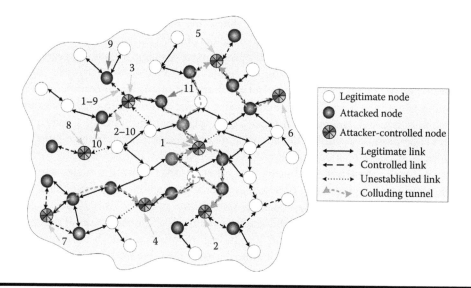

Figure 17.2 An illustration of short-distance collusion attack.

attacker-controlled nodes obtain no added gain. Nevertheless, under collusion, node 3 can achieve the IDs of nodes 1 and 2 and their cryptographic secrets. Using the acquired information, node 3 can establish controlled links with nodes 9 and 10. By this manner, the attacker can expand his attack and eventually gain the control of most of the network.

17.3.2.2 Long-Distance Collusion Attack

In this attack, the distance between any pair of two attacker-controlled nodes is far enough such that it is not reasonable for them to communicate with each other using the existing routing mechanisms. For efficient communications among attacker-controlled nodes, low-latency links can be utilized such as a wired connection, an optic connection, or long-range, out-of-band wireless directional transmission. These links are possible to be set up if the attacker can gain access to the sensor field to deploy extra long-range, low-latency communicating devices (referred to as *helper nodes*) as neighbors of attacker-controlled nodes as illustrated in Figure 17.3a. Alternatively, the attacker can install additional pieces of communications support hardware onto sensor main boards as shown in Figure 17.3b.

Long-distance collusion attack is somehow similar to wormhole attack [46,47] in the sense that some information is obtained at one point of the network, then tunneled to another point of the network via a long-range, low-latency link, and injected back into the network. However, the two attacks are different from each other due to the following points:

- Node compromise is compulsory in long-distance collusion attack while it is not mandatory in wormhole attack.
- Information exchanged in long-distance collusion attack is cryptographic secrets while in wormhole attack the exchanged data are mainly routing information, neighboring information, and location information.
- Objective of the KSCA is at PKPSs while wormhole mainly aims at ruining routing protocols, neighbor discovery mechanisms, and local broadcast protocols such as localization protocols.

17.3.2.3 Mixed-Distance Collusion Attack

This sort of attack might happen when attacker-controlled nodes are distributed unevenly. In other words, there are some regions where attacker-controlled nodes can deploy short-distance collusion attack efficiently, whereas in other regions, long-distance collusion attack might be a more suitable choice. Note that it is possible that two kinds of attacks can exist in one attacker-controlled node. That is, on one side, the node has short-distance colluding tunnels with other attacker-controlled nodes in its close vicinity. On the other side, it also has long-distance colluding tunnels with the other remote ones.

17.4 Countermeasures

Despite the ruinous impacts on a wide variety of processes, surprisingly, countermeasures against the CLEA have received very little attention from researchers thus far. Generally speaking, these countermeasures can be classified into two categories: indirect approaches and direct approaches.

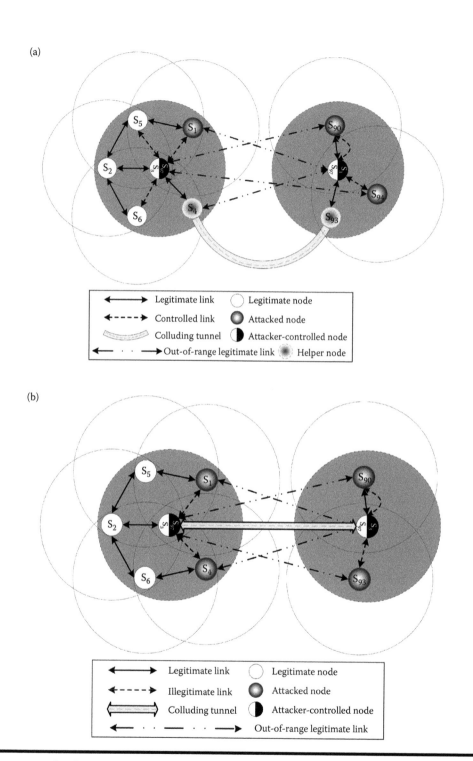

Figure 17.3 The illustrations of long-distance collusion attack.

17.4.1 Indirect Countermeasures

The indirect countermeasures refer to counteractive techniques found out in the literature to address different research problems. As a result, they expose their own limitations when applied to counteract the attack. The very first technique can be confining the usability of predistributed cryptographic secrets to local areas using deployment knowledge and location information as presented in [22,31–33] or location awareness as discussed in [34]. Because compromised cryptographic secrets can only be usable within a geographically specific location, when they are carried in a replica by the NRA or swapped by the KSCA to different locations, they cannot be used to establish illegitimate links. Thus, the CLEA is successfully blocked. However, the existence of deployment knowledge, location information, or location awareness is not always a justifiable assumption. For applications to which this assumption can not be applied, this technique becomes unusable.

The second technique can be key infection [48], which excludes the key predistribution phase by means of weakening the attack model and allowing key transmission in plain-text in the key establishment phase. Hence, despite the fact that the key infection schemes are not vulnerable to the CLEA, they are highly susceptible to powerful on-site attackers.

The third technique can be for nodes to discard unused keys after an initialization phase, but this means once initialization is complete, the network can no longer accommodate new nodes. LEAP+ [6] attempted to address both the problem of network scalability and the CLEA as follows. It is assumed that there exist a lower bound on the time interval T_{min} that is needed for the attacker to compromise a node, and the time T_{est} for a newly deployed node to discover its immediate neighbors such that $T_{min} > T_{est}$. Each new node deployed during the time interval T_i are preloaded with a network-wide initial key K_{IN}^i. The node uses this key to derive pairwise keys with its neighbors within T_{min}. After T_{min}, K_{IN}^i is removed to prevent the attacker from compromising all the established pairwise keys and establishing illegitimate links within T_i. Furthermore, the node is also loaded with master keys derived from the node's ID and initial keys of future intervals. These master keys are used to establish pairwise keys with new nodes deployed in the future intervals. However, if the attacker is powerful enough to compromise a node within T_{est} of each interval, it can launch the CLEA with ease.

Finally, nodes deployed with a uniform density can detect whether they might be under the CLEA by monitoring how many established links they have. If a node finds that the number of the established links is greater than a threshold value, it can consider itself to be attacked by the CLEA. However, it can be very difficult to determine which of its neighbors are lying.

17.4.2 Direct Countermeasures

The direct countermeasures refer to counteractive techniques developed in the literature to defend the KPSs against either the NRA or the KSCA head-on. Since the two attacks are essentially similar, these techniques can be used to counteract the CLEA as well. Therefore, the mention of either the NRA or the KSCA in the following can be regarded as the mention of the CLEA.

17.4.2.1 Witness-Based Detection Schemes

In their pioneering work, Parno et al. [36] presented a wide range of detection protocols against the NRA, including *centralized detection*, *local detection*, *node-to-network broadcasting*, *deterministic multicast*, *randomized multicast*, and *line-selected multicast*. In the centralized detection scheme, each node is required to send a list of its neighbors and their claimed locations to the base station. The base station then scrutinizes every list to find out replicas. Thereafter, it can revoke the detected replicas by flooding the network with authenticated revocation messages. However, several drawbacks render this approach impractical. First, the use of the base station for security purpose apparently introduces a single point of failure. The scheme becomes useless if the attacker can compromise the base station or communication channels around it. Secondly, the nodes nearest to the base station will suffer from a heavy routing load and will become appealing targets for the attacker. The scheme also introduces delays in replica revocation since it takes the base station a long time to look for conflicts in the lists before flooding revocations throughout the networks. Finally, many DSNs are not supported by powerful base stations, thus a distributed approach is unavoidable.

The centralized detection approach can be replaced with the local detection scheme which depends on a node's neighbors to perform the replication detection. The neighbors use a voting mechanism to reach a consensus on the legitimacy of the node. Unluckily, although this scheme achieves detection in a distributed manner, it ensures no success in detecting distributed node replication in disjoint neighborhoods within the network where replicas are at least two hops away from each other.

The node-to-network broadcasting scheme is the very first attempt to detect the NRA in distributed fashion. Accordingly, each node is required to flood the network with its location information in authenticated manner. When receiving the flooded information, a node compares this information with the location information of its neighbors in its memory. If it detects a conflicting claim, it revokes the detected replicas. While this scheme can guarantees 100% detection of duplicate location claims as long as the broadcast messages reach every node, it puts the total communication cost of $O(n^2)$ messages on the network. This cost is not a cause for concern for small networks. But as the size of the network grows, it becomes too costly.

The deterministic multicast scheme is developed to reduce the communication overhead of the previous scheme in such a way that a node's location claim is not sent to all the other nodes in the network, but just to a limited set of nodes, called witnesses. The witnesses are chosen as a function of the node's ID. If a replica exists somewhere in the network, then the witnesses will receive two conflicting location claims for the same node. Assume that the size of the set of witnesses is g and the degree of each node d, then as long as each of the neighbors randomly selects $g \ln g/d$ witnesses from the set, the coupon collector's problem [49] assures that each of the witnesses will receive at least one location claim. Assuming an average network path length of $O(\sqrt{n})$ nodes, the resulting communication cost is $O(g \ln g \sqrt{n})/d$ messages. Unluckily, this decline in the communication cost is achieved at cost to security. Since the function used to select the witnesses is deterministic, the attacker can predict them as well. Therefore, if he is able to capture or jam all of the messages bound for the witnesses, then he can create as many replicas as he wants.

The security weakness in the deterministic multicast scheme can be overcome by randomizing the witnesses as proposed in the randomized multicast scheme. This prevents the attacker from anticipating their identities. Specifically, the scheme requires each node to broadcast its signature-based authenticated location claim. Each of the node's neighbors, with probability p, selects g random locations within the network and uses a routing protocol (e.g., geographic routing [50]) to forward the claim to the nodes closest to the selected locations. The security analysis shows that the probability of choosing the same node as a witness more than once is negligible. In the network of n nodes where each location is testified by $\sqrt{n}$ witnesses, the birthday paradox [49] anticipates that a pair of conflicting location claims will be encountered under the NRA with high probability. For instance, if $n = 10,000$, $g = 100$, $d = 20$, and $p = 0.05$, the probability of detection will be greater than 63% for a single replication and 95% for a double replication. However, this scheme remains inefficient in terms of storage and communication requirements. For example, with the above network setting, each node is required, on average, to store 3700 bytes, which is just under

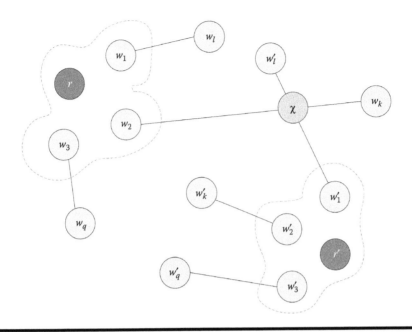

Figure 17.4   Line-selected multicast: The attacker is assumed to have created a replica of *r*, *r'*. The storage of the replicas' location claims at nodes en route to the witnesses (w_is and w_i's) results in an intersection at χ.

91% of the MICA2/MICAz's total RAM. Similarly, as estimated in [36], the total communication cost is $O(n^2)$, equivalent to those of the node-to-network broadcasting scheme.

Owing to the above limitations, a different scheme named line-selected multicast is investigated to improve the performance of the randomized multicast scheme. In this scheme, each node is required to send out its location claim to *r* initial witnesses. The function of witnessing is extended to each of the nodes along the routes from the broadcasting node to the initial witnesses as well. The node en route also stores a copy of the location claim while forwarding the claim to an initial witness. By storing location claims at intermediate nodes, we have effectively drawn line-segments through the network as illustrated in Figure 17.4. Two line segments intersecting correspond to the fact that a witness at the intersection receives two conflicting location claims.

Thereafter, it can flood the network with the undeniable evidence to revoke the detected replicas. This scheme offers a very high detection rate. As pinpointed in [36], the detection rate is 56% with one line segment per node and soars to 95% with five line segments per node. Furthermore, the line-selected multicast scheme also exhibits a reasonable performance feature. Assume that the length of each line segment is $O\left(\sqrt{n}\right)$, then the scheme requires only $O\left(n\sqrt{n}\right)$ messages for the whole network and each node stores $O\left(\sqrt{n}\right)$ location claims.

Table 17.1 summarizes the storage and communication costs for each of the distributed protocols discussed previously. The communication costs are for the whole network while the storage costs are per node.

In their work [37], Conti et al. pointed out the security weakness of the line-selected multicast scheme by introducing a stronger attacker, named *smart attacker*. This attacker aims to anticipate prospective witnesses before each round of the detection scheme to corrupt. If he successfully corrupted these nodes, the scheme would fail to detect the NRA. The anticipation can be carried

Table 17.1   Summary of Scheme Costs

	Communication	Memory
Node-to-network broadcast	$O(n^2)$	$O(d)$
Deterministic multicast	$O\left(\frac{g\ln g\sqrt{n}}{d}\right)$	$O(g)$
Randomized multicast	$O(n^2)$	$O(\sqrt{n})$
Line-selected multicast	$O(n\sqrt{n})$	$O(\sqrt{n})$

out based on nodes' ID information and nodes' location information. To defend against this attacker, the authors proposed a Randomized, Efficient, and Distributed (RED) protocol that keeps changing a node's witness set after each RED iteration to avoid the anticipation. Specifically, the RED executes at fixed time intervals. Each round of the protocol consists of two steps. First, a random value, *rand*, is distributed among all the nodes. Secondly, each node digitally signs and broadcasts its claim, including node ID and geographic location. Each of the node's neighbors, with probability *p*, sends the claim to a set of network locations. These locations are selected using the *PseudoRand* function. The inputs of this function consist of the node's ID, the current *rand* value and the output is the selected network locations. The function also guarantees that the neighbors, who are forwarding the claim, will select the same set of the network locations. The witnesses are the closest nodes to the selection locations.

17.4.2.2 SET: Set Operation-Based Detection Scheme

The witness-based schemes [36,37] described above must rely on public key cryptography that may be impractical for low-end DSNs. To overcome this limitation, Choi et al. [38] proposed a new scheme, named *SET*, to detect the NRA. The key idea is based on computing set operations (intersection and union) of exclusive subsets in a DSN to detect replicas. Correspondingly, the DSN can be perceived as a set of nonoverlapping subregions. Nodes in each subregion form an exclusive subset. Because node IDs are unique throughout the network, the intersection of any two subsets should be empty. If the attacker deploys replicas in the network, the intersection of subsets including these replicas will not be empty, and thus the NRA can be detected.

SET consists of five components: exclusive subset construction, authenticated subset covering, verifiable random member selection, distributed set computation on subset trees, and interleaved authentication on subset trees. The first component uses an exclusive subset maximal independent set (ESMIS) algorithm to form exclusive subsets in a distributed fashion. An exclusive subset consists of one subset leader (SLDR) and a number of subset members within the SLDR's transmission range. The process of forming the exclusive subset in ESMIS happens as follows. The base station generates a random seed and broadcasts it to the network. Upon receiving the seed, every node puts itself in Init state. Each node then computes locally hash values for itself and its neighbors using a hash function $H_1 : (\text{seed}|x) \rightarrow y \in [1, d]$, where d is the average number of neighbors in the network and x is the node's ID or the ID of one of the node's neighbors. The node corresponding to the maximum hash value in its neighborhood becomes an SLDR and switches its state to Ruler. If there is more than one node with the same maximum hash values, the node having the biggest ID will be selected as the SLDR. This SLDR then informs its neighbors of its promotion. Upon receiving the SLDR's message, the other neighbors in the Init state change their state to Ruled. This means that

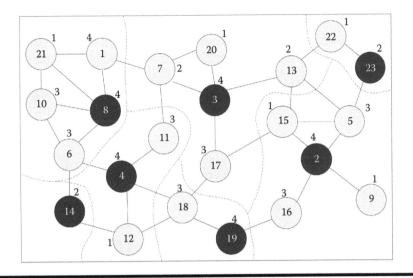

Figure 17.5 The construction of seven subsets in accordance with the ESMIS algorithm. (Data from H. Choi, S. Zhu, and T. F. La Porta, *Proceedings of 3rd International Conference on Security and Privacy in Communications Networks and the Workshops (SecureComm 2007),* **IEEE Communications Society, New York, NY, 2007, pp. 341–350.)**

they become the members of the exclusive subset dominated by the SLDR. Thereafter, the nodes in the ruled state announce their SLDR to their neighbors.

Figure 17.5 illustrates an example of a small network with 23 nodes and $d = 4$. The value outside the circle of each node is the H_1 computation result. The construction of seven subsets in accordance with the ESMIS algorithm is demonstrated. Nodes 2, 3, 4, 8, 14, 19, and 23 are selected as SLDR nodes of the neighborhoods bounded by dashed curves.

The second component of *SET*, authenticated subset covering, is used to cope with the presence of corrupted nodes in hostile environments. These corrupted nodes may attempt to hide their existence by either not following the ESMIS algorithm or announcing their *ruled* state with bogus SLDR identifiers. To address this problem, the component requires each *ruled* node to send the identifiers of all its neighboring SLDR to its SLDR. This SLDR then generates membership authentication of the *ruled* node for each neighboring SLDR. The membership authentication is a message authentication code (MAC) derived from the *ruled* node ID and the SLDR ID using an MAC function and the pairwise key between the SLDR and the neighboring SLDR. This membership MAC is then sent to the neighboring SLDR and used by this node to vouch for the SLDR's covering of the *ruled* node. Because the *ruled* node does not have the pairwise key between the SLDR and the neighboring SLDR, it cannot generate the membership MAC. Hence, the *ruled* node cannot convince its neighboring SLDRs of a bogus SLDR ID.

According to the basic idea, every node ID is reported to the base station through SLDRs. This might introduce significant communication overhead to nodes near the base station. The component of verifiable random member selection is used to reduce the communication cost. The key idea is to allow the base station to randomly select which nodes should be reported, instead of all the nodes. The base station releases addition information which is a string of bits "0" and "1" when broadcasting the *seed*. An SLDR uses this string to pick reported members and then send them to the base station for the attack detection.

Owing to the random selection of SLDRs, the base station might not know which nodes are SLDRs. As a consequence, the attacker can drop reports sent by SLDRs without detection. To tackle this issue, a multiple tree-based approach is employed. This approach builds multiple trees in the network. A tree consists of nodes which are SLDRs of the exclusive subsets. The tree construction starts from selecting an SLDR root randomly. Thereafter, the root discovers the neighboring SLDRs and admits them to be its children. The process keeps continuing until the last SLDR joins the tree. After the tree construction, the leaf SLDR sends its subset report to its parent. The parent collects its children's subsets and computes the intersection of these subsets with its own subset to detect the NRA. If the NRA is not detected, the parent generates a union of its children's subsets and its own subset, and sends this new report to its parent. Each root forwards its final report to the base station. If a node on the tree detects that the computed intersection is not empty, it knows that replicas exist in its subtree. It then notifies the base station of the NRA for further action.

The tree-based approach works well in the absence of corrupted nodes. However, if a parent is corrupted, it may delete replicated identifiers from the computed union to conceal the NRA. Therefore, the set operations (intersection and union of subsets) on the tree should be verified. This verification can be enabled by employing the idea discussed in [51] to develop an interleaved authentication scheme on the tree. This scheme requires that the SLDRs store the path information from the root to themselves during the tree construction. When an SLDR sends a report to its parent, it computes a keyed MAC, for example, HMAC [52], not only for its parent but also for its grandparent (interleaved MAC). The grandparent can detect any changes made by corrupted parents on the results of set operations by computing and checking the interleaved MACs.

17.4.2.3 Bloom Filter-Based Detection Scheme

Approaching the NRA in a different manner, specifically for random key predistribution schemes [8,20,21,25], Brooks et al. [39] introduced the concepts of a Bloom filter [53] and a counting Bloom filter [54] to monitor the use of keys as authentication tokens and detect statistical deviations that indicate the NRA.

The Bloom filter as illustrated in Figure 17.6 is a space-efficient probabilistic data structure that is used to test whether an element is a member of a set via membership queries. Elements can be added to the set, but not removed.

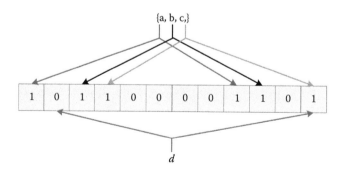

Figure 17.6 An illustration of a Bloom filter, representing the set {a, b, c}. The colored arrows show the positions in the bit array that each set element is mapped to. The element *d* is not in the set {a, b, c}, because it hashes to one bit-array position containing 0. For this figure, *m* = 12 and *k* = 3.

Initially, an empty Bloom filter, which is a bit array of m bits, all set to 0, is created. There must also be k different defined hash functions, each of which maps or hashes a set element to one of the m array positions with a uniform random distribution. The set element is added to the filter by feeding it to each of the k hash function to get k array positions. Bits at all these positions are then set to 1. The membership query about the set element is performed by feeding it to each of the k hash functions to get k array positions. If any of the bits at these positions are 0, the element is not in the set. Otherwise, all the bits would have been set to 1 when it was added. If all are 1, then either the element is in the set, or the bits have been set to 1 during the addition of other elements.

Unfortunately, the removal of an element from the Bloom filter is impossible. Because, to remove the element represented by k bits in the filter, we need to set any one of these k bits to zero. However, this setting may lead to the removal of other elements that are also represented by that bit. A counting filter provides a solution to support a *delete* operation on a Bloom filter. In the counting filter, the array positions (buckets) are extended from being a single bit to being an n-bit counter. The add operation is extended to increment the value of the buckets and the look-up operation checks that each of the required buckets is nonzero. The delete operation essentially consists of decrementing the value of each of the respective buckets.

The main idea of the counting filter-based protocol is as follows. Since each node is preloaded with m randomly selected keys from the pool, the distribution of the number of nodes possessing a given key can be predictable. The distribution of the number of times a key is used for the PKE is directly proportional to the number of nodes owning that key. When the NRA is launched, the key usage distribution is skewed. The reason is that replicated keys appear on a greater number of nodes than normal, and thus are used more frequently than keys that have not been replicated. By gathering key usage statistics, the keys which have been replicated can be specified and revoked.

According to the protocol, the gathering of key usage statistics is conducted as follows. Each node constructs a counting filter from the keys it uses to connect to its neighboring nodes. It appends a random number (nonce) to the created filter and encrypts the result using the base station's public key. This encrypted report is forwarded to the base station. The base station decrypts the received filters and discards duplicate reports by checking nones. The base station performs membership queries on the filter and counts the number of times each key is used in the network. Keys used above a threshold value are considered replicated. The base station creates a Bloom filter from the replicated keys, signs the filter with its private key, and broadcasts this filter to the network using a gossip protocol [55]. Each node verifies the received filter and checks the keys in its key ring for membership of replicated keys in the filter. The node then removes the replicated keys from its key ring and terminates all connections using the replicated keys.

17.4.2.4 One-Way Hash Chain-Based Protection Schemes

Motivated by the KSCA on random PKPSs [8,20,21,25], Tran et al. [56] proposed a light-weight framework for thwarting it. The main idea of this proposal is to minimize the utility of compromised nodes or pairwise keys to the attacker. It is a winning combination of two factors: intermittent deployment strategy and one-way hash chain [57]. In the intermittent deployment strategy as shown in Figure 17.7, the total sensor nodes intended for deployment are classified into generations with identifiers ranging from 1 to t based on a number of deployment times. At one deployment time, either one generation or set of successive generations can be deployed simultaneously complying with application requirements.

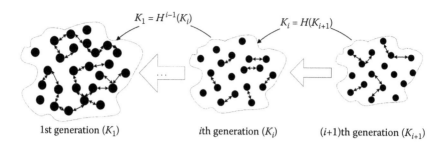

Figure 17.7 An example deployment model.

Besides, an one-way hash chain is generated from a seed K_t. Element i, K_i of the chain is associated with sensor generation i functioning as a key encryption key (KEK). This generation-wide KEK is stored in each node together with its key ring. As illustrated in Figure 17.8, each KEK is also a key recovery key (KRK) for the KEK of the immediate successor generation and can be used to compute the KEKs of predecessor generations.

After deployment, there are two scenarios to be considered: current generation PKE and different generation PKE. For the former, nodes just follow the same procedure as described in a specific random PKRS to establish pairwise keys. After all the PKE activities have been completed, a current ith generation node performs the following operations: encrypting the key ring using K_i, computing the KRK K_{i-1} using the one-way hash function, replacing K_i and the key ring with K_{i-1} and the encrypted key ring, respectively.

What makes the latter different from the former is that in order to participate in the PKE process with a current generation node y, an previous generation node x has to recover its key ring from the encrypted one first. This is done via the aid of y. This node first uses its KEK and the one-way hash function to compute the KEK and KRK for x. It then sends the computed KEK encrypted by the KRK to x. x uses the KRK which is already in its memory to decrypt the received message and obtain its KEK. x uses the KEK to recover its plaintext key ring before joining the PKE process.

The advantage of this approach is threefold. First, it still allows later deployed nodes to integrate with previously deployed nodes. Secondly, it minimizes both the attacker's opportunity to obtain pairwise keys and the quantity of disclosed pairwise keys from compromised nodes. Thirdly, it evades undesirable requirements of functionalities and resources, topological pre-deployment knowledge, or costly location-based detection algorithms.

Unfortunately, this framework suffers from a serious drawback, that is, lack of backward confidentiality for KEKs. If the attacker can compromise one node, he can obtain the compromised node's KRK which is the KEK of the previous generation. Applying the one-way hash function

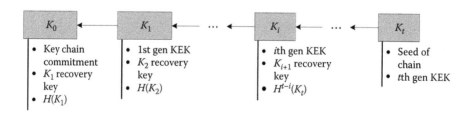

Figure 17.8 Generation and functions of the one-way hash keys.

repeatedly on this key, the attacker can obtain backward KEKs indirectly. Consequently, as long as a group of compromised nodes contains one in the latest generation, the framework almost fails to counter the attack. To provide the backward secrecy for KEKs, Tran et al. [58] devised a scheme against the KSCA which uses a diversified one-way hash chain instead of the simple one. The idea of this scheme is quite similar to the work described previously [56]. In more detail, a one-way hash chain is generated. Each element of the chain is associated with a node generation functioning as a KEK. This key is predistributed in each node of the generation together with KEK-encrypted cryptographic secrets to support the PKE activities. Thereafter, it is diversified to avoid being disclosed due to node compromise. The KEK diversification process happens as follows. Each node selects a random salt, computes a KRK by feeding the KEK and the salt to a hash function. The node then deletes KEK and retains the KRK. The KRK is used to recover plaintext cryptographic secrets for later deployed generation as in [56].

17.5 Future Research Directions

Based on what have been presented in this chapter, future research directions can be classified into three groups. The first is related to improvements on the existing schemes. The second is concerned with the design of new PKE schemes keeping the CLEA in mind. The third is involved with the proposition of new solutions to the attack.

The improvements on the existing schemes can be made by overcoming the limitations of the previous schemes. For instance, all the witness-based schemes use the following costly components:

- Public-key cryptosystem which has very high computational complexity and energy consumption.
- Geographic positions/locations which require localization algorithms.
- Network loose-time synchronization.

A new scheme independent of any of these components would be a significant enhancement.

The design of new PKE schemes which have the capability to foil the CLEA would be a challenging research direction. One example could be LEAP+ [6]. Unluckily, its immunity from the CLEA relies solely on the condition $T_{min} > T_{est}$. Consequently, the powerful attacker, who is able to compromise a node within T_{est} of each interval, can launch the CLEA with ease. More security-efficient schemes need to be developed.

Finally, radical solutions decoupled from any PKE schemes to the attack should be considered as well. One preliminary instance could be one-way hash chain-based approach [56,58]. However, this approach exposes the same limitation as LEAP+ does [6]. In addition, one might come up with ideas that enable nodes to collectively detect early node capture or early illegitimate link establishment.

17.6 Conclusions

This chapter discussed the issue of the CLEA in DSNs and counteractive techniques against the attack. The CLEA is brought into play due to the limited applicability of asymmetric-key-based schemes on resource-constrained nodes and the proliferation of the PKPSs for the PKE in DSNs. It exploits the weakness of the PKPSs which is a destructive combination of network-wide cryptographic secrets and locally communicating nodes to invade the network via the establishment of a large number of illegitimate links. The CLEA represents both the NRA and the KSCA. Therefore, the discussion of the CLEA has been conducted via the discussion of the NRA and the KSCA. We have explained how these attacks work and their impacts upon vast numbers of protocols and mechanisms such as routing, data fusion, distributed voting, resource allocation.

Since research on this attack is still in the early stage, the countermeasures have been found very few in the literature. The first approach relies on key management schemes to reduce the utility of compromised nodes to the attacker. The second approach uses nodes' location information and witnesses to detect replicas. The third approach detects replicas by computing set operations (intersection and union) of exclusive subsets of nodes in the network. Finally, the attack can be foiled using the approach of protecting cryptographic secret.

Unfortunately, each of the above techniques is security inadequate and exposes its own limitations. Hence, we expect to see more advances in counteractive and detection mechanisms against the attack in the near future.

Terminologies

CLEA—Controlled Link Establishment Attack
DSN—Distributed Sensor Network
RSA—An algorithm for public-key cryptography. It is the first algorithm known to be suitable for signing as well as encryption, and one of the first great advances in public key cryptography. The algorithm was publicly described in 1977 by Ron Rivest, Adi Shamir, and Leonard Adleman at MIT; the letters RSA are the initials of their surnames.
ECC—Elliptic Curve Cryptography
MEMS—Micro-electromechanical Systems
PKE—Pairwise Key Establishment
KDC—Key Distribution Center
PKPS—Pairwise Key Pre-distribution Scheme
LDK—Location-Dependent Key
NRA—Node Replication Attack
KSCA—Key-Swapping Collusion Attack
MAC—Medium Access Control; Message Authentication Code
RED—Randomized, Efficient, and Distributed node replication detection protocol
SET—Set operation based node replication detection protocol
ESMIS—Exclusive Subset Maximal Independent Set
SLDR—Subset LeaDeR in an exclusive subset
HMAC—Keyed-Hash Message Authentication Code
KEK—Key Encryption Key
KRK—Key Recovery Key

Questions and Sample Answers

1. Why aren't public-key based algorithms the undesirable approach to the pairwise key establishment in DSNs as opposed to symmetric-key based algorithms?
   Sensor nodes are designed to be compact and therefore are limited by size, energy, computational power, and storage. The limited resources limit the types of security algorithms and protocols that can be implemented. Meanwhile, the public-key-based algorithms are very

costly because of their large code size and data size, intensive computation, and high power consumption. These features make the algorithms impractical to be implemented into sensor nodes. On the contrary, the symmetric-key-based algorithms consume much less computational energy than their public-key-based counterparts. This makes them feasible to be used in the design of PKE schemes.

2. What is the main idea of pairwise key pre-distribution schemes for DSNs?
   A set of the cryptographic secrets is predistributed into each node before deployment. After deployment, neighboring nodes negotiate with each other to agree on common pairwise keys using the preloaded cryptographic secrets.

3. What are the desirable features of PKPSs for DSNs?
   The desirable features of a PKPS for DSNs include low energy consumption, low cost, low memory usage, distributed manner, resilience to large number of node compromises, guaranteed key establishment, direct key establishment, node-to-node identity authentication, and resilience to dynamic topology.

4. Why can the NRA be launched in DSNs?
   After deployment, sensor nodes are usually left unattended. This enables the attacker to gain access to the sensor field and capture nodes. Furthermore, because nodes are typically low-end devices, thus they are not equipped with tamper-resistant hardware. Therefore, the captured nodes can be tampered with and cryptographic secrets can be extracted from the compromised nodes. Having these cryptographic secrets, the attacker can launch the NRA with ease.

5. What is the shortcoming of PKPSs that make the KSCA feasible in DSNs?
   The shortcoming emerges from the fact that although the illegal usage of compromised cryptographic secrets is feasible all over the network via the existing routing mechanisms or out-of-band channels, it is unlikely to be detected as the consequence of the lack of co-operation among locally communicating nodes.

6. Why can countermeasures against either the NRA or the KSCA be applicable to the NIA?
   The eventual goal of these attacks is to gain partial or even full control of DSNs via illegitimate link establishment. Because of the identical attack goal, a countermeasure against one of them can be utilized to the others.

7. How is the KSCA similar to and different from the Sybil attack and the NRA?
   The KSCA is similar to the Sybil attack in that single nodes present multiple identities. However, these identities are not randomly generated but instead are reused according to the pairwise keys available. The attack is similar to the NRA in that copies of a node are inserted into a network. However, the KSCA is unique in that attacker-controlled nodes pretend to be different nodes to different neighbors.

8. What are indirect approaches to the NIA?
   They include confining the usability of preloaded cryptographic secrets to local areas using deployment and location information, excluding the key predistribution phase using the idea of key infection, discarding unused keys after an initialization phase, and counting the number of established links of each node with its neighbors.

9. What is the main idea of the witness-based detection schemes?
   The main idea is that a number of nodes or network locations are selected randomly as witnesses to the NIA. In each round, all the nodes listen to location claims from their neighbors and

with a certain probability then forward the claims to the selected claims. The NIA is detected if a witness receives at least two different claims of the same node identifier.

10. What is the main idea of SET?

    SET divides the network into exclusive subsets of nodes. Nodes in each subset elect one node in the subset as a subset leader. Thereafter, multiple subset trees, each of which consists of all subset leaders, are constructed. Each subset leader reports its subset members to higher subset leaders along the trees. The higher subset leader performs intersection operation of received reports with its report. If the intersection is not empty, then the NRA is detected. Otherwise, it computes the union of the reports and sends it to higher subset leaders. The base station is the last point to compute the intersection for detecting replicas.

11. What is the main idea of Bloom filter-based detection scheme?

    The Bloom filter-based detection scheme is based on the observation that replicated keys that appear on a greater number of nodes than normal are used more frequently than nonreplicated keys. By gathering key usage statistics using a Bloom counting filter, it is not difficult to see the keys which have been used above a threshold value. The scheme treats these keys as replicated keys and removes them from usage.

12. How is the one-way hash chain-based protection approach different from the other approaches?

    The one-way hash chain-based protection approach is different from the witness-based detection approach and the set operation-based detection approach in that the former is aimed at protecting the PKPSs against the NIA specifically while the latter focus on the detection of replicas. It is different from the Bloom filter-based scheme in that the former can be applicable to any PKPSs whereas the latter can be used with random key predistribution schemes only. In other words, the former is more general than the latter.

Author's Biography

Thanh Dai Tran received the BEng degree in information technology from Hanoi University of Technology, Vietnam, in 2005 and MEng in computer engineering from Kyung Hee University, Republic of Korea, in 2007. He joined Networking Laboratory while working on his MEng from 2005 to 2007. Since February 2008, he has been a doctoral candidate at University of Technology, Sydney (UTS) and a member of Centre for Real-time Information Networks, UTS. His current research interests include security in wireless sensor networks (WSNs) and multihop wireless networks. Thus far, he has been the author of many publications that include two book chapters, two international journals, and many international conference papers on key management and attacks in WSNs.

Johnson I. Agbinya received his PhD in 1994 from La Trobe University in Melbourne in electronic communication engineering in remote sensing with ground-probing radar systems, MSc (Electronic Communications) from the University of Strathclyde, Glasgow, Scotland, and BSc from the Obafemi Awolowo University, Nigeria. He is currently a senior lecturer at the University of Technology Sydney and an Adjunct Professor of Communications at the department of Computer Science, the University of the Western Cape and the Alcatel-Lucent Professor of Communications at French South African Institute of Technology (F'SATI) at Tshwane University of Technology, Pretoria, South Africa. He is the author of five technical books, over 140 Journal and conference papers, 13 book chapters, and over 30 classified industrial research reports. Several of his papers have received international awards and formed major contents to patents held by CSIRO where he worked for 8 years.

He was Senior Research Scientist at CSIRO Telecommunications and Industrial Physics, Sydney, Australia, from 1993 to 2000 in the areas of Biometrics and Signal Processing. Between 2000 and 2003 he was Principal Engineer, Vodafone Australia, responsible for managing Vodafone's Industrial Research in Mobile Communications.

His current research interests are in short-range wireless communications and sensing, mobile content development, vehicular networks and networks in uncovered areas, WiMAX, sensor networks, sensor web and biometric security systems. He is a rated researcher by the National Research Foundation (NRF) of South Africa. He is also a core member of Centre for Real-Time Information Networks (CRIN) at UTS and the Telkom Centre of Excellences in Tshwane University of Technology, Pretoria, South Africa, French South African Institute of Technology (F'SATI) and also CoE for IP Communications at UWC South Africa.

He is the current Editor-in-Chief of the *African Journal of Information and Communication Technology* (AJICT) and its pioneer. He has also editor several conference proceedings. He is also the pioneer of and conference chair for AusWireless, BroadCom, and ICCB (2010). He has also been involved in the technical and organizing committees of numerous international conferences, including AusWireless (2006, 2007), BroadCom 2008 and 2009, ICMB, ICT Africa, ICADST and many other conferences.

References

1. I. F. Akyildiz, S. Weilian, Y. Sankarasubramaniam, and E. Cayirci, A survey on sensor networks, *Communications Magazine, IEEE*, 40, 102–114, 2002.
2. M. Ilyas and I. Mahgoub, *Handbook of Sensor Networks: Compact Wireless and Wired Sensing Systems.* 2005, CRC Press, Boca Raton, USA.
3. R. Anderson and M. Kuhn, Tamper resistance—a cautionary note, in *Proceedings of Second USENIX Workshop on Electronic Commerce*, USENIX Association, Berkeley, CA, 1996, pp. 1–11.
4. C. Karlof and D. Wagner, Secure routing in wireless sensor networks: Attacks and countermeasures, in *Proceedings of 1st IEEE International Workshop on Sensor Network Protocols and Applications*, Anchorage, AK, 2003, pp. 113–127.
5. A. Perrig, R. Szewczyk, J. D. Tygar, V. Wen, and D. E. Culler, SPINS: security protocols for sensor networks, *Wireless Network*, 8, 521–534, 2002.
6. S. Zhu, S. Setia, and S. Jajodia, LEAP+: Efficient security mechanisms for large-scale distributed sensor networks, *ACM Transactions on Sensor Networks*, 2, 500–528, 2006.
7. W. Zhang, M. Tran, S. Zhu, and G. Cao, A random perturbation-based scheme for pairwise key establishment in sensor networks, in *Proceedings of 8th ACM Int'l Symposium on Mobile Ad hoc Networking and Computing*, ACM, New York, NY, 2007.
8. W. Du, J. Deng, Y. S. Han, P. K. Varshney, J. Katz, and A. Khalili, A pairwise key pre-distribution scheme for wireless sensor networks, *ACM Transactions on Information and System Security*, 8, 228–258, 2005.
9. W. Yong, G. Attebury, and B. Ramamurthy, A survey of security issues in wireless sensor networks, *IEEE Communications Surveys & Tutorials*, 8, 2–23, 2006.
10. Z. Yun, F. Yuguang, and Z. Yanchao, Securing wireless sensor networks: A survey, *Communications Surveys & Tutorials, IEEE*, 10, 6–28, 2008.
11. W. Diffie and M. Hellman, New directions in cryptography, *IEEE Transactions on Information Theory*, 22, 644–654, 1976.
12. R. L. Rivest, A. Shamir, and L. Adleman, A method for obtaining digital signatures and public-key cryptosystems, *Communications of the ACM*, 21, 120–126, 1978.

13. B. Michael, C. Donny, H. Darrel, H. Julio Lopez, K. Michael, and M. Alfred, PGP in constrained wireless devices, in *Proceedings of the 9th conference on USENIX Security Symposium—Volume 9 Denver*, USENIX Association, Colorado, 2000.

14. D. W. Carman, P. S. Kruus, and B. J. Matt, Constraints and approaches for distributed sensor network security, *DARPA Project report*, Cryptographic Technologies Group, Trusted Information System, NAI Labs, vol. 1, September 2000.

15. A. S. Wander, N. Gura, H. Eberle, V. Gupta, and S. C. Shantz, Energy analysis of public-key cryptography for wireless sensor networks, in *Proceedings of 3rd IEEE International Conference on Pervasive Computing and Communications (PerCom 2005)*, IEEE Computer Society, Washington, DC, 2005, pp. 324–328.

16. N. Gura, A. Patel, A. Wander, H. Eberle, and S. C. Shantz, Comparing elliptic curve cryptography and RSA on 8-bit CPUs, in *Proceedings of 6th International Workshop on Cryptographic Hardware and Embedded Systems (CHES'04)*, Lecture Notes in Computer Science, pp. 119–132, 2004.

17. A. Liu and P. Ning, TinyECC: A configurable library for elliptic curve cryptography in wireless sensor networks, in *Proceedings of 7th International Conference on Information Processing in Sensor Networks*. IEEE Computer Society, Washington, DC, 2008.

18. V. S. Miller, Use of elliptic curves in cryptography, in *Advances in Cryptology (CRYPTO'85)*, Lecture Notes in Computer Science, 1986, pp. 417–426.

19. N. Koblitz, Elliptic curve cryptosystems, *Mathematics of Computation*, 48, 203–209, 1987.

20. L. Eschenauer and V. Gligor, A key-management scheme for distributed sensor networks, in *Proceedings of 9th ACM Conference on Computer and Communications Security (CSS'02)*, ACM, New York, NY, 2002, pp. 41–47.

21. H. Chan, A. Perrig, and D. Song, Random key predistribution schemes for sensor networks, in *Proceedings of IEEE Symposium on Security and Privacy*, IEEE Computer Society, Washington, DC, 2003, pp. 197–213.

22. W. Du, J. Deng, Y. S. Han, S. Chen, and P. K. Varshney, A key management scheme for wireless sensor networks using deployment knowledge, in *Proceedings of IEEE INFOCOM*, IEEE Communications Society, New York, NY, 2004, pp. 586–597.

23. C. Siu-Ping, R. Poovendran, and S. Ming-Ting, A key management scheme in distributed sensor networks using attack probabilities, in *Proceedings of IEEE Global Telecommunications Conference (GLOBECOM'05)*, IEEE Communications Society, New York, NY, 2005.

24. T. D. Tran and C. S. Hong, Efficient ID-based threshold random key pre-distribution scheme for wireless sensor networks, *IEICE Transactions on Communications*, E91-B, 2602–2609, 2008.

25. D. Liu, P. Ning, and R. Li, Establishing pairwise keys in distributed sensor networks, *ACM Trans. Inf. Sys. Sec.*, 8, 41–77, 2005.

26. J. Spencer, *The Strange Logic of Random Graphs*. vol. 22. Springer, Berlin, Germany, 2001.

27. R. D. Pietro, L. V. Mancini, and A. Mei, Random key-assignment for secure wireless sensor networks, in *Proceedings of 1st ACM workshop on Security of ad hoc and sensor networks (SASN '03) Fairfax*, ACM, VA, 2003.

28. R. Blom, An optimal class of symmetric key generation systems, in *Proceedings of the EUROCRYPT 84 workshop on Advances in cryptology: Theory and application of cryptographic techniques,* Springer, Paris, France, New York, 1985.

29. C. Blundo, A. De Santis, U. Vaccaro, A. Hertzberg, S. Kutten, and M. Yong, Perfectly secure key distribution for dynamic conferences, *Information Computing*, 146, 1–23, 1998.

30. T. Matsumoto and H. Imai, On the key predistribution system: A practical solution to the key distribution problem, in *A Conference on the Theory and Applications of Cryptographic Techniques on Advances in Cryptology*. Springer, Berlin/Heidelberg, Germany, 1988.

31. D. Liu and P. Ning, Location-based pairwise key establishments for static sensor networks, in *Proceedings of 1st ACM workshop on Security of ad hoc and sensor networks (SASN '03)*, ACM, Fairfax, VA, 2003.

32. D. Liu and P. Ning, Improving key predistribution with deployment knowledge in static sensor networks, *ACM Transactions on Sensor Networks*, 1, 204–239, 2005.

33. T. Ito, H. Ohta, N. Matsuda, and T. Yoneda, A key pre-distribution scheme for secure sensor networks using probability density function of node deployment, in *Proceedings of 3rd ACM workshop on Security of ad hoc and sensor networks (SASN '05)*, ACM, Alexandria, VA, 2005.

34. F. Anjum, Location dependent key management using random key-predistribution in sensor networks, in *Proceedings of 5th ACM Workshop on Wireless Security*, ACM, Los Angeles, CA, 2006.
35. C. Castelluccia and A. Spognardi, RoK: A robust key pre-distribution protocol for multi-phase wireless sensor networks, in *Proceedings of 3rd International Conference on Security and Privacy in Communications Networks and the Workshops (SecureComm 2007)*, IEEE Communications Society, New York, NY, 2007, pp. 351–360.
36. B. Parno, A. Perrig, and V. Gligor, Distributed detection of node replication attacks in sensor networks, in *Proceedings of IEEE Symposium on Security and Privacy*, IEEE Computer Society, Washington, DC, 2005, pp. 49–63.
37. M. Conti, R. D. Pietro, L. V. Mancini, and A. Mei, A randomized, efficient, and distributed protocol for the detection of node replication attacks in wireless sensor networks, in *Proceedings of 8th ACM International Symposium on Mobile Ad Hoc Networking and Computing*, ACM, New York, NY, 2007.
38. H. Choi, S. Zhu, and T. F. La Porta, SET: Detecting node clones in sensor networks, in *Proceedings of 3rd International Conference on Security and Privacy in Communications Networks and the Workshops (SecureComm 2007)*, IEEE Communications Society, New York, NY, 2007, pp. 341–350.
39. R. Brooks, P. Y. Govindaraju, M. Pirretti, N. Vijaykrishnan, and M. T. Kandemir, On the detection of clones in sensor networks using random key predistribution, *IEEE Transactions on Systems, Man, and Cybernetics, Part C: Applications and Reviews*, 37, 1246–1258, 2007.
40. H. Fu, S. Kawamura, M. Zhang, and L. Zhang, Replication attack on random key pre-distribution schemes for wireless sensor networks, *Computer Communication*, 31, 842–857, 2008.
41. C. Hartung, J. Balasalle, and R. Han, Node compromise in sensor networks: The need for secure systems, *Technical Report CU-CS-990-05*, January 2005.
42. K. W. Kai-Wei Fan, S. Liu, and P. Sinha, Structure-free data aggregation in sensor networks, *IEEE Transactions on Mobile Computing*, 6, 929–942, 2007.
43. H. Cunqing and T. S. P. Yum, Optimal routing and data aggregation for maximizing lifetime of wireless sensor networks, *IEEE/ACM Transactions on Networking*, 16, 892–903, 2008.
44. C. Intanagonwiwat, R. Govindan, D. Estrin, J. Heidemann, and F. Silva, Directed diffusion for wireless sensor networking, *IEEE/ACM Transactions on Networking*, 11, 2–16, 2003.
45. T. Moore, A collusion attack on pairwise key predistribution schemes for distributed sensor networks, in *Proceedings of 4th Annual IEEE International Conference on Pervasive Computing and Communications (PerCom'06)*, IEEE Computer Society, Washington, DC, 2006.
46. Y. C. Hu, A. Perrig, and D. B. Johnson, Packet leashes: A defense against wormhole attacks in wireless networks, in *Proceedings of INFOCOM*, vol. 1973, IEEE Communications Society, New York, NY, 2003, pp. 1976–1986.
47. R. Poovendran and L. Lazos, A graph theoretic framework for preventing the wormhole attack in wireless ad hoc networks, *Wireless Networks*, 13, 27–59, 2007.
48. R. Anderson, H. Chan, and A. Perrig, Key infection: Smart trust for smart dust, in *Proceedings of 12th IEEE International Conference on Network Protocols (ICNP'04)*, IEEE Press, Greater Vancouver A, BC, Canada, 2004, pp. 206–215.
49. T. H. Cormen, C. E. Leiserson, R. L. Rivest, and C. Stein, *Introduction to Algorithms*. MIT Press, 2001.
50. B. Karp and H. T. Kung, GPSR: Greedy perimeter stateless routing for wireless networks, in *Proceedings of 6th Annual International Conference on Mobile Computing and Networking (MobiCom)*, ACM, Boston, MA, 2000.
51. S. Zhu, S. Setia, S. Jajodia, and P. Ning, Interleaved hop-by-hop authentication against false data injection attacks in sensor networks, *ACM Transactions on Sensor Networks*, 3, 14, 2007.
52. M. Bellare, R. Canetti, and H. Krawczyk, Keying hash functions for message authentication, in *Proceedings of 16th Annual International Cryptology Conference on Advances in Cryptology*. Springer, 1996.
53. B. Bloom, Space/time trade-offs in hash coding with allowable errors, *Communications of the ACM*, 13, 422–426, 1970.
54. L. Fan, P. Cao, J. Almeida, and A. Z. Broder, Summary cache: A scalable wide-area web cache sharing protocol, *IEEE/ACM Transactions on Networks*, 8, 281–293, 2000.
55. R. R. Brooks, *Disruptive Security Technologies with Mobile Code and Peer-to-Peer Networks*. CRC Press, Inc., Boca Raton, FL, 2004.

56. T. D. Tran and J. I. Agbinya, A framework for confronting key-swapping collusion attack on random pairwise key pre-distribution schemes for distributed sensor networks, in *Proceedings of 5th IEEE International Conference on Mobile Ad Hoc and Sensor Systems (MASS'08)*, 2008, pp. 815–820.
57. Y. C. Hu, M. Jakobsson, and A. Perrig, Efficient constructions for one-way hash chains, *Applied Cryptography and Network Security*, 2005, pp. 423–441.
58. T. D. Tran and J. I. Agbinya, Combating key-swapping collusion attack on random pairwise key pre-distribution schemes for wireless sensor networks, *Security and Communication Networks*, DOI: 10.1002/sec.106, 2009.

Chapter 18

Proactive Key Variation Owing to Dynamic Clustering (PERIODIC) in Sensor Networks

Gicheol Wang and Gihwan Cho

Contents

18.1 Introduction

Recent advances of technology have made it possible to develop wireless sensor networks consisting of a large number of ultrasmall autonomous devices. Sensor networks are being deployed for a wide variety of applications, such as battlefield surveillance, environment monitoring, patient monitoring and tracking, and health monitoring [1,2].

Because sensor nodes carry out their duties in a hazardous or hostile environment, it is indispensable to deliver data from sensors securely for the wide deployment of sensor networks. To provide the security services in sensor networks, a suitable key management scheme is required [2–5]. Generally, the key management service is provided by a centralized server or by virtue of the help of such a server. For instance, the Base Station (BS) has a key pool of many keys, and distributes a predetermined number of keys to each sensor before the deployment of the duty field. Each sensor establishes communication keys with other sensors using those predistributed keys, after the deployment. Recently, cluster architecture that transforms a network into small groups of nodes has been employed for wireless sensor networks. Under the cluster architecture, sensors belonging to a cluster are served and controlled by a local server, so-called Cluster Head (CH). Also, the responsibility of distributing and managing keys is delegated from the BS to each CH. As a result, the burden laid on the centralized server (that is, the BS) is distributed, and the efficiency of total key management is improved. However, in the cluster architecture, the compromise of CHs is more threatening than that of member sensors. This is because CHs collect data from member sensors and send the collected data to the BS. Therefore, if an attacker compromises a CH, it can obtain all data from the CH's members and even fabricate all data from the members. Furthermore, an attacker can easily identify the CH role nodes and aim at compromising the CH role nodes. If all CHs are compromised by attackers, the attackers can control the network. Assume that a number of sensors are deployed in a mountain to detect a fire. In the network, compromised sensors intentionally send the forged information, indicating that there is no fire to the BS under the occurrence of fire. If all CHs are compromised, the whole network is controlled by attackers and the occurrence of fire is never known to the fire-monitoring center. In this case, the sensor network does not function at all and the mountain will be burnt to ashes. Therefore, in the cluster architecture, it is very important to change the CH role nodes periodically.

The primary problem of key management in sensor networks is that the keys stored in sensors tend to be exposed to attackers. This is because sensor nodes are physically vulnerable so that attackers can easily capture them. The keys from the captured nodes can play a role of lever for obtaining the keys of other sensors. Therefore, the keys stored in sensors should be renewed frequently. Renewing the keys stored in sensors causes a tradeoff between security and efficiency (especially energy efficiency). That is, if the frequency of the key renewal increases, the communication and the computation overhead increase significantly while the security of the keys becomes high.

Key renewal schemes in sensor networks are divided into three classes. The first class has no renewal mechanism. In this scheme, the keys predistributed from a server are never renewed until the extinction of the network [1–4,6]. Therefore, keys from the captured nodes can be continuously employed for uncovering keys inside other sensors until the extinction of the network. We refer to this scheme as nonrenewal scheme hereafter. The second class has an on-demand renewal mechanism where the renewal occurs whenever a compromised node(s) is detected [7–11]. This scheme is available only when a CH or the BS which is responsible for key management can detect the compromised nodes by itself or via the help of pure nodes. We call these schemes as reactive key renewal schemes hereafter. The last class has a periodical renewal mechanism. These schemes are called as proactive key renewal schemes, hereafter. In recent years, a lot of key management schemes for sensor networks have been proposed. However, most of the schemes fall into the category of the nonrenewal schemes or the reactive renewal schemes. Because nonrenewal schemes do not change the keys for communication, they are very vulnerable to the proliferation of compromised nodes. In case of reactive renewal schemes, they change the keys for communication but they do not change the CH role nodes. Therefore, they are vulnerable to the proliferation of compromised CHs. In this chapter, we introduce a proactive key renewal scheme that resolves the above problems that two classes of key renewal schemes have.

18.2 Background

18.2.1 Nonrenewal Schemes

Eschenauer and Gilgor [6] first proposed a key management scheme which is based on the key predistribution. In the scheme, a predefined number of keys are assigned to each node before network deployment. After the network deployment, each node establishes communication keys with neighbor nodes using common assigned keys. If there are no common keys between any two nodes, they can indirectly establish a communication key using neighbors sharing common keys. Eschnauer's scheme is vulnerable to node compromise, since it allows the communication key establishment even when the number of common keys is one. To address this problem, Chan et al. allowed the communication key establishment only if the number of common keys is more than $q(>1)$ keys [1]. Du et al. [2] showed that the predistributed keys can be shrunken, if approximate locations in the deployment area are known to the key predistribution server. The key predistribution server partitions the deployment area into a specific number of subareas and distributes some keys to the key pool of each subarea. Then the key predistribution procedure is adjusted so that neighboring subareas share a lot of common keys. Therefore, although the predistributed keys at a sensor shrink, any two nearby sensors can easily find common keys. Liu et al. proposed a key predistribution scheme in which nodes are deployed in groups [3]. Because nodes belonging to the same group are close to each other, the key predistribution server distributes some keys to each node so that the rate of sharing common keys in the same group is very high. If some nearby nodes belong to different groups, such nodes act as an intergroup gateway to support the establishment of communication keys between nodes belonging to different groups. Traynor et al. proposed a key predistribution scheme for an environment where high and low capability sensors are mixed [4]. High capability sensors hold much more keys than low capability sensors. Also, high capability nodes support communication of low capability nodes, thereby a hierarchical communication model is implemented. Because low capability sensors hold a small number of keys, this scheme saves the memory space of sensors and provides the robustness against node compromise. Figure 18.1 shows the key distribution and employment procedure of nonrenewal schemes.

The above schemes do not have any key renewal mechanisms in the key management process. Therefore, key materials obtained from compromised sensors are valid until the extinction of the network, and the disclosed key materials also exist inside many other sensors with a predefined

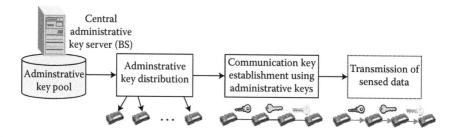

Figure 18.1   Key distribution and employment procedure of nonrenewal schemes.

probability. As a result, the security of the above schemes is paralyzed by attackers as the number of compromised nodes increases.

18.2.2 Reactive Renewal Schemes

Eltoweissy et al. [9] proposed an Exclusion Basis System (EBS) that defends a group key against evicted nodes in a communication group. An EBS server holds $k + m$ administrative keys to secure the group key update, and each group member obtains k administrative keys from the EBS server. If a group member is evicted, updating the current group key needs only m messages. Because the evicted nodes do not have m keys for decrypting m messages, they cannot participate in later communication of the group any more. In [10], Eltoweissy et al. applied the EBS to the group key management in wireless sensor networks. First, each sensor determines its cluster through the training content broadcasted by the BS. Each cluster manages one group key, and EBS is applied to the whole network to update the group keys. Jolly et al. [8] proposed a dynamic key management scheme which is not based on the EBS. The BS is responsible for key generation and assignment while gateways (CHs) are responsible for obtaining some needed keys through communication between neighboring gateways. This scheme reduces memory overhead of sensors (generally, two keys), but increases communication overhead significantly. This is because a key renewal in this scheme causes reformation of clusters and key redistribution. Younis et al. [11] pointed out that Eltoweissy's EBS is vulnerable to collusion attacks launched by a group of attackers. In order to resolve the problem, Younis et al. proposed a scheme, which was called Scalable, Hierarchical, Efficient, Location-aware, and Lightweight (SHELL). SHELL performs location-based key assignment in a cluster to decrease the number of keys revealed by the collusion of attackers. That is, nearby sensors in SHELL share more common administrative keys than distant sensors. SHELL performs the EBS-based key renewal, and the key renewal occurs only within each cluster. Eltoweissy proposed a scheme, so-called Localized Combinatorial Keying (LOCK), where the EBS is employed for key renewals not only between a CH and members but also between the BS and CHs [7]. Figure 18.2 shows the key distribution and renewal procedure of the reactive renewal schemes.

The above schemes perform reactive key renewal in the key management process. A key administrative server detects the compromise of nodes by itself or via the report of a veracious sensor(s), and performs the renewal of the exposed administrative keys. Therefore, a reactive key renewal scheme needs a matured Intrusion Detection System (IDS) for the successful operation. However, a matured IDS for clustered sensor networks was not yet implemented, and it is a challenging problem. In above schemes that employ the cluster architecture, the administrative key server (that is, CH) has a lower size of key pool than the nonrenewal schemes, and each sensor also has a lower size of key ring

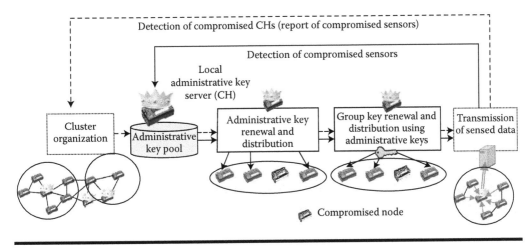

Figure 18.2 Key distribution and renewal procedure of reactive renewal schemes.

than the nonrenewal schemes. Therefore, the above schemes are much more vulnerable to the node compromise. That is, although attackers compromise a small number of nodes, they obtain many administrative keys exploited in the network. Furthermore, because each cluster employs only one key for communication, a compromised node can expose the key to attackers. Most importantly, the above schemes do not rotate the CH role nodes so that a CH is likely to be a compromise target of attackers.

18.2.3 Secure CH Election Schemes

Reactive renewal schemes directly renew the keys which are employed for communication using unexposed keys. However, they do not change the CH role nodes and just redistributes the pure sensors to pure remaining CHs when a compromised CH is detected. If a node keeps playing the CH role, attackers can easily identify their compromise target and shower furious attacks to the node via collusion. This may result in the compromise of all CHs in a short time. To prevent this attack, CH nodes should be changed as frequently as possible. To this aim, the nodes should invoke the CH election procedure in a periodic manner. Even though a proactive key renewal scheme that employs the periodic CH elections was proposed in [12], the CH election method cannot prevent attackers from declaring themselves as CHs. In fact, the CH election method does not need to prevent the misbehavior by virtue of a well-functioning IDS. If such a well-functioning IDS is not available, a secure CH election method is essential in a key renewal scheme because attackers try to become a CH in the periodic CH elections.

Most of the existing CH election schemes elect a CH using a weight value such as ID [13], node degree [13], mobility [14], energy [15], and so on. If a node has a highest weight value among its neighbors, it becomes a CH and its neighbors become the members of the CH. Some nodes cannot determine their role immediately if they have a neighbor whose weight value is higher than them. In those cases, the nodes should wait for the higher weight node to determine its role. If all higher weight nodes determine their role, the waiting nodes then become CHs or members of other CH declaration nodes. The prominent problem of the weight-based CH election is that a compromised node can forge its weight value as if it has a highest weight value among the neighbors. As a result, the compromised node always becomes a CH, and illegally obtains all data from its members and even forges all data from its members toward the BS.

Sirivianos et al. [16] proposed a new type of CH election schemes which is called Secure Aggregator Node Election (SANE). SANE elects a CH randomly in a node group so that a compromised node cannot predict which node will become a CH in the group. This makes the forgery of weight values by attackers meaningless. Therefore, the CH election of SANE is more secure than that of any weight based schemes. In SANE, nodes are pregrouped into sectors and the nodes in each sector elect a CH without interfering other sectors' CH election. First, all nodes in a sector contribute to the generation of a sum. Next, they share the sum and divide the sum by the number of nodes in the sector. The remainder is the index that indicates the position of the CH role node in the sector. Because all nodes have the ordered list of IDs of members in the sector, they can easily agree on a CH role node with each other. In SANE, the generated sum is an important value for determining a CH. If a compromised node changes the generated sum by suppressing its contribution, the CH election result is also changed. This intentional change of CH election result sometimes allows a compromised node to become a CH. If a compromised node transmits its contribution value which is used for the common sum generation to a part of all nodes, the nonreceivers have a different CH election result. This produces many clusters in a sector so that the advantage of cluster architecture is severely impaired. So, a CH election scheme should prevent a compromised node from launching the above attacks.

18.3 Network and Threat Model

18.3.1 Network Model

In the network setting of SANE, sensors are assumed to be pregrouped into sectors and to know their assigned sector ID. It means that sensors are deployed in their assigned sector by some people. This manual deployment causes a lot of management overhead and cost, and it is unrealistic. In our network setting, sensors are deployed by an aircraft so that their deployment position is unpredictable. Once sensors are deployed, they are grouped into some sectors. The reason of grouping is to elect a CH in each sector independently from other sectors. In order to group sensors into some sectors, sensors should invoke a CH election scheme after deployment. After the sector formation, a physical network is transformed into some sectors and each sector has its own local manager, which is called sector manager hereafter. Generally, one sector has only one CH but sometimes one sector has many CHs when they suffer from attackers. Details about the sector formation are described in Section 18.4.1. The sector manager facilitates the pairwise key agreement between sector members when any two members have no common keys. Details about the pairwise key agreement are described in Section 18.4.2.

The network consists of one BS, some CHs, and many member sensors under the CHs as shown in Figure 18.3. All sensors are stationary nodes and can become a CH. Member sensors belong to only one CH and send their data to their CH. The CH collects data from member sensors and transmits the collected data to the BS using a fixed-spreading code and Carrier Sense Multiple Access (CSMA). Namely, a CH first sense the channel to check whether there is a transmission from a different CH or not. If the channel is busy by any other transmission, it should wait to transmit its data. Otherwise, it transmits its data to the BS using the BS spreading code. Note that each cluster exploits a different spreading code for intracluster transmission to avoid the intercluster interference.

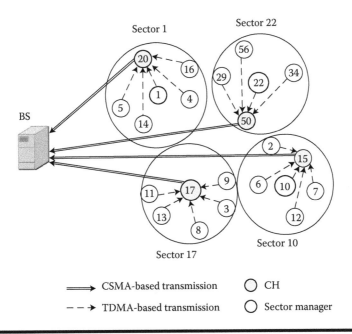

Figure 18.3   Network model of clustered sensor networks.

For the network longevity, each member transmits its data only in allowed time slots and enters to sleep mode during the remaining slots. To implement this energy-efficient communication, each CH broadcasts a Time Division Multiple Access (TDMA) schedule, and members transmit their sensed data to the CH directly according to the schedule. The BS has a large amount of available resources and energy and it is located in a sufficiently safe position which is free from various attacks. In contrast, CHs and member sensors have a small amount of energy and resources, and they are located in a very vulnerable position. In other words, CHs and members are likely to be compromised by attackers at any time.

18.3.2 Threat Model

First, the attackers try to illegally obtain ongoing data from the sensors to the BS. More importantly, they want to fabricate a large amount of data that are transmitted to the BS. Therefore, attackers compromise pure sensors and exploit the compromised sensors to carry out the above attacks. In a clustered sensor network, attackers must try to compromise not member sensors but CHs because they can have a great benefit by compromising the CHs. If a party of attackers compromises all CHs via their cooperation, they can obtain sensed data from all sensors and fabricate all sensed data. In that case, a user who makes a decision with the sensed data will have fabricated information.

Even though the confidentiality and the integrity of sensed data are both important in sensor networks, we need to give a preference to the integrity. This is because the illegitimate fabrication of sensed data makes a wrong decision of a user while the illegal acquisition of sensed data cannot do that. Assuming that pure sensors and compromised sensors coexist in a network, we need to minimize the illegal fabrication of sensed data by periodically and securely electing the CH nodes. Therefore, the aim of our key renewal scheme is to prevent compromised nodes from fabricating a lot of sensed data, even though they share the keys obtained from compromised nodes with each other. Another aim is to minimize the energy consumption for accommodating the key renewal routine.

18.4 Proactive Key Variation Owing to Dynamic Clustering Scheme

We present a novel key renewal scheme that changes the CH role nodes securely and consequently changes the keys employed for intracluster communication. Before the detailed description of our scheme, we assume the followings.

■ A predefined number of administrative keys and an individual key are assigned to all sensors before deployment. If a node identifies a different node's ID, it can also identify the assigned keys of the node. Each node employs the administrative keys to agree pairwise keys with other nodes in its two hop neighborhood at network boot-up time. A CH employs the individual key to communicate with the BS.

■ Nodes complete the sector formation and the pairwise key establishments in a very short time, so that an attacker cannot compromise a node during such a short period.

■ The clocks of BS and all nodes are initially synchronized. When the synchronized timer is expired, all nodes in a sector resynchronize their clock with other nodes in the same sector. Section 18.6 presents the details about the synchronization problem.

Our scheme is called as Proactive Key Variation Owing to Dynamic Clustering (PERIODIC) because it varies keys employed for communication using the periodic CH election instead of renewing the keys directly. PERIODIC consists of five steps; administrative key distribution, sector formation, pairwise key establishments within sectors, secure CH election, and transmission of sensed data. The administrative keys are distributed from the BS with a key pool to sensors before deployment. When the BS determines the IDs of keys which will be assigned to a node, it first generates a predetermined number of integers using a pseudorandom number generator and the node's ID as an input. The output of the pseudorandom number generator is the IDs of keys which will be assigned to the node. Both sector formation and pairwise key establishments are carried out only once after the deployment. The last two steps, such as secure CH election and transmission of sensed data, are performed periodically until the network is extinguished. Figure 18.4 shows the whole picture of PERIODIC. Details of all steps excluding administrative key distribution are described in the following sections.

18.4.1 Sector Formation

After the deployment, each node exchanges its ID with neighbors and makes a neighbor list. Then, each node exchanges the neighbor list with its neighbors again. Consequently, each node identifies the nodes which are at most two hops away and their preassigned keys. If any two nodes share at least one common key, they establish a pairwise key between them using the common key at the pairwise key establishments step.

After identifying two hop away nodes and their assigned keys, nodes should determine their sectors to separately elect a CH in each sector. For example, if Highest Connectivity Clustering Protocol (HCCP) [13] is used for the sector formation, a node first compares its degree (that is, the

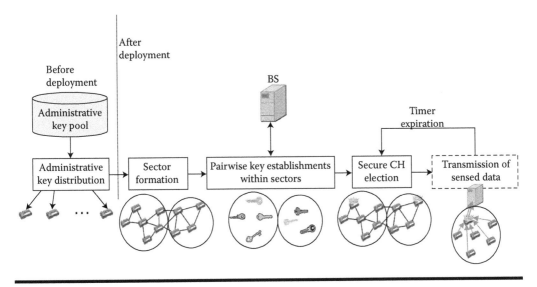

Figure 18.4 **Steps of PERIODIC.**

number of neighbors) with its neighbors. If it has a highest degree among its neighbors, it declares itself as a sector manager and broadcasts the manager declaration message. The neighbors become the members of the manager and transmit a join message to the sector manager. Otherwise, it waits for a node with a higher degree to send a manager declaration message or a join message for any other sector manager. When a node has already joined a sector and it receives a manager declaration message from another manager, it ignores the message and discards it. In general, a weight-based CH election scheme generates some single sectors. A single sector consists of only sector manager itself. Because these single sectors reduce the advantages of clustering, they have to be affiliated into other sectors. So, all single sectors are affiliated into neighboring sectors. Therefore, the hop distance between any two nodes in a sector extends to at most four hops. If the sector formation is completed, the sector managers register themselves into the BS. Each sector employs a unique spreading code to reduce intersector interference. Intrasector communication is performed using the spreading code and the code is assigned when the sector manager registers itself into the BS. For instance, the first sector manager to register is assigned the first code on a predefined list, and the second sector manager to register is assigned the second code, and so on.

18.4.2 *Pairwise Key Establishments within Sectors*

As described before, a randomized CH election is more secure than a weight-based CH election due to unpredictability. If a CH is randomly elected in a sector, any node can play a role of CH. Besides, PERIODIC employs the TDMA Medium Access Control (MAC) protocol in the communication between a CH and its members. Therefore, a CH and its members should communicate directly. This means that all nodes in a sector should establish pairwise keys with each other to protect their data. If any two nodes have common preassigned keys, they easily establish a pairwise key using them. However, some nodes may have no common keys to support the pairwise key establishment. These nodes are hereafter referred to as insecure nodes. In those cases, a proxy node that shares common keys with two insecure nodes can support an indirect pairwise key establishment between

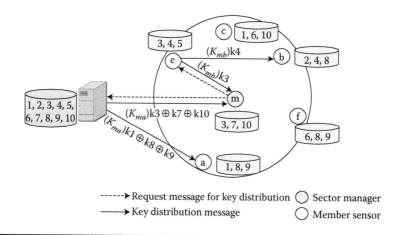

----→ Request message for key distribution ◯ Sector manager

——→ Key distribution message ◯ Member sensor

Figure 18.5   Pairwise key establishments between a sector manager and its members.

them. That is, a proxy node generates a key and sends it to two insecure nodes after encrypting the generated key with common preassigned keys, respectively. Because this indirect pairwise key establishment causes a lot of communication and computation overhead, individual invocation of the indirect pairwise key establishment at each node should be avoided. A sector manager that is directly connected to the most members can make the situation better. First, the sector manager generates pairwise keys with insecure members in its sector using the proxy nodes. For example, in Figure 18.5, the sector manager m establish a pairwise key (K_{mb}) with node b using the proxy node e. Each sector manager has already identified potential proxy nodes through exchanges of the ID and the neighbor list. If the sector manager cannot find a proxy node sharing common keys with an insecure sensor, the sector manager requires the BS to generate and distribute a key. Because the BS has all keys which are preassigned to all nodes in the network, it can fully support the request. For example, in Figure 18.5, the sector manager m requests the BS to generate and distribute a key to it and node a because it cannot find any proxy nodes. The BS generates a pairwise key (K_{ma}) and securely distributes the key to nodes m and a using shared common keys. However, this key establishment using the BS causes a lot of communication overhead because the distance between a sector manager and the BS is fairly long. So, it is desirable for a sector manager to establish pairwise keys with all nodes without the help of the BS. After the pairwise key establishments are completed, the sector manager broadcasts its member list. Each member establishes pairwise keys with its insecure members through the sector manager. That is, when a sector manager is requested to distribute a pairwise key by a member, it generates a pairwise key and distributes it to two members using pre-established pairwise keys.

Consequently, the success of pairwise key establishments within a sector highly relies on the success of pairwise key establishments between the sector manager and the members. For this reason, we need to analyze the success frequency of key establishments between a sector manager and its members as the probability of administrative key assignment and the sector formation scheme vary. The number of proxy nodes highly affects the success of the pairwise key establishments between a sector manager and the members. The number of proxy nodes is highly affected by the probability that an administrative key is assigned to a sensor from the key pool. If this probability is high, the number of proxy nodes rises up and the success frequency of the pairwise key establishments also rises up. Second, a CH election scheme which is used for sector formation varies the membership and

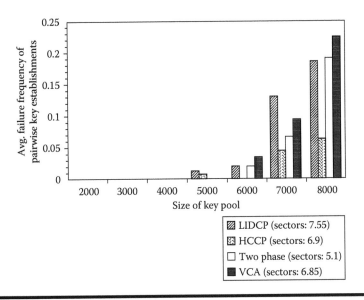

Figure 18.6 Failure frequency of pairwise key establishments versus key pool size.

size of sectors and consequently affects the number of proxy nodes. To prove the above suppositions, we built a simulation environment where 100 nodes with 50 keys were randomly deployed in a simulation area of 100 × 100 m. In the simulation setting, each sector manager tried to establish pairwise keys with its members. We varied the size of key pool to diversify the probability of administrative key assignment and employed four different CH election schemes (that is, LIDCP [13], HCCP [13], two phase [17], and VCA [15]) for sector formation.

Figure 18.6 shows the number of member nodes with which sector managers could not establish pairwise keys even via the proxy nodes. It means the failure frequency of pairwise key establishments between sector managers and their members. Figure 18.6 also shows the variation of the failure frequency which was caused by employing different CH election schemes. As shown in Figure 18.6, all schemes augment the failure frequency as the key pool size increases. This is because the probability that any two nodes share a common key lessens and consequently reduce the number of proxy nodes. Nevertheless, HCCP significantly reduces the failure frequency.

Figure 18.7 shows the variation of energy consumed for sector formation and pairwise key establishments between sector managers and their members as the key pool size increases. As shown in Figure 18.7, VCA consumes much more amounts of energy than three other schemes because all nodes have to exchange three messages during the sector formation. Three other schemes consume almost equal amount of energy for the sector formation and the pairwise key establishments. These simulation results indicate that HCCP is the best candidate for sector formation among various CH election schemes.

18.4.3 Secure CH Election

Before starting the CH election, each node sets its timer interval to a predefined value. The timer interval is long enough to accommodate data transmissions from sensors to the BS as well as the CH election step. Then, nodes should launch a secure CH election scheme to determine a CH role node among the nodes in their sector. Because the CH election is based on a random number

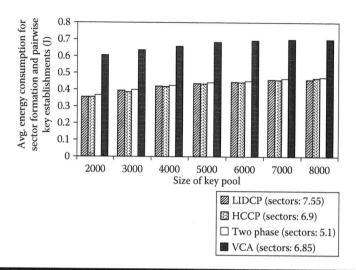

Figure 18.7 Energy consumption for sector formation and pairwise key establishments versus key pool size.

and performed in a periodic manner, the elected CH is likely to be changed with the lapse of time. Therefore, the communication keys between a CH and its members are also periodically changed. For the sake of simplicity, we assume that no MAC layer collisions occur among the nodes. The assumption can be materialized using a broadcast order that prescribes the message transmission order of nodes in a sector. Actually, PERIODIC generates a message transmission order and updates it for secure CH election. So, the order can be reused for the collision avoidance. Refer to Section 18.4.3.2 for details about the message transmission order. As shown in Section 18.2, SANE has some security flaws, that is, a malicious node can change the CH election result illegally. PERIODIC forces all nodes to follow the message transmission order to prevent the attack. Besides, a malicious node can break an agreement on the sum of random values among nodes by selective transmission. For example, a malicious node can easily materialize the selective transmission by lowering the power level of its transmission. In PERIODIC, a node estimates the received signal strength when it receives a message to defeat this agreement prevention attack. Figure 18.8 depicts the flowchart of the secure CH election through the ordered transmissions and the signal strength estimation.

18.4.3.1 Commitment Broadcast

Each node generates a random value and encrypts it with pairwise keys shared with other members in its sector. The encrypted random value is called a commitment and it plays a verifier role for the following fulfillment value. A member node generates the commitments as many as the number of other members for the broadcast transmission. Each member node makes a list of the commitments in the ascending order of all members' ID and transmits the list in a broadcast manner. The distance between any two nodes in a sector is at most four hops. So, a message containing the commitment list should be transmitted with a sufficient power level so as to reach four hops away nodes. Receiving nodes first check whether the message is originated from a member in their sector or not. If the message is not originated from a member in the same sector, the receiving nodes discard the message. Otherwise, the receiving nodes draw their commitment from the commitment list and decrypt it to store with the originator's ID.

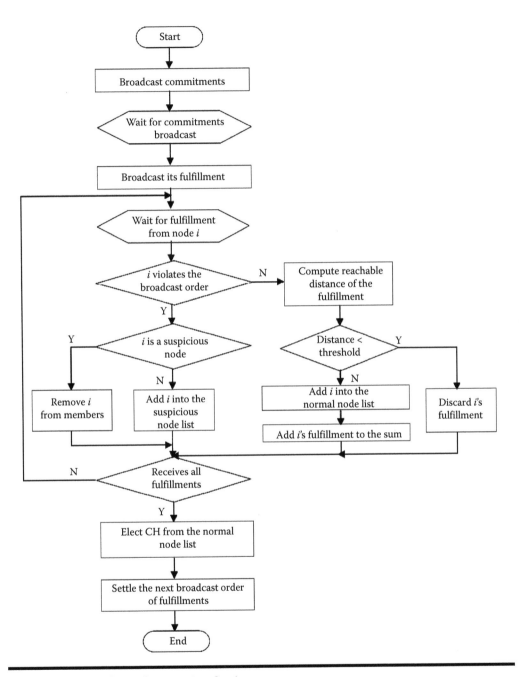

Figure 18.8   Flowchart of secure CH election.

18.4.3.2 Broadcast of Fulfillment Value

A commitment broadcasted by a member can contribute to the sum generation of random values only if the corresponding fulfillment value is received from the sender. For this purpose, each node broadcasts the plain random value that was used for commitment generation to other members. The plain random values are called fulfillment values and they are broadcasted according to the predetermined order. Initially, the order is settled with the ascending order of IDs of members. If a node does not follow the order, the node is recognized as a suspicious node and recorded into the suspicious node list. Besides, receivers drop the fulfillment value from a suspicious node. If a fulfillment value is received in the correct order, receivers verify the fulfillment value using the corresponding commitment. If the verification succeeds, the receivers store the sender into the normal node list. If a suspicious node violates the broadcast order again, the receivers remove it from the member list and the suspicious node list. Therefore, a compromised node can have just one chance to be able to forge the CH election result by passing its fulfillment transmission.

Even if a node transmits its fulfillment value, some nodes cannot receive it when the signal strength of the message is weaker than a specific level. If a fulfillment value is received by only a part of members, other members have a different sum of random values. It results in more CHs than one in a sector, and this result significantly impairs the aim of grouping nodes. Furthermore, some compromised nodes have another chance to become a CH by the generation of redundant CHs. To neutralize this attack in our scheme, a message sender is considered as a compromised node impairing the agreement property when the signal strength of the message is too weak to reach all other member nodes. Receivers can discover this trial by estimating the signal strength of each received message. The estimation of signal strength, however, can vary at each node because of an obstacle, a propagation error, and other reasons. To resolve this problem, a specific level of signal strength is set to a threshold. Transforming the threshold into the physical distance, the distance stands from three hops to four hops and it mostly converges to four hops. Even though this technique cannot prevent the agreement disruption attack perfectly, the number of redundant CHs is significantly reduced by using this technique. In other words, an attacker gets only a trivial benefit from launching the agreement disruption attack. Recall that attackers should transmit a message with a power level that exceeds three hops. Therefore, the number of redundant CHs is reduced. If a message can propagate over the threshold, the receivers preserve the message. Otherwise, legal nodes discard the message. We assume that the energy model in [18] is employed in the energy consumption of transmitters and receivers. Assuming that the two-ray ground reflection model is used for radio propagation, the transmission power of a sender transmitting a fulfillment value (Pt) can be calculated as follows:

$$P_t = \frac{P_r d^4 L}{G_t G_r h_t^2 h_r^2} \tag{18.1}$$

where P_r is the received power, d is the Euclidean distance, and L is the system loss. Besides, G_t and G_r are antenna gains and h_t and h_r are antenna heights.

If the transmission power of a sender is known to a receiving node, the node can transform the power to the maximum reachable distance (d_r) by

$$d_r = \sqrt{\sqrt{\frac{P_t}{E_{\text{two\_ray\_amp}} \times b}}} \tag{18.2}$$

where $E_{two\_ray\_amp}$ is the energy consumed by the amplifier and b is the channel bandwidth. If d_r is shorter than a predetermined threshold, the receivers drop the received fulfillment value. This refusal technique based on the signal strength estimation alleviates the impairment of the agreement property. Besides, the subsidiary threat by the generation of the redundant CHs (that is, increase of compromised CHs) is also reduced.

18.4.3.3 Sum Generation and CH Election

Once all random numbers are received from other members, all members generate a sum of the random numbers. Then they divide the sum by the number of normal nodes to obtain the remainder. Recall that all members keep the list of normal nodes that follow the broadcast order of fulfillment values correctly. The remainder indicates the position of the CH node in the normal node list.

Each node preserves the list of nodes that suppress the transmission of fulfillment value or violate the broadcast order. This suppression may be resulted from an attacker which tries to change a CH election result or a message loss. To deal with this situation flexibly, each node includes these nodes into the suspicious node list at first. Later, if these nodes pass their fulfillment transmission again, each normal node expels them from the member list. So, they cannot join the CH election procedure in the sector any more.

18.4.3.4 Adjustment of Broadcast Order

Once a CH is elected among the normal nodes, each normal node coordinates the broadcast order of fulfillment values. First, each node puts the suspicious nodes to the front of the broadcast order and puts the normal nodes to the tail of the broadcast order. In other words, the broadcast order of the next round is generated by concatenating the suspicious node list and the normal node list. Even if some normal nodes can be considered as suspicious nodes due to message loss, the readjust of broadcast order gives them a chance to escape from the false accusation.

18.4.4 Transmission of Sensed Data

After the completion of CH election, each elected CH generates a TDMA schedule for its members and broadcasts it. All nodes in the jurisdiction of the CH compute their transmission times and rest times according to the schedule. The members send their data to their CH in their allowed time, and each CH gathers data from the members. Then the CHs send the gathered data to the BS. These data transmissions from sensors to the BS are repeated until the timer that was set at the beginning of the secure CH election step expires. If the timer is expired, each sensor re-invokes the secure CH election step.

18.5 Evaluation

We exploited the well-known simulator, namely ns-2 (version 2.27) [19], to evaluate the security and the energy efficiency of PERIODIC. In our simulation setting, 100 nodes were randomly deployed in a 100×100 m area and the BS was located in the position of (50, 175 m). The simulations employed the energy consumption model in [18]. We compared PERIODIC with a nonrenewal scheme and a reactive renewal scheme. We chose the Chan et al.'s [1] scheme as a representative of nonrenewal schemes and the SHELL [11] as a representative of reactive renewal schemes. We ran three different schemes 30 times for each number of compromised nodes and averaged the results to yield a representative value.

The operation of SHELL highly depends on the existence and correctness of an intrusion detection system on the network while PERIODIC does not rely on such a system. So, we executed PERIODIC 30 more times with the intrusion detection system which is applied to SHELL. These additional simulations were performed to know how the intrusion detection system affects the performance of PERIODIC. The simulation parameters and their values are listed in Table 18.1. As an EBS parameter, $k + m$ means the key pool size of cluster and k means the key ring size at each member node. In other words, in each cluster, $k + m$ keys are preserved and k keys are randomly distributed to each member node among them. Besides, m means the number of messages that are employed for a key renewal.

In Chan's scheme, any two neighbors establish a communication key using q common administrative keys, and the communication key is used for secure communication between them. If they share less than q common administrative keys, they have to search for a proxy node that shares common keys with both of them until such a node is found. This search causes a great deal of communication overhead. For this reason, we modified the Chan's scheme not to search such proxy nodes when the number of common keys is less than q. Besides, in Chan's scheme, the Minimum Transmission Energy (MTE) routing protocol [20] was employed for data delivery from sensors to the BS. In the MTE routing, all nodes determine the next hop node, considering the least energy consumption of transmission. In our modified version, the routing protocol takes account of the security as well. Namely, a node first finds out some candidates that share a communication key among its neighbors. Then the node designates the least energy consumption node among the candidates as the next hop node. If there are no candidates among the neighbors, it generates a

Table 18.1   Simulation Parameters

Parameter	Value
Simulation time	3600 seconds
Initial energy	10 Joules/battery
Bandwidth	1 Mbps
Data packet size	500 bytes
Packet header size	25 bytes
Number of compromised nodes (CHs)	10–50 (0–3: SHELL)
Compromise time distribution	Random, 0–900 seconds
Number of clusters	5 (SHELL)
EBS parameters ($k + m$)	7 + 3 (SHELL)
Key renewal period	20 seconds (SHELL)
NEIGHBOR radius	30 meters (PERIODIC)
Expiration time of cluster formation timer	60, 120, 180 seconds (PERIODIC)

communication key with the BS by executing the XOR operation with all preassigned administrative keys. Because the BS has already known all keys assigned to each node, it can easily agree the same key with the node. Then the node directly transmits its data to the BS with the key.

In addition, we modified the compromise detection and key renewal process in SHELL to take place at regular intervals (that is, 20 seconds) to reduce the energy and communication overhead. We considered the following metrics:

1. *Exposure rate:* the rate that represents how many data readings are exposed to compromised nodes. This metric is exploited to evaluate the confidentiality of a key renewal scheme.
2. *Fabrication rate:* the rate that represents how many data readings are fabricated by compromised nodes. This metric is exploited to evaluate the integrity of a key renewal scheme.
3. *Energy consumption:* this metric represents how much amount of energy is consumed for key renewal process per unit time. This metric is exploited to evaluate the energy efficiency of a key renewal scheme.
4. *Network lifetime:* this metric represents the network longevity. This metric is exploited to show how a key renewal scheme affects the availability of a network.

18.5.1 Security Evaluation

Figure 18.9 depicts the exposure rate of the sensor readings as the compromised nodes increase. The increase of compromised nodes also increases the exposure rate of all schemes. As shown in Figure 18.9a, SHELL seems to guarantee better confidentiality than two other schemes. However, the good performance highly relies on the good performance of an IDS. In SHELL, compromised sensors are all evicted from the network by redistributing only pure members to pure CHs. This eviction takes place whenever the BS detects a compromised CH. Therefore, the increase of compromised CHs rather makes the confidentiality better, as shown in Figure 18.9. Actually, assuming such a well-functioning IDS is impractical. If such an IDS is available, PERIODIC can get a great benefit through the IDS. This is because nodes can exclude the compromised nodes by rejecting messages from them in a CH election. In those cases, our scheme guarantees the best confidentiality because our scheme periodically invokes the CH election and expels the compromised nodes, as shown in Figure 18.9b.

Figure 18.10 depicts the fabrication rate of sensor readings as the compromised nodes increase. Chan's scheme shows a much higher fabrication rate than two other schemes. This is because the nodes transmit their data to the BS via multiple relay nodes. As a result, even if only a relay node is compromised, the data from the originator to the BS can be fabricated by the compromised relay node. Moreover, the fabrication rate increases with the increase of compromised nodes. SHELL sharply reduces the fabrication rate as compared to Chan's scheme. Nevertheless, its fabrication rate is so higher than PERIODIC. This is because SHELL excludes the compromised nodes only when a compromised CH is detected by the BS. When a member node is compromised, SHELL just updates the compromised keys without evicting the compromised nodes explicitly. SHELL first renews the administrative keys known to the compromised nodes and then renews the group key using the renewed administrative keys. This eviction scheme is functionally useless, if a CH is compromised after the key renewal process. In other words, compromised CHs can keep fabricating data readings from their members until the timer expiration.

PERIODIC greatly reduces the frequency at which a compromised node is elected as a CH. Therefore, the fabrication rate decreases sharply even if the compromised nodes augment. Note

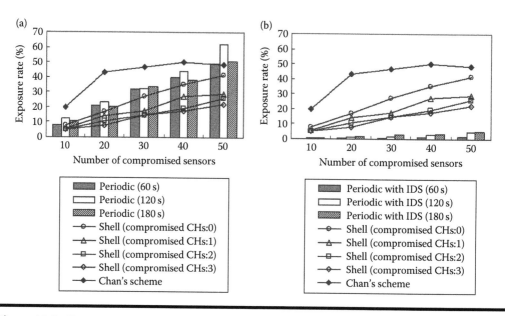

Figure 18.9 Exposure rate versus compromised nodes. (a) Exposure rate of PERIODIC without IDS. (b) Exposure rate of PERIODIC with IDS.

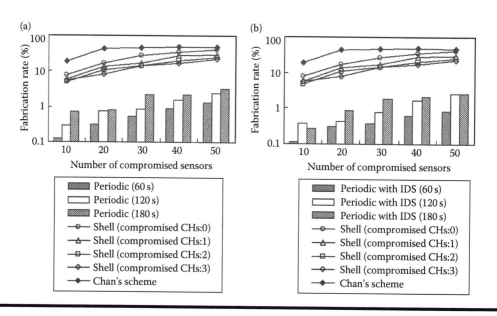

Figure 18.10 Fabrication rate versus compromised nodes. (a) Fabrication rate of PERIODIC without IDS. (b) Fabrication rate of PERIODIC with IDS.

that PERIODIC guarantees the best integrity for sensor readings regardless of IDS employment, as shown in Figure 18.10.

18.5.2 Efficiency Evaluation

Sensors perform their duty, employing a battery with a limited amount of power. Besides, there is no way to recharge them when they are operating in the duty field. This fact indicates that sensor nodes should reduce their energy consumption as significantly as possible when they work in the duty field. Therefore, a key renewal protocol also should be energy efficient. Below, we look into how much amount of energy the three key renewal schemes consume for the key renewal operation.

Figure 18.11 depicts the energy consumption for key renewals as the compromised nodes increase. In Chan's scheme, nodes never invoke a key renewal routine after they generate communication keys with their neighbors. For this reason, nodes consume almost constant amount of energy in spite of the increase of compromised nodes.

Contrarily, SHELL excludes the compromised nodes when the BS recognizes the compromised CHs. Because the excluded nodes do not consume their energy for the network operation, the exclusion produces the variation in the energy consumption. In SHELL, nodes consume more amounts of energy if the compromised nodes proliferate. This problem is caused by the fact that nodes should invoke the key renewal process more often. During the key renewal process, if the BS recognizes a compromised CH, it excludes the compromised CH by redistributing the orphan sensors to the pure CHs. Then the pure CHs distribute their administrative keys to the newly joined nodes before key renewal. This preliminary operation makes sensors consume a large amount of energy. However, if the number of compromised CHs further increases, the energy consumption of sensors rather decreases. This is because more compromised nodes are excluded from the network

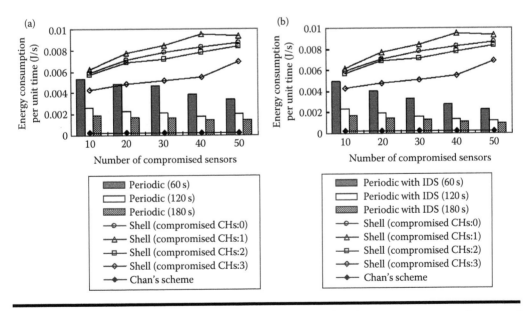

Figure 18.11 Energy consumption rate versus compromised nodes. Energy consumption rate of PERIODIC (a) without IDS and (b) with IDS.

as the BS detects more compromised CHs. Note that the excluded nodes (that is, compromised nodes) have no incentives to consume their energy if they cannot join the network operation.

In PERIODIC, the primary factor of energy consumption is the secure CH election step because sensors repeatedly invoke the step. In the step, all sensors consume their energy for two message transmissions and two message receptions. Energy consumptions for other steps such as the sector formation and the pairwise key establishments is trifling in the viewpoint of total energy consumption. Recall that those steps are launched just one time at network boot-up time. Therefore, PERIODIC dissipates less amount of energy for key renewals, as shown in Figure 18.11. As shown in Figure 18.11b, if PERIODIC employs an IDS, it can save more amount of energy. This is because the compromised nodes are evicted from the network every CH election period and they do not consume their energy for CH elections.

Figure 18.12 shows how a key renewal scheme influences on the network longevity as the compromised nodes increase. Chan's scheme does not have a key renewal mechanism and a way to evict the compromised nodes. As a result, the network lifetime is not greatly affected by the increase of compromised nodes. Contrarily, SHELL and PERIODIC evict the compromised nodes from the network. SHELL evicts all known compromised nodes whenever the BS detects a compromised CH, and therefore the number of evicted nodes is large. In PERIODIC without an IDS, one compromised node is evicted only if it avoids or delays transmitting its fulfillment value more than once. Therefore, the number of evicted nodes is much smaller than SHELL. The eviction of the compromised nodes decreases the active nodes operating in the network. Furthermore, because the active nodes will keep consuming their energy for key renewals and transmission of their data, the active nodes continuously shrink as time goes by. If the number of active nodes is equal to the number of CHs, the network is terminated.

In SHELL, whenever the BS detects the compromised CHs, it evicts the compromised nodes as well as the compromised CHs through redistribution of pure sensors. Therefore, if the compromised CHs increase, the evicted nodes also increase. However, the network lifetime rather increases

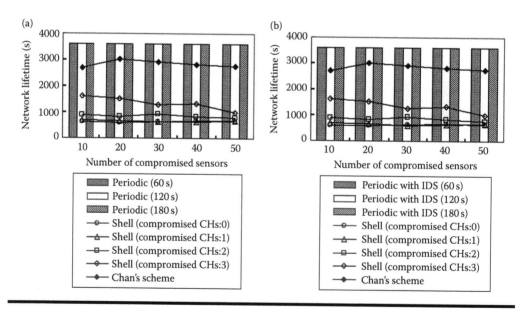

Figure 18.12 Network lifetime versus compromised nodes. (a) Network lifetime of PERIODIC without IDS. (b) Network lifetime of PERIODIC with IDS.

as shown in Figure 18.12. This is because the network lifetime in a clustered sensor network is contingent on not the number of evicted nodes but the number of remaining pure CHs. In other words, if one CH is compromised, then the orphan sensors are redistributed to the remaining pure four CHs. In this case, all orphan sensors are properly divided into four remaining CHs, so that the number of sensors which a pure CH manages is small. Note that the size of clusters determines the length of TDMA schedule of clusters. The smaller the size of a cluster is, smaller is the length of TDMA schedule in the cluster. If a cluster has a shorter TDMA schedule, the sensors in the cluster frequently send their data to the CH. Therefore, the sensors exhaust their energy quickly, and the active nodes rapidly shrink. Contrarily, if three CHs are compromised, two pure CHs should take over the remaining pure sensors. This division makes the TDMA schedule in a cluster very long and the transmission frequency of the sensors in a cluster dwindles significantly. This slows down the energy exhaustion of sensors and the number of active nodes decreases deliberately.

PERIODIC significantly extends the network lifetime by comparison with SHELL regardless of timer intervals. There are two main reasons for the lifetime extension. PERIODIC evicts only a compromised node if and only if it violates the broadcast order of fulfillment values more than once. Therefore, the evicted nodes do not increase significantly with the lapse of time. Besides, in PERIODIC, sensors consume only a small amount of energy for key renewals as compared with SHELL, as shown in Figure 18.11.

In summary, even if SHELL enhances the confidentiality and the integrity of sensed data by comparison with Chan's scheme, it also suffers from the increase of compromised nodes. Especially, as the number of compromised CHs increases, its confidentiality and integrity severely deteriorates. On the contrary, PERIODIC has strong resilience against node compromise. Especially, if it operates with an IDS, it guarantees the best confidentiality and integrity as compared with SHELL and Chan's scheme while it reduces the energy consumption of sensors significantly. Table 18.2 compares three key renewal schemes qualitatively, and summarizes various features of them.

18.6 Synchronization and Scalability

In general, sensor nodes should synchronize their clock with other nodes every CH election round. This is because the nodes should launch the CH election process at the same time to make an election result through a series of message exchanges. In the weight-based CH election schemes, each node should synchronize its clock with all other nodes and it is called as the global synchronization. The global synchronization is essential for a weight-based scheme because a ripple effect in the CH elections happens frequently. The ripple effect means that a CH election result in a region affects other CH election results in other regions. Again, the CH election results may affect the CH elections of other nodes which are waiting for the results. On the contrary, in PERIODIC, nodes do not need to synchronize their clock with all other nodes in the network. PERIODIC requires only a local synchronization among the nodes in the same sector. This is because a CH election result in a sector does not invade the CH elections of other sectors. Up to now, a number of literatures have dealt with the local and global synchronization schemes for sensor networks. Especially, Sun et al. [21] proposed a local synchronization scheme between nodes that share a pairwise key and a global synchronization scheme using μTESLA, which is referred as TinySerSync. PERIODIC can employ the local synchronization scheme of TinySerSync because the nodes in a sector share a pairwise key with each other.

In a clustered sensor network, usual network operation is divided into multiple rounds. Each round consists of cluster formation phase and data transmission phase. To support the addition of

Table 18.2 Qualitative Comparison of Key Renewal Schemes

	Chan's Scheme	SHELL	PERIODIC
Key pool usage	Large sized key pool	Small sized key pool	Large sized key pool
Key distribution	Predistribution before deployment	Redistribution as needed	Predistribution before deployment
Employment of cluster architecture	No need	Large-sized clusters	Medium-sized clusters
Key renewal type	No renewal Although it is available, it causes lots of overhead	Reactive key renewal (on-demand manner)	Proactive key variation (periodical manner)
Key renewal method	Communication from BS to sensors if available	Exclusion Basis System (EBS)	Secure CH election
Additional system for key renewals	No need	A well-functioning intrusion detection system (essential)	A well-functioning intrusion detection system (inessential)
Communication overhead	If key renewal is available, it causes lots of overhead	Overhead for redistribution of pure sensors and reactive key renewals	Overhead for periodic CH election
Memory overhead per node	Predistributed keys and generated pairwise keys	Fewer predistributed keys and group key	Predistributed keys, generated pairwise keys, one individual key
Resilience against compromised nodes	Low	Very low (if the number of compromised nodes exceed a threshold, key renewal is useless)	Medium
Resilience against compromised CHs	N/A	Low (CH role nodes are not changed)	High
Dependence on an IDS	Low	High	Medium

new nodes, all sector managers broadcast a manager advertisement message at the end of every round. A newly added node transmits a join message to the nearest sector manager among them. Then a sector manager which receives the join messages establishes pairwise keys with the newly joined nodes using the method described in Section 18.4. If new nodes are added in a periodic manner, we need to add a new step which contains the above-mentioned actions to PERIODIC. The new step should be located between the pairwise key establishments step and the secure CH election step. Besides, the newly added step should be performed at regular intervals like the following two steps (that is, secure CH election and transmission of sensed data).

18.7 Future Directions for Research

Because the pairwise key establishment method in PERIODIC depends on the administrative keys of sensors, a coalition of some attackers can expose many pairwise keys employed for communication. That is the reason why the confidentiality of PERIODIC deteriorates significantly when an IDS is not in use. One promising solution is that the available time of the administrative keys is shortened and new nodes with new administrative keys should join the network frequently. This resolution again causes the problem of pairwise key establishments between old nodes and new nodes. Therefore, our future research focuses on a pairwise key establishment scheme between different generation sensors with different administrative keys.

If PERIODIC operates with a well-functioning IDS such as SHELL, it is a perfect scheme in respect of security as shown in the simulation results. However, designing such an IDS for clustered sensor networks is still one of challenging problems.

The CH election method in PERIODIC sometimes evicts innocent nodes due to the incorrectness of discriminating compromised nodes from pure ones. This incorrectness is mainly caused by the fact that the discrimination relies on the message reception in the error-prone communication environment. Therefore, we need to devise a CH election method that is immune to the error-prone communication.

We assumed that PERIODIC functions in the network environment where all sensor nodes are stationary. However, in some applications, sensors need to move for their duty fulfillment. So, it is meaningful to design a dynamic key management scheme that can deal with the node mobility in wireless sensor networks.

Some key management schemes [22,23] employ the public key cryptography for key management in *ad hoc* networks. However, the public key operations are so heavy that resource-constrained sensors cannot support such operations for a long time. Furthermore, the schemes require a lot of communication and computation overhead for key management. For this reason, applying the public key cryptography to the key management of real sensor networks is still a challenging problem.

18.8 Conclusions

In this chapter, we have presented a key renewal scheme that varies the keys employed for communication instead of replacing old communication keys with new ones. It implements the periodic key variation through the secure CH elections, so it is referred to as PERIODIC in this chapter. PERIODIC employs two techniques (estimation of received signal strength and adjustment of broadcast order) to elect a CH in a secure manner. These two techniques not only prevent the nodes from electing a compromised node as a CH but also alleviate the partition of a cluster. Simulation results prove that PERIODIC greatly improves the integrity of sensed data as compared to other key renewal schemes. They also show that PERIODIC with an IDS guarantees the best confidentiality over other key renewal schemes. Another simulation results prove that PERIODIC consumes the least amount of energy for key renewals and provides the longest network lifetime as compared to other key renewal schemes.

Terminologies

Key Management
Secure Cluster Head Election
Energy Efficiency
Key Renewal
Confidentiality
Integrity
Intrusion Detection System
Wireless Sensor Network
Availability

Questions and Sample Answers

1. Why do modern sensor networks need to have a key management scheme for their operation? Do you think that a novel key management scheme is sufficient to defend the network from attackers? Otherwise, what kind of system do we need to complement a key management scheme?
   Because sensor nodes sometimes operate in an unattended or even an adversarial environment, a key management scheme which can protect the sensed data using keys is essential. Key management is only a measure for preventing the invasion of attackers toward a network. If all keys employed in the network are exposed to attackers through compromised nodes, it cannot work at all. So, we need an intrusion detection system (IDS) which recognizes the compromised nodes and expels them.

2. What advantages can we have by employing cluster architecture in wireless sensor networks?
   Employing cluster architecture in sensor networks distributes the key management duty from one server (BS) to multiple local servers (CHs) and enables the energy-efficient TDMA communication in sensor networks.

3. We have learned three kinds of key renewal schemes. Explain the key renewal method of them.
   The three types of key renewal schemes are nonrenewal schemes, reactive renewal schemes, and proactive renewal schemes. In nonrenewal schemes, the keys of sensors are employed until the network is terminated. The reactive renewal schemes renew the keys employed for communication whenever a compromised node is detected. They replace the exposed keys with new keys using nonexposed keys. The proactive schemes change the keys employed for communication by periodically changing the CH role nodes instead of replacing old keys with new ones.

4. What types of problems do the reactive key renewal schemes cause in respect of security?
   A local administrative server (CH) has a smaller sized key pool and each sensor in the jurisdiction of the server has a smaller sized key ring than those of nonrenewal schemes. So, they are

more vulnerable to the increase of compromised nodes than the nonrenewal schemes. They employ only one key (group key) for communication in a cluster so that a compromised node can expose the communication key to attackers. Most importantly, they do not change the CH role nodes so that the CH role nodes are likely to be compromise targets of attackers. If all CHs are compromised, the network is laid under the control of attackers.

5. In a CH election, why is a random value-based scheme more secure than a weight-based scheme?

In a weight-based scheme, a CH election result is determined by the comparison of a weight value among nodes. However, the weight value can be forged by attackers that want to become a CH. Furthermore, there is no way to verify the weight value of other nodes. This makes the weight-based CH election vulnerable to various attacks. In a random value scheme, a CH election result is determined by a shared random value among nodes, so an attacker cannot predict which node will become a CH. This unpredictability limits the capability of attackers.

6. What types of attacks are available in the CH election which is based on a random value?

In a random value-based scheme, the random value is created by collecting and summing random numbers from all nodes in a cluster. A compromised node which is expected to lastly transmit its random number can change the random value by suppressing the transmission of its random number. Besides, a compromised node can reduce the power level for its message transmission so as to break an agreement of the sum of the random numbers among nodes. In that case, a cluster is split into multiple ones and the advantage of grouping nodes is weakened.

7. PERIODIC exploits two techniques to defeat the attacks which are delivered to the CH election scheme using a random value. Describe those two techniques.

To prevent attackers from changing a CH election result, PERIODIC keeps the broadcast order of message transmission and forces each node to follow it. If a node violates the order or suppresses its transmission more than once, the node is evicted from the network by all normal nodes. To mitigate the cluster split attack, PERIODIC refuse a message whose transmission power level is too weak to reach all other members.

8. In the evaluation of three key renewal schemes, SHELL showed the best confidentiality over other schemes. What guarantees the best confidentiality for SHELL? Why does PERIODIC expose so many sensor readings to attackers?

The best confidentiality of SHELL is resulted from the good performance of an IDS. If the IDS does not work well or its detection is incorrect, SHELL cannot guarantee such a good performance. PERIODIC employs the preassigned administrative keys for the pairwise key generation. Because the preassigned keys exist in other nodes in the network, they are likely to be exposed when the compromised nodes increase. That is, a coalition of some compromised nodes can expose many preassigned keys and consequently many pairwise keys for communication to attackers. Moreover, PERIODIC basically does not employ an IDS. This makes the confidentiality of PERIODIC deteriorate significantly.

9. Why does PERIODIC require only a local synchronization among nodes?

In PERIODIC, a CH election result in a sector does not affect the neighboring sectors. That is, a CH election in a sector is performed separately from neighboring sectors. Therefore, nodes in a sector need to synchronize their click with other nodes in the same sector.

10. We pointed out that the CH election method in PERIODIC sometimes evict innocent nodes from a network. Even though this is a necessary evil, attackers may employ it for a bad purpose. Suggest an idea to resolve the problem.

In a sector, each node can evaluate all other nodes by giving a grade. That is, each node gives a high grade to a normal nodes and a low grade to a suspicious node. Then each sensor broadcasts the list of grades for other members. According to the average of the grades, the nodes whose grades are lower than a threshold can be excluded from the network.

Author's Biography

Gicheol Wang received the BS degree in computer science from Gwangju University, Gwangju, Korea, in 1997, and the MS degree from Mokpo National University, Mokpo, Korea, in 2000, in computer science and statistics. He received his PhD in computer science and statistics from Chonbuk National University, Jeonju, Korea, in 2005.

He worked for CAIIT (Center for Advanced Image and Information Technology) at Chonbuk National University, Jeonju, Korea, as a postdoctoral research fellow, from January 2006 to Deccember 2007, for the Research Center for Ubiquitous Information Appliances at Chonnam National University, Gwangju, Korea, as a Postdoctoral Research Fellow, from January 2008 to December 2008. From January 2009, he joined to the Supercomputing Center at KISTI (Korea Institute of Science and Technology Information), Daejeon, Korea, and he is currently serving as a senior research scientist. His current research interests include *ad hoc* networks, sensor networks, vehicular networks, security of wireless networks, and mobile computing. He is a member of IEEE, IEICE, KICS, KSII, and IEEK.

Gihwan Cho received his BS degree from Chonnam University, Gwangju, Korea, in 1985, and the MS degree from Seoul National University, Seoul, Korea, in 1987, both in computer science and statistics. He received PhD in computer science from University of Newcastle, Newcastle Upon Tyne, England, in 1996.

He worked for ETRI (Electronics and Telecommunications Research Institute), Daejeon, Korea, as a Senior Member of Technical Staff from September 1987 to August 1997, for the department of computer science at Mokpo National University, Mokpo, Korea, as a full-time lecture from September 1997 to February 1999. From March 1999, he joined to the division of computer science and engineering at Chonbuk National University, Chonju, Korea, and he is currently serving as a professor and chairman of the CH division. His current research interests include mobile computing, computer communication, security of wireless networks, sensor networks, and distributed computing system.

References

1. H. Chan, A. Perrig, and D. Song, Random key predistribution schemes for distributed sensor networks, in *Proceedings of IEEE Symposium Security and Privacy*, Berkeley, CA, May 11–14, 2003, pp. 192–213.
2. W. Du, et al., A key management scheme for wireless sensor networks using deployment knowledge, in *Proceedings of IEEE Infocom '04*, Hong Kong, March 7–11, 2004.
3. D. Liu, P. Ning, and W. Du, Group-based key pre-distribution in wireless sensor networks, in *Proceedings of 2005 ACM Wksp. Wireless Security (WiSe 2005)*, Cologne, Germany, September 2005, pp. 11–20.

4. P. Traynor et al., Establishing pair-wise keys in heterogeneous sensor networks, in *Proceedings of IEEE Infocom '06*, Barcelona, Catalunya, Spain, April 23–29, 2006.

5. S. Zhu, S. Setia, and S. Jajodia, LEAP: Efficient security mechanisms for large-scale distributed sensor networks, in *Proceedings of 10th ACM Conference on Computer and Comm. Security (CCS '03)*, Washington DC, October 2003, pp. 62–72.

6. L. Eschenauer and V. D. Gilgor, A key management scheme for distributed sensor networks, in *Proceedings of 9th ACM Conference on Computer and Communications Security*, November 2002, pp. 41–47.

7. M. Eltoweissy, M. Moharrum, and R. Mukkamala, Dynamic key management in sensor networks, *IEEE Communications Magazine*, 44(4), 122–130, 2006.

8. G. Jolly, et al., A low-energy key management protocol for wireless sensor networks, in *Proceedings of IEEE International Symposium on Computers and Communications (ISCC '03)*, Kemer-Antalya, Turkey, June 2003, pp. 335–340.

9. M. Eltoweissy, et al., Combinatorial optimization of group key management, *Journal of Network and Systems Management*, 12(1), 33–44, 2004.

10. M. Eltoweissy, et al., Group key management scheme for large-scale sensor networks, *Ad hoc Networks*, 3(5), 668–688, 2005.

11. M. Younis, K. Ghumman, and M. Eltoweissy, Location-aware combinatorial key management scheme for clustered sensor networks, *IEEE Transactions on Parallel and Distributed Systems*, 17(8), 865–882, 2006.

12. G. Wang and G. Cho, Clustering-based key renewals for clustered sensor networks, *IEICE Transactions on Communications*, E92.B(2), 612–615, 2009.

13. M. Gerla and C. Chiang, Multicluster, mobile, multimedia radio network, *ACM-Baltzer Journal Wireless Networks*, 1(3), 255–265, 1995.

14. P. Basu, N. Khan, and T.D.C. Little, A mobility based metric for clustering in mobile ad hoc networks, in *Proceedings of IEEE ICDCS 2001 Workshop on Wireless Networks and Mobile Computing*, Valencia, Spain, September 3–7, 2001, pp. 413–418.

15. M. Qin and R. Zimmermann, An energy-efficient voting-based clustering algorithm for sensor networks, in *Proceedings of 6th International Conference on Software Engineering, Artificial Intelligence, Networking and Parallel/Distributed Computing and 1st ACIS International Workshop on Self-Assembling Wireless Networks (SNPD/SAWN '05)*, Towson University, MD, May 23–25, 2005, pp. 444–451.

16. M. Sirivianos, et al., Non-manipulable aggregator node election protocols for wireless sensor networks, in *Proceedings of International Symposium on Modeling and Optimization in Mobile, Ad Hoc, and Wireless Networks (WiOpt '07)*, Cyprus, April 16–20, 2007, pp. 1–10.

17. K. Wang and G. Cho, Two phases based cluster formation scheme for mobile ad hoc networks, in *Proceedings of International Conference on Computational Science 2003 (ICCS 2003)*, LNCS 2657, Melbourne, Australia, and St. Petersburg, Russia, June 2–4, 2003, pp. 194–203.

18. W. Heinzelman, A. P. Chandrakasan, and H. Balakrishnan, An application-specific protocol architecture for wireless microsensor networks, *IEEE Transactions on Wireless Communications*, 1(4), 660–670, 2002.

19. The network simulator-ns-2, http://www.isi.edu/nsnam/ns/.

20. M. Ettus, System capacity, latency, and power consumption in multihop-routed SS-CDMA wireless networks, in *Proceedings of Radio and Wireless Conference (RAWCON)*, Colorado Springs, August 1998, pp. 55–58.

21. K. Sun, et al., TinySerSync: Secure and resilient time synchronization in wireless sensor networks, in *Proceedings of 13th ACM Conference on Communications*, Alexandria, VA, November 2006, pp. 264–277.

22. G. Wang, G. Cho, and S. Bang, A pair-wise key establishment scheme without predistributing keys for ad-hoc networks, in *Proceedings of 2005 IEEE International Conference on Communications (ICC 2005)*, vol. 5, May 2005, pp. 3520–3524.

23. V. Gupta, et al., Sizzle: A standard-based security architecture for the embedded internet, in *Proceedings of 3rd International Conference on Pervasive Computing and Communications (PerCom 2005)*, March 2005, pp. 247–256.

Chapter 19

Secure Routing Architectures Using Cross-Layer Information for Attack Avoidance (with Case Study on Wormhole Attacks)

James Harbin, Paul Mitchell, and Dave Pearce

Contents

19.1 Introduction

19.1.1 Overview of Sensor Networks

A wireless sensor network (WSN) is a distributed, autonomous network of communication and sensory nodes that collaborate to detect and relay information on phenomena under investigation to external coordinators (Figure 19.1). These networks were originally proposed for military applications [1], and have since been proposed and evaluated for deployment in several diverse domains [2], including environmental and habitat monitoring [3], industrial safety and hazardous environments [4], and volcano monitoring [5]. There are numerous economic and resource challenges concerning WSN implementation, and a considerable body of research has concentrated on their energy-efficient routing [6] and media access [7].

Example standardized sensor node platforms have been developed and deployed, and custom hardware has also been used to meet the energy and performance constraints of specific situations. The intent and objectives of a deployment suggest many different organizing paradigms for sensor networks, although a very common architecture for sensor networks, due mainly to the limited range of individual transceivers, requires nodes to collaboratively forward data on behalf of their

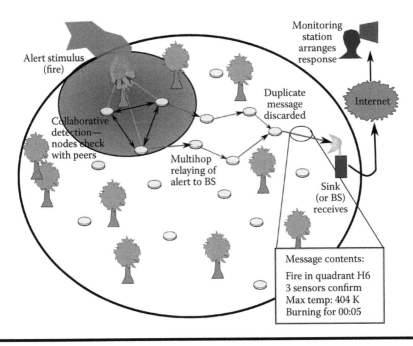

Figure 19.1   An example wireless sensor network scenario intended for an example application of forest fire monitoring.

peers via multihop routing. Some situations in which nodes are static and node failure rates are low concentrate upon initiating long-lived routes with an explicit establishment phase, while others favored in more dynamic and less predictable environments, such as wildlife-monitoring protocol mentioned previously by Juang et al. [3], concentrate upon database merging via flooding to deliver messages progressively to the sink node.

There are differences in the philosophy behind route formation, and the optimal choice is often driven by application-traffic demands. Reactive or on-demand routing, as used in the *Ad-Hoc* On-Demand Distance Vector (AODV) protocol [8], is often favored when traffic demands are unpredictable, despite the higher latency for route acquisition it presents. In contrast, proactive protocols, such as Gradient Broadcast (GRAB) [9] protocol, initiate route discovery centrally for the entire network during setup. The sensor networks considered in this chapter will be of the on-demand philosophy in which there exist explicit route setup and teardown phases initiated by the endpoints. This represents an endpoint of WSN evolution in which there exist not just the low-fidelity message notifications common in the current applications but high-fidelity data reporting streams, perhaps incorporating multimedia data. Under these conditions, it would be suboptimal to proactively allocate capacity evenly to all network nodes in advance of a usage request.

19.1.2 Overview of Sensor Network Security Issues

An often neglected but critical aspect in the deployment of a future-proof sensor network is consideration of the security issues involved. Sensor network protocols depend on distributed interaction to achieve the multihop routing that transports the data, which is dependent on all nodes involved in communication faithfully adhering to the intended protocol. But in the presence of malicious adversaries, such as enemy units in a military reconnaissance network, this assumption may fail to hold. Since nodes are generally unobserved, deployed without hand placement and under severe energy constraints (relative to the devices attackers could introduce), security considerations are of paramount importance in engineering viable future systems. It is not the case that security can be ignored in environments less threatened than a combat theatre; the history of the Internet shows that building into a communication system any dependency on unsecured protocols, or failing to consider security in making a deployment, is inviting failure to meet the underlying purpose of the project. Any situation in which an economic motivation exists for an attack could become a conceivable target; such as a polluting company which may consider subverting a pollution monitoring network to be cheaper than modifying the systems to cease polluting. Furthermore, in a network in which there exist standardized mode platforms that ultimately become so cost-effective as to be universally available and disposable, it is possible that attacks could become based on standard templates and using commodity hardware, mounted with no motive other than mindless vandalism.

Attackers may be able to deploy malicious nodes to subvert or otherwise defeat the design objectives of the network, via mechanisms such as signal jamming, flooding with traffic, advertising malicious routes, dropping traffic, forging application or control packets, or interfering with timing to prevent correct operation. The end result of these activities is that control over the network's stability and performance is transferred to the attacker. These approaches can be ranked as threats based upon the disruption caused relative to the costs in effort and equipment incurred by the attacker. Accordingly, an attack based on scattering a multitude of short-range jamming units that managed to disable half of all network nodes would not be as serious a threat as an attack with only two devices that managed to cut off network communications entirely by exploiting routing dynamics. This illustrates that subtle attacks concentrating on higher layer protocols can prove more destructive, relative to the effort involved, than those focused on the physical layer. Sending out a single one-time forged routing advertisement of the presence of a base station can disable the network-wide connectivity in proactive protocols by directing traffic to a nonexistent sink, while physical layer disruption by signal jamming would require the continuous expenditure of significantly greater energy.

The objectives of security engineering generally amount to protecting some or all of a primary trio of properties: confidentiality, integrity, and availability of all or some network messages [10]. Confidentiality refers to protecting the secrecy of a message and ensuring it is not read by the adversary. Integrity refers to protecting messages from malicious modification while in transit across the system, and the availability refers to ensuring that the attacker cannot deny an intended communication service. A security policy for a system may not demand fulfillment of every one of these objectives, for example, a forest fire detection system may not be designed to ensure message confidentiality as long as integrity is preserved to prevent malicious messages from being injected or delivered.

19.1.3 Techniques for Defending WSN Systems

In many communication systems, cryptography is relied upon to preserve confidentiality and integrity, and protection of availability relies upon defending all dependencies of system architecture (e.g., power, communications spectrum and so on, endpoints and backbone) from attacks that would render them unavailable. It is impossible to achieve a 100% availability in such systems or indeed any system, and particularly so in wireless systems due to the shared nature of the radio spectrum employed. Guaranteeing protection of the availability of the radio spectrum by ensuring the removal of all signal jammers would probably require an economic expense as costly as manual data gathering, or at least the deployment of a wired infrastructure for communications and power, negating the benefits of spontaneous deployment for data gathering in the first place. Therefore, the best that can be planned for in deployment is a graceful degradation in which attacks that disrupt the whole network availability are made resource prohibitive under the worst expected threat considered likely, and that a failure under greater levels of attack is anticipated.

In communication systems, cryptography can be relied upon to authenticate messages and the identity of nodes providing them, provided the algorithms are of sufficient strength and the keys appropriately distributed and protected. Public-key cryptography has widely been deployed on the Internet as it solves the key distribution problem; the need for both endpoints to share an identical key, and the challenges of keeping these keys fresh and safe from interception when individual nodes throughout the system may be compromised. Using the inherent separation of encryption and decryption keys in a public-key framework, the decryption key is held only by the recipient, and it is not possible for decryption keys to be compromised or otherwise extracted from any arbitrary stolen device, helping to ensure that decryption keys remain protected and confidentiality is provided. However, high computational requirements mean that public-key cryptography performed using algorithms other than low-modulus elliptic-curve cryptography is normally energy prohibitive for deployment in sensor networks [11]. Therefore, it is clear that even if they find limited applicability in WSN scenarios, these public-key algorithms must not be relied upon as a panacea and must be restricted to occasional operations, or to help establish symmetric keying schemes.

19.1.4 The Wormhole Attack

Attacks also exist that cannot be countered by cryptographic means, and in this respect a particular challenge in sensor network security comes from the presence of unauthorized elements. A troublesome attack in this regard is the wormhole attack. An operating wormhole of this form is illustrated in Figure 19.2. In the simplest case (the passive case), a malicious attacker deploys one or more pairs of energy-rich external devices that utilize a private (out-of-band) low-latency channel between paired endpoints. Each device listens to the channel in promiscuous mode and intercepts any packets heard locally, tunnels them across the private channel to the other wormhole endpoint, and rebroadcasts them at this remote endpoint. During the update phase of next neighbor discovery, this effectively distorts the network topology by causing the nodes in the regions within the range of the endpoints to consider themselves as direct neighbors. If the wormhole provides a shortcut across a long-range region then, under unsecured protocols which reward shortest paths, the wormhole will feature strongly in the ensemble of routes eventually formed. This transfers the control over network availability to the attacker, who is then free to mount any one of a number of further application-specific attacks. Denial of service can be implemented by suddenly severing the wormhole link, and a partial and a more stealthy denial of service can be implemented via selective traffic dropping or filtering. This is particularly insidious as it may give network operators the impression that the network is still operating properly, due to the continued availability of maintenance traffic or other ordinary traffic over the link.

19.1.4.1 Wormhole Attack Classifications

There is also a distinction between operating wormhole types as considered in this work. The type defined is a passive wormhole, which is generated when an entirely separate device is introduced to the network that merely intercepts all transmissions on a detected frequency and rebroadcasts them without any knowledge of their contents. This is the simplest type of wormhole to implement for

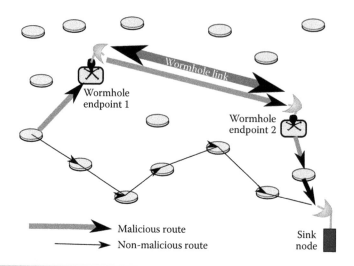

Figure 19.2 A diagram of the wormhole attack (passive) in operation. The malicious wormhole allows distant nodes to form a peer relationship, thus encouraging the formation of malicious routes through this "shortcut" path. The malicious route encourages the source to consider itself only a hop away from the sink, even when a nonmalicious route is much further away in hop count.

an attacker as it can operate using any hardware platform capable of monitoring the radio spectrum and does not require compromise and reprogramming of an existing device. An active wormhole, in contrast, is one in which a device originally introduced by the deployment authority (or a compatible one equipped with stolen network keys, if using a standard platform) has been modified via the addition of the private channel hardware to fulfill the wormhole role.

19.1.4.2 Response to Wormhole Attacks

These attacks may be timed so as to coincide with the phenomena under the control of an attacker which the network may be intended to detect, such as a fire or enemy military attack. In this case, the objectives of the network will not be met, and failure of the overall network-wide security policy will be deemed to have occurred. Obviously, a key objective for network security is to isolate the wormholes to prevent them from participating in routing, or make the wormhole attack resource-prohibitive by ensuring that many devices would be needed to mount a successful attack. Although any system will have failure modes and can never be guaranteed to continue operating under sufficiently intense attack, a proactive approach is to ensure that all envisaged attacks are beyond the anticipated resources of attackers. A key security engineering principle that guides the design here with this is that of defense in depth. The motivation behind this approach is that a single defense layer may be relatively easily overcome, but that a combination of layers provide much more effective protection as an attacker would have to breach all of them simultaneously to succeed in their malicious objectives. Accordingly, this motivates the use of multi-layered security techniques in which encrypted packets, for example, are combined with physical-layer techniques, and finally, routing avoidance aims to detect and avoid suspected regions whose properties suggest a potential threat.

19.1.5 Chapter Overview

The upcoming sections present a technique that attempts to provide a defense against the wormhole attack by routing preferentially around wormholes in the first place. The approach is entirely distributed and does not depend on details of timing, consume excessive system resources, or otherwise exhibit prohibitive computational complexity. It originally emerged from work at the University of York on improvement of network performance via routing around bottlenecks [12] and within this chapter will extensively be explored to discover its security benefits via wormhole mitigation.

19.2 Background

19.2.1 Background Overview

This section will survey the literature on wormhole detection approaches, examining the methods used, their operating principles and drawbacks. The wormhole attack in its WSN context is noted in Karlof and Wagner's paper [13] as part of their taxonomy of routing-level WSN attacks. This section will survey broad classifications and specific examples of solutions that have been proposed for the attack, their underlying philosophy of operation, and implementation considerations or other issues that may underlie them.

19.2.2 Packet Leashing Approaches

An early and logical solution to the wormhole attack is packet leashing [14], which relies on an underlying physical constraint which underlies all communications, namely the speed of light limit. This approach specifies a security module that can be run periodically to bound transmission range between the two nodes via interchange of a pair of specially designed test packets. The logic behind this approach is that the wormhole attempts to distort the topology via use of its private out-of-band channel invisible to network entities. However, unlike a hypothetical wormhole in physics that may enable instantaneous interactions between distant regions, this wormhole is bound by an inescapable physical limit, namely the speed of light delay. Imposing a time limit for communications serves to limit the length of the private channel, bounding the possible distance of interaction between two peers. By using a link range testing process regularly and rejecting links on which these "leashing" test response packets arrive late at their original sender, use of an out-of-band channel can be detected before the pair of endpoints can participate maliciously in any network operations.

The most accurate approach to packet leashing is the temporal synchronization variant. This approach ensures that accurate time synchronization exists via a global clock, and then makes a transmission from one node to another. By timing the return of the acknowledgement, it is possible to estimate a round-trip-time (RTT) and therefore the distance that was travelled by the packet. This allows wormhole links to be rejected, since the distance will be above a predefined threshold.

A major problem with the scheme is its reliance on the exact details of timing. The relatively small distances involved in sensor network communications mean that the time delay must be detected within extremely tight time constraints (e.g., within 67 nanoseconds to bound the remote peer distance to 10 meters). This is difficult to implement on a congested network, given the analog characteristics of low-cost sensor transceivers taking time to switch states, the drift of onboard oscillators and possible heavy contention for the surrounding channel. A further problem with this technique is that it is troublesome to implement in heterogeneous networks, since a maximum transmission threshold distance must be defined, beyond which rejection of the link will occur. Therefore, long-distance bridging links that are under the control of the deployment authority will not deliver their intended benefits to the network as they will be rejected as suspicious.

The geographic form of the packet leashing system verifies peer locations by processing information from a secure localization protocol, and only requires loose time synchronization to accomplish this. However, secure localization is itself a nontrivial problem, which imposes its own hardware requirements; normally, at least a certain proportion of beacon devices, together with tight collaboration to detect devices attempting to forge such localization approaches. Liu et al. [15] presents an approach to assist with this secure localization intent, which is, however, still vulnerable to wormhole attacks. If localization is required by the application, then the coordination overheads of this additional layer of complexity may be acceptable.

19.2.3 Approaches Involving Additional Hardware

Another approach involves the introduction of extra hardware, for example, directional antennas [16]. These antennas can bring an additional performance benefits to the network, and would not have to be manually aligned to an external reference since they can self-align using the Earth's magnetic field. The approach using them can detect the wormhole using the following technique. As shown in Figure 19.3, packets are marked with their transmission direction, and this can be extracted from the header and compared on reception to check that it is sensible. The diagram shows a transmission to the north–east quadrant, which will be accepted by the receiver if it is received from the south to east antenna. Figure 19.4 shows an example security situation in which traffic would be rejected. For example, if a packet is sent in the north–east quadrant and yet arrives at its receiver from a direction other than south–west, it is possible that a wormhole between them has rebroadcasted it improperly.

It is possible that a wormhole could exist in a region and simply adjust the direction markers embedded in packet headers, although presumably encryption would ensure the integrity of packets en-route. However, given that the presence of wormhole may impede even key-exchange, it is possible that a wormhole could take part in a man-in-the-middle attack that sought to interfere with the keys transmitted and thus defeat this approach, altering traffic on rebroadcast to defeat the detection or perform any other malicious traffic alteration.

A major drawback regarding this approach concerns operation in realistic propagation environments. The presence of obstacles in the topology could lead to nondirect signal paths, incorporating

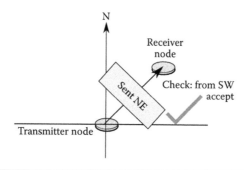

Figure 19.3   Normal operation of a network with directional antennas. With the antennas operating, the receiver can check the packet is received from the expected quadrant.

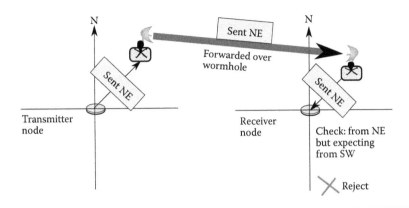

Figure 19.4 Rejection of wormhole traffic using the directional antennas approach. A transmission is made to the north–east but is forwarded over the wormhole and received from the north–east, not the south–west as expected. The receiver can reject the communication links as malicious.

reflection from obstacles, which could lead to the rejection of safe links that do not feature and wormholes and a reduced link density in the final secured topology.

19.2.4 Sink-Based Approaches

Some approaches bring the responsibilities for detection to the network sink or endpoint at the routing level, rather than relying on distributed isolation at the link level. The Statistical Analysis of Multipath (SAM) [17] approach attempts to analyze an ensemble of multipath routes formed, and detect any single links appearing with greater frequency than expected in the distribution. With information on any heterogeneity used being known at the point of detection, this would allow the network to isolate suspicious links and instead choose safer ones that are less likely to be involved in wormhole activity. This would also have the advantage of improving performance by reducing the contention for busy links. This approach does, however, make it difficult for the network to share any long-term information about the traffic levels that exist in the network, as the detection decisions are made entirely on the basis of a single routing request for this endpoint.

19.2.5 Graph-Theory Approaches

Another category of approaches use graph theory to reconstruct the topology and discover unusual properties of regions that could indicate a wormhole. For example, a graph-theoretic approach [18] attempts to discover unusual connectivity properties in a network, for example, direct connectivity between nodes that would otherwise require several hops to reach. This approach does, however, bear a large coordination burden, and may fail in real-world cases in which statistical shadowing means an accurate log-distance loss model is not followed and real multihop connectivity characteristics can be surprising. Similarly, one visualization approach [19] relies on global reconstruction of the entire topology, which carries an unacceptably high (cubic) time complexity. This would likely prove prohibitive for a future-proof algorithm for large-scale networks, but in segmented or hierarchical regions or on current test networks it could be viable.

19.2.6 Spectral Monitoring Approaches

One interesting approach for wormhole detection based on physical layers of the protocol stack is based on spectral properties of received signals [20]. This assumes the transmission of periodic link-state routing announcements under a protocol such as Optimized Link-State Routing (OLSR) [21], and requires low-level physical layer monitoring. The wormhole introduces a characteristic property into the measured power spectral density due to its inherent delay. However, this may not be appropriate for protocols in which widespread link-state flooding is not required, as it proves wasteful of energy and bandwidth in a network in which nodes remain idle until activated as sources. Also, modifications to the transceivers to analyze the spectrum in precise detail would be necessary, which would be beyond the capability of present-day sensor nodes. Unless such spectral monitoring was also a sensory-level goal of the network (e.g., a hostile interference warning network to track and monitor spectral conditions), then it would be unlikely that such high-fidelity monitoring would be realistic.

19.2.7 Unexpected Security Benefits from Wormholes

In engineering a complex system such as a sensor network, unexpected interactions may occur in which the wormhole may offer tangible benefits to the network. The wormhole does offer an immediate performance benefit by providing a high-capacity heterogeneous link across distant regions, which allows it to obtain its particular gains for the attacker at the cost of the security of the network. But interestingly, the presence of one or more wormholes in the network may sometimes offer an unexpected security benefit by mitigating other cruder classes of independent attack. Wormholes may present resilience to other classes of attacks due to their inherent propensity to tunnel signals across the network to geographically remote reasons [22]. For example, if a signal jamming attack exists, then it may be undermined by the presence of a wormhole carrying packets to a remote region free of jammers. This serves as a reminder that in a complex system or in situations in which multiple uncoordinated attacks exist the interactions between components may be unexpected and that a full system analysis must take these cross-phenomena interactions into account to assess security threats fully.

It is instructive to consider the approaches from the literature and how their detection methodologies relate to the proposed work. The scheme to be described in the upcoming section has some similarities to a distributed version of the Statistical Analysis of Multipath (SAM) scheme, building up network state highlighting the wormhole endpoints regardless of the destinations employed. This makes it highly extensible to multisink scenarios, without the burdens of high computational complexity or packet interchange to build up or maintain protocol state.

19.3 Current Research Progress

19.3.1 Introduction to Current Research

This section will describe an original routing-oriented protocol for disturbance-based routing; a class of approaches in which routing decisions take into account the number of peers affected by transmissions and the congestion resulting in order to assist in wormhole avoidance. Performance will be evaluated via simulation in a series of scenarios, presenting a discussion of the conclusions in an appropriate choice of metrics for different classes of scenario. It also provides an approach for deployment authorities and system integrators as to how best to tune a disturbance-based scheme to a particular topology.

19.3.2 Philosophy Behind Disturbance

19.3.2.1 Passive Wormholes and Static Disturbance

There is a notable feature of topologies that assists detection of any passive wormholes that may be operating within them. In the passive wormhole case, a wormhole which forwards neighbor discovery packets can convince nodes surrounding a pair of endpoints that they are all mutually peers of each other. As illustrated in Figure 19.5, this arises as neighbor discovery requests are tunneled across each wormhole endpoint, receiving replies from their recipients on the other site that convince the nodes surrounding the endpoints of their peer relationship. An active wormhole that is a modified network device will not usually have this effect upon connectivity, as it has packet inspection capabilities and can therefore mount a more sophisticated attack by ensuring that it broadcasts only the routing packets. The end result is that in the passive wormhole case the number of peers in the region is inflated, producing what appears as an unusually well-connected set of nodes. This will appear as a static feature of the network topology from the moment routes begin to be formed. This feature motivates an approach that can help in wormhole avoidance, particularly for passive wormholes. Avoiding the overhead of centralized computation to reconstruct this topology structure, discovery of these potential spots of improved connectivity can be integrated with the route discovery phase by routing based on a *static disturbance metric*.

This metric defines the cost of a link by the number of peers reachable from it, and thus rewards routes of low disturbance in which the path of nodes involved is surrounded by the fewest peers. In this view of avoidance, a wormhole is avoided as a consequence of the static disturbance its tunneling function would create to nodes around the endpoint, and this is used as an indicator of

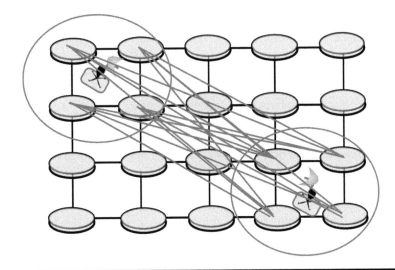

Figure 19.5 The presence of a passive wormhole inflates the number of peers of nodes in a region, appearing to create additional "virtual" malicious links between nodes surrounding the wormhole endpoints.

its potential presence to route away from the affected region. It is notable that in irregular topologies such approaches may not reveal the wormhole perfectly; depending on the degree of irregularity in the topology, regions of high connectivity may be perfectly natural, and wormholes may be missed. This would be a false-positive detection that leads, if the shortest safe path is considered optimal, to a suboptimal routing decision. However, depending on the network objectives it can potentially still be a good idea to route around these regions for the sake of performance, as using these densely connected areas under heavy network load would involve burdening the shared wireless medium in these areas and reducing the capacity available for others. Even if absolute energy consumption is the key criterion, it does not necessarily follow that using longer safe routes would exert an energy cost, as a reduction in collisions or retransmissions on a route occupied by fewer surrounding peers could actually lower the energy burdens. Therefore, the potential false-positive detection effects of this scheme can be mitigated by its performance benefits.

19.3.2.2 Dynamic Disturbance for Detection

The most notable dynamic feature of an ongoing wormhole attack will be the concentration of routes around the active endpoint, as a consequence of its attempts to draw in traffic and rush routing requests across the network. However, since the nonmalicious homogeneous nodes surrounding the endpoint have no extra resources to match the out-of-band channel of the wormhole, then under load there will be a rapid exhaustion of available capacity in the region. This provides a detection framework that can be applied to high-risk regions for routing, by detecting this clustering of traffic that may indicate the presence of a wormhole in the region. It is also the case that this route clustering exposes poor routing choices from a performance point of view, and therefore the network does not suffer extensively from a decision not to route through a suspected wormhole but instead obtains a possible performance benefit. It is logical that as a region becomes heavily loaded with traffic, it becomes a progressively poorer choice for further routing, especially when used with contention-based MAC protocols under which queue build-up and traffic dropping under heavy load is a possibility.

Dynamic disturbance-based routing algorithms serve to distribute traffic evenly throughout the network, leading to an even spread of energy consumption and the reduced chance of early network partition. Again, in some situations, it is possible that the absolute energy consumption which occurs would be increased in this approach, as it preferentially favors longer routes with more intermediaries over shorter routes through congested regions.

To a certain extent, physical constraints of the surrounding medium limit the network's dependence on a wormhole link, and therefore the control it can exert over the network. Under heavy contention, collisions, build-up of MAC layer queues, or slot exhaustion in scheduled schemes would serve to limit the wormhole's success at forwarding route discovery floods. However, this applies only under heavy instantaneous loads as would be seen in synchronized bursts of traffic, and if the duty cycle of these routes is low, a large number of routes may form through the wormhole before it can be detected. A more proactive approach that attempts to detect a concentration of developing routes by observing routing responses is needed, and a dynamic disturbance approach provides it.

The *dynamic disturbance* approach is based upon nodes overhearing a route request from their peers, which serves as an indication of increased route density in a region. This is used to update an activity factor that is used as an expected indication of occupied spectrum in a region. The concentration of these activity requests gives a low-cost approximation of general routing-level activity under shortest path protocols in a region, which is useful to reveal the location of any operating wormholes.

It must be pointed out, however, that if naively constructed an avoidance scheme may negate its own success, reaching a steady state in which there is a stable level of traffic through the wormhole. This arises via a negative feedback loop in which a naive avoidance scheme would negate the disturbance build-up that is integral to its own detection strategy. There are two approaches to solving this problem. The first would be is to separate routing operation network-wide into a set of distinct phases, a shortest path phase, and a disturbance-based phase. During the shortest path phase, noncritical traffic would be submitted for routing, and would be used to accumulate disturbance around the endpoints. It is accepted that during this phase, that network would not be secure against wormhole attacks. On expiry of a timer, the network would switch into disturbance-based routing mode, in which the disturbance metric would be used as the routing metric. On expiry of a further timer, network nodes would switch back into shortest path mode, and thus the cycle would repeat throughout operation of the network. However, this simple phased approach would lead to a number of significant problems. Most notable is the fact that it leads to globally correlated "insecure" phases in which the vulnerability is not challenged, and if secure routing is required during this phase, the service would be unavailable. This would add an unacceptable level of latency or risk in highly secure environments. Other problems relate to the difficulties of scheduling the routing phases globally, namely the possibility of the phases becoming unsynchronized without the intervention of external time synchronization. It would be an advantage if the network could operate without the requirement for an external time synchronization protocol. Therefore, a more sophisticated alternative to this phased approach has been chosen, and is described fully in Section 19.3.4 on protocol implementation (Figure 19.6).

Therefore, these two static and dynamic anomalies point to low-cost mechanisms of potential wormhole avoidance. Although these do not provide hard-line decision thresholds, they can add

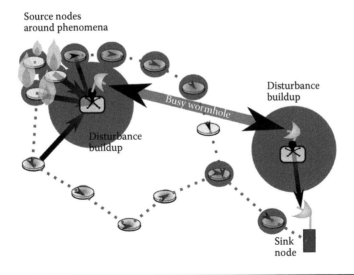

Figure 19.6 Wormhole avoidance by dynamic disturbance. A wormhole experiencing large amounts of dynamic disturbance around its endpoints (intensity of disturbance illustrated as cloud size) encourages routing through longer paths with lower disturbance values on intermediate nodes.

some resilience at the routing layer, contributing to the important security engineering principle of defense in depth. This holds that an effective security infrastructure should be multilayered to protect against failures in any one component, and a strategy that avoids exposure to potential wormholes in the first place is a useful constituent of this.

19.3.3 Metric Definition

19.3.3.1 Static Metric Definition

The *static disturbance* metric $SDYN_{i,j}$ for a link between nodes N_i and N_j is defined in Equation 19.1. The metric is static as it takes into account only the number of peers P_i of the transmitting node N_i. Here, a peer is defined as any adjacent node with which the current node may interchange intelligible messages, negating those that merely interfere destructively. In a real situation, this does not make any connectivity or localization assumptions, and would be obtained empirically from packet exchanges for neighbor discovery. The exponent α is a tuning factor that controls sensitivity of the metric to high node counts, in which high α leads to routing more aggressively around denser regions. In a network of approximately constant density, this will favor routing through the sparser regions at the edges of the network, and as discussed, will penalize the regions of increased peer count.

$$SDYN_{i,j} = P_i^{\alpha} \tag{19.1}$$

SDYN equation for static disturbance of link (i, j), where P_i is the peer count and α a tuneable constant to control sensitivity of avoidance.

19.3.3.2 Dynamic Metric Definition

In the dynamic disturbance case, disturbance imposed is tracked using a routing-level parameter, the *shortest path activity factor*, $SPAF_i$, at node N_i. SPAF is considered as a proportion of a constant maximum activity level and capped to this maximum, that is, $0 < SPAF_i \leq 1$. In a contention-based MAC scheme, $SPAF_i$ is derived by measuring the number of still-active shortest path route responses overheard by node N_i as a proportion of the maximum capacity of the medium in the region (in a homogeneous network, this can be computed by assuming that all nodes are transmitting on a channel at a constant rate). These routes are assumed to stay active for a fixed lifetime determined during setup requests, and upon expiry of a routing timer or explicit teardown message at N_i they are retired and the $SPAF_i$ decremented. This provides only an approximate estimate of capacity and relies upon the route bandwidth requirement being faithfully reported with the outgoing request. However, malicious attempts to insert additional disturbance anywhere else in the network to confuse the dynamic disturbance algorithm increases it around the malicious endpoint, due to the logic of the shortest-path routing protocol.

The metric for dynamic disturbance is defined in Equation 19.2, and uses an exponential function in which $DDYN_{i,j}$ becomes the basis β at full $SPAF_i$. The values of β employed must typically be much larger than the values used for α in the static scheme, as they must influence end-to-end total metrics even when SPAF is accumulating in only a small portion of the network (in fact, β values of several hundreds performed well during the simulations performed). High values of this exponent serve to make the avoidance performed by the scheme more proactive, ensuring it reacts earlier to the build up of congestion and reducing the time for the schemes to stabilize.

$$DDYN_{i,j} = \beta^{SPAF_i} \tag{19.2}$$

DDYN equation for static disturbance of link (i, j), where SPAF$_i$ is the activity factor overheard at node N_i, β is a tuneable constant to control sensitivity of avoidance.

19.3.4 Protocol Logic and Implementation

The static disturbance scheme is intended to build upon an existing protocol such as AODV, in which the static disturbance metric SDYN$_{i,j}$ would replace a unit value as the per-link metric from N_i to N_j. The path that minimizes the aggregate total metric is selected by the protocol as the chosen route.

In this protocol, nodes which did not have a route to the destination would rebroadcast the request, until it reaches either the destination or a node with a sufficiently fresh active route to the destination. Then, a response would be delivered following the reverse path, which would allow the originating node to extract the route and to choose the route with the lowest overall aggregate end-to-end metric. If the wormhole attempts to interfere with this routing logic by dropping or otherwise subverting traffic, then the wormhole impairs only its own participation in routing.

The dynamic disturbance scheme performs a more sophisticated process. The protocol actually finds a pair of routes to a desired endpoint for every route request. First, a shortest path route is discovered using a conventional routing protocol such as AODV to form a sacrificial route. These routes are not designed for the transit of application-level data but during their route response phases register load for the dynamic disturbance metrics. Since the wormhole would concentrate traffic sent under a shortest path protocol, its location will be revealed from its presence in these shortest path routes.

Therefore, in this security-focused scenario, dynamic disturbance technically refers to a hypothetical disturbance that would occur if all application traffic was to be sent over shortest path routes. It may be possible to make profitable use of these sacrificial routes, perhaps for carrying encrypted application traffic, as this would be a source of a redundant multipath flow that exploits some possible heterogeneity provided by the wormhole. Since routes controlled by an attacker are inherently unreliable, if higher guarantees of performance are required, it would be advisable to either ignore them for data transmission or inject disinformation (the exact nature of which would be application-specific) along these routes, thus confusing an attacker who managed to break confidentiality of the messages.

The second stage of dynamic disturbance routing is the stage which obtains the actual dynamic disturbance route. The logic of the protocol sums DDYN$_{i,j}$ along every possible path explored, to find the total cost incurred end-to-end upon a hypothetical route. The protocol finds these routes via a modified AODV protocol, in which an outgoing broadcast accumulating two metrics serves as a route request. A per-hop delay before rebroadcast would be applied to the route request at node N_i in proportion to DDYN$_{i,j}$. The optimal disturbance route is chosen on the basis of delay as well as metric, with rejection of responses after a certain timeout. This delay by the peers of the wormhole endpoint means that a node which tried to artificially lower disturbance values would have its attempts rejected as suspiciously late by the final destination, since it cannot control the delay added by the nonmalicious nodes experiencing high disturbance around it. Replies to both the shortest path route and the optimal disturbance route are unicast back along the reverse path. The unicast reply to the shortest path route updates the SPAF at each node visited to reflect an increased usage. This is the stage in the algorithm at which increases in dynamic disturbance are registered for future routing decisions, since SPAF is used to define the disturbance metrics. Nodes periodically scan their table of the routes that contributed to SPAF and after the endpoint of lifetime defined during setup is exceeded, decrement SPAF to reflect the reduced load upon the network.

A potential security vulnerability of this approach is its reliance upon the wormhole faithfully delivering the reverse unicast replies for the shortest path routes. Since the wormhole would have an interest in attempting to defeat the protocol, it is possible for it to prevent disturbance from accruing along earlier points in a path by dropping the first response seen. This is only an advantage for the attacker's aims if it is known that this disturbance scheme is being used. In its absence, dropping the first (and only) routing response would prevent the wormhole from taking any part whatsoever in routing. But if the attacker knows that the disturbance-based protocol is in use, then an extra security measure can be implemented. This approach involves dispatching a periodic flood from the final routing destination (typically the sink node in a WSN) at the *route update interval*. This route update consists of a single packet containing the largest node identities and their corresponding aggregate increments in SPAF for nodes on all routes found to that destination during that interval. The nodes and their peers apply these disturbance increments for the duration given in the packet.

This *route update flooding* mechanism serves to ensure the propagation of the most significant disturbance changes to all network entities, regardless of attempts by the wormhole to interfere by traffic-dropping. Since there are typically very few destinations in a sink-oriented sensor network relative to nodes themselves, this occasional flooding approach would not impose an excessive burden. This feature could potentially invite spoofing to maliciously adjust and alter disturbance values; however, in a network in only very few potential endpoints for routing exist, it would be possible to authenticate these update flood messages by using a lightweight authentication technique such as the one-way hash chain (OHC) to ensure that all messages are part of an approved series by a trusted authority such as a sink node.

19.3.5 Scenario Description

19.3.5.1 Scenario Introduction

The intent of this section is to provide an application case for a security-critical deployment scenario and specify assumptions about topologies, sources, and detection logic used in investigation of the disturbance-based schemes. The chosen deployment scenario is a military network as might be used in around a base or command station for the monitoring of the movements and behavior of enemy troops in the area.

The sink node represents a base or headquarters. The scenario assumes that the defender has set up a static network over a fixed area of terrain in which detection nodes are deployed. As is required of a sensor network deployment, the detection nodes are equipped with individual sensors capable of picking up suspicious indications of possible enemy presence such as sonic, motion, or thermal phenomena. As a proactive response, these notification warnings trigger the sensor nodes, which then enter a reporting phase, acquire a relatively long-lived route back to the sink node, and generate continuous data on any phenomena in the local region that might be ongoing. This is a useful example case as the routing protocol used fits the traffic dynamics of a security-sensitive scenario: on first detection of a potential anomaly, the nodes establish a persistent route to the base station, proactively scanning for any further threat and reporting continuously along the route all further information that may arise for detailed central analysis at the sink.

As the intent of this case study is to model avoidance strategies for the wormhole attack, the scenario also features a single wormhole installed by the malicious enemy. The enemy has freedom to place either wormhole endpoint anywhere within the topology, but would be anticipated to position its endpoints in a position of maximum advantage. The intent of the defender at their military base is to obtain as many wormhole-free routes to deliver as much accurate information as possible on enemy activity. The intent of the attacker is to ensure their troops can roam as widely as possible over the area and ensure that as many of the reporting flows use the wormhole as possible so they can disrupt or otherwise disable the defender's communications at this single point of attack. The dependence on the wormhole delivers them control over the defender's information on phenomena in the region, which they can use to their advantage.

19.3.5.2 Mobility Parameters

The mobility of the sources used in the scenario is a simple model. The enemy troops that initiate the route discovery at sensors in their region are assigned an initial velocity as if on patrol. Their speed remains constant throughout but their direction changes, in that they "reflect" off the boundary limits of the network with an angle of incidence equal to the angle of reflection. These mobile troops serve to initiate routing at the nearest node within a sensing range that does not have a route active.

19.3.5.3 Geometry Parameters for Deployment Region

The deployment region for the network topology is a square of side 2S, with the military base (sink node) at a constant location midway along the northern edge at $(0, S)$. The region is populated with homogeneous nodes. These nodes are placed upon a basis grid spanning the topology, perturbed by an additional uniform random variation. Accordingly, each has a position $(X + \Delta X, Y + \Delta Y)$ consisting of a contribution from the regular (grid) component (X, Y) and the variation $(\Delta X, \Delta Y)$. The variation is controlled by the *regularity factor* R_f, which produces a uniform distribution within the bound $|\Delta| = (1 - R_f)(S/2)$ for both axes. A situation in which $R_f = 1$ corresponds to a perfectly regular grid placement. It is generally difficult to obtain complete all-pairs connectivity throughout the network without excess peer clustering in some regions and disconnection in others when the regularity factor is low, so during simulation only scenarios in which $R_f > 0.6$ have been explored. Nodes may potentially lie outside of the boundaries of the placement grid if perturbed by a variation in that direction.

This approach allows the simulation to accurately reflect real deployment scenarios, as no matter how useful it would be for the defender to optimize coverage by a regular deployment, in obstructed regions it would be unlikely for perfect grid placement to be attainable due to the presence of physical obstructions, or the distortion in position during aerial deployment.

19.3.5.4 Wormhole Placement Parameters

As mentioned it is assumed that the attacker aims to obtain control over as many routes as possible. This motivates them attempting to place one of their wormhole endpoints close to the sink to maximize wormhole usage, and indeed, the scenarios explored all feature a single wormhole delivery endpoint located close to the sink. However, the options for placing the other endpoint are very varied, and the attacker could place their wormhole in any region they would like to direct their troops toward. This motivates the use of a uniform range strategy for pickup wormhole placement in these simulations, since it is best to attempt to secure the network for an ensemble of all possible wormhole placements when the exact one chosen is uncertain. Another approach is edge positioning, which may be a priority target to defend against as the attacker would have more opportunities to install a wormhole unobserved at the network boundaries out of the defender's control. Accordingly, the scenarios include modeling of wormhole pickup endpoints midway along the West, East and South regions of the network, at coordinates $(-S, 0)$, $(S, 0)$, and $(0, -S)$, respectively.

A sample simulation topology is illustrated in Figure 19.7, which illustrates a representative sample topology under simulation, with the military base as sink node on the midpoint of the northern edge, the wormhole pickup endpoint very close to it, and the other endpoint at the southern edge, with enemy troop clusters in motion in the region.

19.3.5.5 Simulation Logic

To assess the benefits of the disturbance-based scheme, simulation of the given scenario was performed using a custom simulator written in the OCaml language [23]. Input parameters for this simulator consist of a disturbance-based metric (the flag to indicate a STATIC or DYNAMIC scheme and an associated α or β value), a count of topologies to generate, wormhole type and location distribution, and the set of scenario definition parameters governing, among other factors, topology generation, and source behavior. The names and explanations of these parameters are specified in Table 19.1. Upon execution the simulator produces an output metric to indicate the metric's relative success in wormhole avoidance compared to pure shortest path routing. The simulation can be run in both active and passive wormhole modes, modeling the protection provided for different classes of wormhole endpoints.

The simulator generates an ensemble of topologies and connectivity between nodes is modeled according to a standard protocol model, assuming bidirectional binary connectivity within a given peer distance threshold P_R. This assumes that a link is connected if its endpoints lie within the range P_R, and disconnected if they are outside of this. For simplicity, the maximum sensing range S_R (within which the nodes will detect troops) is set equal to P_R.

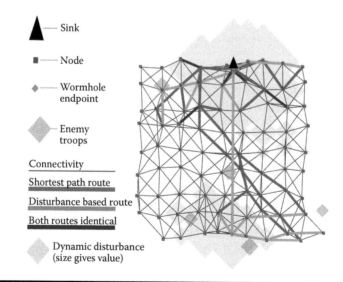

Figure 19.7 The dynamic disturbance routing protocol in operation during simulation. The shortest path routes (gray) are using the wormhole (link between northern and southern edges) but the disturbance-based (black) routes are avoiding it due to the higher disturbance by the wormhole endpoints.

Table 19.1   Default Parameter Values as Used in the Simulations

Parameter	Explanation	Default Value
Topology count	The number of custom topologies generated	200
Simulation runtime	The number of time spent running a particular simulation before it ends	600 seconds
Route interval	The interval at which routing metrics are updated	12 seconds
New flow interval	The interval at which the position of enemy troop clusters is updated and new flows generated	2 seconds
Node count	The number of ordinary data transmission nodes in the network	80
Peer threshold P_R	The distance for communication between nodes, and the maximum range for detection of an approach troop cluster	12 meters
Size limit	The size of the coordinate of the each axis at the edge (half of the perpendicular distance across the topology)	30 meters
Maximum channel rate	The channel data rate of the transceivers used upon the nodes	250 kbits per second
Rate per flow	The date rate assumed on each route that is initiated	5 kbits per second
Regularity factor R_f	The regularity factor that is assumed if no other is provided for the results	0.8

Note: Specific values may be varied in a particular result set, but if not mentioned they are as specified.

Periodically, at the *new flow interval* during simulation, the fresh position of each troop cluster is recomputed using their known velocity, previous position, and the time delay. The simulator checks for idle sources surrounding each troop cluster within S_R, and proceeds to activate the nearest idle source for reporting. This chosen node executes the disturbance-based routing protocol to begin initiating data. Also, at this new flow interval, any expired flows will be removed and the nodes that originated them return to idle. At the *route update interval*, any aggregate statistics on packet flow are updated and the dynamic disturbance metrics, if they are in use, are recomputed, reducing SPAF for any nodes around the completed flows. After a specified time limit, the simulation is suspended.

19.3.5.6 Success Metric Tracked by a Simulator

Upon the end of simulation, the simulator records a success metric to assess the disturbance metric given for analysis. The main avoidance advantage ADV is the additional proportion of all discovered routes throughout the simulation that successfully avoid the wormhole under the disturbance scheme as compared to shortest path. A value of zero corresponds to a state in which the disturbance-based scheme delivers no advantage, as precisely the same proportion of routes used the wormhole as in shortest path routing. In the unlikely case that the disturbance-based routes are actually more susceptible to use the wormhole, the value of ADV will be negative rather than positive. This could occur, for example, with a static disturbance metric in which avoidance preferentially routed on an edge pathway that happened to lead in that particular topology to an active wormhole.

19.3.6 Results

In this section, simulation results will be presented to verify the advantages in wormhole avoidance delivered by the disturbance-based protocols. In the following results, the avoidance advantage ADV is presented as cumulative frequency distributions across an ensemble of topologies, which can incorporate variation in the node placements, wormhole placements, and any other topics which may be specified as nondeterministic in the scenario. The reason for this is to ensure that conclusions obtained are not tailored specifically to a single generated topology, but to the ensemble with the intended variation. This ensures that the results are valid for generated topologies at that regularity R_f and the intended wormhole placement strategy, instead of specific to some particular node arrangement. Table 19.1 shows the values of any default parameters that are assumed in the following result sets.

19.3.6.1 Varying the Static Routing Exponents

Figure 19.8 shows the cumulative frequency distribution of avoidance advantage with passive wormhole pickup endpoints uniformly randomly distributed around the generated topologies, for a variety of static routing exponents. The results show that the avoidance advantage is distributed over the higher values by greater exponents, and therefore that these exponents deliver the greatest avoidance advantages. This is as expected since the design of the metric intended that increasing α indicates that the routing would be more sensitive to suspected wormholes. However, in this case, the clustering of the curves for $\alpha = 5$ toward that of $\alpha = 10$ indicates that a point of diminishing returns is being approached, and especially on power-limited sensor nodes, higher exponents would require additional floating-point multiplications, which are unnecessary. Therefore, there is little point in using higher values than $\alpha = 10$ in a long-lived scenario in which the disturbance metrics have time to reach a steady state which reveals the wormhole (typically at around 90 seconds during simulations).

19.3.6.2 Using Static Disturbance in Topologies of Varying Regularity

The static disturbance technique, although it may not build up knowledge of wormhole locations from traffic flow patterns, was shown in Section 19.3.6.1 to gain some information on passive wormhole locations. Varying the regularity is useful to investigate the performance of static techniques against passive wormholes in a variety of topology characteristics. Figure 19.9 displays the performance of a static disturbance metric with $\alpha = 3.0$ in a range of topology regularities. The close clustering of the CDF plots for $R_f < 0.9$ shows that there is little variation in avoidance success, but in highly regular topologies success increases dramatically with a more consistent avoidance performance. This is explicable from the difficulty of distinguishing the wormhole from natural connectivity variations. In a passive wormhole scenario, the peer inflation effect which reveals the wormhole will be a highly significant anomaly in a perfectly regular topology, in which nearly every node has a constant number of neighbors. However, the connectivity variations brought about by

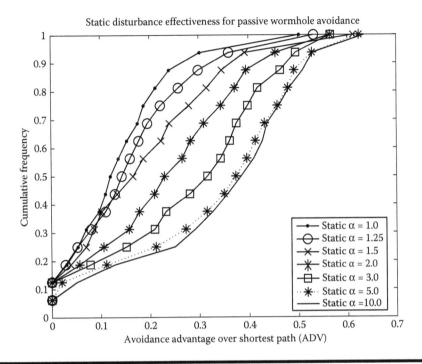

Figure 19.8 Static routing avoidance advantage for passive wormholes, showing better performance for higher exponents.

random placement serve to disguise the wormhole amidst fluctuations in connectivity. This indicates that those deploying a network using static disturbance protocols should assume only protection against passive wormholes if they can ensure this highly homogeneous connectivity by regular placement. It must be pointed out that even if such regular placement is possible in the location, then shadowing and other statistical link properties may disturb the connectivity experienced by the nodes in a real implementation.

19.3.6.3 Relative Performance of Static and Dynamic Disturbance

Figure 19.10 shows a contrast between the effectiveness of static and dynamic metrics at avoidance. The results obtained show that the dynamic metric obtains much better performance under the active wormhole case, whereas static disturbance routing obtains an equivalent performance only under passive wormholes. In this scheme, any performance advantage that exists for static disturbance vanishes under active wormholes, as only the peer inflation effect is present. Also, the wormhole endpoint placement for this scheme was uniform random, and the specific topologies on which static routing performed well were those in which wormhole was located more centrally. Therefore, the static disturbance scheme obtained most of its avoidance gains due to static routing favoring network edges where the wormhole was less likely to be located, rather than knowledge obtained about the wormhole location being reflected in the disturbance metrics. A high value of $\beta = 300$ was employed in the dynamic schemes, which may seem excessive but such a high value is necessary to make the avoidance sensitive to disturbance when it is first building around the wormhole endpoints.

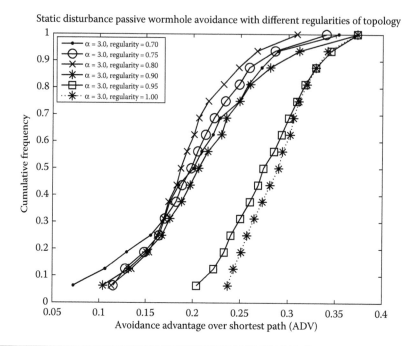

Figure 19.9 Distributions of avoidance advantage for an ensemble of 200 topologies, showing the increased performance of the static metric on perfectly regular topologies due to its sensitivity to irregularities in peer connectivity created by the wormhole endpoint.

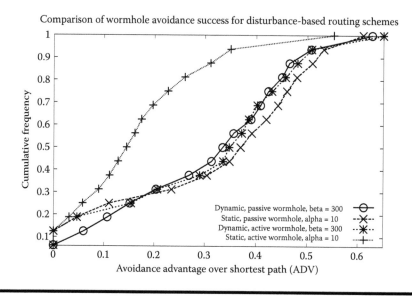

Figure 19.10 Performance comparison of avoidance advantage for the different routing metrics.

19.3.7 Customizing Metrics for Known Topologies

It must be noted that the data sets which generated these results incorporate significant variability in node and wormhole placement, and therefore the metrics will not be optimal for any one particular topology arrangement. If the exact topology is known and particular risk spots for wormhole placement, perhaps via an intended deployment graph, then it would be possible to iterate metric calculation upon a topology to maximize success upon one particular topology. The only variability in the iterations for generated topologies would be in malicious behavior (choice of wormhole placement and source activation), rather than defensive node locations. This algorithm would operate by taking a starting pair of metrics with values of the exponents at the high and low ranges (as found by the general evaluation of the disturbance techniques in previous sections), and repeatedly bisecting or running other interval search strategies until a maximal value of avoidance advantage ADV across the ensemble of possible malicious behaviors is obtained.

19.3.8 Future Directions for Research

19.3.8.1 Overview of Future Directions

The literature review in the background section demonstrated that the general tendency in the literature favors wormhole detection either by means of custom hardware or the fine detail of low-level properties, such as exact timing or spectral signatures, or computationally expensive distributed algorithms. A contrasting principle has been presented here which makes routing sensitive to properties that penalize wormholes, thus harnessing a necessary activity that the network must perform for its routing dynamics and using it to improve security. There is clearly a wide scope for investigating and refining these techniques, further developing the metrics used and applying them in novel ways to other attacks.

19.3.8.2 Application to the Sinkhole Attack

The sinkhole attack [13] is one example of an attack which could benefit from the application of these techniques. In fact, it is similar in dynamic operation to a single endpoint of a wormhole. A sinkhole exists when a node sends out a route advertisement for a high-quality route for the base station or otherwise spoofs a quality route, with the intent of encouraging new routes to travel through it. This may be intended to implement selective forwarding by dropping the incoming traffic (a total or partial denial of service) or simply to cause congestion in a region. The sinkhole is often highly effective in classes of networks operating proactive protocols such as GRAB, since the advertisement propagates outwards to draw traffic into the network, with receiving nodes participating on forwarding the malicious advertisement to draw in further traffic into the sinkhole. Considering an active sinkhole in operation, it is clear that dynamic disturbance-based routing has potential benefits in avoidance here. This occurs as the network's operation brings routes closely together around the sinkhole. Under a dynamic disturbance scheme, these routes would become much less attractive given the clustering of routes and capacity exhaustion as the sinkhole was approached, causing genuine paths to a base station to become more attractive in comparison. More investigation is needed to work out precisely how this effect would manifest and the precise metrics and protocol logic that can exploit it.

It would be interesting to consider the extension of the disturbance-based routing schemes to more exotic topologies, such as multiple-sink and multiple-wormhole topologies. Given that the information about wormhole locations is either inherent in node placement and connectivity (static variants) or in the SPAF values held on nodes (dynamic variants) this state information on likely wormhole locations is implicitly shared and no inter-sink coordination is required to take advantage of it. Therefore, routes to other sinks obtain an avoidance advantage immediately as the avoidance is inherent in the network-wide routing protocol and immediately available regardless of the network endpoint requiring routing service, which gives a security advantage and helps to reduce vulnerable intervals.

A further very interesting approach to investigate would be generalizing the framework into that of a distributed database. The distributed algorithms for topology reconstruction considered in the literature review [18,19] aimed to centralize knowledge of likely wormhole locations, making the sink aware of global topology structure. In contrast, the disturbance-based approaches only focused on locally and passively overheard state to construct their metric values, obtaining this information from the ongoing end-to-end routing process. The distributed database approach would attempt to create a database of network security state informed by cross-layer characteristics, such as interference levels, dynamic and static disturbance from surrounding routes, repeated link failures and incidences of protocol violations or other unusual behavior. Formalizing this as a database allows policies to be defined on how state is shared, finding the best combination of state sharing to detect complex attacks while avoiding unnecessary global transmission of information. Investigating this, it is expected that policies will be defined which allow attacks to be classified using a cross-layer attack signature, and, via the state sharing in the database, give wider regions than the intermediate neighbors possible warning of an ongoing attack before it can affect them adversely.

Our view is that future developments in this field will ultimately explore the gains that can be obtained from cross-layer interaction to reveal security characteristics of a region and ongoing attacks and mitigate them at higher levels such as the routing layer. It is unlikely that mass-market sensor nodes will offer custom hardware to mitigate attacks, or transceivers with stable enough oscillators to successfully implement packet leashing in a close-range WSN, especially in a crowded spectral environment. The economics of deployment will demand security features that add low costs in development and tuning, and also have attractive performance characteristics, as it would be difficult to persuade system integrators to forgo potential performance in the existence of hypothetical threats. Therefore, although a lot of progress can be made via the typical approach of isolating low-level properties of short-range interaction and link layer characteristics for security detection, cross-layer interaction using link layer characteristics to assist secure routing is likely to ultimately form a valuable component of an integrated WSN security suite.

19.4 Conclusions

It has been demonstrated that the disturbance-based routing schemes can provide benefits for wormhole avoidance in sensor networks. The dynamic disturbance metrics have shown the ability to deliver avoidance advantages through building up state to indicate wormhole locations for both active and passive wormholes, and not merely through any artifacts of topology structure or wormhole placement. Although a variety of schemes exist to provide wormhole detection, they suffer many drawbacks, such as requirements for centralized communication, unreasonable algorithmic complexity, inability to integrate with realistic MAC schemes, and requirements for nonstandard hardware. The disturbance-based approaches, in contrast, provide a way to embed awareness of possible wormhole locations into decisions made by a distributed routing protocol, which provides a valuable compliment to the existing detection schemes and the provision of defense in depth for the overall network.

These schemes can also provide significant performance benefits in networks which are capacity limited rather than energy-limited, for example, short-term monitoring networks which are richly resourced. By giving an incentive to route through sparsely deployed areas of the network (static disturbance schemes) and through regions with the lowest level of activity (dynamic disturbance schemes) the network can reduce contention and early exhaustion of critical nodes by spreading out the burdens of routing more evenly.

The results have demonstrated the benefits of the disturbance schemes in proactively avoiding wormholes, and provided a strategy for tuning metrics optimally to predetermined network topologies, as well as approximate values demonstrated to be a good choice for general deployments.

Future directions have been discussed and extensions of the previous scheme considered incorporating the sinkhole attack, as well as considering extension to other security attacks and richer frameworks for detection in the form of the distributed security database idea. However, as was discussed, even though the simulations have shown good performance in a wide variety of cases, it is important to consider thoroughly the unique characteristics of a particular system and how the proposed measures would assist or hamper overall network security in the presence of other threats.

Terminologies

AODV—*Ad hoc* On-Demand Distance Vector Routing
GRAB—Gradient Broadcast
OLSR—Optimised Link-State Routing
Activity factor
Capacity allocation
Commodity hardware
Contention
Wormhole

Questions and Sample Answers

1. How can routing dynamics and overall date flow in a network be compromised?
   If the topology is distorted by introduction of malicious code or external hardware, then routing dynamics and overall data flow may be compromised.

2. What is a WSN?
   A wireless sensor network (WSN) is a distributed, autonomous network of communication and sensory nodes which collaborate to detect and relay information on phenomena under investigation to external coordinators.

3. How does GRAB protocol initiate route discovery?
   Gradient Broadcast (GRAB) protocol initiates route discovery centrally for the entire network during setup.

4. What are the main objectives of security engineering?
   The objectives of security engineering generally amount to protecting some or all of a primary trio of properties: confidentiality, integrity, and availability of all or some network messages. Confidentiality refers to protecting the secrecy of a message and ensuring it is not read by the

adversary. Integrity refers to protecting messages from malicious modification while in transit across the system, and availability refers to ensuring that the attacker cannot deny an intended communication service. A security policy for a system may not demand fulfillment of every one of these objectives, for example, a forest fire detection system may not be designed to ensure message confidentiality as long as integrity is preserved to prevent malicious messages from being injected or delivered.

5. Draw a figure which shows how wormhole attack takes place in a wireless sensor network.

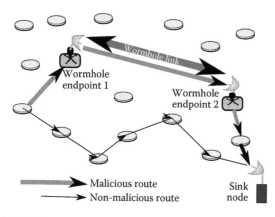

The wormhole attack occurs when an attack introduces a pair of malicious devices. These relay traffic over a private low-latency channel, allowing a short-cut compared to standard routes in the network.

The wormhole allows the attacker to defeat the redundancy that is normally present in a WSN. Since many routes feature the wormhole link, after routing begins to use the wormhole it is likely that disconnecting it will severely affect network availability.

Wormhole attack example scenario.

6. How does dynamic disturbance approach work?

The *dynamic disturbance* approach is based upon nodes overhearing a route request from their peers, which serves as an indication of increased route density in a region. This is used to update an activity factor which is used as an expected indication of occupied spectrum in a region. The concentration of these activity requests gives a low-cost approximation of general routing-level activity under shortest path protocols in a region, which is useful to reveal the location of any operating wormholes.

7. What is route update flooding mechanism?

This route update flooding mechanism serves to ensure the propagation of the most significant disturbance changes to all network entities, regardless of attempts by the wormhole to interfere by traffic-dropping.

8. When does a sinkhole exist?

A sinkhole exists when a node sends out a route advertisement for a high quality route for the base station or otherwise spoofs a quality route, with the intent of encouraging new routes to travel through it.

9. When a sinkhole is highly effective?

The sinkhole is often highly effective in classes of networks operating proactive protocols such as gradient broadcast, since the advertisement propagates outwards to draw traffic into the network, with receiving nodes participating on forwarding the malicious advertisement to draw in further traffic into the sinkhole.

10. What could be expected in the future in this field?

Future developments in this field will ultimately explore the gains that can be obtained from cross-layer interaction to reveal security characteristics of a region and ongoing attacks and mitigate them at higher levels such as the routing layer. It is unlikely that mass-market sensor nodes will offer custom hardware to mitigate attacks, or transceivers with stable enough oscillators to successfully implement packet leashing in a close-range WSN, especially in a crowded spectral environment. The economics of deployment will demand security features that add low costs in development and tuning, and also have attractive performance characteristics, as it would be difficult to persuade system integrators to forgo potential performance in the existence of hypothetical threats. Therefore, although a lot of progress can be made via the typical approach of isolating low-level properties of short-range interaction and link layer characteristics for security detection, cross-layer interaction using link layer characteristics to assist secure routing is likely to ultimately form a valuable component of an integrated WSN security suite.

Author's Biography

Mr James Harbin is a PhD candidate in the Communications Research Group at the University of York, York, United Kingdom. He received his BA in computer science from the University of Cambridge, Cambridge, United Kingdom in 2005 and his MSc in communications engineering from York in 2007. His research interests include wireless sensor and *ad-hoc* networks, security issues in energy-constrained networks, interference and capacity issues in sensor networks and cross-layer optimization ideas.

Dr. Paul Mitchell is lecturer in the Communications Research Group at the University of York. He received his MEng and PhD degrees from York in 1999 and 2003, respectively. Doctoral research on medium access control for satellite systems was supported by BT, with other industrial experience gained at QinetiQ. Dr. Mitchell is an author and reviewer of refereed journal papers, has served on international conference programme committees, and has been an invited speaker at the IET and the University of Bradford. He is an executive committee member of the IET Knowledge Network on Satellite Systems and Applications. His research expertise and interests include medium access control, sensor networks, queuing theory, traffic modelling and system level simulation in which he has over nine years experience.

Dr. Dave Pearce is lecturer in the Communications Research Group at the University of York. He graduated from the University of Cambridge in 1985, and worked in industry for Plessey and Madge Networks researching and developing communication equipment before commencing his PhD at the University of York in 1995. He is now responsible for the teaching of communications at York, and is programme leader for the taught MSc in communication engineering. His research interests are in the interaction of Internet protocols with wireless access-layer protocols and spectrum management.

References

1. *Proceedings of the Distributed Sensor Nets Workshop*. Department of Computer Science, Carnegie Mellon University, 1978.

2. K. Romer and F. Mattern, The design space of wireless sensor networks, *Wireless Communications, IEEE*, 11(6), 54–61, 2004.

3. P. Juang, H. Oki, Y. Wang, M. Martonosi, L. S. Peh, and D. Rubenstein, Energy efficient computing for wildlife tracking: Design tradeoffs and early experiences with zebranet, *SIGPLAN Not.*, 37(10), 96–107, 2002.

4. A technology review of smart sensors with wireless networks for applications in hazardous work environments, Department of Health and Human Services, Centers for Disease Control and Prevention, National Institute for Occupational Safety and Health, Pittsburgh Research Laboratory Pittsburgh, PA, Technical Report, April 2007. http://www.cdc.gov/niosh/mining/pubs/pdfs/2007-114.pdf.

5. G. Werner-Allen, J. Johnson, M. Ruiz, J. Lees, and M. Welsh, Monitoring volcanic eruptions with a wireless sensor network, in *Second European Workshop on Wireless Sensor Networks (EWSN'05)*, Istanbul, Turkey, January 2005.

6. J. N. Al-Karaki, Routing techniques in wireless sensor networks: A survey, *Wireless Communications*, 11(6), 6–28, 2004.

7. I. Demirkol, C. Ersoy, and F. Alagoz, MAC protocols for wireless sensor networks: A survey, *IEEE Communications Magazine*, 44(4), 115–121, 2006.

8. C. Perkins and E. M. Royer, Ad-hoc on-demand distance vector (AODV) routing, in *IEEE Workshop on Mobile Computer Systems and Applications*, New Orleans, LA, 100, pp. 90–100, February 1999.

9. F. Ye, G. Zhong, S. Lu, and L. Zhang, Gradient broadcast: A robust data delivery protocol for large scale sensor networks, *Wireless Networks*, 11(3), 285–298, 2005.

10. R. Anderson, *Security Engineering*. Wiley, New York, 2001.

11. D. Malan, M. Welsh, and M. Smith, A public-key infrastructure for key distribution in TinyOS based on elliptic curve cryptography, in *2004 First Annual IEEE Communications Society Conference on Sensor and Ad Hoc Communications and Networks, IEEE SECON 2004*, Santa Clara, CA, pp. 71–80, 2004.

12. B. Han and D. Grace, Using cognitive interference routing to avoid congested areas in wireless ad hoc networks, in *International Workshop on Cognitive Networks and Communications (COGCOM2009)*, August 2009.

13. C. Karlof and D. Wagner, Secure routing in wireless sensor networks: Attacks and countermeasures, *Ad Hoc Networks*, 1(2–3), 293–315, 2003.

14. Y. C. Hu, A. Perrig, and D. B. Johnson, Packet leashes: A defense against wormhole attacks in wireless networks, *IEEE Computer and Communications Societies. IEEE*, 3, 1976–1986, 2003.

15. D. Liu, P. Ning, A. Liu, C. Wang, and W. K. Du, Attack-resistant location estimation in wireless sensor networks, *ACM Transactions on Information and System Security*, 11(4), 1–39, 2008.

16. L. Hu and D. Evans, Using directional antennas to prevent wormhole attacks, in *Network and Distributed System Security Symposium (NDSS)*, San Diego, CA, pp. 131–141, 2004.

17. L. Qian, N. Song, and X. Li, Detection of wormhole attacks in multi-path routed wireless ad hoc networks: A statistical analysis approach, *Journal of Network and Computer Applications*, 30(1), 308–330, 2007.

18. R. Poovendran and L. Lazos, A graph theoretic framework for preventing the wormhole attack in wireless ad hoc networks, *Wireless Networks*, 13(1), 27–59, 2007.

19. W. Wang and B. Bhargava, Visualization of wormholes in sensor networks, in *Proceedings of the 3rd ACM Workshop on Wireless Security*. ACM Press, New York, NY, USA, 2004, pp. 51–60.

20. M. A. Gorlatova, M. Kelly, R. Liscano, and P. C. Mason, Enhancing frequency based wormhole attack detection with novel jitter waveforms, in *Security and Privacy in Communications Networks and the Workshops*, pp. 304–309, 2007.

21. P. Jacquet, P. Muhlethaler, T. Clausen, A. Laouiti, A. Qayyum, and L. Viennot, Optimized link state routing protocol for ad hoc networks, *IEEE INMIC*, pp. 62–68, 2001.

22. M. Cagalj, S. Capkun, and J.-P. Hubaux, Wormhole-based antijamming techniques in sensor networks, *IEEE Transactions on Mobile Computing*, 6(1), 100–114, 2007.

23. J. Harrop, *OCaml For Scientists*, Flying Frog Consultancy, Cambridge, UK, 2007. Online: http://www.ffconsultancy.com/products/ocaml_for_scientists/index.html

Chapter 20

Reputation-Based Trust Systems in Wireless Sensor Networks

Hani Alzaid

Contents

20.1 Introduction

Wireless Sensor Network (WSN) is a highly distributed network of small, lightweight wireless nodes, deployed in large numbers to monitor the environment or other systems by the measurement of physical parameters such as temperature, pressure, or relative humidity [1, p. 647]. Sensor nodes collaborate to form an *ad hoc* network capable of reporting network activities to a data collection sink. Recently, WSNs have been used in many promising applications, including habitat monitoring [2], military target tracking [3,4], natural disaster relief [5], and health monitoring [6]. However, WSNs are resource constrained with a limited energy lifetime, slow computation, small memory, and limited communication capabilities [7].

The current version of sensor nodes such as mica2 uses a 16-bit, 8-MHz Texas Instruments MSP430 microcontroller with only 10-KB RAM, 128-KB program space, 512-KB external flash memory to store measurement data, and is powered by two AA batteries [8]. Therefore, the energy impact of adding security features should be considered. For example, data authentication in TinyOS increases the consumed energy by almost 3%, while data authentication and encryption increases by 14% [9]. Furthermore, the processing capabilities in sensor nodes are generally not as powerful as those in the nodes of wired networks. Complex cryptographic algorithms are consequently impractical for WSNs. Not only do the resource limitations affect the WSN performance, but also the deployment nature does as well. Most WSNs are deployed in remote or hostile environments where nodes are exposed to physical attacks since anyone can access the deployment area and these sensors lack tamper-resistance property. The adversary can easily compromise one or more sensor nodes, extract secrets, and then affect the overall performance of the network. This attack is referred to as the node compromise attack and is sometimes referred to as the supervision attack or the physical attack [10,11].

Many existing security protocols in WSNs, such as in [12], rely on pure cryptography. However, cryptographic mechanisms alone are insufficient to protect WSNs, because sensor nodes are deployed for long periods in hostile environments where it is possible for an adversary to physically take over a sensor node and obtain access to the cryptographic keys used to provide security within the network [11]. Protocol designers, therefore, should consider these challenges and make their protocols able to function even in the presence of a small fraction of compromised nodes. The wireless security community has consequently developed a suite of mechanisms to complement cryptographic techniques such as reputation-based trust systems that can be defined as a system that collects, processes, and disseminates feedback about the history of the sensors' behaviors.

Trust has become an important topic of research in many fields, including sociology, psychology, philosophy, economics, business, law, and information technology. The most cited definition of trust has been presented by Dasgupta as "*the expectation of one person about the actions of others that affects the first person's choice, when an action must be taken before the actions of others are known*" [13]. This definition captures both the purpose of trust and its nature in a form that can be reasoned. Another definition for trust by Gambetta is also often quoted in the literature: "*trust (or, symmetrically, distrust) is a particular level of the subjective probability with which an agent assesses that another agent or group of agents will perform a particular action, both before he can monitor such action (or independently of his capacity ever to be able to monitor it) and in a context in which it affects his own action*" [14]. Mui et al. [15] define reputation as "*a perception that an agent creates through past actions about its intentions and norms.*" A similar definition given by Abdul-Rahman et al. is "*a reputation is an expectation about an agent's behavior based on information about or observations of its past behavior*" [16]. Although many of the proposed definitions are available in the literature, a complete formal unambiguous definition of trust is rare because trust is a complex term with multiple dimensions.

One way of computing the trust level is based on social networks (or reputation systems). The Feedback Forum on eBay is the most prominent example of online reputation systems [17] in which the basic idea is to let parties rate each other. After the completion of a transaction, each party is allowed to leave feedback about their experience of the other party. Then, the aggregated ratings about a given party are used to derive a reputation score, which can assist other parties in deciding whether or not to deal with that party in the future. Reputation is generally defined as the opinion or view of one party about the character of another entity. In the WSNs context, an entity could be a sensor node, cluster head, or an activity.

Different reputation-based trust systems have been designed for WSNs, each of which is used in a specific context. This has led to different trust system architectures and different attack-resilient levels.

Our contributions in this chapter include the following:

■ The security concerns in reputation-based trust systems designed for WSNs are discussed.
■ A survey of the "state-of-the-art" in reputation-based trust systems for WSNs is presented and then these systems are classified according to the context they were designed for.
■ Finally, the security analysis for current reputation-based trust protocols is discussed in order to establish common ground (or test-bed) to compare different reputation-based trust protocols. This will help draw a road map for the future design of attack resistant reputation-based trust systems.

20.2 Security Concerns

One of the potential vulnerabilities in WSNs is the security compromise of sensor nodes, given the lack of tamper-resistant packaging [10]. An adversary can gain control of one or more sensor nodes and readily access sensitive information. Even worse, the adversary may also inject their own commodity nodes into the network by fooling nodes into believing that these commodity nodes are legitimate members of the network, especially if there is no proper authentication scheme in place. Another adversary activity is launching a selective forwarding attack where a node, which is under the control of an adversary, selectively drops legitimate packets in order to affect the overall performance of the system [18].

Integrating reputation system capabilities within WSNs can strengthen the performance and security levels of the WSNs by providing continuous monitoring, calculating the reputation metrics based on the quality of different activities, and reporting un-trusted behaviors to neighboring nodes. However, vulnerabilities related to WSNs can affect the reputation system functionality. For example, a selective forwarding attack can affect the reputation component by selectively dropping direct observations propagation of neighboring nodes. In the following sections, two types of attack are discussed: WSN-related attacks that might affect the reputation system functionalities, once they are integrated, and reputation-related attacks that might affect the overall performance and security of WSNs.

20.2.1 WSN Attacks

WSNs are vulnerable to different types of attacks due to the nature of the transmission medium (broadcast), remote and hostile deployment location, and the lack of physical security in each node [19]. The damage caused by these attacks varies from one protocol to another according to the adversarial model which is assumed by the designers of the protocol. This section studies how attacks related to WSNs can affect the reputation system component, which is the security primitive used to improve the security level in WSNs. These attacks are as follows:

Sybil attack (SY): The SY is a type of attack where the adversary is able to present more than one identity (node) within the network to deceive other nodes. A node that wishes to conduct the SY attack can create a new identity or impersonate the identity of an existing node. The adversary can create multiple identities to affect the reputation values of legitimate nodes in reputation-based applications by falsely degrading their reputation values.

Let us consider the example shown in Figure 20.1 where the adversary creates fake IDs in order to affect the overall performance of the network. Figure 20.1a shows a sketch of the real scenario with no existence of any adversary. The real path starts from node $A(D)$ and ends at node $D(A)$. Nodes B and C are adjacent neighbors. Once the adversary has succeeded in launching an SY attack, node B' will be added to the network by manipulating route discovery messages within the routing activities. The adversary thus communicates with node A using node B and communicates with node C using node B'. Any detection of misbehaved activities can be caused by either node B or node B' from the perspectives of nodes A and C. The adversary can trickily blame node B' (or node B) for those misbehaviors and leave the reputation value of node B (or node B') untouched.

Selective Forwarding (SF) attack: With no consideration of security, it is assumed in WSNs that each node will accurately forward the received messages. However, a compromised node may refuse to do so. It is up to the adversary, which is controlling the compromised node, whether to forward received messages or not. To put it another way, the process of stopping the propagation of certain

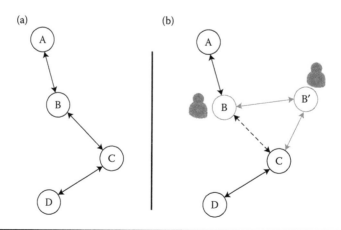

Figure 20.1 Sybil attack. (a) Real scenario and (b) modified scenario.

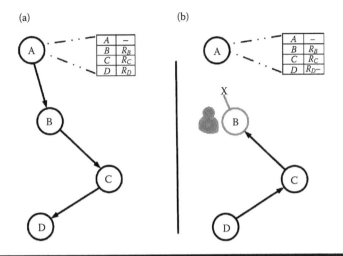

Figure 20.2   Selective forwarding attack. (a) Request path and (b) reply path.

messages at the compromised node is under the control of the adversary. Once the adversary has succeeded in launching an SF attack, it can affect the propagation of the reputation information such as direct and indirect observations across the network.

Figure 20.2 depicts a simplified scenario where an SF attack can affect the reputation systems. The scenario follows the single aggregator model [20], where node A acts as an aggregator. In Figure 20.2a, the adversary succeeded in compromising node B but behaved well and forwarded the request message sent by node A. Later on, node B, which is still under the adversary control, dropped the response from D. Subsequently, the aggregator has not received any reply for its recent request. Consequently, node A updates its reputation table (RT) and reduces the reputation value of node D as in Figure 20.2b.

Replay attack (RE): Some applications that are destined for WSNs are vulnerable to RE where the adversary is able to eavesdrop on the traffic and replay old messages. This type of attack is the easiest one because the adversary does not need to physically capture a sensor node or get access to its internal memory. In the reputation-based applications context, the adversary can record some reputation information from the network without even understanding its content and then replay them (with no changes) to mislead other nodes.

Figure 20.3 describes a simplified scenario of an RE attack where the adversary is able to capture the update message for reputation values at a certain time t_1 (see Figure 20.3a), and then re-inject it at time t_2 where $t_2 > t_1$ (see Figure 20.3b). With no proper verification, nodes B, C, and D will accept this re-injection and consequently end up with incorrect reputation values.

Spoofed Data (SD) attack: In this type of attack, the adversary alters intercepted data in order to inject false data into the network and affects the reputation values. This attack cannot be launched alone; the adversary needs to combine either an RE attack or node compromise attack with an SD attack. In the former, the adversary first eavesdrops on the traffic, captures some reputation information in understandable format, performs some changes in the captured information, and then re-injects it into the network. In the latter, the adversary first needs to overtake a sensor node and then it can affect the reputation calculation by falsely claiming that his direct observation for node N_i is R'_i but in fact it is R_i. Other nodes are misled by the received R'_i and thus their calculations for the reputation value of N_i are affected.

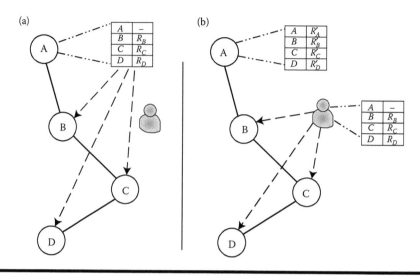

Figure 20.3 Replay attack. (a) Reputation update at t_1 and (b) reputation update at t_2.

Figure 20.4 presents a simplified scenario of an SD attack once the adversary has succeeded in compromising node B. The adversary, in Figure 20.4b, during the reputation update phase, claims that the reputation value for node A is R'_A not R_A and then sends it to the neighboring nodes C and D. Therefore, nodes C and D will use R'_A as a second-hand information when they calculate the reputation for node A.

20.2.2 Reputation Attacks

The reputation system itself is threatened by different types of attack [21]. Understanding these attacks is crucial in order to ensure that the integration between reputation systems and WSNs does not open doors to more threats. Attacks that are applicable only to the reputation system are discussed in this section as follows:

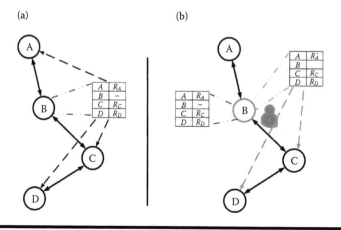

Figure 20.4 Spoofed data attack. (a) Real scenario and (b) modified scenario.

Bad Mouthing (BM) attack: Once the adversary has compromised a sensor node, it can affect the reputation system by reporting invalid direct observations and assigning negative feedback to well behaving nodes. When these direct observations are propagated to other neighboring nodes, they will be considered as indirect observations or sometimes as second-hand information. With no proper verification, these indirect observations are considered by neighboring nodes at the reputation calculation phase as will be discussed in Section 20.3. In other words, the BM attack happens when the adversary has the ability to assign negative feedback for well-behaved nodes in order to reduce the trustworthiness in those nodes. This attack is possible in scenarios where the indirect observations are taken into consideration and parties are allowed to share their negative feedback with nodes in the neighborhood.

Figure 20.5 depicts a simplified scenario where the BM attack can take place. The adversary has succeeded in compromising node *B*. Later on, it assigned a negative reputation value for a well-behaved node *C* in order to mislead node *A* with its calculation of the reputation value of node *C* (see Figure 20.5b).

Ballot attack (BA): A BA is similar to the BM attack, but the adversary tries to perform exactly the opposite. The trustworthiness of well-behaved nodes, in this attack, is not affected as in the BM attack; however, the trustworthiness of the bad-behaved nodes is affected by assigning positive feedback to malicious nodes. This attack is possible in scenarios where the indirect observations are taken into consideration and parties are allowed to share their positive feedback with their neighboring nodes.

Figure 20.6a shows that nodes *B* and *C* are compromised and their reputation values are low. These compromised nodes colluded with each other and assigned higher reputation values to each other as in Figure 20.6b. Generally speaking, the adversary can replace low reputation values with high reputation values.

On/Off attack (OO): In this type of attack, the adversary aims to distribute the system's overall performance with the hope that it will not be detected or excluded from the network. The adversary behaves badly and correctly alternatively in order to extend the detection time required to recognize its misbehaviors. This attack can be launched against either the reputation activities or general activities in WSNs.

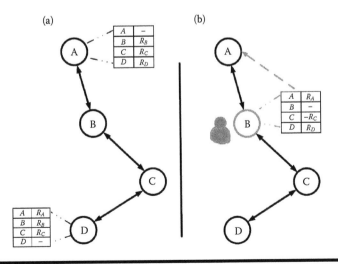

Figure 20.5 Bad mouthing attack. (a) Real reputation update and (b) altered reputation update.

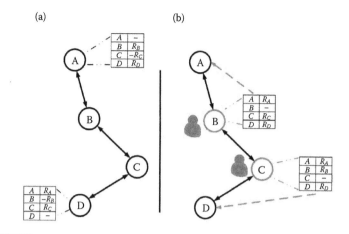

Figure 20.6 Ballot attack. (a) Real reputation update and (b) altered reputation update.

A simplified scenario where the adversary is able to perform some OO attack activities is shown in Figure 20.7. The adversary had compromised node B and later on, it behaved badly and correctly when with nodes A, C, and D. It behaved badly when it had dealt with nodes C and D and claimed that the reputation value for node A is R'_A instead of R_A. On the other hand, it behaved correctly when it had dealt with node A and disseminated the real reputation values for nodes C and D. Another form of OO attack happens when node B misbehaves once every t well-behaved transactions, which means nodes A, C, and D are uncertain about the behavior of node B. They are not sure whether this misbehavior was intended by node B or whether it was due to some other factors such as the wireless medium.

Newcomer attack (NC): As soon as the adversary's reputation value drops below the threshold value, which moves the node from the trusted mode into an untrusted mode, the adversary will consider other ways to increase its reputation value. One way to do so is to rejoin the network with a new ID and wipe out all its bad history. This attack is referred to as the newcomer attack and is sometimes referred to as the identity attack or white washing attack. If the adversary has the ability

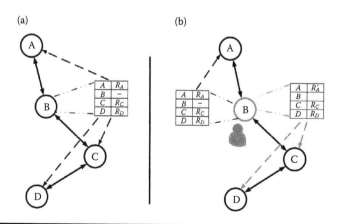

Figure 20.7 On/off attack. (a) Real reputation update and (b) altered reputation update.

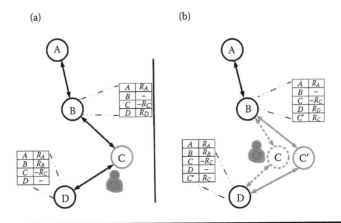

Figure 20.8 Newcomer attack. (a) Reputation update at t_1 and (b) reputation update at t_2.

to launch this attack, then detecting the adversary's misbehaviors is not an issue from the adversary's perspective due to the fact that all the old history can be wiped out at any stage.

A sketch of a simplified scenario for an NC attack is shown in Figure 20.8. The reputation value of node C fell below the predefined threshold value as a result of its previous misbehaviors. Therefore, the adversary decided to rejoin the network with another identity C' and neutral reputation value.

20.3 Analysis Framework for Reputation Systems

Reputation systems often share similar structural patterns due to their common purposes such as enhancing the system's overall performance by monitoring the network activities. They consist of three main phases/stages: information gathering and sharing, information modeling (or reputation calculation), and decision-making (see Figure 20.9). These three stages are discussed in the following sections.

20.3.1 Information Gathering and Sharing

The main task of this stage is to determine what type of information should be collected about other neighboring nodes and how. And what reputation metrics are used and how are they distributed. For example, a reputation system may accept positive, negative feedback information, or both. It is believed that this stage is the core component of any reputation system, because it gathers current activities and the available information about the system and then hands it to the next stage, which is the information modeling stage. Then, the information modeling stage transforms this gathered information into some usable reputation metrics. This stage is composed of four components: information source, information type, information gathering approach, and gathering scope.

Information Source: The information source in any reputation system can be either manual or automatic. The manual information source is obtained in the form of user ratings for other entities as a result of being involved in a single transaction such as in the eBay rating system [17]. This type of source is not available in WSNs due to the lack of user interaction with the network. The only user interaction with WSNs usually occurs at the base station (BS) while the reputation system

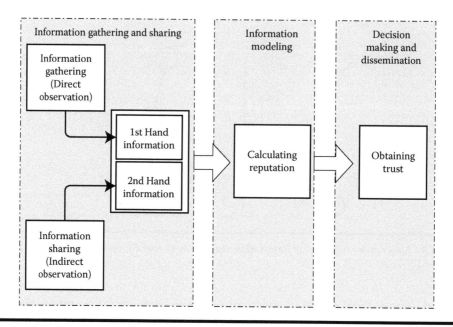

Figure 20.9   Reputation system stages.

gathers information from every device within the WSNs. The automatic information resource does not involve user interaction and can be either direct or indirection observations. Direct observations (sometimes called first-hand information) are computed based on the node's observations and experience about neighboring nodes such as the success and failure of forwarding aggregated data within an error rate. In some reputation systems, the first-hand information needs to be propagated to other nodes in the neighborhood and then this propagated information is called second-hand information (or indirect observations) at the receiving nodes. In other words, second-hand information for one node is the propagated first-hand information of another node. Second-hand information helps build up the reputation system more quickly than using only first-hand information since nodes will be able to know about the other nodes' behaviors even though no direct communications (observations) have occurred. However, propagating reputation information between nodes makes the system vulnerable to different attacks such as BM and BA attacks as discussed in Section 20.2.

Information Type: The information type can be negative, positive, or both. The choice of the information type is up to the system designer, but designers should be aware of the consequences of any choice. Considering only positive feedback on the one hand, the BM attack can be prevented because malicious nodes would not be able to affect the trust level of well-behaved nodes by propagating negative reputation ratings. However, malicious nodes can collude and falsely praise misbehaved nodes to launch a BA attack. Propagating positive feedback also exhausts the network's limited resources since the number of nodes that behave correctly, in general, is supposed to be larger than those which do not. Thus, the number of transmissions required to update reputation values is high.

On the other hand, considering only negative feedback helps prevent malicious nodes from colluding and praising misbehaving nodes (BA attack), because they could not propagate positive feedback. It also helps to minimize the number of transmission required to update the reputation values. However, malicious nodes can assign negative reputation ratings/feedback for well-behaved nodes in order to affect their trust level (BM attack).

Information Gathering Approach: As discussed earlier, the main task of this stage is to collect information about other sensor nodes in the neighborhood. This information is gathered by a sensor node based on its observations and experience about other nodes. The output of this component is called first-hand information. Most current reputation-based trust systems in WSNs use monitoring mechanisms such as Watchdog mechanism (WDM) [22].

When a node forwards a packet, the node's WDM verifies that the next node in the path also forwards the packet. The WDM is implemented by maintaining a buffer of recently sent packets and comparing each overheard packets with the packet in the buffer in order to see whether they match. Once there is a match, the packet is removed from the buffer. If the packet has remained in the buffer for longer than a certain timeout, the WDM increments a failure tally for the node which is responsible for forwarding the packet. If the tally exceeds a certain threshold, then the node attached to this tally is considered a misbehaving node.

Reputation System Scope: In the current literature, each reputation-based trust system focuses on specific functions. For example, CORE [22], CONFIDANT [23], and PLUS [24] focus on detecting misbehaviors related to routing functionalities, whereas DRBTS [25] focuses on enforcing cooperation between beacon nodes by motivating them to provide correct location information. It is important to know that reputation-based trust systems with different scopes make the comparison between these systems invisible because trust systems with scope being limited to a specific function do not suffer from issues related to another scope. For example, a reputation system limited to ensuring correct location information in WSNs does not cover issues related to routing activities. To the best of our knowledge, scopes where reputation-based trust systems were designed to increase the trustworthiness between sensor nodes, include routing, aggregation, mobility, and localization.

20.3.2 Information Modeling

The main task of this stage is to calculate the reputation values for such a node from the available information (direct and indirect observations), which are provided by the previous phase—the information gathering and sharing phase. This phase is composed of two components: the information modeling structure and the information modeling approach.

Information Modeling Structure: Reputation systems can be designed to calculate the reputation values via a centralized entity, distributed entities, or a hybrid approach. In the centralized approach, observations about a node's performance are propagated to a central authority that collects these observations, derives reputation values for each node, and subsequently updates nodes with new reputation values. This approach relies on some assumptions, namely nodes completely trust the centralized authority which in turn must be correct and always available. If the centralized approach is not carefully designed, it can become a single point of failure for the whole reputation-based trust system. Centralized systems suffer from the lack of scalability, especially if the information is obtained from high-latency sources. In the domain of WSNs, most recent applications were designed with a central robust authority (BS) in place. However, propagating observations across the network to the central point is impractical due to the scalability issue and the huge energy consumption. Hence, minimizing the energy consumption is important in such environment where end nodes are operated with two AA batteries such as Mica2 sensor nodes [8]. One way to minimize the energy consumption is by considering the distributed structure for information modeling.

In the distributed approach, each node propagates its observations to neighboring nodes and then these nodes calculate the reputation values individually. In other words, each node is responsible for collecting first- and second-hand information (if they exist), calculating reputation values of other nodes in the neighborhood. Although the distributed structure of the information modeling is inherently more complex, it scales well, avoids single points of failure in the system, and balances load across multiple nodes.

Finally, the reputation values in the hybrid approach are calculated by more than one entity. For example, Shaikh et al.'s scheme [26] follows the distributed approach for calculating reputation values for nodes within the cluster, but it follows a centralized approach when the BS calculates reputation values for cluster heads.

Information Modeling Approach: The information modeling approach can be either deterministic or probabilistic. In the former, the output is uniquely determined by the input with no existence for randomness while the output, in the latter, can be predicted only within certain errors due to some randomness resources added to the input. Examples for the probabilistic and deterministic information modeling are a simple Bayesian model and a summation model, respectively.

20.3.3 Decision Making and Dissemination

The main task of this stage is to decide based on the calculated reputation values, which results from the information modeling stage, whether or not the trustworthiness of a specific node is enough to cooperate with. This stage is composed of three components: decision and dissemination structure, decision metric, and dissemination approach.

Decision and Dissemination Structure: Calculated reputation values are distributed within the trust systems according to the dissemination structure, which can be either a distributed or centralized structure. In the former, each sensor node calculates reputation values of other nodes in the neighborhood, stores them locally, and then shares them with its neighbors. This type of structure helps sensor nodes being updated about other nodes by quickly filling their RTs. However, redundancy in this reported reputation information exists, which affects the limited energy source in sensor nodes. Unfortunately, the distributed structure opens doors for attackers to affect the reputation values by launching BA or BM attacks. Consequently, system designers should pay careful attention when they follow this structure. In the latter, the calculated reputation values are stored and distributed by a single entity, which can be a cluster-head or a BS. However, this single entity has to have greater resources (such as enough memory space to store reputation information for other nodes, and enough energy and processing capabilities to ensure the availability of this single entity) to manage the dissemination activities. It is worth mentioning that there is an overlap between the information modeling structure component and this component as will be discussed in Section 20.5.2.

Decision Metric: The decision metric can be either binary or discrete. In the former, the decisions (cooperate and do not cooperate notions) are represented by two symbols, 1 and 0, respectively. This is usually based on a threshold policy, which is common in most reputation-based trust systems for WSNs. If the reputation value of a sensor node is above a predefined threshold, then cooperation with this node is preferable. In the latter, reputation systems have various predefined levels of trust which allow for more flexibility since different actions can be taken, depending on different levels of reputation.

Dissemination Approach: The dissemination approach can be either proactive or reactive. In the former, reputation values are broadcasted periodically, although there are no changes in the reputation values since last update. In the latter, reputation values are broadcasted only when there are sufficient changes to these reputation values or when a specific event occurs. Periodic dissemination, on the one hand, is suitable for resource-constraint devices in busy networks, because reputation values are updated regularly for more than one activity. This helps reduce the number of transmissions required to update reputation values. The reactive dissemination approach, on the other hand, is suitable in networks with light traffic. This helps minimize the number of transmissions in cases where there are no sufficient changes in the reputation values. It also covers designs where reputation values are piggy-backed on reply messages such as in CORE [27].

20.4 The State-of-the-Art of Reputation-Based Trust Systems in WSNs

Research on reputation-based trust systems for WSNs is at a very early stage. To the best of our knowledge, there is only one survey in which current reputation-based trust systems for WSNs have been studied. Roman et al. gave "the state-of-the-art" in trust management systems for WSNs and they also tried to identify the main features of the architectures of these systems [28]. The main two features, according to Roman et al.'s study, are information gathering and information modeling.

This chapter extends the work in [28] by considering more features of the architecture of reputation-based trust systems and analyzing more trust systems. It also provides insights into the reputation components, or features, and the visibility of attacks discussed in Section 20.2 for each system. Discussion on nine representative reputation-based trust systems is given in the following sections.

20.4.1 A Trust-Based Security System for Ubiquitous and Pervasive Computing Environments [29]

Boukerche and Ren designed a security system based on trust management that involves developing a trust model, assigning credentials to nodes, updating private keys, managing trust values of each node, and making appropriate decisions about the nodes' access rights [29]. They stressed that a reputation-based trust system can track the behavior of nodes and thereby proceed by rewarding well-behaved nodes and punishing misbehaving nodes. They verified, through the presentation of a formal security analysis, that the stated goals are achieved and that malicious nodes can be effectively excluded from the network.

Their security system, TOMS, introduces the concept of community. The authors of TOMS defined a node that is a central node and its entire one-hop neighboring nodes as a community. Let us consider the community model example shown in Figure 20.10. In the community of the central node C, C has six neighbors, but neighbors D and G are malicious and they are excluded from C's community. Thus, the community of the central node C consists of nodes A, B, E, and F, as well as the central node C.

TOMS is composed of a trust model and trust management phases (or information modeling phase, and the decision-making and dissemination phase according to the discussion in Section 20.3). In the former, the trust metric is affected by two factors: the time that the node stays in the community and the past activity record of the node. The recent trust rt of the node n_i can reflect the past behavior of n_i. The trust in TOMS is defined as a function that depends on the time a node has stayed in the community and on the past trust to which this node has belonged in recent periods [30]. First, Boukerche and Ren defined the trust as a function that depends on the time that a node has stayed in the community and on the past trust to which this node has gained,

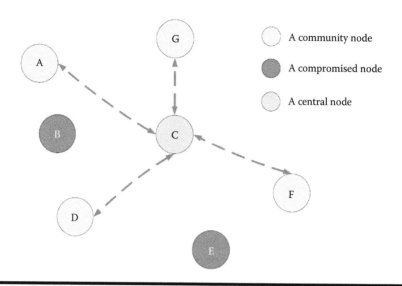

Figure 20.10 An example for a community in TOMS. (Data from A. Boukerche and Y. Ren. *Computer Communications*, 31(18):4343–4351, 2008.)

as follows:

$$N = 0.51 + rt$$

This will yield a value very close to 1 for nodes with a moderate trust ($rt = 0.5$), a value below 1 for nodes that have lower trust ($rt < 0.5$), and a value above 1 for nodes that have a higher trust ($rt > 0.5$). Then, they defined the time factor (W) as follows:

$$W = K^{\text{time}} + ra$$

where K is a discount factor between 0 and 1 and ra is the node's recent activities, which can include a successful forwarding or a deliberate exaggeration. Finally, the trust metric is evaluated as follows:

$$T = \gamma \frac{1 - N^{1+W}}{1 - N}$$

where γ is a scaling factor to keep the trust T at a value between 0 and 1. On the other hand, the trust management phase covers the trust assistant policy (TAP), the trust query, the key revocation, the black list, and the trust route selection. The TAP helps the central node in better evaluating its neighboring nodes' trusts. When the central node wants to evaluate a neighboring node's trust, it will query its trust assistants about this neighboring node x. Then, these trust assistants will provide the node's trust in their individual community to the central node. Subsequently, the node's final trust can be calculated as follows:

$$T_x^{\text{final}} \frac{T_{(C,x)} + [T_{(A_1,x)} + \cdots + T_{(A_i,x)} + T_{(A_n,x)}]}{n + 1}$$

where $T_{(C,x)}$ is the trust value of the central node C to a certain node x, $T_{(A_i,x)}$ is the trust value of the trust assistant i to the same node x, and n is the number of trust assistants in the community.

In TOMS, the trust assistants are allowed to provide the central node with either positive or negative feedback. In other words, the trust assistant can answer any trust query, which is sent by the central node, with either a high or low trust value.

Generally speaking, reputation information in TOMS is obtained automatically due to the lack of interaction between the users and the network. TOMS uses the direct information source (first-hand information) and indirect information source (second-hand information) in order to calculate the trust value of a specific node, say node x. The direct information represents the experience of a central node with node x, whereas the indirect information represents the experience of trust assistants with the same node, node x.

TOMS uses a binary decision metric whenever it evaluates other nodes in its community. In other words, any node within the central node's community is assigned either a trust or distrust notion. The decision is made based on a threshold policy in which the central node will independently set a trust threshold for its community, and the neighboring nodes that cannot meet the trust requirement will be taken out of its community. The central node will keep a black list to record all malicious nodes that have been excluded from its community due to their malicious behaviors in recent periods. This means that TOMS is a centralized system and thus, any node in the community needs to send a trust query to the central node in order to obtain trust information for another node in the community.

20.4.2 Reputation-Based Secure Data Aggregation in WSNs [31]

Alzaid et al. proposed a Reputation-based Secure Data Aggregation for WSNs (RSDA) that integrates aggregation functionalities with the advantages provided by a reputation system to enhance the network lifetime and the accuracy of the aggregated data [31]. The target terrain where RSDA is implemented is divided into smaller cells of equal size (see Figure 20.11). Each cell has T nodes where only one of them is selected, based on its reputation value, to be the cell representative (C^{rep}). Each node has a monitoring mechanism similar to the WDM [22] in order to compare its result with the results of its neighbors. Each node in a cell performs redundant operations to monitor the cell representative performance.

RSDA follows a request–response paradigm where the BS initiates the aggregation process by flooding a query message into the network. The transformation from this paradigm to the periodic paradigm, however, is straight forward by letting the representatives periodically report their data without the need to wait for the BS's query. RSDA focuses on the multiple-aggregator model that was identified by Alzaid et al. [20] where the aggregation is performed at each cell. Each node monitors the behavior of the other nodes within the same cell and then calculates the reputation value for them based on participation in some cell operations such as sensing, forwarding, and aggregation. RSDA is composed of two types of identities: a BS and normal sensor nodes. The BS is entrusted with the task of initiating queries to the network, processing received answers for these queries, and deriving meaningful information that reflects the events in the target field. The normal sensors are grouped into cells and in each cell one of sensors is selected to be the cell representative C^{rep}.

Initially, C^{rep} is chosen randomly since all nodes start with same reputation value. The cell can be an intermediate cell, which receives data from downstream cells and performs sensing, aggregation, and forwarding operations, or a leaf cell that does not receive data from downstream cells and does not perform aggregation functions (see Figure 20.11). The C^{rep} is responsible for confirming its cell reading C^{read} (reported by other cell member), aggregating it with other readings (if the cell is an intermediate cell), and forwarding it to an upper stream cell. In addition to reporting C^{read} to C^{rep}, other nodes evaluate the behavior of their C^{rep} and other nodes in the same cell. The

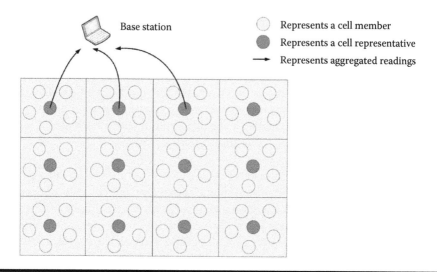

Figure 20.11 A simplified version of the deployment area in RSDA.

node's behavior is represented in the form (α, β) where α and β represent the amount of positive and negative ratings that are calculated by each node for other nodes in its cell and then stored in the reputation table (RT) (see Table 20.1). If x has behaved well for a specific function, α for node x is incremented by one. Otherwise, β is incremented. The nodes' behaviors are examined for three functions: data sensing, data forwarding, and data aggregation (if x is the C^{rep} for an intermediate cell). Each node, therefore, maintains an RT for its cell members and keeps recording α and β for these functions separately as in Table 20.1. To fill this RT, each sensor node evaluates the sensing, forwarding, and aggregation (if it is in an intermediate cell) functionalities of each cell members and computes the amount of positive and negative (α, β) ratings for each function. RSDA uses the beta probability density function (PDF) to update the reputation value of each sensor node due to its flexibility, strong foundation on the statistics, and simplicity that meets the needs of the resource constraint nodes [32]. The reputation value which factors in sensing, aggregation, and forwarding $R_{S/F/A}$ can be expressed as follows:

$$R_{S/F/A} \frac{\alpha_{S/F/A}}{\alpha_{S/F/A} + \beta_{S/F/A}}$$

when nothing is known, the *a priori* distribution is the uniform beta PDF with $\alpha = 1$ and $\beta = 1$.

Table 20.1   Reputation Table Format in RSDA

	R_S		R_F		R_A	
Node ID	α_S	β_S	α_F	β_F	α_A	β_A
x_1	10	4	8	6	—	—
x_2	13	1	14	—	14	—
.	—	—	—	—	—	—
x_n	—	—	—	—	—	—

When a leaf cell receives the query, C^{rep} randomly selects a sensor node x from its cell to send back an answer to the received query, r_x. As a response, x senses some physical phenomena (as requested) and then sends them back to C^{rep}. Other nodes in the cell are listening to the on-going traffic between the elected node and C^{rep} since they share the same cell key. A neighboring node y does not agree on the reading r_x if $|r_y - r_x| < Thr_S$. If they agree on r_x, they update α_S^x and α_F^x of node x and consider r_x as the C^{read}. They also update α_S for all other nodes because of their agreements on the C^{read}, which means their readings were within Thr_S. If they do not agree, they

- Update β_S^x (if the reading was unacceptable) or β_F^x (if the destination was not the cell representative or no reply was sent).
- Collaborate with other nodes in the cell to supply C^{rep} with the correct C^{read}.

Each disagreeing node, say node y, sends its reading to the C^{rep} and thus other nodes are updating α_S^y and β_S^y. After receiving claims from $n > t$ eligible nodes regarding the reported reading, C^{rep} computes the cell reading by using exogenous discounting of unfair ratings where the reputation values of these n nodes are used to determine the weight given to the information as follows:

$$C_i^{read} = \frac{\sum_{i=1}^{n}(r_i * R_S^i)}{\sum_{i=1}^{n}(R_S^i)}$$

It is based on the assumption that sensors with low reputation are likely to give unfair information and vice versa. Besides what a leaf cell does, an intermediate cell C_j also performs the aggregation activities. The cell representative of an intermediate cell waits until receiving readings from its cell, which is done in the same way as the leaf cell does, and other children cells in order to apply the aggregation function on them as follows:

$$AD_{C_j} = F(C_1^{read}, C_2^{read}, \ldots, C_i^{read}, C_j^{read})$$

Other nodes in cell C_j are still able to keep an eye on the aggregation and forwarding behaviors of C_j^{rep}. They recalculate the aggregation function $AD_{C_j}^*$ and match the result with AD_{C_j} and then update their RTs. If they are bound by small value such as $|AD_{C_j}^* - AD_{C_j}| < Thr_S$, $\alpha_A^{C_j^{rep}}$ is increased by one. Otherwise, $\beta_A^{C_j^{rep}}$ is increased by one. Also, the $\alpha_F^{C_j^{rep}}$ is increased by one if C_j^{rep} forwards the packet to right C^{rep} that is not in the black list and is one-cell closer to the BS. It is worth mentioning that the RT is kept locally at each node in the cell. All nodes in the cell, except the cell representative, are responsible for monitoring the behavior of their cell representative, taking it out, and then black listing it once it has been detected as a misbehaving representative.

In conclusion, RSDA uses a direct information source (first-hand information) to update the reputation information within a cell. The negative feedback of the cell representative is forwarded to the next cell only if the revocation mechanism is needed to replace the misbehaving cell representative. In other words, negative feedback is only forwarded to neighboring cells once the current cell representative is needed to be revoked. If the reputation value of the current cell representative is within an acceptable range, no reputation information is forwarded to the neighboring cells. In RSDA, reputation values are distributed only when there is a need. For example, when the reputation value for the cell representative falls below a predefined threshold, a revocation mechanism is initiated. In the revocation mechanism, the current reputation value of the cell representative is attached with a request to change the misbehaving representative.

20.4.3 The Trust Management Problem in Distributed WSNs [26]

Shaikh et al. argued that traditional trust management schemes used for wired and wireless *ad hoc* networks are not suitable for WSNs due to higher computational costs and large memory overheads [26]. Shaikh et al.'s scheme is based on hybrid trust management architecture in which the trust value is calculated at three phases/places: at each sensor node, at each cluster head (or group leader), and at the BS. At the first phase, a trust value is calculated by using first- and second-hand information, or time-based past interactions and peer recommendations according to the designers' terminology. The reputation value is calculated by node x for node y as follows:

$$R_{x,y} = \frac{(R_{x,y})_\mathrm{D} + (R_{x,y})_\mathrm{IND}}{2}$$

The first-hand information $(R_{x,y})_\mathrm{D}$, which is the node's direct observations, is calculated by node x for node y as follows:

$$(R_{x,y})_\mathrm{D} = 1 - \frac{1}{\max[\{w_s\alpha_{x,y} - w_u\beta_{x,y}\}, 0] + 1}$$

where $\alpha_{x,y}$, $\beta_{x,y}$, w_s, and w_u represent the number of positive feedback (or successful interactions), the number of negative feedback (or unsuccessful interactions), corresponding weight of α, and corresponding weight of β, respectively. The scheme assigns higher weight if the difference between the current time and the time of the last successful/unsuccessful interaction is small and vice versa. The second-hand information is calculated as:

$$(R_{x,y})_\mathrm{IND} = \frac{\sum_{i=1}^{m-1}(R_{x,i})_\mathrm{D}(R_{i,y})_\mathrm{IND}}{m - 1}$$

where m denotes the number of sensor nodes in the group. Each cluster head (ch), in the second phase, broadcasts request-messages to its group members in order to update their reputation values. In response, each group member sends to ch its reputation values of other members. After that, ch calculates the reputation vector and then sends it to the BS. The reputation vector is calculated as:

$$\overrightarrow{R_{ch}} = (R_{ch,1}, R_{ch,2}, \ldots, R_{ch,m})$$

where $R_{ch,i}$ represents the reputation value of node i, which is calculated by ch as follows:

$$R_{ch,i} = \frac{\sum_{k=1}^{m-1} R_{k,i}}{m - 1}$$

The cluster head calculates the reputation value of another cluster head (ch_j) in the same way as a sensor node does in the first phase. However, the indirect information for ch_j is provided by the BS, which can be obtained from $\overrightarrow{R_{ch_j}}$. In the last phase, the BS calculates the reputation values for each cluster head ch_i from received reputation vectors $\overrightarrow{R_{ch}}$ as follows:

$$R_{B,ch_i} = \frac{\sum_{k=1}^{m} R_{k,ch_i}}{m}$$

Then, the BS calculates the trust decision based on the threshold policy as follows:

$$R_{B,ch_i} = \begin{cases} \text{trusted,} & 0.6 \leq R_{B,ch_i} \leq 1 \\ \text{uncertain,} & 0.4 \leq R_{B,ch_i} \leq 0.6 \\ \text{untrusted,} & 0 \leq R_{B,ch_i} \leq 0.4 \end{cases}$$

20.4.4 A Collaborative Reputation Mechanism to Enforce Node Cooperation in Mobile Ad Hoc Networks [22]

Michiardi and Molva [22] developed a collaborative reputation mechanism to enforce node cooperation in mobile *ad hoc* networks (CORE) where each node computes a reputation value for every neighbor, based on its observations that are collected in the same way as Watchdog. The reputation mechanism differs between subjective reputation, indirect reputation, and functional reputation. Subjective reputation, or direct reputation/first-hand information according to the discussion in Section 20.3, is calculated directly from neighbors' observations where the past observation is given more weight than the present observation in order to minimize false detection influence. Indirect reputation is the information collected through interaction and exchanged with other nodes. Functional reputation is the global reputation value associated with every node. CORE prohibits the spread of negative rating and allows only the positive ratings to be distributed in order to resist attacks such as BM. Unfortunately, giving greater weight to the past observations enables a malicious node to misbehave temporarily if it has accumulated a high reputation value. Moreover, combining the reputation values for various functions into a single global value is another problem since it helps a malicious node to hide its misbehavior with respect to certain functions by behaving cooperatively with respect to the remaining functions.

CORE is a distributed reputation system, which means that the indirect (second-hand) information is propagated to other neighboring nodes. However, only positive feedback is propagated when an event of interest occurs (reactive dissemination) to prevent BM attacks.

According to CORE specifications, three information sources are available: subjective, indirect, and functional reputations. The subjective reputation can be directly observed by using the WDM, which is the same as first-hand information. The designers give more weight to previous observations in order to reduce the influence of any misbehavior in a recent observation. The indirect reputation is the subjective reputation of one node that has been propagated and received by other nodes, which is the same as second-hand information. Moreover, the functional reputation is the combination of indirect and subjective reputation with respect to a specific function. It is also possible to assign more weight for a specific function using the following formula:

$$R_{s_i}(s_j) = \sum_{k=1}^{n} W_k \{R_{s_i}(s_j | f_k) + IR_{s_i}(s_j | f_k)\}$$

where $R_{s_i}(s_j)$ represents the global reputation value, W_k represents the weight associated with a specific function f_k, $R_{s_i}(s_j|f_k)$ represents the subjective reputation value (first-hand information) calculated by s_i on s_j with respect to the function f_k, and $IR_{s_i}(s_j|f_k)$ represents the indirect reputation of s_j collected by s_i for the function f_k. The reputation value varies between $[-1,1]$. A node is considered trusted if its reputation value is positive, otherwise it is distrusted.

20.4.5 Performance Analysis of the Confidant Protocol [23]

Buchegger and Boudec [23] introduced the CONFIDANT system, which works as an extension to a reactive source routing protocol for mobile and *ad hoc* networks and its implementation assumes that the network layer is based on the dynamic source routing [33]. Each node is composed of four modules in order to accomplish the CONFIDANT functionality. These modules are: monitor, reputation system, path manager, and trust manager. The monitor module is responsible for performing neighborhood watch by monitoring the behavior of its one-hop neighborhood. The monitor module registers any deviation from the normal behavior and then the reputation system module is called once a deviation occurs.

The reputation system module then updates the rating of the node that caused this deviation. There are three information sources available to the reputation system module in order to calculate the reputation value for a specific node. These sources are: personal experience, direct observation, and second-hand information. CONFIDANT weights these types of information differently. It assigns the greatest weight for personal experience, a smaller weight for observations in the neighborhood and the smallest weight for the reported experience (second-hand information). This is because nodes trust their own experiences and observations more than those of other nodes. Unfortunately, the authors did not explain how these three types of information are calculated together to give the reputation value of a node.

After that, the path manager and the trust manager modules are called. The path manager module detects all routes containing the misbehaving node from the path cache, whereas the trust manager module sends ALARMs to warn other nodes in the neighborhood, which are stored in its friends list, about malicious nodes. This means that only negative trust values are propagated by the node's trust manager module to other nodes.

20.4.6 A Distributed Reputation-Based Beacon Trust [25]

Srinivasan et al. [25] developed a distributed reputation-based beacon trust system for WSNs called DRBTS, which excludes malicious beacon nodes that provide false location information. Beacon nodes are special sensor nodes that have the capability of knowing their location through a GPS receiver, manual configuration, and so on. DRBTS helps sensor nodes to validate whether given location information is correct or not. It differs from previous reputation-based trust systems in calculating the trust values for beacon nodes not for normal sensor nodes.

The network topology consists of three types of devices: sensor nodes, beacon nodes, and a BS. The BS is not involved in any function of DRBTS because of the distributed architecture of the system. The information gathering stage is done at two points: the sensor node and the beacon node. Each beacon node runs an adaptive version of the WDM in order to monitor other beacon nodes within one-hop of its neighborhood. When a sensor node broadcasts a query asking about its location, each beacon that is able to hear this broadcast should respond with the sensor's location information. Other beacon nodes, which are able to hear this query and replies, compare their calculations of the sensor's location information with others' calculations. If the difference is within a certain range, the reputation values of the other beacons are increased. Otherwise, they are decreased.

On the other hand, sensor nodes are not concerned with gathering direct observations. They are only interested in indirect observations or second-hand information, which are propagated from the beacon nodes. DRBTS is a distributed reputation system which requires direct observation (first-hand information) to be propagated to other nodes to speed up spreading the reputation values within the network. The scheme designers have chosen to couple the answer to the location's query (received from a sensor node) with the dissemination of the reputation values of other beacon nodes. Thus, when a beacon node answers the sensor query, it broadcasts the sensor's estimated location with the beacon's neighbor RT (NRT). This table contains positive and negative feedback about the other beacon nodes.

The reputation value for each beacon node is updated after obtaining first- or second-hand information. Suppose a beacon node B_i overhears location information transmitted by another beacon node B_j, it first compares B_j's location information with its estimation. If B_j's location information is acceptable, then $\tau = 1$; otherwise, $\tau = 0$. The reputation value of B_j is calculated by B_i as follows:

$$R_{i,j}^n = \mu_1 R_{i,j}^c + (1 - \mu_1)\tau$$

where $R_{i,j}^n$, $R_{i,j}^c$ represents the new and current reputation values calculated by B_i for B_j, respectively, and μ_1 denotes a factor that is used to weight previous experience against the current information.

Furthermore, B_i considers the overhead NRT_j for updating its NRT_i. Suppose NRT_j has a reputation value for another beacon node B_k which also exists in NRT_i. Beacon node B_i performs a deviate test on these two reputation values as follows:

$$|R_{i,k}^c - R_{j,k}^c| \le d$$

If the result of the deviation test is positive, then the published information by B_j is considered to be compatible with B_i's first-hand information. Then, B_i accepts this published information and updates $R_{i,k}$ in its NRT_i as follows:

$$R_{i,k}^n = \mu_2 R_{i,k}^c + (1 - \mu_2)R_{j,k}^c$$

However, if the result is negative, then the published information by B_j is considered to deviate too much from its own first-hand experience, and is disregarded as incompatible information. Moreover, the beacon node B_j has to be punished by reducing its reputation value as follows:

$$R_{i,j}^n = \mu_3 R_{i,j}^c$$

The sensor node that produced the location request will receive NRTs and location information from its beacon neighbors. It counts the number of positive and negative votes, and then stores them in the trusted beacon neighbor (TBN) table. A positive vote for a beacon node B_j is given when B_i reports a reputation value for B_j greater than a predefined trust value threshold in a sensor node. DRBTS uses majority votes to decide the final reputation value of the beacon node B_j.

20.4.7 Reputation-Based Framework for High Integrity Sensor Networks [34]

Ganeriwal and Srivastava [34] designed a reputation-based framework for high-integrity sensor networks (RFSN) in order to create a community of trustworthy sensor nodes. RFSN is a distributed, symmetric reputation-based model that uses both first- and second-hand information for updating reputation values. In RFSN, nodes maintain the reputation and trust values for only nodes in their neighborhood. RFSN is the first reputation system work that designed a reputation system specifically for WSNs. It uses the WDM in order to perform two operations: collecting observations and making decision on these observations. Each node stores RT that is maintained by the node itself. Each entry of the table is built over time through the WDM and the output of the WDM is used to update the reputation value $R_{i,j}$ which is calculated by node i for node j.

If this reputation value does not fall below a threshold, which is a positive reputation value, then it is propagated to other nodes. The reputation value (R) of each node consists of direct reputation $(R)_D$, which is built up using direct observation through the WDM, and indirect reputation $(R)_{ID}$, which is built up using second-hand information received from a highly reputed node. Finally, the decision on cooperating with other nodes is made based on a threshold-based policy. If $R_{i,j}$ is greater than a threshold value, then node i cooperates with node j; otherwise, it does not.

20.4.8 Trust-Based Security for Wireless Ad Hoc and Sensor Networks [35]

Boukerche et al. [35] introduced the agent-based trust and reputation management scheme (ATRM) in order to manage trust and reputation in a clustered WSN with mobile backbones. In mobile backbone cluster WSNs, cluster heads are required to be predeployed backbone nodes, which are capable of communicating through radios of different ranges and are connected directly through long-range radios instead of message relay. Nodes in the same level share the same communication channel, but radios in different levels use different frequencies and channel resources. Nodes communicate only with other nodes that belong to the same cluster, and inter-cluster communication has to be done through cluster heads in the higher level backbone network. Consequently, each cluster can be considered an independent small-sized wireless network, and the higher level backbone network is another wireless network linking the clusters together.

Different from traditional trust and reputation management systems, ATRM requires that nodes trust and reputation information to be stored, respectively, in the forms of t-instrument and r-certificate by the node itself. ATRM also requires that every node locally holds a mobile agent that is in charge of administrating the trust and reputation of its hosting node. In this sense, mobile agents provide nodes with a one-to-one trust and reputation management service.

The execution of ATRM involves two phases: the network initialization and the service-offering. The purpose of the network initialization phase is to distribute a mobile agent (TRA) to each node. The authors assumed that there is a trusted authority that is responsible for generating and launching mobile agents.

The second phase, the service-offering phase, starts as soon as the TRA has been distributed. It is composed of four modules: r-certificate acquisition, t-instrument issuance, r-certificate issuance, and trust management routine. The objective of the r-certificate acquisition module is to obtain the reputation of the service provider to the requester. The reputation of the provider is stored in the r-certificate which is located at the provider's mobile agent. Therefore, the requester asks its mobile agent to communicate directly with the provider's mobile agent to obtain the provider's reputation value. The second module is the *t*-instrument issuance, which allows the requester to evaluate the offered service from the provider. The third module is *r*-certificate issuance, which involves two types of operation: computing reputation and generating *r*-certificate. For any node, its mobile agent computes the reputation of the node based on the old *r*-certificate value and the *t*-instruments that resulted from previous transactions with other nodes. Then, the node's mobile agent generates an *r*-certificate with time stamp and replaces the old *r*-certificate value. Unfortunately, no details have been given on how this can be computed. The final module is the trust management routine where the mobile agent keeps recording the trust evaluation values of other nodes.

In ATRM, a transaction is defined as the process of interaction between two nodes, the requester and the provider, and it is triggered by the requester and may be accepted/rejected by the provider. Before starting any transaction, the requester asks its local mobile agent to obtain the *r*-certificate of the provider by directly querying the provider's local mobile agent. On the basis of the provider's *r*-certificate, the requester decides whether or not to start the transaction. After a transaction is finished, the requester makes a trust evaluation on the provider based on the quality of service it gets from the provider during the transaction, and then submits the evaluation to its local mobile agent which then accordingly generates a *t*-instrument for the provider and sends the *t*-instrument to the provider's local mobile agent. On the basis of the collected *t*-instruments, a mobile agent periodically issues its hosting node updated *r*-certificates.

20.4.9 Formal Reputation System for Trusting WSNs [36]

Xiao et al. [36] argued that cryptography-based security alone is not enough in WSNs due to unique characteristics and resource constraints. They, therefore, proposed a reputation-based trust system that uses both direct and indirect reputation information as sources of reputation values. Direct reputation information $(R)_D$ is built using direct observations about neighboring sensor nodes, whereas indirect reputation information $(R)_{IND}$ is built using second-hand information received from neighboring sensor nodes. The final reputation value $(R_{x,y})$ is calculated by a sensor node x for its neighbor y as:

$$R_{x,y} = (R_{x,y})_D + (R_{x,y})_{IND}$$

where $(R_{x,y})_{IND}$ represents the reputation value for node y reported by other neighbors (but not node x) and multiplied by the available reputation value of the reporting node at node x. For example, the reputation value of node y, which is calculated by node x with the help of second-hand information reported by nodes a and b can be calculated as:

$$R_{x,y} = (R_{x,y})_D + (R_{x,y})_{IND}$$
$$= (R_{x,y})_D + [(R_{x,a})_D(R_{a,y})_{IND} + (R_{x,b})_D(R_{b,y})_{IND}]$$

After that, trust is obtained by taking the statistical expectation of reputation value as:

$$T_{x,y} = E[R_{x,y}]$$

Xiao et al.'s scheme uses a binary decision metric whenever a sensor node cooperates with another. The scheme designers used a simple threshold-based policy to decide the node's decision as:

$$D_{x,y} = \begin{cases} \text{accept}, & \forall T_{x,y} \geq \theta \\ \text{reject}, & \forall T_{x,y} < \theta \end{cases}$$

20.5 Comparison of Current Reputation-Based Systems in WSNs

This section provides the security and performance analysis of current reputation-based trust systems in WSNs. This analysis can be difficult for the following reasons:

■ The designers of each system solved the trustworthiness problem in WSNs from different angles and for different activities. For example, some designers solved the problem by considering only routing misbehaviors as in [22–24,37]. It is believed that each activity, such as routing or data aggregation, has its own challenges that need to be considered carefully.
■ There is no standard adversarial model where current reputation-based trust systems compete to provide a higher level of security, or resilience to attacks discussed in Section 20.2.
■ There is not enough information about different reputation components discussed in Section 20.3, which sometimes makes the comparison invisible.

Existing reputation-based trust systems, consequently, are compared in a number of different ways: the system scope they follow, reputation components their systems are composed of, and resilience against attacks described in Section 20.2.

20.5.1 Classification Model

On the basis of the discussion in Section 20.4, the current reputation-based trust systems in WSNs were designed in order to enhance the trustworthiness between sensor nodes. These systems fall under one of five categories: generic, localization, mobility, routing, and aggregation.

Figure 20.12 concludes the discussion in Section 20.4 and classifies the current reputation-based trust systems, depending on what activity attracts most the system designers. Ganeriwal and Srivastava [34], Chen [38], Yao et al. [39], Xiao et al. [36], and Boukerche et al. [35] designed generic reputation-based trust systems, which do not consider a specific activity. They argued that their systems can be tailored to do any sort of activity. Moreover, Boukerche and Ren [29] introduced the concept of community and then they proposed a reputation-based system that considers the mobility of nodes in the community, which is also addressed by Srinivasan et al. [40]. Furthermore, Srinivasan et al. [25] designed a reputation-based system that enforces cooperation between beacon nodes by motivating them to provide correct location information. Moreover, Michiardi and Molva [22], Buchegger and Boudec [23], Chen et al. [37], and Yao et al. [24] considered only the routing misbehaviors when a node evaluates another one. Finally, Alzaid et al. [31] and Özdemir [41] integrated the aggregation functionalities with the advantages provided by a reputation component in order to enhance the accuracy of the aggregated values.

20.5.2 Reputation Components Visibility

According to the discussion in Section 20.3, reputation-based trust systems often share similar structural pattern. They consist of three main phases: information gathering and sharing, information modeling (or reputation calculation), and decision-making (see Figure 20.9). This section investigates the visibility of these phases (and the internal components of each phase) in the current reputation-based trust systems. Table 20.2 incorporates the discussion on Section 20.3 and then analyzes trust systems for WSNs discussed in this chapter. It also depicts the information related to each phase (and its components) covered by the designers of each trust system, which helps

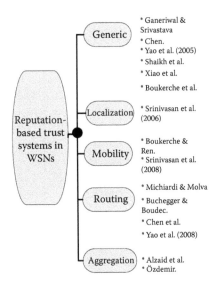

Figure 20.12 The classification model for reputation-based trust systems in WSNs.

understanding the differences between the reputation-based trust systems in the current literature. We believe that Table 20.2 is self-explanatory and hence we do not provide a discussion on it.

20.5.3 Attack Visibility

This section investigates the current reputation-based trust systems for WSNs to determine whether or not these systems are vulnerable to attacks discussed in Section 20.2. Table 20.3 shows that all systems except Boukerche et al.'s [35] system are vulnerable to these attacks. Damage caused by these attacks varies from one system to another, depending on whether these attacks were considered at the design time or not. Importantly, attacks are not visible in the trust system in [35] due to the use of mobile agents, which is assumed to be generated and launched by a trusted authority. We believe that if this assumption is relaxed, some of the attacks may threaten their system.

The SF attack occurs when an adversary, which is controlling a compromised node, selectively forwards received messages. The damage caused to the reputation component by the SF attack varies from no damage to maximum damage, as shown in Table 20.3. The SF attack causes partial damage in systems [22,23,29,31,34,37–39,41] although they monitor the forwarding activity. This is because these systems use a binary decision method when they evaluate the trust level of a specific node. This method is based on a threshold policy, and once the node's reputation is above this threshold value, then the node is considered trusted. For example, an adversary can launch the SF attack once every t transactions to be above the threshold value and then to keep its trust level. The damage is considered partial because adjusting the threshold value or applying mechanisms such as ageing factor and weighting can help defeating this attack. Unfortunately, some systems designers did not consider forwarding misbehaving in their systems such as in [25,26,36,40] and therefore, the damage caused by the SF attack is maximum.

Moreover, Table 20.3 shows that there is a link between the adversary capability of launching SY and NC attacks. According to the discussion in Section 20.2, the adversary can launch the SY attack by presenting more than one identity, which means that the adversary is able to launch the NC attack once it has succeeded in presenting another identity beside its original identity. Interestingly,

Table 20.2 Classification of Existing Reputation-Based Trust Systems

Scheme	Gathering & Sharing				Calculation		Dissemination		
	Source	WDM	Type	Scope	Structure	Approach	Structure	Approach	Decision Metric
Michiardi and Molva [22]	D/I	Y	+	R	Di	?	Di	Re	B
Buchegger and Boudec [23]	D/I	Y	−	R	Di	?	Di	Re	B
Ganeriwal and Srivastava [34]	D/I	Y	+	G	Di	Pr	Di	P	B
Srinivasan et al. [40]	D/I	Y	+,−	M	C	De	C	Re	B
Boukerche et al. [35]	D	N	+,−	G	P2P	?	P2P	Re	?
Alzaid et al. [31]	D	Y	−	A	Di	Pr	Di	Re	B
Yao et al. [24]	D/I	Y	+,−	R	C	De	Di	P	Disc
Shaikh et al. [26]	D	?	+,−	G	H	De	H	P,Re	Disc
Özdemir [41]	D/I	Y	+,−	A	Di	Pr	Di	P	B
Bouckerche and Ren [29]	D/I	Y	+,−	M	C	De	Di	P,Re	B
Yao et al. [39]	D/I	?	+,−	G	Di	De	Di	P	Disc
Chen et al. [37]	D/I	?	+,−	R	Di	Pr	Di	P	B
Chen [38]	D/I	Y	+,−	G	Di	Pr	Di	P	B
Xiao et al. [36]	D	?	+,−	G	Di	Pr	Di	P	B
Srinivasan et al. [25]	D	Y	+,−	L	Di	De	Di	Re	B

Note: +, Positive feedback; −, Negative feedback; ?, Not available; A, Aggregation misbehavior; B, Binary; C, Centralized; D, Automatic direct; De, Deterministic; Di, Distributed; Disc, Discrete; G, Generic misbehavior–module dependent; H, Hybrid; I, Automatic indirect; L, Localization misbehavior; M, Mobility; P, Proactive; Pr, Probabilistic; P2P, Peer to peer; R, Routing misbehavior; Re, Reactive; WDM, Watchdog mechanism.

Table 20.3 Attack Visibility in the Existing Reputation-Based Trust Systems

Scheme	WSNs Attacks				Reputation Attacks			
	SF	SY	SD	RE	BM	BA	OO	NC
Michiardi and Molva [22]	·		··	··	··		··	
Buchegger and Boudec [23]	·	··	·	··		··	··	··
Ganeriwal and Srivastava [34]	·		··	··	·		··	·
Srinivasan et al. [40]	··	··	··	··	··	··	··	··
Boukerche et al. [35]								
Alzaid et al. [31]	·		·			·	·	
Yao et al. [24]	·	··		··	··	··	··	
Shaikh et al. [26]	··		··	··	··	··	··	
Özdemir [41]	·		·	··	··	··	··	
Bouckerche and Ren [29]	·		·	·	··	··	··	
Yao et al. [39]	·	··			··	··	··	··
Chen et al. [37]	·	··		··			··	··
Chen [38]	·	··		··			··	··
Xiao et al. [36]	··	··	··	··	··	··	··	··
Srinivasan et al. [25]	··		··	··	··	··	··	

Note: " " denotes robust, "·" denotes partial visibility, and "··" denotes maximum visibility.

reputation-based trust systems such as [23,24,36–40] are vulnerable to SY and NC attacks. This is due to the lack of discussion on an authentication process used between sensor nodes in these systems.

Furthermore, the RE attack occurs when the adversary has the ability to replay old messages into the network. Surprisingly, this attack is visible in reputation-based trust systems such as [22,23,25,26,29,34,36–38,40]. This can harm these systems, especially if the adversary is able to replay old reputation information, which is not valid anymore. We argue that systems with vulnerability to the RE attack are also vulnerable to the SD attack because the adversary can first capture some reputation information and then replay it into the network, even without changing it, in order to affect the performance of the reputation component—which is one form of the SD attack.

Moreover, BM and MA attacks are visible in systems that use second-hand information in the reputation calculation. On the one hand, the BA attack is visible in reputation-based trust systems that allow sensor nodes to exchange their negative feedback such as in [23–26,29,31,36,39–41]. However, the damage caused by this attack in limited/partial in [31] because the negative feedback is accepted to revoke the cell representative only if it is supported by $t - 1$ witnesses. On the other hand, the BM attack is visible in systems that allow sensor nodes to propagate their positive feedback such as in [22,24–26,29,34,36,39–41]. Finally, the OO attack occurs when the adversary tries to launch a mixture of attacks discussed in Section 20.2 in an irregular basis in order to keep its reputation value within an acceptable trust value. Importantly, Table 20.3 shows that all reputation-based trust systems are vulnerable to this attack. The damage caused by this attack varies, depending on how many attacks these systems are vulnerable to.

20.6 Conclusion

This chapter provides a detailed review of reputation-based trust systems in WSNs. It first explains the motivation behind adding the reputation system capabilities into WSNs, which in brief helps to enhance the trustworthiness between sensor nodes. It then discusses how the integration between WSNs and reputation systems can open doors for an adversary to threaten reputation-based trust systems destined for WSNs, and affect the entire performance. After that, the "state-of-the-art" in reputation-based trust systems is surveyed and classified into five categories (generic, localization, mobility, routing, and aggregation) depending on what activity attracts most the system designers. Subsequently, the current reputation-based trust systems in WSNs are compared in a number of different ways: the reputation components they are composed of, and the attacks they secure against.

Terminologies

ATRM—Agent-based Trust and Reputation Management Scheme Security
NRT—Neighbor Reputation Table
RSDA—Reputation-based Secure Data Aggregation
TAP—Trust Assistant Policy
TBN—Trusted Beacon Neighbor table
WDM—WatchDog Mechanism
WSN—Wireless Sensor Network

Questions and Sample Answers

1. Why are WSNs vulnerable to different types of attacks?
   WSNs are vulnerable to different types of attacks due to the nature of the transmission medium (broadcast), remote and hostile deployment location, and the lack of physical security in each node.

2. What is Sybil attack?
   The Sybil attack is a type of attack where the adversary is able to present more than one identity (node) within the network to deceive other nodes. A node that wishes to conduct the SY attack can create a new identity or impersonate the identity of an existing node. The adversary can create multiple identities to affect the reputation values of legitimate nodes in reputation-based applications by falsely degrading their reputation values.

3. What is a replay attack?
   Some applications that are destined for WSNs are vulnerable to replay attack where the adversary is able to eavesdrop on the traffic and replay old messages. This type of attack is the easiest one because the adversary does not need to physically capture a sensor node or get access to its internal memory.

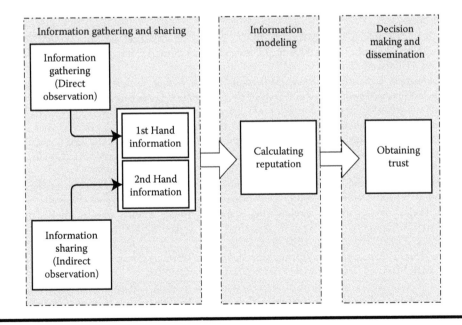

On/off attack. (a) Real reputation update and (b) altered reputation update.

4. Describe OO attack. Draw figures of OO attack.

   In this type of attack, the adversary aims to distribute the system's overall performance with the hope that it will not be detected or excluded from the network. The adversary behaves badly and correctly alternatively in order to extend the detection time required to recognize its misbehaviors. This attack can be launched against either the reputation activities or general activities in WSNs.

5. Show "Reputation System Stages" with a diagram.

Reputation system stages.

Author's Biography

Hani Alzaid received his BEng in computer engineering from King Saud University (Riyadh, Kingdom of Saudi Arabia) in 2000 and his MS from University of New South Wales (Sydney, Australia) in 2005. He is currently working toward his PhD in computer science in Queensland University of Technology, Australia. His research interests include secure data aggregation, key management, and reputation-based solutions for wireless sensor networks (WSNs).

References

1. C. Siva Ram Murthy and B.S. Manoj. *Ad Hoc Wireless Sensor Networks Architectures and Protocols*. Prentice Hall PTR, Upper Saddle River, NJ, USA, 2004.
2. A.M. Mainwaring, D.E. Culler, J. Polastre, R. Szewczyk, and J. Anderson. Wireless sensor networks for habitat monitoring, in *Proceedings of the 1st ACM International Workshop on Wireless Sensor Networks and Applications (WSNA'02)*, Atlanta, GA, USA, pp. 88–97, September 28, 2002.
3. T. He, P. Vicaire, T. Yan, L. Luo, L. Gu, G. Zhou, R. Stoleru, Q. Cao, J.A. Stankovic, and T.F. Abdelzaher. Achieving real-time target tracking using wireless sensor networks, in *Proceedings of the IEEE Real Time Technology and Applications Symposium*, pp. 37–48. IEEE Computer Society, Los Alamitos, CA, USA, 2006.
4. G. Simon, M. Maróti, Á. Lédeczi, G. Balogh, B. Kusy, A. Nádas, G. Pap, J. Sallai, and K. Frampton. Sensor network-based countersniper system, in *Proceedings of the 2nd International Conference on Embedded Networked Sensor Systems (SenSys'04)*, ACM Press, New York, NY, USA, pp. 1–12, November 3–5, 2004.
5. E. Cayirci and T. Coplu. Sendrom: Sensor networks for disaster relief operations management. *Wireless Networks*, 13(3):409–423, 2007.
6. A. Milenkovic, C. Otto, and E. Jovanov. Wireless sensor networks for personal health monitoring: Issues and an implementation. *Computer Communications*, 29(13–14):2521–2533, 2006.
7. J. Yick, B. Mukherjee, and D. Ghosal. Wireless sensor network survey. *Computer Networks*, 52(12):2292–2330, 2008.
8. Crossbow Technology, Inc. Mica2 datasheet, 2006. The datasheet of Mica2 mote platform is retrieved on November 10, 2009, http://www.xbow.com/Products/Product_pdf_files/Wireless_pdf/MICA2_Datasheet.pdf.
9. G. Guimarães, E. Souto, D. Fawzi Hadj Sadok, and J. Kelner. Evaluation of security mechanisms in wireless sensor networks, in *Proceedings of the International Conference on Wireless Technologies/High Speed Networks/Multimedia Communications Systems/ Sensor Networks, ICW/ICHSN/ICMCS/SENET*, Montreal, Canada, pp. 428–433, August 14–17, 2005.
10. C. Hartung, J. Balasalle, and R. Han. Node compromise in sensor networks: The need for secure systems. *Technical Report CU-CS-990-05*, Department of Computer Science, University of Colorado at Boulder, January 2005.
11. Z. Yan, P. Zhang, and T. Virtanen. Trust evaluation based security solution in *ad hoc* networks. Technical report, December 2003. This technical report is retrieved on November 12, 2009, http://research.nokia.com/publications/trust_evaluation_based_security_solution_ad_hoc_networks.
12. A. Perrig, R. Szewczyk, J.D. Tygar, V. Wen, and D.E. Culler. SPINS: Security protocols for sensor networks. *Wireless Networks*, 8(5):521–534, 2002.
13. P. Dasgupta. Trust as a commodity, in *Trust: Making and Breaking Cooperative Relations*, D. Gambetta (ed.), Basil Blackwell, University of Oxford, UK, pp. 49–72, 2000.
14. D. Gambetta. Can we trust trust? in *Trust: Making and Breaking Cooperative Relations*, D. Gambetta (ed.), Blackwell Publishing Ltd., Oxford, UK, pp. 213–237, 1988.
15. L. Mui, M. Mohtashemi, and A. Halberstadt. A computational model of trust and reputation for e-businesses, in *Proceedings of the 35th Annual Hawaii International Conference on System Sciences (HICSS'02)*, Hilton Waikoloa Village Island of Hawaii, USA, vol. 7, p. 188, January 7–10, 2002.

16. A. Abdul-Rahman and S. Hailes. Supporting trust in virtual communities, in *Proceedings of the 33th Annual Hawaii International Conference on System Sciences (HICSS'00)*, Maui, Island of Hawaii, USA, vol. 6, January 4–7, 2000.

17. C. Keser. Experimental games for the design of reputation management systems. *IBM Systems Journal*, 42(3):498–506, 2003.

18. C. Karlof and D. Wagner. Secure routing in wireless sensor networks: Attacks and countermeasures. *Ad Hoc Networks*, 1(2–3):293–315, 2003.

19. T. Roosta, S. Shieh, and S. Sastry. Taxonomy of security attacks in sensor networks and countermeasures, in *Proceedings of the 1st IEEE International Conference on System Integration and Reliability Improvements*, IEEE International, Washington, DC, USA, 2006.

20. H. Alzaid, E. Foo, and J. Manuel González Nieto. Secure data aggregation in wireless sensor network: A survey, in *Proceedings of the 6th Australasian conference on Information security (AISC'08)*, Wollongong, NSW, Australia, pp. 93–105, January, 2008.

21. R. Ismail. Security of Reputation Systems. Ph.D. in Computer Science, Queensland University of Technology, Brisbane, Australia, 2004. The thesis is retrieved on November 24, 2009, http://eprints.qut.edu.au/15964/.

22. P. Michiardi and R. Molva. CORE: a collaborative reputation mechanism to enforce node cooperation in mobile *ad hoc* networks, in *Communications and Multimedia Security*, vol. 228 of IFIP Conference Proceedings, pp. 107–121, September 26–27, 2002.

23. S. Buchegger and J.-Y. Le Boudec. Performance analysis of the confidant protocol, in *Proceedings of the 3rd ACM International Symposium on Mobile Ad Hoc Networking and Computing (MobiHoc'02)*, Lausanne, Switzerland, pp. 226–236, June 9–11, 2002.

24. Z. Yao, D. Kim, and Y. Doh. PLUS: Parameterized localized trust management-based security framework for sensor networks. *IJSNET*, 3(4):224–236, 2008.

25. A. Srinivasan, J. Teitelbaum, and J. Wu. DRBTS: Distributed reputation-based beacon trust system, in *Proceedings of the 2nd International Symposium on Dependable Autonomic and Secure Computing (DASC'06)*, Indianapolis, IN, USA, pp. 277–283, 29 September–1 October 2006.

26. R. Ahmed Shaikh, H. Jameel, S. Lee, S. Rajput, and Y. Jae Song. Trust management problem in distributed wireless sensor networks, in *Proceedings of the 12th IEEE Conference on Embedded and Real-Time Computing Systems and Applications (RTCSA'06)*, Sydney, Australia, pp. 411–414, August 16–18, 2006.

27. S. Marti, T.J. Giuli, K. Lai, and M. Baker. Mitigating routing misbehavior in mobile *ad hoc* networks, in *Proceedings of the 6th annual international conference on Mobile computing and networking (MOBICOM)*, Boston, MA, USA, pp. 255–265, 2000.

28. R. Roman, M. Carmen Fernandez-Gago, J. Lopez, and H.-H. Chen. Trust and reputation systems for wireless sensor networks, in C. Skianis Stefanos Gritzalis, T. Karygiannis, eds. *Security and Privacy in Mobile and Wireless Networking*, pp. 223–233. Troubador Publishing Ltd., Leicester, 2008.

29. A. Boukerche and Y. Ren. A trust-based security system for ubiquitous and pervasive computing environments. *Computer Communications*, 31(18):4343–4351, 2008.

30. S. Buffett, N. Scott, B. Spencer, M. Richter, and M.W. Fleming. Determining internet users' values for private information, in *Proceedings of the 2nd Annual Conference on Privacy, Security and Trust (PST)*, Wu Centre, University of New Brunswick, Fredericton, New Brunswick, Canada, pp. 79–88, October 13–15, 2004.

31. H. Alzaid, E. Foo, and J. Gonzalez Nieto. RSDA: Reputation-based secure data aggregation in wireless sensor networks, in *Proceedings of the 9th International Conference on Parallel and Distributed Computing, Applications and Technologies (PDCAT'08)*, Dunedin, New Zealand, pp. 419–424, December 1–4, 2008.

32. R. Ismail and A. Jøsang. The beta reputation system, in *Proceedings of the 15th Bled Conference on Electronic Commerce*, Bled, Slovenia, 2002.

33. D.B. Johnson, D.A. Maltz, and J. Broch. DSR: The dynamic source routing protocol for multihop wireless Ad Hoc networks. http://www.monarch.cs.rice.edu/monarch-papers/dsr-chapter00.pdf (retrieved on November 10, 2009).

34. S. Ganeriwal and M.B. Srivastava. Reputation-based framework for high integrity sensor networks, in *Proceedings of the 2nd ACM Workshop on Security of Ad Hoc and Sensor Networks (SASN'04)*, Washington, DC, USA, pp. 66–77, October 25, 2004.

35. A. Boukerche, L. Xu, and K. El-Khatib. Trust-based security for wireless *ad hoc* and sensor networks. *Computer Communications*, 30(11–12):2413–2427, 2007.

36. D. Xiao, J. Feng, and H. Zhang. A formal reputation system for trusting wireless sensor network. *Wuhan University Journal of Natural Sciences*, 13(2):173–179, 2008.

37. H. Chen, H. Wu, J. Hu, and C. Gao. Agent-based trust management model for wireless sensor networks, in *Proceedings of the International Conference on Multimedia and Ubiquitous Engineering (MUE'08)*, Busan, Korea, pp. 150–154, April 24–26, 2008.

38. H. Chen. Task-based trust management for wireless sensor networks. *International Journal of Security and Its Applications*, 3(2):21–26, 2009.

39. Z. Yao, D. Kim, I. Lee, K. Kim, and J. Jang. A security framework with trust management for sensor networks, in *Proceedings of the 1st International Conference on Security and Privacy for Emerging Areas in Communication Networks*, Athens, Greece, pp. 190–198, September 5–9, 2005.

40. A. Srinivasan, F. Li, and J. Wu. A novel CDS-based reputation monitoring system for wireless sensor networks, in *Proceedings of the 28th IEEE International Conference on Distributed Computing Systems Workshops (ICDCS'08)*, Beijing, China, pp. 364–369, June 17–20, 2008.

41. S. Özdemir. Functional reputation based reliable data aggregation and transmission for wireless sensor networks. *Computer Communications*, 31(17):3941–3953, 2008.

Chapter 21

Major Works on the Necessity and Implementations of PKC in WSNs

A Beginner's Note

Al-Sakib Khan Pathan

Contents

21.1 Introduction

Security is an indispensable issue in many types of wireless sensor networks (WSNs). Although all types of applications of WSNs may not require security, most of the applications need a minimum level of data authenticity, privacy, and integrity. The applications may range from simple monitoring services to high-level security-demanding critical services such as military reconnaissance or disaster warning and response system. In many WSN-based applications, crucial decisions are taken based on the data supplied by the tiny devices dispersed over the target areas. Hence, it is imperative to ensure the legitimacy of the devices participating in the network and that of the data they exchange within the network. Another important aspect of WSNs is that such type of network is usually left unattended in the deployment area and the base station (BS) is often placed in remote location. This particular feature along with the broadcast nature of wireless communication opens the door of many types of logical and physical attacks against WSNs. Strong cryptography can be beneficial to defend against many of the logical attacks on sensor data. Though all types of physical attacks cannot be directly defended using cryptographic mechanisms, the method of operation of Public-Key Cryptography (PKC) can provide some types of protection (we explain this issue in case 5 presented later).

There are mainly two types of cryptographies: symmetric and asymmetric. In Symmetric Key Cryptography (SKC), the sender and the receiver of a message share a single, common key for encryption and decryption of the message, whereas in asymmetric key cryptography (also known as PKC), the key (public key) used to encrypt a message differs from the key (private key) used to decrypt it. These two keys are mathematically related to each other. For managing security issues in WSNs, among the two types of cryptographies, SKC has been preferred over PKC from the very beginning stage because of its low-resource requirements. The use of PKC in WSNs was totally ruled out for a long time with the idea that PKC is energy-inefficient, slow, and computationally complex; thus not at all suitable for resource-constrained devices such as sensors. More or less this was the commonly held belief of the researchers until late 2004 when some courageous researchers came up with positive results about the feasibility of PKC in WSNs. From that time up to today a good number of works have proved the suitability of PKC with practical and experimental validations. The intent of this chapter is to present a survey on these notable implemented works. Our focus is to get a lucid picture of the current status of practical achievements, challenges, and future hopes of harnessing the benefits of PKC in low-resource sensor network environment. With the survey of various notable works, we also present an overview of PKC, analyze its necessity for WSNs, and the hurdles to implementing it in WSNs. We conclude the chapter with a summary of proven facts and future expectations of exploiting PKC in WSNs.

21.2 PKC in WSNs

In PKC, a pseudo-random private key and a corresponding public key are generated by an entity. In a network, the public key of a participating entity is made open to all but the private key is kept secret only to its owner. All other entities in the network can use the public key for encrypting the plaintexts while sending them to the public-key owner. *Plaintext* is a term used in communication technology to refer to the original message sent by the sending entity. When the receiver receives an encrypted plaintext, it can decrypt it using its private key.

A typical scenario of using PKC in a WSN is shown in Figure 21.1. In this particular scenario, a deployed sensor node only needs to encrypt its plaintext (M) before sending it to the BS. The encrypted message is termed *ciphertext* (shown as C), which is transmitted through an unsecured channel. Only the entity that knows the corresponding private key (in this figure, the BS) is able to decipher the encrypted message. So, any other entity or intermediate node cannot decipher C without knowing the secret private key (PRI) of the BS. This method allows the ciphertext to travel through multiple hops without public disclosure of the plaintext. In the figure, $E_{PUB}(M)$ stands for

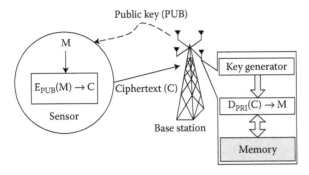

Figure 21.1   An example of using PKC in WSNs.

encryption of message "M" with the public key of BS, "PUB" (that becomes the ciphertext C) and $D_{PRI}(C)$ stands for decryption of the ciphertext C using the private key of the BS, "PRI."

In almost all types of applications of WSNs, the task of the sensors is to sense their surrounding environments and send the readings of certain parameters (e.g., light, temperature, sound, magnetism) to the BS or sink. Usually, BS is considered to be fully secure and cannot be compromised in any way. Often, a routing tree is formed over the entire network putting BS as its root. For collecting data from sensors, the important concern of the BS is to ensure authenticity, privacy, and integrity of the sensor readings as well as the authenticity of the source of data. As these networks are often deployed in physically accessible open environments, the sensors are exposed to potential physical intruders who can even capture and tamper the sensors, extract secret information, add rogue nodes in the network, or physically damage the sensors. Apart from other logical attacks within the network (e.g., Sybil attack, Sinkhole attack, Hello Flood attack, Wormhole attack), some of these physical attacks can lead to various types of logical attacks on sensor data such as eavesdropping, altering sensor readings, pilfering information, and injecting false readings. To deal with a wide range of physical and logical attacks in WSNs, using strong cryptographic mechanism is required. Given these characteristics and nature of operations of WSNs, sometimes the use of PKC can be advantageous than that of its symmetric counterpart. In this section, we will explore these cases for WSNs.

Case 1. Complexity of key management and storage requirement in some network settings: PKC allows flexible key management in WSNs. In some symmetric key based security solutions for WSNs, complex broadcast authentication and delayed key disclosure are necessary, which complicate the task of key management. Delayed key disclosure is a method where a key associated with a particular time slot (i.e., key is valid only for a particular time slot) is revealed after some interval. As the symmetric keys are revealed sequentially over time, the nodes might have to store multiple (or, large volume) messages before they can be authenticated. This type of broadcast authentication scheme also means that the BS needs to broadcast the updated keys to all nodes in regular intervals. As WSN is a multihop network and often the sensors cannot directly reach the BS because of limited wireless communication range, the updated keys need to be forwarded from node to node (hop by hop). Moreover, proper time synchronization in the network is necessary. This type of complex key management and high storage requirements for multiple keys and messages consume considerable amount of resources of the nodes. PKC does not have such type of complexity in its operational method and at the same time it reduces the memory usage as only the public key of the BS needs to be stored in the sensor memories.

Case 2. Absence of pairwise key: In many WSN applications, the sensor nodes just need to send their readings all the way to BS and in a particular network setting, pairwise keys may not exist (or may not be used) between the nodes and the BS. In such cases, PKC can be useful as the reporting nodes can encrypt their readings with the public key of BS and send them to BS using multiple hops. Other mechanisms can work side by side for authenticating the source of data. An example case is a healthcare sensor network that is protected from intruder intervention in some way (physically secure environment, the ids of the sensors along with their placements are well-recorded in the BS), but which needs to keep the patient-data confidential while transmitting through the network using wireless communications. In this case, using PKC is beneficial as no secret key needs to be established rather a node only needs to know the publicly available key of the BS. Any new sensor associated with a new patient, doctor, or nurse can join the network anytime and send data securely to BS if it knows the public key of the BS.

Case 3. Offline processing and number of keys in some network settings: In some key predistribution-based symmetric key mechanisms, huge number of keys or keying information need to be stored in sensor memories prior to their deployment (e.g., probabilistic key establishment methods). It increases the burden of offline processing and often requires considerable amount of storage resource. On the other hand, in case of PKC, only the public key of the BS needs to be embedded into the sensors, which is much easier and memory-efficient.

Case 4. Establishing symmetric session key in some networks: A great advantage of PKC over SKC in WSNs is that it can be used to establish symmetric session keys, which can then be used with a private key cryptographic algorithm to encrypt and decrypt messages.

Case 5. Better protection under physical capture scenario: As WSNs are often deployed in physically accessible open environments, physical intruders could capture or tamper nodes, extract secret keys, and use those for injecting false data in the network. As in many key predistribution-based symmetric key solutions, the deployed nodes carry almost all the important keys or keying information, the damage probability from such attack is very high. For some symmetric key-based schemes, it can jeopardize the security of a large portion of the network if multiple nodes use the same secret key. PKC can be very helpful in this scenario. For instance, PKC can be used to establish symmetric session keys that could be valid for short time periods. Capturing a node in this case can reveal the key of a particular session (which would be invalid after some time). Again, in other general cases of physical capture of nodes, revealing the public-key materials of the BS would not cause any harm.

Case 6. Scalability of a network: PKC does not have scalability problem like that of some of the symmetric key solutions for WSNs. Once the network starts running, in case of key predistribution-based symmetric key schemes, assigning new keys to newly deployed sensors is a cumbersome task.

Case 7. An ad hoc *deployment scenario:* As in usual case, a WSN is deployed in an ad hoc fashion, often a sensor does not know in advance who its neighbor node will be in the network. In such an *ad hoc* deployment case, if the neighboring nodes know the public keys of each other (by exchanging their public-key information within one-hop), they can use those to derive secret key and then use that for node-to-node secure communications (within one hop).

In spite of having these advantageous cases, there are some critical factors in WSNs that deter the frequent use of PKC. We will learn about these factors in the next section.

21.3 Major Challenges to Implementing PKC in WSNs

For implementing PKC in WSNs, several aspects are taken into consideration, such as energy consumption, communication costs, computational complexity, time complexity, and the level of security. However, the major barriers to implementing it are:

- *Constrained resources of the sensors:* The sensors that build up the network are usually of an inadequate memory, processing, and communication capabilities and they cannot support the execution of a large amount of codes. Their energy sources are also very limited. For instance, very popular Crossbow MICA2 mote is equipped with an ATmega128L 8-bit processor at 8 MHz, 128 kB program memory (flash), 512 kB additional data flash memory, 433, 868/916, or 310 MHz multichannel radio transceiver, 38.4 kbps radio, 500–1000 ft. outdoor range (depending on versions) with a size of only $58 \times 32 \times 7$ (mm). Usually, it is run by TinyOS operating system and powered by two AA-sized batteries. Clearly, a device with this configuration cannot support security mechanisms that require executing a large number of instructions. If a node is busy relatively longer than other nodes in the network for performing huge calculations related to security mechanisms, it might lose its energy rapidly and can fail much sooner than the other less active nodes.
- *Throughput rate:* Usually, the throughput rates of well-known PK encryption methods are several orders of magnitude slower than that of the major symmetric key schemes.
- *Need for supporting hardware:* If not accelerated by extra cryptographic hardware, strong PKC is resource-hungry and thus may not be suitable for resource-constrained devices like the sensors.

So far we have explored the advantageous cases and challenges of implementing PKC in WSNs. Despite the initial hesitance of working on this area, in the recent years, some implementations have been done to verify how PKC performs if tested with real test-beds. In the next section, we present a survey of these prominent works. We have selected only the innovative/ground-breaking works for our survey (as well as the most recent notable works).

21.3.1 Survey on the Notable Implementations of PKC in WSNs

21.3.1.1 Types of Implementations

Table 21.1 gives a summary of the significant implementations of PKC in WSNs. As it can be noticed from the table that various implementations of PKC-based mechanisms can be categorized mainly into three approaches:

- a. *Software implementation:* Writing customized software or program that can support PKC operation or using lightweight versions of the public-key-based schemes.
- b. *Hardware implementation:* Design and development of customized hardware so that they can be used in the sensor boards for supporting PKC-based operations and computations.
- c. *Hardware/software blended implementation:* A third approach is the combination of hardware support and software optimization that also works effectively in some cases.

In the following sections, let us explore a bit how different researchers have tackled various facets of PKC in their implementations from different viewing angles.

21.3.1.2 Software Implementations

Gura et al. [1] present the first implemented work that gives the idea about the applicability of PKC in small resource-constrained devices. They present a comparative analysis of Elliptic Curve

Table 21.1 Comparative Features and Summary of the Major Implementations of PKC in WSNs

Work, Year	Type of Device Used	Mechanism(s) Used	Type of Implementation	Achieved Results in Brief
Gura et al. [1], 2004	8-bit CPUs (Chipcon CC1010 8-bit microcontroller and Atmel Atmega128 processor)	Optimized SECG-standardized elliptic curves, Rivest, Shamir and Adleman (RSA)-1024 on both processors. Optimized RSA-2048 on ATmega128	Software implementation (in assembly code)	ECC outperforms RSA
Malan et al. [2], 2004	8-bit, 7.3828 MHz MICA2 mote	ECC over F_{2p}	Software implementation	ECC is viable on MICA2 mote platform
Gaubatz et al. [19], 2004	Regular ASIC standard cell library (not optimized)	Rabin's Scheme and NtruEncrypt	Software/hardware implementation	Designing PKC architecture with low power consumption is possible
Watro et al. [3], 2004	UC Berkeley MICA1, MICA2 motes	RSA, Diffie-Hellman	Software implementation	PKC is feasible for lightweight sensor networks
Wander et al. [4], 2005	Atmel Atmega128L microcontroller (MICA2dot)	RSA and ECC (RSA-1024, ECDSA-160, RSA-2048, and ECDSA-224)	Software implementation (nesC, C, assembly language)	ECC shows better performance than RSA
Gaubatz et al. [15], 2005	TSMC 0.13µ CMOS standard cell technology	Rabin's scheme, NtruEncrypt, and ECC	Hardware implementation	Special purpose ultra-low power hardware implementation is feasible
Blaß and Zitterbart [5], 2005	8-bit Atmega128 microcontroller	Lightweight versions of ECC-based algorithms	Software implementation	Acceptable speed of PKC operations for WSN
Gupta et al. [6], 2005	Berkeley/Crossbow Motes	ECC	Software implementation	Embedded secure web server on motes

Murphy et al. [20], 2006	Tyndall National Institute Mote	Rabin's scheme-based PKC	Hardware/software codesign approach	HW implementation of enc. algorithm is much faster than SW implementation
Piotrowski et al. [7], 2006	MICA2dot, MICA2, MICAz, TelosB	RSA and ECC	Software implementation	Energy consumption of the computation of PKC in WSNs is feasible
Yang et al. [8], 2006	Tmote Sky mote	TTS, enTTS	Software implementation	Optimized multivariate schemes may show better performance than established schemes
Batina et al. [16], 2006	Customized low-cost ECC processor	ECC	Hardware implementation	Low-cost ECC with hardware assisted approach
Bertoni et al. [17], 2006	Customized dedicated co-processor	ECC	Hardware implementation	Low energy, reduced silicon area, speed-up compared to software solutions
Arazi et al. [9], 2006	Intel Mote 2	ECC-based method	Software implementation	PKC in WSN is promising
Ugus et al. [10], 2007	MICAz	EC-ElGamal	Software implementation	EC-ElGamal is suitable for WSN
Großschädl et al. [11], 2007	Rockwell WINS node	Lightweight Kerberos with 128-bit AES encryption and ECMQV	Software implementation	Energy requirements for supporting PKC operation is manageable
Murphy et al. [18], 2008	Tyndall Mote	ECC and RSA	Hardware implementation	PKC processor suitable for low-resource devices

Cryptography (ECC) and RSA on 8-bit CPUs. Their experimental findings show that ECC point multiplication is comparable in performance to RSA public-key operations in low-resource devices. Besides their implementations and evaluations, they also propose a novel multiplication algorithm that leads to 25% performance increase for ECC point multiplication on Atmel ATmega128 platform. Malan et al. [2] presents the first implementation of ECC over F_{2^p} for sensor networks (on 8-bit ATmega128 chip [MICA2 mote]). Although their primary implementation of a TinyOS module (EccM 1.0, that implements ECC) was a failure, they got success in their second implementation (EccM 2.0). Their work proves that public keys can be generated and distributed among the sensor nodes within a short time and using very limited resource that can be afforded by the state-of-the-art sensor technologies. TinyPK system presented in [3] shows that a public-key-based protocol is feasible even for an extremely lightweight sensor network. TinyPK security scheme provides authentication and key exchange between a sensor network and an external party. It is mainly based on RSA cryptosystem. In order to adapt TinyPK for sensors, the authors in this work design their protocols that require only the PK operations (e.g., data encryption, signature verification) in the sensor network. They show in this work that TinyPK can be used together with other symmetric encryption services for mutual authentication and secure communications between a sensor and a third party.

Although Gura et al. [1] show the computational feasibility of PKC on 8-bit CPUs, their work does not provide any energy cost analysis or the applications of PK operations in various security protocols. In order to fill that void, Wander et al. [4] analyze and compare the energy costs of mutual authentication and key exchange between two nodes in a WSN, based on RSA and ECC using MICA2dot motes. The findings from their implementations are mainly two: (a) the energy costs for authentication and key exchange protocols with optimized software implementations of PKC are feasible on sensors and (b) ECC outperforms RSA in saving energy as it reduces computation time and communication costs.

As the implementation efforts of elliptic curves on 8 Bit ATmega128 chips by Malan et al. [2] reached poor results, Blaß and Zitterbart [5] investigate several ECC-based schemes: ECDH, El-Gamal encryption algorithm, and ECDSA algorithm in their work. They test the mechanisms on the same 8-bit ATmega128 microcontroller platform. For making the entire implementation easier and lightweight, they also consider some other optimizations for memory savings, point multiplications, handcrafting a source to the target platform, and sophisticated loop-unrolling. On the basis of the achieved results, the authors conclude that PKC is feasible in WSNs and lightweight versions perform relatively better.

In [6], the authors present the smallest known secure web server named Sizzle, which is implemented on several versions of Berkeley/Crossbow motes. Sizzle can run efficiently with very low resources and can take advantage of higher performance of ECC when communicating with an ECC-enabled version of Mozilla. Sizzle uses highly optimized implementations of PKC for offering scalable key management and end-to-end security. By the implementation and successful operation of Sizzle, the authors demonstrate in this work that ECC as a public-key cryptosystem is not only suitable for resource-constrained sensors but also it allows running a complete secure web server stack under very limited amount of resources.

Piotrowski et al. [7] analyze the influence of PKC operations on the lifetimes of wireless sensors. They provide the results of their implementations about the costs of cryptographic operations in software and energy costs for data transmission. They implement RSA and ECC operations on two groups of sensor nodes. The first group was the MICA motes (MICA2dot, MICA2, MICAz) based on ATmega128L microcontroller and the second group was the sensors based on the MSP430F1611 microcontroller (TelosB and Tmote Sky). With all the evaluated cost charts, they calculate the amount of PKC operations the sensors can perform with their available amount of energies. Their findings indicate that the energy consumption for the computation of PKC in WSNs is feasible because the operations related to PKC are often performed intermittently. Side-by-side, if dedicated hardware is used for data transmission operations, it can improve the energy efficiency of the sensors and thus can extend their lifetimes.

In [8], the authors implement minimized multivariate PKC on low-resource embedded systems. They illustrate most of their minimization techniques on a current variant in the family of multivariate PKC called the Enhanced TTS (enTTS). Tame Transformation Signatures (TTS) is a consequence of the public-key cryptosystem Tame Transformation Method (TTM) and shares many of its superior properties, resulting in low signature delays, fast verification, and high complexity. To evaluate the performance of enTTS on modern sensor nodes, they benchmark enTTS on the Tmote Sky mote. The results from this work show that multivariate schemes can be better candidates against the established PKC schemes if they are a lot better customized and optimized for use in WSNs. They also present a partial hardware implementation (ASIC implementation) of enTTS.

Arazi et al. [9] present an efficient ECC-based method for self-certified public-key generation in resource-constrained sensor nodes. Self-certified public keys reduce the amount of storage and computations in public-key schemes, whereas secret keys could still be chosen by the user and could be kept unknown to the certifying authority. This work shows that PKC in WSNs is in fact a reality; however, full-scale utilizations of PKC might take some more time. In [10], Ugus et al. implement EC-ElGamal cryptosystem on MICAz mote, which is equipped with an 8-bit processor. With the test results, the authors compare the performance of point multiplication with other previously devised solutions. Implementation results show that the EC-ElGamal encryption operation is fairly usable in the resource-constrained sensors with some careful design decisions at finite field level and at elliptic curve level. In [11], the authors implement two protocols and evaluate the energy requirements. The first protocol employs a lightweight variant of the Kerberos key transport mechanism with 128-bit AES symmetric encryption. The second protocol is based on ECMQV and uses a 256-bit prime field GF(p) as underlying algebraic structure. The reason for choosing these parameters is that a properly selected 256-bit elliptic curve system provides the same level of security as a symmetric algorithm like AES with a 128-bit key. The authors evaluate the energy costs of both protocols on a Rockwell WINS node equipped with a 133 MHz StrongARM processor and a 100 kbit/second radio module. The evaluation considers both the processors' energy consumptions for calculating cryptographic primitives and the energy cost of radio communication for different transmit power levels. The results demonstrate that the key establishment protocols using PKC are affordable in sensor nodes considering energy consumption.

Other than these major software implementations, there are many optimized ECC-based implementations like [12], TinyECC [13], or NanoECC [14], which demonstrate the feasibility of optimized software implementations of PKC in various sensor network platforms.

21.3.1.3 Hardware Implementations

In [15], the authors show that special purpose ultralow power hardware implementations of PK algorithms can be used on sensor nodes. The authors in this work select three low-complexity PKC schemes (Rabin's scheme, NtruEncrypt, and Elliptic Curve) and for each of the schemes they develop three basic encryption architectures in TSMC 0.13μ Complementary Metal-Oxide-Semiconductor (CMOS) standard cell technology. For comparing the inherently different algorithms and their feasibility for ultralow power implementations, they choose algorithm-specific parameter sets to provide approximately the same level of security. This work shows that PKC greatly simplifies the implementation of many typical security services and additionally reduces transmission power due to less protocol overhead. Batina et al. [16] propose a custom hardware-assisted approach of implementing ECC with the goal of obtaining stronger cryptography and minimizing the energy consumption. The results from their experiments indicate that this low power processor called ECP can efficiently be used for supporting ECC-based operations in resource-constrained WSNs.

In [17], the authors present a hardware implementation of PKC for elliptic curve over binary extension fields. The authors in this work design two different hardware devices in Very-High-Speed Integrated Circuit Hardware Description Language (VHDL) and synthesize them using the 0.18 μm CMOS technology library by ST Microelectronics. By using their coprocessor, the authors show that meeting the goals of low energy consumption, reduced silicon area requirements, and significant speed-up compared to software solutions is possible. Other than these hardware implementations, in a recent work, Murphy et al. [18] present an area-efficient processor for PKC operations in WSNs. A hardware chip area is very stringent. Area efficiency in this case ensures that the chip area is utilized in the best possible way (e.g., trading speed for area) so that the required services (related to PKC operation) could be provided with the minimum use of the physical area.

21.3.1.4 Hardware/Software-Blended Implementations

In [19], the authors propose a custom hardware-supported approach for implementing PKC. In order to validate their claim of success, they present proof of concept implementations of two schemes; Rabin's Scheme and NtruEncrypt. As these two schemes are inherently different, the authors carefully choose the specific parameters for making them comparable in sensor platform. They analyze the viabilities of the two schemes based on chip area, the level of security, delay, average power, energy per bit, and throughput. Their work shows that it is possible to design public-key encryption architectures with manageable power consumptions using the right selection of algorithms and associated parameters, optimization, and low-power techniques.

Another significant hardware/software codesign approach is presented by Murphy et al. [20] which successfully maps a public-key cryptosystem based on Rabin's scheme onto the motes developed by Tyndall National Institute. Their implementation mainly focuses on efficient architectures that execute the PK algorithms using minimal resources. The achieved results show that hardware implementation of encryption algorithm is much faster than software implementation. Software implementations of the algorithms are also possible and have the benefit of low cost and high flexibility. However, the time necessary to perform encryption and decryption is significantly increased by using a software-only approach.

21.3.2 Summary of Implementations

Having learnt all these information, before wrapping up the chapter, let us have a quick look at the comparative achievements. Here, Tables 21.2 and 21.3 present the comparative results of the most significant and recent implementations of PKC in WSNs (notable software and hardware implementations).

In Figure 21.2, we show the year-wise number of innovative published works on the implementations of PKC in WSNs. Here, we have only considered the significant implementations in this area (other than these, numerous optimization-related implementations are also available). As we can see from the figure, the number of novel implemented works in this area has significantly reduced within the last couple of years. It indicates that at some stage between the year 2004 to

Table 21.2 Comparative Features of Significant Hardware Implementations

Works	Scheme	Area	Freq.	Point Mult.	Enc.	Dec.	Signing	Verif.
Gaubatz et al. [15]	Rabin	17,000 Gates	500 kHz	–	2.88 ms	1.089 s	1.089 s	2.88 ms
	NTRU	3000 Gates	500 kHz	–	58 ms	116.9 ms	233.8 ms	58.45 ms
	ECMV	18,720 Gates	500 kHz		817.7 ms	411.5 ms	–	–
	ECDSA	18,720 Gates	500 kHz	~ 400 ms	–	–	410.45 ms	822.5 ms
Batina et al. [16]	ECC	12,000 Gates	500 kHz	115 ms	–	–	–	–
Bertoni et al. [17]	ECC	12,000 Gates	8 MHz	17 ms	–	–	–	–
Murphy et al. [18]	ECC	858 (LUTs) [ALU Radix 16]	65.3 MHz	228.7 ms	–	–	–	–
	RSA	858 (LUTs) [ALU Radix 16]	65.3 MHz	–	–	–	–	68 ms

Table 21.3   Comparisons of Software Implementations (Best Results)

	Malan et al. [2]	Watro et al. [3]		Wander et al. [4]		Blaß and Zitterbart [5]	Yang et al. [8]
Mechanism	ECC	RSA	D-H	ECC	RSA	ECC	enTTS
Platform/Device	MICA2	MICA1	MICA2	MICA2dot		8-bit Atmega128	Tmote Sky
Test ROM size	34 kB (req.)	128 kB	12,408 bytes	128 kB		73 kB (req.)	48 kB
Test SRAM size	1 kB (req.)	4 kB	1167 bytes	4 kB		208 bytes (req.)	10 kB
Time	34.161 s (Pub. Key Gen.)	14.5 s (Small Exp. Op.)	–	–	–	6.74 s (key gen.) 6.88 s (sign) 24.17 s (verification)	71 ms (sign) 726 ms (verifi.)
CPU utilization	2.512×108 cycles	–	–	–	–	–	–
Energy	0.816 J	–	–	61.54 mJ (sign) 121.98³ mJ (verifi.)	2302.7 mJ (sign) 53.7 mJ (verify.)	–	–

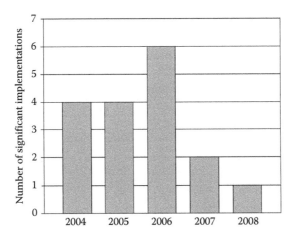

Figure 21.2   Number of works per year (significant and innovative implementations only).

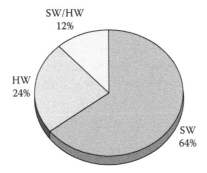

Figure 21.3   Percentage of various types of implementations (major implementations only).

2006, many researchers were curious to find out the answer of the critical question regarding the practicability of PKC in WSNs. However, once some experimental results showed the viability, more concentrations have been given toward optimization issues, applications, and theoretical aspects of PKC (in resource-constrained devices).

Among the major works surveyed in this chapter, around 64% deal with software implementations, 24% deal with hardware implementations, and the rest 12% are about the software–hardware-blended approach. Figure 21.3 shows this with a pie chart. As shown in the figure, mainly software-based implementations are used to verify the performance of PKC in WSNs. This is in general true if we consider even a bigger picture taking all the implemented works in this area.

21.4   Conclusions and Future Expectations

The works surveyed in this chapter demonstrate that most of them have implemented various mechanisms based on ECC. Other popular public-key cryptosystems have also been tested. However, encouraging performances have been gained mainly from ECC-based implementations. So far, the device platforms used for all these implementations are Chipcon CC1010 8-bit microcontroller, Atmel ATmega128 processor, 8-bit, 7.3828 MHz MICA2 mote, Regular ASIC standard cell library,

MICA2dot mote, TSMC 0.13μ CMOS standard cell technology, Tyndall National Institute Mote, MICAz, TelosB, Tmote Sky mote, Intel Mote 2, and other customized low-cost coprocessors or hardware. Other than these implementations, a lot of theoretical works are also available that propose various types of frameworks and security architectures using PKC mechanisms.

On the basis of proven results and facts presented in this chapter, it can be realized that it is possible to utilize PKC in a sensor network environment, if a particular WSN application needs it. It cannot be flatly said whether PKC in WSNs is preferable than SKC or not. We have previously seen with various cases that different scenarios demand different course of actions. Hence, the verdict depends on the application at hand, specific network settings, and given requirements. As for the use of specific mechanism for a particular platform, ECC-based mechanisms prove to be the best choice. However, depending on the application and type of mote, multivariate PKC schemes can be chosen. Again, if hardware support could be afforded, hardware–software-blended approaches can fasten PKC operations in low-resource devices such as sensors.

Another important point is that any PKC-based scheme cannot be applied alone. To ensure holistic security in WSNs, efficient security architectures and management policies such as efficient deployment policies, proper mapping of the attackers, tamper-resistant methods, node identity verification at the bootstrapping stage of the network, and so on should also be employed alongside PKC mechanisms. At this stage, it is comprehensible that due to the design and developments of various types of supporting hardware and software optimizations of well-known PKC schemes, the notion of PKC's unsuitability has partially been diminished. With the sophistication of sensor network technologies, future might see abundant use of PKC-based schemes in WSNs.

Terminologies

ECC—Elliptic Curve Cryptography
Implementation
Lightweight
PKC—Public Key Cryptography
Security
Sensor
SKC—Symmetric-key Cryptography

Questions and Sample Answers

1. How many types of cryptographies are there? What are SKC and PKC?
   There are mainly two types of cryptographies: symmetric and asymmetric. In Symmetric Key Cryptography (SKC), the sender and the receiver of a message share a single, common key for encryption and decryption of the message, whereas in asymmetric key cryptography (also known as Public Key Cryptography or PKC), the key (public key) used to encrypt a message differs from the key (private key) used to decrypt it. These two keys are mathematically related with each other.

2. Why PKC was thought to be not suitable for wireless sensor networks in the initial stage?
   For managing security issues in WSNs, among the two types of cryptographies, SKC has been preferred over PKC from the very beginning stage because of its low-resource requirements. The use of PKC in WSNs was totally ruled out for a long time with the idea that PKC is

energy-inefficient, slow, and computationally complex; thus, not at all suitable for resource-constrained devices such as sensors.

3. Describe the use of PKC in a WSN with an example figure.

A typical scenario of using PKC in wireless sensor network is shown in Figure 21.1. In this particular scenario, a deployed sensor node only needs to encrypt its plaintext (M) before sending it to the base station (BS). The encrypted message is termed as *ciphertext* (shown as C) which is transmitted through an unsecured channel. Only the entity that knows the corresponding private key (in this figure, the BS) is able to decipher the encrypted message. So, any other entity or intermediate node cannot decipher C without knowing the secret private key (PRI) of the BS. This method allows the ciphertext to travel through multiple hops without public disclosure of the plaintext. In the figure, $E_{PUB}(M)$ stands for encryption of message "M" with the public key of BS, "PUB" (that becomes the ciphertext C) and $D_{PRI}(C)$ stands for decryption of the ciphertext C using the private key of the BS, "PRI."

4. How does PKC provide better protection against physical capture attack?

As WSNs are often deployed in physically accessible open environments, physical intruders could capture or tamper nodes, extract secret keys, and use those for injecting false data in the network. As in many key predistribution-based symmetric key solutions, the deployed nodes carry almost all the important keys or keying information, the damage probability from such an attack is very high. For some symmetric key-based schemes, it can jeopardize the security of a large portion of the network if multiple nodes use the same secret key. PKC can be very helpful in this scenario. For instance, PKC can be used to establish symmetric session keys that could be valid for short time periods. Capturing a node in this case can reveal the key of a particular session (which would be invalid after some time). Again, in other general cases of physical capture of nodes, revealing the public-key materials of the BS would not cause any harm.

5. Describe the major approaches of implementations of PKC in WSNs.

Various implementations of PKC-based mechanisms can be categorized mainly into three approaches:

a. *Software implementation:* Writing customized software or program that can support PKC operation or using lightweight versions of the public-key-based schemes.

b. *Hardware implementation:* Design and development of customized hardware so that they can be used in the sensor boards for supporting PKC-based operations and computations.

c. *Hardware/software-blended implementation:* A third approach is the combination of hardware support and software optimization that also works effectively in some cases.

6. Which types of implementations have been mostly done?

Software implementation.

Author's Biography

Al-Sakib Khan Pathan is an assistant professor at computer science and engineering department in BRAC University, Bangladesh. He worked as a researcher at Networking Lab, Department of Computer Engineering in Kyung Hee University, South Korea, where he received his PhD in 2009. He received his BSc degree in computer science and information technology from Islamic University of Technology (IUT), Bangladesh, in 2003. He has served as a Chair, Organizing Committee

Member, and Technical Program Committee member in some international conferences/workshops such as HPCS 2010, ICA3PP 2010, WiMob'09 and 08, HPCC'09, IDCS'09 and 08, and so on. He is currently serving as an Area Editor of IJCNIS, Associate Editor of IASTED/ACTA Press *IJCA*, Guest Editor of several international journals, including Elsevier's *Mathematical and Computer Modelling*, and editor of a book. He also serves as a referee of a few renowned journals such as *IEEE Transactions on Dependable and Secure Computing* (IEEE TDSC), *IEEE Transactions on Vehicular Technology* (IEEE TVT), *IEEE Communications Letters*, Elsevier's *Computer Communications, Computer Standards and Interfaces, Computers & Electrical Engineering* journal, *Journal of High Speed Networks* (JHSN, IOS Press), *EURASIP Journal on Wireless Communications and Networking* (EURASIP JWCN), *International Journal of Communication Systems* (IJCS, Wiley), and so on. He is a member of IEEE and several other international organizations. His research interest includes wireless sensor networks, network security, and e-services technologies.

References

1. Gura, N., Patel, A., Wander, A., Eberle, H., and Shantz, S. C. Comparing elliptic curve cryptography and RSA on 8-bit CPUs, *CHES 2004, LNCS 3156*, Springer, Berlin/Heidelberg, Germany, 2004, pp. 119–132.
2. Malan, D. J., Welsh, M., and Smith, M. D. A public-key infrastructure for key distribution in TinyOS based on elliptic curve cryptography, in *Proceedings of IEEE SECON 2004*, Santa Clara, CA, USA, October 4–7, 2004, pp. 71–80.
3. Watro, R., Kong, D., Cuti, S.-F., Gardiner, C., Lynn, C., and Kruus, P. TinyPK: Securing sensor networks with public key technology, in *Proceedings of ACM SASN'04*, Washington, DC, USA, 2004, pp. 59–64.
4. Wander, A. S., Gura, N., Eberle, H., Gupta, V., and Shantz, S. C. Energy analysis of public-key cryptography for wireless sensor networks, in *Proceedings of PerCom'05*, Hawaii, USA, 2005, pp. 324–328.
5. Blaß, E. O. and Zitterbart, M. Towards acceptable public-key encryption in sensor networks, in *Proceedings of ACM 2nd International Workshop on Ubiquitous Computing*, Miami, USA, May 2005, pp. 88–93.
6. Gupta, V., Wurm, M., Zhu, Y., Millard, M., Fung, S., Gura, N., Eberle, H., and Shantz, S. C. Sizzle: A standards-based end-to-end security architecture for the embedded internet, *Technical Report*, SMLI TR-2005-145, Sun Microsystems, Inc., Menlo Park, CA, USA, June 2005.
7. Piotrowski, K., Langendoerfer, P., and Peter, S. How public key cryptography influences wireless sensor node lifetime, in *Proceedings of ACM SASN 2006*, VA, USA, 2006, pp. 169–176.
8. Yang, B.-Y., Cheng, C.-M., Chen, B.-R., and Chen, J.-M. Implementing minimized multivariate PKC on low-resource embedded systems, *SPC 2006, LNCS 3934*, Springer, Berlin/Heidelberg, Germany, 2006, pp. 73–88.
9. Arazi, O., Elhanany, I., Rose, D., Qi, H., and Arazi, B. Self-certified public key generation on the Intel Mote 2 sensor network platform, in *Proceedings of 2nd IEEE Workshop on Wireless Mesh Networks 2006 (WiMesh 2006)*, VA, USA, pp. 118–120.
10. Ugus, O., Hessler, A., and Westhoff, D. Performance of additive homomorphic EC-ElGamal encryption for TinyPEDS, *Technical Report*, 6. Fachgespräch "Drahtlose Sensornetze," July 2007, http://www.ist-ubisecsens.org/publications/EcElgamal-UgHesWest.pdf
11. Großschädl, J., Szekely, A., and Tillich, S. The energy cost of cryptographic key establishment in wireless sensor networks, in *Proceedings of the 2nd ACM Symposium on Information, Computer and Communications Security, ASIACCS 2007*, March 20–22, Singapore, pp. 380–382.
12. Uhsadel, L., Poschmann, A., and Paar, C. Enabling full-size public-key algorithms on 8-bit sensor nodes, *ESAS 2007, LNCS 4572*, Springer, Berlin/Heidelberg, Germany, 2007, pp. 73–86.
13. Liu, A. and Ning, P. TinyECC: A configurable library for elliptic curve cryptography in wireless sensor networks, in *Proceedings of IPSN'08*, Missouri, USA, April 22–24, 2008, pp. 245–256.

14. Szczechowiak, P., Oliveira, L. B., Scott, M., Collier, M., and Dahab, R. NanoECC: Testing the limits of elliptic curve cryptography in sensor networks, *EWSN 2008, LNCS 4913*, Springer, Berlin/Heidelberg, Germany, 2008, pp. 305–320.
15. Gaubatz, G., Kaps, J.-P., Ozturk, E., and Sunar, B. State of the art in ultra-low power public key crytography for wireless sensor networks, in *Proceedings of the Third IEEE International Conference on Pervasive Computing and Communications Workshops*, Hawaii, USA, 2005, pp. 146–150.
16. Batina, L., Mentens, N., Sakiyama, K., Preneel, B., and Verbauwhede, I. Low-cost elliptic curve cryptography for wireless sensor networks, *ESAS 2006, LNCS 4357*, Springer, Berlin/Heidelberg, Germany, 2006, pp. 6–17.
17. Bertoni, G., Breveglieri, L., and Venturi, M. Power aware design of an elliptic curve coprocessor for 8 bit platforms, in *Proceedings of the Fourth Annual IEEE International Conference on Pervasive Computing and Communications Workshops (PERCOMW'06)*, Pisa, Italy, 2006, p. 337.
18. Murphy, G. D., Popovici, E. M., and Marnane, W. P. Area-efficient processor for public-key cryptography in wireless sensor networks, in *Proceedings of SENSORCOMM 2008*, Cap Esterel, France, August 25–31, 2008, pp. 667–672.
19. Gaubatz, G., Kaps, J.-P., and Sunar, B. Public key cryptography in sensor networks-eevisited, *ESAS 2004, LNCS 3313*, Springer, Berlin/Heidelberg, Germany, 2005, pp. 2–18.
20. Murphy, G., Keeshan, A., Agarwal, R., and Popovici, E. Hardware-software implementation of public-key cryptography for wireless sensor networks, in *Proceedings of Irish Signals and Systems Conference*, Dublin, Ireland, June 28–30, 2006, pp. 463–468.

WIRELESS MESH NETWORK SECURITY

Chapter 22

Secure Access Control and Authentication in Wireless Mesh Networks

Bing He, Bin Xie, David Zhao, and Ranga Reddy

Contents

22.1 Introduction

In the Wireless Local Area Network (IEEE802.11a/b/g/n-based WLAN), each access point is standalone for Internet accessibility by operating single channel and single radio, resulting in limited efficiency. Differently, a Wireless Mesh Network (WMN), for example, IEEE 802.11a/b/g/n-based or IEEE 802.16 based, promotes the increased network capacity and coverage using multihop, multichannel, and multiradio technologies (e.g., multiple input and multiple output). It has been recently emerging as a novel communication paradigm to offer high-speed Internet connection for mobile users with secure and ubiquitous mobility [1]. A WMN consists of wireless entities organized in an arbitrary mesh topology where a number of Mesh Routers (MRs) are interconnected by wireless links and form a wireless backbone to provide ubiquitous high-speed Internet connectivity for mobile clients (MCs). An MR can be equipped with multiple radios to simultaneously transmit or receive with orthogonal wireless channels (i.e., nonoverlapping channels), which significantly increases the network capacity. The wireless backbone is tightly integrated with the Internet by a few special MRs called Internet Gateways (IGWs), which have an additional high-speed wired connection to the Internet or other wired networks. An MR is then associated with one or more IGWs by single- or multihop paths by employing multiple radios. These IGWs are able to collaboratively deploy in such a way that the Internet traffic is well balanced for all the MRs in the same domain. The superiority of WMN network not only allows high-bandwidth Internet access but also offers a low cost and flexible deployment as compared to traditional networks such as WLAN. One of its promising applications is to provide a cost-effective alternative to high-speed Internet connectivity for mobile users, in place of conventional expensive cables or digital subscriber line (DSL). On the other hand, a WMN can cover a large area such as a city and a town to support the applications that require high bandwidth communications. These applications, for example, include online video broadcast, video conference, and other multimedia services.

Before practical deployment and use of a WMN, secure authentication with access control is the critical part to deliver reliable applications for mobile users. The secure authentication should enable two entities to validate the authenticity of each other and generate the shared common session keys which can be used for subsequent cryptographic algorithms (e.g., symmetric key cryptosystems and message authentication codes). These further keys enable two entities (e.g., an MC and an MR, or two MCs) to transmit/receive data packets in an authentic way over open wireless links between any two communication parties. As other wireless networks, the authentication can easily be compromised due to several factors [2] such as distributed network architecture, the vulnerability of channels and network nodes in the shared wireless medium, and the dynamic change of network topology. From the network side, authentication should be able to protect its network infrastructure (e.g., IGWs, MRs) and the services provided by the network. If the network is accessed by illegal users, the service of the innocent users cannot be degraded due to limited network bandwidth. This means that the network should guarantee only the legitimate users to access the network for any services. Furthermore, the service can be interrupted if the adversary launches security attacks on the network. The security attacks, for example, include unauthorized network access, replay attack, spoof attack, denial of service (DoS) attacks, and compromised or forged MR attacks. Furthermore, the authentication should ensure the authenticity of access points (e.g., MRs in a WMN) from the viewpoint of the users. The secure authentication is also critical for Internet

Service Providers (ISPs; e.g., WMN operators), and they want to ensure that the mobile users are authorized customers and that the payment for the service has or will be received. Therefore, they need to verify the user identity or authorization before granting a network access request. Network operators are traditionally very keen on preventing unauthorized access. For them, unpaid seconds or bits equal lost revenue.

A WMN can span through a large area such as in a city or a town and contains multiple network domains owned and maintained by the same or different ISPs. Therefore, the authentication is not only needed for accessing the WMN at one time, it is also required for mobile handoff, that is, accessing from one MR to another or from an ISP-managed network to the network managed by another ISP. To perform a handoff, the MR should collaboratively to support reauthentication in a fast way. In addition, session keys for data protection have to be fast established without rendering high delay on it. Otherwise, the Quality of Service (QoS) for users cannot be guaranteed for the services. Packets may be significantly delayed in redirecting to the newly accessing MR and packets may be dropped. This is unacceptable for delay sensitive applications such as real-time services. However, the 802.1X security framework is highly limited in offering advanced performance for mobile handoff. As a result, many efforts are focused to improve the authentication schemes for providing the better performance. For example to decrease the authentication delay, several attempts have been reported to achieve localized authentication for both intra- and interdomain roaming, including the reactive and proactive authentication.

In this chapter, we investigate the WMN authentication to provide a comprehensive understanding on it. Before discussing the security requirements on the security and performance, we first illustrate the general network architecture and identify its security attacks that are related to authentication. The background of the security authentication is also depicted by using the WLAN. Then, a number of secure access control and authentication approaches are illustrated to show the current progress in developing robust and high-performance authentication protocols.

22.2 Background

22.2.1 Wireless Mesh Network

The popularity of WMNs originates from its rapid proliferation with a low deployment cost as well as its advanced performance. It recently allures considerable interest in the commercial and academic spheres owning to their high bandwidth per node, extended coverage, self-configurability, scalability and self-healing ability. It popularizes the age old concept of mobile ad hoc networks (MANETs) to form a multihop wireless backbone extended from the Internet infrastructure. Figure 22.1 represents a generic WMN model that has a hierarchical architecture consisting of three layers as shown in Figure 22.1. IGWs, located at the top of the hierarchy, form the backbone infrastructure for providing the Internet service to the second-level entities. As the IGW, all the Internet traffic flows to and from the WMN have to travel through the IGW.

The second level consists of a mesh of interconnected MRs that collaboratively forward the network traffic in a multihop fashion toward the IGW. It is noted that different technologies can be used to implement the MR functionalities. Different spectrum technologies can be used to implement the MR functionalities, depending on the application scenarios. In the radio interface, IEEE 802.16 adapts Orthogonal Frequency–Division Multiplexing (OFDM) and Orthogonal Frequency–Division Multiple Access (OFDMA) in accordance with standard IEEE 802.16-2004 and IEEE 802.16e-2005 for static and mobile WiMAX, respectively. Differently, IEEE 802.11s

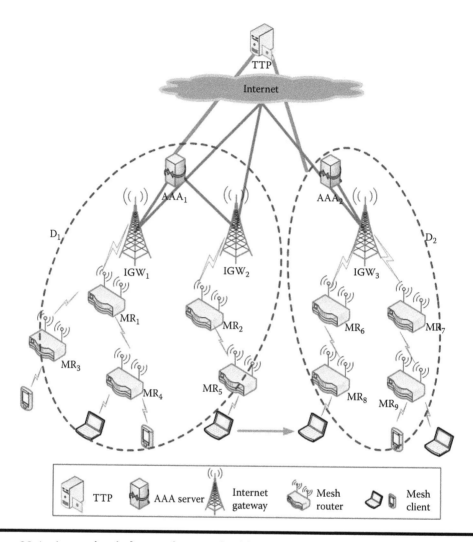

Figure 22.1 A generic wireless mesh network with two ISPs.

uses Frequency Hopping Spread Spectrum (FHSS) or Direct Sequence Spread Spectrum (DSSS) over the unlicensed spectrum to provide network access to users. In the link layer, IEEE 802.11s uses the contention-based Media Access Control (MAC) such as CSMA/CA (Carrier Sense Multiple Access/Collision Avoidance) that is connectionless based whereas WiMAX runs a connection-oriented MAC to establish a point-to-point wireless connection for the user.

On the lowest level, the MCs can be static or mobile while connecting to the nearest MR in a single- or multihop fashion. As shown in Figure 22.1, every MR has the functionality of aggregating traffic from MCs or distributing traffic to MCs. Unlike a pure ad hoc network where traffic is randomly generated between peer nodes, the traffic in the WMN is almost IGW oriented, that is, the traffic of an MR is predominantly directed either toward the IGW or from the IGW to the MR.

In spite of different MR implementation on the radio and link layers, an IGW provides the Internet access for a couple of MRs located in its nearby area, resulting in similar network structures and common security issues for applications. In a large area covered by MRs, the WMN could have more than one ISPs to ensure the Internet services in different parts of the area. Each ISP could maintain and manage some IGWs as well as the associated MRs, which again is denoted as the administrative mesh domain of an ISP (i.e., ISP domain in brief). To receive the Internet services form the WMN, MCs will be initially registered into an ISP domain that is called the home domain. Similar to Remote Authentication Dial-In User Service (RADIUS) [3] authentication protocol used in wireless LAN, each ISP maintains an administrative center with the function of authentication, authorization, and accounting (AAA) [4] to validate the authenticity of a variety of network entities (i.e., for customers like MCs, network devices like MRs and IGWs) in its domain. We use the AAA server here to denote such a central administrative center. Upon the request from the authenticated MC, the AAA server assigns authorization policies for different entities, defines their roles, and issues the permissions. These operations may also be used for accounting. We denote an ISP domain as D_i, which includes multiple IGWs, MRs, and registered MCs.

If a WMN fails in protecting itself, the open wireless channel and the multihop nature of communication pave many ways to malicious intruders. A malicious intruder can disrupt the network activities by conducting a number of security attacks. By an authentication hacking attack, a malicious MC illegally access the WMN and enjoy free network services by using single or multihop communications. The ISPs can be further accessed by the unauthorized MC due to improper ISP configurations and inadequate security measures at the network devices. An attacker can further gain the access to the network infrastructure using some techniques like MAC address spoofing. An adversary can conduct the impersonation attack that a forged MR claims itself as authenticated node by sending forged/replayed network messages (e.g., beacons) to entice MCs who assume that they are connected to the correct IGW. A rogue MR can conduct the attack that causes other MRs to redirect their traffic toward itself by advertising to them a higher rate link/less congested link to the Internet.

Authentication and secure access control is one of the fundamental issues in preventing these attacks and they should be enforced across multiple ISPs to support MC's mobility (e.g., secure handoff from one ISP to another). In a federated WMN, different ISPs may decide to couple initial separated mesh domains for mutual benefit (e.g., global roaming of MCs). Different ISP domains cooperate as a "federation" to provide seamless Internet access for MCs belong to the federated domains. For example, WMN in Figure 22.1 consists of two federated ISP domains operated by ISP_1 and ISP_2, with two AAA servers AAA_1 and AAA_2, respectively. AAA_1 provides authentication, authorization and accounting for mesh domain of ISP_1 (i.e., D_1), which includes two IGWs (IGW_1 and IGW_2) and five MRs. AAA_2 implements the security access for IGW_3 and four MRs of ISP_2. All the ISPs register with a Trusted Third Party (TTP), which is a trusted security center or a public CA. When an MC is moving across multiple ISP, a reauthentication scheme is necessary to prevent the security attacks caused by handoff between ISP domains.

22.2.2 Authentication Schemes in WLANs

In this section, we illustrate the authentication schemes for wireless access in the WLAN and illustrate the IEEE 802.1X security framework that is also the basic scheme for WMN also.

22.2.2.1 IEEE 802.1X Authentication

As the most widely used broadband wireless network, WLAN provide high-speed Internet access for mobile terminals such as laptops, PDA which configured with WiFi network interface card. IEEE

802.1X [5], which is an authentication protocol included in the 802.11i [6], is a current solution for the WiFi authentication. IEEE 802.1X authenticates the access request from the IEEE 802.11 media from the link layer. IEEE 802.1X performs authentication and session encryption at the higher layer, and thus can be applied for different PHY layers (Ethernet, IEEE 802.11, 802.16). This achieves multiventor interoperability and different hardware platforms. Specifically, IEEE 802.1X contains a Port-based Network Access Control scheme as well as an Extensible Authentication Protocol (EAP) as below.

Port-based Network Access Control: The IEEE 802.1X port-based access control scheme as shown in Figure 22.2 has the basic functional components:

- *Supplicant:* A supplicant at the end of a point-to-point WLAN segment represents an MC that wishes to join the network.
- *Authenticator:* It is an AP or an MR that is directly connected to the MC seeking network services.
- *Authentication server* (AS)*:* AS is the backend central server which acts as AAA server and maintains all user credentials such as secret keys, public-key certificates, and passwords. It determines, from the credentials provided by the supplicant, whether the supplicant can be authorized to access the services provided by the authenticator.

Port access entity (PAE) is the protocol entity associated with a port. A given PAE can support the protocol functionality associated with the authenticator, the supplicant, or both. In IEEE 802.1X, there are two types of network ports: the controlled port and the uncontrolled port. The uncontrolled port is used for transmission of the authentication messages such as "access request" and "access reply" messages. On the other hand, the controlled port is used for data transmission [7]. An MC can obtain access to the controlled port only after performing the user authentication and receiving session and WEP keys. In other words, when an MC attempts to connect to an AP, it is initially restricted to send only authentication messages from the uncontrolled port. The AP acts as a proxy between the MC and the backend AS, and relays the authentication requests to the AS. The AS and the client mutually authenticate each other and generate a secret Pairwise Master Key (PMK).

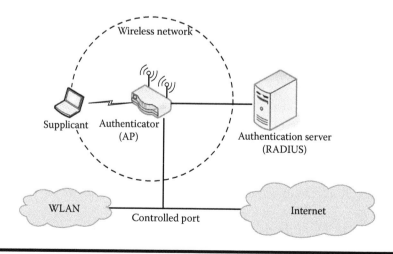

Figure 22.2 802.1X authenticating components.

Once authenticated by the AS, the MC will gain access to multiple APs in the visiting region using multiple session and WEP keys.

EAP-based Authentication and Key Distribution: The EAP is an IETF standard that provides an infrastructure for authentications between clients and authentication servers. IEEE 802.11 WLAN uses 802.1X for EAP-based access control standard to provide authentication for network access control and encryption with key distribution. It then is enhanced for 802.16e to support the EAP-based authentication and key distribution for mobile WiMAX networks. However, EAP is quite complex and the communication and computational overhead are expensive, and thus lacks the ability to support mobility or to deal with the need for an MC to quickly extend its ongoing session before initially authorized session expires [8]. According to the latency/jitter guideline identified by WiMAX Forum, a quality user experience can be ensured if the latency does exceed about 50 milliseconds in VoIP or video conference, and 100 milliseconds in streaming media [9]. As pointed out by Aboba et al. [10], every full EAP-based 802.1X authentication leads to a 1000-millisecond latency and 250 milliseconds in fast resume mode. The authentication delay will be even worse in the multihop scenario where the authentication message will have to travel multiple hops between supplicant and authentication server. Thus, fast and reliable authentication schemes need to be designed to meet the requirements of security and QoS in a WMN.

22.2.2.2 IEEE 802.1X Limitations

The IEEE 802.1X are extensible for adapting different authentication methods such as public-key certificate based (EAP-TLS: EAP Transport Layer Security), one-time password based (EAP-MS-CHAP: EAP Microsoft Challenge-Handshake Authentication Protocol), secret key based (EAP-TTLS: EAP Tunneled Transport Layer Security), smartcards, and so on, which have widely been used in the commercial wireless products. For example, EAP-TLS is based on the secure socket layer protocol used for secure Web traffic, and it has become the defacto standard by its inclusion in Windows XP. However, these authentication methods suffer from a common limitation of delay for using in the WLAN. A complete EAP-TLS hand shake, including RADIUS messages, causes on the order of 1 second that are not affordable for any form of streaming media [11]. To facilitate fast authentication, IEEE group again recommends preauthentication which allows the MC to be authenticated by a potential AP as the next Internet attachment. The preauthentication occurs before associating with the potential AP and before disassociating from the previous AP. Unfortunately, such a preauthentication mechanism has several shortcomings. The preauthentication works only when the MC can guess which AP it will next use and when the old and new APs have to be connected on the same wired link [12]. Consequently, an MC can only preauthenticate to another AP on the same local area network and cannot authenticate beyond the nearest accessing APs. The primary hindering factor in extending the IEEE 802.1X framework to WMNs is that it is operated at the link layer. This mandates the MC to be directly connected to the IGW (i.e., single hop extension). Without any multihop security protection, the 802.1X framework fails to provide the secure authentication among the components of the WMN backbone. Therefore, two MRs in a multihop path cannot be mutually authenticated and an MC/MR on the WMN backbone cannot be authenticated by the ISP when the multihop communication is required for the MC to connect with the IGW.

In addition, the IEEE 802.1X suffers from the limitation of supporting MC handoff where the collaboration of multiple APs is required. Due to mobility, an MC may move out the coverage of its current AP. In this case, the received signal strength (RSS) and the signal-to-noise ratio seen by the MC will degrade significantly and the MC will loss its connectivity with its current AP. Thus, the MC needs to initiate a handoff to the roaming AP. The handoff procedure of the IEEE 802.11 with 802.11i enhancements includes two main phases [13]:

- Probe-and-decision
- Execution

In the first Probe-and-decision phase, the moving MC scans the channels to find potential APs nearby. Based on the scanning result and some decision metrics such as RSS, it will choose a target AP as its new AP. After that, the MS will start the Execution phase to attach to the selected target AP. In an IEEE 802.11i-supported WLAN, the Execution phase involves three steps:

- Reassociation
- 802.1X Authentication
- Four-way handshake

In the reassociation, the MC exchanges reassociation request and reassociation response messages with its new associated AP. In 802.1X authentication step, the MC is finally authenticated to a backend authentication (AAA) server, such as RADIUS server, via the new AP. If the authentication is accomplished, both the MS and authentication server will derive the same PMK. The authentication server would transmit the PMK to the new AP, and triggers the new AP to start the four-way handshake, which is between the AP and MC. In the four-way handshake, some temporal keys will be derived, including a data encryption key and a data message integrity code (MIC) key, to protect wireless data communication between the MC and the new AP. And then, the IEEE 802.11 handoff completes. All these phases may cause considerable delay. Experiences by Aboba [10] show latencies of 300–400 milliseconds for the Probe-and-decision phase, 800 milliseconds for the 802.1X authentication, and 40 milliseconds for the four-way handshake. It is obvious that the total handoff delay is not acceptable for real-time applications. Also, in the current 802.1X authentication, users are not allowed to transmit/receive normal data via the new AP before the authentication is accomplished.

22.3 Access Control and Authentication in WMN

Authentication and authorization is the first step against fraudulent accesses from illegal users. As WLAN, the network authentication provides a secure way such that an MC and the corresponding MR can mutually validate their legislations with each other before the MC accesses the WMN for network services. In this section, we investigate the security attacks in the WMN and basic assumptions in implementing the access control and authentication.

22.3.1 Authentication-Related Security Attacks

The attackers in the WMN may come from three sources: External Attackers, Dishonest Customers, and Dishonest operators. From these attacking sources, the attacker may gain illegal and unaccountable network access, intrude the privacy of legitimate network users, or launch DoS attacks against network resource availability. There are several types of attacks that are related to authentication as below.

- *Unauthorized Access:* It is the attacker access to the service provided by the WMN (i.e., the Internet access). Further, the attacker may access to customer data and private information.
- *Replay Attack:* In a replay attack, a malicious attacker caches a legitimate authentication request (by eavesdropping or interception) and replays it at a later time. Thus, if an MR fails to recognize the authenticity of the request, it could be granting access to malicious attackers.
- *Spoof Attack:* IP address spoofing is a rampant problem in WMNs due to the multihop nature of communication. Spoofing is the act of forging a false IP address in the packets originating from an attacker. Using such a spoofed address, a malicious attacker can perform a man-in-the-middle attack by cleverly intercepting a termination request and thereby hijacking the session.
- *DoS:* A malicious attacker can conduct a DoS attack by sending a flood of packets to the MR. This results in DoS to legitimate mesh clients. Another well-known security flaw that can be exploited by an attacker is that the deauthentication/termination message from the 802.11-based wireless clients is left unauthenticated. Thus, a malicious attacker can send false termination messages on behalf of a legitimate MC.
- *Comprised or Forged MR:* The adversary can compromise and control a number of users and MRs subject to his choices. For example, the attackers may comprise the existing MRs by physical tampering or logical break-in. The adversary may also set up rogue MRs to conduct a variety of attacks. The fake or comprised MRs can be arbitrarily manipulated by the attacker to perform arbitrary attacks. The attacks may be implemented by attacking on wireless links (e.g., eavesdropping, jamming, replay and injection of message and traffic analysis). Advanced attacks can also be implemented such as the attacker can advertise itself as a genuine MR, using some forged messages or duplicate beacons procured by eavesdropping on a genuine MR. When a multihop MC hears these fraudulent beacons from a malicious MR, it assumes that it is within the radio coverage of a genuine MR and initiates a registration procedure. After registering, the MR assumes that it has obtained the Internet connection and disconnects its communication from the genuine MR. Slowly, the forged MR could entice a number of MCs to disconnect from the genuine MRs. This attack is possible in situations where the MR is not authenticated by the MC and breaks a genuine Internet connection or causes unwarranted registration delay. The forged MR that acts as the relay can seize the data packets or capture sensitive personal (e.g., password) of MCs and other MRs.

22.3.2 Secure Authentication Assumptions

Most of the current works on WMN access control and authentication consider MRs and IGW are static and connected with one or multiple hop high-speed wireless connection. In a cooperative WMN where multiple ISPs exists and each maintain individual service domains, WMN assumes a TTP which can serve as a certificate authority (CA) and issue every legitimate ISP with its corresponding certificate such that each ISP can check the validity of another. They assume that the ISP is trustable due to following reasons [14]:

- A legitimate ISP does not intentionally misbehave, which is reasonable since the attacks on its serviced MCs will decrease the satisfactory of end users on the ISP, and will lead to reduction of its long-term revenue.

■ The attacks launched by an ISP can be easily detected by the CA, and the malicious ISP will be deprived of its ISP qualification with subsequent penalties.
■ Furthermore, since the number of revoked ISPs should be small for most of the time, it is feasible to real-time update and distributed the Certificate Revocation List (CRL) of ISPs.

Therefore, ISPs are assumed to be the initial trustable points in designing an access control and authentication scheme for WMNs. A practice way for establishing a trust relationship among different ISPs considers a centralized TTP trusted by all the ISPs [14]. Under this framework, when an MC roams into a foreign network domain, the foreign ISP simply forwards the corresponding AAA session of the MC to the home ISP of the MC for authorization via the TTP. A more elaborated approach can be devised on top of the centralized TTP architecture by taking advantages of the Public-Key Infrastructure (PKI), where the TTP serves as not only a TTP, but also a CA which issues public-key certificates to the ISPs and MCs. The trust relationship among ISPs, or between an ISP and MCs, can be easily established by validating the public-key certificates issued by the TTP. The TTP architecture can effectively solve the interdomain roaming and billing problem; unfortunately, the TTP may become a performance bottleneck for the interdomain handoff authentication. In addition, the long signaling propagation latency of every transaction may not be tolerable to the real-time services in the interdomain roaming events. Thus, it is desirable to develop a new framework in meeting with the stringent requirements on authentication latency and scalability without losing the security assurance.

Similarly, the IGWs are also assumed to be trustable and reliable since the IGW is directly managed by an ISP and each ISP maintains only a few IGWs with have high-speed wired connection. With the above assumptions, the authentication between mesh clients and the mesh back bone become the focus of the access and authentication of WMN.

22.3.3 Requirements for Authentication in WMNs

On the basis of whether a central authentication server is available, there are two types of implementation of access control enforcement: central access control enforcement and distributed access control enforcement. For both schemes, the access control enforcement should be at the border of the mesh network. In the distributed access control enforcement, the access points could work as the distributed authentication servers. The authentication could also be performed in three different places:

■ An remote central authentication center
■ Local entities such as IGWs or some MRs that playing the role of authentication sever
■ Local accessed MRs

The main benefit of central authentication server is the easy of management and maintenance, while suffering the danger of single point failure. Owing to the round trip time and authentication delay in the multihop mesh network, a solely centralized authentication scheme is not desirable and the implementation of the authentication should be allowed at some local nodes such as IGW or MRs. In many cases, the physical protection cannot be assured for the special MRs with usually need to store sensitive data. Thus, the additional solution is that to delegate the authentication power to a group of MRs. Wherever ways used, the access control and authentication has to achieve both security requirement and performance requirement.

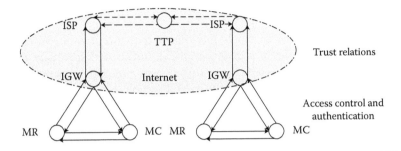

Figure 22.3 Mutual authentication in a WMN with multiple ISPs.

22.3.3.1 Security Requirements

An elaborate authentication mechanism is to guarantee only legitimate users to have access to the network service. Figure 22.3 illustrates the mutual authentication requirement for WMN with multiple ISPs. As shown in Figure 22.3, each network component and the MC have to be trustable by a trust chain. On the one hand, the provider (e.g., ISP) has to be ensured that users get the necessary permissions before access to the network resource. On the other hand, from the client's point of view, the authentication and key establishment should occur as fast as possible in order to provide seamless handoffs. The source of the incoming messages should be specifically identified to prevent malicious nodes gaining network access. The authentication of network devices is also important for customers to ensure the authenticity of access points (e.g., MRs in a WMN) and avoid any subscription fraud. Once the user has accessed into the network, a key establishment protocol enables two entities to share common session keys which can be used for subsequent cryptographic algorithms (e.g., symmetric key cryptosystems and message authentication codes). An authenticated key establishment scheme in a WMN enables two entities (e.g., an MC and an MR, or two MCs) to share common session keys in an authentic way over open wireless links while providing mutual identity authentication between these two parties. The basic objective of the access control and authentication is to ensure that the WMN functions properly and securely.

■ *Mutual authentication:* Any network elements (IGW, MR, and MC) have to authenticate each other mutually if required. An MR and MC should be able to mutually authenticate each other to prevent unauthorized network access and the security attacks as listed in Section 22.3. Figure 22.3 shows the possible trust relations in the WMN. In addition, the whole mutual authentication system should be scalable in terms of the number of MRs and MCs.

■ *Security Keys between MC and MR:* The MCs and MRs should be able to establish a share pair-wise communication session key to encrypt corresponding traffic. It should support dynamical key distribution and the negotiated keys should be independent between each other. Therefore, the authentication should be maintained fresh. Furthermore, the connection key should be controlled by each participant.

■ *Standardization:* The characteristic of multioperator environment renders specific requirement. The existing standards should remain useable, and it is unacceptable to change on any part of the current standard. This is beneficial to avoid using single entity that all the operators have to trust.

In addition to preventing attackers, the authentication and access control schemes for WMN have to satisfy performance requirements imposed by networking environment as below.

22.3.3.2 Performance Requirements

The access control and authentication should meet the performance from the aspects of delay and packet loss, computational complexity, signaling overhead, and network scalability. Fast authentication with less delay and packet loss is required to support user mobility. As analyzed by Askoxylakis [15], the main challenge in WMN security is that the authentication method has to support seamless or fast handoff caused by user mobility. Note that due to limited coverage of an MR (e.g., IEEE 802.11b/g generally covers only 35 meters indoor and 110 meters outdoor), it is undesirable for a roaming MC to authenticate at every MR from the ISP. Also, key generation mechanism for data encryption should be scalable to achieve continuous data services for a moving MC in WMNs. When an MC moves into a new access MR area, it should perform a new user authentication procedure and receive a new session key, which encrypts the transmitted data in the wireless link. If an MC conducts a nontrivial process in reauthentication in every MR, not only the migration of the MC from an MR to another will be delayed, the network services will also be significantly interrupted. This problem becomes worse when an MR roams across two MRs that are managed by two different ISPs. In general, an MC subscribes to an ISP, called home ISP, and signs up its account in order to access the wireless internet services at the hotspots managed by the ISP. However, the MC cannot access to the hotspots operated by any other ISP (called foreign ISP) unless the foreign ISP and the home ISP have a cooperative roaming agreement which stipulates how each others' users can have access to the hotspots managed by these two ISPs when a handoff event occurs. Therefore, the delay for a handoff between two MRs managed two ISPs will be much higher than that between two MRs that are managed by the same ISP. It is a critical issue in access control and authentication to provide a secure, light-weight, and cost-effective approach in authenticating the MCs' login requests, along with the consideration on the limitations of wireless communications environments, such as the limited computation power, memory, battery capacity of the MCs, and the ping-pong movement phenomena.

Fast authentication with less delay is essential to support voice and multimedia with continuous mobility, in response to the handoff time-scale required in the real-time services. The delay of a secure handoff comes from two sources: handoff and authentication. Given a handoff either between networks belonging to different operators or between different service domains of one ISP, some data packets usually get lost or take a significantly longer time to arrive at the receiver. During handoff, the location of the MC is updated in a corresponding network path of the current service, resulting in handoff delay that may significantly degrade the performance of real-time services. The authentication causes an additional delay in the network access, resulting in a much longer delay than the typical handoff delay without security. In other words, the access control and authentication process can be one of the major sources of latency and jitter. In multihop WMNs, AAA usually are located far from the access MRs, the authentication delay in the WMN will be longer than that in the WLAN. In a real-time service such as voice, the total latency (layers 2 and 3) for a secure handoff between access points must be small, and the overall latency should not exceed 50 milliseconds to prevent the excessive jitter [16] on the service. Owing to this constraint, the delay in secure handoff should be carefully design to satisfy for real-time multimedia applications. The handoff protocol has to perform location updates as locally as possible and avoid message exchanges with remote nodes that might not be available at the time. On the other hand, the authentication should be performed as quickly as possible to reduce the authentication delay while satisfy the security requirement, for example, reducing the delay in contacting a remote server such as RADIUS server or AAA server.

In addition to delay, there are other performance requirements. At first, the access control and authentication should achieve low communication overhead. In the case of authentication and authorization, messages are exchanged with a trusted on-line authority or a certificate revocation database. The message exchange for handoff and authentication should be reduced to a low level to save the network bandwidth, especially in the mutihop scenarios in WMNs. Secondly, the access control and authentication should achieve low computational overhead. The authentication and security protocols that require the computation of complex cryptographic functions are not feasible for the MC which has limited computational capability and power resource. Efficient cryptographic computations are highly preferred for cheap and battery-based mobile devices. Thirdly, the access control and authentication should achieve high network scalability. The authentication and access control should be able to support a large cooperative WMNs in which multiple ISPs cooperate to provide a large area as well as a small community access mesh network maintained by a single ISP.

22.4 Access Control and Authentication Schemes

The current access control and authentication schemes for WMNs can be divided into localized authentication, preauthentication, EAP-based authentication with improved roaming support, and ID-based authentication. In this section, we present the works of these access control and authentication schemes.

22.4.1 Localized Authentication Based on Public Certificate

In the localized authentication, the TTP serves as the trusted CA to issue certificates to both ISPs and MCs. The certificates issued to an MC or an ISP is a digital signature signed by the TTP on its public key, and the certificates establishes an secure association between the public key and the MC or ISP identity. By preloading each MR with necessary cryptographic information, some required security capability can be achieved such that the roaming/handoff authentication and billing can be performed in a localized manner. Owing to localized authentication at MR, the performance can be improved in support of MC's mobility. The localized authentication and billing scheme is expected to effectively solve the scalability problem due to the centralized TTP and dramatically reduce the interdomain roaming latency by avoiding any intervention of the TTP.

The approach proposed by Buttyan et al. [17] achieves the localized authentication by a public certificate scheme. In this approach, the authentication is performed locally between the access point (e.g., MR) and the MC. To enable the localized authentication between the MC and MR in a multioperator maintained WMN, each operator maintains its own CA service. Each CA is responsible for issuing certificates to their subscribers. The CA also maintains the CRL. To cooperate between each other, the operators issue cross-certificates for each other's CA, which enable entities (subscribers or MRs) to perform certificate-based authentication and key exchange even if they belong to different operators. In the process of authentication between MC and MR, each MC exchange with MR their certificates with an nonce or timestamps in a three-step handshake, therefore an connection key is setup. To minimize the authentication delay, the Blake–Wilson–Menezes (BWM) Provably Secure Key Transport Protocol [18] is used, which is claimed having the minimum number of public key-based computations.

The authentication scheme proposed by Buttyan [17] is integrated into the EAP framework. Since their scheme is based on the public encryption, how to choose the efficient public-key algorithms and key parameters is the major concern in the implementation of their scheme. Two public-key sets are proposed: one for a powerful MC and one for a capacity-constrained MC. In the former set, the computationally intensive operations are shifted to the MC, while in the latter certificate set it suggests the usage of weak keys and short-term certificates for digital signature. In other words, this scheme supports a powerful MC which can perform expensive public-key computation. In this scheme, each MR of a hotspot is assumed to have unlimited power and sufficient computational capacity while the MCs could only have relatively limited resources. Owing to expensive public-key operation, this scheme could not perform well for resource-constraint MCs, especially in the scenario of the ping-pong movement phenomenon when handoff across different hotspots occurs.

Considering the authentication for interworking of multiple domains in a metropolitan area operated by different ISPs, Lin et al. [19] proposed another localized authentication scheme. In this scheme, an embedded two-factor authentication mechanism is proposed to determine whether a roaming MC is authentic or not without the needs of the intervention from the MC's home ISP and the authentication key exchanging. The two-factor authentication mechanism includes two methods of authentication, namely password and smart card. The security performance is considered as much stronger since two independent authentication methods are adopted to ensure an MC to be a legitimate user. To mitigate the negative impact due to the ping-pong movement phenomena, the session key is cached in the current network domain. Whenever the MC requests a handoff into a neighboring MR which has a nonexpired shared session key with the MC, a user-authenticated key agreement protocol with secret-key cryptography will be performed instead of the full authentication procedure based on an asymmetric key encryption. The proposed authenticated key agreement protocol uses only the symmetric-key encryption and keyed-hash message authentication codes, which are very fast compared with the asymmetric-key encryption.

These localized authentication schemes are based on the assumption that the accessed MRs are trusted and fully protected by certificates. In practical, MRs are most probably low-cost devices without expensive and wholesome protection, thus such an authentication scheme will be at the expense of reduced security level of the system due to the compromise-prone. In a compromise event, the cryptographic secrets, such as the public/secret key pairs, could be deprived by the attackers, who may launch some serious attacks by manipulating the secret information. To overcome this shortcoming, Zhu et al. [14] proposed secure localized authentication and billing scheme. To protect the attacks due to comprised MRs, the local voting strategy is adopted based on the threshold digital signature mechanism [20] to enhance the system security level. In the local voting strategy, certificates of an MC are issued by a group of neighboring MRs instead of a single serving MR. In the threshold digital signature technique, each neighboring MRs carries a piece of the digital certificate. To jointly generate a digital certificate of a visiting MC, more than k-neighboring MRs are required to send the consistent copies. The proposed digital signature issuing and localized interdomain handoff authentication scheme can be applied to not only the interdomain handoffs between the home ISP domain and a foreign ISP domain but also the ones between two foreign ISPs.

22.4.2 Predictive Authentication and Preauthentication

To reduce the handoff latency for the moving MCs, several efforts have been focused on decreasing the negative impact of the user mobility and several preauthentication schemes have been developed. Pack et al. [7] proposed a fast handoff scheme based on mobility prediction that can be used for WMNs. In this scheme, an MC entering the radio area covered by an access point performs authentication procedures for multiple neighboring MRs (or APs), rather than just the current MR. When an MC sends an authentication request, the AAA server authenticates not only the currently attached AP, but also multiple MRs, and sends multiple session keys to the MC. Therefore, the authentication to a neighboring MR can be fast conducted once the MC is moving to it. Such an approach could be very effective in decreasing the handoff delay if it is adapted to a mobile WMN (e.g., mobile IEEE 802.16j WiMax) while the mobility is much higher since MRs are also moving. In this approach, a prediction method is also developed to further reduce the handoff delay. The prediction method is called the Frequent Handoff Region (FHR) selection algorithm that multiple MRs are selected as the frequent MR for a specific MC. FHR takes into account users' mobility patterns, user characteristics, service classes, and so on, which are collected and managed by the centralized system. In this way, an MC is authenticated and registered in an FHR in advance, thus the handoff delay resulting from reauthentication can be significantly reduced.

To increase the accuracy of the prediction of user mobility, Mishra et al. [11] have proposed a proactive key distribution approach to reduce the latency of the authentication phase. This approach predistributes key material ahead of the accomplishment of the authentication. A new data structure, neighbor graphs, is used to determine the candidate set of MRs that a roaming MC could potentially reassociate to. They have demonstrated their approach achieves the same level of security as a 1.1 seconds full EAP/TLS authentication. The experiments further show a lower latency of 20 milliseconds as shown by simulation and of 50 milliseconds as shown by test-bed experiments.

Aura et al. [12] have proposed a reauthentication approach. In this approach, the MR issues a credential for its serving MC, and the credential later can be used to certify MC's authenticity for the next MR. This mechanism requires the MC to trust MRs belonging to other operators when issuing credentials. The credential which the MC receives from an MR is a proof of past honest behavior to the newly associating MR. Suppose an MC receives the credential from an MR (e.g., MR1) where it has been successfully authenticated or paid for the access. MR1 gives the MC the credential as a proof of its verified honest behavior. When the MC arrives at a new MR (MR2), it presents this credential to MR2. In a sequence, the MC could be credited by a series of MRs. The credential is protected by keyed one-way functions that result in low computation and communication overhead at both the MC and the MR. An MC whose honesty has previously been proven will be given a weak authentication. After the weak protocol, the server has some level of assurance that the client is honest and can allocate some resources to it. As a result, this protocol rewards frequent handoff and well-behaving customers. The authentication delay is relatively high only when the MC visits an MR or a group of co-operating MRs for the first time, and the delay will be reduced with the increase of the visiting times. A risk of the credential is that the MC may use the pirated credentials that were issued to other mobiles. To prevent an MR from the use of pirated credentials, this approach uses a cryptographic mechanism to prove that it is indeed the same MC to which the credential was issued by the corresponding MCs. A challenge–response protocol is used for the proof. The authors have demonstrated its effectiveness in supporting real-time or multimedia applications. The results show that it has the great potential to significantly reduce the authentication delay experienced by MC during movement.

Chen et al. [13] have proposed an approach that allows a roaming MS to execute the 802.1X authentication and enjoy its service simultaneously for a short period of time. To prevent the possible security loophole, the MC is enforced to access the wireless network via its previous MR before handoff complete. Since the previous AP has already authenticated the MS, thus it can check whether the MC is legal and use their prior data encryption key to get the wireless access. The proposed seamless authentication scheme is composed of a Dynamic Tunnel Establishing procedure and a seamless handoff process. Dynamic Tunnel Establishing is used for each MR to construct tunnels with its neighbor MRs that verifies if the neighboring MRs can be trusted or not. In the seamless handoff process, the roaming MC is allowed to access the network via its previous MR by the tunnel between the new MR and the previous one during handoff. During the execution phase of a handoff to a new MR, the MC temporally uses the trusted tunnel established between the previous MR and the new MR. Before the authentication between the new AP and MS is completed, the MS will use the old key to encrypt data and communicate via the old MR. The new MR tunnels only the MC's data to the old MR or forwards the encrypted data to the MC. This relaying process will stop once a new data encryption key is derived between the new MR and the MC, or a specific timer is expired.

Soltwisch et al. [21] proposed a fast authentication and key exchange mechanism to support seamless domain handoffs. Context Transfer Protocol (CTP) [22] is utilized to forward session key from previous router to the new access router. In the proposed Interdomain Key Exchange (IDKE) mechanism, the security associations (SAs) between two access routers are established. Then the credentials (session key) could be forwarded securely between these access routers using CTP. Without contacting the home network, an MC, the CTP-based IDKE mechanism, is fast in forwarding setting and credentials from previous access router to the new access router, thus provides seamless handoff between different access MRs. Their scheme is shown with low-bandwidth consumptions compared with GSM authentication and EAP-type authentication.

22.4.3 EAP-Based Authentication Schemes for WMNs

As discussed in Section 22.2, EAP-based authentication suffers from the limitation of high latency. The authentication delay increases in supporting handoff and multihop WMNs. In order to decrease the authentication delay, Hur et al. [8] recently proposed an EAP-enhanced preauthentication scheme for mobile WMN (IEEE 802.16e) in the link layer. In this approach, the PKMv2 (public-key management version 2) has been slightly modified based on the key hierarchy in a way that the communication key can be established between the MC and the target MR before handoff in a proactive way. The modification allows the master session key generated by the authentication server to bind the MR identification (i.e., base station identification) and the MAC address of the MC. In the preauthentication, the authentication server generates and delivers the unique public session keys for the neighbor MRs of the MC. The neighboring MRs are the access points that the MC potentially moves to. These MRs can use the public session key to derive an authorization key of the corresponding MC. In the same way, the MC can derive the public session key and the authorization key for its neighbor MRs with the MR identification. Upon the handoff, the MC is only need to perform a three-way handshake and update the encryption key since the MC and MR already have the authentication key. This avoids the reauthentication from the authentication server and reduces the delay.

Khan and Akbar [23] proposed a symmetric authentication scheme for WMNs, which is based on EAP-Tunneled Transport Layer Security (EAP-TTLS) [24] over Protocol for carrying Authentication for Network Access (PANA) [25]. As IEEE 802.1X, PANA is an authentication framework that is independent of the underlying access technologies. On the basis of this framework, the approach in [23] tries to provide a quick reauthentication using the EAP-TTLS over PANA. For reauthentication efficiency, the proposed EAP–TTLS authentication passes only the session keys without performing the entire TTLS that is required in the standard EAP-TTLS.

Fitzek et al. [26] describe an application scheme of IEEE 802.1X in combination with UMTS–AKA to enable authentication and security in IP-based multihop networks. In their scheme, each authenticated clients can work as the authenticator, and is able to forward the authentication data from the new client to the real AAA through a secure communication channel, thus authenticating newly joined MC using UMTS-AKA-over-EAP. While this scheme is based on the assumption that the authenticated clients will work as the authenticator and the communication overhead is high, and is not bandwidth efficient, especially for the mutual communication between two MCs.

Lee et al. [27] proposed a distributed authentication method which is aimed at decreasing the authentication delay in a wireless LAN-based mesh network. In the proposed Dynamic Distributed Authentication (DDA) method, multiple trusted nodes are distributed over the WMN to server the role of an authentication server. This significantly eases the management burden and reduces the storage space on mesh points. Since the number of trusted nodes which can play the role of authentication server is limited compared with the access routers, the performance of their scheme will substantially decrease when multiple MCs send out authentication request.

On the basis of a 4-way-handshake mechanism introduced by 802.11i, Lukas et al. [28] proposed an improved security protocol for WMNs, which is called WMNSec. In WMNSec, a dedicated station, Mesh Key Distributor (MKD), generates one single dynamically generated key for the whole network—Global Key (GK). GK is propagated from MKD to authenticated stations (MRs) using the 4-way-handshake from 802.11i. A newly joined MR would become another authenticator after it is authenticated and become the authenticated part of the WMN. Thus, the iterative authentication forms a spanning tree starting at MKD and expanding to the whole network. To provide the high-level security, the duration that a key can be actually used is limited using a key validity field. Periodic rekeying ensures the key material that is used in all stations is up-to-date. With these, WMNSec provides a much higher security barrier than WEP.

22.4.4 Identity-Based Cryptography-Based Authentication

22.4.4.1 Identity-Based Cryptography

Identity-based cryptography (IBC) is an emerging public-key cryptography in which the public key of a user could be derived from some public unique identity information about the user (e.g., SSN, email address, or unique domain name, etc.). The concept of identity-based cryptography was first proposed by Shamir [29] in 1984. Shamir also constructed an identity-based signature (IBS) scheme based on existing RSA to simplify the certificate management in e-mail system. In this approach, the digital signatures can be verified by using the user's unique identity, such as the e-mail address, as the public key. While the full functional identity-based encryption scheme is not established until Boneh and Franklin [30] applied Weil pairing to construct a bilinear map as described below. Let G_1 denote an additive group and G_2 a multiplicative group of order q for some large prime q. Let $P, Q \in G_1$ be generators. A admissible bilinear map $e{:}G_1 \times G_1 \rightarrow G_2$ should have the following properties:

1. *Bilinearity:* $\forall P, Q \in G_1$, and $a, b \in Z_q^* : e(aP, bQ) = e(P, Q)^{ab}$
2. *Non-degenerate:* $\forall P \in G_1$, $P \neq \infty : e(P, P) \neq 1$, that is, the map e does not send all pairs in $G_1 \times G_1$ to the identity in G_2
3. *Computable:* There is an efficient algorithm to compute $e(P, Q)$ for any $P, Q \in G_1$

Beside Boneh/Franklin's Weil pairing-based encryption scheme [30], Cocks et al. [31] also implemented an IBE scheme based on quadratic residues, which is a variant of interfere factorization problem [32]. In a typical IBC system, a trusted authority, which is called the Private Key Generator (PKG), is responsible to manage the public key cryptography system. The PKG first generates the system parameters including a master private/public key pair. Given the unique identity of an entity, its corresponding public key could be computed by combining the master public key of PKG. The authenticated entity could also request its private key from the PKG. After the authenticity of the entity is proved, the PKG will generate the corresponding private key using its master private key and the unique identity of the entity. Figure 22.4 illustrates the operation steps of the ID-based encryption and decryption.

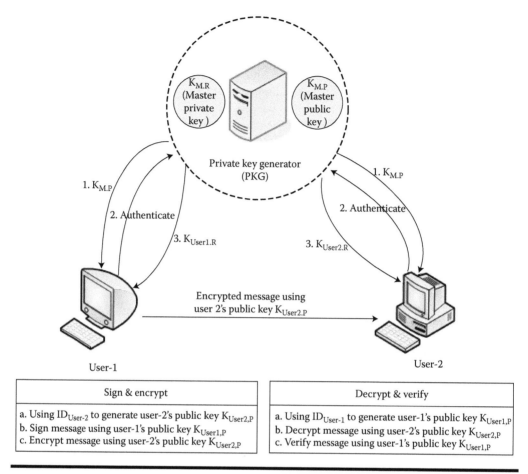

Sign & encrypt
a. Using ID_{User-2} to generate user-2's public key $K_{User2,P}$ b. Sign message using user-1's public key $K_{User1,P}$ c. Encrypt message using user-2's public key $K_{User2,P}$

Decrypt & verify
a. Using ID_{User-1} to generate user-1's public key $K_{User1,P}$ b. Decrypt message using user-2's public key $K_{User2,P}$ c. Verify message using user-1's public key $K_{User1,P}$

Figure 22.4 ID-based encryption and signature.

As a powerful alternative to the traditional certificate-based cryptography, the advantage of IBE is that parties may encrypt messages with no prior distribution of key distribution, which will be very useful in the conditions where predistribution of authenticated keys are inconvenient or infeasible. In the case of WMN, the deployment of MRs may be before the registration of MCs, thus the predistribution of authenticated keys between MRs and MCs will be infeasible. Thus, IBE is suitable for the application in the WMN. On the other hand, to encrypt or sign messages, the authorized user must obtain the appropriate private key form the PKG. Thus, the PKG is the key element in this scheme and must be highly trustable.

22.4.4.2 ID-Based Authentication in WMNs

Considering the IBC, Zang and Fang [33] present an attack-resilient security architecture, called ARSA, for large-scale WMNs where different MRs are operated by different parties, imposing another set of security vectors and issues like billing. The authors consider that the relationships among three parties in ARSA, that is, brokers, users, and WMN operators, are analogous to that among a bank, a credit-card holder, and a merchant. Each user acquires a universal pass from a broker which can be regarded as the TTP. Once the authenticated user receives a secure pass where the ID is enveloped, the WMN operator then grants the access of the pass holder. Compared to the conventional cellular-like authentication, ARSA is a homeless solution in which each client is not being bound to any specific WMN operator, and can get ubiquitous network access by a universal pass issued by a third-party broker. ARSA does not require a WMN operator to establish pairwise bilateral service-level agreements and interact in real-time with potentially other WMN operators. Another benefit brought by ARSA is that the mutual authentication and key agreement (AKA) between a mesh client and the serving WMN domain just involve local interactions without the real-time involvement of the corresponding broker, which is particularly beneficial for reducing the authentication signaling overhead and latency. ARSA also provides an efficient mutual Authentication and Key Agreement (AKA) between a user and a serving WMN domain or between users served by the same WMN domain.

In order for providing anonymity and privacy for the users, Zhang et al. [34] further proposed an improved a security solution, where the authors exploit that the public–private key pairs can be used both for authentication and for key agreement with an off-line central authority. It also provides mutual authentication and key establishment between MCs and MRs, and between MCs and MRs themselves. By maintaining a long-term relationship between MC and TTP, which is a trusted broker in the proposed scheme, the MC may take advantage of being able to get network access by any WMN operators on demand without being bound to any specific operator.

Another solution based on the ID-based encryption is PEACE, a sophisticated privacy-Enhanced yet Accountable seCurity framEwork, was proposed by Ren et al. [35]. PEACE, a user authenticates only himself as a legitimate service subscriber without disclosing any of his identity information by utilizing the group signature technique. Neither the adversary nor the user group managers can tell which particular user generates a given signature. The adversary, even by compromising MRs and other network users who may know a number of group private keys in addition to the group public key, still cannot derive any information regarding the particular group private key used for signature generation. In these solutions, the handoff is only considered within the same operators, and thus fast handoff cannot be guaranteed when the handoff is performed between two APs belonging to different operators.

22.5 Future Directions for Research

Most of the available authentication schemes are focused on the authentication between MC and mesh backbone, that is, the access level. The authentication on backbone level is highly ignored in the current research. While due to the ease-of-deployment, the WMN can be interconnected by many of small domains that are operated by different operators, and the operators can be many types. Unlike a cellular operator that often manage a nationwide or larger scale, a WMN operator may be the owner of on a community, school, enterprise, and many other small privates. IEEE 802.11-based WMN can even consists of many MRs deployed by individuals. These small WMNs are then wirelessly interconnected as a large network to cover a large area. In such a case, heterogeneous MRs and IGWs are administrated by many independent operators. Consequently, the number of WMN operators will be much larger than that of cellular operators. The authentication between MRs and IGWs from different WMN operators becomes one of the main challenges in building up such a large-scale cooperative WMN. The secure collaboration protocol among ISPs should be effectives across different WMN domains. The failure of the interoperation causes the interruption of services and any reselection of another WMN operator (if available) will also make the service unacceptable for users. The frequent handoffs among different operators will significantly increases the authentication delay if they could not be effectively collaborated.

The current works also ignore the authentication and key distribution on the mobile WMN such as the mobile WiMAX where the IEEE 802.16e/j MR (i.e., BS and relaying stations) are mobile to form a mobile wireless backbone for many applications such as disaster rescue. In the static WMN, all the MRs and the IGWs are fixed and their authentication can be maintained for a long time. Owing to dynamicity of the MRs in the mobile WMNs, the dynamical authentication should be performed once the link is broken and new link is opportunistically established. The reauthentication of an MR may causes the group handoff for all attached MC on the corresponding MR and group reauthentication should be conducted to mitigate the handoff delay of the users. Furthermore, an effective and robust key negotiation should be available in the dynamic network environment.

The authentication for high-mobility users is the other issue that should be further explored in the research. Owing to very limited coverage of IEEE 802.11-based MRs (e.g., 100 meters), the high-mobility user (e.g., a user on the car) will migrate from an MR to another. It is not acceptable for the user to authenticate and negotiate the key each MR with a long process, and the group key may be the solution that a user can access a couple of MRs by using the same key.

Anonymity and privacy for users are also important for providing a user-friendly WMN. In addition to the traffic information and data packets, the privacy of the user also includes its location, identification, personal preference, and other environmental data. These personal sensitive data should be protected with no disclosure due to authentication.

22.6 Conclusions

WMN is emerging as a promising networking technology to offer users with high-bandwidth Internet accessibility with a lower cost, as compared to traditional wireless networks. To support practical use of it, robust authentication and access control schemes are required in many application scenarios to prevent security attacks to offer reliable network services for the authentic users. All the network resources such as radios and channels have to be properly protected from any malicious access and attacks. In this chapter, we investigate these issues, elaborate the security challenges in WMNs, and illustrate the possible types of security attacks in the networks. The discussion on the unique network characteristics allows us to identify the security requirement for designing access control and authentication schemes. The current authentication mechanisms such as EAP-based IEEE 802.11x access control standards are illustrated to show their limitations in support of secure mobility in a WMN. The recently developed accessing and authentication schemes such as localized authentication, preauthentication scheme, and access control scheme based on IBC are further introduced. Finally, we discuss the future directions in the WMN authentication.

Terminologies

WMN—Wireless Mesh Network
EAP—Extensible Authentication Protocol
IBC—Identity-based cryptography
IGW—Internet Gateway
ISP—Internet Service Provider
IEEE 802.11s
IEEE 802.16j
IEEE 802.1X

Questions and Sample Answers

1. Why authentication is needed in WMNs?

   A WMN can span through a large area such as in a city or a town and contains multiple network domains owned and maintained by the same or different ISPs. Therefore, the authentication is not only needed for accessing the WMN at one time, it is also required for mobile handoff, that is, accessing from one Mesh Router (MR) to another or from an ISP managed network to the network managed by another ISP.

2. Discuss a malicious intruder's activities in brief.

   A malicious intruder can disrupt the network activities by conducting a number of security attacks. By an authentication hacking attack, a malicious MC illegally access the WMN and enjoy free network services by using single- or multihop communications. The ISPs can be further accessed by the unauthorized MC due to improper ISP configurations and inadequate security measures at the network devices. An attacker can further gain the access to the network infrastructure using some techniques such as MAC address spoofing. An adversary can conduct the impersonation attack that a forged MR claims itself as a authenticated node by sending forged/replayed network messages (e.g., beacons) to entice MCs who assume that they are connected to the correct IGW. A rogue MR can conduct the attack that causes other MRs to redirect their traffic toward itself by advertising them to a higher rate link/less congested link to the Internet.

3. Mention the steps in an execution phase in an IEEE 802.11i-supported WLAN.

   In an IEEE 802.11i-supported WLAN, the Execution phase involves three steps:
   - Reassociation
   - 802.1X authentication
   - Four-way handshake

4. What are the authentication-related security attacks against a WMN? Describe each of them.

   Unauthorized access: It is the attacker access to the service provided by the wireless mesh network (i.e., Internet access). Further, the attacker may access to customer data and private information.

   Replay attack: In a replay attack, a malicious attacker caches a legitimate authentication request (by eavesdropping or interception) and replays it at a later time. Thus, if an MR fails to recognize the authenticity of the request, it could be granting access to malicious attackers.

   Spoof attack: IP address spoofing is a rampant problem in WMNs due to the multihop nature of communication. Spoofing is the act of forging a false IP address in the packets originating from an attacker. Using such a spoofed address, a malicious attacker can perform a man-in-the-middle attack by cleverly intercepting a termination request and thereby hijacking the session.

   DoS: A malicious attacker can conduct a DoS attack by sending a flood of packets to the MR. This results in DoS to legitimate mesh clients. Another well-known security flaw that can be exploited by an attacker is that the deauthentication/termination message from the 802.11-based wireless clients is left unauthenticated. Thus, a malicious attacker can send false termination messages on behalf of a legitimate MC.

   Comprised or forged MR: The adversary can compromise and control a number of users and MRs subject to his choices. For example, the attackers may comprise the existing MRs by physical tampering or logical break-in. The adversary may also set up rogue MRs to conduct a variety of attacks. The fake or comprised MRs can be arbitrarily manipulated by the attacker to perform arbitrary attacks. The attacks may be implemented by attacking on wireless links (e.g., eavesdropping, jamming, replay and injection of message, and traffic analysis). Advanced attacks can also be implemented such as the attacker can advertise itself as a genuine MR, using some forged messages or duplicate beacons procured by eavesdropping on a genuine MR. When a multihop MC hears these fraudulent beacons from a malicious MR, it assumes that it is within the radio coverage of a genuine MR and initiates a registration procedure. After registering, the MR assumes that it has obtained the Internet connection and disconnects its communication from the genuine MR. Slowly, the forged MR could entice a number of MCs to disconnect from the genuine MRs. This attack is possible in situations where the MR is not authenticated by the MC and breaks a genuine Internet connection or causes unwarranted registration delay. The forged MR that acts as the relay can seize the data packets or capture sensitive personal (e.g., password) of MCs and other MRs.

5. Draw a diagram showing the mutual authentication in a WMN with multiple ISPs.

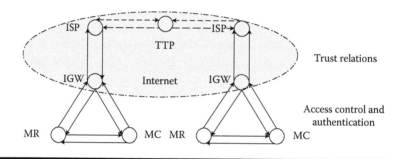

Mutual authentication in a WMN with multiple ISPs.

6. What is identity-based cryptography?

Identity-based cryptography (IBC) is an emerging public-key cryptography in which the public key of a user is some unique information about the identity of the user (e.g., SSN, email address, or unique domain name). The first conception of an email-address-based PKI was described by Adi Shamir in 1984. In this approach, users can verify the digital signatures by using only public information such as the user's identifier. IBC completely eliminates the need for public-key distribution via conventional public-key certificates.

Author's Biography

Bing He is a PhD candidate in computer science at the Center for Distributed and Mobile Computing in the Department of computer science, University of Cincinnati. His current research interests include architecture design, performance optimization and security of wireless mesh networks, and resource allocation in 802.16 WiMAX network. He received his BS degree in communication engineering and MS degree in signal and information processing from Northern Jiaotong University of China. He has been an engineer with Honeywell Technology Solutions Lab in China.

Dr. Bin Xie received his MSc and PhD degrees (with honors) in computer science and computer engineering from the University of Louisville, Kentucky, 2003 and 2006, respectively. He is currently the president of the InfoBeyond Technology LLC. He is the author of the books titled *Heterogeneous Wireless Networks-Networking Protocol to Security (VDM Publishing House: ISBN: 3836419270)* and *Handbook/Encyclopedia of Ad Hoc and Ubiquitous Computing (World Scientific: ISBN-10: 981283348X)*. He has published 50+ papers in the IEEE conferences and journals. His research interests are focused on mobile computing, embedded wireless sensor networks, network security, in particular, on the fundamental aspects of mobility management, performance evaluation, Internet/wireless infrastructure security, and network capacity. He is an IEEE senior member and the vice chair of the IEEE Computer Society Technical Committee on Simulation (TCSIM). He serves as the program cochair of IEEE HPCC-09 and received the Outstanding Leadership award. He also is the program cochair of CyberC 2009, ICICTA 2009, and CPSC2009, and the TPC of many international conferences, and was the keynote of ICICTA 2008. He also is the editor of *International Journal of Information Technology, Communications and Convergence* (IJITCC).

David Zhao completed his bachelor degree in electric engineer in the City College of City University of New York and master degree in Electronics Engineer from Stevens Institute of Technology, New Jersey. He is currently a senior engineer at US ARMY Communications and Electronics Research and Development Center (CERDEC) at Fort Monmouth, New Jersey. His work is to research, examine, and evaluate various commercial wireless technologies and products, and also to determine its adaptability for use in the military tactical environment.

Ranga Reddy received his MS in electrical engineering from Stevens Institute of Technology in 2004. Ranga is involved in wireless communication standardization efforts within IEEE 802. He is a member in the 802.16 working group, and actively participates in the 802.16j and 802.16m task groups. Ranga is also a participant in the 802.22 working group activities. Since 2001 he has been working US Army Communications/Electronics Research & Development Engineering Center (CERDEC) at Fort Monmouth on various projects involving systems engineering, modeling and simulation efforts surrounding development and engineering of several communication technologies MANET, IPv6, and broadband wireless communications.

References

1. N. Nandiraju, D. Nandiraju, L. Santhanam, B. He, J. Wang, and D. P. Agrawal, Wireless mesh network: Current challenges and future directions of web-in-the-sky, *IEEE Wireless Communications*, 14(4), 79–89, 2007.
2. I. F. Akyildiz, X. Wang, and W. Wang, Wireless mesh networks: a survey, *Computer Networks and ISDN Systems*, 47(4), 445–487, 2005.
3. C. Rigney, S. Willens, A. Rubens, and W. Simpson, Remote authentication dial in user service (RADIUS), IETF RFC 2865, IETF, June 2000.
4. S. Baek, S. Pack, T. Kwon, and Y. Choi, A localized authentication, authorization, and accounting (AAA) protocol for mobile hotspots, in *Proceedings of the Third Annual Conference on Wireless on Demand Network Systems and Services*, Les Ménuires, France, pp.144–153, January 2006.
5. IEEE Standard 802.1X-2004, *IEEE Standard for Local and Metropolitan Area Networks—Port-based Network Access Control*, IEEE Computer Society, 2004.

6. IEEE Computer Security, IEEE 802.11i Standard, *IEEE Standard for Information Technology—Telecommunications and Information Exchange between Systems—Local and Metropolitan Area Networks—Specific Requirements Part 11: Wireless LAN Medium Access Control (MAC) and Physical Layer (PHY) Specifications Amendment 6: Medium Access Control (MAC) Security Enhancements*, July, 2004.

7. S. Pack and Y. Choi, Fast handoff scheme based on mobility prediction in public wireless LAN systems, *IEE Communications*, 151(5), 489–495, 2004.

8. J. Hur, H. Shim, P. Kim, H. Yoon, and N.-O. Song, Security considerations for handover schemes in mobile WiMAX networks, in *Proceedings of IEEE Wireless Communications and Networking Conference, WCNC 2008*, Las Vegas, NV, March 2008.

9. WiMAX Forum, *Mobile WiMAX—Part I: A Technical Overview and Performance Evaluation*, February 2006.

10. B. Aboba, Fast handoff issues, IEEE-03-155r0-I. IEEE 802.11 Working Group, March 2003.

11. A. Mishra, M. H. Shin, N. L. Petroni, J. T. Clancy, and W. A. Arbauch, Proactive key distribution using neighbor graphs, *IEEE Wireless Communication*, 11(1), 26–36, 2004.

12. T. Aura and M. Roe, Reducing reauthentication delay in wireless networks, in *Proceedings of IEEE SecureComm'05*, Athens, Greece, 2005.

13. J.-J. Chen, Y.-C. Tseng, and H.-W. Lee, A seamless handoff mechanism for IEEE 802.11 WLANs supporting IEEE 802.11i security enhancements, in *Proceedings of IEEE APWCS*, Hsinchu, Taiwan, 2007.

14. H. Zhu, X. Lin, R. Lu, P.-H. Ho, and X. Shen, SLAB: A secure localized authentication and billing scheme for wireless mesh networks, *IEEE Transactions on Wireless Communications,* 7(10): 3858–3868 2008.

15. I. Askoxylakis, B. Bencsáth, L. Buttyán, L. Dóra,V. Siris, D. Szili, and I. Vajda, Securing multi-operator based QoS-aware mesh networks: requirements and design options, *Wireless Communications and Mobile Computing* (Special Issue on QoS and Security in Wireless Networks), 10(5), pp. 622–646, 2010.

16. International Telecommunication Union, General characteristics of international telephone connections and international telephone circuits. ITU-TG.114, 1988.

17. L. Buttyán and L. Dóra, "An authentication scheme for QoS-aware multi-operator maintained Wireless Mesh Networks", in *Proceedings of the First IEEE WoWMoM Workshop on Hot Topics in Mesh Networking (HotMESH'09)*, Kos, Greece, June, 2009.

18. S. Blake-Wilson and A. Menezes, Entity authentication and authenticated key transport protocols employing asymmetric techniques, in *Proceedings of the 5th International Workshop on Security Protocols*, London, UK, pp. 137–158, 1998.

19. X. Lin, X. Ling, H. Zhu, P.-H. Ho, and X. Shen, A novel localized authentication scheme in IEEE 802.11 based wireless mesh networks, *International Journal of Security and Networks*, 3(2), 122–132, 2008.

20. Z. Cao, H. Zhu, and R. Lu, Provably secure robust threshold partial blind signature, *Science in China Series F: Information Sciences*, 49(5), 604–615, 2006.

21. R. Soltwisch, X. Fu, D. Hogrefe, and S. Narayanan, A method for authentication and key exchange for seamless inter-domain handovers, in *Proceedings of 12th IEEE International Conference on Networks (ICON 2004)*, Singapore, pp. 463–469, November 2004.

22. J. Loughney, M. Nakhjiri, C. Perkins, and R. Koodli, Context transfer protocol (CXTP), IETF RFC 4067, IETF, July 2005.

23. K. Khan and M. Akbar, Authentication in multi-hop wireless mesh networks, in *Proceedings of World Academic of Science, Engineering and Technology*, November 2006.

24. P. Funk and S. Blake-Wilson, Extensible Authentication Protocol Tunneled Transport Layer Security Authenticated Protocol Version 0 (EAP-TTLSv0), IETF RFC 5281, IETF, August 2008.

25. D. Forsberg, Y. Ohba, B. Patil, and H. Tschofenig, Protocol for carrying authentication and network access (PANA), IEFT RFC 5191, IETF, May 2008.

26. F. Fitzek, A. Kopsel, and P. Seeling, Authentication and security in IP based multi-hop networks, in *Proceedings of Wireless World Research Forum 7 (WWRF7)*, Eindnoven, NL, December 2002.

27. I. Lee, J. Lee, W. Arbaugh, and D. Kim, Dynamic distributed authentication scheme for wireless LAN-based mesh networks, in *Information Networking, Towards Ubiquitous Networking and Services:*

international Conference, ICOIN 2007, Estoril, Portugal, January 23–25, 2007. Revised selected papers, T. Vazão, M. M. Freire, and I. Chong, eds. Lecture Notes in Computer Science, vol. 5200. Springer, Berlin, Heidelberg, pp. 649–658, 2008.

28. G. Lukas, and C. Fackroth, WMNSec: security for wireless mesh networks, in *Proceedings of the 2009 International Conference on Wireless Communications and Mobile Computing: Connecting the World Wirelessly, IWCMC '09*, Leipzig, Germany, June 21–24, 2009. ACM Press, New York, NY, pp. 90–95.
29. A. Shamir, Identity based cryptosystems and signature schemes, Lecture Notes in Computer Science. Springer-Verlag, Berlin, Germany, 1984, vol. 196, Proc. CRYPTO, pp. 47–53.
30. D. Boneh and M. Franklin, Identify-based encryption from the Weil pairing, in Lecture Notes in Computer Science. Springer, Berlin, Germany, 2001, vol. 2139, Proc. CRYPTO, pp. 213–229.
31. C. Cocks, An identity based encryption scheme based on quadratic residues, in *Proceedings of the 8th IMA international Conference on Cryptography and Coding*, December 17–19, 2001. B. Honary, Ed. Lecture Notes in Computer Science, vol. 2260. Springer, London, pp. 360–363.
32. J. Baek, J. Newmarch, R. Safavi-Naini, and W. Susilo, A Survey of Identity-based Cryptography, in *Proceedings of the 10th Annual Conference for Australian Unix and Open Systems User Group (AUUG 2004)*, pp. 95–102, 2004.
33. Y. Zhang and Y. Fang, ARSA: an attack-resilient security architecture for multihop wireless mesh networks, *IEEE Journal on Selected Areas in Communications*, 24(10), 1916–1928, 2006.
34. Y. Zhang and Y. Fang, A secure authentication and billing architecture for wireless mesh networks, *Wireless Networks*, 13(5), 663–678, 2007.
35. K. Ren and W. Lou, A sophisticated privacy-enhanced yet accountable security framework for metropolitan wireless mesh networks, in *Proceedings of the 2008 The 28th International Conference on Distributed Computing Systems*, IEEE Computer Society, Washington, DC, USA, pp. 286–294, 2008.

Chapter 23

Misbehavior Detection in Wireless Mesh Networks

Md. Abdul Hamid and Md. Shariful Islam

Contents

23.1 Introduction

The continuing driving force in the development of wireless mesh networks (WMNs) comes from their envisioned advantages, including extended coverage, robustness, self-configuration, easy maintenance, and low cost. The term "wireless mesh network" is often used interchangeably with an *ad hoc* or a multihop wireless network, although all the terms have slightly different nuances. WMNs attempt to change the economics of wireless networking by aggregating traffic for Internet backhaul. As such, wireless mesh networking has garnered widespread interest in its applications to enterprise, military, sensor, community, and public safety networks [1–3]. WMNs [4,5] are gaining in importance as an alternative to cable and Digital Subscriber Line (DSL), and are envisioned to provide fixed, nomadic, portable, and eventually mobile-wireless broadband connectivity. These aspects attract the interests toward a wide variety of potential applications and usage scenarios for the mesh networking domain. WMNs have the potential to bring diverse advantages to wireless communication services, allowing clients to exchange information in a decentralized manner and also to extend coverage of cellular and other networks by allowing relay-based networking at the edge terminals. Similar to the paradigm shift experienced in wired networks during the late 1960s and early 1970s that led to a hugely successful and distributed wired network, WMNs are promising directions in the future of wireless networks [6–8].

The architecture of WMN exhibits some unique characteristics (e.g., use of a shared wireless medium, stationary wireless nodes, heterogeneous traffic pattern, both access point and relay functionalities, and so on) that differentiate a WMN from other networks. A WMN architecture is composed of four different entities or network elements: Mesh Point Portal (MPP), Mesh Access Point (MAP), Mesh Point (MP) [9], and legacy nodes which are also called Mesh Client (MC). An MPP is a gateway node that connects the WMN to a wired infrastructure, possibly the Internet or other local networks. There can be multiple MPPs deployed in a WMN. An MP is a relay node responsible for relaying traffic to other MPs or MAPs, whereas an MAP performs both the functions of an MP and an access point. Therefore, a WMN is an emerging two-tier architecture, where WGs, MRs form the backbone (infrastructure) based on wireless multihop transmission, and MCs are connected to the MRs using the existing wireless LAN technologies; for example, 802.11 (WiFi). The backbone entities offer connectivity to the MCs by acting like access points, forming at the same time a self-organized wireless backbone. This backbone has two possible roles. It can be either a standalone network simply offering interclient connectivity or a local extension for the wired Internet if there are available connections between one or more MPPs. In both cases, the WMN's backbone is in charge of relaying all the traffic from/to MCs.

The desirable features make WMNs an appealing solution for a plethora of applications, such as broadband home networking, community networking, and so on. However, there are still several challenges and issues preventing WMNs to be widely deployed in large scales. The first major issue is the performance (throughput, delay, or packet loss rate) of WMNs, which drops sharply with an increasing number of wireless hops the packets traverse through. Throughput maximizing routing metrics [10–12] and the multiradio, multichannel techniques [13–15] are being researched to overcome this problem. The second major issue is the lack of an integrated solution to provide security in WMNs, which has received meager attention in the literature. Clearly, without a well-designed security solution, WMNs are vulnerable to various types of internal and external attacks that may cause significant inconvenience to the users and operators. The potential of wireless mesh networking cannot be exploited without considering and adequately addressing the involved security issues. The existing security schemes proposed for *ad hoc* networks can be adopted for WMNs, but several issues exist [1]: (i) most security solutions for *ad hoc* networks are still not mature enough to be practically implemented and (ii) the network architecture of WMNs is different from a conventional *ad hoc* network, which causes differences in security mechanisms. As a consequence, new security schemes ranging from encryption algorithms to security key distribution, secure MAC and routing protocols, intrusion detection, and security monitoring need to be developed. To further ensure network survivability and reliability, a new security mechanism needs to be developed, which can detect the presence of malicious node(s) in the network and also prevent malicious use of network resources.

In the context of mesh networks, information about the behavior of mesh clients can become readily available to their immediate mesh routers through direct observation measurements. If these measurements are compared with their counterparts for normal protocol operation, it is then contingent upon the detection rule to decide whether the protocol is normally executed or not. In this chapter, we describe two detection mechanisms to distinguish the clients gaining higher share of bandwidth (by deliberately misusing the access mechanism, for example, smaller contention window) than other clients. We consider that all the clients have similar application running and they have same data rates, so clients attached to a router are expected to have equal share of bandwidth. Therefore, we have designed the detection mechanism mainly by measuring the throughput to identify malicious clients. In the first mechanism, we use correlation coefficient-based detection via common set of mesh routers. In the second scenario, detection mechanism is independent of the common set of routers, that is, each mesh router can directly identify the misbehaving client(s) in the network.

The rest of the chapter is organized as follows. In Section 23.2, we introduce misbehavior in WMNs and review the related works of misbehaviors in wireless networks and their counter measures. In Section 23.3, we discuss the detection mechanisms along with the algorithms that identify the malicious clients in WMNs. Also, we evaluate and discuss the performance of the detection mechanisms through simulations. In Section 23.4, we discuss the avenues of potential research directions. Finally, we conclude the chapter in Section 23.5 with a brief summary.

23.2 Background

Deviation from legitimate protocol operation in wireless networks has received considerable attention from the research community in recent years. The pervasive nature of wireless networks with devices, which are gradually becoming essential components in our life-style, justifies the rising interest on that issue. In addition, the architectural organization of wireless networks in distributed secluded user communities raises issues of compliance with protocol rules. More often than not, users are clustered in communities that are defined on the basis of proximity, common service, or some other common interest. Since such communities are bound to operate without a central supervising entity, no notion of trust can be presupposed. Furthermore, the increased level of sophistication in the design of protocol components, together with the requirement for flexible and readily reconfigurable protocols has led to the extreme where wireless network devices have become easily programmable. As a result, it is feasible for a network peer to tamper with software and firmware, modify its wireless interface and network parameters and ultimately abuse the protocol. This situation is referred to as protocol misbehavior.

23.2.1 Misbehavior in WMNs

Contention-based MAC (medium access control) protocols are usually adopted in WMNs for wireless users to share a common wireless channel. Therefore, a greedy mesh client (MC) can substantially increase his share of bandwidth, at the expense of the other clients, by misusing the access protocols, which unfairly occupies wireless channel and resources [16]. This can become a serious problem in Internet access hotspots where individual clients have to pay for network usage and hence may be motivated to cheat in order to increase their share of the medium. For example, a small backoff interval gives the corresponding client the advantages of gaining access to the wireless channel quickly.

Apart from small backoff values, a malicious client can disobey the MAC protocol in other ways as well. It can choose a smaller contention window size or it may reserve the channel for larger interval than the maximum allowed network allocation vector (NAV) duration. As a result, it may gain higher share of bandwidth by sending more packets in the network over regularly behaving honest clients. This can be dangerous, since the devices themselves control their random deferment. Indeed, with the higher programmability of the network adapters, the temptation to tamper with the software or firmware is likely to grow. Therefore, in the presence of malicious clients (even though they are legitimate in the sense that they use cryptographic keys and obey the underlying security protocol) that disobey the standard, bandwidth share of the well-behaved clients may significantly degrade [16,17]. Since all the wireless stations use the similar IEEE 802.11 CSMA/CA MAC protocol, malicious use of the medium is always possible to gain higher share of bandwidth at the expense of other users in the network. Even though the cryptographic solution may achieve authentication, data confidentiality and other security issues, this mechanism may not detect/restrict the MAC layer greedy behavior since wireless devices (e.g., mesh clients) directly deal with the wireless medium [16]. For network survivability and reliability, it is necessary to develop an efficient technique to detect misbehaving client(s) to defend the network being crippled. Therefore, along with the cryptographic solution, a detection mechanism, as a second line of defense, needs to be developed to defend/restrict malicious clients. Since the goals of a misbehaving peer range from exploitation of available network resources for its own benefit up to network disruption, the solution to the problem is the timely and reliable detection of such misbehavior instances, which would eventually lead to network defense and response mechanisms and isolation of the misbehaving peer and thus ensuring efficient and fair use of network resources and minimizing performance losses. Furthermore, the basic feature of attack and misbehavior strategies is that they are entirely unpredictable. In the presence of such uncertainty, it is meaningful to seek models and decision rules that are robust, namely they perform well for a wide range of uncertainty conditions.

23.2.2 Related Works on Misbehavior Detection

Misbehavior in wireless networks has been studied at different layers, for example, network, MAC layer, and so on. However, solutions to the misbehavior were mostly developed in *ad hoc* or WLAN [16,18–22]. In fact very few works [10,23] have been done for misbehavior detection so far in the domain of WMN. For IEEE 802.11 equipments, Kyasanur and Vaidya [21] have proposed that the receiver assigns the backoff value to be used by the sender, so the former can detect any misbehavior of the latter. If the sender deviates from the assigned value, it will be assigned high backoff values on the next round to compensate its deviation. As mentioned by the authors, this mechanism has several limitations such as the possible collusion between sender and receiver, and the fundamental change to the protocol. Konorski [22] has proposed a misbehavior-resilient backoff algorithm; both [21,22] exhibit the same drawback: they require to changing the current protocol. The authors in [11] focus on MAC layer misbehavior in wireless hot-spot communities. They have proposed a sequence of conditions on some available observations for testing the extent to which MAC protocol parameters have been manipulated. The advantage of the scheme is its simplicity and easiness of implementation, although in some cases the method can be deceived by cheating peers, as the authors pointed out. In the context of WMN, an active cache-based mechanism [23] is proposed to defend DoS attack caused by flooding a large volume of traffic in the network by malicious intruders. On the basis of the detection method of *ad hoc* networks [16], an MAC layer selfish behavior detection model is presented in [10] for WMN and different detection mechanisms are used for router and client selfish attacks.

Misbehavior detection has been studied at the network layer [12,18–20,24–26] for routing protocols as well. The work in [18] presents the watchdog mechanism, which detects nodes that do not forward packets destined for other nodes. The pathrater mechanism evaluates the paths in terms of trustworthiness and helps in avoiding paths with untrusted nodes. The technique presented in [19] aims at detecting malicious nodes by means of neighborhood behavior monitoring and reporting from other nodes. A trust manager, a reputation manager, and a path manager aid in information circulation throughout the network, evaluation of appropriateness of paths and establishment of routes that avoid misbehaving nodes. Detection, isolation, and penalization of misbehaving nodes are also attained by the above technique.

In SORI [12], all nodes maintain a confidence-level table for them to exchange information with each other and penalize the bad reputation selfish node. They use one-way hashing to ensure the selfish node could not impersonate other nodes in improving its own reputation. However, a malicious node can always fake the information and keep condemning other innocent nodes and eventually causing a chaos in the network.

SMDP [24] is a session-based detection protocol and it uses the principle of data flow conversation where the data flow in and flow out from a node should always be equal. At the end of each data session, all the nodes along the path will send the total packet they received to the previous hop and the total packet they transmitted to the next hop. After gathering all these transmission reports, all the nodes will rebroadcast the sum of the packets to the surrounding nodes. A node will be suspected if the total transmission is much different from the total reception. Digital signature has been used to ensure no one can fake the integrity of the report. However, the source can defame the next forwarder by reporting an incorrect number of total transmitted packets.

Tan and Bose [25] discussed a reward-based scheme that relies on the secured module where it must be tamper resistance and protected from illegal manipulation. The secured module is only feasible under a controlled environment. Thus, it is not suitable for real-world implementation as the availability and the robustness of the modules are not guaranteed. URSA [26] is a robust network access control that based on the ticket certification service through multiple node consensus and fully localized instantiation. It is a protocol that relies on the multisignature (threshold cryptography) to achieve the group trust model where each of the legitimate node holds a portion of the secret key (SK) and k portion of SKs are needed to renew the signature of the ticket. URSA is affected by the connectivity of the network. In other words, insufficient k nodes in an area of network will cause an innocent node be excluded from the network. Besides, the robustness of URSA relies on the strength of the threshold cryptography and it needs to periodically refresh the network SK to avoid the malicious node from obtaining k SK illegally by joining the network repeatedly.

In general, the detection approaches are developed from two directions, namely active and passive. Active approaches usually deal with the correction of the underlying protocol to thwart/remove the existing attacks, and passive approaches use statistical means. Note that it is unrealistic and difficult in practice to change the existing protocols due to equipments' compatibility problems. Moreover, the existing mechanisms focus on specific protocol's misbehavior, which are not applicable for a wide variety of attacks. Therefore, we base our detection approach using passive monitoring since no modification of the existing protocols need to be done, and our mechanism identifies misbehaving mesh clients those aim at gaining higher share of the wireless medium. Generally speaking, in mesh networks, mesh clients should have fair share of bandwidth to ensure the uninterrupted and unfair service provided by the network. Fare share of the wireless medium can be damaged by easily exploiting the access mechanisms and/or other techniques. Compared with the usual network traffic pattern, greedy behavior can dramatically change the fare share of bandwidth among the mesh clients, in which the abnormal traffic pattern may interrupt, damage, or disable the normal network functionality. Therefore, in the access network, service disruption of other legitimate clients would totally be crippled. Along with the security challenges, this sort of vulnerabilities require a comprehensive network client-monitoring framework to achieve real-time awareness, immediate response, and even traceback to malicious users. Our endeavor in this chapter is to design a comprehensive monitoring framework to detect malicious mesh clients to prevent their greedy behavior in WMNs. The detection system is implemented only on the mesh routers and thus it is inherently distributed. With our techniques, we believe that the detection system can become an integral part of the security scheme to expedite the rapid deployment of this promising wireless technology.

23.3 Misbehaving Client Detection

As stated earlier, in this chapter, two mechanisms are described, namely correlation coefficient-based detection using the trust relationships of the mesh clients with the common set of routers (CSRs) [27], and the Wilcoxon signed rank test (WSRT)-based detection [28]. We describe each of the detection mechanisms in the sequel.

23.3.1 Correlation Coefficient-Based Detection via CSRs

We develop a preventive solution to deal with the colluding actions taken by the malicious intruder (i.e., mesh clients) based on the correlation coefficient between the communicating clients. Let us consider the communication scenario in Figure 23.1, where two clients have the common set of routers through which their messages traverse in the network. Figure 23.1 shows the communication scenario between two MCs, p and q via a common set of routers $MR_1, MR_2, \ldots, MR_m$. Common set is chosen based on the close relationship with the two communicating clients. For example, all the past messages between client p and q traverse through this set of routers and/or both clients have individual communications with those routers and therefore, they have an existing trust history with the routers. If we judge in the statistical way, their behaviors will be linearly correlative with each other. In other words, bandwidth share (under saturated condition) will be almost equal for all the clients. On the basis of this trust relationship, an algorithm is developed to calculate the correlation coefficient between client p and q, and a decision whether client q is malicious or not is compared against client p and/or vice-versa. Correlation is calculated by the associated router using the trust relationship of the client with the common set of routers. In the following, we describe the protocol in details.

23.3.1.1 Protocol Description

Suppose $M = \{R1, R2, R3, \ldots, R_m\}$ is a common set of routers through which clients MC_p and MC_q exchange messages. Let us define two set of trust values $T_p = \{t_1^p, t_2^p, \ldots, t_m^p\}$ and $T_q = \{t_1^q, t_2^q, \ldots, t_m^q\}$ for the two clients MC_p and MC_q, respectively. Then, individual trust between

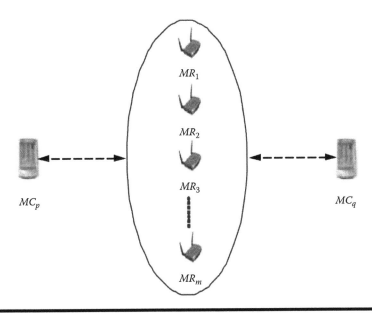

Figure 23.1 A communication scenario between mesh clients MC_p and MC_q via common set of mesh routers $R_1, R_2, \ldots, R_m$.

MC_p and its common communicating set M is evaluated as

$$t_1^p = \frac{\left(\omega_{R1}^p - \mu_{R1}^p\right)}{\left(\omega_{R1}^p + \mu_{R1}^p\right)}, \quad t_2^p = \frac{\left(\omega_{R2}^p - \mu_{R2}^p\right)}{\left(\omega_{R2}^p + \mu_{R2}^p\right)}, \ldots, t_m^p = \frac{\left(\omega_{Rm}^p - \mu_{Rm}^p\right)}{\left(\omega_{Rm}^p + \mu_{Rm}^p\right)}.$$

Similarly, individual trust between MC_q and M is evaluated as

$$t_1^q = \frac{\left(\omega_{R1}^q - \mu_{R1}^q\right)}{\left(\omega_{R1}^q + \mu_{R1}^q\right)}, \quad t_2^q = \frac{\left(\omega_{R2}^q - \mu_{R2}^q\right)}{\left(\omega_{R2}^q + \mu_{R2}^q\right)}, \ldots, t_m^q = \frac{\left(\omega_{Rm}^q - \mu_{Rm}^q\right)}{\left(\omega_{Rm}^q + \mu_{Rm}^q\right)}.$$

Here, ω and μ denote the number of times an MC have normal behavior (throughput does not exceed beyond a threshold) and malicious behavior (throughput exceeds the threshold), respectively, recorded by the common set M. Then, the correlation coefficient [29], denoted by ρ, is calculated according to the following equation:

$$\rho(T_p, T_q) = \frac{\text{cov}(T_p, T_q)}{\sigma_{T_p} \sigma_{T_q}} \tag{23.1}$$

where σ_{T_p} and σ_{T_q} are the standard deviations of client p and q, respectively. On the basis of this correlation coefficient value, a decision is made whether a client is malicious or not.

Algorithm 23.1: **Correlation coefficient-based malicious client detection via common set of Routers**

1. Calculate the trust values T_p and T_q with the common set M
2. Divide the routers into $g(g \geq 1)$ groups, and accordingly divide the trust values T_p and T_q into g groups as $\{\{T_{p1}\}, \{T_{p2}\}, \ldots, \{T_{pg}\}\}$ and $\{\{T_{q1}\}, \{T_{q2}\}, \ldots, \{T_{qg}\}\}$
3. Arrange the trust values according to groups as $\{\{T_{p1}, T_{q1}\}, \{T_{p2}, T_{q2}\}, \ldots, \{T_{pg}, T_{qg}\}\}$
4. Calculate the correlation coefficient according to Equation 23.1
5. Calculate the average correlation coefficient, $\rho_{\text{avg}} = \sum_{i=1}^{g} \frac{\rho_i}{g}$
6. Compare ρ_{avg} with a predefined threshold *thresh*
7. **If** $\rho_{\text{avg}} \leq$ *thresh* **then**
8. **return** *true* {i.e., presence of malicious client in the network}
9. **else**
10. **return** *false* {i.e., network is not under attack}
11. **end if** ■

The value of the parameters ω and μ are calculated by the router(s) in the following way. Under the normal operation of the network, the average throughput is calculated by each router in the time interval T [T is the monitoring interval described in Section 4.3.1.2 (Performance Evaluation)]. Then, while the detection mechanism is executed by each router, clients' current throughput is measured within the monitoring interval T and compared against the reference value (i.e., average throughput). If the throughput exceeds the reference value, the mesh router increments the counter and records the value as μ. Otherwise, if the throughput is less or equal to the reference value, then the mesh router increments the counter and records the value as ω. Therefore, ω refers to the counter that records the number of times a client accesses the medium which is considered to be a normal behavior. Similarly, μ refers to the counter that records the number of times a client accesses the medium which is considered to be a malicious behavior.

Algorithm 23.1 depicts the detection procedure. Each MR collects the trust values of same clients from other MRs (i.e., for those clients who move from one router to another) as well as for the clients attached to this MR. Then, each MR calculates the correlation coefficient according to Equation 23.1, and then calculates the average correlation coefficient value. After that, the average value is compared against the reference threshold. The presence of the malicious client is detected by the MR if the average correlation coefficient is less than the threshold.

23.3.1.2 Performance Evaluation

In this section, we evaluate the performance of the proposed detection algorithm through simulations using NS-2 [30]. We examine the effectiveness of the proposed detection algorithm with the number of malicious clients varying from 5% to 30% of the total clients in terms of the following performance metrics:

Detection efficiency: The detection efficiency ε is defined as $\varepsilon = z/Z$, where z denotes the number of malicious client detected and Z denotes the total number of malicious clients in the network.

False positive: The false-positive rate γ is defined as $\gamma = y/Y$, where y denotes the number of legitimate clients detected as malicious ones and Y denotes the total number of clients in the network.

We consider two scenarios in the simulation. In the first scenario, we fix the total number of MRs to 10 and MCs to 80 in the network. In the second scenario, we fix the total number of MRs to 125 and MCs to 1000 in the network. In each case, there are eight MCs attached to each MR. Matrix location is uniformly placed in the terrain. Transmission range is taken as 50 meters for both MR and MC, and MCs are attached to their nearest MRs. We consider that all the clients always have packets (of same size) to send (i.e., all the mesh clients in the network are backlogged). The simulation parameters are listed in Table 23.1.

To avoid the false alarm and increase the detection accuracy, monitoring interval, denoted by T, is particularly important so that the data required for detection are collected during configurable intervals of time; at the end of each interval, the detection mechanism is run. This provides the ability to collect more statistical data and hence increase the accuracy. In addition, it is shown in [31–34] that the binary exponential backoff algorithm of IEEE 802.11 is unfair in the short term. This would result in false positives if stations were monitored over short-term periods (even in the absence of misbehavior). Therefore, the monitoring period has to be large enough to rely on long-term backoff fairness. To get the reference record (throughput or number of packets) at each mesh router, the value of T should be chosen carefully, so that the usual variation of the short-term unfairness of the 802.11 MAC protocol is taken into account. We derive the value of T to be 11 seconds for the first scenario from our simulation in ns-2 as follows. The raw bandwidth is set to 2 Mbps. Due to the overhead of IEEE 802.11, in ideal case, a maximum throughput of around 1.4 Mbps can be achieved. In other words, considering a packet size equals 512 bytes, a mesh router can receive a total of around 341 packets/second from all of its clients. Considering eight clients

Table 23.1 Simulation Parameters

Parameter	*Value(s)*
Num. of mesh routers	10,125
Num. of mesh clients	80,1000
Node distribution	Uniform
Transmission range	50 meters
Channel bandwidth	2 Mbps
Packet size	512 Bytes
Traffic	CBR (Constant Bit Rate)
MAC	CSMA/CA (IEEE 802.11 DCF)
Minimum contention window (CW_{min})	32
Monitoring interval, T	11 and 16.35 seconds
Simulation time	300 seconds

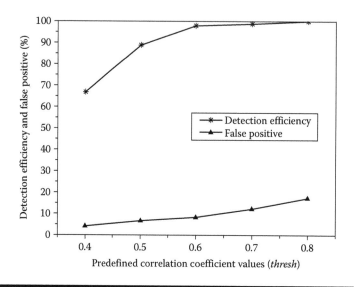

Figure 23.2 Detection efficiency (ε) and false-positive rate (γ) with variable threshold values. There are total of 10 MRs and 80 MCs in the network.

under each router, an individual client may send 42 packets/second at best. We set the source rate higher (50 packets/second) than the medium capacity to saturate the channel. Without introducing malicious actions taken by the clients in the network, we observe the number of packets sent by all the clients to their corresponding routers for different time interval T. We observe that all the clients get almost equal share of bandwidth when T equals 11 seconds. So, we set the monitoring period $T = 11$ seconds in our detection mechanism. Similarly, we get $T = 16.35$ seconds for the second scenario. To implement the proposed detection mechanism in NS-2, we randomly vary contention window sizes for 5% to 30% of the total clients while setting that to the default value (of 802.11 MAC layer implementation in NS-2) for the rest of the clients. Since clients with smaller contention window size access the medium quicker than the clients with default contention window size, they send more packets to the routers. We have used the well-known random way-point mobility model as the client mobility pattern by modifying the default implementation in NS-2 [30]. Each client is initially attached with a mesh router within the simulation terrain. As the simulation progresses, each client pauses at its current location for a period, and then randomly chooses a new location to move to and velocity between 0.2 and 0.6 meters per second at which to move there. Each client continues this behavior, alternately pausing and moving to a new location (and attaches to its nearby mesh router), for the duration of the simulation. Results are averaged over 10 simulation runs.

Figure 23.2 shows the detection efficiency and false-positive rate when we use the number of MRs and MCs 10 and 80 (i.e., there are eight MCs associated with each MR), respectively. The results are plotted as a function of the predefined correlation coefficient values. We observe that the choice of threshold (i.e., correlation coefficient) is an important factor to design an efficient detection mechanism. From Figure 23.2, the optimum threshold value (i.e., high detection efficiency with low false positive) may be found between 0.5 and 0.6. More specifically, at the threshold value 0.55, the detection efficiency can be achieved more than 95% while keeping the false-positive rate below 8%. Therefore, 0.55 has been chosen to be the optimal threshold.

Figure 23.3 depicts the simulation results about the detection efficiency and false-positive rate as a function of the percentage of malicious clients. The threshold 0.55 is considered as the optimal

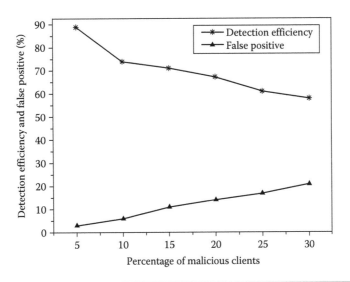

Figure 23.3 Detection efficiency (ε) and false-positive rate (γ) with different percentage of malicious clients (*thresh* is set to 0.55). There are total of 10 MRs and 80 MCs in the network.

value in the simulation. The mechanism performs better when the percentage of malicious clients is smaller. About 90% to 70% malicious nodes can be detected while keeping the false-positive rate below 3% to 7%, for 5% to 10% malicious clients. However, as the number of these clients increases, the detection efficiency reduces and false-positive increases quickly. So, the algorithm has its limitation as the detection efficiency gets confined by the number of misbehaving clients.

Figure 23.4 depicts the detection efficiency and false-positive rate when we use the number of MRs and MCs 125 and 1000 (i.e., there are eight MCs associated with each MR), respectively. The

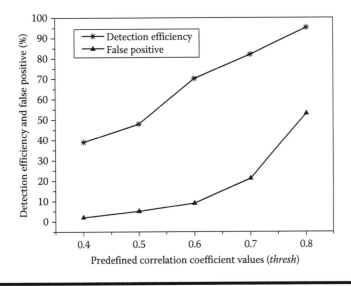

Figure 23.4 Detection efficiency (ε) and false-positive rate (γ) with variable threshold values. There are total of 125 MRs and 1000 MCs in the network.

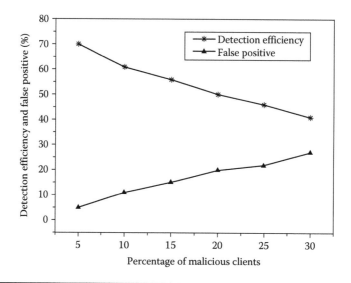

Figure 23.5 Detection efficiency (ε) and false-positive rate (γ) with different percentage of malicious clients (*thresh* is set to 0.65). There are total of 125 MRs and 1000 MCs in the network.

results are plotted as a function of the predefined correlation coefficient values. Optimal threshold value (i.e., detection efficiency is high with minimum false positive) may be found between 0.6 and 0.65 from Figure 23.4.

Figure 23.5 shows the efficiency and false-positive rate under optimal threshold 0.65 and we conclude that our algorithm performs better when the percentage of malicious clients is smaller. So, the correlation coefficient-based mechanism has its limitation as the detection efficiency gets confined by the number of misbehaving clients.

23.3.1.3 Discussion

The correlation coefficient-based detection algorithm works well when a client behaves maliciously (i.e., obtaining higher share of bandwidth) with all the routers. However, if the client strategically behaves maliciously with selective routers, then it might maintain the average correlation (with common set of routers) higher than a detectable threshold value. Furthermore, to detect whether client q is malicious or not, mesh router has to request each of the common routers between client p and q about their ratings (i.e., trust values), and this introduces delay and communication overhead. Therefore, for large-scale attacks (i.e., when the number of malicious clients is high), correlation coefficient based detection mechanism poses fundamental performance limits in terms of accuracy or detection delay for misbehavior detection.

23.3.2 WSRT-Based Detection

To overcome the limitations of the correlation coefficient-based detection technique, we propose a lightweight mechanism to detect the presence of malicious client(s) in the network based on the WSRT technique [35]. The proposed detection algorithm relies on the observation that the greedy behavior of malicious client(s) may significantly decrease the bandwidth share for the other well-behaved clients in the network, as described in Section 23.4.1. In our approach, during the

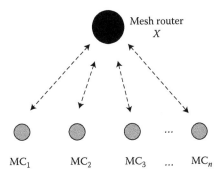

Figure 23.6 **Malicious client detection: there are *n* mesh clients under a mesh router *X*. Router *X* periodically updates its clients' information to identify malicious client(s).**

normal functioning of the network (i.e., no client is malicious), each router measures the number of packets received from each of its clients for a monitoring period T and stores them as reference values. Then, the mesh router periodically compares the current values against the reference values by executing the distribution-free statistical test based on the WSRT technique to identify the malicious clients. We have selected the Wilcoxon test because it attains good detection rate even with a small sample set, uses paired measurements (i.e., measurements from the same samples before and after an experiment), and does not assume any underlying distribution on the measurement (e.g., normal distribution). In one hand, using a small sample set adds little computation overhead. On the other hand, the ability to use paired measurements allows mesh routers to use the number of packets of the clients, which is achieved by means of a statistical passive approach based on traffic monitoring. Our detection technique is distributed since mesh routers act as local aids in the detection of misbehaving clients. We describe the detection mechanism in the following sections.

23.3.2.1 Protocol Description

Let n be the number of client nodes whose packets are received by the router X, as shown in Figure 23.6. Traffic traces of clients are collected (by the mesh router X) periodically during the monitoring period T and are compared against the reference value (threshold) to observe the deviation if there is any. Typically, 0.05 is used for the threshold, which results in the commonly used confidence level of 95% for the test results. Intuitively, all the MCs should enjoy equal share of bandwidth under normal network environment. However, if one or more MCs maliciously gain higher share of bandwidth (by manipulating the wireless access mechanism), bandwidth share degrades significantly for the other legitimate MCs. Therefore, router X can identify the misbehaving client by running the algorithm. The detection mechanism performed by the router X is presented in WSRT-based detection algorithm, as depicted in Algorithm 23.2.

Algorithm 23.2: WSRT-based malicious client detection

1. During the normal functioning of the network (i.e., no client is malicious), router X records the number of packets it received from n mesh clients as $(b_1, b_2, \ldots, b_n)$ for a monitoring period T as the reference

2. Router X periodically compares the current packet counts recorded at it as $(c_1, c_2, \ldots, c_n)$ against the reference records $(b_1, b_2, \ldots, b_n)$ using the WSRT procedure, stated as follows
   a. Compute $d_i = b_i - c_i$. Exclude d_i that is 0 and order non-zero absolute values $|d_i|$ to obtain the rank r_i for each ordered $|d_i|$
   b. Let g be the number of non-zero d_i's and I be the indicator function with value 1 and -1, compute $\mathbb{R} = \sum_{i=1}^{g} I(d_i) r_i$
   c. If there is no malicious client, the statistic $\mathbb{R}$ follows normal distribution with mean 0 and standard deviation $\sigma_{\mathbb{R}} = \left(\sqrt{g(g+1)(2g+1)}\right)/6$. Then, the p value is obtained from the calculated $\mathbb{R}$ based on the expected normal distribution d
   d. Return p
3. **if** $p \leq thresh$ **then**
4. **return** *true* {i.e., presence of malicious client in the network}
5. **else**
6. **return** *false* {i.e., network is not under attack}
7. **end if**   ■

Since the WSRT-based algorithm exploits only the readily available information of the number of packets counted by the router X; it does not incur any bandwidth (communication) overhead. The computation overhead involves $O(n)$ operations by the router, which lie mainly in the WSRT procedure. In real network environment, this algorithm may raise false alarm due to unfairness in accessing the wireless medium. To overcome this limitation, the reference records are stored in the router for an appropriate value of monitoring interval T, which takes into account the usual variations or unfairness in accessing the wireless medium by the MCs. Derivation of the value of T is described in Section 23.3.1.2.

A natural question that arises after misbehavior is detected concerns notifying the network about the attack. This is an essential step that needs to be accomplished so that the network learns about the attack and can initiate a response or isolate the attacker. It will also prevent further propagation of misbehavior in the network. In the context of mesh network, routers may maintain a separate control channel, in which case the notification message can be transmitted in that channel.

After observing any malicious behavior, a router can make a notification about the client to the WMEN operator. Multihop relaying method can be used by the mesh routers via router-to-router wireless links, finally reaching the information to the network controller (i.e., WMEN operator). However, if the multihop relaying method is employed, the collection process might interfere with normal network traffic transport. In this situation, resources must be carefully allocated to balance the transmission of normal packets and notification messages. Upon receiving the notification, the operator can decide how to react to malicious client(s). For example, the operator can charge a penalty bill, reduce the service quality, or even completely stop the service, depending on the extent of the observed behavior and the responsiveness of the client. As a result, a malicious client will be restricted from taking the advantage at the expense of other well-behaved clients in the network.

23.3.2.2 Performance Evaluation

In this section, we evaluate the performance of the proposed WSRT-based detection algorithm and compare the results with the correlation coefficient-based algorithm. As before, we have considered two scenarios in the simulation. In the first scenario, we fix the total number of MRs to 10 and MCs to 80 in the network. In the second scenario, we fix the total number of MRs to 125 and MCs to 1000 in the network. In each case, there are eight MCs attached to each MR. Matrix location

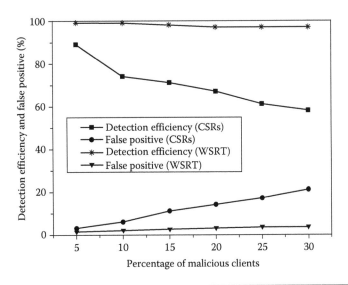

Figure 23.7 Detection efficiency (ε) and false-positive rate (γ) with different percentage of malicious clients. Results are compared with the previous malicious client detection algorithm (CSRs). There are total of 10 MRs and 80 MCs in the network.

is uniformly placed in the terrain. Transmission range is taken as 50 meters for both MR and MC, and MCs are attached to their nearest MRs. We consider that all the clients always have packets (of same size) to send (i.e., all the mesh clients in the network are backlogged).

Figure 23.7 shows the detection and false-positive rates of our algorithm for different values of malicious clients. We observe that the WSRT algorithm produces a high detection rate, even when the number of attackers is large, compared to CSRs algorithm. When the number of misbehaving MCs is small, for example, with 5% to 15% MCs being malicious, detection efficiency is around 99% while keeping the false-positive rate below 2%. CSR mechanism based on correlation coefficient can achieve up to 70% detection efficiency while keeping the false-positive rate more than 10%. In the worst case, with 30% malicious MCs, CSRs can achieve only up to 60% detection efficiency while the false-positive rate is more than 20%. On the other hand, in WSRT-based detection, efficiency is more than 90% while keeping the false-positive below 4%, as can be observed from Figure 23.7.

Figure 23.8 shows the detection and false-positive rates of two algorithms for different values of malicious clients. We observe that the WSRT algorithm produces a high detection rate, even when the number of attackers is large, compared to CSRs algorithm. For example, the detection rate is over 98% when there are only 5% to 10% malicious clients (50–100 of 1000 clients). The detection rate is more than 95% when the number of attackers is smaller than 30% (300 clients out of 1000). The false-positive rate of the WSRT algorithm also outperforms the CSR-based algorithm as shown in Figure 23.8. As can be seen, the WSRT reduces the false-positive rate below 4% while it is more than 30% in CSRs-based algorithm when the number of malicious clients is 30%. Since malicious clients are allowed to vary the contention window size, they may exploit the medium access mechanism with some routers and behave normally with other routers. Consequently, malicious clients may maintain average correlation (calculated from the common set of routers) higher than the detectable threshold as explained in Section 23.3.1.3. On the contrary, with the proposed algorithm, each router independently detects misbehaving client(s) regardless of how they behave

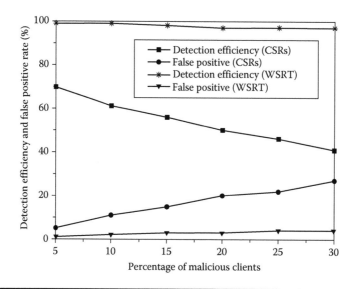

Figure 23.8 Detection efficiency (ε) and false-positive rate (γ) with different percentage of malicious clients. The results are compared with the previous malicious client detection algorithm (CSRs). There are total of 125 MRs and 1000 MCs in the network.

with other routers. Thus, the WSRT algorithm improves the detection efficiency compared to the correlation coefficient-based technique presented in the previous section. Furthermore, unlike the correlation coefficient-based algorithm presented in the previous section, detection efficiency using WSRT technique is not confined by the number of malicious clients.

23.3.2.3 Discussion

Our detection technique identifies the misbehaving client(s) using the number packets as the metric with the assumption that all clients have the same application running and they have same data rates. There might be different applications with different data rates [36,37] and/or different packet lengths [38] in the network and hence some clients may have more traffic compared to others. For example, if two stations have different data rates and delays, such as VoIP versus streaming video sources, the throughput of the latter will be naturally much larger than that of VoIP. However, our detection algorithm can easily handle such cases as described subsequently.

There can be different traffic classes (i.e., different applications) in the network and hence traffic of a certain class might generate data at a rate higher than the others. Suppose, there are C traffic classes and data rate associated with the ith class (where, $i = 1, 2, \ldots, C$) is R_i packets/second. We define the normalized data rate, $\bar{d}$ (in packets/second) of the ith class as

$$\bar{d} = \frac{R_i}{w_i} \tag{23.2}$$

where w_i is the weight associated with the traffic class i. A higher value of the weight indicates the importance of the traffic class and hence, higher rate of the traffic class. We assume that weights are assigned in a way so that normalized rates are equal for all the traffic classes. In this case, the normalized data rate $\bar{d}$ in Equation 23.2 is expected to be equal for the clients attached to an MR.

So, instead of number of packets, using the normalized data rate as the metric, router can distinguish a client that accesses the medium higher than the other clients.

In another case, there can be different traffic classes with different packet lengths. Suppose, there are C traffic classes and data rate associated with the ith class (where, $i = 1, 2, \ldots, C$) is R_i packets/second. We define the normalized data rate, $\bar{d}$ (in packets/second), of the ith class as

$$\bar{d} = \left(\frac{\text{packetsize}_i}{\text{Max(packetsize)}} \times \frac{R_i}{w_i} \right) \tag{23.3}$$

where, packetsize$_i$ is the length (e.g., in Bytes) of a data packet of the ith class, Max(packetsize) is the maximum length (in Bytes) of a data packet in the network, and w_i is the weight associated with the traffic class i. Again, instead of number of packets, using the normalized data rate as the metric, router can detect a client that gets higher share of bandwidth at the expense of the other clients.

23.4 Future Directions of Research

WMNs along with *ad hoc*, sensor, and ubiquitous networks have recently witnessed their fastest growth period ever in history, and this trend is likely to continue for the foreseeable future. However, as mentioned earlier, due to the unique characteristics of WMNs, they are vulnerable to security attacks compared to wired networks or other wireless networks. Developing a robust security solution will play crucial part in their design process to ensure their successful deployment and exploitation in practice. Nevertheless, there are several future directions in which this work might be augmented. An interesting research direction might be to consider addressing in more detail the case of adaptive cheating. Adaptive cheating might be defined as the set of misbehavior techniques that exploit some knowledge about the way detection system works. For example, a malicious client may switch frequently enough between several techniques in such a way that the system fails to collect enough data to detect misbehavior. We have considered the greedy behavior of mesh clients and detect a client as a misbehaving one if it gains higher share of bandwidth compared to other clients in the network. However, a client can misbehave in other ways too. As a matter of fact, an intelligent adaptive client, after knowing the detection procedure, can change its strategy. For example, a misbehaving client might find a way to increase collisions in the neighborhood. Thus, the misbehavior detection mechanism might be enhanced by considering other metrics for detecting misbehaving clients.

23.5 Conclusions

In this chapter, we have presented misbehaving client detection mechanisms for the network to be survivable and robust against malicious clients. First solution concerns the exploitation of observations from several mesh routers (i.e., common set of routers) in order to identify the malicious clients. This amounts to the scenario where observers (i.e., mesh routers) pass their measurements (trust values) to each other which then combines them appropriately and derives a decision as to the occurrence or not of an attack. Second solution concerns the detection approach to be independent of common set of routers. Each mesh router individually detects the malicious client(s) exploiting the WSRT technique. The results of our second approach eliminate the problem of the performance limits in terms of detection accuracy and communication overhead posed by the first solution.

In our approach, we have assumed continuously backlogged nodes and we have used bandwidth share as a means of measuring the benefit of the attacker and corresponding performance loss of legitimate nodes. Implicitly, we assumed that fair sharing of the medium is reflected by this measure. However, there might be different applications with different data rates and/or different packet lengths in the network and hence some clients may have more traffic compared to others. Our detection algorithm can easily be extended to handle such cases as described in the previous section (Section 23.3.2.3). Therefore, our approach can encompass different cases that can be present in the network.

Terminologies

Wireless mesh networks
Misbehavior
Malicious mesh clients
Intrusion detection
Correlation coefficient
Signed ranked test
Detection efficiency
False positive

Questions and Sample Answers

1. Draw an architectural diagram of a Wireless Mesh Network that reflects the description in Section 23.1.

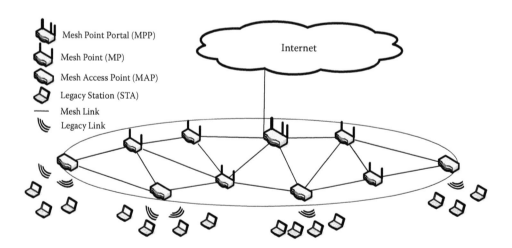

2. Please specify the characteristics that make WMN different from other wireless networks. Although both WMN and mobile *Ad Hoc* networks are multihop wireless networks, there exist some key characteristics in WMN, which make it different from *ad hoc* networks. First, the WMN is composed of static wireless nodes that form a wireless backbone for providing Internet services to the end users. In contrast, in MANET, nodes are usually mobile and the network itself is not designed for providing backbone services. Secondly, the traffic pattern

in WMN is mostly gateway centric, such as traffic flows between mesh routers and gateway. On the other hand, in MANET, traffic flows between any two nodes. Finally, energy is not considered an issue with WMN where nodes are static and assumed to be energy efficient. Whereas, the use of small battery power devices (like PDA) in MANET, make it a more energy constraint network.

3. List some applications of WMN.
   a. Broadband home networking
   b. Community networking
   c. Enterprise networking
   d. Public safety, rescue and recovery operations
   e. Military operations

4. Why security in WMN is more critical than other wireless networks?
   WMN is a collection of static wireless nodes (i.e., mesh routers) that form the backhaul network to provide services to mobile users attached with the static mesh routers. Providing security for such type of network is critical and different from the existing solutions for wireless networks. First, the multihop architecture of WMN renders general security schemes as insufficient for a WMN. Secondly, WMN exhibits a two-tier architecture which requires different security requirements for wireless access of mesh clients to router and mesh router to router. Finally, WMN may incorporate multiple wireless networks with different security architecture and schemes. So, there remains a challenge to develop a scheme to provide secure and reliable inter-network communication.

5. What is misbehavior in wireless networks?
   Deviation from legitimate protocol operation in wireless networks has received considerable attention from the research community in recent years. The increased level of sophistication in the design of protocol components, together with the requirement for flexible and readily reconfigurable protocols, has led to the extreme where wireless network devices have become easily programmable. As a result, it is feasible for a network peer to tamper with software and firmware, modify its wireless interface and network parameters and ultimately abuse the protocol. This situation is referred to as protocol misbehavior.
   Since all the wireless stations use the similar IEEE 802.11 CSMA/CA Medium Access Control (MAC) protocol, malicious use of the medium is always possible to gain higher share of bandwidth at the expense of other clients/users in the network. Even though the cryptographic solution may achieve authentication, data confidentiality, and other security issues, this mechanism may not detect/restrict the MAC layer greedy behavior since wireless devices (e.g., mesh clients) directly deal with the shared wireless medium.

6. Give an example of misbehavior in WMN.
   In a typical WMN, mesh clients usually access the medium using contention-based medium access protocols. For example, if the mesh clients use generic IEEE 802.11 DCF, then it is possible to act like a misbehaving client by altering the value of the backoff parameter. A small backoff interval gives the corresponding client the advantages of gaining access to the wireless channel quickly. This, in turn, will force adjacent nodes to defer their transmission.

7. How correlation coefficient-based detection mechanism works?
   In the statistical way, behaviors of all the mesh clients (MCs) will be linearly correlative with each other. In other words, bandwidth share (under saturated condition) will be almost equal

for all the clients. For example, let us consider that two MCs, p and q, via a common set of mesh routers (MRs) $MR_1, MR_2, \ldots, MR_m$. Common set is chosen based on the close relationship with the two communicating clients. For example, all the past messages between client p and q traverse through this set of routers and/or both clients have individual communications with those routers and therefore, they have an existing trust history with the routers. Each MR collects the trust values of same clients from other MRs (i.e., for those clients who move from one router to another) as well as for the clients attached to this MR. Then, each MR calculates the correlation coefficient, and then calculates the average correlation coefficient value. After that, the average value is compared against the reference threshold. The presence of the malicious client is detected by the MR if the average correlation coefficient is less than the threshold.

8. What are the limitations of correlation coefficient-based detection mechanism?
   The correlation coefficient-based detection algorithm works well when a client behaves maliciously (i.e., obtaining higher share of bandwidth) with all the routers. However, if the client strategically behaves maliciously with selective routers, then it might maintain the average correlation (with common set of routers) higher than a detectable threshold value. Furthermore, to detect whether client q is malicious or not, mesh router has to request each of the common routers between client p and q about their ratings (i.e., trust values), and this introduces delay and communication overhead. Therefore, for large-scale attacks (i.e., when the number of malicious clients is high), correlation coefficient-based detection mechanism poses fundamental performance limits in terms of accuracy or detection delay for misbehavior detection.

9. How WSRT-based detection mechanism works?
   To overcome the limitations of the correlation coefficient-based detection technique, a lightweight mechanism to detect the presence of malicious client(s) in the network based on the Wilcoxon signed rank test (WSRT) technique is developed. The detection algorithm relies on the observation that the greedy behavior of malicious client(s) may significantly decrease the bandwidth share for the other well-behaved clients in the network. In this approach, during the normal functioning of the network (i.e., no client is malicious), each router measures the number of packets received from each of its clients for a monitoring period T and stores them as reference values. Then, the mesh router periodically compares the current values against the reference values by executing the distribution-free statistical test based on the WSRT technique to identify the malicious clients. Intuitively, all the MCs should enjoy equal share of bandwidth under normal network environment. However, if one or more MCs maliciously gain higher share of bandwidth (by manipulating the wireless access mechanism), bandwidth share degrades significantly for the other legitimate MCs. Therefore, a router can identify the misbehaving client by running the algorithm.

10. What are limitations of WSRT-based detection mechanism?
    WSRT-based detection technique identifies the misbehaving client(s) using the number packets as the metric with the assumption that all clients have the same application running and they have same data rates. There might be different applications with different data rates and/or different packet lengths in the network and hence some clients may have more traffic compared to others. For example, if two stations have different data rates and delays, such as VoIP versus streaming video sources, the throughput of the latter will be naturally much larger than that of VoIP.

Author's Biography

Md. Abdul Hamid, PhD received his BEng in computer and information engineering in 2001 from International Islamic University Malaysia (IIUM). In 2002, he joined as a lecturer in the Computer Science & Engineering Department, Asian University of Bangladesh, Dhaka. He received the PhD degree at Kyung Hee University, South Korea, in August 2009. Since September 2009, he has been working as a professor of the Department of Information and Communication Engineering, Hankuk University of Foreign Studies, South Korea. His research interests include wireless sensor networks, wireless mesh networks, and *ad hoc* networks with particular emphasis on the network security, the design of routing protocol/metric, enhanced MAC protocols, fairness and channel assignment.

Md. Shariful Islam received his BSc and MSc degree in computer science from the University of Dhaka, Bangladesh, in the year 2000 and 2002, respectively. He completed his MS degree in Information Technology from the Royal Institute of Technology (KTH), Sweden, in 2005. He worked as a faculty member in the Institute of Information Technology (IIT), University of Dhaka. His current research interests include the design of routing protocol/metric and enhanced MAC protocols for wireless mesh networks. He also worked on security issues-related wireless *ad hoc* and mesh networks. He is now a PhD student in the Department of Computer Engineering, Kyung Hee University, Korea.

References

1. I. F. Akyildiz, X. Wang, and W. Wang, Wireless mesh networks: A survey, *Computer Networks*, 47(4), 445–487, 2005.
2. I. Akyildiz and X. Wang, A survey on wireless mesh networks, *Communications Magazine, IEEE*, 43(9), S23–S30, 2005.
3. R. Bruno, M. Conti, and E. Gregori, Mesh networks: Commodity multihop ad hoc networks, *Communications Magazine, IEEE*, 43(3), 123–131, 2005.
4. Microsoft research. Mesh networking. http://research.microsoft.com/mesh/.
5. Intel. Wimax. http://www.intel.com/netcomms/technologies/wimax/.
6. Y. Zhang, J. Luo, and H. Hu, eds., Wireless mesh networking: Architectures, protocols and standards. New York: Auerbach Publications, 2007. http://www.auerbach-publications.com.
7. E. Hossain and K. Leung, eds., Wireless mesh networks: Architectures and protocols. New York: Springer Science + Business Media, LLC, 2008. http://www.springer.com.
8. G. Held, *Wireless Mesh Networks*. Boston, MA, USA: Auerbach Publications, 2005.
9. IEEE 802.11s Task Group, draft amendment to standard for information technology telecommunications and information exchange between systems—LAN/MAN specific requirements—part 11: Wireless medium access control (MAC) and physical layer (PHY) specifications: Amendment: ESS mesh networking, IEEE p802.11s/d1.06, September 2008.
10. H. Li, M. Xu, and Y. Li, Selfish MAC layer misbehavior detection model for the IEEE 802.11-based wireless mesh networks, in *Proceedings of Advanced Parallel Programming Technologies (APPT07)*, vol. 4847. Springer, Berlin/Heidelberg, 2007, pp. 382–391.
11. M. Raya, J.-P. Hubaux, and I. Aad, DOMINO: A system to detect greedy behavior in ieee 802.11 hotspots, in *MobiSys '04: Proceedings of the 2nd International Conference on Mobile Systems, Applications, and Services*. New York, NY, USA: ACM Press, 2004, pp. 84–97.
12. Q. He, D. Wu, and P. Khosla, SORI: A secure and objective reputation-based incentive scheme for ad-hoc networks, in *Wireless Communications and Networking Conference, 2004*. WCNC. 2004 IEEE, vol. 2, March 2004, pp. 825–830.
13. M. Alicherry, R. Bhatia, and L. E. Li, Joint channel assignment and routing for throughput optimization in multi-radio wireless mesh networks, in *MobiCom '05: Proceedings of the 11th Annual*

International Conference on Mobile Computing and Networking. New York, NY, USA: ACM Press, 2005, pp. 58–72.

14. A. Raniwala and T. Chiueh, Architecture and algorithms for an IEEE 802.11-based multi-channel wireless mesh network, INFOCOM 2005, in *24th Annual Joint Conference of the IEEE Computer and Communications Societies.* Proceedings IEEE, vol. 3, pp. 2223–2234, March 2005.

15. M. Kodialam and T. Nandagopal, Characterizing the capacity region in multi-radio multi-channel wireless mesh networks, in *ACM MobiCom.* New York, NY, USA: ACM Press, 2005, pp. 73–87.

16. D. S. Radosavac, J. S. Baras, and I. Koutsopoulos, A framework for MAC protocol misbehavior detection in wireless networks, in *WiSe '05: Proceedings of the 4th ACM Workshop on Wireless Security.* New York, NY, USA: ACM Press, 2005, pp. 33–42.

17. M. Cagalj, S. Ganeriwal, I. Aad, and J.-P. Hubaux, On selfish behavior in CSMA/CA networks, in *INFOCOM*, 2005, pp. 2513–2524.

18. S. Marti, T. J. Giuli, K. Lai, and M. Baker, Mitigating routing misbehavior in mobile ad hoc networks, in *MobiCom'00: Proceedings of the 6th Annual International Conference on Mobile Computing and Networking.* New York, NY, USA: ACM Press, 2000, pp. 255–265.

19. S. Buchegger and J. yves Le Boudec, Performance analysis of the CONFIDANT protocol (Cooperation Of Nodes: Fairness In Dynamic Ad-hoc NeTworks), in *MOBIHOC02*, 2002, pp. 226–236.

20. N. B. Salem, L. Buttyan, J.-P. Hubaux, and M. Jakobsson, A charging and rewarding scheme for packet forwarding in multi-hop cellular networks, in *MobiHoc '03: Proceedings of the 4th ACM International Symposium on Mobile Ad Hoc Networking & Computing.* New York, NY, USA: ACM Press, 2003, pp. 13–24.

21. P. Kyasanur and N. Vaidya, Detection and handling of MAC layer misbehavior in wireless networks, in *Proceedings 2003 International Conference on Dependable Systems and Networks,* June 2003. pp. 173–182.

22. J. Konorski, Multiple access in ad-hoc wireless LANs with noncooperative stations, in *NETWORKING '02: Proceedings of the Second International IFIP-TC6 Networking Conference on Networking Technologies, Services, and Protocols; Performance of Computer and Communication Networks; and Mobile and Wireless Communications.* London, UK: Springer, 2002, pp. 1141–1146.

23. L. Santhanam, D. Nandiraju, N. Nandiraju, and D. Agrawal, Active cache based defense against dos attacks in wireless mesh network, in *2nd International Symposium on Wireless Pervasive Computing, 2007. ISWPC '07*, pp. 419–424, February 2007.

24. T. Fahad, D. Djenouri, R. Askwith, and M. Merabti, A new low cost sessions-based misbehaviour detection protocol (SMDP) for MANET, in *AINAW '07: Proceedings of the 21st International Conference on Advanced Information Networking and Applications Workshops.* Washington, DC, USA: IEEE Computer Society, 2007, pp. 882–887.

25. C.-W. Tan and S. K. Bose, Enforcing cooperation in an ad hoc network using a cost-credit based forwarding and routing approach, in *Wireless Communications and Networking Conference, 2007. WCNC 2007. IEEE*, March 2007, pp. 2935–2939.

26. H. Luo, J. Kong, P. Zerfos, S. Lu, and L. Zhang, URSA: Ubiquitous and robust access control for mobile ad-hoc networks, *IEEE/ACM Transactions on Networking*, 12, 1049–1063, 2004.

27. M. A. Hamid, M. S. Islam, and C. S. Hong, Developing security solutions for wireless mesh enterprise networks, in *Wireless Communications and Networking Conference, 2008. WCNC 2008. IEEE*, 31 March–3 April 2008, pp. 2549–2554.

28. M. A. Hamid, M. Abdullah-Al-Wadud, C. S. Hong, O. Chae, and S. Lee, A robust security scheme for wireless mesh enterprise networks, *Annals of Telecommunications*, 64(5–6), pp. 401–413, 2009.

29. S. M. Ross, *Introduction to Probability Models*, 9th edition. San Diego, CA, USA: Academic Press, 2007.

30. The Network Simulator – ns-2, http://www.isi.edu/nsnam/ns/index.html.

31. Z. Li, S. Nandi, and A. K. Gupta, Modeling the short-term unfairness of IEEE 802.11 in presence of hidden terminals, *Performance Evaluation*, 63(4), 441–462, 2006.

32. C. L. Barrett, M. V. Marathe, D. C. Engelhart, and A. Sivasubramaniam, Analyzing the short-term fairness of IEEE 802.11 in wireless multi-hop radio networks, in *MASCOTS'02: Proceedings of the 10th IEEE International Symposium on Modeling, Analysis, and Simulation of Computer and Telecommunications Systems.* Washington, D.C., USA: IEEE Computer Society, 2002, p. 137.

33. C. E. Koksal, H. Kassab, and H. Balakrishnan, An analysis of short-term fairness in wireless media access protocols, in *Proceedings of ACM Sigmetrics*, New York, NY, USA, 2000, pp. 118–119.

34. T. Nandagopal, T.-E. Kim, X. Gao, and V. Bharghavan, Achieving MAC layer fairness in wireless packet networks, in *MobiCom'00: Proceedings of the 6th Annual International Conference on Mobile Computing and Networking*. New York, NY, USA: ACM Press, 2000, pp. 87–98.
35. R. Lowry, *Concepts and Applications of Inferential Statistics*. Vassar College, Poughkeepsie, NY, USA, 2006.
36. X. Cheng, P. Mohapatra, S.-J. Lee, and S. Banerjee, Performance evaluation of video streaming in multihop wireless mesh networks, in *NOSSDAV '08: Proceedings of the 18th International Workshop on Network and Operating Systems Support for Digital Audio and Video*. New York, NY, USA: ACM Press, 2008, pp. 57–62.
37. D. Niculescu, S. Ganguly, K. Kim, and R. Izmailov, Performance of VoIP in a 802.11 wireless mesh network, in *INFOCOM 2006. 25th IEEE International Conference on Computer Communications. Proceedings*, Barcelona, Catalunya, Spain, April 23–29, 2006, pp. 1–11.
38. G. Xylomenos and G. Polyzos, TCP and UDP performance over a wireless LAN, in *INFOCOM '99. Eighteenth Annual Joint Conference of the IEEE Computer and Communications Societies. Proceedings*. IEEE, vol. 2, New York, NY, USA, March 1999, pp. 439–446.

Index

Milton Keynes UK
Ingram Content Group UK Ltd.
UKHW052029071024
449327UK00027B/2493

9 780367 383527